STATISTICAL ANALYSIS: A DECISION-MAKING APPROACH

STATISTICAL ANALYSIS: A DECISION-MAKING APPROACH

Robert Parsons
NORTHEASTERN UNIVERSITY

Harper & Row, Publishers
New York, Evanston, San Francisco, London

Sponsoring Editor: John Greenman
Project Editor: Eleanor Castellano
Designer: Rita Naughton
Production Supervisor: Stefania J. Taflinska

STATISTICAL ANALYSIS: A Decision-Making Approach

Library of Congress Cataloging in Publication Data
Parsons, Robert.
    Statistical analysis: a decision-making approach.
      1.   Statistics.   2.   Statistical decision.
I.   Title.
HA29.P326   519.5'4   73-13204
ISBN 0-06-045022-3

To Ginger
the spice of my life

CONTENTS

Bayesian material is clearly indicated by screens in the Table of Contents.

x CONTENTS

PREFACE

Statistical Analysis: A Decision-Making Approach is a textbook for beginning statistics courses in business administration and economics at both the under-graduate and graduate levels. It presumes no calculus, only a knowledge of basic algebra. It has been designed to provide breadth of topic coverage; yet, it minimizes interdependence among chapters. For example, the concept of least-squares curve fitting has been developed in a technical note that can be referred to equally well from the regression chapter or from the chapter on secular trend. Similarly, time series analysis precedes regression analysis, but the order may be reversed, according to the instructor's preference. This kind of flexibility enables the text to serve the needs of a one-semester, a two-semester, a one-quarter, or a two-quarter course. Several alternative course outlines are presented in the Instructor's Solutions Manual.

The question of incorporating Bayesian analysis has been considered through-out. Accordingly, Bayesian materials have been presented in a fashion that enables the instructor to emphasize or deemphasize them at many points. These topics may be integrated or omitted as the user sees fit, and they have been clearly demarked in the Table of Contents by gray panels like this one. One might elect to omit *all*

Bayesian materials. Instructors preferring such an approach will find the text equally well suited to presenting only classical material.

I would like to express appreciation to the many individuals who made helpful suggestions, comments, and criticisms at various stages of the manuscript. I am indebted to the Literary Executor of the late Sir Ronald A. Fisher, F.R.S., to Dr. Frank Yates, F.R.S., and to Oliver and Boyd, Edinburgh, for their permission to reprint Table V-B from *Statistical Methods for Research Workers* and Table III from *Statistical Tables for Biological, Agricultural and Medical Research*. I would particularly like to thank Professors Cecil H. Meyers and David W. Miller, who reviewed all or portions of the manuscript, and Warren Briggs for his contributions to the end of Chapter 22.

A particular debt of gratitude is owed to instructors who used preliminary versions of the text. These include Jules Borack, Steve Brenner, Warren Briggs, Vin Giovinazzo, Steve Grossman, Ken Moore, Barry Shore, and Fred Wiseman. Their reactions and suggestions resulted in many revisions and improvements in the text. I would also like to thank Connie Fernandes, Classia Simmons, and Sue Skalder for the tireless efforts they expended on the typing of the manuscript.

ROBERT A. PARSONS

STATISTICAL ANALYSIS: A DECISION-MAKING APPROACH

INTRODUCTION

The need for decision making is inescapable. Virtually every waking moment involves decision making to a greater or lesser degree. If you are in a managerial position, these decisions may have far-ranging implications for the future well-being of the firm. Fortunately, most decision making is associated with situations that continually reoccur, and one's subconscious can be relied on to generate the tried and true response. When decisions are of a nonroutine, nonrepetitive nature, conscious deliberation on the part of the decision maker is required. Ideally, after carefully structuring the alternatives, the decision maker should amass all data that could conceivably exert an influence on the consequences of these actions. The likelihood of each influence and the relative desirability of each possible package of consequences should be assessed. Finally, after careful consideration of all available evidence, an effective strategy or course of action should be formulated.

Can you imagine the reaction of a decision maker to the above suggestion? Undoubtedly the response would be: "Impossible! Do you have any idea how much data is out there? The file cabinet population is growing at a faster rate than the human population! Even if the physical location of all relevant data were known, the sheer physical volume would be totally overwhelming. Moreover, data gathering is an expensive proposition. There is a point at which the cost of gathering additional data (even though it is relevant) outweighs the benefits to be derived from acquisition of that data.[1] If this were not enough, the time available for data gathering and analysis places a further restriction on the data-gathering process." Unfortunately, such reactions on the part of the decision maker are quite valid. The decision maker must operate under the constraints of limited time, limited money, and the responsibility for arriving at effective decisions. Properly gathered and evaluated statistics can help. One of the principal objectives of this book is to generate a solid understanding of what statistical analysis is all about and how its use can aid effective decision making.

It is appropriate at this time to define the word statistics formally. Current usage finds it used a number of ways. The two most common are as follows:

1. Information communicated in a numerical form, for example, the number of highway fatalities last Labor Day Weekend, the number of checks cleared this week through the local Federal Reserve Bank, the sales volume at the Louisville branch last week, batting averages for members of the New York Yankee organization for the 1970 season, the number of defective parts in the last shipment, or the total accounts receivable balance outstanding over 60 days. Statistics, thus defined, are certainly useful to the decision maker. Statistical data serve as informational inputs in decision analysis. Statistical data can be thought of as the raw material on which the structure of analysis is built.

2. The word statistics is also employed to refer to the body of analytical procedures or techniques employed in the manipulation of numerical data. To generate information in as useful a form as possible, it is often necessary to subject data to statistical analysis prior to its incorporation in the decision-making process. Furthermore, techniques have recently been developed to utilize information more effectively *after* it has been incorporated into the decision-making process. These techniques are referred to collectively as *statistical decision theory*.

Our concern with statistics will be with the way data and techniques can be used to improve the quality of decision making. The next few pages are devoted to a brief overview of the techniques and procedures treated in the text.

1.1 PREPARATION OF INFORMATION AS AN INPUT TO THE DECISION-MAKING PROCESS

After careful definition of the problem and structuring of the decision alternatives, the decision maker should determine the various factors likely to influence the outcomes associated with the decision alternatives under consideration. The next

---

[1] Techniques that indicate whether or not acquisition of additional information is warranted on a cost-benefit basis will be examined in Chapter 9.

step is generation of information to project the nature of the influence these factors are likely to exert.

Consider an urban university faced with a parking problem. One possible solution is to build parking garages. One factor that would exert considerable influence on the financial feasibility of the project is the amount faculty, students, and staff are willing to pay for the use of such a facility. Another factor to consider is the percentage of current automobile commuters who would shift to a car pool or an alternative mode of transportation (such as rapid transit) as the charge for university-sponsored parking increased.

The total set of observations relevant to a study under investigation is referred to as the *universe* or *population.* It is extremely important to consider carefully how the data is to be used prior to actual data collection. This helps ensure identification of the relevant universe (or population). In the above illustration, the universe consists of all individuals currently commuting to the university as operators of automobiles. Characteristics of interest of this universe include the dollar amount each commuter is willing to pay for the privilege of using the facility and the percentage who would discontinue commuting as an operator at various parking fees. The actual process of data gathering can be an expensive proposition. Occasionally, the data may previously have been acquired in a study taken for another purpose. The analyst should attempt to determine if a previous study exists before undertaking an expensive original study. For example, another urban university with a student body of similar income composition, transit alternatives, and commuting pattern may have recently conducted a similar study. Before using data gathered in another survey, one should, of course, carefully verify the applicability of the data gathered to the problem currently under investigation.

Originally collected data is usually relatively unstructured. As a result, once relevant data has been identified and collected, it is necessary to restructure it in such a fashion that the salient information contained therein is effectively communicated. The body of statistical techniques that has evolved to perform this function is referred to under the general heading of *techniques of statistical description.* Chapters 2, 3, and 4 of the text are devoted to this area of statistical analysis. Chapter 2 discusses the frequency distribution, a technique which enables the analyst to restructure the data into either a tabular or graphic format. Often, however, it is desirable to go beyond the tabular restructuring involved in the construction of a frequency distribution and capture the salient characteristics of the universe in a few succinct numerical values.

To illustrate, the values observed for individual items frequently tend to cluster around a particular numerical value. Such a value is a measure of central tendency and can be used as a representative response for the entire population. In the case of the parking study, a measure of central tendency could be used to indicate the representative amount respondents would be willing to pay for the use of the proposed parking facility. Chapter 3 discusses measures of central tendency, their construction, and interpretation.

A measure of dispersion is designed to indicate the extent to which individual responses tend to deviate from the value of the measure of central tendency. A low value for a measure of dispersion indicates that the values of the individual observations are closely clustered around the measure of central tendency. A large value for the measure of dispersion indicates that they are widely scattered. When the measure of dispersion is quite small, the measure of central tendency can be said to serve effectively as a representative value. Chapter 4 examines measures of dispersion, their construction, and interpretation.

1.2 STATISTICAL INFERENCE

Descriptive measures of a universe (or population) are referred to as *parameters* of the universe. When one computes a measure of central tendency for a universe, the resultant value is a parameter (or descriptive measure of the universe). In the same fashion, the computation of a measure of dispersion for the universe provides another parameter (another descriptive measure of the universe). To compute parameters of a universe, it is necessary to collect observations on every single item comprising the totality of items defined as the universe. Gathering of information on every member of a universe is referred to as a *census*. In some cases such an activity is feasible. Usually it is not. Either insufficient time or the cost prohibits such extensive data gathering. In situations where a census is not feasible, the statistical analyst must resort to a *sample*. A sample is a collection of observations on only a few members of the universe.[2] Descriptive measures for samples (for example, measures of central tendency and dispersion) can be computed in a fashion similar to their determination for a universe. A descriptive measure of a sample is referred to as a *statistic*.[3]

It is, of course, not the statistic (descriptive measure of the sample) but rather the parameter (descriptive measure of the universe) that is of interest to the decision maker. A body of knowledge relating to statistical analysis referred to as *statistical inference* has evolved that is concerned with the development of procedures whereby inferences are drawn with respect to the parameters of the universe on the basis of statistics observed in a sample selected from that universe. Estimation of universe parameters is examined in Chapter 13. The inability or infeasibility of dealing with a total universe often results in one working with only a portion of the entire universe. In this case it is necessary to generate estimates of the values of parameters. Since we do not have the entire universe, these estimates will be subject to error (where error is defined as the difference between the value of the parameter of the universe and the value of the statistic of the sample used as the estimate). Techniques have been developed that enable the decision maker to measure the probable amount of error introduced. Some sampling plans are more subject to error than others. The merits of alternative sampling plans are examined in Chapter 12.

Another major branch of statistical inference deals with tests of hypotheses. In tests of hypotheses, the analyst states a hypothesis with respect to the value of a universe parameter. Sample evidence is then collected. If the evidence is consistent with the hypothesis, the hypothesis is accepted; otherwise it is rejected. Chapter 15 is devoted to an examination of the techniques associated with hypothesis testing.

Up to this point, discussion has centered on the form of uncertainty associated with estimation of characteristics of the whole when information is restricted to

[2] It is interesting to note that the same set of data could serve as a population in one situation and as a sample in another. To illustrate, consider a shop steward. If he is concerned with the attitude of the members of the local toward a proposed pension plan, the members of the local would constitute the universe of concern. On the other hand, if the National Union is conducting a study to determine the attitude of the membership at large, then the responses of the local would constitute only a sample (or a segment of the total universe under investigation).

[3] A third use of the term *statistic* is introduced. In this context, it performs the important function of differentiating between situations where the analyst is referring to descriptive measures of the universe (designated *parameters*) and situations where the analyst desires to refer to the descriptive measures of a sample (designated *statistics*).

characteristics of the part. The decision maker must deal with another type of uncertainty as well. It is a fact of life that decision making takes place in an uncertain probabilistic environment. The decision maker is constantly faced with the necessity of deciding on a course of action, knowing full well that the consequences associated with the selection of any given act depend on future conditions and events, the outcomes of which are unknown at the time the decision has to be made. To illustrate: Suppose Babylon Steel is in the process of deciding whether to initiate a price increase and if so how great an increase. A number of factors must be considered in weighing the possible consequences associated with each of the action alternatives under consideration. What will White House reaction be to any one of a number of price increases under consideration? Will White House reaction take the form of retaliation by raising foreign import quotas? How will U.S. Steel react? Will it ignore the price increase? Match it? Will U.S. Steel raise its prices by some lesser amount? If U.S. Steel raises its prices by some lesser amount and Babylon subsequently cuts its price increase back to the U.S. Steel figure, will the White House accept the price increase in its reduced form? If Babylon Steel were in a position to anticipate the correct answer to these and associated questions, the best or optimal action would be clear. Unfortunately, this is seldom the case. Invariably, the decision maker is faced with a number of uncertainties with respect to the nature of the environment in which he must operate. The existence of uncertainty among several chance outcomes clearly complicates the decision-making process. Nevertheless, decisions must be made. A number of techniques have evolved that are designed to aid the decision maker in arriving at effective decisions under such conditions. Decision analysis is a relative newcomer to the area of statistical analysis, and some of the techniques discussed in this text have recently been the source of a great deal of controversy. The major bone of contention involves the validity of incorporating the *feelings* of the decision maker in assigning probabilities describing the likelihood of occurrence of various events. Two major schools of thought have emerged. One school, referred to as the *classicist school*, believes that only hard, objective evidence should be taken into consideration when assessing probabilities. The other school, referred to as the *Bayesian school*,[4] believes that the intuitive feelings of the decision maker should be incorporated in the analysis of a decision problem. Bayesians believe that the probability assignments to various chance events should reflect the strength of conviction of the decision maker as to the likelihood of occurrence. The position of the author is basically Bayesian, although the bulk of the statistical techniques examined in the text will be initially developed within a classical framework. The justification for this approach is that an understanding of the classical techniques provides a useful basis for extension of the techniques to incorporate the philosophy of the Bayesian school. Moreover, an understanding of the classical approach is a necessary prerequisite to an intelligent evaluation of the merits of the two alternative points of view. Both schools of thought will be discussed. Which of the two approaches is the legitimate one to use is just another decision you, the reader, will have to make.

---

[4] The label Bayesian emanates from Sir Thomas Bayes, the initial proponent of the incorporation of subjective probabilities into the decision-making process, in "An Essay Toward Solving a Problem in the Doctrine Of Chance," published in *Philosophical Translations of the Royal Society* in 1763.

1.3 TIME-SERIES ANALYSIS

A time series is a set of chronological observations on the varying value of a statistical series. An example of a time series is the retail sales volume by month for the past 20 years for the Honson Johnson Ice Cream Chain. The statistical analyst is concerned with analyzing such collections of historical data in an effort to understand why a given statistical series has been characterized by a particular historical behavior pattern. The analyst is not restricted to analyzing the past behavior of a series; he is equally if not more concerned with the prediction of future behavior. Such projections serve as important informational inputs in the structuring of the analysis of decision problems. Exploration of the techniques available to the statistical analyst designed to generate answers to questions of this type constitute the subject matter of Chapters 24, 25, and 26.

1.4 REGRESSION AND CORRELATION ANALYSIS

Regression analysis is concerned with determination of the nature of the relationship between a variable of interest and a number of explanatory variables. For example, a farming cooperative may be interested in predicting yield per acre of an agricultural crop. Using regression analysis, the relationship would be determined between yield per acre and such explanatory variables as fertilizer application, fertility of the soil, rainfall, and degree of mechanization. Given the relationship and the value of the explanatory variables, an estimate could then be generated for the variable of interest, yield per acre. The degree to which variation in the variable of interest is associated with variation in the explanatory variables is measured by correlation analysis. Correlation analysis provides an index that indicates the strength of the relationship between the explanatory variables and the variable being explained. Regression and correlation analysis are treated in Chapters 27 and 28.

1.5 A WORD OF CAUTION

Before delving into the techniques of statistical analysis, a word of warning is appropriate. An incorrect application of a technique typically results in an incorrect and misleading conclusion. The misuse of statistics may even be deliberate. Certainly, this occasionally occurs. In fact, it is a wise policy when using the results of statistical analyses performed by others to maintain a questioning attitude rather than to accept the conclusions at face value.[5] All too often, however, statistical misuse is unintentional. This can be attributed to insufficient understanding of the techniques involved. A firm grounding in the underlying concepts of statistical analysis places one in a position to evaluate the legitimacy of the statistical analyses and conclusions of others. It also serves to eliminate the possibility of self-inflicted careless statistical analysis. With this in mind, attention will be drawn throughout the text to the common pitfalls that await the careless statistical analyst and business decision maker.

[5] Darrell Huff has written a most entertaining book entitled *How to Lie with Statistics*, New York: Norton, 1954, in which he describes a number of statistical deceptions that await the unwary.

THE FREQUENCY DISTRIBUTION

2

def.

A pile of sales slips representing last month's sales
activity contains a great deal of useful information. The
problem is that the data is not structured in its existing
form. Thus, various inherent data characteristics are
not discernible. The major emphasis of statistical
descriptive techniques is to take a large mass of useful
but poorly structured data and condense or restructure
it so that the basic characteristics of the data are
clearly evident. A statistical technique frequently
employed for this purpose is the frequency distribution.
A *frequency distribution* is a table in which observed
values of a variable are grouped or classified according
to numerical magnitude with respect to a particular
trait under consideration. For example, if total dollar
sales is the trait of interest, a particular sales slip
would be assigned to a given category on the basis of
the dollar value of the sale reflected on the sales slip.

7

The construction of a frequency distribution to describe a set of data is not a haphazard undertaking. The procedure followed in construction should be such that the resulting frequency distribution will be appropriate for the purpose intended. Typical objectives and the procedures that lead to their attainment will be examined, but first it will be useful to acquire an understanding of the vocabulary commonly employed in discussions pertaining to frequency distributions. A frequency distribution is illustrated in Table 2.1 and will serve as a reference whereby terms and their definitions can be related to their physical counterparts in an actual frequency distribution.

Table 2.1 Frequency distribution—L. Gozman and Sons: Distribution of sales by dollar magnitude, March, 1971

Dollar sales as indicated on sales slip ($)	Number of sales slips
0–2	832
3–5	1507
6–10	631
11–15	238
16–20	74
21 and over	6
Total	3288

2.1 DEFINITIONS

Frequency Distribution

A frequency distribution is a table in which observed values of a variable are grouped or classified according to their numerical magnitude with respect to the trait under consideration.

Variable

The variable is the trait on the basis of which classification is to be made. In Table 2.1, the variable is "Dollar sales as indicated on the sales slip."

Class Frequency

Class frequency is the number of observations for which the value of the variable falls within the confines of a particular class. In Table 2.1, the class frequency for the class $6–$10 is 631.

Total Frequency

The total frequency is the total number of observations to be classified. Clearly, the total frequency must equal the total of the class frequencies. In Table 2.1, the total frequency for the frequency distribution is 3288.

Class Limits Versus Class Boundaries

There is an important technical distinction between class limits and class boundaries. Class limits are the stated class limits, and class boundaries are the real class limits. The lower class boundary is actually the lowest possible value that can be assigned to the given class. The upper class boundary is the highest possible value that can be assigned to the given class. Unfortunately, the stated class limits do not always coincide with the real class limits. Referring to Table 2.1, to which class would the dollar sales figure $10.15 be assigned? $10.15 clearly exceeds the upper class limit of the third class and lies below the lower class limit of the fourth class. In actuality, it is necessary to know the *rule of rounding*, which in turn indicates the real class limits or class boundaries.

EXAMPLE 2.1

Suppose the rule of rounding was

Anything below $0.50, round down.
Anything above $0.50, round up.
Exactly $0.50, round to the even number.

Then:

Stated class limits ($)	Real class limits or class boundaries ($)
6–10	5.50–10.50
11–15	10.51–15.49
16–20	15.50–20.50

EXAMPLE 2.2

Suppose the rule of rounding was to round everything down to the nearest dollar.
Then:

Stated class limits ($)	Real class limits or class boundaries ($)
6–10	6.00–10.99
11–15	11.00–15.99
16–20	16.00–20.99

In Example 2.2, a more effective way of designating the stated class limits would have been:

$6 up to but not including $11.
$11 up to but not including $16.
$16 up to but not including $21.

An effective way to avoid misunderstandings as to where the real class limits (or class boundaries) lie is to use one more significant digit in defining the class limits than is to be found in the actual data under observation.

Another element to be considered in selecting class limits is whether the data are *continuous* or *discrete*. If the data are discrete, only certain numbers at given intervals along the number scale can occur. Discrete data is generated by a counting process. An example of discrete data is a survey of the number of children in a family. The possible values of the variable would be restricted to whole numbers (0, 1, 2, 3, etc.). The *real* class limits in the case of discrete data would be the smallest and largest discrete values falling within a given class. *Continuous* data can take on any value over a range of feasible values. To illustrate, the weights of all males in the population is a continuously distributed variable. No matter what two weights are selected within the range of possible values, another possible value can always be determined halfway between. This subdividing process could be continued indefinitely. Values for a continuous variable are obtained by measurement. Therefore, the number of significant digits one is able to obtain in the original data is a function of the precision of the measuring device. Confusion can be avoided in assigning values of the variable to a particular class by simply defining class limits one significant digit greater than the measuring device used is capable of generating.

Class Interval

The class interval is the difference between the *real* lower limit and the *real* upper limit of a given class (i.e., the spread between the class boundaries).

In Table 2.1, using the rounding alternative referred to as Example 2.1, the class intervals would be as follows:

Class number	Stated class limits ($)	Real class limits or boundaries ($)	Class intervals ($)
I	0–2	0–2.50	2.50
II	3–5	2.51–5.49	2.98
III	6–10	5.50–10.50	5.00
IV	11–15	10.51–15.49	4.98
V	16–20	15.50–20.50	5.00
VI	21 and over	20.51–	—

The last class in Table 2.1 is what is referred to as an *open-end class*. It is impossible to establish a class interval unless the class in question has both an upper and lower limit.

Class Mark or Class Midpoint

A class mark is the value of the variable that lies halfway between the value of the variable assigned to the *real* upper and the *real* lower class limits (i.e., the class boundaries). The value of the midpoint can be computed by either averaging the values of the two class boundaries or by adding one-half the value of the given class interval to the value of the lower class boundary.

2.2 PROCEDURE FOR CONSTRUCTING A FREQUENCY DISTRIBUTION

The set of data in Table 2.2 will be used as a vehicle for discussion in developing a procedural framework for the construction of a frequency distribution. The data represents information culled from the time cards of 112 workers in an assembly plant in Richmond, Virginia. The figures are presented in the order in which they were encountered on successive time cards.

Table 2.2 Daily wage rates for 112 workers in an assembly plant in Richmond, Virginia

$30.00	$15.80	$12.00	$20.90	$20.00
28.00	17.00	14.40	16.50	23.40
20.00	19.50	11.20	20.80	31.00
22.80	18.10	18.20	18.00	32.00
20.80	19.20	22.00	16.80	15.80
18.60	22.00	25.20	26.50	24.00
20.00	12.00	15.20	24.00	25.60
24.00	20.10	12.40	16.90	28.00
24.00	13.20	18.00	22.40	16.00
16.00	20.00	14.00	15.30	28.00
16.00	28.40	18.00	20.20	27.50
11.60	18.40	16.40	18.00	16.00
22.00	20.00	20.00	20.00	16.00
16.00	16.40	10.00	18.40	16.00
25.00	13.00	18.00	19.00	19.60
20.00	27.00	18.00	18.00	16.00
16.00	10.00	20.50	20.00	22.00
19.00	18.80	12.20	23.90	16.00
19.50	15.60	20.00	21.20	24.00
25.80	10.00	21.60	32.00	12.00
23.60	16.00	16.00	13.50	
24.80	14.60	16.80	10.40	
16.00	18.00	16.40	12.00	

Although a quick survey of the data reveals that the lowest daily wage is $10.00 and the highest daily wage is $32.00, it is difficult to obtain a clear picture of the overall pattern of the data. (Around what value of the variable do the observations tend to cluster? To what extent is the data dispersed within the range of values?) Had the data consisted of 1112 observations rather than 112, the attempt to perceive a discernible pattern in the data in its present form would have been an even greater exercise in frustration.

The first stage in the development of a frequency distribution is to rearrange the data in the form of an array. An *array* is an ordering of the values of the variable in order of magnitude, usually from smallest to largest. In constructing the array, all values in the range of possible values should be represented, whether or not they have actually occurred. This enables the analyst to visualize the extent to which the observed values are scattered over the range of possible values. Construction of a frequency distribution is facilitated by bringing the underlying structure of the distribution clearly into view. It will be assumed that the firm to which the example refers has a wage structure such that daily wages are always some multiple of $0.10.

In constructing the array, one should not begin at the lowest observed numerical value and end at the highest. It is preferable to expand the range somewhat in both directions in order to facilitate the development of a feel for the spread that exists in the data. In accordance with this policy, the array to be constructed will extend from a low of $9.00 in successive $0.10 increments to a high of $32.90. Once the array has been constructed, the next step is to tally for each of the listed values the actual number of observations possessing the associated value. The result is depicted in Table 2.3. When presented in this form, the data is referred to as a *tally sheet*.

Table 2.3 Tally sheet: daily wage rates for 112 workers in an assembly plant in Richmond, Virginia

9.00	13.00 I	17.00 I	21.00	25.00 I	29.00
9.10	13.10	17.10	21.10	25.10	29.10
9.20	13.20 I	17.20	21.20 I	25.20 I	29.20
9.30	13.30	17.30	21.30	25.30	29.30
9.40	13.40	17.40	21.40	25.40	29.40
9.50	13.50 I	17.50	21.50	25.50	29.50
9.60	13.60	17.60	21.60 I	25.60 I	29.60
9.70	13.70	17.70	21.70	25.70	29.70
9.80	13.80	17.80	21.80	25.80 I	29.80
9.90	13.90	17.90	21.90	25.90	29.90
10.00 III	14.00 I	18.00 ⦀ III	22.00 IIII	26.00	30.00 I
10.10	14.10	18.10 I	22.10	26.10	30.10
10.20	14.20	18.20 I	22.20	26.20	30.20
10.30	14.30	18.30	22.30	26.30	30.30
10.40 I	14.40 I	18.40 II	22.40 I	26.40	30.40
10.50	14.50	18.50	22.50	26.50 I	30.50
10.60	14.60 I	18.60 I	22.60	26.60	30.60
10.70	14.70	18.70	22.70	26.70	30.70
10.80	14.80	18.80 I	22.80 I	26.80	30.80
10.90	14.90	18.90	22.90	26.90	30.90
11.00	15.00	19.00 II	23.00	27.00 I	31.00 I
11.10	15.10	19.10	23.10	27.10	31.10
11.20 I	15.20 I	19.20 I	23.20	27.20	31.20
11.30	15.30 I	19.30	23.30	27.30	31.30
11.40	15.40	19.40	23.40 I	27.40	31.40
11.50	15.50	19.50 II	23.50	27.50 I	31.50
11.60 I	15.60 I	19.60 I	23.60 I	27.60	31.60
11.70	15.70	19.70	23.70	27.70	31.70
11.80	15.80 II	19.80	23.80	27.80	31.80
11.90	15.90	19.90	23.90 I	27.90	31.90
12.00 IIII	16.00 ⦀⦀ III	20.00 ⦀⦀	24.00 ⦀	28.00 III	32.00 II
12.10	16.10	20.10 I	24.10	28.10	32.10
12.20 I	16.20	20.20 I	24.20	28.20	32.20
12.30	16.30	20.30	24.30	28.30	32.30
12.40 I	16.40 III	20.40	24.40	28.40 I	32.40
12.50	16.50 I	20.50 I	24.50	28.50	32.50
12.60	16.60	20.60	24.60	28.60	32.60
12.70	16.70	20.70	24.70	28.70	32.70
12.80	16.80 II	20.80 II	24.80 I	28.80	32.80
12.90	16.90 I	20.90 I	24.90	28.90	32.90

By presenting the data in a format such as Table 2.3, the analyst is able to obtain a rough approximation of the underlying pattern that exists in the data—the extent to which it tends to cluster as well as the nature of the dispersion in the data.

Suppose that in constructing the tally sheet, allowance was made only for those values of the variable that actually occurred. Column 1 of Table 2.3 would then have appeared as follows:

$10.00 |||
10.40 |
11.20 |
11.60 |
12.00 ||||
12.20 |
12.40 |

If the data had been summarized in the above format, the underlying pattern in the data, particularly the scatter or dispersion in the data, would have been much more difficult to detect. For this reason, when setting up an array or tally sheet, it is important that space be allowed to reflect those values of the variable that did not occur as well as those that did occur.

In practice, it is not necessary and frequently cumbersome to allow for each and every conceivable value. In such a situation, it is useful to assign a number of values to a representative value when constructing the tally sheet. To illustrate, if the assumption that daily wages are always some multiple of $0.10 is released, it can alternatively be assumed that the analyst developed Table 2.3 by pursuing a policy of assigning all wage rates from $14.55 up to but not including $14.65 to the value $14.60; all wage rates from $14.65 up to but not including $14.75 to the value $14.70; and so on. Pregrouping in this manner greatly facilitates the construction of the tally sheet. The margin of error thereby introduced is usually insignificant with respect to the use the frequency distribution is intended to serve.

Construction of the tally sheet normally represents only a preliminary grouping. The next step involves compressing the data into *6 to 25* classes, depending on the nature of the data and the objectives the frequency distribution is to serve.[1] It is important to recognize that there is no such thing as *the* correct frequency distribution for a set of data. There are usually several alternative possible groupings, any one of which would be satisfactory. By the same token, there also exists a category of unsatisfactory frequency distributions. About the most one can do is to become familiar with the characteristics that satisfactory frequency distributions possess (and *understand* why they are desirable characteristics). Then attempt to construct a frequency distribution in a given situation that best meets these criteria.

The following represent important relationships for which to strive when attempting to establish the class intervals and class limits in a frequency distribution:

1. Be careful that the width of the class interval selected is not so wide that it causes the resulting frequency distribution to fail to reflect the basic pattern in the underlying data. It is equally undesirable for the class intervals to be excessively narrow. In the latter case, one would be no better off than when dealing with the tally sheet. The actual size of the intervals—the number of classes combination must be arrived at in each situation on a trial-and-error basis.

---

[1] A useful guide for beginners is to use *Sturges' rule* to generate the approximate number of classes to construct:

$K = 1 + 3.3(\log N)$

where K = the number of classes, and N = the number of observations.

If the frequency distribution is to be depicted in graphic form to facilitate visualization of the underlying pattern of the data, the number of classes will be fewer (say 6–15) than if the function of the frequency distribution is to provide a more convenient framework for the performance of further computations (in which case the number of classes will tend to range from, say, 15 to 25).

2. Unless absolutely necessary to reflect correctly the underlying pattern of the data, classes with zero frequencies should be avoided. The principal reason is the attendant awkwardness in subsequent graphic and computational procedures.

3. If the distribution is to be compared with another, the class intervals in the two distributions should be identical. Although this characteristic is not essential, it is desirable because it greatly facilitates comparison. If the two distributions are not the same, the process of comparison is similar to trying to compare the shooting abilities of two basketball players, one of whom shot at a basket with a diameter of 15 in. and the other at a basket with a diameter of 24 in. In comparing their abilities, one would not only have to compare the number of baskets made but would also have to make allowance simultaneously for the difference in diameter of the baskets at which they are shooting. Similar allowances must be made if the class intervals in two distributions being compared are not identical, thereby making the comparison process more difficult.

4. All class intervals should be equal if possible. There are three reasons why this is desirable:
   a. It facilitates comparison of the frequency of occurrence between classes by eliminating the necessity of allowing for differences in class interval size.
   b. If the frequency distribution is to be presented graphically, the process of construction and interpretation will be greatly simplified. This will be demonstrated in section 2.4.
   c. If the frequency distribution is to serve as a framework for further computations, short-cut formulas are available for distributions that have equal class intervals.

5. Class midpoints should be established, if possible, so that the midpoint is approximately equal to the arithmetic average of the items in the class. If the frequency distribution is to be used for further computations, this characteristic is extremely important. Once the individual items have been placed in a frequency distribution, they lose their individual identity. For example, the observer would know that there are ten items in a certain class but would be unable to determine the individual values of the variable for the ten items. For purposes of further computation, the ten items would have to be assigned a representative value. The value assigned is the midpoint of the class. If the class midpoint is to be truly representative of the items in the class, then its value should closely coincide with the arithmetic mean of those values. In constructing the classes of the frequency distribution, every effort should be made to cause this assumption to conform to reality. Otherwise, computations based on these assumptions will be distorted. That this is indeed the case will be illustrated in the discussion of measures of central tendency (Chapter 3).

6. If possible, class limits should be established so that the class marks or midpoints fall on whole numbers. This is desirable simply because computations with whole numbers are simpler and less subject to error.

7. If at all possible, open-end distributions should be avoided. Table 2.1 is an example of an open-end distribution. An open-end distribution is a distribution

that contains an open-end class such as "under $20.00" or "over $150.00." It is desirable to avoid classes of this type because the analyst often will be unable to determine a midpoint for such a class—a prerequisite for certain computations. This problem can be overcome, however, if the actual values of the individual items in the open-end class are known. The arithmetic average of the items in that class can then be computed and used as the representative value in further computations.

Keeping the above criteria in mind, let us attempt to develop a frequency distribution from the data base presented in Table 2.2. Recall that in a preliminary stage of the analysis, the data was restructured in the form of a tally sheet (Table 2.3). This was to provide a format in which the basic pattern in the data could be more readily discerned. One possibility for a frequency distribution for the data in Table 2.3 is six classes, each with a class interval of $4, as presented in Table 2.4.

Table 2.4 Daily wages

$9 up to but not including $13
$13 up to but not including $17
$17 up to but not including $21
$21 up to but not including $25
$25 up to but not including $29
$29 up to but not including $33

Examination of the tally sheet presented in Table 2.3 discloses that each of the classes in Table 2.4 corresponds to one of the columns in the tally sheet. In this example, the objective of equal class intervals has been achieved. A further question to be raised, however, is how this class breakdown has fared with respect to meeting the criterion that the midpoint of a class should fall at about the arithmetic average of the items in the class if the midpoint is to function effectively as a representative value. Strange as it may seem, a principle of physics can be drawn upon as an aid in evaluating how effectively the above criterion has been met. Think of the range between the lower class limit and the upper class limit as a bar and each of the tallies as weights placed upon that bar (Figure 2.1).

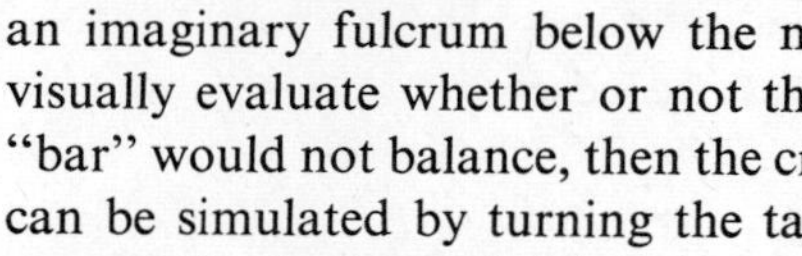

lower
class
limit
      fulcrum
           upper
           class
           limit

Figure 2.1

In order for the bar to balance, the fulcrum must be placed directly below the arithmetic average of the values represented by the tally marks. Thus, by placing an imaginary fulcrum below the midpoint of a proposed class, the analyst can visually evaluate whether or not the "bar" would balance. If it appears that the "bar" would not balance, then the criterion has been violated. The above procedure can be simulated by turning the tally sheet (Table 2.3) so that the page number appears in the lower left-hand corner. A fulcrum could then be placed below the midpoints of each of the "bars," that is, below $11.00, $15.00, $19.00, $23.00, $27.00, and $31.00. Recall that each column constituted a proposed class. A visual evaluation could then be made as to whether or not the bar representing each of the classes would balance.

In Figure 2.2, a fulcrum placed below $11.00 appears to come close to balancing the first "bar" or class. In actuality, the true arithmetic average is $11.30, so that some understatement does take place when the midpoint of $11.00 is used as the representative value for the class. The classes $17–$21, $21–$25, $25–$29, and $29–$33 also appear as if they would come close to balancing if a fulcrum were to be placed below their respective midpoints. The only serious misrepresentation arises with respect to the class $13–$17.

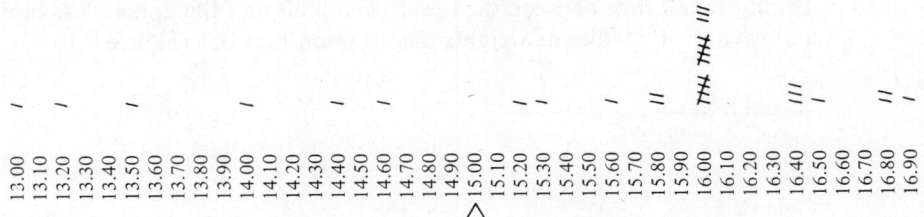

Figure 2.2

As can be seen in Figure 2.3, a fulcrum placed below the midpoint of $15.00 would not come close to balancing the "bar" representing the second class. The fulcrum would have to be placed below the value $15.63 in order for the bar to balance. Thus, use of $15.00 as the representative value for the 31 items in the class results in a considerable understatement of the total value of the items in the class "$13 up to but not including $17." The projected total value of the 31 items, using the midpoint as the representative value, would be 31 times $15.00 or $465, whereas the true total value of the items in the class is $484.60. This points out that the impact on further computations of failing to meet this criterion is a function not only of the difference between the midpoint and the true arithmetic average of the items in the class but also of the number of items represented by the non-representative midpoint. The greater the number of items in the class, the more the error is compounded. Although there clearly are some undesirable features associated with the above representation, it nevertheless represents one of the alternatives available, and as such will be referred to as Frequency Distribution Construction Alternative A (Table 2.5).

Figure 2.3

Table 2.5 Frequency Distribution Construction Alternative A

Daily wages	Class frequency
$9 up to but not including $13	12
$13 up to but not including $17	31
$17 up to but not including $21	37
$21 up to but not including $25	17
$25 up to but not including $29	11
$29 up to but not including $33	4
	112

The class frequencies are obtained by simply counting the number of items that possess a value falling within the limits of each class on the tally sheet and then recording the total number of such values for each class in the distribution.

Another alternative frequency distribution construction from the basic data as presented in Table 2.3 is 12 classes, each with a class interval of $2 as depicted in Table 2.6.

Table 2.6

Daily wages ($)

9–11	The dash is to be interpreted as
11–13	"up to but not including." The
13–15	fact that $11 appears as both the
15–17	upper limit of the first class and
17–19	the lower limit of the second class
19–21	is interpreted as indicating that
21–23	everything up to $11.00 falls in
23–25	the first class but that $11.00 itself
25–27	falls in the second class, etc.
27–29	
29–31	
31–33	

Referring again to Table 2.3, it can be seen that each of the classes in Table 2.6 corresponds to either the first half or the second half of a column on the tally sheet. Verification of the extent to which the midpoint falls at or about the arithmetic average of the items can be obtained by placing an imaginary fulcrum below the midpoint of the class and observing the tendency or lack of tendency for the "bar" representing the class interval to balance [Figures 2.4(a) and 2.4(b)].

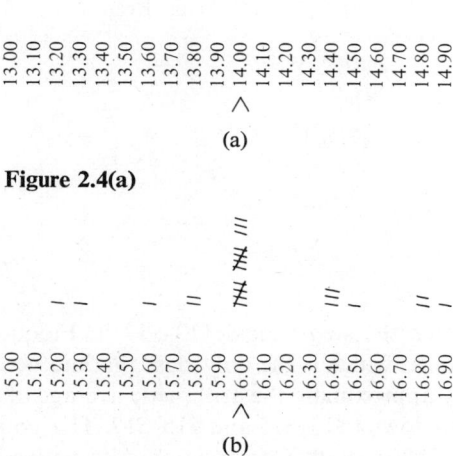

(a)

Figure 2.4(a)

(b)

Figure 2.4(b)

Recall that under Frequency Distribution Construction Alternative A, the midpoint of the class representing the entire second column (i.e., $13.00–$17.00) violates the criterion significantly (the midpoint, $15.00 is considerably below the arithmetic average of the items in the class, $15.63). However, when the second column is broken down into two separate classes, as reflected in Figures 2.4(a) and 2.4(b), the criterion appears to be effectively met. The only class that appears to be poorly represented of the 12 classes in the second alternative frequency distribution construction is the class $17.00–$19.00. Even in this case the difference is not significant, since the true arithmetic average is only $18.10 (see Figure 2.5).

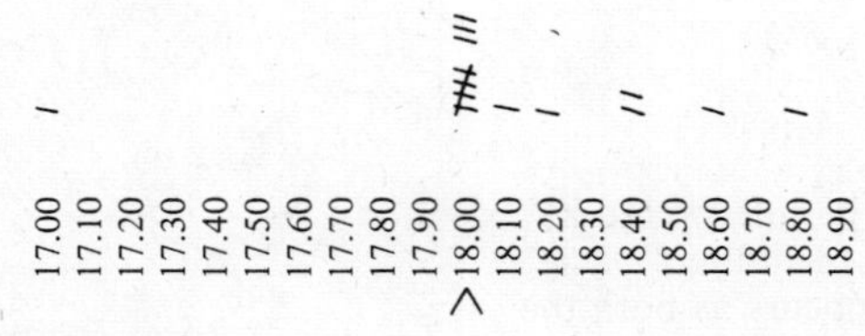

Figure 2.5

For reference purposes, the second frequency distribution will be referred to as Frequency Distribution Construction Alternative B (Table 2.7).

Table 2.7 Frequency Distribution Construction Alternative B

Daily wages ($)	Frequency
9–11	4
11–13	8
13–15	6
15–17	25
17–19	15
19–21	22
21–23	8
23–25	9
25–27	5
27–29	6
29–31	1
31–33	3
	112

A third possibility is to replace the single class $13–$17 in Frequency Distribution Construction Alternative A, which is the only class that did not meet the criterion that the midpoint closely approximate the arithmetic average of the items in the class, by the two class breakdowns $13–$15 and $15–$17. This combination will be referred to as Frequency Distribution Construction Alternative C (Table 2.8).

Later in the discussion of measures of central tendency (Chapter 3), the computations will be performed with all three frequency distribution alternatives. The difference in results will be observed, and the differences traced back to the varying degree of effectiveness with which they met the criterion that the midpoint of a class should fall at about the arithmetic average of the items in that class.

Table 2.8 Frequency Distribution Construction Alternative C

Daily wages ($)	Frequency
9–13	12
13–15	6
15–17	25
17–21	37
21–25	17
25–29	11
29–33	4
	112

2.3 PERCENTAGE FREQUENCY DISTRIBUTIONS

On many occasions the analyst is interested in determining the percentage of the total number of items falling into each of the class categories rather than the absolute frequency of occurrence. It is a simple matter to make such a transformation. Simply express each of the class frequencies as a percentage of the total frequency for the distribution in question. To illustrate, Frequency Distribution Construction Alternative A will be converted into this format (Table 2.9). When

Table 2.9 Computation of the class percentages: Frequency Distribution Construction Alternative A

Daily wages ($)	Class frequency	Class percentage
9–13	12	12/112 = 11
13–17	31	31/112 = 28
17–21	37	37/112 = 33
21–25	17	17/112 = 15
25–29	11	11/112 = 10
29–33	4	4/112 = 3
	112	100

expressed in this form, the data is referred to as a *percentage* distribution (Table 2.10).

Table 2.10 Percentage distribution: Frequency Distribution Construction Alternative A

Daily wages ($)	Class percentage
9–13	11
13–17	28
17–21	33
21–25	15
25–29	10
29–33	3
	100

One advantage of presenting the frequency distribution in percentage form is that it facilitates evaluating the relative importance of each of the classes, because the analyst relates to a familiar 100 percent base instead of the total frequency of 112. It is more awkward to compare 37 to 112 than to compare 33 percent to 100 percent, a framework of comparison with which one is used to dealing.[2]

Percentage frequency distributions are particularly useful where comparison is to be made between two different frequency distributions that are similar with respect to class breakdowns but differ with respect to the total frequency allocated among their respective classes. For instance, when one compares the two distributions in the form shown in Table 2.11, one must not only compare the two fre-

Table 2.11

Richmond		Chicago	
Daily wages ($)	Class frequency	Daily wages ($)	Class frequency
9–13	12	9–13	20
13–17	31	13–17	40
17–21	37	17–21	85
21–25	17	21–25	60
25–29	11	25–29	35
29–33	4	29–33	10
	112		250

quencies for the class $17–$21 (i.e., 37 versus 85), but in addition one must allow for the fact that 37 occurred out of a total possible of 112 while the 85 occurred out of a total possible of 250. Such comparisons are awkward.

The data in Table 2.11, reexpressed in percentage form, are reproduced as Table 2.12. The simplification in comparison is self-evident. The 33 percent can

Table 2.12

Richmond		Chicago	
Daily wages ($)	Class frequency (%)	Daily wages ($)	Class frequency (%)
9–13	11	9–13	8
13–17	28	13–17	16
17–21	33	17–21	34
21–25	15	21–25	24
25–29	10	25–29	14
29–33	3	29–33	4
	100		100

be compared directly to the 34 percent for the class $17–$21 without having to allow for a difference in base. By converting both classes to percentage form, they have been restructured with a common base of 100 percent, thereby eliminating the need to take the base into consideration.

[2] The disadvantage of percentage distributions is that they preclude the use of chi-square calculations to determine the significance of the differences that they highlight (see Chapter 19).

2.4 GRAPHIC REPRESENTATIONS OF FREQUENCY DISTRIBUTIONS

When the objective is to present the data in a format designed to facilitate the visualization of the underlying pattern that exists in the data, a graphic representation of the frequency distribution is appropriate. There are two alternative graphical constructions from which to choose. One is a bar-chart representation of the frequency distribution called a *histogram*. The other is a line-chart representation called a *frequency polygon*.

The Histogram

A histogram is defined more formally as a bar chart of a frequency distribution in which the relative area assigned to the bars representing each of the classes is proportional to the relative size of the class frequencies. This is an extremely important relationship since the human eye makes visual comparisons on the basis of area not height.

Figure 2.6

In Figure 2.6, the block to the left represents the number of A grades in section 1 and the block to the right represents the number of A's in section 2. If an individual were asked to indicate the ratio of the number of A's in section 2 to the number of A's in section 1, he would undoubtedly answer approximately 4 to 1. He would be comparing the area of the two blocks, not the height. The area of the block to the right is four times the area of the block to the left, although the linear dimensions are only in a ratio of 2 to 1. It is because of this natural tendency to compare areas, not height above the base line in visually assigning relative importance, that it becomes so important that the areas assigned to each of the classes be in the same proportion as their respective class frequencies. Otherwise a visual distortion will result, leading to a misinterpretation of the data presented in the histogram. The proper visual representation is easily attained for frequency distributions with equal class intervals by simply locating the height of the bar representing a class a distance above the base line equal to its class frequency. To illustrate, the histogram for Frequency Distribution Construction Alternative A, which has equal class intervals, is developed in Figure 2.8. Frequency of occurrence is measured on the vertical axis and values of the variable are designated on the horizontal axis. Demarcations on the horizontal axis serve to indicate the range of values served by each of the classes.

The bars representing the frequency of occurrence for each class are obtained by raising a perpendicular at the lower and upper limits of each class up to the value of the class frequency and connecting the tops of the two lines to close off the bar (Figure 2.8).

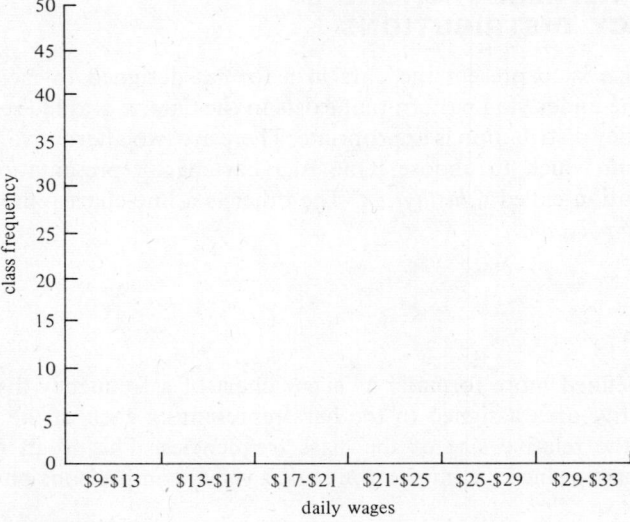

Figure 2.7

The same procedure is followed in graphing a percentage frequency distribution, except that the height of the bar is determined by the percentage falling in a particular class. Graphically, the distribution will look the same, except that the percentage vertical axis at the right of the diagram is used instead of the absolute value axis depicted on the vertical axis to the left.

It is a simple matter to demonstrate that the relative areas assigned to each of the bars is in direct proportion to the class frequency for the classes the bars represent. The desired relationship can be shown as follows:

$$\frac{\text{area (class 1)}}{\text{area (class 2)}} = \frac{\text{frequency (class 1)}}{\text{frequency (class 2)}}$$

The area for any of the bars is equal to the height times the base of that bar. The height is equal to the class frequency, and the base is equal to the class interval,

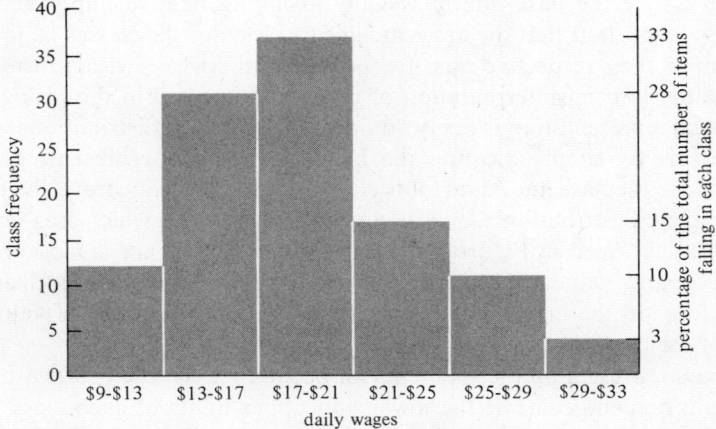

Figure 2.8 Histogram: Frequency Distribution Construction Alternative A.

which is identical for all classes if the distribution has equal class intervals. Thus:

$$\frac{\text{area (class 1)}}{\text{area (class 2)}} = \frac{(f_1)\ \text{(base 1)}}{(f_2)\ \text{(base 2)}} = \frac{f_1}{f_2}$$

Base 1 and base 2 must cancel, because they are identical with equal class intervals. Thus, quite clearly the areas assigned to each of the bars is directly proportional to the relative class frequency of those bars.

Frequency Distribution Construction Alternative B also has equal class intervals, and would be constructed in a similar fashion (see Figure 2.9). Note that the labels on the horizontal axis in Figure 2.9 are the class midpoints. In the histogram for Frequency Distribution A (Figure 2.8), the class limits were used as the designations. Either technique may be used to provide reference to the values of the variable at various benchmark points on the horizontal axis.

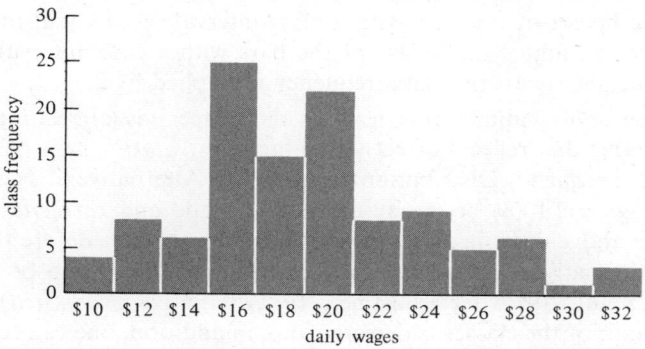

Figure 2.9 Histogram: Frequency Distribution Construction Alternative B.

By comparing the histograms of Alternative A and Alternative B, one can observe how the effect of the clustering of the values in the distribution in the area \$15–\$17 (midpoint \$16) is lost when that interval is combined with the interval \$13–\$15 into the single class interval \$13–\$17 in Frequency Distribution Alternative A. Graphically, therefore, Frequency Distribution Alternative B is preferable to Frequency Distribution Alternative A, because it captures an important pattern in the data—one that is not reflected in Frequency Distribution Alternative A. Frequency Distribution Alternative A actually indicates the data peaking to the right (around \$19) of where it actually does peak (around \$16).

Unequal Class Intervals

When the objective is to graph Frequency Distribution Alternative C, a new problem arises—unequal class intervals. There are five classes with an interval of \$4 and two classes with an interval of \$2. If the bars for the histogram for Frequency Distribution Construction Alternative C were constructed by the same procedure used for frequency distributions with equal class intervals, the resulting histogram would no longer reflect the true relative importance of the frequency of occurrence by the areas of the bars assigned to the classes.

$$\frac{\text{area of the class \$17-\$21}}{\text{area of the class \$15-\$17}} = \frac{\text{(class frequency)(base or class interval)}}{\text{(class frequency)(base or class interval)}}$$

$$= \frac{(37)(\not{4})}{(25)(\not{2})} = \frac{(37)(2)}{25} = \frac{74}{25}$$

The ratio of the areas of the two bars would be 74/25, while the ratio of the frequencies of the two classes is only 37/25. Clearly, an adjustment is necessary if the histogram is to reflect the true relationship visually. The reason for the problem, of course, is that the bases of the bars are no longer equal. The area of the bars cannot be adjusted by varying the base, because the class interval is given. Thus, the procedure for adjustment must involve the other dimension—namely, the height of the bar. There are two alternative adjustment procedures available:

1. Leave the height of the bars with a class interval of $2 equal to the class frequency and adjust the height of the bars with a class interval of $4 so that the height equals the class frequency divided by 2.
2. Leave the height of the bars with a class interval of $4 equal to the class frequency and adjust the height of the bars with a class interval of $2 so that the height equals the class frequency multiplied by 2.

Either of the above adjustments restores the proper visual relationship and provides histograms that reflect correct visual images.

To illustrate, Frequency Distribution Construction Alternative C is constructed *incorrectly* in Figure 2.10(a) before adjustment is made and *correctly* in Figure 2.10(b) after the above-described adjustments have been made. Before the adjustment (i.e., in Alternative C *unadjusted*) the distribution appears to be clustering around the midpoint $19. After adjustment (in Alternative C *adjusted*), the true relative importance of the classes is depicted and, in addition, one can see that the greatest clustering occurs around $16, the midpoint of the class with the highest bar. If the histogram is constructed correctly, the class with the *highest* bar is the class with the greatest *density* of occurrence. The greatest density of occurrence signifies the greatest clustering.

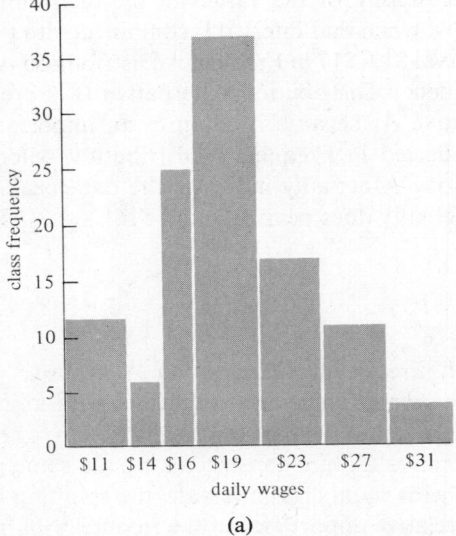

(a)

Figure 2.10(a) Histogram: Alternative C unadjusted. An *incorrect* representation.

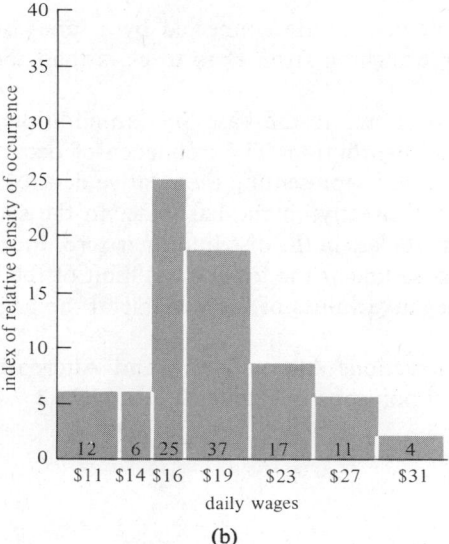

Figure 2.10(b) Histogram: Alternative C adjusted. A *correct* representation.

It is important to note that one can no longer determine the frequency of occurrence in a class by reading off the value on the vertical axis that corresponds to the height of the bar. The vertical axis no longer represents class frequency. It now refers to the value of the density function associated with each of the classes (an index of the relative density of occurrence with reference to some standard interval). It will equal class frequency for some classes but not for others. Because of this, the vertical axis, once plotting has taken place, is often eliminated, and instead the class frequency is indicated by inserting that value in the appropriate bar of the histogram as illustrated in Alternative C adjusted. The problems created in graphically depicting and interpreting histograms of frequency distributions with unequal class intervals provide a strong argument for using equal class intervals whenever the data and the other constraints enable one to do so.

The Frequency Polygon

The frequency distribution can also be presented graphically in the form of a line chart called a frequency polygon. Once again, relative density of occurrence is measured on the vertical axis and values of the variable are designated on the horizontal axis. As before, demarcations on the horizontal axis serve to indicate the range of values served by each of the classes.

The significant difference with respect to construction of the frequency polygon is that the frequency of occurrence in a class is represented by a single point rather than by a bar. The point is plotted directly over the class midpoint. This is consistent with the role of the midpoint as the representative value for the items in the class. If the frequency distribution has equal class intervals, the point is plotted a distance above the base line equal to the class frequency. If the distribution contains unequal class intervals, an adjustment similar to that required in the construction of the histogram must be introduced (for the same reason). After adjustment, the height of the point would then be located a distance above the base line equal to the value of the density function associated with that particular class. After all the points are plotted, they are connected by a series of straight lines, which provides a smoother transition from class to class than the abrupt, steplike jumps of the histogram.

To bring the frequency polygon down to the base line, an additional class is usually added at either end of the distribution. The frequency of occurrence for these classes is zero, and thus the point representing the relative density of occurrence for the classes will be plotted directly on the base line. In the special case where the lower class limit of the first class in the distribution is zero, the frequency polygon is brought down to the base line at the lower class limit of the first class rather than adding a class with negative values of the variable at the beginning of the distribution.

Frequency Distribution Constructions Alternative A and Alternative C are presented in the form of frequency polygons in Figures 2.11 and 2.12.

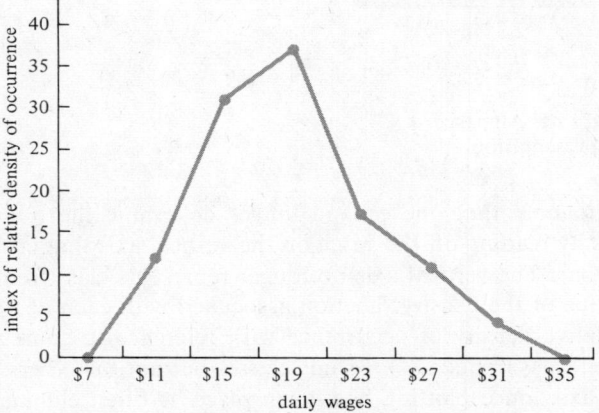

Figure 2.11 Frequency Polygon: Frequency Distribution Construction Alternative A.

The histogram is ordinarily preferred to the frequency polygon when the objective is to emphasize the differences between the frequency of occurrence of each of the classes. The frequency polygon is favored when the objective is to depict visually the basic sweep or underlying pattern of the data.

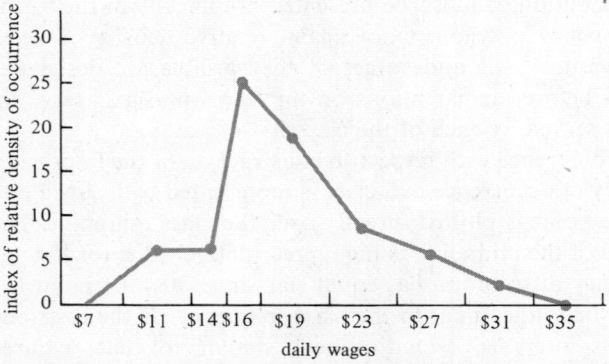

Figure 2.12 Frequency Polygon: Frequency Distribution Construction Alternative C.

An advantage of the histogram over the frequency polygon is that the area under the graphic representation for each class is directly proportional to the number of items falling in that particular class. This relationship is violated somewhat in the case of the frequency polygon because of the smoothing influence of the gradual transition from one class to the next. For example, if the areas of bar A and bar B in Figure 2.13 were in proportion to their relative class frequencies, the relationship would no longer hold when the data was presented in the form of a frequency polygon. The area assigned to class A would remain about the same (solid area lost is approximately equal to the striped area gained), while the area assigned to class B would be significantly reduced (the two solid areas would be taken away with no compensating areas added).

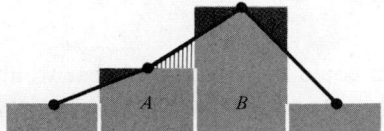

Figure 2.13

Another advantage of the histogram is that with discrete data a gap may be left between the bars in the histogram, thereby reinforcing visually the discrete nature of the data as well as emphasizing the differences between the classes.

The frequency polygon is the preferable graphic representation when two or more frequency distributions are to be compared by simultaneous presentation on the same graph. Presentation of two or more histograms on the same graph is an heroic undertaking except in the instance where the height of the bars in one histogram exceeds the height of the bars in the other for all classes.

2.5 CUMULATIVE AND DECUMULATIVE FREQUENCY DISTRIBUTIONS

Sometimes the analyst is interested, not in the number of items falling in each class, but rather in how many of the items in the distribution have a value equal to or greater than certain benchmark values (or how many have a value less than certain benchmark values). The data in the basic frequency distribution can easily be restructured to provide this information. Let us first direct our attention to the "more than" or decumulative frequency distribution. The series of questions asked in this case is how many of the items in the distribution have a value greater than or equal to the value of the *lower* limit of the first class; greater than or equal to the value of the *lower* limit of the second class; greater than or equal to the value of the *lower* limit of the third class, and so on. Using Frequency Distribution Construction Alternative A, a "more than" or decumulative frequency distribution will be constructed. Clearly, all 112 of the items have a value greater than or equal to the lower class limit of the first class. Similarly, all but the 12 items in the first class (or 100 of the items) have a value greater than or equal to the lower class limit of the second class. All but the (12 + 31) = 43 items in the first and second classes (or 69 of the items) have a value greater than or equal to the lower class limit of the third class, and so on (see Table 2.13). The term decumulative is used to describe the "more than" frequency distribution because movement through the distribution is accompanied by a decumulation in frequency.

Table 2.13 Decumulative frequency distribution ("more than" frequency distribution) based on Frequency Distribution Construction Alternative A

Daily wages	Number of employees earning more than the indicated daily wage (decumulative frequency)
Greater than or equal to $9	112
Greater than or equal to $13	100
Greater than or equal to $17	69
Greater than or equal to $21	32
Greater than or equal to $25	15
Greater than or equal to $29	4

In a similar fashion, a distribution could be constructed that would indicate the number of items in the frequency distribution having a value less than the various upper class limits. A distribution of this type is referred to as a "less than" or cumulative frequency distribution. The series of questions that would be asked is how many of the items in the distribution have a value less than the upper class limit of the first class; less than the upper class limit of the second class, and so on. Clearly, only 12 items in the first class have a value less than the upper class limit of the first class; only the 43 items in the first and second classes have a value less than the upper class limit of the second class, and so on (see Table 2.14).

Table 2.14 Cumulative frequency distribution ("less than" frequency distribution) based on Frequency Distribution Construction Alternative A

Daily wages	Number of employees earning an amount less than the indicated daily wage (cumulative frequency)
Less than $13	12
Less than $17	43
Less than $21	80
Less than $25	97
Less than $29	108
Less than $33	112

Decumulative and cumulative frequency distributions can also be expressed in percentage form by dividing the decumulative frequencies, or the cumulative frequencies as the case may be, by the total number of frequencies in the distribution in question.

Ogive

Cumulative and decumulative frequency distributions may also be presented graphically in the form of a line chart. When presented in this format, the graphical representations are referred to as a cumulative ogive and a decumulative ogive, respectively.

Cumulative Ogive

A cumulative ogive is drawn by plotting a point representing the cumulative frequency (or the cumulative percentage) above each of the *upper* class limits referred to in the frequency distribution being described. The ogive for the cumulative frequency distribution of Table 2.14 is illustrated in Figure 2.14.

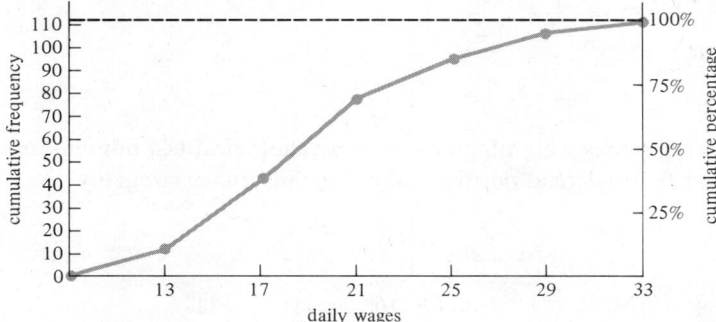

Figure 2.14 Cumulative ogive.

Decumulative Ogive

A decumulative ogive is drawn by plotting a point representing the decumulative frequency (or the decumulative percentage) above each of the *lower* class limits. The ogive for the decumulative frequency distribution in Table 2.13 is illustrated in Figure 2.15.

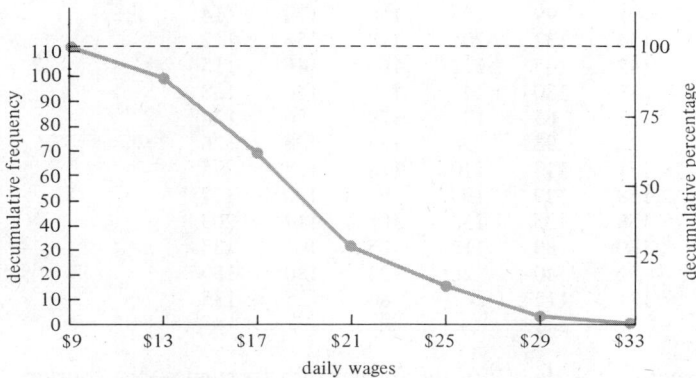

Figure 2.15 Decumulative ogive.

If the decumulative ogive and the cumulative ogive were plotted on the same graph, they would intersect above the value on the horizontal axis that is exactly halfway through the distribution; i.e., 50 percent of the items would have a value greater than that value and 50 percent of the items in the distribution would have a value less than that value. This value is referred to as the *median*. The greater the concentration of the values, the steeper the slope of the ogive. The class with the greatest concentration of values can be located by determining for which class the slope of the ogive is steepest. It should be noted that the construction and interpretation of ogives is in no way affected by equality or nonequality of the class intervals in the underlying distribution. No adjustment is necessary for distributions with unequal class intervals.

Each of the frequency distribution structures discussed has merit. The analyst must decide what his objective is and which of the available formats discussed will most effectively enable him to achieve that goal.

PROBLEMS

Problem 2.1

The following scores were obtained on a psychological test administered to some 200 applicants for clerical positions at a large insurance company.

			Test scores				
148	150	106	172	98	106	117	142
188	125	87	163	152	166	115	145
136	167	70	106	92	151	98	132
102	132	123	143	160	149	76	96
101	172	139	128	164	110	195	150
169	72	184	127	172	121	98	138
113	154	69	41	118	123	98	94
183	101	128	169	118	135	129	100
79	113	105	148	115	202	113	103
144	122	134	145	129	115	83	211
105	144	106	67	171	123	187	122
116	141	120	115	56	99	80	42
138	59	112	169	135	142	183	82
158	115	191	99	86	150	179	114
107	160	154	132	204	118	156	112
135	162	142	105	173	164	146	115
107	102	89	130	143	140	150	138
86	83	123	65	136	179	86	124
154	138	122	98	98	138	108	126
113	131	171	113	110	128	118	95
158	123	158	129	199	78	198	127
103	182	158	135	130	103	114	203
156	106	160	89	113	105	101	133
94	122	52	140	126	123	180	150
154	219	131	112	207	86	135	135

1. Restructure the data into the format of a frequency distribution.
2. Construct a percentage frequency distribution, a cumulative frequency distribution, and a decumulative frequency distribution.
3. Graph as a histogram and as a frequency polygon.
4. Construct cumulative and decumulative ogives.

Problem 2.2

The coach of a major league baseball team was extremely interested in a statistical study being conducted on the batting averages of his players (including farm teams) during spring training. It was suggested to him that this study could be used to show upper management that his decision to increase batting practice time was correct, contrary to the belief of some that it was not.

Batting averages for 100 players prior to increase in practice time				Batting averages for 100 players after increase in practice time			
0.240	0.330	0.430	0.130	0.250	0.380	0.250	0.240
0.450	0.290	0.150	0.260	0.360	0.440	0.340	0.380
0.550	0.240	0.250	0.130	0.330	0.260	0.370	0.320
0.330	0.440	0.230	0.250	0.440	0.380	0.350	0.440
0.240	0.250	0.290	0.150	0.380	0.370	0.450	0.350
0.450	0.450	0.130	0.210	0.250	0.330	0.370	0.370
0.260	0.550	0.230	0.170	0.360	0.440	0.230	0.140
0.550	0.260	0.260	0.250	0.350	0.350	0.330	0.320
0.450	0.150	0.290	0.210	0.460	0.260	0.320	0.250
0.150	0.350	0.130	0.330	0.260	0.450	0.340	0.330
0.560	0.250	0.350	0.150	0.360	0.350	0.150	0.240
0.330	0.450	0.290	0.370	0.250	0.320	0.370	0.270
0.250	0.130	0.230	0.250	0.350	0.360	0.340	0.350
0.350	0.330	0.370	0.350	0.240	0.350	0.230	0.250
0.150	0.250	0.170	0.290	0.360	0.150	0.270	0.350
0.370	0.160	0.450	0.250	0.320	0.240	0.250	0.340
0.240	0.370	0.210	0.140	0.250	0.460	0.350	0.350
0.150	0.210	0.350	0.350	0.150	0.350	0.240	0.240
0.350	0.350	0.140	0.160	0.350	0.550	0.350	0.270
0.210	0.270	0.370	0.260	0.560	0.250	0.160	0.230
0.170	0.140	0.250	0.150	0.260	0.350	0.350	0.540
0.540	0.350	0.350	0.230	0.460	0.460	0.250	0.330
0.170	0.170	0.230	0.250	0.350	0.230	0.380	0.550
0.470	0.270	0.270	0.150	0.550	0.350	0.350	0.340
0.270	0.150	0.460	0.270	0.350	0.260	0.260	0.270

1. Restructure each of the two sets of data into the format of a frequency distribution.
2. Construct a percentage frequency distribution, a cumulative frequency distribution, and a decumulative frequency distribution for each case.
3. Graph each as a histogram and as a frequency polygon.
4. Construct cumulative and decumulative ogives.

Problem 2.3

Dollar sales per order for a small five-and-ten-cent store are presented in the tabular format below.

				Dollar sales per order						
1.67	8.17	20.08	3.78	11.34	7.72	1.72	26.33	2.87	8.01	3.35
2.24	6.89	1.77	4.64	4.35	2.36	1.71	6.12	4.64	26.50	22.36
1.67	10.07	4.89	6.33	6.16	4.19	24.14	1.94	3.56	4.97	18.93
1.82	9.71	1.57	13.76	2.06	7.18	9.93	2.57	2.78	8.48	10.33
10.34	5.30	7.05	8.57	2.63	8.81	2.81	5.34	3.03	9.99	26.71
6.04	5.52	4.70	6.62	1.96	7.71	5.46	23.84	1.81	8.28	11.00
5.54	7.79	9.92	8.37	2.26	6.84	1.25	1.69	4.91	21.65	1.61
3.49	22.93	3.41	6.92	5.73	17.44	4.56	13.52	7.18	1.82	1.32
11.92	10.65	4.29	1.62	4.81	14.33	6.92	7.05	14.39	1.92	26.72
19.68	6.42	15.82	16.56	1.86	1.97	2.12	4.91	9.14	13.62	23.84
8.43	5.84	19.28	9.43	1.96	12.14	1.72	2.57	29.52	9.56	11.40
1.75	4.83	1.23	7.61	7.82	1.11	3.92	20.49	7.91	8.22	18.37
2.51	8.66	1.70	4.08	5.29	7.81	4.58	6.48	21.34	21.22	4.83
9.11	8.60	5.22	20.36	2.73	7.31	2.71	24.42	9.06	9.81	9.86
6.81	10.55	26.95	25.83	13.43	6.81	21.21	20.19	6.01	7.12	1.37
3.11	5.65	20.63	19.10	25.54	2.21	13.80	6.96	3.77	1.27	13.44
9.39	14.20	12.81	11.38	14.31	3.15	9.73	2.67	4.70	5.34	1.23
23.14	11.43	2.10	19.99	12.38	1.36	3.04	2.21	7.54	7.83	16.36
19.58	2.74	3.64	2.29	11.25	28.82	2.53	15.47	5.96	2.34	9.25
6.88	7.27	4.46	4.87	9.07	7.91	8.46	4.24	25.77	23.76	4.61
18.69	12.17	20.52	1.94	1.97	1.83	13.52	9.67	2.58	2.28	3.49
5.90	22.57	4.29	1.42	28.11	20.97	1.89	5.09	8.74	5.72	15.67
17.73	7.11	1.06	26.99	8.47	15.32	2.78	14.31	9.70	8.46	7.33
3.38	12.69	25.12	1.42	29.41	3.80	4.67	13.53	14.05	8.04	5.81
1.57	9.41	9.27	7.23	1.52	6.81	6.87	6.05	17.83	8.87	2.04
1.98	1.99	4.68	4.42	5.69	6.72	6.68	8.04	7.56	14.70	8.15
5.91	6.64	22.35	21.40	3.56	2.11	3.08	6.97	10.61	27.14	1.76
20.30	9.32	5.13								

1. Restructure the data into the format of a frequency distribution.
2. Construct a percentage frequency distribution, a cumulative frequency distribution, and a decumulative frequency distribution.
3. Graph as a histogram and as a frequency polygon.
4. Construct cumulative and decumulative ogives.

Problem 2.4

The production manager of a food-processing plant wishes to determine the effectiveness of a costly retraining program which is to be started in a few weeks in order to try to reduce the number of batches of food rejected daily due to cooking errors. An initial study has been conducted with respect to the number of batches rejected daily prior to the retraining program. A similar study will be conducted after the retraining program. The two studies will then be compared to determine if any significant improvement occurred.

Number of batches rejected daily prior to retraining program

1	34	31	34	23	21	41	40
34	13	36	13	39	30	39	41
14	24	6	44	3	69	23	15
32	1	39	24	41	15	43	36
13	23	10	2	30	33	24	21
16	36	44	33	13	43	21	33
27	2	16	21	44	18	44	4
12	31	55	18	22	62	25	35
37	14	4	42	12	4	14	24
26	33	22	28	27	55	36	18
5	19	15	11	14	27	55	
36	29	43	37	43	61	29	
27	5	38	54	12	19	61	
10	39	14	21	65	36	38	
32	31	44	53	35	65	18	
40	17	28	15	52	45	55	
14	42	45	62	25	6	45	
54	32	13	27	51	58	59	
53	49	54	63	50	37	29	
31	9	26	7	35	29	46	
56	64	35	64	25	49	21	
36	26	8	21	49	47	57	
32	56	67	56	47	21	49	
66	21	35	46	68	46	50	
37	65	66	37	46	57	47	

1. Restructure the data into the format of a frequency distribution.
2. Construct a percentage frequency distribution, a cumulative frequency distribution, and a decumulative frequency distribution.
3. Graph as a histogram and as a frequency polygon.
4. Construct cumulative and decumulative ogives.

Problem 2.5

Data gathered on the number of batches rejected daily after completion of an extensive retraining program is recorded below (see Problem 2.4).

Number of batches rejected daily after
completion of retraining program

13	43	41	34	17	34	39
48	22	13	49	39	8	13
9	12	42	15	11	35	25
35	35	3	26	35	15	17
28	26	28	2	12	25	7
15	8	19	28	28	34	
24	27	15	14	34	13	
23	17	27	24	4	36	
4	15	2	12	32	19	
26	38	33	32	21	7	
15	25	24	22	31	23	
35	2	38	9	5	36	
9	33	12	18	38	25	
33	18	25	27	29	21	
21	35	24	5	36	17	
47	12	14	29	16	45	
17	34	29	25	37	27	
43	6	21	44	25	21	
23	23	37	18	47	25	
5	43	29	45	27	23	
26	14	9	23	6	34	
16	33	23	34	34	16	
28	9	36	16	36	45	
26	37	15	33	25	25	
14	29	45	21	48	23	

---

1. Restructure the data into the format of a frequency distribution.
2. Construct a percentage frequency distribution, a cumulative frequency distribution, and a decumulative frequency distribution.
3. Graph as a histogram and as a frequency polygon.
4. Construct cumulative and decumulative ogives.

Problem 2.6

A real estate agency conducted a study on the selling prices of homes sold in the Boston area. The study was limited to homes sold under $50,000. Data compiled on 120 sales consummated within the past 6 months is summarized below:

34,500	42,150	17,500	39,500	31,950	9,500
31,750	26,500	14,950	37,500	16,900	12,000
21,950	21,900	22,000	16,000	21,500	17,500
41,350	17,950	23,500	42,500	25,000	17,500
22,500	19,250	28,500	27,350	10,000	19,950
16,500	16,200	36,500	15,500	28,750	40,500
18,900	22,950	25,500	18,500	43,900	42,500
23,475	27,000	34,000	24,900	29,950	31,900
20,975	18,500	36,000	27,500	35,950	26,950
18,500	18,500	27,500	23,900	21,950	34,500
16,500	18,900	19,950	22,100	41,800	27,500
7,500	19,750	30,000	15,000	17,500	21,500
28,500	35,000	46,000	24,200	17,750	22,950
17,500	28,900	15,000	15,800	25,000	16,500
21,500	22,500	35,000	23,000	19,800	26,950
22,000	25,500	41,900	13,250	16,250	13,000
18,950	22,950	17,200	16,950	15,950	32,000
29,000	24,850	14,500	30,000	15,500	11,750
16,900	29,950	45,000	14,900	26,100	14,950
29,000	27,650	26,500	22,000	20,000	18,500

1. Restructure the data into the format of a frequency distribution.
2. Construct a percentage frequency distribution, a cumulative frequency distribution, and a decumulative frequency distribution.
3. Graph as a histogram and as a frequency polygon.
4. Construct cumulative and decumulative ogives.

Problem 2.7

The following grades were compiled on a year-end exam given by the statistics department to all students presently taking statistics.

57	82	83	89	77
70	63	75	63	89
62	71	50	82	59
66	51	70	75	88
68	65	61	58	97
62	78	69	68	86
76	61	80	84	75
53	81	57	62	96
74	66	81	76	64
75	73	65	52	95
61	51	72	92	94
86	75	59	66	86
65	64	72	74	75
95	67	84	72	73
71	54	62	56	87
78	85	71	85	64
51	83	64	78	94
83	72	73	67	79
95	79	54	93	65
77	55	79	94	86

1. Restructure the data into the format of a frequency distribution.
2. Construct a percentage frequency distribution, a cumulative frequency distribution, and a decumulative frequency distribution.
3. Graph as a histogram and as a frequency polygon.
4. Construct cumulative and decumulative ogives.

Problem 2.8

In order to be better able to bill customers for services rendered, a telephone answering service decided that it needed to know how long it takes to complete various incoming calls. Accordingly, 100 random calls were monitored and timed. The data gathered is reproduced below.

3.6	4.5	10.0	8.4	6.8
8.2	10.0	6.3	3.6	7.8
5.6	2.1	7.4	11.7	6.6
10.2	8.3	4.2	4.4	11.9
4.4	4.2	5.4	8.0	9.8
6.3	6.1	8.1	2.5	3.0
5.0	6.6	5.2	6.0	9.8
7.0	2.7	8.5	5.6	6.8
5.7	7.3	3.4	4.3	3.2
7.2	5.5	8.2	7.2	7.9
2.3	7.0	5.8	2.3	9.7
9.0	3.8	5.0	7.6	7.8
6.5	8.7	7.6	6.5	10.8
9.0	4.6	6.5	4.8	4.8
3.9	4.2	2.9	10.2	11.8
11.7	10.4	9.3	7.0	3.2
5.9	7.0	9.9	9.0	11.0
11.6	3.0	6.6	5.0	9.6
4.6	11.4	10.7	9.5	7.8
11.9	9.3	7.0	7.7	9.7

1. Restructure the data into the format of a frequency distribution.
2. Construct a percentage frequency distribution, a cumulative frequency distribution, and a decumulative frequency distribution.
3. Graph as a histogram and as a frequency polygon.
4. Construct cumulative and decumulative ogives.

Problem 2.9

A quality control statistician has obtained the following measurements on 90 parts produced by a machining operation. The measurements refer to the diameter of the machined part and are recorded to the nearest ten-thousandth of an inch.

0.9985	0.9976	1.0005	0.9978	1.0018	0.9993
0.9994	1.0003	0.9971	1.0014	1.0006	1.0015
0.9996	1.0011	1.0021	1.0006	0.9981	0.9983
1.0015	0.9992	1.0007	0.9987	1.0024	1.0021
1.0006	1.0012	1.0017	1.0025	1.0013	1.0011
0.9987	1.0019	0.9988	0.9995	1.0004	0.9990
0.9998	1.0032	1.0022	1.0010	0.9985	1.0035
1.0007	0.9988	0.9982	1.0013	0.9999	1.0004
0.9999	1.0023	1.0005	1.0005	1.0016	1.0009
1.0027	1.0011	1.0016	0.9997	1.0002	0.9993
1.0014	1.0025	0.9984	1.0001	0.9997	0.9995
1.0004	0.9994	1.0018	1.0014	1.0010	1.0000
0.9986	0.9973	1.0037	0.9986	1.0003	0.9996
1.0022	1.0002	1.0000	0.9999	0.9984	1.0005
1.0008	1.0016	0.9997	1.0007	0.9995	0.9998

1. Restructure the data into the format of a frequency distribution.
2. Graph as a histogram.
3. Graph as a frequency polygon.
4. What conclusions can you draw with respect to the nature of the variability in the machining process?

Problem 2.10

In constructing a frequency distribution, why is it desirable for the class limits to be established in such a fashion that the midpoint of the class will fall at approximately the arithmetic average of the items in the class?

Problem 2.11

Both the height *and* the area of the bars in a histogram often reflect the relative frequency of occurrence of the associated classes. Under what conditions does this relationship not hold?

Problem 2.12

When designing a tally sheet for the purpose of facilitating the construction of a frequency distribution, why is it desirable to leave spaces for those values of the variable that do not occur as well as for those that do occur? That is,

why 31	instead of 32 II
32 II	34 III
33	36 I
34 III	38 ℕℍ
35	39 I
36 I	
37	
38 ℕℍ	
39 I	

Problem 2.13

Under what circumstances would a histogram be preferable to a frequency polygon as a method of graphic presentation and vice versa?

Problem 2.14

It is considered desirable, if possible, when constructing a frequency distribution to have equal class intervals. Why are equal class intervals desirable?

Problem 2.15

The following were listed as important criteria to keep in mind when constructing a frequency distribution:
   a. Capture the discernable pattern in the data.
   b. All class intervals should be equal if possible.
   c. Class midpoints should be established if possible so that the midpoint is approximately equal to the arithmetic average of the items in the group.
   d. If possible, class limits should be set so that the class mark or midpoints fall on whole numbers.

Assume that you are constructing a frequency distribution to serve as a base for further computations. It is often difficult to achieve all the criteria listed.
   1. Rank the above criteria in order of importance, i.e., a, b, c, d, or d, c, a, b, and so on. Justify your ranking.
   2. Would you have varied the ranking if the frequency distribution were to be constructed for purposes of graphic representation. If so, how and why?

Problem 2.16

Describe the precise adjustment you would make in graphing the following frequency distribution in the form of a histogram and explain *why* the adjustment should be made.

Wage rates ($)	f
5–10	2
10–15	4
15–20	9
20–35	36
35–40	7
40–45	3
	61

MEASURES OF CENTRAL TENDENCY AND LOCATION

3

Although on many occasions restructuring the data in the form of a frequency distribution is sufficient for the purpose at hand, it is often useful to carry the analysis one step further and summarize the underlying characteristics of the data with a few well-chosen numerical values. Numerical measures designed to play such a role fall into four categories: (1) measures of central tendency and location, (2) measures of dispersion, (3) measures of skewness, and (4) measures of kurtosis.

In any study, there is usually a range of possible values for the variable of interest. For example, in Chapter 2 the variable of interest was the daily wage rate, and the range of possible values consisted of all daily wage rates that could occur. Geometrically, this concept can be represented by a number line along which all possible values of the variable are recorded. In

Chapter 2, a histogram was constructed in association with a segment of such a number line. The histogram is reproduced as Figure 3.1.

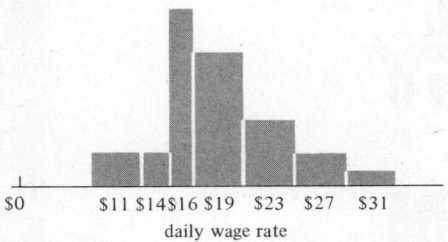

Figure 3.1 Histogram: daily wage rate distribution for an assembly plant in Richmond, Virginia.

Suppose that in the city of Mudville, environmental conditions are particularly disadvantageous such that a $30 premium must be paid to each and every worker for the same kind of work as is performed in the Richmond plant. Also assume that the pattern of wage rates is the same. The distribution of wages for the plant in Mudville, when reflected by a histogram superimposed on the same number line as used to represent the distribution of earnings in the Richmond plant, appears as depicted in Figure 3.2.

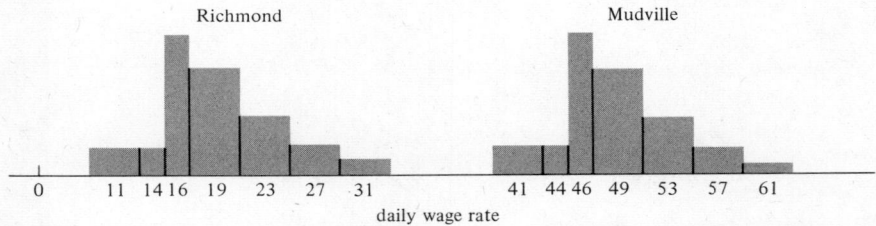

Figure 3.2 Histograms: daily wage rate distributions for assembly plants in Richmond and Mudville.

Notice how the distribution of daily wage rates for the Richmond plant appears to cluster or center around the daily wage rate of $15 and how that for Mudville clusters or centers around $45. In fact, the only difference between the two distributions is the point on the number line around which the daily wage rate patterns tend to cluster. There is no need to describe the distributions in the form of histograms to communicate this information—it is communicated more efficiently by simply indicating the values on the number line around which the distributions tend to cluster ($15 for the Richmond plant, $45 for the Mudville plant). This type of information is conveyed by measures of central tendency and location.

Of course, there are other ways in which distributions differ other than where they are located along a number line. To illustrate, two distributions could be centered on exactly the same value on the number line but differ in the extent to which individual observations in the distribution cluster around the measure of central tendency. Consider a firm attempting to decide from which of two potential suppliers to purchase subassemblies. Both suppliers promise that over time the percent of defectives shipped will approximate 8 percent. For planning purposes and in the interest of avoiding work stoppages due to an inadequate supply of subassemblies, the firm will also desire information with respect to the variation in percent of defectives likely to occur from lot to lot. Investigation reveals the distributions of percent defectives from lot to lot for each of the suppliers shown in Figures 3.3(a) and 3.3(b).

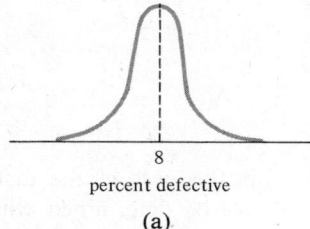

percent defective

(a)

Figure 3.3(a) Supplier A: distribution of defective rates observed.

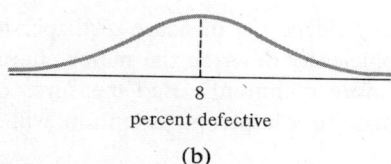

percent defective

(b)

Figure 3.3(b) Supplier B: distribution of defective rates observed.

Clearly, the superior stability of the defective rate from lot to lot for supplier A would facilitate planning. All other things being equal, supplier A would be the preferable choice. Measures of dispersion communicate information of this type. Measures of dispersion indicate the extent to which individual observations cluster closely or are widely dispersed. Other things being equal, the greater the measure of dispersion, the less representative the measure of central tendency will be of the distribution as a whole.

Measures of skewness indicate the direction and degree of departure from symmetry.

1. If a distribution is perfectly symmetrical, the measure of skewness will be equal to zero.

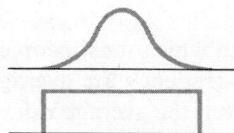

2. If the distribution is asymmetrical (i.e., not symmetrical) and the tail of the distribution extends in the direction of the positive values along the number line, the distribution is referred to as positively skewed. The measure of skewness will have a positive sign, and the magnitude of the value will increase as the extent of departure from symmetry increases.

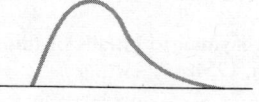

3. If the distribution is asymmetrical and the tail of the distribution extends in the direction of the negative numbers along the number line, the distribution is referred to as negatively skewed. The measure of skewness will have a negative sign, and the magnitude of the value will increase as the extent of departure from symmetry increases.

Measures of kurtosis indicate the extent of peakedness in the distribution. By observing the magnitude of the measure, it can be determined whether the distribution is[1]

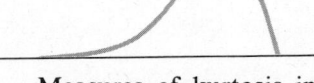

peaked   , normal   , or quite flat

By computing a measure of central tendency, a measure of dispersion and a measure of skewness, an analyst can efficiently describe the major characteristics of a set of data. In this chapter, the more commonly used measures of central tendency and location will be examined. In Chapter 4, attention will focus on measures of dispersion and skewness.

The following commonly used measures of central tendency and location will be examined: (1) the arithmetic mean, (2) the median, (3) the mode, (4) quartiles, and (5) percentiles.

Three other measures that are appropriate under specialized sets of circumstances are (1) the geometric mean, (2) the quadratic mean, and (3) the harmonic mean. The quadratic and harmonic means are specialized measures and will not be treated in this text. The geometric mean is discussed in a technical note at the end of the chapter.

As the above lists clearly illustrate, there are a number of different measures of central tendency. To select the appropriate measure to use in a given situation, one should be familiar with the characteristics of each of the measures, the type of situations in which they perform well, and perhaps more important, those situations in which they perform inadequately. With this orientation, the arithmetic mean will be examined.

3.1 THE ARITHMETIC MEAN

The arithmetic mean is a fancy name for a measure to which most people typically refer as the simple average. All measures of central tendency are averages in one sense of the word or another. Thus, when one states that the average value of something is such and such, it is not clear whether one is referring to the arithmetic mean, median, mode, geometric mean, and so on. For this reason, it is desirable to use the label the arithmetic mean when referring to the simple average (unless it is your objective to mislead your audience).[2]

---

[1] Measures of kurtosis are seldom employed by the business analyst. Hence, discussion of kurtosis will be left to other texts.

[2] A very entertaining primer devoted to alerting the layman to pitfalls of this type is *How to Lie with Statistics* by Darrell Huff, New York: Norton, 1954.

By definition, the arithmetic mean of a group of values is equal to the sum of the values in the group divided by the number of values comprising the group. To illustrate: Suppose a firm stocked the following five models of sump pumps:

Model	Cost to the firm ($)
A	11.85
B	11.75
C	11.80
D	13.15
E	12.45

The arithmetic average of the cost of the sump pumps to the firm is

$$\frac{\$11.85 + \$11.75 + \$11.80 + \$13.15 + \$12.45}{5} = \frac{\$61.00}{5} \quad \text{or} \quad \$12.20$$

Algebraically, the relationship is expressed by the following formula:

$$\overline{X} = \frac{\sum_{i=1}^{n} X_i}{n}$$

Explanation of the symbols in the formula:

X_i is a subscripted variable. The variable of interest in this case is the cost to the firm of each sump pump model. The function of the subscript is to indicate which of the five models is being referenced.

X_1 refers to the cost to the firm of sump pump model A or $11.85.

X_2 refers to the cost to the firm of sump pump model B or $11.75.

X_i is a general subscript referring to the cost to the firm of sump pump model i.

Accordingly:

$$X_1 = \$11.85$$
$$X_2 = \$11.75$$
$$X_3 = \$11.80$$
$$X_4 = \$13.15$$
$$X_5 = \$12.45$$

$\overline{X}$ is read as X-bar, and the "bar" is interpreted as an instruction to take the arithmetic mean of the variable indicated below the bar. $\overline{Y}$, for example, is an instruction to take the arithmetic mean of the Y values where Y_i might refer to the retail price of the ith sump pump model.

n refers to the number of values.

$\sum$ is the Greek capital letter sigma and constitutes a generalized instruction to "take the sum of." To indicate specifically the values of the variable that are to be summed, the symbol is "dressed up" as follows:

$$\sum_{i=1}^{n} X_i$$

This is interpreted as an instruction to take the sum of the X_i values for $X_i = X_1$ through $X_i = X_n$. In the illustrative example, the instruction would be $\sum_{i=1}^{5} X_i$.

$$X_1 = \$11.85$$
$$X_2 = \$11.75$$
$$X_3 = \$11.80$$
$$X_4 = \$13.15$$
$$X_5 = \$12.45$$
$$\overline{}$$
$$\$61.00$$

Thus $\sum_{i=1}^{5} X_i = \$61.00$.

The notation $\sum_{i=1}^{n} X_i$ can be replaced with the simpler instruction $\sum X$, when the specified values to be summed are self-evident. An effort will be made to maintain notational simplicity in all formulas developed. In line with this policy, the formula for the arithmetic mean will be expressed as follows:

$$\overline{X} = \frac{\sum X}{n}$$

Applying this formula to the illustrative example, we obtain

$$\overline{X} = \frac{\sum_{i=1}^{n} X_i}{n} = \frac{\sum X}{n} = \frac{\$61.00}{5} = \$12.20$$

The above formula is appropriate only when dealing with individual items (i.e., data not structured in the format of a frequency distribution) and where each X_i value is considered to be of equal importance.[3] The computational procedures to be followed when the above restrictions are removed will be examined. First, however, it is useful to investigate the properties of the arithmetic mean as a measure of central tendency. An understanding of the properties of a measure facilitates rational choice when deciding which measure to use in a particular application.

3.2 PROPERTIES OF THE ARITHMETIC MEAN

1. Every item is included in the computation. The value of each item is taken into consideration and is an integral part of the computation. This is not true for all measures of central tendency.

2. An extreme value can exert an influence on the arithmetic mean that is out of proportion to its true importance, thereby destroying the representativeness of the arithmetic mean for the group of values from which it was computed.

For example, if X_5 had been $112.45 instead of $12.45, then $\sum X = \$161.00$

---

[3] $\overline{X}$ is typically used to refer to the arithmetic mean of a sample of items that has been selected out of a much larger group of values called the universe. Another formula the analyst will have occasion to employ in the computation of the arithmetic mean is $\mu_x = \sum X/N$. The symbol μ (pronounced mu) is used to refer to the arithmetic mean of all the values in the universe. A descriptive measure for a universe is called a parameter, and the corresponding descriptive measure for a sample drawn from the universe is called a statistic. Thus, μ_x is a parameter and $\overline{X}$ is a statistic. Different symbols facilitate communicating whether one is referring to the parameter or a statistic. The role of parameters and statistics in statistical analysis will be explored in Chapter 13 of the text.

and $\overline{X} = \sum X/n = \$161/5 = \$32.20$. Clearly, \$32.20 is not typical of the five values that entered into its computation. The arithmetic mean should not be utilized as a representative value when there are a few extremely large or a few extremely small values. Fortunately, the median, an alternative measure of central tendency, often serves quite well in this situation.

3. The arithmetic mean can take on a value that is not among the values of the variable from which it was computed.

To illustrate, a study yielded the conclusion that the mean number of cars per family is 1.7 cars. For any one family, the value of the variable can take on only the discrete values 0, 1, 2, and so on. The arithmetic mean for a set of discrete values is not subject to such a limitation, however, and could conceivably take on any value on the number line within the range of feasible values of the variable. Thus, the arithmetic mean for a set of values could conceivably take on a value that could never occur for an individual observation of the variable being analyzed.

4. $\overline{X} = \sum X_i/n$. Multiplying both sides by n, one obtains $n\overline{X} = \sum X_i$. Thus, if the value of the arithmetic mean is multiplied by the number of items involved in its computation, the product will *always* be equal to the sum of the X_i values. This relationship does not hold for the other measures of central tendency. This characteristic explains why it is desirable to have the midpoint of the classes in a frequency distribution approximate the value of the arithmetic mean of the items comprising the classes. If this is the case, then when the class frequency is multiplied by the midpoint, the product closely approximates the sum of the items comprising the class.

Given that

$$n = f$$
$$\overline{X} = M$$

it follows that

$$fM = \sum X_i$$

Even though the individual items have lost their identity, the sum of the items can still be obtained.

5. The arithmetic mean is rigidly determined. This property refers to the fact that there is one and only one arithmetic mean for a given set of data. Alternative computational formulas exist, but all are based on the same relationship and generate the same numerical value. Some other measures, such as the mode, generate different numerical values depending on the computational procedure selected. This complicates the interpretation of such measures.

6. The sum of the deviations of the values of the individual items from the arithmetic mean is equal to zero.

$$
\begin{aligned}
X_1 - \overline{X} &= \$11.85 - \$12.20 = -\$0.35 \\
X_2 - \overline{X} &= \$11.75 - \$12.20 = -\$0.45 \\
X_3 - \overline{X} &= \$11.80 - \$12.20 = -\$0.40 \\
X_4 - \overline{X} &= \$13.15 - \$12.20 = +\$0.95 \\
X_5 - \overline{X} &= \$12.45 - \$12.20 = +\$0.25
\end{aligned}
$$

$$
\begin{aligned}
&-\$1.20 \\
&+\$1.20 \\
\hline
&0
\end{aligned}
$$

One of the short-cut techniques for computing the arithmetic mean from a frequency distribution is based on the relationship that the sum of the deviations of the individual values from the mean is equal to zero. It involves finding the value on the number line from which the sum of the deviations is equal to zero and in so doing simultaneously identifying the value of the arithmetic mean. This approach is presented in Technical Note 1 at the end of the chapter.

7. If one squares the deviations of the values of the individual items from the arithmetic mean and obtains their sum, the sum will be less than it would be if it were to be computed from any other point along the number line.

		$X_i - \overline{X}$	$(X_i - \overline{X})^2$
X_1	11.85	$11.85 - 12.20 = -0.35$	0.1225
X_2	11.75	$11.75 - 12.20 = -0.45$	0.2025
X_3	11.80	$11.80 - 12.20 = -0.40$	0.1600
X_4	13.15	$13.15 - 12.20 = +0.95$	0.9025
X_5	12.45	$12.45 - 12.20 = +0.25$	0.0625
		-1.20	1.4500
		$+1.20$	
		0	

least-squares = minimum value

For the five values 11.85, 11.75, 11.80, 13.15, and 12.45, a challenge could be issued. Choose any value along the number line other than 12.20 and compute the sum of the squared deviations from that value. No matter what value is selected, the sum of the squared deviations obtained will be greater than 1.4500. The above is referred to as the *least-squares criterion*, which states that the most representative value is the one from which the sum of the squared deviations is minimized. According to the least-squares criterion, the arithmetic mean is the most representative value for a given set of data.[4]

The above characteristics should be considered when deciding whether or not the arithmetic mean is the appropriate measure of central tendency. In particular, the analyst should attempt to evaluate the validity of the information communicated by the arithmetic mean with respect to what he wants to know about the underlying distribution.

3.3 THE WEIGHTED ARITHMETIC MEAN

Suppose the firm is interested in determining the average cost to the firm of sump pumps purchased for resale during the last fiscal year. Assume the quantity purchased varies depending on the model. This means that the cost of the various models would no longer be of equal importance and, as a result, the formula for the simple arithmetic average would no longer be applicable.

[4] The least-squares criterion is also used to determine the line of best fit when describing functional relationships. Its use for this purpose is demonstrated in the Chapter on simple regression and correlation.

Let W_i represent the number of units of sump pump model i purchased for resale.

model	W_i	X_i
A	150	11.85
B	200	11.75
C	150	11.80
D	100	13.15
E	100	12.45
	700	

The total number of units of all models purchased was 700. The true mean cost per sump pump purchased for resale is the sum of the 700 individual cost figures divided by the number of units purchased. In effect, one would have a list of 700 numerical values, the first 150 of which would be $11.85, the next 200, $11.75, and so on. Instead of summing the 700 values directly, the sum could alternatively be generated by multiplying $11.85 by the number of times it occurred, $11.75 by the number of times it occurred, and so on for each of the five models and then summing the five subproducts:

$$\$11.85 \times 150 = \$1777.50$$
$$\$11.75 \times 200 = \$2350.00$$
$$\$11.80 \times 150 = \$1770.00$$
$$\$13.15 \times 100 = \$1315.00$$
$$\$12.45 \times 100 = \$1245.00$$

$$\$8457.50$$

The sum would then be divided by the number of values it represents, 700. The resulting value is the arithmetic mean.

$$\frac{\$8457.50}{700} = \$12.08$$

Actually, the above procedure is a description of the computation of a weighted arithmetic mean where each X_i is multiplied by a W_i (a weight indicating its relative importance as compared to the other X_i's, in this instance the weight is the number of units purchased of the model under consideration). The $W_i X_i$ subproducts are then summed and divided by the sum of the weights (i.e., the sum of the W_i values).

The procedure is formalized as follows:

$$\bar{X} = \frac{\sum (W_i X_i)}{\sum (W_i)}$$

where W_i is the weight assigned to the X_ith item (see Table 3.1).

Table 3.1 Computation of the weighted arithmetic mean

Sump pump model	Number of units purchased during the last fiscal year, W_i	Per unit cost to the firm, X_i	Weighted X_i values, $W_i X_i$
A	150	$11.85	$W_1 X_1 = 150 \times 11.85 = \$1777.50$
B	200	$11.75	$W_2 X_2 = 200 \times 11.75 = \$2350.00$
C	150	$11.80	$W_3 X_3 = 150 \times 11.80 = \$1770.00$
D	100	$13.15	$W_4 X_4 = 100 \times 13.15 = \$1315.00$
E	100	$12.45	$W_5 X_5 = 100 \times 12.45 = \$1245.00$
	700		$8457.50

In line with the policy of keeping formulas notationally simple, $\overline{X} = \sum (WX)/\sum (W)$ will be used in place of $\overline{X} = \sum (W_i X_i)/\sum (W_i)$ whenever the values to be summed are self-evident. Thus:

$$\overline{X} = \frac{\sum (WX)}{\sum (W)} = \frac{\$8457.50}{700} = \$12.08$$

The function of the weights is to give each X_i value its proper relative importance. The absolute values of the weights are of no importance provided that the proper relative relationship is retained. Whenever possible, therefore, the weights should be reduced by a common factor in order to simplify the computational procedure. In the above example, the weights can be reduced by the common element of 50. The computation performed with the reduced weights appears in Table 3.2.

Table 3.2 Computation of the weighted arithmetic mean (weights reduced by the common element of 50)

Sump pump model	Number of units purchased during the last fiscal year (in units of 50), W_i	Per unit cost to the firm, X_i	Weighted X_i values, $W_i X_i$
A	3	$11.85	$35.55
B	4	$11.75	$47.00
C	3	$11.80	$35.40
D	2	$13.15	$26.30
E	2	$12.45	$24.90
	14		$169.15

$$\overline{X} = \frac{\sum (WX)}{\sum (W)} = \frac{\$169.15}{14} = \$12.08$$

As the computations in Table 3.2 clearly demonstrate, it is the relative relationship, not the absolute size, of the weights that is important. In actuality, there is no such thing as an unweighted arithmetic mean. If weights are not explicitly introduced, it indicates that the analyst has concluded that the values are all of equal importance and are to be assigned a common weight of 1.

3.4 CALCULATION OF THE ARITHMETIC MEAN FROM A FREQUENCY DISTRIBUTION (GROUPED DATA)

Let us now examine the computational procedures appropriate for computing the arithmetic mean for data that have been restructured into the format of a frequency distribution, thereby causing the individual items to lose their identity. Three alternative computational procedures exist: (1) the long method, (2) the actual deviations method, and (3) the step deviations method. The actual deviations method and the step deviations method are short-cut formulas and are treated in Technical Note 1 at the end of the chapter. The discussion of the short-cut formulas can be ignored without loss of continuity. If one's objective is simply to gain a conceptual understanding, mastery of the formulas for ungrouped data and the long method for frequency distributions is sufficient.

The Long Method

FORMULA

$$\overline{X} = \frac{\sum (f_i \cdot M_i)}{n}$$

where

$\overline{X}$ = arithmetic mean
$\sum$ = the sum of
f_i = frequency of the ith class
M_i = midpoint of the ith class
n = total number of frequencies in the frequency distribution (i.e., $\sum f_i$)

PROCEDURE

FIRST STEP. Multiply the midpoint of each class by its respective class frequency:

$$f_i \cdot M_i$$

This method is based on the assumption that the frequency distribution has been constructed in such a way that the midpoint of each class has a value approximately equal to the value of the arithmetic mean of the items in the class. Therefore, $f_i \cdot M_i$ should approximate the sum of the X values for the ith class, since f_i equals the number of items in the ith class and M_i should be equal to the arithmetic mean of the items in the ith class. To the extent that the assumption "the midpoint of a class is equal to the arithmetic mean of the items in the class" is violated, distortion is introduced into the computations.

SECOND STEP. Sum the products obtained in step 1:

$$\sum (f_i \cdot M_i)$$

If one obtains the sum of all the items in each class and then takes a grand sum of all these values, the sum of the values in the entire distribution will be obtained.

THIRD STEP. Divide by the total frequency:

$$\frac{\sum (f_i \cdot M_i)}{n}$$

If the sum of a group of values is divided by the number of items comprising the group, the arithmetic mean of the group is generated.

The computation of the arithmetic mean by the long method is illustrated in Tables 3.3(a), 3.3(b), and 3.3(c) for the three frequency distribution alternatives generated in Chapter 2.

Table 3.3(a) Computation of the arithmetic mean by the long method: Frequency Distribution Alternative A

Daily wages ($)	M	f	fM	
9–13	$11	12	$ 132	$\bar{X} = \dfrac{\sum (fM)}{n}$
13–17	15	31	465	
17–21	19	37	703	
21–25	23	17	391	$= \dfrac{2112}{112}$
25–29	27	11	297	
29–33	31	4	124	$= \$18.86$
		112	2112	

Table 3.3(b) Computation of the arithmetic mean by the long method: Frequency Distribution Alternative B

Daily wages ($)	M	f	fM	
9–11	$10	4	$ 40	$\bar{X} = \dfrac{\sum (fM)}{n}$
11–13	12	8	96	
13–15	14	6	84	
15–17	16	25	400	$= \dfrac{2146}{112}$
17–19	18	15	270	
19–21	20	22	440	$= \$19.16$
21–23	22	8	176	
23–25	24	9	216	
25–27	26	5	130	
27–29	28	6	168	
29–31	30	1	30	
31–33	32	3	96	
		112	2146	

Table 3.3(c) Computation of the arithmetic mean by the long method: Frequency Distribution Alternative C

Daily wages ($)	M	f	fM	
9–13	$11	12	$ 132	$\bar{X} = \dfrac{\sum (fM)}{n}$
13–15	14	6	84	
15–17	16	25	400	
17–21	19	37	703	$= \dfrac{2131}{112}$
21–25	23	17	391	
25–29	27	11	297	$= \$19.03$
29–33	31	4	124	
		112	2131	

The actual value of the arithmetic mean for the 112 items is $19.13. Frequency Distribution Alternatives A, B, and C provide different estimates of this value. Drawing on the discussion of the issues involved in the construction of a frequency distribution in Chapter 2, an explanation of the source of the differences serves to reinforce some of the statements made.

1. The best estimate is provided by Frequency Distribution Alternative B. This is in accordance with expectations for two reasons:

    a. The requirement that "the midpoint should be approximately equal to the value of the arithmetic mean of the items in the class" is effectively met for all classes.

    b. If the frequency distribution is to be used for computational purposes, it is desirable to have more classes than if graphic representation is all that is desired. In general, the more classes, the narrower the interval, the less the range for error, and the more accurate the resulting computations.

2. The most inferior estimate of the value of the arithmetic mean is provided by Frequency Distribution Alternative A which violates the assumption that "the midpoint is approximately equal to the arithmetic mean of the items in the class" quite severely with respect to the class $13–$17.

Recalling the discussion in Chapter 2 on frequency distributions, the true sum of the values in the class $13–$17 is $484.60. The true arithmetic mean for the items in that class is $484.60/31 = $15.63. In Frequency Distribution Alternative A, $15.00 is used as the midpoint. The estimated sum of the values in that class is $f_i M_i = 31 \times \$15.00$ or $465 (an understatement by $19.60), which in turn is reflected as an understatement in the estimated sum of the $f_i M_i$ values. When divided by n, this leads to an understatement of the value of the arithmetic mean.

In Frequency Distribution Alternative B, the class $13–$17 is subdivided into two classes, $13–$15 and $15–$17, for each of which the assumption that "the midpoint is approximately equal to the arithmetic mean of the items in the class" is met. As a result, the estimated sum of the values in the intervals $13–$15 and $15–$17 closely approximates the true sum of the values between $13–$17.

	M	f	fM
$13–$15	14	6	84
$15–$17	16	25	400
			484 $\cong$ 484.60

The impact on the above computations serves to stress the importance of "the midpoints of the classes approximating the arithmetic means of the items in the classes" if accurate computations are to result.

Actually, the formula $\overline{X} = \sum (fM)/n$ is simply the formula for a weighted arithmetic mean using symbols that are more appropriate when referencing a frequency distribution. The individual items lose their identity when placed in a frequency distribution and henceforth are represented by the midpoints of the classes to which they have been assigned. M_i is the grouped-data equivalent to X_i for ungrouped data. The relative importance of each of the midpoints is signified by the number of items it represents or the class frequency. The class frequency, f, is thus the appropriate weight to use to indicate the proper relative importance of each of the midpoints. The sum of the class frequencies, n, refers to the sum of the weights.

3.5 THE MEDIAN

The median is defined as being equal to the value of the middle item in a distribution when the items have been arranged in order of magnitude. The median is a positional measure in the sense that its value is determined by the value of the item occupying a particular position in the distribution.

3.6 COMPUTATION OF THE MEDIAN FOR A SET OF INDIVIDUAL OBSERVATIONS

An industrial engineer designing a piece-work incentive system desires to determine the representative amount of time a worker should require to perform an indicated operation. The industrial engineer decides to obtain seven observations of a worker's performance and use the median time of the performances as the representative length of time required to perform the operation.

Let n equal the number of observations and X_i equal the length of time the worker requires to perform the operation the ith time.

Observations

$X_1 = 3$ min, 12 sec
$X_2 = 3$ min, 26 sec
$X_3 = 2$ min, 54 sec
$X_4 = 3$ min, 9 sec
$X_5 = 3$ min, 28 sec
$X_6 = 4$ min, 17 sec
$X_7 = 3$ min, 18 sec

1. The first step is to rearrange the data in the form of an array. This involves arranging the observations by order of magnitude from smallest to largest.

Observations
Rearranged in
an Array

$X_3 = 2$ min, 54 sec
$X_4 = 3$ min, 9 sec
$X_1 = 3$ min, 12 sec
$X_7 = 3$ min, 18 sec
$X_2 = 3$ min, 26 sec
$X_5 = 3$ min, 28 sec
$X_6 = 4$ min, 17 sec

2. The second step is to locate the *position* in the array occupied by the median item. The following formula generates the position in the array occupied by the median item when dealing with individual items:

$$\text{position occupied by the median item} = \frac{n + 1}{2}$$

For the industrial engineer's seven observations, the formula generates

$$\frac{n + 1}{2} = \frac{7 + 1}{2} = \frac{8}{2} = \text{fourth position}$$

In other words, when the items have been arranged in order of magnitude, the median value will be in the fourth position.

Note: If there had been eight observations, the formula would have generated 4.5 as the position occupied by the median item. It is customary to interpret this as an instruction to take the arithmetic mean of the fourth and fifth items and use the resulting value as the median or representative value for the group. Thus, whenever there is an even number of items, the median is interpreted as the mean of the middle two items in the array.

3. The third step is to determine the *value* of the item occupying the median position. When dealing with individual items, this is simply a matter of observation. X_7 is in the fourth position. Therefore, the median value is 3 min, 18 sec.

3.7 PROPERTIES OF THE MEDIAN

1. The median is not affected by extreme values. If X_6, the largest value, had been 44 min, 17 sec instead of 4 min, 17 sec, the value of the median would have been unaffected and would have remained at 3 min, 18 sec. On the other hand, if the arithmetic mean was being used as the representative value, its value would have been influenced considerably by the extreme observation. The arithmetic mean would have changed from 3 min, 23 sec to 9 min, 6 sec. Thus, the median is particularly suitable for skewed distributions where there are a few very large or a few very small values.

2. The value obtained for the median may be nonrepresentative if the individual items do not tend to cluster at the center of the distribution.

In the distribution depicted in Figure 3.4, the median would not serve very effectively as a representative measure. For that matter, neither would the arithmetic mean. Another measure of central tendency called the mode, which will be examined shortly, is the most suitable measure to use in this situation.

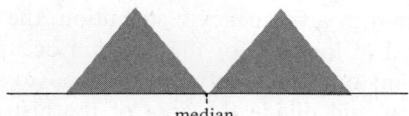
median

Figure 3.4

3. The median is the measure of central tendency from which the sum of the *absolute* deviations is minimized. Referring to the seven observations by the industrial engineer:

X_i values	Absolute deviations $\lvert X_i - \text{Me} \rvert$
X_1 = 3 min, 12 sec or 192 sec	$\lvert 192 - 198 \rvert = 6$
X_2 = 3 min, 26 sec or 206 sec	$\lvert 206 - 198 \rvert = 8$
X_3 = 2 min, 54 sec or 174 sec	$\lvert 174 - 198 \rvert = 24$
X_4 = 3 min, 9 sec or 189 sec	$\lvert 189 - 198 \rvert = 9$
X_5 = 3 min, 28 sec or 208 sec	$\lvert 208 - 198 \rvert = 10$
X_6 = 4 min, 17 sec or 257 sec	$\lvert 257 - 198 \rvert = 59$
X_7 = 3 min, 18 sec or 198 sec	$\lvert 198 - 198 \rvert = 0$

A challenge could be issued. Select a value other than 198 sec and compute the sum of the absolute deviations from it. Regardless of the value selected, the sum of the absolute deviations obtained would exceed 116.

The measure of central tendency selected as the representative value is often used as a proxy for the value of the individual observations it represents. If the magnitude of the cost associated with an estimating error is proportional to the absolute amount of deviation, the sum of these costs incurred over time can be minimized by using the median as the representative value. Similarly, if the cost associated with an estimating error is proportional to the square of the deviation, the arithmetic mean is the preferable measure of central tendency.

3.8 COMPUTATION OF THE MEDIAN FOR A FREQUENCY DISTRIBUTION

The procedure for determining the median for a frequency distribution is quite similar to that for determining the value of the median for a set of individual observations.

1. The first step, involving the placing of the data in the form of an array, would already have been accomplished in the process of constructing the frequency distribution. The individual items, however, will have lost their identity. Thus, the analyst ends up with an array of midpoints and the necessity of introducing an assumption about the manner in which the individual items are arrayed within each class. The convention is to assume that the individual items within a class are evenly distributed over the interval of that class.

2. The second step is to determine the *position* occupied by the median item. Because the individual items have lost their identity, it is necessary to use a different formula to locate the position occupied by the median item in a frequency distribution. The formula $(n + 1)/2$ is used when working with individual items because the analyst is searching for the position occupied by a particular item. When working with data in the format of a frequency distribution, the individual items have lost their identity. Instead of looking for the position occupied by an item, the analyst searches for a point on the horizontal axis above which, if a perpendicular were to be raised, it would divide the area of the histogram representing the frequency distribution exactly in half. Recall that area assigned under a histogram is proportional to relative frequency of occurrence.[5] Thus, if 50 percent of the area is to the left of a point on the number line, 50 percent of the items will lie to the left of that value as well (Figure 3.5). A perpendicular raised at the value of the item that is in the position defined by the formula $n/2$ will achieve this result. After computing the position occupied by the median, the next step is to determine the class in the frequency distribution containing the median item.

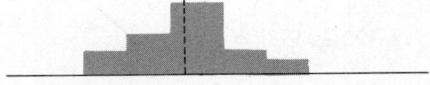

Figure 3.5

The procedure for the computation of the median for Frequency Distribution Alternative B is described in the following section and summarized in Table 3.4.

[5] See p. 21.

For Frequency Distribution Alternative **B**:

$$\text{position occupied by the median item} = \frac{n}{2} = \frac{112}{2} = 56\text{th position}$$

The median class would be the class that contains the 56th item when the values are arranged in order of magnitude.

Daily wages ($)	Frequency	Cumulative frequency
9–11	4	4
11–13	8	12
13–15	6	18
15–17	25	43
17–19	15	58
19–21	22	
21–23	8	
23–25	9	
25–27	5	
27–29	6	
29–31	1	
31–33	3	
	112	

Create a cumulative frequency column. Continue moving down the column until the class in which the item that is 56th in order of magnitude is located. For example, by the upper limit of the class $13–$15, the first 18 items in order of magnitude have been passed. By the time the upper limit of the next class ($15–$17) is reached, the value of another 25 items have been exceeded. Thus, the first 43 items in order of magnitude have been passed. By the upper limit of the next class ($17–$19), the value of 58 of the items have been exceeded. Clearly, the item in the 56th position lies between the lower limit of the class ($17–$19) at which point 43 positions had been cumulated and the upper limit of that class where the number of items whose value has been exceeded is 58. Thus, the class $17–$19 contains the median item and is identified as the median class. The convention is to assume that the 15 items in the median class are evenly distributed over the $2 interval from $17–$19 or, in other words, an observation occurs every $0.133 through the class ($\frac{2}{15}$ = $0.133).

The 44th item in order of magnitude would be located $\frac{1}{15}$ of the distance through the class or at $17.133. The 45th item would be located $\frac{2}{15}$ of the distance through the class or at $17.267, and so on. At $17.00, 43 items had been cumulated. Another 13 items are needed to reach the 56th position. Thus, the 56th position will be located $\frac{13}{15}$ of the distance through the class or at $18.733. See Figure 3.6.

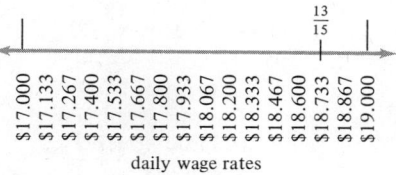

daily wage rates

Figure 3.6

3. The value of the item occupying the median position derived intuitively above is computed algebraically by the formula

$$Me = L + \frac{a}{f} C$$

where

Me = value of the median
L = lower limit of the median class
a = number of additional items needed in the median class in order to reach the position occupied by the median item
f = frequency of the median class
C = class interval of the median class

Therefore, a/f represents the proportion of the distance the median item lies through the median class (assuming that the items are distributed evenly throughout the class). Multiplying this proportion by the class interval provides the amount to add to the lower limit of the class to generate the value of the median item

To illustrate, the value of the median is computed in Table 3.4 for Frequency Distribution Alternative B. The value $18.733 is an estimate based on the assumption that the items in the median class are evenly distributed throughout the class. Distortion in the estimate of the median will occur to the extent that the assumption does not correspond to reality.

Table 3.4 Computation of the median: Frequency Distribution Alternative B

Daily wages ($)	Frequency	Cumulative frequency
9–11	4	4
11–13	8	12
13–15	6	18
15–17	25	43
17–19	15	58
19–21	22	
21–23	8	
23–25	9	
25–27	5	
27–29	6	
29–31	1	
31–33	3	
	112	

$a = $ Position occupied by the median item $-$ Number of positions cumulated prior to entering the median class

$= 56 - 43$
$= 13$

$$Me = L + \frac{a}{f} C$$
$$= 17.00 + \frac{13}{15} (2)$$
$$= 17.00 + \frac{26}{15}$$
$$= 17.00 + 1.733$$
$$= \$18.733$$

3.9 THE MODIFIED MEAN

In situations where a representative value is desired and it is clear that extreme values are included in the data, rather than totally reject the arithmetic mean and rely on the median, which incorporates only the value of a single item in the distribution, a compromise measure, the modified mean, is occasionally utilized. The analyst simply omits the extreme values (usually eliminating an equal number from each end of the distribution although this is not absolutely necessary). The arithmetic mean of the remaining items is then computed. It is important to note that the judgment of the analyst as to the legitimacy of the eliminations is crucial.

3.10 THE MODE

The mode is defined as the most commonly occurring value—that value of the variable that occurs with the greatest frequency. The mode is the appropriate statistic for communicating such information as the model of sump pump most frequently purchased, the primary cause of industrial accidents, or the most prevalent level of family income in a given geographical area. The mode is a meaningful statistic only in those instances where a particular value of the variable clearly occurs more frequently than any of the others.

When dealing with individual items, the mode is determined by observation. In determining the modal value for a frequency distribution, the analyst attempts to determine the value on the number line that is associated with the greatest clustering or density of occurrence. Graphically, the modal value is the value on the number line associated with the maximum value of the density function (Figure 3.7).

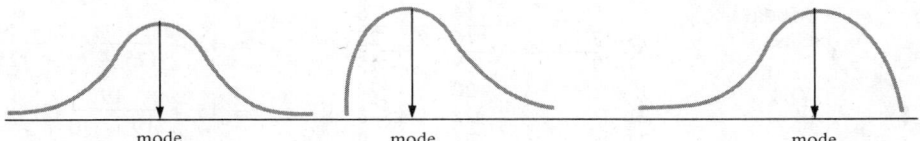

Figure 3.7

The use for which the measure is intended determines the degree of accuracy required. Often, it is sufficient simply to identify the modal class. The modal class is the class with the greatest density of occurrence per unit interval on the horizontal axis. (If the classes have equal class intervals, the modal class is the class with the largest class frequency—otherwise an adjustment is necessary to determine the greatest density of occurrence per unit interval.)

A number of different formulas exist for generating a single-value estimate of the mode. Unfortunately, the mode is not a rigidly defined measure. This creates a problem in interpretation. The different methods generate different estimates for the same set of data. Knowledge of the method by which a particular estimate of the mode is obtained aids in the interpretation of the resulting value. Three common methods employed to estimate the mode are discussed below.

The Crude Mode

The midpoint of the modal class is used as a point estimate of the modal value. For Frequency Distribution Alternative B, the value of the crude mode is $16.00.

The Interpolation Formula

$$Mo = L + \frac{d_1}{d_1 + d_2} C$$

The interpolation formula takes into consideration the density of occurrence in the class preceding the modal class relative to the density of occurrence in the class following the modal class. The relative density of occurrences are then utilized as a weighting factor influencing the extent to which the estimate of the mode is shifted away from the midpoint of the modal class. The estimate will shift toward the class limit associated with the greater density of occurrence.

L = lower limit of the modal class
d_1 = difference between the density of occurrence in the modal class and the density of occurrence in the class immediately preceding the modal class
d_2 = difference between the density of occurrence in the modal class and the density of occurrence in the class immediately following the modal class
C = class interval of the modal class

The computation for Frequency Distribution Alternative B is illustrated in Table 3.5.

Table 3.5 Computation of the mode by the interpolation formula: Frequency Distribution Alternative B

Daily wages ($)	M	f	
9–11	$10	4	$Mo = L + \dfrac{d_1}{d_1 + d_2} C$
11–13	12	8	
13–15	14	6	
15–17	16	25	$= 15 + \dfrac{(25 - 6)}{(25 - 6) + (25 - 15)}$ (2)
17–19	18	15	
19–21	20	22	$= 15 + \dfrac{19}{19 + 10}$ (2)
21–23	22	8	
23–25	24	9	$= 15 + \frac{19}{29}$ (2)
25–27	26	5	$= 15 + \frac{38}{29}$
27–29	28	6	$= 15 + 1.31$
29–31	30	1	$= \$16.31$
31–33	32	3	
		112	

Modal class: 15–17

The Empirical Mode

In a perfectly symmetrical unimodal (single-peaked) distribution, the values of the mean, median, and mode coincide (Figure 3.8). As the distribution departs from symmetry, however, the values of the mean, median, and mode move apart in a predictable manner (provided that the departure from symmetry is moderate). The mode remains under the peak of the curve (the maximum value of the density function). The value of the arithmetic mean, being influenced by extremes, will be pulled away from the mode in the direction of the extreme values (i.e., in the direction of the tail of the distribution). The median is influenced by the number of extremes, although not by their values. Thus, the median will also be pulled in the direction of the extreme values, although to a lesser extent than the mean.

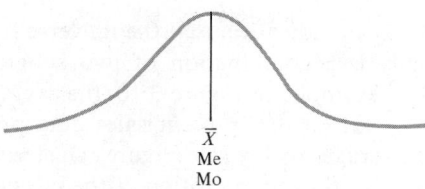

Figure 3.8

Provided the departure from symmetry is moderate, if the arithmetic mean is pulled in the direction of the extreme values by an amount designated as *d*, the median can be said to be pulled in the same direction by an amount equal to approximately two-thirds of *d* (see Figure 3.9). Thus, the relationship between the

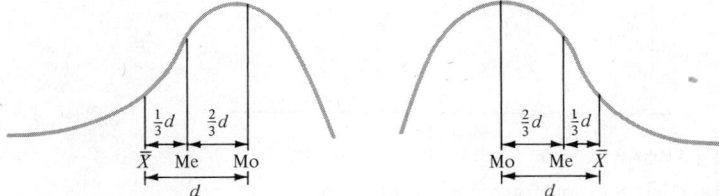

Figure 3.9

values of the mean, median, and mode can be described by the following algebraic relationship[6]:

$$Mo = \overline{X} - 3(\overline{X} - Me)$$

Given a previous determination of the values of the mean and median, the above relationship can be used to generate an estimate of the value of the mode. The computation is illustrated below for Frequency Distribution Alternative B.

$$\overline{X} = \$19.16$$

$$Me = 18.73$$

$$Mo = \overline{X} - 3(\overline{X} - Me)$$

$$= \$19.16 - 3(\$19.16 - \$18.73)$$

$$= \$19.16 - 3(\$0.43)$$

$$= \$19.16 - \$1.29$$

$$= \$17.87$$

This estimate is referred to as the empirical mode because the relationship on which its computation is based is derived from observed empirical relationships.

Occasionally, a distribution will be bimodal. A bimodal distribution contains two distinct values that occur much more frequently than other values of the

---

[6] Assume that the largest value in the distribution is $100 and that the distribution is moderately skewed. The value of the mode could then be estimated by the relationship $Mo = \overline{X} - 3(\overline{X} - Me)$. Then, holding everything else equal, suppose the value $100 is replaced by the value $1,000,000. In the new situation, the mode would be unchanged, the median would be unchanged, but the arithmetic mean would be pulled significantly to the right; thus, the relationship $Mo = \overline{X} - 3(\overline{X} - Me)$ would break down.

variable. When this occurs, it usually is an indication that the universe from which the observations were drawn actually is a combination of two subgroups with distinctly different characteristics. For example, in Figure 3.10, the average weekly sales figures clustering around mode₁ ($10,000) represent sales data prior to the opening of a new branch, while the average weekly sales figures clustering around mode₂ ($15,000) represent sales data after the incorporation of the sales of the new branch.

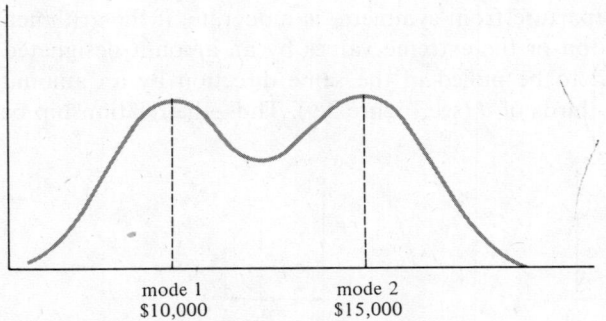

mode 1 $10,000   mode 2 $15,000

Figure 3.10 Distribution of average weekly sales.

3.11 QUARTILES AND PERCENTILES

Quartiles

[handwritten: Position + value]

The quartiles are positional measures, their values being determined by the values of the items occupying particular positions within the distribution. The three quartiles approximately divide the distribution into four equal parts. Graphically, the quartiles can be considered as the three values along the number line above which the construction of perpendiculars causes the histogram of the distribution to be disected into four equal parts (Figure 3.11).

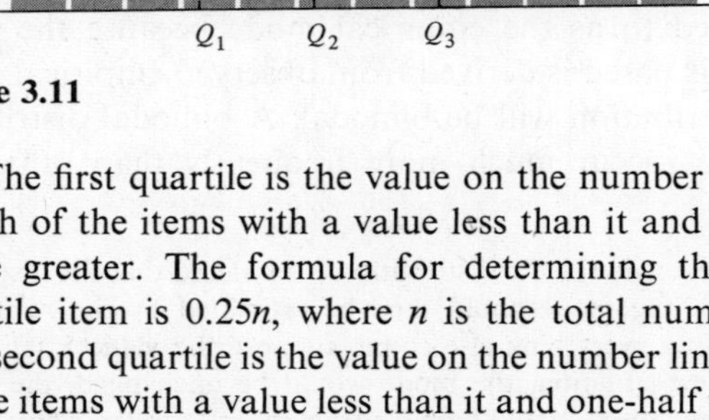

[handwritten in left margin: Q₂ = Median; .25n; n = Total # of items; for position; for mode distribution]

Q_1 Q_2 Q_3

Figure 3.11

The first quartile is the value on the number line that has approximately one-fourth of the items with a value less than it and three-fourths of the items with a value greater. The formula for determining the *position* occupied by the first quartile item is $0.25n$, where n is the total number of items in the distribution. The second quartile is the value on the number line that has approximately one-half of the items with a value less than it and one-half with a value greater. As is evident from its definition, the second quartile is none other than an old friend, the median.

The formula for determining the *position* occupied by the second quartile item is $0.50n$. The third quartile is the value that has approximately three-fourths of the items with a value less than it and one-fourth of the items with a value greater. The formula for determining the position in the distribution occupied by the third quartile item is $0.75n$.

Computationally, the formula used to determine the value of the quartiles is identical to the formula used to compute the median (the second quartile). The only difference is that the symbols now refer to the particular quartile class with which the investigator is concerned. For example, if it is desired to compute the first quartile, the symbols in the formula have the following interpretation:

$$Q_1 = L + \frac{a}{f} C$$

where

L = lower limit of the first quartile class
a = number of additional items needed in the first quartile class in order to reach the position occupied by the first quartile item
f = frequency of the first quartile class
C = class interval of the first quartile class

Computation of the first, second, and third quartiles for Frequency Distribution Alternative B is illustrated below.

Daily wages ($)	Frequency	Cumulative frequency	
9–11	4	4	
11–13	8	12	
13–15	6	18	
15–17	25	43	First quartile class
17–19	15	58	Second quartile class
19–21	22	80	
21–23	8	88	Third quartile class
23–25	9	97	
25–27	5	102	
27–29	6	108	
29–31	1	109	
31–33	3	112	
	112		

(handwritten: $C = 2$; $28 - 18 = a$; $\frac{28-18}{25} = 2$; $15 + \frac{10}{25}(2)$; $15 + .4(2) = 15.8$)

POSITIONS

first quartile $0.25n = (0.25)(112) = 28\text{th position}$

second quartile $0.50n = (0.50)(112) = 56\text{th position}$

third quartile $0.75n = (0.75)(112) = 84\text{th position}$

VALUES

first quartile $Q_1 = L + \dfrac{a}{f} C = 15 + \frac{10}{25}(\$2) = 15 + \frac{20}{25}$

$$= 15 + 0.80 = \$15.80$$

(handwritten: $Q_2 = L + \frac{a}{f}C = 17 + \frac{56-43}{15} \cdot 2 = 17 + \frac{26}{15}$)

(handwritten: $Q3 = 21 + \frac{84-80}{8} \cdot 2 = 21 + \frac{4}{8}(2) = 22$)

second quartile $\quad Q_2 = L + \dfrac{a}{f} C = 17 + \tfrac{13}{15}(\$2) = 17 + \tfrac{26}{15}$

$$= 17 + 1.733 = \$18.733$$

third quartile $\quad Q_3 = L + \dfrac{a}{f} C = 21 + \tfrac{4}{8}(\$2) = 21 + \tfrac{8}{8}$

$$= 21 + 1.00 = \$22$$

INTERPRETATION

25 percent of the workers earn a daily wage of $15.80 or less

50 percent of the workers earn a daily wage of $18.733 or less

75 percent of the workers earn a daily wage of $22 or less

Percentiles

A similar line of reasoning enables the analyst to utilize the formula to generate other positional measures. There are 99 percentiles. They possess the characteristic of dividing the distribution into 100 equal parts. For example, approximately 37 percent of the items have a value less than the 37th percentile, and 63 percent of the items have a value greater than the 37th percentile.

The quartiles are special cases of the percentile. The 25th percentile is the first quartile. The 50th percentile is the second quartile. The 75th percentile is the third quartile.

When the formula $L + (a/f)C$ is used to generate the values of the percentiles, the symbols relate to the particular percentile class of interest. The computation of the 83rd percentile for Frequency Distribution Alternative B is illustrated below.

position $\quad 0.83(n) = (0.83)(112) = 92.96$

value $\quad P_{83} = L + \dfrac{a}{f} C$

$$= 23 + \dfrac{4.96}{9}(2)$$

$$= 23 + \dfrac{9.92}{9}$$

$$= 23 + 1.102$$

$$= \$24.102$$

INTERPRETATION

83 percent of the workers earn a daily wage of $24.102 or less

17 percent of the workers earn a daily wage of more than $24.102

TECHNICAL NOTE 1

Short-cut formulas for the computation of the arithmetic mean for a frequency distribution

3.12 THE ACTUAL DEVIATIONS METHOD

The actual deviations method is based on the characteristic that the sum of the deviations from the arithmetic mean is equal to zero. The method involves determining the value on the number line from which the sum of the deviations is equal to zero and in so doing simultaneously identifying the value of the arithmetic mean.

$$\bar{X} = A + \frac{\sum (f_i d_i)}{n}$$

where

$\bar{X}$ = arithmetic mean
A = assumed mean

   The assumed mean can be the midpoint of any of the classes. To facilitate computation, however, it is preferable to choose the midpoint of a class close to the center of the distribution and associated with a large class frequency.

$\sum$ = the sum of
d_i = the difference between the value of the assumed mean and the value of the midpoint of the ith class
   $M_i - A = d_i$
f_i = the class frequency of the ith class (and the number of times the deviation d_i occurs)
n = the total frequency for the frequency distribution
$\sum (f_i d_i)$ = the sum of the deviations from the assumed mean. If A, the assumed mean, actually is the true mean, then $\sum (f_i d_i)$ is equal to zero and the second term disappears:

$$\bar{X} = A + \frac{\sum (f_i d_i)}{n}$$

$$= A + \frac{0}{n}$$

$$= A$$

If the assumed mean is not the true mean, the second term $\sum (f_i d_i)/n$ acts as a correction factor and indicates by its sign, the direction, and its magnitude the distance along the number line that the assumed mean lies away from the value from which the sum of the deviations will be equal to zero.

   The supporting computations associated with computing the arithmetic mean by the actual deviations method for Frequency Distribution Alternative B are illustrated in Table 3.6. Any of the midpoints could serve as the assumed mean, but $20 is a logical choice. It is located near the center of distribution and is associated with a large class frequency.

   The next step is to compute the d values or the difference between the value

Table 3.6 Computation of the arithmetic mean by the actual deviations method: Frequency Distribution Alternative B

Daily wages ($)	M	f	d	fd
9–11	$10	4	−10	−40
11–13	12	8	−8	−64
13–15	14	6	−6	−36
15–17	16	25	−4	−100
17–19	18	15	−2	−30
19–21	20	22	0	0
21–23	22	8	+2	+16
23–25	24	9	+4	+36
25–27	26	5	+6	+30
27–29	28	6	+8	+48
29–31	30	1	+10	+10
31–33	32	3	+12	+36
		112		−270
				+176
				−94

of the assumed mean and the value of the midpoint for each of the classes in the distribution.

It would be incorrect just to add up the values in the d column to get the sum of the deviations from the assumed mean, because the deviations or d values do not occur just once but rather once for every item in the class. Thus, each of the four items in the class $9–$11 is treated as having a deviation of −10, generating a total deviation for the class $9–$11 of −40. Similarly, for the class $11–$13, the deviation of −8 occurs not once but rather once for each of the eight items in the class, generating a total deviation for the items in the class $11–$13 of −64, and so on. By creating an fd column, the total deviation for each of the classes in the distribution can be generated. Summing the fd column, the total deviation from the assumed mean for the entire frequency distribution is obtained.

The positive and negative deviations must be netted off to obtain the net sum of the deviations from the assumed mean. Substituting into the formula, one obtains

$$\overline{X} = A + \frac{\sum (fd)}{n}$$

$$= 20 + \frac{-94}{112}$$

$$= 20 - 0.84$$

$$= \$19.16$$

Note that the subscripts have been dropped in the interest of notational simplicity.

The correction factor of −0.84 is an instruction to move away from the assumed mean a distance of 0.84 along the number line in a negative direction in order to reach the value from which the sum of the deviations is equal to zero.

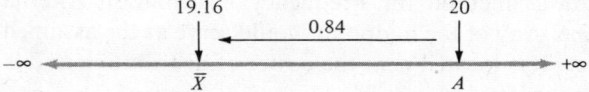

Note that it is important that deviations for values less than the assumed mean be assigned negative values and that deviations for values greater than the assumed mean be assigned positive values. If the pattern of signs were to be reversed, the correction factor would indicate a movement of 0.84 in a positive direction along the number line, which would result in a movement away from the mean rather than toward it. The value of the arithmetic mean computed for Frequency Distribution Alternative B by the actual deviations method is identical to that computed by the long method. This is in accordance with expectations, since the arithmetic mean is rigidly defined.

Inspection of the d column reveals that the d's or deviations from the assumed mean are all multiples of the class interval or $2. This will always be the case for a frequency distribution with equal class intervals. The third technique examined, the step deviations method, uses this relationship to advantage by expressing the deviations in class interval units rather than in actual units. By utilizing this technique, the magnitude of the figures used in the computation are reduced. Computations are performed with the reduced step deviation units. The results are converted back into actual deviation units in the last step by multiplying by the value of the class interval.

3.13 THE STEP DEVIATIONS METHOD

$$\overline{X} = A + \frac{\sum (f_i d_{c_i})}{n} \cdot C$$

where

$\overline{X}$ = arithmetic mean
A = assumed mean
f = frequency of occurrence in the ith class
d_{c_i} = the difference between the value of the assumed mean and the value of the midpoint of the ith class in the distribution *expressed in class interval units*:

$$d_{c_i} = \frac{M_i - A}{c}$$

Thus, the d value for the class $27–$29 of $+$8 would now be expressed as a d_c of $+4$ (i.e., four class interval units of $2 apiece).

C = the class interval, which should be the same for all classes if this method is to be used.

Note that it is necessary to multiply each d_c value by the frequency of the class with which it is associated. Each d_c value or deviation in class interval units occurs not once but rather once for every item in the class it represents.

$$\overline{X} = A + \frac{\sum (f d_c)}{n} \cdot C$$

$$= 20 + \frac{-47}{112} (2)$$

$$= 20 + \frac{-94}{112}$$

$$= 20 - 0.84$$

$$= \$19.16$$

Note that the subscripts have been dropped in the interest of notational simplicity.

Table 3.7 Computation of the arithmetic mean by the step deviations method: Frequency Distribution Alternative B

Daily wages ($)	M	f	d_c	fd_c
9–11	$10	4	−5	−20
11–13	12	8	−4	−32
13–15	14	6	−3	−18
15–17	16	25	−2	−50
17–19	18	15	−1	−15
19–21	20	22	0	0
21–23	22	8	+1	+8
23–25	24	9	+2	+18
25–27	26	5	+3	+15
27–29	28	6	+4	+24
29–31	30	1	+5	+5
31–33	32	3	+6	+18
		112		−135
				+88
				−47

TECHNICAL NOTE 2

The geometric mean

The concept of the geometric mean will be developed within the framework of the following example:

A machine with an expected life of 5 years is to be depreciated at a uniform rate of diminishing book value. The amount charged to depreciation is to be some constant percentage of book value. Book value is the portion of original cost still undepreciated and, of course, declines over the life of the asset. Assume that the original cost of the machine was $10,000 and that it is estimated it will be worth $200 at the end of the 5-year period. Determine the constant rate of declining balance that would reduce book value to the desired $200 level at the end of the fifth year.

Let X_i represent the percent decline in book value in the ith year. Then:

$$\$10,000X_1 = \text{book value at end of first year}$$

$$(\$10,000X_1)X_2 = \text{book value at end of second year}$$

$$(\$10,000X_1 \cdot X_2)X_3 = \text{book value at end of third year}$$

$$(\$10,000X_1 \cdot X_2 \cdot X_3)X_4 = \text{book value at end of fourth year}$$

$$(\$10,000X_1 \cdot X_2 \cdot X_3 \cdot X_4)X_5 = \text{book value at end of fifth year}$$

By definition:

$$X_1 = X_2 = X_3 = X_4 = X_5$$

and

$$\$10,000X_1 \cdot X_2 \cdot X_3 \cdot X_4 \cdot X_5 = \$200$$

Substituting X for X_1, X_2, X_3, X_4, and X_5, we obtain

$$\$10{,}000X \cdot X \cdot X \cdot X \cdot X = \$200$$

$$\$10{,}000X^5 = \$200$$

$$X^5 = \frac{\$200}{\$10{,}000}$$

$$X^5 = 0.02$$

$$X = \sqrt[5]{0.02}$$

The easiest approach for determining the fifth root of a number involves the use of logarithms. Table A in the Appendix contains a brief explanation of the concept of a logarithm and a table of common logarithms. In logarithmic form, $X = \sqrt[5]{0.02}$ becomes $\log X = (\log 0.02)/5$.

$$\log X = \frac{\log 0.02}{5} = \frac{8.3010 - 10}{5} = \frac{48.3010 - 50}{5}$$

$$= 9.6602 - 10$$

Finding the antilog:

$$\text{antilog } 9.6602 - 10 = X$$

$$0.4573 = X$$

Thus:

$$X = 45.73 \text{ percent}$$

The constant percent decline in book value is 45.73 percent. It follows that 54.27 percent of the preceding year-end book value will be the depreciation charge.

End of year	Book value		Depreciation charge	
0	$10,000			
1	(0.4573)($10,000)	= $4,573	(0.5427)($10,000)	= $5,427
2	(0.4573)($4,573)	= $2,091	(0.5427)($4,573)	= $2,482
3	(0.4573)($2,091)	= $956	(0.5427)($2,091)	= $1,135
4	(0.4573)($956)	= $437	(0.5427)($956)	= $519
5	(0.4573)($437)	= $200	(0.5427)($437)	= $237

Technically, the preceding computation involved finding a geometric mean. A geometric mean (GM) is a mean rate of change or mean percentage change. In general, it involves finding the nth root of the product of the values X_1, X_2, ..., X_n where X_1, X_2, ..., X_n are not necessarily equal (in fact, usually not equal):

$$\text{GM} = \sqrt[n]{X_1 \cdot X_2 \cdot \ \cdots \ \cdot X_n}$$

The nth root is most efficiently found by using logarithms. The geometric mean, therefore, is usually expressed in the convenient computational form

$$\log \text{GM} = \frac{\sum\limits_{i=1}^{n} \log X_i}{n}$$

In words: The geometric mean is the antilog of the sum of the logarithms of the values of the factors divided by the number of factors comprising the group.

Characteristics of the geometric mean:

1. It is rigidly defined.
2. It is a difficult concept to understand.
3. The geometric mean is always less than the arithmetic mean of the same set of variables unless all the variables are equal to some constant amount. This is a reflection of the fact that the geometric mean assigns less weight to extremes. To illustrate:

arithmetic value	logarithm
10,000	4
100	2
10	1

$$\bar{x} = \frac{\sum x}{n} \qquad\qquad \log \text{GM} = \frac{\sum \log x}{n}$$

$$\bar{x} = \frac{10,110}{3} \qquad\qquad \log \text{GM} = \tfrac{7}{3} = 2.3333$$

$$\bar{x} = 3370 \qquad\qquad \text{antilog } 2.3333 = 215.4$$

A simple survey of the above example demonstrates why the mean of the logarithms of a set of numbers will be significantly less influenced by extremes than the mean of the original arithmetic values. Occasionally, an analyst will elect to use the geometric mean in order to achieve precisely this effect.

The geometric mean is typically used in averaging index numbers, rates of change, ratios, and other sets of data expressed in ratio or percentage form.

EXAMPLE 3.1: Computation of an Index Number

Product	Ratio of this year's price to last year's price X_i
1	1.07
2	0.98
3	1.26

Assume that the products are of equal importance, thus eliminating the necessity of introducing weights. Determine the average rate of change in price.

$$\log \text{GM} = \frac{\sum \log X_i}{n}$$

Product	X_i	$\log X_i$
1	1.07	0.0294
2	0.98	9.9912 − 10
3	1.26	0.1004
		10.1210 − 10

$$\sum \log X_i = 10.1210 - 10 \text{ or } 0.1210$$

$$\log \text{GM} = \frac{\sum \log X_i}{n} = \frac{0.1210}{3} = 0.0403$$

antilog $0.0403 = 1.097$

The average rate of change in price is 1.10 or a 10 percent increase.

EXAMPLE 3.2

Suppose a piece of property was purchased for $20,000 and sold 10 years later for $32,600. What is the average annual rate of return on the original $20,000 investment?

$$\frac{\$32,600}{\$20,000} = 1.63$$

Alternatively, this could be written as

$$\$20,000 X^{10} = \$32,600$$

$$X^{10} = 1.63$$

Thus

$$X = \sqrt[10]{1.63}$$

$$\log \text{GM} = \frac{\log 1.63}{10}$$

$$= \frac{0.2122}{10}$$

$$= 0.0212$$

antilog $0.0212 = 1.05$

Thus, the investment yielded a mean rate of return of 5 percent over the 10-year period.

PROBLEMS

Problem 3.1

Given the following frequency distribution as one of the possible solutions for Problem 2.1:

Test scores	Number of applicants
40–60	5
60–80	8
80–100	24
100–120	46
120–140	47
140–160	33
160–180	20
180–200	11
200–220	6
	200

Compute:
1. The arithmetic mean
2. The median
3. The crude mode
4. The interpolation formula for the mode
5. The empirical mode
6. The first quartile
7. The third quartile
8. The 80th percentile

Problem 3.2

Given the following frequency distributions as one of the possible solutions for Problem 2.2:

Batting averages	Prior to increase in practice time	After increase in practice time
0.100–0.200	25	5
0.200–0.300	40	30
0.300–0.400	20	50
0.400–0.500	10	10
0.500–0.600	5	5
	100	100

Compute for each of the distributions:
1. The arithmetic mean
2. The median
3. The crude mode
4. The interpolation formula for the mode
5. The empirical mode

Problem 3.3

Given the following frequency distribution:

Dollar sales	Number of sales slips
15–25	9
25–35	21
35–45	42
45–55	17
55–65	6
65–75	5
	100

Compute:
  1. The arithmetic mean
  2. The median
  3. The crude mode
  4. The interpolation formula for the mode
  5. The empirical mode
  6. The first quartile
  7. The third quartile
  8. The 90th percentile

Problem 3.4

The computer center at Motheaten University provides an over-the-counter quick batch service. An instructor who desired to evaluate the effectiveness of the operation had students in his class record the time elapsed between submission of their programs and receipt of the printed outputs. The pattern that emerged for 200 observations is presented in the frequency distribution below:

Time elapsed (min)	Number of occasions
0–2	40
2–4	80
4–6	50
6–8	27
Over 8	3
	200

  1. Explain why the arithmetic mean cannot be computed from the above distribution.
  2. What additional information would be needed to render the arithmetic mean capable of computation?
  3. Compute the median.
  4. Interpret the value obtained for the median in part 3.
  5. Compute the value of the 90th percentile.
  6. Interpret the value obtained for the 90th percentile.

Problem 3.5

The following observations were drawn from the experience of Grossman's, Inc., with respect to reorder times for a machine part from Dependable Distributors:

first observation	7 days
second observation	3 days
third observation	6 days
fourth observation	5 days
fifth observation	7 days
sixth observation	4 days
seventh observation	8 days

1. Determine the modal reorder time.
2. Comment on the meaningfulness of the value obtained in part 1.
3. Determine the median reorder time.
4. Determine the mean reorder time.

Problem 3.6

A sample of size 7 selected from the sophomore class revealed the following commuting distances: 12, 15, 7, 13, 9, 11, and 3 miles.

1. Determine the arithmetic mean of the sample.
2. Determine the median of the sample.

Problem 3.7

The following frequency distribution represents the distribution of distances a number of claimants for unemployment insurance indicated they would be willing to travel to a place of employment.

Round-trip travel restrictions (min)	Percent of respondents
Under 30	2
30–60	34
60–90	20
90–120	37
Over 120	7
	100

1. Explain why the arithmetic mean cannot be computed from the above distribution.
2. What additional information would be needed to render the arithmetic mean capable of computation?
3. Compute the median.
4. Interpret the value obtained for the median in part 3.
5. Compute the value of the 90th percentile.
6. Interpret the value obtained for the 90th percentile.
7. Explain why the mode cannot be computed for the above distribution by the interpolation formula for the mode. What additional information would be needed? Is this the same information that would be required for the computation of the arithmetic mean, or is it of a different nature? Be specific.

Problem 3.8

Suppose the instructor in an advanced elective containing 15 students observed the following grades on the final (and only) exam: 15, 15, 16, 17, 17, 17, 25, 27, 29, 29, 31, 32, 33, 33, and 100.

1. Determine:
   a. the modal grade
   b. the median grade
   c. the mean grade
2. What information (if any) is conveyed by the following:
   a. the modal grade
   b. the median grade
   c. the mean grade
3. The course is taken on a pass–fail basis. Those students attaining less than average performance are to fail. Which of the three measures referred to should be used as the "average" in this situation? Justify your answer.
4. If the distribution of grades had been different, would your answer have changed? Why or why not?

Problem 3.9

One of the world's most urgent problems is the continual growth of population and the pressure increased population places on scarce resources. A measure of the extent to which various countries contribute to this problem is the length of time it would take a country to double its population given its present growth rate. Given the following distribution, compute the median length of time required to double a country's population.

Number of years required to double its population	Number of countries
20–40	30
40–60	28
60–100	22
100–150	16
150–200	4
	100

Problem 3.10

The arithmetic mean is described as being "rigidly defined." What does this mean? Is such a trait desirable?

Problem 3.11

In computing representative family income for a geographic locality, median family income rather than mean family income is typically used. Why?

Problem 3.12

In general terms, describe the type of situation where the median functions more effectively than the arithmetic mean as a representative value.

Problem 3.13

Given the following frequency distribution:

Part-time hours of work per week	Number of students
5–10	11
10–15	23
15–25	40
25–30	20
30–35	6
	100

Compute:
1. The mean
2. The median
3. The crude mode
4. The mode by the interpolation formula
5. The empirical mode

A word of caution: The distribution contains unequal class intervals.

Problem 3.14

Given:

Income of persons purchasing ordinary life insurance

Income ($)	Number of policies
Under 2,000	175
2,000–3,000	641
3,000–4,000	1225
4,000–5,000	1225
5,000–7,500	1458
7,500–10,000	408
10,000 and over	699
	5831

1. Compute:
   a. the first quartile
   b. the second quartile
   c. the third quartile
2. What information is conveyed by each of the above three measures with respect to the income of persons purchasing ordinary life insurance?

Problem 3.15

What underlying assumption is reflected in the formula used to calculate the median from a frequency distribution?

MEASURES OF DISPERSION AND SKEWNESS

4

Handwritten notes:

1. Central Tendency: $\bar{X}$, Md, Mo

2. Meas. of Dispersion

3. Measures of Skewness

4. Meas. of Kurtosis
   - leptokurtic
   - mesokurtic
   - platykurtic

Measures of dispersion play an extremely important role in statistical analysis. Specifically:

1. Measures of dispersion provide information with respect to the extent of scatter or, conversely, the degree of clustering in a set of data.
2. Measures of dispersion are useful in evaluating the representativeness of a measure of central tendency. The significance of an average as a representative value is a function of the degree of dispersion of the individual items around it. The more clustered the data, the more effectively the measure of central tendency represents the distribution as a whole.
3. Measures of dispersion play a vital role in the analytics of statistical inference. The process of sampling is designed to provide an estimate of a universe parameter of interest to the analyst. For

example, the sample arithmetic mean is used as an estimate of the value of the universe mean. Naturally, a certain amount of error (sample value not equal to the universe value) is expected. Measures of dispersion enable the analyst to generate statements about the magnitude of such errors likely to be incurred in association with a particular sampling scheme. Although this is an extremely important application, treatment of this topic will be postponed until the subject arises in the natural progression of understanding of the concepts of statistical inference.[1] For the present, emphasis will be restricted to the computational techniques associated with the various measures of dispersion and the role the measures play in statistical description.

Measures to be examined include: (1) the range, (2) the interquartile range, (3) the average deviation, and (4) the standard deviation.

4.1 THE RANGE

The range is the simplest measure of dispersion to compute. It refers to the distance along the number line encompassed by the smallest and largest values in the distribution. When computing the range for a frequency distribution, it is customary to take the distance between the lower limit of the first class and the upper limit of the last class as the value of the range.

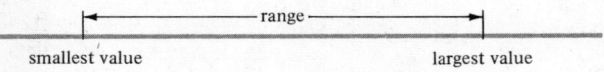

smallest value   largest value

The major disadvantage associated with the range as a measure of dispersion is that it is based solely on the dispersion of the two extreme values. Thus, the range fails to communicate any information about the clustering or lack of clustering of the values in the distribution located between the two extremes.

4.2 THE INTERQUARTILE RANGE

A measure occasionally introduced as an alternative to the range is the interquartile range. The interquartile range refers to the difference in value between the first and third quartile values. This measure of dispersion is also based on the values of only two of the items in the distribution (in the case of the interquartile range, the two values are the first and third quartiles). Although the interquartile range overcomes a major criticism of the range (the interquartile range is not influenced by extremes), it fails to convey any information with respect to the dispersion of the items located either within or outside the interquartile range. As a result, the interquartile range does not generate a great deal of information about the nature of dispersion for the distribution as a whole.

4.3 THE AVERAGE DEVIATION

The average deviation is an improvement over the range and the interquartile range in the sense that the values of all the items in the distribution are included in

[1] The role of measures of dispersion in the analytics of statistical inference is discussed in Chapters 12 and 13.

the computation. It is advisable to specify whether the average deviation has been computed from the mean or the median since the average deviation will always be smaller when computed from the median, provided the values of the mean and median are not the same.[2]

Computation of the average deviation consists of determining the amount each item in the distribution differs from the mean (or median), summing the absolute values of the deviations (ignoring plus and minus signs), and dividing by the number of deviations. Recall that a characteristic of the arithmetic mean is that the sum of the deviations from the mean always totals zero when the signs are taken into consideration. Thus, it is necessary to work with the absolute values if the measure is to make any sense.

The Formulas

AVERAGE DEVIATION COMPUTED FROM THE MEAN: INDIVIDUAL ITEMS

$$AD_{\bar{x}} = \frac{\sum |X_i - \bar{X}|}{n}$$

where

$AD_{\bar{x}}$ = average deviation when computed from the arithmetic mean

$X_i - \bar{X}$ = deviation of the ith item from the mean

$| \quad |$ is an operator that is interpreted as an instruction to use the absolute value (i.e., ignore the plus and minus signs)

n = total number of observations (consequently the total number of deviations)

AVERAGE DEVIATION COMPUTED FROM THE MEDIAN: INDIVIDUAL ITEMS

$$AD_{Me} = \frac{\sum |X_i - Me|}{n}$$

where

AD_{Me} = average deviation when computed from the median

$X_i - Me$ = deviation of the ith item from the median

The example of the five different models of sump pumps and their respective costs introduced in Chapter 3 will be used to illustrate the computational procedure (see Tables 4.1 and 4.2).

4.4 THE STANDARD DEVIATION

When the analyst's intention is to use the measure of dispersion to provide a description of the scatter in the data, the average deviation performs quite effectively. When the measure of dispersion is to be used as an input in further statistical

---

[2] Recall that the sum of the absolute deviations is minimized when it is computed from the median.

Table 4.1 Illustration of the computation of the average deviation from the mean

Model	Cost to the firm, X_i ($)	$X_i - \bar{X}$	$|X_i - \bar{X}|$
A	11.85	−0.35	0.35
B	11.75	−0.45	0.45
C	11.80	−0.40	0.40
D	13.15	+0.95	0.95
E	12.45	+0.25	0.25
	61.00		2.40

(handwritten annotations: "Mean" with arrow; "12.20", "12.20", "12.20", "12.20", "12.20" beside the cost column)

$$\bar{X} = \frac{\$61.00}{5} = \$12.20$$

$$AD_{\bar{x}} = \frac{\sum |X_i - \bar{X}|}{n} = \frac{\$2.40}{5} = \$0.48$$

Note that the subscripts in the formula have been dropped in the interest of notational simplicity.

Table 4.2 Illustration of the computation of the average deviation from the median

Model	Cost to the firm, X_i ($)	$X_i - \text{Me}$	$|X_i - \text{Me}|$
A	11.85	0	0
B	11.75	−0.10	0.10
C	11.80	−0.05	0.05
D	13.15	+1.30	1.30
E	12.45	+0.60	0.60
			$2.05

(handwritten annotation: "Median" with arrow)

$$\text{Me} = \$11.85$$

$$AD_{\text{Me}} = \frac{\sum |X_i - \text{Me}|}{n} = \frac{\$2.05}{5} = \$0.41$$

(handwritten annotation: "Av. Sqd Dev. from mean = variance")

analysis, however, problems arise. The computation of the average deviation necessitates the violation of a mathematical relationship (namely, ignoring the plus and minus signs). On the other hand, consideration of the plus and minus signs causes the deviations to cancel to zero when computed from the arithmetic mean. A need exists for a measure of dispersion that takes all values of the variable in the distribution into consideration yet overcomes the problem of the signs canceling out in a mathematically acceptable manner. The standard deviation is such a measure. The computational process consists of determining the deviations of each of the individual items from the mean, squaring the deviations (thereby converting both negative and positive values to positive values in a mathematically acceptable manner), summing the squared deviations, and then dividing by the number of deviations. The result is the average squared deviation from the mean and is called the *variance*. As will be demonstrated in later chapters, the variance possesses properties that make it a very useful measure in further statistical analysis. In many situations, however, what is needed is not a measure of dispersion expressed

[handwritten: Std Deviation = σ = $\sqrt{Variance}$]

in squared units but a measure of dispersion expressed in original units. Such a measure is obtained by taking the square root of the variance. The resulting measure is referred to as the *standard deviation.*[3]

4.5 FORMULAS FOR THE COMPUTATION OF THE VARIANCE AND STANDARD DEVIATION FOR INDIVIDUAL ITEMS

The Universe Variance and Standard Deviation

universe variance
$$\sigma_x^2 = \frac{\sum (X - \mu_x^2)}{N}$$

universe standard deviation
$$\sigma_x = \sqrt{\frac{\sum (X - \mu_x)^2}{N}}$$

[handwritten table and notes:
$(X-\mu)^2$

X	μ	$X-\mu$	$(X-\mu)^2$
1	3	-2	4
2	3	-1	1
3	3	0	0
4	3	1	1
6	3	2	4

Variance = $\frac{10}{5}$

Variance = 2 *]*

where

     σ is the Greek small letter sigma and is used to represent the standard deviation of the universe. The function of the x subscript is to indicate that it is the standard deviation of the X values.

$X - \mu_x$ refers to the difference between a given value and the arithmetic mean of all the X_i values.

     N is the total number of items comprising the universe.

The Sample Variance and Sample Standard Deviation

The formulas below are based on the assumption that the sample standard deviation is not to be used as an estimate of the universe standard deviation.[4]

sample variance
$$S_x^2 = \frac{\sum (X - \overline{X})^2}{n}$$

sample standard deviation
$$S_x = \sqrt{\frac{\sum (X - \overline{X})^2}{n}}$$

where

     S_x = standard deviation of the X values contained in the sample
     n = sample size
$X - \overline{X}$ refers to the difference between a given X_i value and the arithmetic mean of all the X_i values comprising the sample

It is the dispersion of the universe, not the dispersion of the sample, that is of concern to the analyst. Thus, the standard deviation of the sample is seldom computed for its own sake, but rather is computed to provide an estimate of the

---

[3] The standard deviation is also referred to as the root-mean-square deviation.

[4] If the sample variance and/or sample standard deviation are desired not for their own values but rather as estimates of the corresponding universe parameters, $n - 1$ is used in the formulas in place of n. The symbols $\hat{\sigma}^2$ and $\hat{\sigma}_x$ are used to indicate that the values obtained are estimates of the universe parameters. The "hat" symbol will be employed to indicate an estimated value throughout the text.

standard deviation of the universe from which the sample has been selected. A desirable characteristic for a sample statistic to possess if it is to provide representative estimates of the universe parameter is that it be unbiased.

4.6 THE CONCEPT OF AN UNBIASED ESTIMATOR

An estimator is said to be unbiased if the average value of the sample statistic is equal to the universe parameter it is used to estimate. In effect, if one were to take all possible samples of a given size that could be selected from a universe and computed the given sample statistic for each, the average value of these sample statistics should equal the value of the universe parameter. The sample arithmetic mean is an unbiased estimator of the universe mean. The sample variance is *not* an unbiased estimator of the universe variance. It can be demonstrated that the average value of the sample variance provides an estimate of the universe variance that is biased downward in a manner described by the following relationship:

$$\begin{array}{l} \text{average value of the} \\ \text{sample variance} \end{array} = \left(\frac{n-1}{n}\right)\left(\begin{array}{l}\text{value of the} \\ \text{universe variance}\end{array}\right)$$

$$\text{Ave}\left[\frac{\sum (X - \overline{X})^2}{n}\right] = \frac{n-1}{n}(\sigma_x^{\,2})$$

Multiplying both sides by $n/(n-1)$, the following relationship emerges:

$$\left(\frac{n}{n-1}\right)\text{Ave}\left[\frac{\sum (X - \overline{X})^2}{n}\right] = \left(\frac{n}{n}\right)\sigma^2 \cdot \left[\frac{n}{n-1}\right]$$

$$\text{Ave}\left[\frac{\sum (X - \overline{X})^2}{n-1}\right] = \sigma^2$$

The average value of the sample variance is equal to $(n-1)/n$ times the value of the universe variance when the sample variance is computed by dividing by n. Manipulation of the relationship, however, demonstrates that the average value of the sample variance will be exactly equal to the universe variance (i.e., be an unbiased estimator) if the sample variance is computed by dividing by $n-1$. Thus, whenever the sample measure of dispersion is generated for the purpose of providing an estimate of the universe measure, one should divide by $n-1$ rather than n. The symbol n is used to designate sample size, and $n-1$ is referred to as the number of degrees of freedom. The number of degrees of freedom is interpreted as the number of useful items of information generated by a sample of a given size with respect to the estimation of a given universe parameter. A sample of size 1 generates an estimate of the universe mean but no information with respect to the universe variance. Thus, a sample of size 1 generates one piece of useful information if one is estimating the universe mean, but none if one is estimating the universe variance. If the sample size is increased to 2, more information is generated with respect to the mean, and an estimate of dispersion can now be generated. In the sample of size 2, two pieces of information are generated with respect to the estimation of the universe mean. The sample of size 2 provides only one useful input in the estimate of dispersion. The variance is always computed from the arithmetic mean; therefore, before any deviations can be generated, the arithmetic mean of the sample must be computed. Once the value of the first deviation is computed, the other is automatically and simultaneously determined because of the constraint that the sum of the deviations from the mean is always zero. Thus, although there are two items in the sample, there is only one independent observation with respect to dispersion in the sample. (The other is automatically determined with the assignment of the first.)[5] In general, every item in a sample generates information with respect to the estimation of the universe mean, but only $n - 1$ of the items in the sample provide information with respect to the estimation of the universe variance.

Actually, the general expression for degrees of freedom is $n - m$, where n is the number of observations and m is the number of degrees of freedom lost (m is equal to the number of constants that have to be estimated from the original data in order to compute the deviations—in this case there is only one, the arithmetic mean).

An empirical demonstration of the relationship between the universe parameters μ_x and σ_x^2 and the average values of the corresponding sample statistics is demonstrated employing the artificial universe consisting of the following four values:

$$X_1 = 1$$
$$X_2 = 2$$
$$X_3 = 3$$
$$X_4 = 4$$

Table 4.3 Computation of the universe mean and variance

X	$X - \mu_x$	$(X - \mu_x)^2$
1	-1.5	2.25
2	-0.5	0.25
3	$+0.5$	0.25
4	$+1.5$	2.25
10	0	5.00

Universe mean:

$$\mu_x = \frac{\sum X}{N} = \frac{10}{4} = 2.5$$

Universe variance:

$$\sigma_x^2 = \frac{\sum (X - \mu_x)^2}{N} = \frac{5.00}{4} = 1.25$$

The values of the universe parameters are $\mu_x = 2.5$ and $\sigma_x^2 = 1.25$. Next, a listing of all possible samples of size 2 is generated from the universe, assuming that the item drawn on the first draw is replaced prior to the second item being drawn. In association with each of the four values in the universe that could be selected on the first draw, any one of the four values in the universe could be selected on the second draw, generating 16 different ways in which a sample of size 2 could be obtained. The various possibilities are shown in Figure 4.1.

---

[5] It should be noted that if the universe mean were known and did not have to be estimated from the sample data, a sample of size n would provide n pieces of useful information, not $n - 1$, in the estimation of the value of the universe variance. This is because the sum of the deviations of the individual items in the *sample* from the value of the *universe mean* would not have to equal zero.

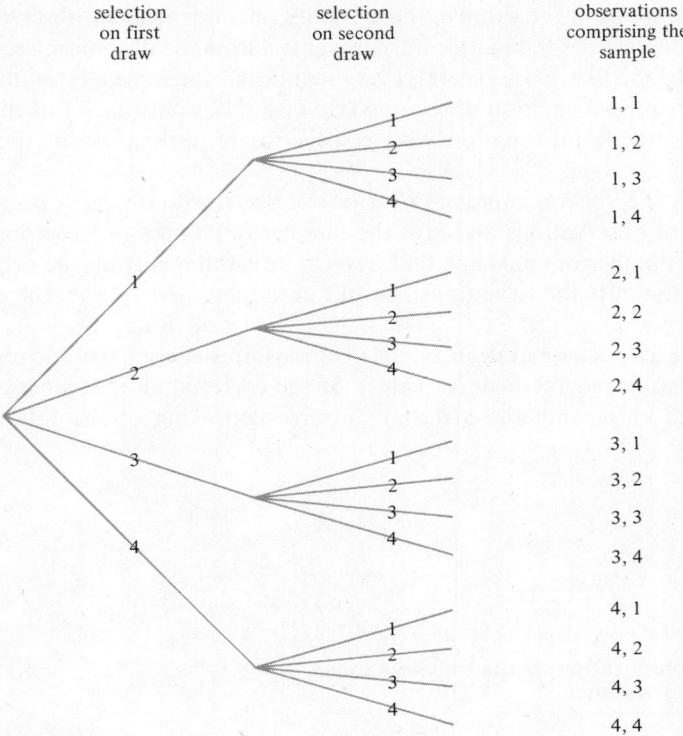

selection on first draw	selection on second draw	observations comprising the sample

Figure 4.1

Column 1 of Table 4.4 contains an enumeration of the composition of each of the samples of size 2 that could be generated. The arithmetic mean ($\bar{X}$) for each of the 16 samples was then computed and recorded in column 2. The average value of the sample means is 2.5 ($\bar{\bar{X}} = \sum \bar{X}/K = \frac{40}{16} = 2.5$, where K equals the number of sample means). The average value of the sample statistic (sample mean) is equal to the universe parameter (universe mean) that it is used to estimate.

Thus, the sample mean is an unbiased estimator of the universe mean. For each of the 16 samples, the deviations of the individual items from the mean of the sample were obtained, squared, and summed. These values are recorded in column 3. The entries in column 3, $\sum (X - \bar{X})^2$, are obtained in the following fashion:

For the sample comprised of the items 1, 2				For the samples comprised of the items 1, 4		
X	$X - \bar{X}$	$(X - \bar{X})^2$		X	$X - \bar{X}$	$(X - \bar{X})^2$
1	-0.5	0.25		1	-1.5	2.25
2	$+0.5$	0.25		4	$+1.5$	2.25
3		0.50		5		4.50

$$\bar{X} = \tfrac{3}{2} = 1.5$$
$$\sum (X - \bar{X})^2 = 0.50$$

$$\bar{X} = \tfrac{5}{2} = 2.5$$
$$\sum (X - \bar{X})^2 = 4.50$$

The sample variance was then computed by dividing the entries in column 3 by n, the sample size. The resulting values are recorded in column 4. The sample

Table 4.4 Computation of the sample means and variances

Column 1	Column 2	Column 3	Column 4	Column 5
			$\dfrac{\sum (X - \bar{X})^2}{n}$	$\dfrac{\sum (X - \bar{X})^2}{n - 1}$
All possible ways a sample of size 2 can be generated	Associated sample mean, $\bar{X}$	Associated sum of the squared deviations from the mean, $\sum (X - \bar{X})^2$	or $\dfrac{\sum (X - \bar{X})^2}{2}$	or $\dfrac{\sum (X - \bar{X})^2}{1}$
1, 1	1	0	0	0
1, 2	1.5	0.50	0.25	0.50
1, 3	2.0	2.00	1.00	2.00
1, 4	2.5	4.50	2.25	4.50
2, 1	1.5	0.50	0.25	0.50
2, 2	2.0	0.00	0.00	0.00
2, 3	2.5	0.50	0.25	0.50
2, 4	3.0	2.00	1.00	2.00
3, 1	2.0	2.00	1.00	2.00
3, 2	2.5	0.50	0.25	0.50
3, 3	3.0	0.00	0.00	0.00
3, 4	3.5	0.50	0.25	0.50
4, 1	2.5	4.50	2.25	4.50
4, 2	3.0	2.00	1.00	2.00
4, 3	3.5	0.50	0.25	0.50
4, 4	4.0	0.00	0.00	0.00
	$\sum \bar{X} = 40.00$		$\sum \left[\dfrac{\sum (X - \bar{X})^2}{n} \right]$ $= 10.00$	$\sum \left[\dfrac{\sum (X - \bar{X})^2}{n - 1} \right]$ $= 20.00$

variance was also computed dividing by $n - 1$, the number of degrees of freedom. The $\sum (X - \bar{X})^2/(n - 1)$ values so obtained are recorded in column 5.

When the sample variance is computed using the formula $\sum (X - \bar{X})^2/n$, the average value of the sample variances equals

$$\frac{\sum [\sum (X - \bar{X})^2/n]}{K} = \frac{10.00}{16} = 0.625$$

where K equals the number of sample variances. The average value of the sample variances has a value equivalent to 50 percent of the value of the universe variance, $\sigma_x^2 = 1.25$. This is in accordance with expectations as described by the relationship

$$\text{Ave} \left[\frac{\sum (X - \bar{X})^2}{n} \right] = \frac{n - 1}{n} \sigma^2 = \frac{2 - 1}{2} \sigma^2 = (0.50)\sigma^2$$

On the other hand, when each sample variance is computed using the formula $\sum (X - \bar{X})^2/(n - 1)$, the average value of the sample variances equals

$$\frac{\sum [\sum (X - \bar{X})^2/(n - 1)]}{K} = \frac{20.00}{16} = 1.25$$

Provided the sample variances are computed by dividing by $n - 1$, the average value of the sample variances will be equal to the value of the universe variance.

Thus, $\sum (X - \bar{X})^2/(n - 1)$ is an unbiased estimator of the universe variance σ_x^2. The above relationship holds regardless of the sample size selected, provided that drawings are made *with replacement*.

When sampling without replacement, the expected value of the sample variance using the formula $\sum (X - \bar{X})^2/n$ is equal to $[N/(N - 1)][(n - 1)/n]\sigma^2$. Using the formula $\sum (X - \bar{X})^2/(n - 1)$, the expected value of the sample variance is equal to $[N/(N - 1)]\sigma^2$. N is the size of the universe. The proof of the above relationships will be left to more advanced texts. When the universe is quite small and samples are drawn without replacement, the factor $N/(N - 1)$ must be taken into consideration. When the universe is large, the expression $N/(N - 1)$ approximates 1 and is typically dropped. Problem 4.1 provides an opportunity to demonstrate the above relationship for samples of size 2 drawn *without replacement*. Problem 4.2 provides an opportunity to demonstrate this for samples of size 3 drawn *with replacement*, and Problem 4.3 for samples of size 3 drawn *without replacement*.

Working under the assumption that the reason for computing sample measures of dispersion is to provide estimates of universe measures of dispersion, $n - 1$ rather than n will be used in the formulas for the computation of the sample variance and the sample standard deviation. The computation of the standard deviation of costs of the five sump pump models is illustrated in Table 4.5. It is presumed in this analysis that the five observations constitute a sample drawn from a universe of values.

Table 4.5 Illustration of the computation of the standard deviation by the basic formula

Model	Cost to the firm, X_i ($)	$X_i - \bar{X}$	$(X_i - \bar{X})^2$
A	11.85	−0.35	0.1225
B	11.75	−0.45	0.2025
C	11.80	−0.40	0.1600
D	13.15	+0.95	0.9025
E	12.45	+0.25	0.0625
	61.00		1.4500

$$\bar{X} = \frac{\sum X}{n} = \frac{61.00}{5} = \$12.20$$

$$\hat{\sigma}_x = \sqrt{\frac{\sum (X - \bar{X})^2}{n - 1}} = \sqrt{\frac{1.4500}{4}} = \sqrt{0.3625} = \$0.602$$

Provided the analyst knows the shape of the distribution, statements can often be generated with respect to the nature of the dispersion of the values comprising the distribution. A distribution that occurs quite frequently in statistical analysis is the normal distribution. The normal distribution is characterized by a bell-shaped curve such as that depicted in Figure 4.2. The normal distribution is examined in detail in Chapter 11. The standard deviation as an indicator of the extent of dispersion in a distribution can be examined in terms of its significance within the framework of a normal curve. Perpendiculars raised on the number line at the values $\mu + \sigma$ and $\mu - \sigma$ encompass 68.27 percent of the area under the curve (hence, 68.27 percent of the values of the variable under examination fall in this range). The above statement holds only if the distribution in question is characterized by a normal curve. The value of the standard deviation communicates information with respect to the extent of clustering in the distribution by indicating the span required to encompass the middle 68.27 percent of the area under the curve. Clearly, the shorter the span (i.e., the smaller the value of the standard deviation), the greater the degree of clustering in the distribution.[6]

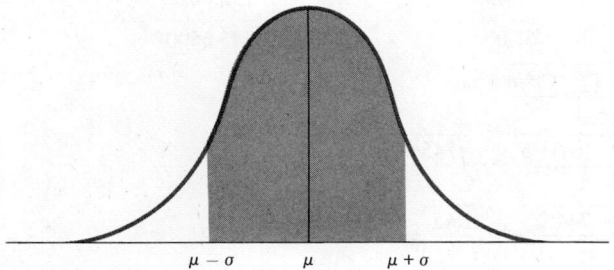

Figure 4.2

4.7 SHORT-CUT FORMULAS

The universe standard deviation:

$$\sigma_x = \sqrt{\frac{\sum X^2 - (\sum X)^2/N}{N}}$$

where X equals the value of the x_ith item and N equals the universe size.

The sample standard deviation:

$$\hat{\sigma}_x = \sqrt{\frac{\sum X^2 - (\sum X)^2/n}{n-1}}$$

where X equals the value of the X_ith item and n equals the sample size.

Note: Division by $n-1$, the number of degrees of freedom, rather than by n, the sample size, is necessary in order to generate an unbiased estimator.

In effect, the short-cut formula computes the sum of the squared deviations from *zero* (rather than from the arithmetic mean), obtaining $\sum X^2$. A correction factor, $(\sum X)^2/n$, is then subtracted from $\sum X^2$ to obtain the sum of the squared deviations from the *arithmetic mean*.

The step of determining the deviations from the mean is replaced by the step of finding the deviations from the arbitrary origin of zero. The deviations of the original values from zero are, of course, equal to the original values themselves, and thus a computational step is eliminated. Computation of the standard deviation by the zero-origin formula is illustrated in Table 4.6.

The value of the standard deviation is $0.602 for both the basic formula and the zero-origin method. This is in accordance with expectations since the standard

---

[6] A similar statement can be generated with respect to the average deviation. $\mu_x \pm 1$ AD encompasses the middle 57.5 percent of the area under the normal curve.

Shert-cut

Table 4.6 Illustration of the computation of the standard deviation by the zero-origin method

Model	Cost to the firm, X_i ($)	X_i^2
A	11.85	140.4225
B	11.75	138.0625
C	11.80	139.2400
D	13.15	172.9225
E	12.45	155.0025
	61.00	745.6500

$$\hat{\sigma}_x = \sqrt{\frac{\sum X^2 - (\sum X)^2/n}{n - 1}}$$

$$= \sqrt{\frac{745.65 - (61)^2/5}{5 - 1}} = \sqrt{\frac{745.65 - 3721/5}{4}}$$

$$= \sqrt{\frac{745.65 - 744.2}{4}} = \sqrt{\frac{1.45}{4}} = \sqrt{0.3625} = \$0.602$$

deviation is rigidly defined. Although the formulas appear to be different, both evolve from the same underlying relationship.[7]

4.8 COMPUTATION OF THE STANDARD DEVIATION FROM A FREQUENCY DISTRIBUTION

Two formulas will be discussed for the computation of a standard deviation from a frequency distribution. They are counterparts to the formulas examined in the discussion of the standard deviation for individual observations. A short-cut formula based on the use of an assumed mean and related to the step deviations method for the computation of an arithmetic mean is presented in Technical Note 4.

The Basic Formula

$$\hat{\sigma}_x = \sqrt{\frac{\sum [f(M - \bar{X})^2]}{n - 1}}$$

which is the counterpart of

$$\hat{\sigma}_x = \sqrt{\frac{\sum (X - \bar{X})^2}{n - 1}}$$

when dealing with individual items.

Once in a frequency distribution format, the individual items lose their identity and henceforth are represented by the midpoints of the classes to which they are assigned. Thus, the analyst finds the deviations of the midpoints from the mean and squares these values to obtain the representative squared deviation for each class. The representative squared deviation occurs not once but rather once for every item in the class. To obtain the total squared deviation for a class, the representative squared deviation is multiplied by the number of items it represents (the class frequency). The sum of these products provides the sum of the squared deviations from the mean for the distribution as a whole.

---

[7] The derivation of the short-cut formula from the relationship $\sum (X - \mu_x)^2/N$ is presented in Technical Note 3.

The Zero-Origin Method

$$\hat{\sigma}_x = \sqrt{\frac{\sum (fM^2) - (\sum fM)^2/n}{n - 1}}$$

which is the counterpart of

$$\hat{\sigma}_x = \sqrt{\frac{\sum X^2 - (\sum X)^2/n}{n - 1}}$$

when dealing with individual items.

The estimated sum of the values in a distribution, after they have been grouped into the format of a frequency distribution, is $\sum (fM)$. In a frequency distribution, the deviation of a midpoint from zero is the value of the midpoint itself. The deviation squared is the midpoint squared. Recalling that the squared deviation occurs not once but rather once for every item in the class, the midpoint squared must be multiplied by the class frequency to generate the total squared deviation for each class from the origin of zero. Summing these products, $\sum (fM^2)$, generates an estimate of the sum of the squared deviations from zero for the distribution as a whole (i.e., generates an estimate of $\sum X^2$). A correction factor, $(\sum fM)^2/n$, is then subtracted from $\sum (fM^2)$ to obtain the sum of the squared deviations as computed from the arithmetic mean.

The computation of the standard deviation for Frequency Distribution Alternative B by the basic method and by the zero-origin method is illustrated in Tables 4.7 and 4.8, respectively. *Basic Formula for f.d.*

Table 4.7 Computation of the standard deviation by the basic formula (long method): Frequency Distribution Alternative B (arithmetic mean = $19.16)

Daily wages ($)	M	f	$(M - \bar{X})$	$(M - \bar{X})^2$	$f(M - \bar{X})^2$
9–11	10	4	−9.16	83.9056	335.6224
11–13	12	8	−7.16	51.2656	410.1248
13–15	14	6	−5.16	26.6256	159.7536
15–17	16	25	−3.16	9.9856	249.6400
17–19	18	15	−1.16	1.3456	20.1840
19–21	20	22	+0.84	0.7056	15.5232
21–23	22	8	+2.84	8.0656	64.5248
23–25	24	9	+4.84	23.4256	210.8304
25–27	26	5	+6.84	46.7856	233.9280
27–29	28	6	+8.84	78.1456	468.8736
29–31	30	1	+10.84	117.5056	117.5056
31–33	32	3	+12.84	164.8656	494.5968
		112			2,781.1072

$$\hat{\sigma}_x = \sqrt{\frac{\sum [f(M - \bar{X})^2]}{n - 1}}$$

$$= \sqrt{\frac{2781.1072}{112 - 1}}$$

$$= \sqrt{\frac{2781.1072}{111}} = \sqrt{25.055} = 5.005$$

Short-cut for F.D.

Table 4.8 Computation of the standard deviation by the zero-origin method: Frequency Distribution Alternative B

Daily wages ($)	M	f	fM	fM^2
9–11	10	4	40	400
11–13	12	8	96	1152
13–15	14	6	84	1176
15–17	16	25	400	6400
17–19	18	15	270	4860
19–21	20	22	440	8800
21–23	22	8	176	3872
23–25	24	9	216	5184
25–27	26	5	130	3380
27–29	28	6	168	4704
29–31	30	1	30	900
31–33	32	3	96	3072
		112	2146	43900

$$\hat{\sigma}_x = \sqrt{\frac{\sum (fM^2) - (\sum fM)^2/n}{n-1}}$$

$$= \sqrt{\frac{43900 - (2146)^2/112}{111}}$$

$$= \sqrt{\frac{43900 - 4,605,316/112}{111}}$$

$$= \sqrt{\frac{43900 - 41118.89}{111}} = \sqrt{\frac{2781.11}{111}} = \sqrt{25.055} = 5.005$$

4.9 SKEWNESS

Measures of skewness are designed to indicate the extent and direction of lopsidedness or asymmetry (departure from symmetry) in a distribution.

Visually:

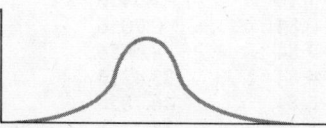

A symmetrical or nonskewed distribution.

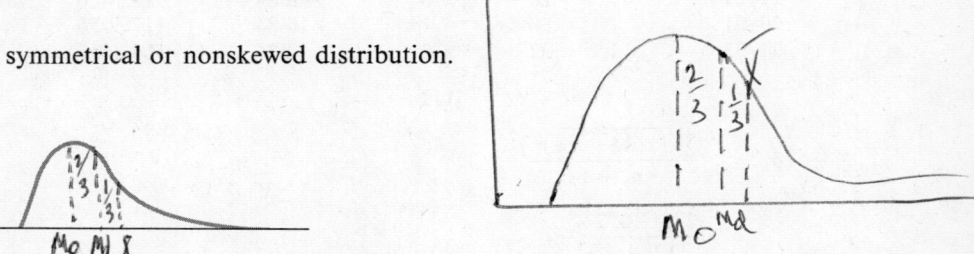

A positively skewed distribution: The "tail" of the distribution extends toward the larger values (indicating that there are a few relatively large values relative to the other items in the distribution).

A negatively skewed distribution: The "tail" of the distribution extends toward the smaller values (indicating that there are a few relatively small values relative to the other items in the distribution).

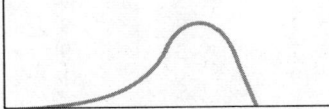

Desirably, a measure of skewness should indicate two things:

1. The direction of skewness (positive or negative) $+$ $-$
2. The extent of skewness, that is, the extent of the departure from symmetry

The measure of skewness presented below, *the third-moment method*, meets these qualifications. An alternative approach, occasionally used, called the Pearsonian coefficient of skewness, is presented in Technical Note 5.

The third moment around the mean is defined as the mean of the cubed deviations (deviations to the third power) from the arithmetic mean and is used as a measure of skewness.

Recall that in the computation of the standard deviation, the deviations were squared in order to eliminate the negative signs. When the deviations are cubed, the signs return.

Table 4.9

X	$X - \mu_x$	$(X - \mu_x)^2$	$(X - \mu_x)^3$
7	-1	1	-1
7	-1	1	-1
7	-1	1	-1
11	$+3$	9	$+27$
			-3
32	0	12	$+27$
			$+24$

$$\bar{X} = \frac{\sum X}{N} = \frac{32}{4} = 8$$

If, as in Table 4.9, the positive deviations are of larger magnitude than the negative deviations, the positive deviations will be given much greater weight when cubed, causing the positive subtotal to exceed the negative subtotal. If the positive deviations are larger than the negative, then $\sum (X - \mu_x)^3$ will be positive; if the negative deviations are larger than the positive, then $\sum (X - \mu_x)^3$ will be negative. Thus, the net sum of the cubed deviations column automatically generates the correct sign.

In addition, the more extreme the deviations, the greater the amount by which the positive subtotal exceeds the negative subtotal (or vice versa depending on the direction in which the extreme deviations occur). Thus, by evaluating the magnitude of the net total of the $(X - \mu_x)^3$ column, a determination of the extent of skewness can be obtained.

To illustrate, compare the two distributions shown in Tables 4.10(a) and 4.10(b).

Table 4.10(a)

X ($)	$X - \mu_x$	$(X - \mu_x)^2$	$(X - \mu_x)^3$
7	-1	1	-1
7	-1	1	-1
7	-1	1	-1
11	$+3$	9	$+27$
32	0	12	-3
			$+27$
			$+24$

$\mu_x = \$8$

$\sigma_x = \sqrt{\frac{12}{4}} = \sqrt{3} = 1.732$

M_3 = third moment
= average of the cubed
deviations from the mean

$= \dfrac{\sum (X - \mu_x)^3}{N} = \frac{24}{4} = \$6$

Table 4.10(b)

X (lb)	$X - \mu_x$	$(X - \mu_x)^2$	$(X - \mu_x)^3$
7	-3	9	-27
7	-3	9	-27
7	-3	9	-27
19	$+9$	81	$+729$
40	0	108	$+729$
			-81
			$+648$

$\mu_x = 10$ lb

$\sigma_x = \sqrt{\frac{108}{4}} = \sqrt{27} = 5.196$

$M_3 = \dfrac{\sum (X - \mu_x)^3}{N} = \frac{648}{4} = 162$ lb

M_3, the average amount of cubed deviation from the mean, has two disadvantages when expressed in its existing form:

1. *The unit of measurement* presents a problem because it is desirable to have a *common denominator* in which skewness is expressed in order to facilitate comparison of the degree of skewness between two different distributions that are expressed in different units.
2. *The variance of the given distribution* presents a problem because for two distributions *of the same shape*, the larger the variance, the larger the value of M_3.

To remove the effect of the units under consideration and to remove the effect of the variance, the third moment is standardized by dividing through by σ^3 (the standard deviation cubed). The resultant standardized third moment is referred to as α_3 (alpha three).

$$\alpha_3 = \frac{M_3}{\sigma^3}$$

Note: Dividing M_3 by σ^3 is equivalent to standardizing the original deviations before cubing, namely,

$$\alpha_3 = \frac{\sum [(X - \mu)/\sigma]^3}{N}$$

Thus, α_3 is the average of the cubed *standardized* deviations from the mean.
Referring to Table 4.10:

Table 4.10(a) $\alpha_3 = \dfrac{M_3}{\sigma^3} = \dfrac{\$6}{(\$1.732)^3} = \dfrac{6}{5.196} = 1.15$

Table 4.10(b) $\alpha_3 = \dfrac{M}{\sigma^3} = \dfrac{162 \text{ lb}}{(5.196 \text{ lb})^3} = \dfrac{162}{140.29} = 1.15$

The effect of standardization is to obtain values of α_3 that are directly comparable from one distribution to another regardless of the units of measurement or the size of the variables involved. If the value of $\alpha_3 > 0.5$, there is considerable skewness present.

TECHNICAL NOTE 3

Derivation of the short-cut formula for the computation of the standard deviation for a set of individual items

Recalling that the variance of the universe is defined as the average value of the squared deviations from the mean, one can write

$$\text{variance} = \frac{\sum (X - \mu)^2}{N}$$

$$= \frac{1}{N} \cdot \sum (X - \mu)^2$$

Expanding the binomial, the following equivalent expression is obtained:

$$\text{variance} = \frac{1}{N} \cdot \sum (X^2 - 2\mu X + \mu^2)$$

The expression $(X^2 - 2\mu X + \mu^2)$ occurs once for each X value. Thus, if there are four items in the universe:

item 1 $X_1^2 - 2\mu X_1 + \mu^2$

item 2 $X_2^2 - 2\mu X_2 + \mu^2$

item 3 $X_3^2 - 2\mu X_3 + \mu^2$

item 4 $X_4^2 - 2\mu X_4 + \mu^2$

The expression $\sum (X^2 - 2\mu X + \mu^2)$ refers to the sum of these values and can be expressed alternatively by taking the sum for each of the terms in the expression separately.

Note the following rules:

1. The sum of a group of terms involving a constant multiplied by a variable equals the constant times the sum of the values of the variable.

$$-2\mu = \text{constant}$$

$$X_i = \text{variable}$$

Therefore,

$$\sum (-2\mu X_i) = -2\mu \cdot \sum X_i$$

2. When a constant appears once for each expression, the sum of the constants can be expressed as the product of the number of times the constant appears and the value of the constant.

$$\mu^2 = \text{constant}$$

$$N = \text{number of items in the universe}$$

$$\sum (\mu^2) = N \cdot \mu^2$$

Thus:

item 1	$X_1{}^2 - 2\mu X_1 + \mu^2$
item 2	$X_2{}^2 - 2\mu X_2 + \mu^2$
item 3	$X_3{}^2 - 2\mu X_3 + \mu^2$
item 4	$X_4{}^2 - 2\mu X_4 + \mu^2$

$$\sum (X_i{}^2) - 2\mu \cdot (\sum X_i) + N\mu^2$$

Dropping the subscripts, the sum of the terms becomes

$$\sum (X^2) - 2\mu \cdot (\sum X) + N\mu^2$$

Therefore

$$\text{variance} = \frac{1}{N} \cdot \sum (X^2 - 2\mu X + \mu^2)$$

can alternatively be expressed as

$$\text{variance} = \frac{1}{N} \left[\sum (X^2) - 2\mu \cdot (\sum X) + N\mu^2 \right]$$

Dividing each term by N,

$$\text{variance} = \frac{\sum (X^2)}{N} - 2\mu \cdot \left(\frac{\sum X}{N} \right) + \frac{N\mu^2}{N}$$

Canceling, we obtain

$$\text{variance} = \frac{\sum (X^2)}{N} - 2\mu \cdot \left(\frac{\sum X}{N} \right) + \mu^2$$

Substituting $\mu = \sum X / N$,

$$\text{variance} = \frac{\sum (X^2)}{N} - 2\mu \cdot \mu + \mu^2$$

$$= \frac{\sum (X^2)}{N} - 2\mu^2 + \mu^2$$

Combining similar terms,

$$\text{variance} = \frac{\sum (X^2)}{N} - \mu^2$$

Substituting $\sum X/N = \mu$,

$$\text{variance} = \frac{\sum (X^2)}{N} - \left(\frac{\sum X}{N}\right)^2$$

$$= \frac{\sum (X^2)}{N} - \frac{(\sum X)^2}{N^2}$$

Factoring out $(1/N)$,

$$\text{variance} = \frac{1}{N}\left[\sum X^2 - \frac{(\sum X)^2}{N}\right]$$

$$= \frac{\sum X^2 - (\sum X)^2/N}{N}$$

$$\frac{\sum X^2}{N} - \left(\frac{\sum X}{N}\right)^2 = \text{Variance}$$

The formula for the standard deviation of the universe is simply the square root of this expression,

$$\text{Variance} = \sqrt{}$$

$$\sigma_x = \sqrt{\sigma_x^2} = \sqrt{\frac{\sum X^2 - (\sum X)^2/N}{N}}$$

The corresponding formula for the sample standard deviation (when it is to be used to generate an estimate of the universe value) is

$$\hat{\sigma}_x = \sqrt{\hat{\sigma}_x^2} = \sqrt{\frac{\sum X^2 - (\sum X)^2/n}{n - 1}}$$

TECHNICAL NOTE 4

*The short-cut or step deviations method
for the computation of the standard
deviation from a frequency distribution*

Formula:

$$\hat{\sigma}_x = C\sqrt{\frac{\sum (fd_c^2) - (\sum fd_c)^2/n}{n - 1}}$$

The computation of the standard deviation by the step deviations method involves first calculating the sum of the squared deviations from an assumed mean in class interval units, namely $\sum (fd_c^2)$. A correction factor is then introduced, $\sum (fd_c)^2/n$, which converts this figure to the sum of the squared deviations in class interval units associated with the value from which the sum of the deviations is equal to zero (namely, the true mean). Performance of the calculations contained within the radical sign generates the value of the variance in class interval units. Taking the square root, the value of the standard deviation in class interval units is obtained. The value of the standard deviation in original units is obtained in the last step by multiplying by the value of C, the class interval. Note that the step deviations method can be used only in conjunction with equal class intervals.

The calculation of the standard deviation by the step deviations method for Frequency Distribution Alternative B is illustrated in Table 4.11.

Table 4.11 Illustration of the computation of the standard deviation by the step deviations method: Frequency Distribution Alternative B

Daily wages ($)	M	f	d_c	fd_c	fd_c^2
9–11	10	4	−5	−20	100
11–13	12	8	−4	−32	128
13–15	14	6	−3	−18	54
15–17	16	25	−2	−50	100
17–19	18	15	−1	−15	15
19–21	20	22	0	0	0
21–23	22	8	+1	+8	8
23–25	24	9	+2	+18	36
25–27	26	5	+3	+15	45
27–29	28	6	+4	+24	96
29–31	30	1	+5	+5	25
31–33	32	3	+6	+18	108
		112		−135	715
				+88	
				−47	

$$\hat{\sigma}_x = C \sqrt{\frac{\sum (fd_c^2) - (\sum fd_c)^2/n}{n-1}}$$

$$= 2 \sqrt{\frac{715 - (-47)^2/112}{112 - 1}}$$

$$= 2 \sqrt{\frac{715 - 2209/112}{111}}$$

$$= 2 \sqrt{\frac{695.28}{111}} = 2\sqrt{6.263783} = 2(2.5025) = 5.005$$

A survey of the formula reveals the need to generate the sum of the deviations from the assumed mean in class interval units $\sum (fd_c)$ and the sum of the squared deviations from the assumed mean in class interval units $\sum (fd_c^2)$. For the computations in Table 4.11, the midpoint of the class $19–$21 is selected as the assumed mean. Next the deviations of the midpoints of each of the classes from the assumed mean are determined, expressed in class interval units, and recorded in the d_c column. The sum of the deviations in class interval units for each class is obtained by multiplying the d_c value for the class by the corresponding class frequency. The resulting products are entered in the fd_c column and summed (netting off the positive and negative subtotals) to obtain the sum of the deviations from the assumed mean in class interval units, $\sum (fd_c)$. The sum of the squared deviations from the assumed mean for each class could be obtained by squaring each of the d_c values and then multiplying by the associated class frequencies. However, $(f)(d_c^2) = f(d_c)(d_c)$. Thus, the fd_c^2 products can be obtained by multiplying the previously recorded fd_c products by the associated d_c values. The latter approach has the advantage of eliminating a step in the computational process (it is not necessary to generate the d_c^2 values as a separate computation). Moreover, the tabular arrangement of the d_c and fd_c columns as adjacent entries facilitates the computational process. Summing the fd_c^2 column provides the sum of the squared deviations from the assumed mean, $\sum (fd_c^2)$. All computations inside the radical sign are performed in class interval units. In the last step it is necessary to multiply by C, the class interval, to convert to original units.

A comparison of Tables 3.7 and 4.11 reveals the closeness of the relationship between the step deviations method for the computation of the arithmetic mean and the step deviations method for computing the value of the standard deviation. The value of $\sum (fd_c)$ is required for both formulas.

$$\overline{X} = A + \frac{\sum (fd_c)}{n} (C)$$

$$\hat{\sigma}_x = C \sqrt{\frac{\sum (fd_c^2) - (\sum fd_c)^2/n}{n - 1}}$$

The formula for the computation of the standard deviation requires the additional input, $\sum (fd_c^2)$. To generate the fd_c^2 column (the only difference between Tables 3.7 and 4.11), one need simply multiply the already computed fd_c values by the associated values for d_c. Thus, very little additional computational effort is required to generate the inputs necessary for the computation of the standard deviation when the arithmetic mean has previously been computed by the step deviations method.

TECHNICAL NOTE 5

The Pearsonian coefficient of skewness

The *Pearsonian coefficient of skewness* is based on the same relationship as the formula for the *empirical mode*. Figure 3.9 illustrates the relationship. First, the direction of skewness is determined by observing whether the mean is greater than the mode (positive skewness) or less than the mode (negative skewness). The extent of departure from symmetry is ascertained by observing the extent to which the mean is pulled away from the mode. The extent of departure is expressed in standard deviation units in order to obtain a measure that is *independent* of the unit of measurement. Either of the following formulas provide a measure of skewness based on the relationship of the median, mode, and mean in a skewed distribution:

$$S_k = \frac{\overline{X} - \text{mode}}{\hat{\sigma}_x}$$

$$= \frac{3(\overline{X} - \text{median})}{\hat{\sigma}_x}$$

The formulas automatically generate the correct sign (minus if negatively skewed; positive if positively skewed). The value of the coefficient itself *can* range from $+3$ to -3, but *usually* will range only from $+1$ to -1.

The procedure is illustrated below with data derived from Frequency Distribution Alternative B.

$$\overline{X} = \$19.16$$

$$\text{median} = 18.733$$

$$\hat{\sigma}_x = 5.008$$

$$S_k = \frac{3(\overline{X} - \text{median})}{\hat{\sigma}_x} = \frac{3(19.16 - 18.733)}{5.008} = \frac{3(0.427)}{5.008} = \frac{1.281}{5.008} = 0.256$$

Interpretation: The mean is greater than the mode by an amount equal to about 25.6 percent of the value of a standard deviation. This is considered negligible skewness.

Qualification: As the departure from symmetry becomes substantial, the relationship on which the Pearsonian coefficient formula is based breaks down and the Pearsonian coefficient no longer provides reliable results (see footnote 6 in Chapter 3, p. 61). The numerical value of skewness derived from the third-moment approach is not interchangeable with the numerical value that is obtained from the Pearsonian coefficient. They are computed on the basis of different relationships. However, it is true that the measure of skewness will be equal to zero in both approaches if the distribution is perfectly symmetrical.

PROBLEMS

Problem 4.1

Referring to the universe in Table 4.3 consisting of the four values 1, 2, 3, and 4:

1. Determine all possible samples of size 2 drawn *without replacement* (without replacement indicates that the first item selected is not replaced in the universe before the second item is drawn).

2. Compute the mean for each of the 12 possible samples and demonstrate that Ave $(\overline{X}) = \mu_x$. (Note that if drawings such as 1, 2 and 2, 1 are treated as the same sample, there are six different samples, each of which can occur two ways.)

3. It is stated without proof that the expected value of the sample variance using the formula $\sum (X - \overline{X})^2/n$ is equal to $[N/(N - 1)][(n - 1)/n].\sigma^2$ When the universe is quite small and samples are drawn without replacement, the factor $N/(N - 1)$ must be taken into consideration. For large universes, the expression $N/(N - 1)$ is virtually equal to 1 and is typically dropped from the expression. Sampling with replacement is equivalent to sampling from an infinite universe. Thus, the factor $N/(N - 1)$ can be ignored when sampling with replacement. Compute the sum of the squared deviations for each of the 12 possible samples.

   a. For each possible sample, divide the sum of the squared deviations by n and then demonstrate that

   $$\text{Ave}\left[\frac{\sum (X - \overline{X})^2}{n}\right] = \left(\frac{N}{N - 1}\right)\left(\frac{n - 1}{n}\right)\sigma^2$$

b. For each possible sample, divide the sum of the squared deviation by $n - 1$ and then demonstrate that

$$\text{Ave} \left[\frac{\sum (X - \bar{X})^2}{n - 1} \right] = \left(\frac{N}{N - 1} \right) \sigma^2$$

Problem 4.2

Referring to the universe in Table 4.3 consisting of the four values 1, 2, 3, and 4:

1. Determine all possible samples of size 3 drawn *with replacement*. (With replacement indicates that the first item selected is replaced in the universe before the second item is drawn.) Figure 13.1 (p. 316) is a tree diagram depicting all possible ways a sample of size 3 could be drawn with replacement from the universe consisting of the four values 1, 2, 3, and 4. Treating drawings such as 3, 2, 1; 1, 2, 3; 2, 1, 3; and so on as alternative ways of generating the same sample, the resulting possibilities are captured in Table 13.3 (p. 317). Note that some of these possibilities can occur only one way, whereas others can occur as many as six ways. Verify this in Figure 13.1.

2. The sample mean for each possible sample has been computed and recorded in Table 13.4. Using the formula for a weighted arithmetic mean (where the weight refers to the number of ways a particular sample could have been generated), compute the mean of the sample means. Demonstrate that

   $$\text{Ave} (\bar{X}) = \mu_x$$

3. Compute the sum of the squared deviations for each of the possible samples in Table 13.4.

   a. Obtain $\sum (X - \bar{X})^2 / n$ for each possible sample. Compute

   $$\text{Ave} \left[\frac{\sum (X - \bar{X})^2}{n} \right]$$

   using the formula for a weighted mean where the weight refers to the number of ways a particular sample could have been generated. Demonstrate that

   $$\text{Ave} \left[\frac{\sum (X - \bar{X})^2}{n} \right] = \frac{n - 1}{n} \sigma^2$$

   b. Obtain $\sum (X - \bar{X})^2 / (n - 1)$ for each possible sample. Compute

   $$\text{Ave} \left[\frac{\sum (X - \bar{X})^2}{n - 1} \right]$$

   using the formula for a weighted mean where the weight refers to the number of ways a particular sample could have been generated. Demonstrate that

   $$\text{Ave} \left[\frac{\sum (X - \bar{X})^2}{n - 1} \right] = \sigma^2$$

Problem 4.3

Referring to the universe in Table 4.3 consisting of the four values 1, 2, 3, and 4:
1. Determine all possible samples of size 3 drawn *without replacement*. (Without replacement indicates that the items selected on the first two drawings are not replaced prior to the selection of the third item.) Table 13.6 (p. 321) contains a tree diagram depicting all possible ways a sample of size 3 could be drawn *without* replacement from the universe consisting of the four values 1, 2, 3, and 4. Treating drawings such as 2, 1, 3 and 3, 1, 2 as alternative ways of generating the same sample, the resulting possibilities are captured in Table 13.6(b). (Note that when drawings are made without replacement, the four possible samples are equally likely.) Verify this in Table 13.6(a).
2. The sample mean for each possible sample has been computed and recorded in Table 13.6(b). Compute the mean of the sample means.
    a. Should you use the formula for a weighted arithmetic mean or the formula for the simple arithmetic average?
    b. Demonstrate that Ave $(\overline{X}) = \mu_x$.
3. It is stated without proof that the expected value of the sample variance [using the formula $\sum (X - \overline{X})^2/n$] is equal to $[N/(N-1)][(n-1)/n]\sigma^2$. When the universe is quite small and samples are drawn without replacement, the factor $N/(N-1)$ must be taken into consideration. For large universes, the expression $N/(N-1)$ is virtually equal to 1 and is typically dropped from the expression. Sampling with replacement is equivalent to sampling from an infinite universe. Thus, the factor $N/(N-1)$ can be ignored when sampling with replacement. Compute the sum of the squared deviations for each of the possible samples in Table 13.6(b).
    a. Obtain $\sum (X - \overline{X})^2/n$ for each possible sample. Compute

    $$\text{Ave}\left[\frac{\sum (X - \overline{X})^2}{n}\right]$$

    Demonstrate that

    $$\text{Ave}\left[\frac{\sum (X - \overline{X})^2}{n}\right] = \left[\frac{N}{N-1}\right]\left[\frac{n-1}{n}\right]\sigma^2$$

    b. Obtain $\sum (X - \overline{X})^2/(n-1)$ for each possible sample. Compute

    $$\text{Ave}\left[\frac{\sum (X - \overline{X})^2}{n-1}\right]$$

    Demonstrate that

    $$\text{Ave}\left[\frac{\sum (X - \overline{X})^2}{n-1}\right] = \left(\frac{N}{N-1}\right)\sigma^2$$

Problem 4.4

It is necessary to divide by $n - 1$ rather than n when computing the sample variance if the sample variance so obtained is to be used as an unbiased estimator. Recall that the sum of the squared deviations for a set of values will be minimized when the deviations are measured from the arithmetic mean of that set of values. Thus, for each sample, the sum of the squared deviations when computed from $\overline{X}$, the sample mean, will always be *less* than they would have been had they been computed from μ_x, the universe mean. Dividing by $n - 1$ is the adjustment required to correct for this understatement. In Problem 4.2, it was demonstrated that Ave $[\sum (X - \bar{X})^2/(n - 1)] = \sigma_x^2$. Repeat the procedure described in Problem 4.2, but this time compute the squared deviations from μ_x for each of the samples. Then demonstrate that Ave $[\sum (X - \mu_x)^2/n] = \sigma_x^2$ (that is, that the need for the adjustment no longer exists).

Problem 4.5

The procedure for computing the average deviation for a set of data described by a *frequency distribution* is summarized below:

1. Determine whether the average deviation is to be computed from the arithmetic mean or the median.
2. For each class, determine the *absolute* difference between the value of the mean (or median) and the midpoint of the class.

$$|M - \bar{X}| \quad \text{or} \quad |M - \text{Me}|$$

3. For each class, multiply the representative *absolute* deviation for the class by the number of items it represents (the class frequency) to obtain the total *absolute* deviation for the class.

$$f|M - \bar{X}| \quad \text{or} \quad f|M - \text{Me}|$$

4. Sum the products obtained in step 3. This provides the total *absolute* deviation from the mean (or median) for the entire frequency distribution.

$$\sum (f|M - \bar{X}|) \quad \text{or} \quad \sum (f|M - \text{Me}|)$$

5. Dividing the sum obtained in step 4 by the total frequency or n. The result is the average deviation from the mean (or median).

$$AD_{\bar{x}} = \frac{\sum (f \,|\, M - \bar{X}|)}{n} \quad \text{or} \quad AD_{\text{Me}} = \frac{\sum (f \,|\, M - \text{Me}|)}{n}$$

Required:
1. For the frequency distribution below, compute the average deviation from the mean.
2. For the frequency distribution below, compute the average deviation from the median.

Exam score	Number of students
45–55	5
55–65	15
65–75	40
75–85	30
85–95	10
	100

Problem 4.6

Referring to the frequency distribution provided in Problem 3.1, compute the following measures:
1. The range
2. The interquartile range
3. The average deviation from the mean (see Problem 4.5)
4. The standard deviation (presume that the resulting value is to be used as an estimate of the unknown universe standard deviation)
5. The Pearsonian coefficient of skewness
6. α_3, the standardized third moment

Problem 4.7

For each of the frequency distributions provided in Problem 3.2, compute the following measures:
1. The range
2. The interquartile range
3. The average deviation from the mean (see Problem 4.5)
4. The standard deviation (presume that the resulting values are to be used as estimates of the unknown universe standard deviations)
5. The Pearsonian coefficient of skewness
6. α_3, the standardized third moment

Problem 4.8

Referring to the frequency distribution provided in Problem 3.3, compute the following measures:
1. The range
2. The interquartile range
3. The average deviation from the mean (see Problem 4.5)
4. The standard deviation (presume that the resulting value is to be used as an estimate of the unknown universe standard deviation)
5. The Pearsonian coefficient of skewness
6. α_3, the standardized third moment

Problem 4.9

Referring to the frequency distribution provided in Problem 3.4, indicate why the standard deviation cannot be determined. What measure of dispersion can be computed from this distribution? Compute this measure of dispersion.

Problem 4.10

Referring to the data provided in Problem 3.5, compute the values of the following measures:
1. The range
2. The average deviation from the mean
3. The average deviation from the median
4. The standard deviation (presume that the result obtained is to be used as an estimate of the unknown universe standard deviation)

Problem 4.11

Referring to the data provided in Problem 3.6, compute the values of the following measures:
1. The range
2. The average deviation from the mean
3. The average deviation from the median
4. The standard deviation (presume that the result obtained is to be used as an estimate of the unknown universe standard deviation)

Problem 4.12

Referring to Problem 3.7, only one of the following three measures of dispersion can be computed:
1. The interquartile range
2. The average deviation
3. The standard deviation
    a. Why can only one of the three be computed?
    b. What additional information would be needed to allow the computation of all three?
    c. Compute the one capable of computation in the problem as given.

Problem 4.13

Referring to Problem 3.8, compute the following measures:
1. The range
2. The average deviation from the mean
3. The average deviation from the median
4. The standard deviation (presume that the result obtained is to be used as an estimate of the unknown universe standard deviation)

Problem 4.14

Referring to Problem 3.13, compute the following measures:
1. The range
2. The average deviation from the median
3. The standard deviation (presume that the result obtained is to be used as an estimate of the unknown universe standard deviation)
A word of caution: The distribution contains unequal class intervals.

Problem 4.15

You are currently employed by a professional football team that is in desperate need of a punter (Right! the New England Patriots). You have been assigned the task of examining the statistics describing the current crop of college punters. It is your responsibility to determine those who should be given further consideration.
1. For any given candidate, what information would the following measures convey?
    a. modal punt
    b. mean punt
    c. median punt
    d. standard deviation
    e. α_3, the standardized third moment

2. Which of the above (none, some, or all) should you examine? Justify the inclusion of those measures you feel should be included in the analysis.

Problem 4.16

The following set of distributions reflects the pattern of activity on a representative group of 1000 of Giovinazzo Office Supply, Inc.'s accounts. Distribution A shows the distribution of the mean dollar magnitude per transaction for the 1000 customers. Distribution B depicts the manner in which the standard deviations of the 120 "large-order" customers are distributed. Distribution C depicts the manner in which the standard deviations of the 880 "small-order" customers are distributed.

Distribution A		Distribution B		Distribution C	
Dollar magnitude of mean order (for individual customers)	Number of customers	Dollar magnitude of standard deviation for the 120 "large customers," i.e., mean $\geq$ \$60	Number of customers	Dollar magnitude of standard deviation for the 800 "small customers," i.e., mean < \$60	Number of customers
0–20	532	0–10	24	0–5	79
20–40	233	10–20	32	5–10	140
40–60	115	20–30	40	10–15	350
60–80	87	30–40	18	15–20	223
80–100	33	40–50	6	20–25	88
	1000		120		880

1. Compute the mean and standard deviation for distribution A.
2. Compute the mean and standard deviation for distribution B.
3. Compute the mean and standard deviation for distribution C.

Problem 4.17

Giovinazzo Office Supply, Inc. (see Problem 4.16) is concerned about the freight costs and clerical costs incurred on small orders. In an effort to reduce expenditures in this area, they have decided to introduce a discount policy rewarding large orders (over \$60). It is hoped that this will have the effect of causing customers to consolidate a number of small orders into large orders.

1. If the policy is successful:
   a. Will the mean of distribution A increase, decrease, or be unaffected?
   b. Will the standard deviation of distribution A increase, decrease, or be unaffected?
2. If the policy is successful:
   a. Will the mean of distribution B increase, decrease, or be unaffected?
   b. Will the standard deviation of distribution B increase, decrease, or be unaffected?
3. If the policy is successful:
   a. Will the mean of distribution C increase, decrease, or be unaffected?
   b. Will the standard deviation of distribution C increase, decrease, or be unaffected?

Problem 4.18

When computing the standard deviation for a sample, the formula $\hat{\sigma}_x = \sqrt{\sum (X - \overline{X})^2/(n - 1)}$ is generally used rather than $S_x = \sqrt{\sum (X - \overline{X})^2/n}$. What is the reason for dividing by $n - 1$ rather than dividing by n? Explain fully.

Problem 4.19

Carefully explain the concept of an *unbiased estimator*.

Problem 4.20

What information is conveyed by a measure of skewness?

Problem 4.21

Given:

$$\overline{X} = 30.83$$

$$\text{Me} = 22.67$$

Is the distribution positively or negatively skewed?

Problem 4.22

You are considering the purchase of a small restaurant. Among other things, you are interested in the average number of customers per day and the extent of variation from day to day. To capture this information in a pair of concise figures, compute the mean and standard deviation for the frequency distribution below. Assume that the resulting values are to be used as estimates of the unknown universe mean and standard deviation.

Number of customers	Number of days
0–20	5
20–40	15
40–60	40
60–80	30
80–100	10
	100

THE PROBABILITY CALCULUS

5

Two basic schools of thought have emerged with respect to the interpretation of the concept of probability. The earlier of the two, the *classical* or *objectivist school*, interprets probability as the relative frequency of occurrence of some event or outcome over the long-run performance or repetition of an experiment. The word experiment, as used here, is defined broadly as an activity that could lead to any one of a number of outcomes, the particular outcome observed on any one performance of the experiment being determined by chance. The objectivist school developed out of the interest of a number of seventeenth-century mathematicians in the mathematics underlying games of chance. Their major concern was the determination of the likelihood of success over many plays of the game. Out of the correspondence of a number of these parties (among them Blaise Pascal and Pierre De Fermet) emerged the *basic rule of probability*.

5.1 THE BASIC RULE OF PROBABILITY

Rule

Given that all possible outcomes to an experiment are equally likely, the probability of occurrence of any specified outcome (or group of outcomes) on a given performance of the experiment is equal to the ratio of the number of ways that particular outcome (or group of outcomes) can occur to the total number of ways an outcome to the experiment can be generated.

To illustrate: The probability of observing a head on any one flip of a fair coin is equal to the ratio of the number of ways that event can occur (one: a head) to the total number of ways an outcome to the experiment can occur (two: either a head or a tail). Thus, the probability of observing a head is $\frac{1}{2}$ or 0.5. It is important to note that the probability obtained using this rule refers to the long-run frequency of occurrence of heads if the experiment were to be performed many times. This relationship is depicted in Figure 5.1.

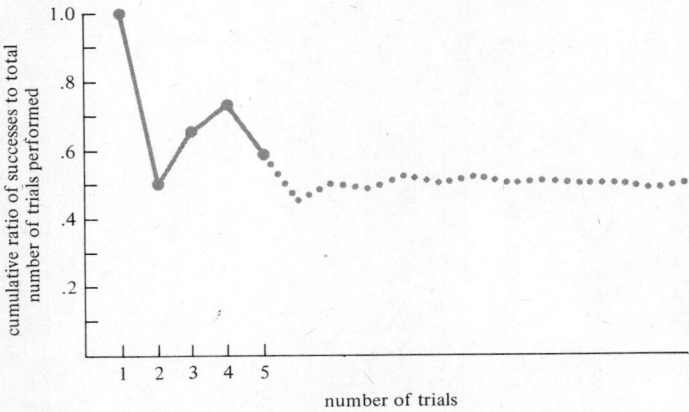

Figure 5.1

In Figure 5.1, the diagram reflects that the outcome on the first trial was a head; therefore, the cumulative ratio equalled $\frac{1}{1}$ or 1.0. If a tail had been observed, the ratio would have been $\frac{0}{1}$ or 0.00. The second trial resulted in a tail, causing the cumulative ratio to become $\frac{1}{2}$ or 0.5. On the third and fourth trials a head was observed, causing the cumulative ratio to change to $\frac{2}{3}$ or 0.67 and $\frac{3}{4}$ or 0.75, respectively. On the fifth trial a tail was observed, causing the ratio to decline to $\frac{3}{5}$ or 0.6, and so on. Given the continuation of the experiment, the long-run cumulative ratio of total number of heads observed to total number of trials would eventually approach the ratio $\frac{1}{2}$ or 0.5. It bears repeating that when the outcomes to an experiment are *equally probable*, the basic rule of probability can be used to generate the limiting ratio that the probability of success will approach in the long run. Probabilities generated by the basic rule of probability are referred to as *a priori probabilities* because the probability assignments can be made *prior* to generating observations on the outcomes of the experiment. *A priori probabilities* are determinable once the nature of the experiment is known. Three additional illustrations of *a priori probabilities* appear below:

1. The probability of obtaining a 3 on a single roll of a fair die. With a fair die, all six "faces" are equally likely to appear. There is one way the experiment can generate a success out of six ways the experiment can generate an out-

come. Therefore, the probability of obtaining a 3 on a single roll of a fair die is $\frac{1}{6}$.

2. The probability of drawing an ace out of a standard deck of playing cards in a single draw. There are four ways the experiment can generate a success out of 52 ways the experiment can generate an outcome. Therefore, the probability is $\frac{4}{52}$.

3. The probability of obtaining a sum of 10 on two rolls of a fair die. As an initial step, it is useful to enumerate all the possibilities by constructing a "Two-Die Addition Table" (see Table 5.1). The column headings represent the outcome on the first roll of the die and the row headings refer to the outcome on the second roll of the die. The entries in the cells are the sums of the row and the column headings. The 36 cells in the table represent all possible outcomes to the experiment. Success is a sum of 10. Inspection of Table 5.1 discloses that there are three ways this can occur. Therefore, the probability of obtaining a sum of 10 is $\frac{3}{36}$.

Table 5.1 Two-die addition table

Outcome on the second roll	Outcome on the first roll					
	1	2	3	4	5	6
1	2	3	4	5	6	7
2	3	4	5	6	7	8
3	4	5	6	7	8	9
4	5	6	7	8	9	10
5	6	7	8	9	10	11
6	7	8	9	10	11	12

An alternative way of setting up this problem is to use a tree diagram (see Figure 5.2). The first set of branches illustrate the six possible outcomes on the first roll of the die. Emanating from each of the first set of branches is a second set of six branches. Each branch in the second set represents the six possible outcomes on the second roll of the die. Each path from the origin to an endpoint represents a possible outcome to the experiment. Summing the numbers along the paths provides the total observed on the two dies associated with each sequence. Observation discloses that three of the 36 possible outcomes have a total of 10. It follows that the probability that a sum of 10 will be observed is $\frac{3}{36}$.

The basic rule will generate the correct answer in all situations given *equiprobable outcomes*. However, it can be an extremely tedious and roundabout way to arrive at the probability of an event. First, all possible outcomes to the experiment must be determined. Next, it must be determined which of those outcomes constitute a success. Given the number of outcomes in each of these categories, the ratio can then be computed. This process is called enumeration.

Suppose the probability of obtaining a sum of 35 on nine rolls of a fair die is desired. If a tree diagram describing this experiment were to be drawn, it would contain nine sets of branches and there would be 6^9 or 10,077,696 endpoints. It would then be necessary to sum up the nine values along each of the 10,077,696 paths to determine the total for each path. One could then determine by inspection the number of paths with a total of 35 and compute the following ratio:

$$\frac{\text{number of outcomes totaling 35}}{\text{total number of possible outcomes}}$$

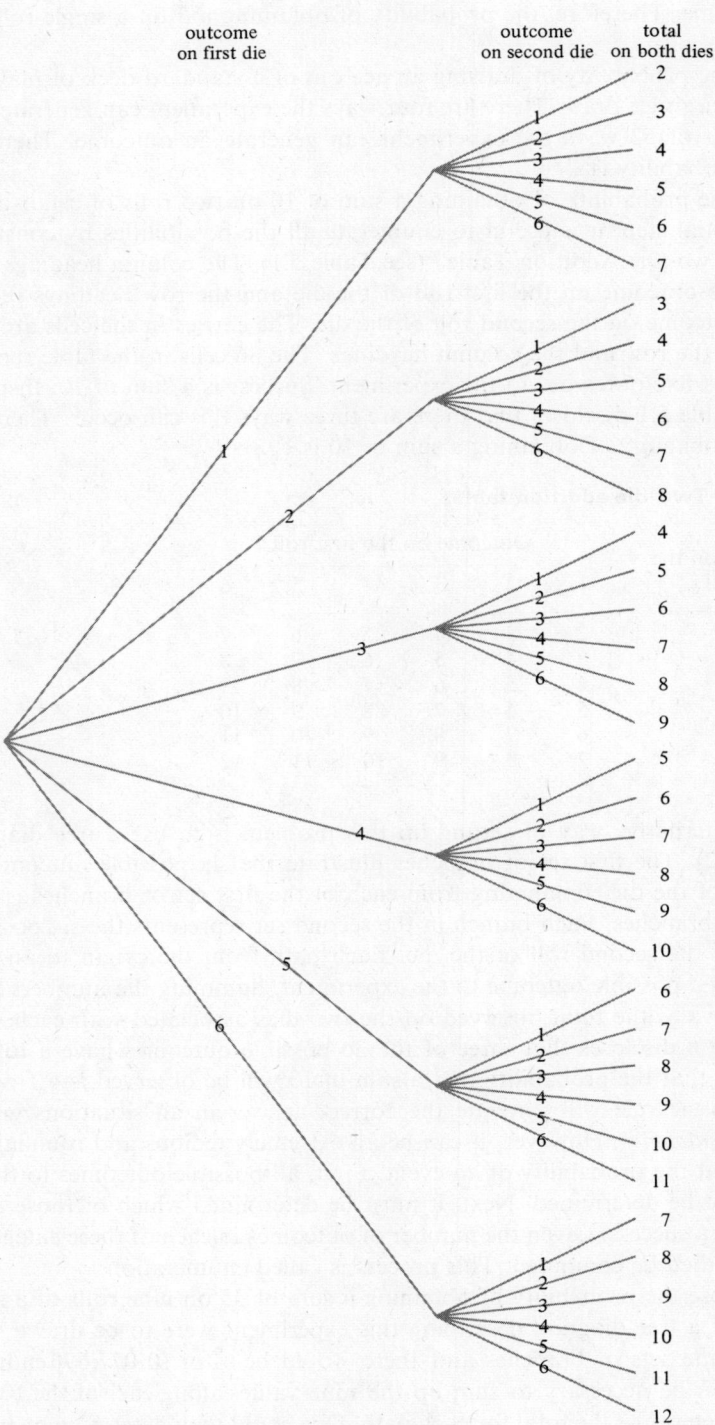

Figure 5.2 Two-die addition tree.

At this point you must be saying that there has to be a better way. There is. Fortunately, shortcuts exist that enable the analyst to obtain such probabilities with a minimum of work. These techniques (the addition rule and the multiplication rule) will be examined later in the chapter. Before introducing the rules, however, the situation where the outcomes to an experiment are not equiprobable will be examined.

5.2 NONEQUIPROBABLE EVENTS

If one were to flip an unfair coin weighted such that heads turn up more frequently than tails, the basic rule of probability would no longer be appropriate. The two outcomes to the experiment would no longer be equally likely. The objectivist school, however, can still handle this situation. The probability of a head can be generated by performing the experiment many times and observing the limiting value that the ratio of total number of successes to total number of trials approaches. The results of such an undertaking are depicted in Figure 5.3.

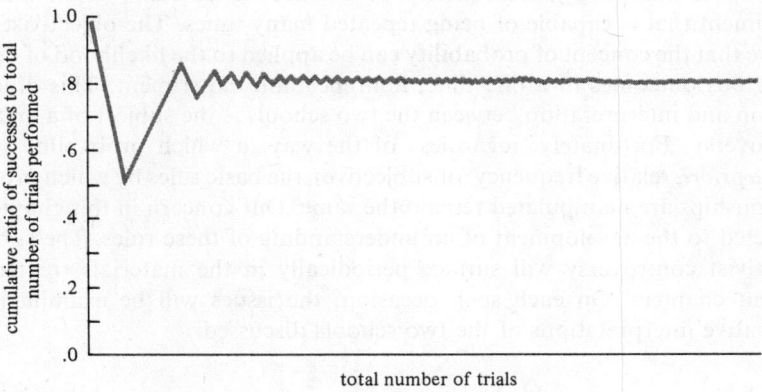

Figure 5.3

The limiting value, in this case 0.8, would be assigned as the probability that a head will be observed on any one flip of the coin. Probabilities generated in this fashion are referred to as *relative frequency probabilities* or *a posteriori probabilities*. The label *a posteriori* refers to the fact that probability assignments cannot be generated until *after* evidence on the relative frequency of occurrence has been accumulated. Note that the relative frequency approach can also be used to generate the long-run probability of success for an experiment in which the outcomes are equally likely (see Figure 5.1). The difference is that with experiments in which outcomes are not equally likely, the relative frequency approach *must* be used, whereas for experiments characterized by equally likely outcomes, the long-run probability of occurrence of a given outcome *can* be determined *a priori*.

5.3 SUBJECTIVE PROBABILITIES

The *subjectivist school* believes that probability should be interpreted as a measure of the strength of conviction of an individual as to the likelihood of occurrence of some specified event. Such probabilities are arrived at by considering such objective evidence as is available and, in addition, incorporating the subjective feelings of the individual. The subjectivist approach allows the individual to assign probabilities to unique events (i.e., possible outcomes of experiments that are conducted on a single-time basis). Some examples are the assignment of the probability that a given horse will win the Kentucky Derby, assignment of a probability to the event "the heavyweight champion will successfully defend his crown," assignment of a probability to the event "your proposal in a sealed-bid competition will be accepted," assignment of a probability to the event "a new product, if added to the line, will sell more than 100,000 units"; assignment of probabilities to the events "research and development costs on a new component will be less than $5000," "research and development costs will be between $5000 and $10,000," and "research and development costs will exceed $10,000." The subjectivist school is also referred to as the Bayesian school. In this context, the label Bayesian signifies a propensity or willingness to incorporate subjective probabilities in statistical analysis. Individuals identifying with the objectivist school are unwilling to incorporate subjectively arrived at probabilities into the formal structure of statistical analysis. Objectivists believe that the concept of probability is only appropriate when referring to the long-run frequency of occurrence of some given event in an experiment that is capable of being repeated many times. The objectivist does not believe that the concept of probability can be applied to the likelihood of occurrence of various outcomes in a one time, nonrepeatable experiment. This difference of opinion and interpretation between the two schools is the subject of a great deal of controversy. Fortunately, regardless of the way in which probability is defined (be it *a priori*, relative frequency, or subjective), the basic rules by which probabilistic relationships are manipulated remain the same. Our concern in this chapter will be restricted to the development of an understanding of these rules. The subjectivist–objectivist controversy will surface periodically in the materials treated in subsequent chapters. On each such occasion, the issues will be examined and the alternative interpretations of the two schools discussed.

Table 5.2

Statistics grade		Dean's list, event 6	Not on dean's list, event 7	Total
Event 1	A	20	5	25
Event 2	B	15	50	65
Event 3	C	5	90	95
Event 4	D	0	80	80
Event 5	F	0	35	35
Total		40	260	300

A useful base on which to build an understanding of probability is an ability to recognize the symbols that are encountered and the relationships to which they refer. To aid in perceiving what the symbols represent, they will be referenced to the presentation in Table 5.2, which describes the academic performance of the 300 students who completed the statistics course last semester. Observe that the 300 students have been cross classified on the basis of their grade in the statistics course *and* as to whether or not they made the dean's list. The body of the table contains ten cells. The categories are so defined that each student is assigned to one and only one cell. For example, 20 of the students have been assigned to the cell associated with receiving a grade of A in statistics *and* also making the dean's list. The total number of possible outcomes in the experiment corresponds to the total number of items in the universal set, where the universal set is defined as the set of all possible outcomes. A set is defined as a collection of objects bound together by a common characteristic or set of characteristics that sets them apart from other objects. In this case, the universal set consists of those individuals who completed the statistics course last semester.

A number of subsets can be identified within the universal set. To illustrate, the subset consisting of those students who made the dean's list consists of the 40 students in the first column; the subset of students who received C's consists of the 95 students in the third row; the subset of students who received a C *and* also made the dean's list consists of the five students who are members of both the first column and the third row (i.e., members of the cell at the intersection of the first column and the third row).

5.4 SYMBOLS

$P(E_1)$

E_1 refers to event 1. Event 1 is defined as "the student selected received a grade of A in statistics." $P(E_1)$ refers to the probability of occurrence of event 1. $P(E_1)$ can also be interpreted as the probability of success, where success is defined as the individual selected from the universal set also is a member of the subset designated as E_1. Event 1 is said to occur if any of the members of subset E_1 are selected. Thus, event 1 would be said to occur if any of the 25 individuals who received an A in statistics should be selected. Given that each of the 300 individuals in the universal set has an equal opportunity of being selected, $P(E_1)$ is equal to the ratio of the number of elements in subset E_1 to the total number of elements in the universal set.

$$P(E_1) = \frac{\text{number of elements in subset } E_1}{\text{number of elements in the universal set}} = \frac{25}{300}$$

Probabilities such as $P(E_1)$ are referred to as *marginal probabilities*. The word marginal is derived from the fact that the number of elements in the subset in question can be determined by finding the number of elements in a particular row of the table or the number of elements in a particular column. This information is conveyed in the *margins* of the table. For example, the number of elements in subset E_1 is equal to the sum of the elements in the first row or 25. Similarly, the number of elements in subset E_6 is equal to the sum of the elements in the first column or 40.

$P(E_1 \cup E_6)$

The symbol $\cup$ in $P(E_1 \cup E_6)$ is a symbol used in set theory to represent the *union* of two sets.[1] The union of two sets refers to the new set formed by combining two existing sets where the new set consists of those elements that can be found in *at least one* of the two existing sets.

---

[1] $P(E_1 \cup E_6)$ is also written $P(E_1 \text{ or } E_6)$.

EXAMPLE 5.1

Given: set A consists of the four numbers $\{1, 3, 5, 7\}$ and set B consists of the four numbers $\{1, 3, 9, 11\}$.

Then: set $A \cup B$ would consist of the six elements $\{1, 3, 5, 7, 9, 11\}$.

A set can be described either by definition, for example, "all the students who received an A in statistics" or by enumeration, that is, a pair of brackets containing a list of the names of the 25 students who received an A in statistics last term. Using enumeration, the set E_1 would contain a list of 25 names and the set E_6 would contain a list of 40 names.

set $E_1 = \{$name 1, name 2, . . . , name 25$\}$

set $E_6 = \{$name 1, name 2, . . . , name 40$\}$

$E_1 \cup E_6$ refers to the new subset formed by those individuals who belong to either subset E_1 or to subset E_6, one or the other being sufficient. $E_1 \cup E_6$ is composed of all 25 individuals in subset E_1 plus the 20 individuals in subset E_6 not already mentioned in subset E_1.

$E_1 \cup E_6 = \{$name 1, name 2, . . . , name 44, name 45$\}$

$P(E_1 \cup E_6)$ refers to the probability of success, where success is defined as the student drawn out of the universal set is also a member of subset $E_1 \cup E_6$. Event E_1 is defined as "the student selected obtained an A in statistics," and event E_6 is defined as "the student selected made the dean's list." Note that in order for an outcome to be considered a success, it need be associated with *either* event E_1 or event E_6. Both events could occur (the selected student could have received an A *and* made the dean's list), but one or the other is sufficient. Observe that success can occur 45 ways—there are 45 students who fall into at least one of the two categories. It would be incorrect, however, simply to sum together the number of students who made the dean's list (40) and the number of students who received an A (25) to obtain the total number of ways success can occur. The resulting total of 65 would overstate the number of ways success can occur. The 20 students who possess both characteristics would have been included twice.

Given that each of the 300 individuals in the universal set has an equal opportunity of being selected, $P(E_1 \cup E_6)$ is equal to the ratio of the number of elements in subset $E_1 \cup E_6$ to the total number of elements in the universal set.

$$P(E_1 \cup E_6) = \frac{\text{number of elements in subset } E_1 \cup E_6}{\text{number of elements in the universal set}} = \frac{45}{300}$$

$P(E_1 \cap E_6)$

The symbol $\cap$ in $P(E_1 \cap E_6)$ is a symbol used in set theory to represent the *intersection* of two sets.[2] The intersection of two sets refers to the new set formed by extracting from the two existing sets only those elements that the two existing sets share in common.

---

[2] $P(E_1 \cap E_6)$ is also written $P(E_1$ and $E_6)$.

EXAMPLE 5.2

Given: set A consists of the four numbers $\{1, 3, 5, 7\}$ and set B consists of the four numbers $\{1, 3, 9, 11\}$.

Then: set $A \cap B$ consists of the two elements the sets share in common or $\{1, 3\}$.

The intersection of subsets E_1 and E_6 refers to the new set composed of the 20 individuals whose names appear in both the enumeration of the 25 individuals in set E_1 and the enumeration of the 40 individuals in set E_6.

$$E_1 \cap E_6 = \{\text{name } 1, \text{name } 2, \ldots, \text{name } 20\}$$

All 20 of the above names have the common characteristic of appearing in both the separate enumeration of E_1 and the separate enumeration of E_2.

If two existing sets share no elements in common, then there is no intersection and we have what is known as an empty set. Such a set (i.e., $\{\ \}$) is known as a *null set*.

EXAMPLE 5.3

Given

$$A = \{0, 1, 2, 3\}$$

$$B = \{0, 7, 8\}$$

$$A \cap B = \{0\}$$

null $\neq$ 0 so

$A \cap B$ is *not* a null set. It contains the element 0.

Given:

$$C = \{4, 5, 6\}$$

$$D = \{0, 1, 2\}$$

$$C \cap D = \{\ \}$$

$C \cap D$ is a null set. The null set is typically represented by the symbol $\varnothing$.

Referring to Table 5.2, $E_5 \cap E_6$ is a null set or $\varnothing$.

$P(E_1 \cap E_6)$ refers to the probability that an individual selected at random out of the universal set will also belong to the subset formed by the intersection of subsets E_1 and E_6. Note that the occurrence of *both* events E_1 and E_6 (*either simultaneously or sequentially depending on the nature of the experiment*) is required in order for an outcome to the experiment to be considered a success. Both must occur—one or the other is not sufficient.

Given that each of the 300 individuals in the universal set has an equal opportunity of being selected, $P(E_1 \cap E_6)$ is equal to the ratio of the number of elements in subset $E_1 \cap E_6$ to the total number of elements in the universal set.

$$P(E_1 \cap E_6) = \frac{\text{number of elements in subset } E_1 \cap E_6}{\text{number of elements in the universal set}} = \frac{20}{300}$$

Probabilities such as $P(E_1 \cap E_6)$ are referred to as *joint probabilities* because they refer to the probability of the joint occurrence of two events.

$P(E_1 \mid E_6)$

$P(E_1 \mid E_6)$ is an example of a conditional probability and is identified by the slash separating E_1 and E_6. The slash is a symbol used to represent the expression "given that." In words, $P(E_1 \mid E_6)$ refers to the probability that event E_1 will occur, *conditioned* by the information that event E_6 has already occurred. The basic difference between conditional probabilities and those previously examined is that in the case of conditional probabilities, selections are made out of the subset indicated to the right of the slash, whereas for $P(E_1)$, $P(E_1 \cup E_6)$, and $P(E_1 \cap E_6)$, selections are made from the universal set.

$$P(E_1) = P(E_1 \mid \text{universe})$$

$$P(E_1 \cup E_6) = P(E_1 \cup E_6 \mid \text{universe})$$

$$P(E_1 \cap E_6) = P(E_1 \cap E_6 \mid \text{universe})$$

Occasionally, $P(E_1)$ is written as $P(E_1 \mid \text{universe})$, which is read as "Given that the drawing is to be made out of the universal set, what is the probability that the element selected will also be a member of subset E_1." Usually, it is written as $P(E_1)$. Unless otherwise stated, it is presumed that the drawing is to be made from the universal set. The conditional format typically is reserved for those occasions where it is necessary to draw attention to the fact that drawings are to take place out of a subset of the universe rather than from the universal set itself.

$P(E_1 \mid E_6)$ refers to the probability that an individual selected at random out of the subset E_6 is also a member of subset E_1. The experiment consists of selecting an element out of subset E_6 that consists of the 40 individuals who made the dean's list. Success is defined as the individual selected is also a member of subset E_1 (i.e., the individual selected received an A).

$$P(E_1 \mid E_6) = \frac{\text{number of elements in the subset } E_6 \text{ that are also members of subset } E_1}{\text{number of elements in the subset } E_6} = \frac{20}{40}$$

Similarly,

$$P(E_6 \mid E_1) = \frac{\text{number of elements in the subset } E_1 \text{ that are also members of subset } E_6}{\text{number of elements in the subset } E_1} = \frac{20}{25}$$

5.5 VENN DIAGRAMS

The relationships referred to above can also be represented visually by Venn diagrams.

A rectangular shape is typically used to represent the universal set.

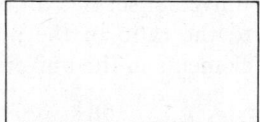

Subsets of the universal set are represented by geometric configurations placed within the universal set. Subsets are represented by circles in the following Venn diagram.

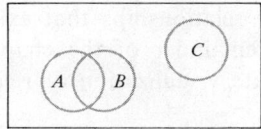

If the circles for two subsets intersect, it indicates that they share common elements. If the circles for two subsets do not intersect, it indicates that they have no common elements (i.e., any elements that are in one subset are not among the elements included in the other subset). A number of subsets are listed below. To the right of each subset is a Venn diagram that depicts the designated subset as the shaded area.

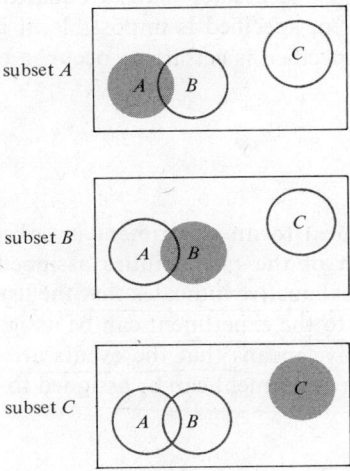

Unions: The union of two sets is represented by the total area encompassed by the circles representing the subsets.

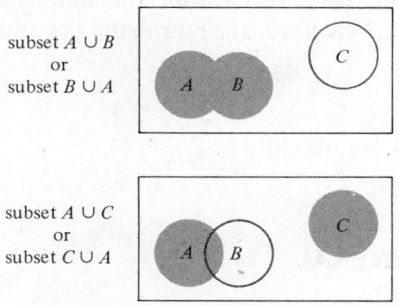

Intersections: The intersection of two sets is represented by the area shared by the two circles representing the subsets.

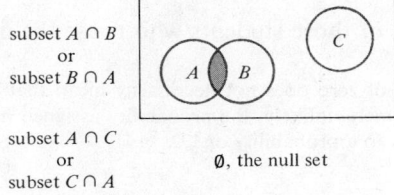

Venn diagrams are useful for depicting the relationships that exist among various subsets and the universal set. In the remainder of the chapter, Venn diagrams will frequently be employed to facilitate visualization of relationships under discussion.

5.6 AXIOMS

In assigning probabilities to the outcomes of an experiment, the following axioms must be observed.

$O \leq P(\text{event}) \leq 1$

In words, the probability of any event must be greater than or equal to zero and less than or equal to one. If the given event specified is impossible, it is assigned a probability of zero. If the given event specified is certain to occur, a probability of one is assigned.[3]

$\sum_{i=1}^{n} P(\text{event}_i) = 1$

Given that the list of events (or outcomes) to an experiment is *collectively exhaustive and mutually exclusive*, the sum of the probabilities assigned to the n outcomes must total to one. Collectively exhaustive indicates that the list is defined in such a way that any possible outcome to the experiment can be assigned to one of the events in the list. Mutually exclusive means that the events are defined in such a fashion that no one outcome to the experiment can be assigned to more than one event.

$1 - P(\text{event}) = P(\overline{\text{event}})$, where $\overline{\text{event}}$ means the event does *not* occur

In words, the probability of any given event not occurring is equal to one minus the probability that the event will occur. These two possible outcomes are mutually exclusive and collectively exhaustive. Therefore, their probabilities must sum to one.

$$P(\text{event}) + P(\overline{\text{event}}) = 1$$

Subtracting $P(\text{event})$ from both sides, we obtain

$$P(\overline{\text{event}}) = 1 - P(\text{event})$$

5.7 DERIVATION OF THE GENERAL RULE OF ADDITION

Consider $P(E_1 \cup E_6)$. Referring again to Table 5.2,

E_1 refers to the subset comprised of those students who obtained an A in statistics.

E_6 refers to the subset comprised of those students who made the dean's list.

---

[3] It should be noted that a probability of zero does not necessarily mean that an event is impossible. In cases where an event is extremely unlikely, the probability assigned may be zero due to rounding. A similar comment applies to a probability of 1.0.

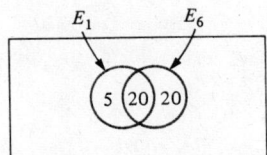

The number of ways success can occur is equal to the sum of the members of the following three subsets:

$$\text{subset } (E_6 \cap \bar{E}_1) + \text{subset } (\bar{E}_6 \cap E_1) + \text{subset } (E_6 \cap E_1) = \text{subset } (E_6 \cup E_1)$$

the 20 students who made the dean's list but did not get an A in statistics	+ the 5 students who received an A in statistics but did not make the dean's list	+ the 20 students who both made the dean's list and also received an A in statistics	= the 45 students who *either* received an A *or* made the dean's list

In the language of Venn diagrams,

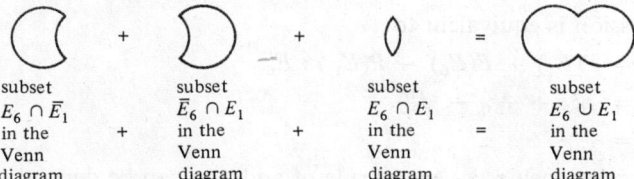

subset $E_6 \cap \bar{E}_1$ in the Venn diagram	+	subset $\bar{E}_6 \cap E_1$ in the Venn diagram	+	subset $E_6 \cap E_1$ in the Venn diagram	=	subset $E_6 \cup E_1$ in the Venn diagram

An alternative (and easier) approach for determining the number of elements in the subset $E_6 \cup E_1$ is as follows:

$$\text{subset } E_1 \quad + \text{subset } E_6 \quad - \text{subset } (E_1 \cap E_6) = \text{subset } (E_6 \cup E_1)$$

the 25 students who received an A in statistics	+ the 40 students who made the dean's list	− the 20 students who both received an A and made the dean's list	= the 45 students who *either* received an A *or* made the dean's list

In the language of Venn diagrams,

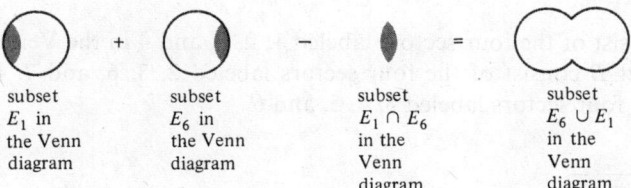

subset E_1 in the Venn diagram	+	subset E_6 in the Venn diagram	−	subset $E_1 \cap E_6$ in the Venn diagram	=	subset $E_6 \cup E_1$ in the Venn diagram

Dividing both sides of the expression by $\boxed{N}$ (i.e., the number of elements in the universal set) maintains the following equality.

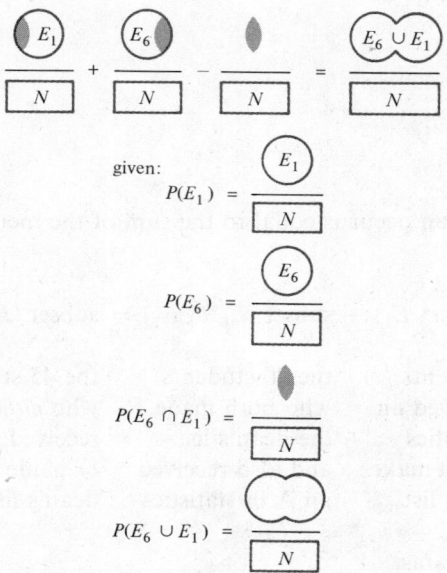

given:

$$P(E_1) = \dfrac{\boxed{E_1}}{\boxed{N}}$$

$$P(E_6) = \dfrac{\boxed{E_6}}{\boxed{N}}$$

$$P(E_6 \cap E_1) = \dfrac{}{\boxed{N}}$$

$$P(E_6 \cup E_1) = \dfrac{}{\boxed{N}}$$

The above expression is equivalent to

$$P(E_1 \cup E_6) = P(E_1) + P(E_6) - P(E_1 \cap E_6)$$
$$= \tfrac{25}{300} + \tfrac{40}{300} - \tfrac{20}{300}$$
$$= \tfrac{45}{300}$$

From the above relationship, a general rule of addition can be derived.

5.8 THE GENERAL RULE OF ADDITION

$$P(A \cup B) = P(A) + P(B) - P(A \cap B)$$

The probability of the union of two events is equal to the marginal probability of the first plus the marginal probability of the second minus the probability of their intersection.

The function of the term $P(A \cap B)$ is to rectify the double counting that takes place when $P(A)$ and $P(B)$ are summed. The general rule of addition can be extended to the probability of the union of three events. The general rule of addition for three events is

$$P(A \cup B \cup C) = P(A) + P(B) + P(C) - P(A \cap B)$$
$$- P(A \cap C) - P(B \cap C) + P(A \cap B \cap C)$$

EXAMPLE 5.4

Let set A consist of the four sectors labeled 1, 2, 3, and 4 in the Venn diagram below. Let set B consist of the four sectors labeled 2, 3, 6, and 7. Let set C consist of the four sectors labeled 3, 4, 5, and 6.

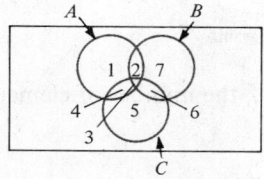

Observe the effect of combining the subsets referred to in the general rule of addition:

$$\text{set } A = 1 + 2 + 3 + 4$$
$$\text{plus set } B = 2 + 3 + 6 + 7$$
$$\text{plus set } C = 3 + 4 + 5 + 6$$
$$\text{minus set } (A \cap B) = - 2 - 3$$
$$\text{minus set } (A \cap C) = - 3 - 4$$
$$\text{minus set } (B \cap C) = - 3 - 6$$
$$\text{plus set } (A \cap B \cap C) = + 3$$
$$\overline{}$$
$$1 + 2 + 3 + 4 + 5 + 6 + 7$$

Thus,

$$(A \cup B \cup C) = A + B + C - (A \cap B) - (A \cap C)$$
$$- (B \cap C) + (A \cap B \cap C)$$

Therefore,

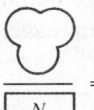 $=$ $\dfrac{\text{number of elements in the new subset formed by the union of the three subsets } A, B, \text{ and } C}{\text{total number of elements contained in the universal set}}$ $= P(A \cup B \cup C)$

5.9 THE MEANING OF THE TERM MUTUALLY EXCLUSIVE

Two or more events are considered to be mutually exclusive if the occurrence of one automatically eliminates the possibility of the occurrence of the others. In set theory if two or more events are mutually exclusive, there can be no intersection of their respective subsets because the joint occurrence of the events is impossible. The Venn diagram below depicts the subsets for events A and B given that the two events are mutually exclusive.

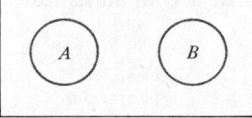

Examples:

1. The two events "making the dean's list" and "receiving an F in statistics" are mutually exclusive events.
2. The two events "drawing a 4 on the first draw from a deck of cards" and "drawing a 5 on the first draw from a deck of cards" are mutually exclusive events.

3. The two events "drawing a 4 on the *first* draw from a deck of cards" and "drawing a 5 on the *second* draw from a deck of cards" are *not* mutually exclusive events. This differs from Example 2 in that Example 2 referred to outcomes on a single draw from a deck of cards and Example 3 refers to outcomes on successive draws from a deck of cards.

4. The two events "observing a 2 on the first roll of a die" and "observing a 6 on the first roll of a die" are mutually exclusive events.

5. The two events "drawing a red card on the first draw from a deck of cards" and "drawing a queen on the first draw from a deck of cards" are *not* mutually exclusive events. One could draw a red queen and satisfy both conditions. The occurrence of one event does not prevent the occurrence of the other.

5.10 THE SPECIAL RULE OF ADDITION (OR THE ADDITION RULE FOR MUTUALLY EXCLUSIVE EVENTS)

When two events are mutually exclusive, the general rule of addition simplifies to the following:

$$P(A \cup B) = P(A) + P(B)$$

In words, if two events A and B are mutually exclusive, the probability that *either A or B* will occur is equal to the sum of their individual probabilities.

Referring to Table 5.2 and $P(E_1 \cup E_2)$, in Venn diagram language,

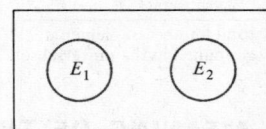

$$P(E_1 \cup E_2) = \frac{\boxed{E_1}}{\boxed{N}} + \frac{\boxed{E_2}}{\boxed{N}} = \frac{\boxed{E_1} + \boxed{E_2}}{\boxed{N}}$$

where $\boxed{N}$ represents the universal set

E_1 refers to the subset of students who obtained an A in statistics.

E_2 refers to the subset of students who obtained a B in statistics.

$$P(E_1) = \frac{\boxed{E_1}}{\boxed{N}} = \frac{\text{number of elements in subset } E_1}{\text{number of elements in the universe}} = \frac{25}{300}$$

$$P(E_2) = \frac{\boxed{E_2}}{\boxed{N}} = \frac{\text{number of elements in subset } E_2}{\text{number of elements in the universe}} = \frac{65}{300}$$

Recall:

The general rule of addition:

$$P(A \cup B) = P(A) + P(B) - P(A \cap B)$$

The special rule of addition:

$$P(A \cup B) = P(A) + P(B)$$

The function of the term $P(A \cap B)$ is to rectify the double counting that takes place when $P(A)$ and $P(B)$ are summed. The special rule of addition does not contain the term $P(A \cap B)$ because it applies only to mutually exclusive events. When events are mutually exclusive, there is no intersection and thus no double counting. If two events are mutually exclusive, the probability of their joint occurrence is zero. The occurrence of one automatically precludes the occurrence of the other. Thus, for mutually exclusive events, $P(A \cap B) = 0$ and the general rule, $P(A \cup B) = P(A) + P(B) - P(A \cap B)$, simplifies to $P(A \cup B) = P(A) + P(B)$.

Similarly, for mutually exclusive events, the general rule of addition for three events simplifies to $P(A \cup B \cup C) = P(A) + P(B) + P(C)$ because there are no intersections and the terms $P(A \cap B)$, $P(A \cap C)$, $P(B \cap C)$, and $P(A \cap B \cap C)$ all become equal to zero.

5.11 DERIVATION OF THE GENERAL RULE OF MULTIPLICATION

Recalling the definition of conditional probability, it was stated that $P(E_1 \mid E_6)$ refers to the probability that an individual selected out of subset E_6 will also be a member of subset E_1. Another way of stating this is to say: Of the elements that are members of subset E_6, what percentage are also members of subset E_1?

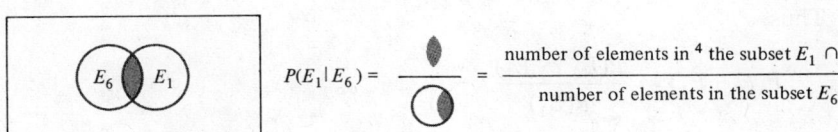

$$P(E_1 | E_6) = \frac{\text{shaded}}{\text{circle}} = \frac{\text{number of elements in }[4]\text{ the subset } E_1 \cap E_6}{\text{number of elements in the subset } E_6}$$

A ratio is unaffected when numerator and denominator are divided by the same positive value. Dividing both numerator and denominator by the number of elements comprising the universal set, we obtain

$$P(E_1 \mid E_6) = \frac{\text{shaded} / N}{E_6 / N} = \frac{\text{number of elements in the subset } E_1 \cap E_6 / \text{number of elements in the universal set}}{\text{number of elements in subset } E_6 / \text{number of elements in the universal set}}$$

By definition,

$$\frac{\text{shaded}}{N} = P(E_1 \cap E_6)$$

$$\frac{E_6}{N} = P(E_6)$$

---

[4] The symbol E_6 is interpreted as representing the entire subset E_6. The shaded portion is simply intended to emphasize the portion of subset E_6 associated with the intersection.

Thus, it follows that[5]

$$P(E_1 \mid E_6) = \frac{P(E_1 \cap E_6)}{P(E_6)}$$

Multiplying both sides of $P(E_1 \mid E_6) = [P(E_1 \cap E_6)/P(E_6)]$ by $P(E_6)$ generates the following equality:

$$P(E_6) \cdot P(E_1 \mid E_6) = \frac{P(E_1 \cap E_6)}{P(E_6)} \cdot P(E_6)$$

Canceling, we obtain

$$P(E_6) \cdot P(E_1 \mid E_6) = P(E_1 \cap E_6)$$

It follows that

$$P(E_1 \cap E_6) = P(E_6) \cdot P(E_1 \mid E_6) \tag{1}$$

A similar line of reasoning applied to $P(E_6 \mid E_1)$ leads to the following relationship:

Thus,

$$P(E_6 \mid E_1) = \frac{P(E_6 \cap E_1)}{P(E_1)}$$

Multiplying both sides by $P(E_1)$ and canceling, we obtain

$$P(E_6 \cap E_1) = P(E_1) \cdot P(E_6 \mid E_1) \tag{2}$$

The two derivations result in the following two equalities:

$$P(E_1 \cap E_6) = P(E_6) \cdot P(E_1 \mid E_6) \tag{1}$$

$$P(E_6 \cap E_1) = P(E_1) \cdot P(E_6 \mid E_1) \tag{2}$$

Since $P(E_1 \cap E_6)$ and $P(E_6 \cap E_1)$ refer to the same thing, the following equivalency exists:

$$P(E_1 \cap E_6) = P(E_6) \cdot P(E_1 \mid E_6) \quad \text{or} \quad P(E_1) \cdot P(E_6 \mid E_1)$$

The above relationship is known as the *general rule of multiplication* and is used to compute the probability of occurrence of joint events.

---

[5] Note that any conditional probability is equal to the ratio formed by a numerator equal to the probability of the joint occurrence of the events on either side of the slash and a denominator equal to the probability of occurrence of the event indicated to the right of the slash. Bayes' rule, an extremely useful technique for revising probabilities in the light of new information, is based on this relationship. Bayes' rule is discussed in detail in Chapter 6.

5.12 THE GENERAL RULE OF MULTIPLICATION

$$P(A \cap B) = P(A) \cdot P(B \mid A) \quad \text{or} \quad P(B) \cdot P(A \mid B)$$

The probability of the joint occurrence of two events is equal to the marginal probability of one of the events multiplied by the conditional probability of the other event, given that the first event has occurred. Note that it makes no difference which of the joint events is depicted as occurring first.

Referring once again to Table 5.2,

$$P(E_1 \cap E_6) = P(E_1) \cdot P(E_6 \mid E_1)$$

$$= \frac{25}{300} \times \frac{20}{25} = \frac{20}{300}$$

or

$$P(E_1 \cap E_6) = P(E_6) \cdot P(E_1 \mid E_6)$$

$$= \frac{40}{300} \times \frac{20}{40} = \frac{20}{300}$$

Note that with $P(E_1)$, which is a marginal probability, the objective is to determine the percentage of the elements in the universe that possess the characteristic "received an A in statistics" or, in other words, are in subset E_1. In the second factor, $P(E_6 \mid E_1)$, which is a conditional probability, drawings are to take place out of subset E_1 instead of the universe. The ratio of interest in the second factor is the proportion of the 25 elements in subset E_1 that also belong to subset E_6.

5.13 THE MEANING OF THE TERM INDEPENDENCE

Two or more events are said to be independent if the occurrence or nonoccurrence of one event in no way affects the probability of occurrence (or nonoccurrence) of the other events.

EXAMPLE 5.5

Given a deck of 52 playing cards, let

A = ace on the first draw

a = not an ace on the first draw

B = ace on the second draw

b = not an ace on the second draw

Consider the case where drawings are made *with replacement*. With replacement simply means that the card removed on the first draw is replaced in the deck prior to selection of the second card.

With replacement:

$$P(A) = \tfrac{4}{52} \qquad P(B \mid A) = \tfrac{4}{52}$$

$$P(a) = \tfrac{48}{52} \qquad P(B \mid a) = \tfrac{4}{52}$$

$P(A)$ is the ratio of the number of ways an ace could be drawn to the total number of different cards that could be drawn or $\tfrac{4}{52}$.

With replacement, there are still 4 aces and 52 cards left in the deck, so the probability of drawing an ace on the second draw remains at $\frac{4}{52}$ regardless of what transpired on the first draw (i.e., ace or not ace). Thus, $P(B \mid A) = P(B \mid a) = \frac{4}{52}$. Thus, events A and B are independent. The probability of event B occurring remains the same regardless of whether an ace was drawn or not drawn on the first draw.

Consider the same situation, only this time *without replacement*; that is, the card drawn on the first draw is *not* returned to the deck prior to the second drawing.

Without replacement:

$$P(A) = \tfrac{4}{52} \qquad P(B \mid A) = \tfrac{3}{51}$$

$$P(a) = \tfrac{48}{52} \qquad P(B \mid a) = \tfrac{4}{51}$$

The probability of obtaining or not obtaining an ace on the first draw is unaffected by the introduction of drawing without replacement. This is not the case, however, on the second draw. If an ace were to be obtained on the first draw, there would be 51 cards remaining for the second draw. Only 3 would be aces, so $P(B \mid A)$ would be $\frac{3}{51}$. If other than an ace were obtained on the first draw there would again be 51 cards left, but this time 4 would be aces so $P(B \mid a) = \frac{4}{51}$. Clearly, the probability of event B occurring varies depending on whether or not an ace was selected on the first draw. Thus, in this example, when drawing under conditions of nonreplacement, events A and B are *not independent*. The probability of B (and b) occurring *depends* on whether A or a occurred. The probability of event B occurring is conditional upon what has previously transpired.

For events to be independent, the following relationship must hold:

$$P(B) = P(B \mid A) = P(B \mid a)$$

The above requirement is met for this example when drawings are made *with replacement*.

$$P(B) = \tfrac{4}{52}$$

$$P(B \mid A) = \tfrac{4}{52}$$

$$P(B \mid a) = \tfrac{4}{52}$$

The above requirement is not met for this example when drawings are made under conditions of *nonreplacement*. Therefore, for this example, under conditions of *nonreplacement* events A and B are not independent.

$$P(B) = \tfrac{4}{52}$$

$$P(B \mid A) = \tfrac{3}{51}$$

$$P(B \mid a) = \tfrac{4}{51}$$

In the above two comparisons, it is stated without proof that $P(B) = \frac{4}{52}$ under both conditions of replacement and nonreplacement. The truth of this assertion will be demonstrated in Section 5.16 in the discussion of *the principle of insufficient reason*.

5.14 THE SPECIAL RULE OF MULTIPLICATION (OR THE MULTIPLICATION RULE FOR INDEPENDENT EVENTS)

$$P(A \cap B) = P(A) \cdot P(B)$$

$$P(A \cap B \cap C) = P(A) \cdot P(B) \cdot P(C)$$

*The probability of the joint occurrence of two or more independent events is
equal to the product of their marginal probabilities.*

If events A and B are independent, then $P(B \mid A) = P(B)$, and the general
rule of multiplication simplifies from $P(A) \cdot P(B \mid A)$ to $P(A) \cdot P(B)$. Similarly,
if events A, B, and C are independent, $P(A) \cdot P(B \mid A) \cdot P(C \mid A \cap B)$ simplifies
to $P(A) \cdot P(B) \cdot P(C)$.

EXAMPLE 5.6

Consider $P(C \cap D)$, where

$C =$ ace on first draw
$D =$ ace on second draw

and drawings are performed with replacement, which as previously demonstrated makes events C and D independent.

$$P(C) = \tfrac{4}{52}$$

$$P(D) = \tfrac{4}{52}$$

$$P(C \cap D) = P(C) \cdot P(D)$$

$$= \tfrac{4}{52} \cdot \tfrac{4}{52} = \tfrac{16}{2704}$$

If you tried to solve by enumeration, the tree diagram would have 2704
endpoints and of the 2704 paths from the origin to an endpoint, 16 would
consist of an ace on the first draw and an ace on the second draw.

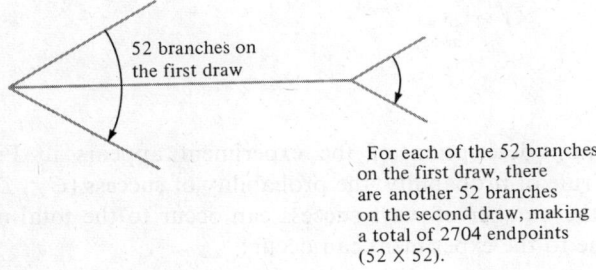

52 branches on
the first draw

For each of the 52 branches
on the first draw, there
are another 52 branches
on the second draw, making
a total of 2704 endpoints
(52 × 52).

Consider the same example *without replacement.*

$$P(C \cap D)$$

where

$C =$ ace on the first draw
$D =$ ace on the second draw

and drawings are made *without replacement.*

Note: When drawings are made without replacement, the events C and D will not be independent. When events are not independent, the *general rule of multiplication* must be used.

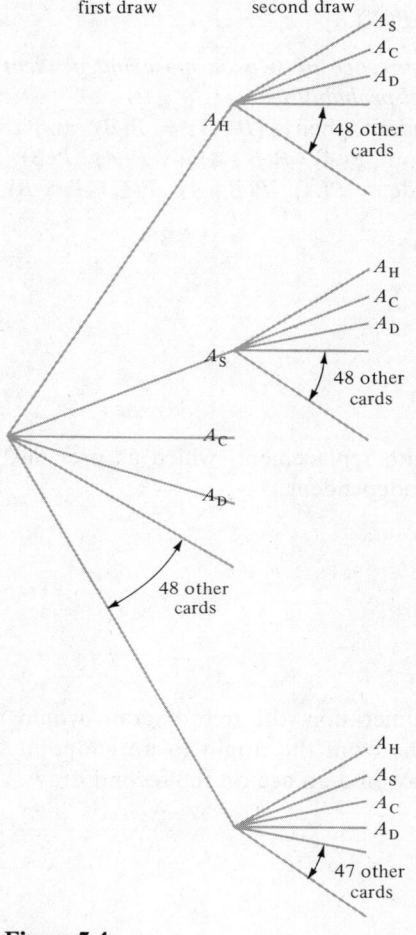

There are 52 possibilities and therefore 52 branches associated with the first draw. There are only 51 possibilities left for the second draw, and therefore there are 51 possible paths to an endpoint in the diagram associated with each of the 52 branches on the first draw.

Therefore, the total number of endpoints or total number of paths from the origin to an endpoint is 52 × 51 = 2652.

Examination of the 2652 endpoints reveals that there are only four ways an ace could have been selected on the first draw and, associated with each of these, only three ways an ace could be drawn on the second draw; therefore, the total number of ways success could occur (ace on the first draw *and* ace on the second draw) is equal to (4) (3) = 12.

Figure 5.4

A tree-diagram description of the experiment appears in Figure 5.4. Using the basic rule of probability, the probability of success ($C \cap D$) is equal to the ratio of the number of ways success can occur to the total number of ways an outcome to the experiment can occur:

$$P(C \cap D) = \tfrac{12}{2652}$$

Alternatively, one can turn to the general rule of multiplication (which applies whether events are independent or not) and generate the probability directly:

$$P(C \cap D) = P(C) \cdot P(D \mid C)$$
$$= \tfrac{4}{52} \cdot \tfrac{3}{51}$$
$$= \tfrac{12}{2652}$$

EXAMPLE 5.7

Find $P(A \cap B \cap C \cap D)$
where

A = ace on the first draw
B = ace on the second draw
C = ace on the third draw
D = ace on the fourth draw

and drawings are made *without replacement*.

$$P(A \cap B \cap C \cap D) = P(A) \cdot P(B \mid A) \cdot P(C \mid A \cap B) \cdot P(D \mid A \cap B \cap C)$$

$$= \frac{4}{52} \cdot \frac{3}{51} \cdot \frac{2}{50} \cdot \frac{1}{49}$$

$$= \frac{24}{6,497,400}$$

The specific outcome, four aces, would be expected to turn up on the average 24 out of every 6,497,400 times the experiment is performed.

In practice, the fractions are reduced before being multiplied out. This was deliberately not done in the preceding examples to enable the reader to observe the actual *number* of ways an event can succeed as compared to the actual *number* of ways an outcome can occur. When fractions are reduced, one obtains the *ratio* of the number of ways an event can succeed to the number of ways an outcome could occur.

$$P(A \cap B \cap C \cap D) = \frac{\overset{1}{\cancel{4}}}{\underset{13}{\cancel{52}}} \cdot \frac{\overset{1}{\cancel{3}}}{\underset{17}{\cancel{51}}} \cdot \frac{\overset{1}{\cancel{2}}}{\underset{25}{\cancel{50}}} \cdot \frac{1}{49} = \frac{1}{13 \cdot 17 \cdot 25 \cdot 49} = \frac{1}{270,725}$$

The result (1/270,725) is interpreted as indicating that the relative frequency with which the outcome "four aces in four draws (without replacement)" would be expected to be observed is once out of every 270,725 times the experiment is performed.

EXAMPLE 5.8

Determine

$$P(A \cap B \cap C \cap D \cap E)$$

where

A = head on first flip
B = head on second flip
C = head on third flip
D = head on fourth flip
E = head on fifth flip

and it is a fair coin.
The general rule of multiplication is

$$P(A \cap B \cap C \cap D \cap E) = P(A) \cdot P(B \mid A) \cdot P(C \mid A \cap B)$$
$$\cdot P(D \mid A \cap B \cap C) \cdot P(E \mid A \cap B \cap C \cap D)$$

However, the events are independent, so it simplifies to

$$P(A \cap B \cap C \cap D \cap E) = P(A) \cdot P(B) \cdot P(C) \cdot P(D) \cdot P(E)$$

$$= \tfrac{1}{2} \cdot \tfrac{1}{2} \cdot \tfrac{1}{2} \cdot \tfrac{1}{2} \cdot \tfrac{1}{2}$$

$$= \tfrac{1}{32}$$

The following discussion should help resolve a point of confusion for some: Experiment 1 (Table 5.3): Let

A = A queen is selected on the first draw
B = A red card is selected on the first draw

Table 5.3 Experiment 1

Face value of card	Color of card		Total
	Red	Black	
2	2	2	4
3	2	2	4
4	2	2	4
5	2	2	4
6	2	2	4
7	2	2	4
8	2	2	4
9	2	2	4
10	2	2	4
J	2	2	4
Q	2	2	4
K	2	2	4
Ace	2	2	4
Total	26	26	52

In Experiment 1, there is only one trial, but it is multidimensional—we are concerned with whether the card is red or black and also whether it is a 2, 3, 4, 5, 6, 7, 8, 9, 10, J, Q, K, or ace.

$P(A \mid B)$ = probability of event A occurring given that event B occurred. If it is specified that event B has occurred, this means the card drawn was a red card and possible outcomes are restricted to the first column.

How many cards in the first column? 26. Of the cards in the first column (red cards), how many are queens? 2. Therefore, $P(A \mid B) = \tfrac{2}{26}$.

You should have had little difficulty following the reasoning associated with Experiment 1.

Experiment 2 (Table 5.4). Let

C = ace on the first draw
D = ace on the second draw

Drawings are made without replacement.

$$P(D \mid C) = \frac{\text{number of elements in subset } C \text{ that are also in subset } D}{\text{number of elements in subset } C}$$

Table 5.4 Experiment 2

Outcome on first draw	Outcomes on second draw								
	A_H	A_S	A_C	A_D	K_H	K_S	K_C	$\cdots$	
A_H	0	1	1	1	1	1	1	$\cdots$	51
A_S	1	0	1	1	1	1	1	$\cdots$	51
A_C	1	1	0	1	1	1	1	$\cdots$	51
A_D	1	1	1	0	1	1	1	$\cdots$	51
K_H	1	1	1	1	0	1	1	$\cdots$	51
K_S	1	1	1	1	1	0	1	$\cdots$	51
K_C	1	1	1	1	1	1	0	$\cdots$	51
K_D	1	1	1	1	1	1	1	$\cdots$	51
Q_H	1	1	1	1	1	1	1	$\cdots$	51
$\vdots$	$\vdots$	$\vdots$	$\vdots$	$\vdots$	$\vdots$	$\vdots$	$\vdots$		$\vdots$
Total	51	51	51	51	51	51	51		2652

> This step is where some students become confused. They are unable to visualize how "ace on the second draw" is part of the subset associated with "ace on the first draw."

Let us attempt to clarify the confusion. Experiment 2, unlike Experiment 1, deals with an experiment that has a sequence of trials. Again it is multidimensional, but now each dimension is associated with a different trial. Note that there are 52 rows each with 51 entries for a total of 2652 entries in the universal set.

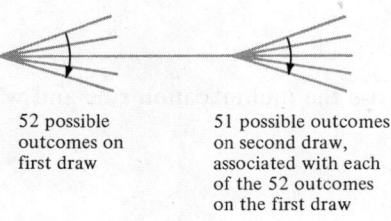

52 possible outcomes on first draw

51 possible outcomes on second draw, associated with each of the 52 outcomes on the first draw

$52(51) = 2652$ endpoints

Note that if an ace of hearts is selected on the first draw, there is no way an ace of hearts can be selected on the second draw given that drawings are made without replacement. Each of the other 51 cards, however, has an equal chance of being selected on the second draw. The remainder of the table can be interpreted in a similar fashion.

$$P(D \mid C) = \frac{\text{number of elements in subset } C \text{ that are also in subset } D}{\text{number of elements in subset } C}$$

Subset C refers to the event "an ace is obtained on the first draw" and restricts possible outcomes to the first four rows of the table (51 in each row or a total of 204). Subset D refers to the event "an ace is obtained on the second draw" and restricts possible outcomes to the first four columns (again a total of 204 elements).

$P(D \mid C)$ inquires: Of the elements in the first four rows (i.e., in subset C), how many are also in the first four columns (i.e., in subset D)? The table reproduced below represents the subset $(C \cap D)$. Clearly, subset $(C \cap D)$ contains 12 elements.

	A_H	A_S	A_C	A_D
A_H	0	1	1	1
A_S	1	0	1	1
A_C	1	1	0	1
A_D	1	1	1	0

Thus

$$P(D \mid C) = \tfrac{12}{204} = \tfrac{3}{51}$$

If drawings had been made with replacement, all the entries in the body of the table would have been 1's, and the total number of outcomes in the universal set would have been $52 \times 52 = 2704$; the number of elements in subset C would have been $4 \times 52 = 208$; the number of elements that are in both subset C and subset D would now be 16.

	A_H	A_S	A_C	A_D
A_H	1	1	1	1
A_S	1	1	1	1
A_C	1	1	1	1
A_D	1	1	1	1

and

$$P(D \mid C) = \tfrac{16}{208} = \tfrac{4}{52}$$

It is important to recognize when to use the multiplication rule and when to use the addition rule.

Addition Rule $P(A \cup B)$

The addition rule should be used whenever success can occur more than one way, that is, when the connector *or* is used [e.g., $P(A \text{ or } B)$]. The probability of success in this instance is equal to the sum of the probabilities of the various ways success can occur (eliminating double counting where necessary).

Multiplication Rule $P(A \cap B)$

The multiplication rule should be used whenever success requires the occurrence of a joint event, that is, when the connector *and* is used [e.g., $P(A \text{ and } B)$].

Occasions arise when both rules are used in combination. To illustrate: Let

A = a red card is selected on the first draw with replacement from a standard deck of 52 playing cards
B = a black jack is selected on the second draw
C = a red queen is selected on the second draw

Let success be defined as a red card is selected on the first draw and either a black jack or a red queen is selected on the second draw.

$$P(A \cap (B \cup C)) = [P(A)][P(B) + P(C)] = P(A) \cdot P(B) + P(A) \cdot P(C)$$

addition
rule

multiplication
rule

5.15 A DIVERSION

Can two events be mutually exclusive and independent at the same time?

If events A and B are independent, it follows that $P(B) = P(B \mid A)$. In words, this means that if the number of elements in subset B is equal to 10 percent of the number of elements in the universal set, then the number of elements in subset $(A \cap B)$ must also be equal to 10 percent of the number of elements in subset A. It follows that $(A \cap B)$ must contain some elements. $P(B) = P(B \mid A)$ implies that

where ⬥ represents $A \cap B$

On the other hand, if events A and B are mutually exclusive, they cannot occur jointly and the subset $(A \cap B)$ must be a null set (empty). Clearly, both of these situations cannot occur at the same time; therefore, two events cannot be mutually exclusive and independent at the same time.[6]

Another inconsistency is that when events A and B are independent, $P(B \mid A) = P(B)$, whereas when events A and B are mutually exclusive, $P(B \mid A) = 0$.

Still another way of perceiving the relationship is to think in terms of a table depicting the types of relationship that can exist between two events (Table 5.5).

Table 5.5 Types of relationships that can exist between two events

	Dependent		Independent
Mutually exclusive	Nonmutually exclusive		Independent
Occurrence of one event automatically causes the conditional probability of the second event to go to zero.	Occurrence or nonoccurrence of one event affects the conditional probability of the other event occurring, although it does not preclude the other events occurrence; i.e., it does not cause the conditional probability of the other event to go to zero.		Occurrence or nonoccurrence of one event in no way affects the conditional probability of occurrence of the other.
$P(B \mid A) = 0$	$P(B) \neq P(B \mid A) \neq P(B \mid \bar{A}) \neq 0$		$P(B) = P(B \mid A) = P(B \mid \bar{A})$

[6] There is one exception: when $P(B) = 0$.

Consider a universal set consisting of 100 equally likely elements and two subsets: subset A with 20 elements and subset B with 40 elements.

1. If A and B are independent:

$$P(A) = P(A \mid B) = P(A \mid \bar{B})$$

$$\tfrac{20}{100} = \tfrac{8}{40} = \tfrac{12}{60}$$

The Venn diagram would appear as follows:

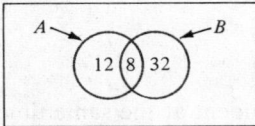

2. If A and B are mutually exclusive:

$$P(A) = \tfrac{20}{100} \neq P(A \mid B) = \tfrac{0}{40} = 0$$

The Venn diagram would appear as follows:

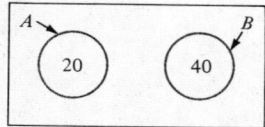

3. If A and B are dependent events but not mutually exclusive,

$$P(A) = \tfrac{20}{100} \neq P(A \mid B)$$

There would be an intersection, and the number of elements contained in $A \cap B$ could be any nonnegative integer up to 20 except 0 (which would mean no intersection and that the events are mutually exclusive) and 8 (which would mean that the events are independent). Clearly, the number of elements in the intersection can never exceed the number of elements in the smaller of the two subsets, in this case 20.

5.16 THE PRINCIPLE OF INSUFFICIENT REASON[7]

On p. 126, it was stated without proof that $P(B) = \tfrac{4}{52}$ under conditions of both replacement and nonreplacement, where B was defined as an ace being selected on the second draw. It is now appropriate to demonstrate the validity of this statement. This will be accomplished by drawing on the multiplication rule. In the process of generating the proof, an opportunity also exists to demonstrate the principle of insufficient reason.

Previously it was indicated that where

B = ace on the second draw
A = ace on the first draw

---

[7] The principle of insufficient reason is also known as the *equivalence law*.

then

$P(B \mid A)$ without replacement $= \frac{3}{51}$

$P(B \mid a)$ without replacement $= \frac{4}{51}$

It was also stated that $P(B) = \frac{4}{52}$ with or without replacement.

The principle of insufficient reason states that with no information as to the outcome of the first draw (i.e., as to whether A or a occurred), there is insufficient reason to assign a probability of other than $\frac{4}{52}$ to the event "an ace will be drawn on the second draw." There are 52 unassigned cards—each has an equal opportunity of being the second one drawn, just as each has an equal opportunity of being the first or the last one drawn. Of the 52 cards, 4 are aces; therefore, $P(B) = \frac{4}{52}$.

Given that event A has occurred, the analyst has more information and would conclude that $P(B \mid A) = \frac{3}{51}$. Of the 51 cards still unassigned, 3 are aces. Similarly, given that event a (not an ace on the first draw) occurred, the analyst would conclude that $P(B \mid a) = \frac{4}{51}$.

The claim that $P(B) = \frac{4}{52}$ can easily be substantiated. There are two alternative paths that lead to the occurrence of event B.

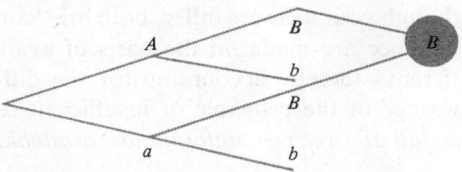

$A \longrightarrow B$ Event A occurs on the first draw and then event B occurs on the second draw (i.e., ace on the first and ace on the second).

$a \longrightarrow B$ Event a occurs on the first draw and then event B occurs on the second draw (i.e., no ace on the first draw and then an ace on the second draw).

The probability that event B will occur is equal to the sum of the probabilities of either of the two paths being followed. Thus:

With replacement:

$P(B) = P(A \cap B) + P(a \cap B)$

$P(B) = P(A) \cdot P(B \mid A) + P(a) \cdot P(B \mid a)$

$$= \frac{4}{52} \cdot \frac{4}{52} + \frac{48}{52} \cdot \frac{4}{52}$$

$$= \frac{4 \cdot 4 + 48 \cdot 4}{52 \cdot 52} = \frac{4(4 + 48)}{52 \cdot 52} = \frac{4(52)}{(52)(52)} = \frac{4}{52}$$

and, as indicated, $P(B) = \frac{4}{52}$.

Without replacement:

$P(B) = P(A) \cdot P(B \mid A) + P(a) \cdot P(B \mid a)$

$$= \frac{4}{52} \cdot \frac{3}{51} + \frac{48}{52} \cdot \frac{4}{51}$$

$$= \frac{4 \cdot 3 + 48 \cdot 4}{52 \cdot 51} = \frac{4(3 + 48)}{52 \cdot 51} = \frac{4(51)}{(52)(51)} = \frac{4}{52}$$

and, as indicated, $P(B) = \frac{4}{52}$.

EXAMPLE 5.9

Suppose a bin of parts contains 100 parts, 6 of which are defective. Let G_n represent a good part selected on the nth draw. Let D_n represent a defective part selected on the nth draw.

$$P(G_1) = \tfrac{94}{100}$$
$$P(D_1) = \tfrac{6}{100}$$
$$P(D_{13}) = \tfrac{6}{100}$$
$$P(D_{100}) = \tfrac{6}{100}$$
$$P(D_2 \mid G_1) = \tfrac{6}{99}$$
$$P(D_1 \mid D_7) = \tfrac{5}{99}$$
$$P(D_3 \mid G_1 \cap G_2) = \tfrac{6}{98}$$

Suppose you observe your instructor flip a fair coin. He then requests you to write down the probability that it is a head. You write down $\tfrac{1}{2}$. Then your instructor inspects the coin and records what he feels is the probability that it is a head. He writes down 1. Although your answers differ, both are "correct." Judgments on the probability of occurrence are made on the basis of available information. Your information is different—thereby accounting for the different evaluations. This, in effect, is the message of the principle of insufficient reason: *Every probability is correct if it uses all the relevant information available but only that which is available.*

In Chapter 6, a technique is introduced that enables the decision maker to revise the probabilities he would assign to the likelihood of occurrence of given events as the state of his information changes.

5.17 THE FUNDAMENTAL RULES OF PROBABILITY

The basic rule of probability: Given that all possible outcomes to an experiment are *equally likely*, the probability of occurrence of any specified outcome (or group outcomes) on a given performance of the experiment is equal to the ratio of the number of ways that particular outcome (or group of outcomes) can occur to the total number of ways an outcome to the experiment could be generated.

The general rule of addition:

$$P(A \cup B) = P(A) + P(B) - P(A \cap B)$$

The special rule of addition (applicable only to mutually exclusive events):

$$P(A \cup B) = P(A) + P(B)$$

The general rule of multiplication:

$$P(A \cap B) = P(A) \cdot P(B \mid A) \quad \text{or} \quad P(B) \cdot P(A \mid B)$$

The special rule of multiplication (applicable only to independent events):

$$P(A \cap B) = P(A) \cdot P(B)$$

Conditional probability (also known as Bayes' rule):

$$P(A \mid B) = \frac{P(A \cap B)}{P(B)}$$

Bayes' rule is often expressed formally as

$$P(A_1 \mid B) = \frac{P(A_1) \cdot P(B \mid A_1)}{\sum\limits_{i=1}^{n} [P(A_i) \cdot P(B \mid A_i)]}$$

but basically it is equivalent to the relationship describing the conditional probability above.

PROBLEMS

Problem 5.1

Mark Shark has recently been spending Saturday and Sunday afternoons at the local pool hall. Each weekend, Mark engages in one game apiece with each of two brothers, Jack and Bill Pigeon. Mark believes that his chance of winning on any given occasion is .80 regardless of which Pigeon he plays. He also believes that winning or lack of winning on any one occasion in no way influences his chances of success on subsequent outings.
1. Determine the probability that Mark will lose to both Pigeons this weekend.
2. Determine the probability that Mark will beat only one of the two Pigeons.
3. Determine the probability that Mark will beat Jack but lose to Bill.
4. Determine the probability that Mark will beat both Pigeons.
5. Determine the probability that Mark will beat *at least* one of the Pigeons.
6. Determine the probability that Mark will beat both Pigeons two weeks in a row.

Problem 5.2

Differentiate between $P(A \mid B)$ and $P(A \cap B)$.

Problem 5.3

Given: $P(A) = 0.20$; $P(B) = 0.15$.
1. What is $P(A \mid B)$ if events A and B are independent?
2. What is $P(A \mid B)$ if events A and B are mutually exclusive?
3. What is $P(A \cap B)$ if events A and B are mutually exclusive?
4. What is $P(A \cap B)$ if events A and B are independent?
5. What is $P(A \cup B)$ if events A and B are mutually exclusive?
6. What is $P(A \cup B)$ if events A and B are independent?

Problem 5.4

Given: $P(A)$ is 0.25 and $P(B)$ is 0.35. Is it possible for the above two events to be both mutually exclusive and independent at the same time? Justify your answer.

Problem 5.5

The general rule of addition is $P(A \cup B) = P(A) + P(B) - P(A \cap B)$. The special rule of addition is $P(A \cup B) = P(A) + P(B)$. Explain why the general rule of addition contains the term $P(A \cap B)$ while the special rule does not.

Problem 5.6

Dan's Print Shop has three model A-4 presses. Model A-4 presses are down for repairs 10 percent of the time. Assuming that the likelihood of any one press having to be shut down is in no way influenced by the current operating condition of the other two, determine the probability that:
1. All three presses will be shut down.
2. Number 1 will be shut down but numbers 2 and 3 will be operating.
3. Numbers 1 and 2 will be shut down but number 3 will be operating.
4. One of the three will be shut down.
5. Two of the three will be shut down.
6. At least one press will be shut down.
7. At least two presses will be shut down.

Let
X = number 1 model A-4 press is operating
Y = number 2 model A-4 press is operating
Z = number 3 model A-4 press is operating

Solve the above using the addition and/or multiplication rules (in combination where necessary).

Problem 5.7

Referring to Problem 5.6, if, due to old age, the model A-4 down time due to repairs increased to 20 percent:
1. By how much will the probability increase that at least one press will be shut down?
2. By how much will the probability increase that two or more presses would be shut down?

Problem 5.8

Spacecraft are designed in such a way that back-up systems exist for many of the operations vital to the safety of the crew. Suppose that one system consists of three major components. As long as two of the three components are operative, the system will function effectively. The three components of the system will be designated A, B, and C. The probability that the components will be damaged during liftoff are 8 percent for component A, 3 percent for component B, and 2 percent for component C. The components are subject to totally different stresses and strains. As a result, the fact that any one component is damaged will in no way influence the likelihood that damage has been suffered by the other two.
1. Draw a Venn diagram showing the interrelationship of the three subsets of interest.
2. Shade in the portion of the Venn diagram corresponding to the inability of the system to function.
3. Compute the probability that the system will be unable to function.

Problem 5.9

Referring to the two-die addition table (Table 5.1):
1. What is the probability
   a. of rolling a total of 8?
   b. of rolling a total of 3 or 4?
   c. of rolling a total of 6 or more?

2. What is the probability that die number 1 will be a 4 *and* that the total of the two dice will be 6, 7, or 8?
3. What is the probability of a 5 appearing on at least one of the two dice?

Problem 5.10

Immediately after the November 1968 election, one noted political scientist assigned the following probabilities to the 1972 presidential race:

Nixon   = 0.50
Muskie   = 0.15
Kennedy   = 0.10
Humphrey = 0.03
Agnew   = 0.02

1. What probability would he assign to either Nixon, Agnew, or Kennedy being elected?
2. What probability would he assign to Nixon and Agnew and Kennedy being elected?
3. Why doesn't the total of the probabilities in the list total to 1?

Problem 5.11

There are 12 balls in an urn; three red, five white, and four green. Balls drawn are not replaced.

1. What is the probability of a red ball on the first draw?
2. What is the probability of a red ball on the fifth draw?
3. What is the probability of at least one red ball in two draws?
4. What is the probability of exactly one red ball in two draws?
5. What is the probability of a red ball on the fifth draw, given that the first ball drawn was green?
6. What is the probability of a red ball on the fifth draw, given that the twelfth ball drawn was green?

Problem 5.12

1. What is the probability of drawing a 2 or a 4 out of a standard deck of 52 playing cards on a single draw?
2. What is the probability of drawing a red card or a jack on a single draw from a standard deck of 52 playing cards?
3. In drawing one card at a time without replacement from a standard deck of 52 playing cards, what is the probability that the first spade will be the third card drawn?
4. Given drawings *with replacement*, what is the probability that the third card drawn will be a spade?
5. Given drawings *without replacement*, what is the probability that the third card drawn will be a spade?

Problem 5.13

Assume that a firm has determined the following relationship between income level and current status of accounts receivable balances:

Family income	Less than 30 days	30 to 60 days	60 to 90 days	Over 90 days	Total
Less than $5000	50	125	100	25	300
$5000–$10,000	200	150	75	25	450
Greater than $10,000	70	125	50	5	250
Total	320	400	225	55	1000

1. Indicate the marginal probabilities.
2. Determine the conditional probability that the customer has an income over $10,000 given that his balance is over 90 days.
3. Determine the probability that a customer selected at random will have a family income level less than $5000 *and* have an account balance between 30–60 days. Determine this probability using the table above.
4. Find the probability referred to in part 3, but use the general rule of multiplication to do so. Show all work.
5. Is the age of the accounts receivable balance independent of family income? Justify your answer.

Problem 5.14

Given:
  $P(A) = 0.3$
  $P(B) = 0.7$
  A and B are independent
Find:
  1. $P(A \cap B)$
  2. $P(A \mid B)$
  3. $P(A \cup B)$
  4. $P[\bar{A} \cap B) \cap (\bar{B} \cap A)]$
  5. $P(\bar{A} \cap B)$
  6. $P(\bar{B} \cap A)$
  7. $P(\bar{A} \cap \bar{B})$
Draw Venn diagrams if you have difficulty visualizing the relationship specified.

BAYES' RULE— REVISION OF PROBABILITIES IN THE LIGHT OF NEW INFORMATION

6

Consider three urns with the following compositions: urn I contains 10 red marbles and 2 green for a total of 12; urn II contains 4 red marbles and 6 green for a total of 10; urn III contains 2 red marbles and 7 green for a total of 9.

urn I	*urn II*	*urn III*
10 red	4 red	2 red
2 green	6 green	7 green
12 marbles	10 marbles	9 marbles

The urns are placed on a table and are indistinguishable one from another, except for a label attached to the base of the urn. A marble is to be selected from one of the urns, and the urn from which the selection is to be made will be determined by the following procedure.

A fair die will be tossed once.

1. If the die turns up a 1, the individual will make a selection from urn I.
2. If the die turns up a 2 or a 3, the individual will make a random selection from urn II.
3. If the die turns up a 4, 5, or 6, the individual will make a random selection from urn III.

Given the preceding as the state of information, what probability should be assigned to the individual selecting urn I? Selecting urn II? Selecting urn III?

Applying the basic rule of probability (the probability of success is equal to the ratio of the number of ways success can occur to the total number of ways an outcome to the experiment can be generated), one obtains

$$P(\text{I}) = \tfrac{1}{6} = 0.167$$
$$P(\text{II}) = \tfrac{2}{6} = 0.333$$
$$P(\text{III}) = \tfrac{3}{6} = 0.500$$
$$\overline{\phantom{P(\text{III})} 1.000}$$

Urns I, II, and III are the three possible states of the universe. $P(\text{I}) = \tfrac{1}{6}$, $P(\text{II}) = \tfrac{2}{6}$, and $P(\text{III}) = \tfrac{3}{6}$ represent the probability of occurrence of each of the possible states of the universe given no additional information. As such, they are referred to as *prior probabilities*. They are the probabilities assigned to the various states of the universe *prior* to receiving new information.

Now let us change the state of information. Suppose it is observed, without knowing the outcome of the roll of the die (and therefore without knowing whether the selected urn was urn I, II, or III), that the selected marble is a red marble. The symbol K is introduced to represent the additional information acquired. In this problem, K, the new information, is "one red marble out of one observed." The new information that a red marble has been drawn will change the probabilities assigned to the occurrence of the three alternative states of the universe. In effect, the following questions are now being asked:

How does the new information that a red marble has been drawn change the probability one should assign to the occurrence of the event "the selected urn is urn I"?

How does the new information that a red marble has been drawn change the probability one should assign to the occurrence of the event "the selected urn is urn II"?

How does the new information that a red marble has been drawn change the probability one should assign to the occurrence of the event "the selected urn is urn III"?

The above three questions, when expressed in the language of probability symbols, appear as follows:

$$P(\text{I} \mid K) = \frac{P(\text{I} \cap K)}{P(K)}$$

$$P(\text{II} \mid K) = \frac{P(\text{II} \cap K)}{P(K)}$$

$$P(\text{III} \mid K) = \frac{P(\text{III} \cap K)}{P(K)}$$

$P(K)$ refers to the probability of event K occurring. The probability of event K occurring is equal to the sum of the probabilities of the various ways event K can occur. In general, there will be one way associated with each state of the universe. In this problem, there are three states of the universe and therefore three ways in which event K can occur.

I $\cap$ K urn I can be selected *and* then a red marble drawn

II $\cap$ K urn II can be selected *and* then a red marble drawn

III $\cap$ K urn III can be selected *and* then a red marble drawn

The probability of event K occurring is equal to the sum of the probabilities of the various ways event K can occur. Therefore,

$$P(K) = P(\text{I} \cap K) + P(\text{II} \cap K) + P(\text{III} \cap K)$$

Note that the three terms in the right-hand member of the above expression coincide with the numerators in the three "Bayesian questions."

Expanding the terms in the right-hand member by the general rule of multiplication, the expression becomes

$$P(K) = P(\text{I}) \cdot P(K \mid \text{I}) + P(\text{II}) \cdot P(K \mid \text{II}) + P(\text{III}) \cdot P(K \mid \text{III})$$

In words:

$$P(K) = \begin{pmatrix}\text{prior probability} \\ \text{of urn I}\end{pmatrix} \begin{pmatrix}\text{conditional probability that} \\ \text{event } K \text{ will occur given that} \\ \text{the urn selected is urn I}\end{pmatrix}$$

plus

$$\begin{pmatrix}\text{prior probability} \\ \text{of urn II}\end{pmatrix} \begin{pmatrix}\text{conditional probability that} \\ \text{event } K \text{ will occur given that} \\ \text{the selected urn is urn II}\end{pmatrix}$$

plus

$$\begin{pmatrix}\text{prior probability} \\ \text{of urn III}\end{pmatrix} \begin{pmatrix}\text{conditional probability that} \\ \text{event } K \text{ will occur given that} \\ \text{urn III is the selected urn}\end{pmatrix}$$

Thus, we have

$$P(K) = (\text{prior})(\text{associated conditional}) + (\text{prior})(\text{associated conditional})$$
$$+ (\text{prior})(\text{associated conditional})$$

Recall that

$P(K \mid \text{I})$ = probability of drawing a red marble given that the draw is from urn I (i.e., $\frac{10}{12}$)

$P(K \mid \text{II})$ = probability of drawing a red marble given that the draw is from urn II (i.e., $\frac{4}{10}$)

$P(K \mid \text{III})$ = probability of drawing a red marble given that the draw is from urn III (i.e., $\frac{2}{9}$)

The above three probabilities are *conditional probabilities*. They refer to the probability that event K will occur given the existence of a particular state of the universe. Clearly, event K is more likely to occur in association with some states of the universe than with others.

Recall the relationship

$$P(K) = P(\text{I}) \cdot P(K \mid \text{I}) + P(\text{II}) \cdot P(K \mid \text{II}) + P(\text{III}) \cdot P(K \mid \text{III})$$

The prior probabilities of the three states of the universe are

$$P(\text{I}) = \tfrac{1}{6}$$
$$P(\text{II}) = \tfrac{2}{6}$$
$$P(\text{III}) = \tfrac{3}{6}$$

Substituting, we obtain

$$
\begin{aligned}
P(K) &= (\tfrac{1}{6})(\tfrac{10}{12}) + (\tfrac{2}{6})(\tfrac{4}{10}) + (\tfrac{3}{6})(\tfrac{2}{9}) \\
&= \tfrac{10}{72} + \tfrac{8}{60} + \tfrac{6}{54} \\
&= 0.139 + 0.133 + 0.111 \\
&= \tfrac{139}{1000} + \tfrac{133}{1000} + \tfrac{111}{1000} \\
&= \tfrac{383}{1000}
\end{aligned}
$$

Interpretation: Suppose the experiment is to be performed 1000 times; that is, a die is rolled and then a marble selected from the indicated urn. It would be expected that of the 1000 times the experiment is performed, a 1 would turn up one-sixth of the time or approximately 167 times, a 2 or a 3 would appear two-sixths of the time or approximately 333 times, and a 4, 5, or 6 would occur three-sixths of the time or approximately 500 times.

Referring to Figure 6.1, if one were to start out at the origin 1000 times, it would be expected that the roll of the die would result in the outcome (path) labeled urn I, 167 times, the outcome labeled urn II, 333 times, and the outcome labeled urn III, 500 times.

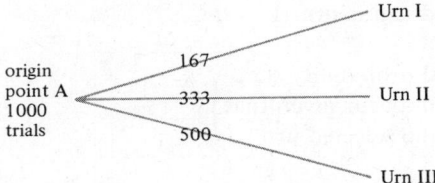

Figure 6.1

The second stage of the experiment consists of selecting a marble from the indicated urn. Referring to Figure 6.2, of the 167 times the roll of the die resulted in the selection of urn I, one would subsequently expect a red marble to be selected $\tfrac{10}{12}$ of the time or 139 times. The other 28 times or $\tfrac{2}{12}$ of the time, a green marble would be expected to be selected.

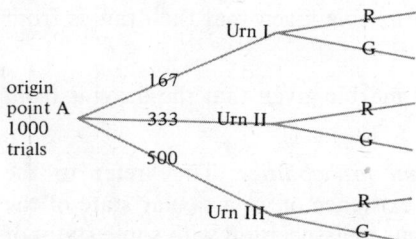

Figure 6.2

A similar line of reasoning provides the following information: Of the 333 times urn II is selected, it would subsequently be expected that a red marble would be selected $\frac{4}{10}$ of the time or 133 times and a green marble 6 out of every 10 times or 216 times. Of the 500 occasions urn III is selected, it would subsequently be expected that a red marble would be selected $\frac{2}{9}$ of the time or 111 times. A green marble would be expected to be selected $\frac{7}{9}$ of 500 or 389 times.

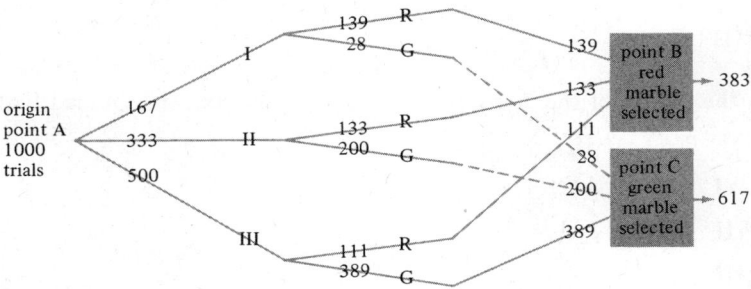

Figure 6.3

Starting out at the origin 1000 times, one would expect to observe the specific result of a red marble selected 383 times (Figure 6.3). Of the 383 times a red marble is selected, it would be expected that 139 would be associated with urn I, 133 with urn II, and 111 with urn III.

$P(I \mid K) = $ ratio of the number of elements in subset $[I \cap K]$ (i.e., 139) to the number of elements in the subset $[K]$ (i.e., 383)

$$= \tfrac{139}{383} = 0.363$$

Once again, this can be related to the basic rule of probability: "The ratio of the number of ways the event can succeed (139) to the total number of ways the event can occur (383)."

Similarly,

$$P(II \mid K) = \tfrac{133}{383} = 0.347$$

$$P(III \mid K) = \tfrac{111}{383} = 0.290$$

Thus, we have

$$P(I \mid K) = 0.363$$

$$P(II \mid K) = 0.347$$

$$P(III \mid K) = 0.290$$

$$\overline{1.000}$$

Returning to the formulas, recall the following "Bayesian questions":

$$P(I \mid K)$$

$$P(II \mid K)$$

$$P(III \mid K)$$

Also recall that any conditional probability is equal to the ratio of the probability of the intersection of the symbols to the left and to the right of the slash

to the marginal probability of the symbol indicated to the right of the slash.
Thus:

$$P(\text{I} \mid K) = \frac{P(\text{I} \cap K)}{P(K)}$$

$$P(\text{II} \mid K) = \frac{P(\text{II} \cap K)}{P(K)}$$

$$P(\text{III} \mid K) = \frac{P(\text{III} \cap K)}{P(K)}$$

In the course of the analysis, it has previously been determined that

$$P(K) = \frac{383}{1000}$$

$$P(\text{I} \cap K) = \frac{139}{1000}$$

$$P(\text{II} \cap K) = \frac{133}{1000}$$

$$P(\text{III} \cap K) = \frac{111}{1000}$$

Substituting:

$$P(\text{I} \mid K) = \frac{P(\text{I} \cap K)}{P(K)} = \frac{\frac{139}{1000}}{\frac{383}{1000}} = \frac{139}{383} = 0.363$$

$$P(\text{II} \mid K) = \frac{P(\text{II} \cap K)}{P(K)} = \frac{\frac{133}{1000}}{\frac{383}{1000}} = \frac{133}{383} = 0.347$$

$$P(\text{III} \mid K) = \frac{P(\text{III} \cap K)}{P(K)} = \frac{\frac{111}{1000}}{\frac{383}{1000}} = \frac{111}{383} = 0.290$$

The last set of probabilities, $P(\text{I} \mid K)$, $P(\text{II} \mid K)$, and $P(\text{III} \mid K)$, are called
posterior or *revised probabilities*. The term posterior refers to the fact that the
probabilities are assigned to the various states of the universe after or posterior to
the acquisition of new information. Note that the probabilities arrived at intuitively
with the tree-diagram approach are identical to those obtained by manipulation
of the formulas. This is in accordance with expectations, since both approaches
are based on the same relationship.

Once the underlying rationale behind the computations is understood, the
tabular format, illustrated in Table 6.1, should be used. The tabular format greatly

Table 6.1

Column 1	Column 2	Column 3	Column 4	Column 5
States of the universe	Priors	Conditionals	Joint probabilities	Posterior probabilities
I	$\frac{1}{6}$	$\frac{10}{12}$	$\frac{1}{6} \times \frac{10}{12} = \frac{10}{72}$ or 0.139	$\frac{0.139}{0.383} = 0.363$
II	$\frac{2}{6}$	$\frac{4}{10}$	$\frac{2}{6} \times \frac{4}{10} = \frac{8}{60}$ or 0.133	$\frac{0.133}{0.383} = 0.347$
III	$\frac{3}{6}$	$\frac{2}{9}$	$\frac{3}{6} \times \frac{2}{9} = \frac{6}{54}$ or 0.111	$\frac{0.111}{0.383} = 0.290$
			0.383	$\frac{0.383}{0.383} = 1.000$

facilitates the computational process by substituting location in the table for writing out the formulas. The computations in the joint probability column correspond to the computation of $P(K)$ in the formula $P(K) = P(I \cap K) + P(II \cap K) + P(III \cap K)$.

$P(I \cap K)$ is the first term in column 4, namely $P(I) \cdot P(K \mid I)$.

$P(II \cap K)$ is the second term in column 4, namely $P(II) \cdot P(K \mid II)$.

$P(III \cap K)$ is the third term in column 4, namely $P(III) \cdot P(K \mid III)$.

The sum of the three terms in the joint probability column generates $P(K)$.

Dividing the first figure in the joint probability column by the total of the joint probability column generates the first revised or posterior probability.

$$\frac{0.139}{0.383} = \frac{P(I \cap K)}{P(K)} \quad \text{or} \quad P(I \mid K)$$

Similarly,

$$\frac{0.133}{0.383} = \frac{P(II \cap K)}{P(K)} \quad \text{or} \quad P(II \mid K)$$

$$\frac{0.111}{0.383} = \frac{P(III \cap K)}{P(K)} \quad \text{or} \quad P(III \mid K)$$

6.1 AN EXTENSION

Suppose a second marble is drawn from the same urn (without the first marble being replaced) and is also observed to be a red marble. Once again, *new information* is obtained and the probabilities assigned to the various states of the universe must be revised.

Visually, the change in the state of information can be depicted as shown in Figure 6.4.

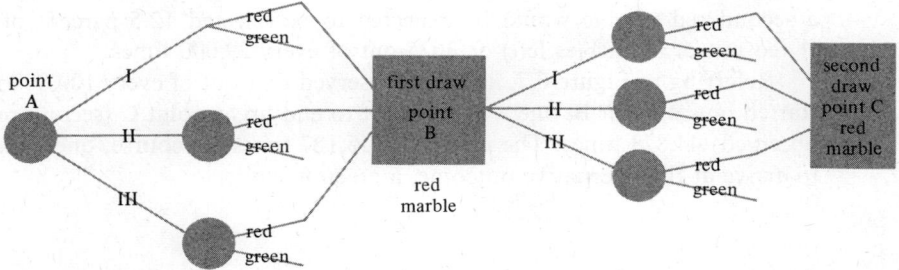

Figure 6.4

Recall that after the first marble drawn was observed to be red, the state of your information placed you at point B, and as a result the probabilities assigned to urn I, urn II, and urn III were revised to 0.363, 0.347, and 0.290, respectively. The posterior or revised probabilities as of the first stage become, in turn, the prior probabilities assigned to the states urn I, urn II, and urn III at point B, that is, at the beginning of stage 2 before any further additional information is made available.

The state of your information after the first draw places you at point B. Recall that point B implies that a red marble has been removed from the selected urn.

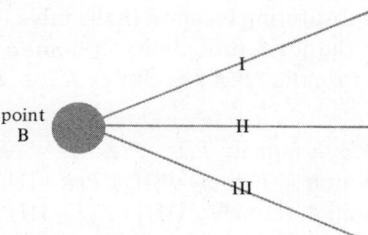

Figure 6.5

It would be expected that out of every 100,000 times you proceeded beyond point B, that is, continued to experiment (Figure 6.6), you would be operating out of urn I 36.3 percent of the time or 36,300 times, out of urn II 34.7 percent of the time or 34,700 times, and out of urn III 29 percent of the time or 29,000 times.

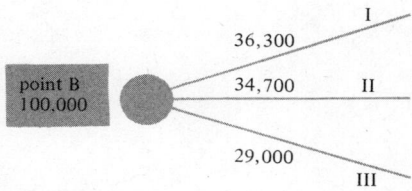

Figure 6.6

Of the 36,300 times the selected urn is urn I (Figure 6.7) one would expect a second red marble to be selected 81.8 percent of the time (9 red marbles out of 11 marbles left). This is approximately 29,693 times (36,300 × 0.818). Of the 34,700 occasions on which the selected urn is expected to be urn II, it is anticipated that a second red marble would be selected 33.3 percent of the time (3 red out of 9 marbles left) or 11,555 times. Similarly, if urn III is the actual state of the universe, a second red marble would be expected to be selected 12.5 percent of the time (1 red out of 8 marbles left) or 3625 out of every 29,000 times.

Referring to Figure 6.7, it can be observed that out of every 100,000 times one started out at point B, one would expect to end up at point C (second red marble observed) 44,873 times. The remaining 55,127 times, of course, one would expect to arrive at the alternative outcome, a green marble.

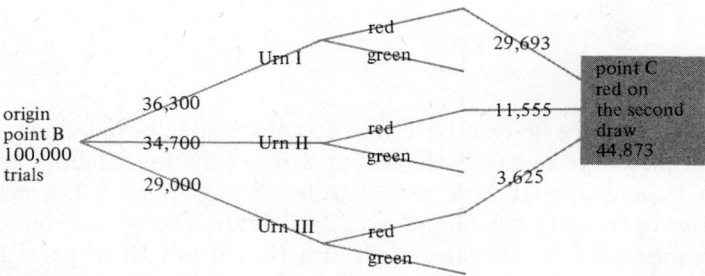

Figure 6.7

Given an arrival at point C, the probability assigned to the state "urn I is the selected urn" is equal to the ratio of the number of ways one could arrive at the outcome "red marble on the second draw given that a red marble was selected on the first draw and not replaced associated with urn I" to the total number of ways that the sequence of events, "red marble on the second draw given that a red marble was selected on the first draw and not replaced" could have been expected to occur, that is, 29,693/44,873 or 66 percent. Similarly, the probability assigned to the state "urn II is the selected urn given the latest information" would be 11,555/44,873 or 26 percent and to the state "urn III is the selected urn" 3625/44,873 or 8 percent. Thus, after the second revision, the posterior probabilities assigned to the states of the universe are as follows:

Posterior probability

urn I	0.66
urn II	0.26
urn III	0.08
	———
	1.00

If developed in the tabular format (by far the most efficient approach), the computations would appear as in Table 6.2. Note that the sum of the joint probability column equals $P(K)$.

Table 6.2

K = New information = red marble selected on the second draw given that a red marble was removed and not replaced on the first draw				
States of the universe	Prior probabilities	Conditional probabilities	Joint probabilities	Revised or posterior probabilities
I	0.363	0.818	0.29693	$\dfrac{0.29693}{0.44873} = 0.66$
II	0.347	0.333	0.11555	$\dfrac{0.11555}{0.44873} = 0.26$
III	0.290	0.125	0.03625	$\dfrac{0.03625}{0.44873} = 0.08$
	———		———	———
	1.000		0.44873	1.00

Table 6.3

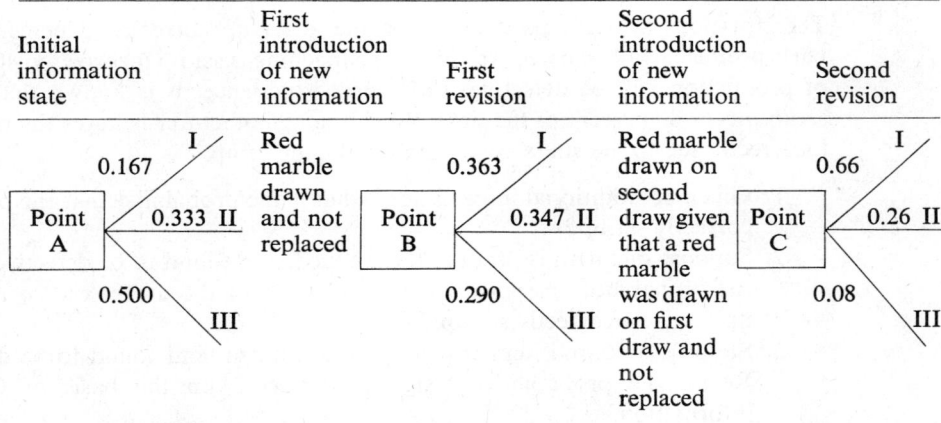

Notice, as the state of information changed, how the probabilities assigned to the likelihood of the various states of the universe are adjusted to reflect the new information (Table 6.3).

If the information obtained had been pooled and treated as one observation, the subsequent revision of the probabilities assigned to the three states of the universe would have been identical to that obtained at the final stage of the two-stage revision process where each observation was treated separately and the priors were revised after each observation (see Table 6.4).

Table 6.4

K = New information = two red marbles out of two observed where the first marble is not replaced in the urn before the second marble is drawn

States of the universe	Prior probabilities	Conditional probabilities	Joint probabilities	Revised or posterior probabilities
I	0.167	$\frac{10}{12} \times \frac{9}{11} = \frac{90}{132} = 0.682$	0.113894	$\frac{0.113894}{0.172183} = 0.66$
II	0.333	$\frac{4}{10} \times \frac{3}{9} = \frac{12}{90} = 0.133$	0.044289	$\frac{0.044289}{0.172183} = 0.26$
III	0.500	$\frac{2}{9} \times \frac{1}{8} = \frac{2}{72} = 0.028$	0.014000	$\frac{0.014000}{0.172183} = 0.08$
	1.000		0.172183	1.00

It does not matter whether the information is treated as one pooled observation and one revision made or as a series of observations and a series of revisions made—in the end, the new information totals up to the same. As the computations bear out, the revised probabilities assigned to the various states of the universe will prove identical after the final revision under either approach.

6.2 REINFORCEMENT

A Second Example

The situation: When a particular machine is set up correctly, 3 percent of the parts produced will be defective. When the machine is set up incorrectly, 40 percent of production will be defective. From past experience, it is known that correct setups occur 80 percent of the time. When the manufacturer believes the process is incorrectly set up, he shuts down and checks the setup.

1. Given no additional information, what is the probability that the process is correctly set up?
2. Suppose one item is produced, examined, and found to be defective. Revise the probabilities assigned to the two states of the universe: "correctly set up" and "incorrectly set up."
3. Suppose a second item is produced, examined, and found to be defective. Revise the probabilities assigned in part 2 on the basis of this new information.

4. Treat the information in parts 2 and 3 as one pooled observation of two defectives observed out of two parts produced. Revise the probabilities assigned to the two states of the universe in part 1 to reflect the pooled information.

Part 1

States of the universe	Prior probabilities
Correct setup	0.80
Incorrect setup	0.20
	——
	1.00

Part 2

States of the universe	Prior probabilities	Conditional probabilities	Joint probabilities	Revised or posterior probabilities
Correct setup	0.80	0.03	0.024	$\dfrac{0.024}{0.104} = 0.231$
Incorrect setup	0.20	0.40	0.080	$\dfrac{0.080}{0.104} = 0.769$
	——		——	——
	1.00		0.104	1.000

The probability that the process is correctly set up has been revised from 0.80 to 0.231 on the basis of the new information.

Part 3

$K =$ Second part produced defective given that the first part produced was also defective

States of the universe	Prior probabilities	Conditional probabilities	Joint probabilities	Revised or posterior probabilities
Correct setup	0.231	0.03	0.00693	$\dfrac{0.00693}{0.31453} = 0.022$
Incorrect setup	0.769	0.40	0.30760	$\dfrac{0.30760}{0.31453} = 0.978$
	——		——	——
	1.000		0.31453	1.000

An interesting issue is raised in part 3: Is it legitimate to use 0.03 as the probability that the second part produced will be defective given that the first part produced was defective? Suppose the process could produce 100,000 parts—3000 of them would be defective.

The probability of the first part being defective is equal to the ratio of defective parts to the total number of parts produced:

$$\frac{3000}{100,000} = 0.03$$

The probability of the second part being defective given that the first was defective would be equal to the ratio of the 2999 defectives left to the total number of parts left 99,999:

$$\frac{2999}{99,999} = 0.02999$$

Clearly, although the probability is not exactly 0.03 that the second part would be defective given that the first was defective, it is so close that to treat it as such results in an infinitesimal difference in the computations performed and the resultant conclusions. Such an assumption is only legitimate when the population from which the sample is to be selected is quite large relative to the size of the sample.

Part 4

$K = $ Two defectives out of two observed

States of the universe	Prior probabilities	Conditional probabilities	Joint probabilities	Revised or posterior probabilities
Correct setup	0.80	$(0.03)(0.03) = 0.0009$	0.00072	$\frac{0.00072}{0.03272} = 0.022$
Incorrect setup	0.20	$(0.40)(0.40) = 0.1600$	0.03200	$\frac{0.03200}{0.03272} = 0.978$
	1.00		0.03272	1.000

Note that the revised probabilities assigned to the states of the universe when the information is pooled are identical to those arrived at when the information is treated as a series of observations and a two-stage revision undertaken.

6.3 BAYESIAN VERSUS CLASSICIST APPLICATION OF BAYES' RULE

It is the author's contention that the basic concept of revising probabilities assigned to states of the universe in the light of new information should be equally acceptable to the Bayesian (subjectivist) and the classical (objectivist) schools as long as these probabilities are interpreted as representing the long-run relative frequency of occurrence of a given state of the universe over many repetitions of the "experiment." The Bayesian school would go one step further and also use the priors to represent the probability of occurrence of the various states of the universe on a particular performance of the experiment.

As a further qualification, the classical (or objectivist) school would be expected to accept this interpretation only when the "prior" probabilities can be generated by one of the following procedures[1]:

1. *A priori* knowledge of the way the prior probability distribution will emerge given a knowledge of the nature of the underlying "experiment" to be performed. The urn illustration is an example of a case where the prior probabilities are capable of being assigned *a priori*.
2. *A posteriori* knowledge of the prior probability distribution based on observations of the relative frequency of occurrence of alternative outcomes over many past performances of the experiment. The machine setup illustration provides an example of a case where the prior probabilities are capable of being assigned *a posteriori*. Use of *a posteriori* probabilities implies acceptance of the belief that forces operating in the past causing a particular probability distribution pattern to emerge will continue to operate in a similar fashion over the current decision period.

The Bayesian Position

Suppose the decision maker believes that underlying conditions have changed. It would follow that the *a posteriori* probabilities of the classical school would no longer provide a relevant description of the relative likelihood of occurrence of possible outcomes to the experiment. To illustrate: Suppose a new mechanic has been hired to make machine setups and no objective evidence exists as to the quality of his performance. The classicist (or objectivist) would be unwilling to assign prior probabilities to the states of the universe. The classicist would reserve judgment until objective evidence on the quality of a given setup had been obtained. Upon receipt of such information, the classicist would assign a probability to the process being correctly set up solely on the basis of the observed sample evidence. What is often not realized is that proceeding in this fashion is equivalent to assigning equal prior probabilities to the alternative states of the universe. By considering only sample evidence, the classicist would be implicitly concluding prior to the acquisition of sample evidence that it is just as likely the new mechanic will make a correct setup as it is that he will make an incorrect one.

The Bayesian (or subjectivist), on the other hand, would draw upon the attitude, experience, and convictions of the decision maker in this situation. Prior probabilities would be assigned that would reflect the decision maker's opinion as to the relative frequency with which the newly hired mechanic would make correct setups.[2] Certainly the decision maker would have reached some conclusions on this issue in the process of deciding whether or not to hire the new mechanic. It is the author's contention that such evidence should not be ignored but rather incorporated into the decision-making procedure. One of the key abilities decision makers should possess is the ability to assess the likelihood of occurrence of events that are liable to exert an influence on the outcome of various actions under consideration. To fail to incorporate such judgments into the decision-making process is to waste a valuable managerial resource. Over time, objective evidence

---

[1] One would expect that the classicist would also rebel at the use of the word "prior" because of its Bayesian connotation. The classicist interprets these values as "long-run relative frequencies of occurrence given the existing state of information."

[2] Problems 6.14 through 6.17 provide illustrative examples of the approach utilized by a subjectivist to convert his information and attitudes into the form of subjective probabilities.

on the performance of the new mechanic would accumulate. The Bayesian would incorporate such evidence in association with other insights in assigning prior probabilities to the likelihood of a correct setup or the likelihood of an incorrect setup on subsequent setups. As more and more objective evidence accumulated, a closer and closer correspondence would occur between the prior probability assignments of the subjectivist (or Bayesian) school and the long-run relative frequencies of occurrence assigned by members of the objectivist school.[3] It should be noted that different individuals would tend to assign different subjective probabilities to the same event because no two individuals possess the same package of experience or attitudes. As objective evidence with respect to the relative frequency of occurrence of the event in question builds up, the subjective probability assignments of different individuals tend to converge on the long-run frequency of occurrence of the event (and in so doing converge on the *a posteriori* probability assignment of the classicist).

The major point is that the Bayesian can proceed, with minimal objective evidence, to assign probabilities reflecting strength of conviction as to the likelihood of occurrence of various states of the universe. As sample evidence on the actual state of the universe is generated, the Bayesian then revises his probability assignments using Bayes' rule. The classicist would be unwilling to assign explicit long-run frequencies of occurrence unless (1) he has sufficient historical objective evidence to justify *a posteriori* assignments to the prior probabilities, or (2) the nature of the underlying experiment is such that prior probabilities can be assigned *a priori* (i.e., the limiting relative frequencies the probabilities would approach over many performances of the experiment can be predicted from knowledge of the characteristics of the experiment to be performed).

If neither of the above cases hold, the classicist would hold off judgment until sample evidence has been generated. Only this sample evidence would be considered in assigning probabilities to the various states of the universe. As previously mentioned, unwillingness to assign prior probabilities is equivalent to assigning equal probabilities to all possible states of the universe. It is clear that such a posture is much less realistic than the Bayesian position of subjective assignments. Less frequently, but occasionally, this issue arises in the assignment of conditional probabilities. (See Problem 6.17.) Examples of *a priori* prior probability assignments, *a posteriori* prior probability assignments, and subjective prior probability assignments are reflected in the problems at the end of the chapter.

PROBLEMS

Problem 6.1

Peter Berenson always uses the following technique to determine whether or not to study for an exam. He goes to the gym and takes ten foul shots. If he gets more than seven (i.e., eight, nine, or ten), he will not study. If he gets seven or less, he will study. In the past he has made more than seven out of ten only 20 percent of the time. The probabilities of Peter obtaining various grades on the statistics exam obviously vary depending on whether or not he studies and are as follows:

---

[3] Even members of the objectivist school could arrive at different long-run relative frequency of occurrence assignments as a result of different decisions with respect to the relevant period of historical evidence to consider in the generation of *a posteriori* probabilities.

If he studies		If he does not study	
Grade	Probability	Grade	Probability
A	0.3	A	0
B	0.4	B	0
C	0.2	C	0.2
D	0.1	D	0.3
F	0	F	0.5
	1.0		1.0

1. The last time you saw Peter he was on his way to the gym. What is the probability that he studied for the statistics exam?
2. Assume that when Peter gets his exam back, his grade is a D. Revise your estimate of the probability that he studied in the light of this new information.

Problem 6.2

A buddy of yours has recently left for Australia, leaving in your care two black books (A and B). He swears to you that in book A, 90 percent of the girls are fabulous, while book B contains only 30 percent fabulous girls. The books are identical and not marked A or B but strictly contain a list of names. (He also mentioned that the names are randomly good and bad.) You select one book and start to date.

1. What prior probability would you assign to the selected book being book A?
2. What revised probability would you assign to the selected book being book A given the information that the first girl dated turned out to be fabulous?
3. Revise the probability assigned in part 2 on the basis of the additional information that the second girl dated (out of the same book) was terrible.
4. Instead of the two-stage revision reflected in parts 2 and 3, make a one-stage revision of the probabilities established in part 1 on the basis of the information "the first girl dated was fabulous and the second girl was terrible."
5. Explain intuitively why part 3 and part 4 generate the same revised probabilities.

Problem 6.3

Distinguish between the terms prior and posterior as used in Chapter 6.

Problem 6.4

Why does $P(K)$ equal less than 1, where K refers to the new information generated?

Problem 6.5

Three clerks handle mail orders in the exotic foods section. Lotta Errors handles 40 percent of the orders, Will Botchit handles 35 percent of the orders, and Philya

Order handles 25 percent. The likelihood that each of these clerks will make an error is summarized below:

Clerk	Likelihood of an error (%)
Lotta Errors	15
Will Botchit	10
Philya Order	2

1. If you placed an order for exotic foods, what is the probability that an error would occur in handling your order?
2. Given that an error occurs in the handling of your order, what probability should be assigned to the order having been handled by Lotta; by Will; by Philya?

Problem 6.6

A test for detecting a certain rare disease has been perfected that is capable of discovering the disease in 97 percent of all afflicted individuals. When the same test is tried on healthy individuals, 5 percent of them are incorrectly diagnosed as having the disease. When it is tried on individuals who have certain other milder diseases, 10 percent of them are incorrectly diagnosed. It is known that the percentage of individuals of the three types in the population at large are 1 percent, 96 percent, and 3 percent, respectively. Calculate the probability that an individual selected at random from the population at large and tested for the rare disease actually has the disease if the test indicates he is so afflicted. Does the answer you obtain seem correct? Justify it.

Problem 6.7

Al Mello, "the thinking man's quarterback," is faced with selecting a play to call on third down with big yardage to go. His major concern is whether Bridgeport (opponent for the day) will "red dog" (rush the passer). If they red dog, the optimal play to have called would be X-7b. Otherwise, Mello's best call in this situation would be play X-12. Based on observation of Bridgeport game films, Mello believes the probability that they will "red dog" is 30 percent.
Mello's decision rule is:

If the probability of a "red dog" $\geq$ 60 percent, call X-7b.
If the probability of a "red dog" $<$ 60 percent, call X-12.

When Mello brings the team up to the line of scrimmage, he notices the linebacker adjusting his forearm guard. From Bridgeport game films, Mello knows that the linebacker always does this when he is going to "red dog" and does this only 20 percent of the time when he is not going to "red dog." Having observed this behavior on the part of the linebacker (and given his decision rule), should Mello change to play X-7b by calling an audible at the line of scrimmage or should he stay with his original call, X-12? Your answer must be supported by computations.

Problem 6.8

Suppose that at the beginning of the term your instructor, being the optimist that he is, felt that you had a 50–50 chance of passing. That is, there was a 50–50 chance that you were that "type" of student who would work hard enough to pass. However, through past experience, he is aware that students who work hard enough to pass miss a *scheduled* exam only 5 percent of the time, whereas students who do not work hard enough to pass miss these exams 20 percent of the time. If you had missed the only scheduled exam during the term, what would your instructor consider your probability of passing?

Problem 6.9

Machine A produces 90 percent acceptable parts, while machine B produces 75 percent acceptable parts. Before testing a part, the manager of the department assesses a probability of 0.3 that the part was produced on machine A. Therefore, he feels that there is a 0.7 probability that it was produced on machine B. A test procedure was carried out. *The part was good.* Now what probability should the manager assess that the part was produced on machine A? On machine B?

Problem 6.10

The new products manager for Imagination, Inc., in assessing the market potential of a new product, has rated the probability of wide acceptance at 20 percent, of low acceptance at 30 percent, and of medium acceptance at 50 percent. A prototype is developed and a test is conducted to evaluate market response in a representative geographical area. Thirty-seven percent of the clientele in the test area react favorably to the product. The probability of this response given that the true state of the market place is wide acceptance is 0.42. Given that the true state of the market place is medium acceptance, the probability of observing this response is 0.29. The probability of observing this response given that the true state of the market place is low acceptance is 0.22. Given the new information (market response in the test area), determine the revised probabilities the new products manager should assign to market states "wide acceptance," "medium acceptance," and "low acceptance."

Problem 6.11

When a machine is set up correctly by an operator, 0.2 of the items produced are defective. For an incorrect machine setup, 0.6 of production is defective. Further, since the manufacturer knows from past experience that the probability of a machine being set up correctly is 0.5, he is uncertain whether or not to proceed with production. To reduce the uncertainty, he decides to obtain additional information on the quality of this particular setup by producing and inspecting a sample item.
1. Suppose this piece proves to be defective. What is the probability (posterior) that this particular setup is correct?
2. Suppose the manufacturer takes a sample of size 2, and both sample pieces are defective. What is the probability that the sample came from a correct setup?

Problem 6.12

The probability of getting a tail with a fair coin is 0.5. The probability of getting a tail with a particular unfair coin is 0.2. Assume a probability of 0.9 that the coin is fair. If we toss the coin once and get a head, what is the probability that it is a fair coin?

Problem 6.13

A restaurant offers its patrons a choice of either steak or chicken and either red
or white wine with the main course. It is known from experience that the probabilities that the customer will order steak or chicken are 0.70 and 0.30, respectively.
The probabilities that red or white wine will be ordered are 0.80 and 0.20, respectively. Further, the probability that a customer will leave a good tip is 0.90
if he has steak *and* red wine, 0.40 if he has steak and white wine, 0.60 if he has
chicken and red wine, and 0.10 if he has chicken and white wine. (The white wine
is so bad that it is a poor match even with poultry.)
1. What are the probabilities that the customer will order:
   a. steak and red wine?
   b. steak and white wine?
   c. chicken and red wine?
   d. chicken and white wine?
2. What is the probability that the waiter will get a good tip (no knowledge
   of food ordered)?
3. Given that the waiter receives a good tip, what is the revised probability
   that chicken and red wine was ordered?

Problem 6.14

Select two professional basketball teams, one of which in your opinion is definitely
superior to the other. The objective of this problem is to develop a probability
distribution describing the point differential likely to be observed at the end of
regulation play between the two teams in any one given game. Overtime is omitted
from consideration, so a zero differential is possible.

Approach:
1. Calibrate the horizontal axis to represent the point differentials between
   the largest amount by which you believe the inferior team could win any
   given game (a negative sign indicates the inferior team has the higher
   score) and the largest amount by which you believe the superior team
   could win any given game.

point
differential

2. Calibrate the vertical axis from 0.0 to 1.00 to represent the probability
   that the point differential (including sign) will be equal to the corresponding
   value on the horizontal axis or less. Plot a point equal to $P(PD \leq X) =$
   0.00 above the largest negative differential and a point equal to
   $P(PD \leq X) = 1.00$ above the largest positive differential.

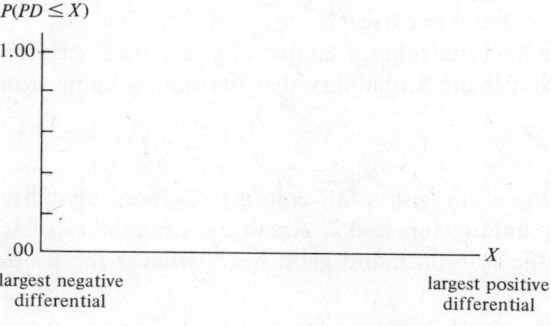

$P(PD \leq X)$

1.00

.00

X

largest negative
differential

largest positive
differential

3. What value of the point differential do you feel would be exceeded 50 percent of the time? Plot a value of $P(PD \leq X) = 0.50$ above this value on the horizontal axis.

4. What value would you expect the point differential to exceed 10 percent of the time? Above this value on the horizontal axis, plot a value of $P(PD \leq X) = 0.90$.

5. What value would you expect the point differential to exceed 90 percent of the time? Above this value on the horizontal axis, plot a value of $P(PD \leq X) = 0.10$.

6. What value would you expect the point differential to exceed 25 percent of the time? Above this value on the horizontal axis, plot a value of $P(PD \leq X) = 0.75$.

7. What value would you expect the point differential to exceed 75 percent of the time? Above this value on the horizontal axis, plot a value of $P(PD \leq X) = 0.25$.

8. Fit a smooth curve to the plotted points. The completed curve will be a cumulative percentage distribution (see p. 29). To verify the representativeness of the curve, read some values off the curve to determine if they correctly reflect your posture as to the likelihood of occurrence of the events in question. If they do not, make appropriate adjustments in the curve.

9. Divide the horizontal axis into classes characterized by a common class interval. Select the number of classes in such a fashion that the underlying pattern in the data will be reflected in the resulting distribution. Raise perpendiculars at the class limits. Read from the cumulative probability ogive the probability of observing a point differential falling within the limits of each class. To illustrate:

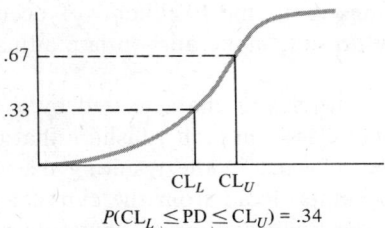

$$P(CL_L \leq PD \leq CL_U) = .34$$

10. Present the information obtained in part 9 in the format of a percentage frequency distribution.

11. Graph the distribution obtained in part 10 as a percentage frequency polygon.

12. What would be the impact on the percentage frequency polygon if the number of classes were to be increased (with a corresponding decrease in the width of the common class interval)?

Problem 6.15

The time required to commute from your place of residence (or place of work) to the university varies from one occasion to the next. The objective of this problem is to develop a probability distribution describing the pattern of times likely to be observed.

1. The amount of time required to make the trip is recorded on the horizontal axis. The vertical axis is calibrated from 0.0 to 1.00 and indicates the probability that the time required will be equal to or less than the corresponding value on the horizontal axis. Let t represent the time required to commute and X a stipulated value on the horizontal axis.

2. Plot a point equal to $P(t \leq X) = 0.00$ above the value on the horizontal axis that represents the least amount of time you believe would be required to make the trip. Plot a point equal to $P(t \leq X) = 1.00$ above the value on the horizontal axis that you believe will virtually never be exceeded (the situation where your car becomes inoperable is excluded from consideration).

3. Half of the trips would be expected to exceed what length of time? Above this value on the horizontal axis, plot a value of $P(t \leq X) = 0.50$.

4. What value would you expect the time to commute to exceed 10 percent of the time? Above this value on the horizontal axis, plot a value of $P(t \leq X) = 0.90$.

5. You would expect the time required to commute to be less than what value, 10 percent of the time? Above this value on the horizontal axis, plot a value of $P(t \leq X) = 0.10$.

6. What value would you expect the time to commute to exceed 25 percent of the time? Above this value on the horizontal axis, plot a value of $P(t \leq X) = 0.75$.

7. You would expect the time required to commute to be less than what value, 25 percent of the time? Plot a point at $P(t \leq X) = 0.25$ above this value.

8. Fit a smooth curve to the plotted points. The completed curve will be a cumulative percentage distribution (see p. 29). To verify the representativeness of the curve, read some values off the curve to determine if they correctly reflect your feelings as to the likelihood of occurrence of the events in question. If they do not, make appropriate adjustments in the curve.

9. Divide the horizontal axis into classes characterized by a common class interval. Select the number of classes in such a fashion that the underlying pattern in the data will be reflected in the resulting distribution. Raise perpendiculars at the class limits. Read from the cumulative probability ogive the probability that the time required to commute will fall within the various class intervals. To illustrate:

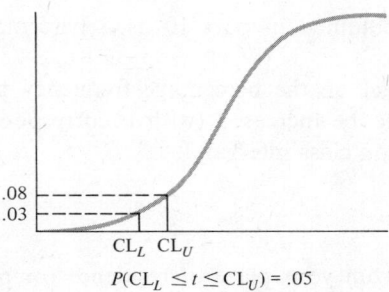

$$P(CL_L \leq t \leq CL_U) = .05$$

10. Present the information obtained in part 9 in the format of a percentage frequency distribution.

11. Graph the distribution obtained in part 10 as a percentage frequency polygon.
12. Is the percentage frequency polygon skewed or symmetrical? Is this in accordance with your expectations? Justify your answer.
13. Compute the mean and standard deviation from the percentage frequency polygon obtained in step 10. Use the following formulas:

$$\mu_x = \frac{\sum [P(M) \cdot M]}{\sum P(M)} = \sum [P(M) \cdot M]$$

where

$$M = \text{class midpoint}$$

$$P(M) = \text{probability an observation will fall in the class with the midpoint, } M$$

$$\sum P(M) = 1.0$$

$$\sigma_x = \sqrt{\frac{\sum [P(M) \cdot M^2] - \{\sum [P(M) \cdot M]^2 / \sum P(M)\}}{\sum P(M)}}$$

Substituting $\sum P(M) = 1.0$, the formula simplifies to

$$\sigma_x = \sqrt{\sum [P(M) \cdot M^2] - \sum [P(M) \cdot M]^2}$$

14. Interpret the result obtained in part 13.
15. Compare the distribution you generated in part 8 with a similar distribution prepared by someone in your car pool. Are they different? If so, why? If they are different, get together and thrash out your differences. Arrive at a common curve with which you both can live.

Problem 6.16

The following cumulative probability ogive reflects the attitude of the new products manager with respect to the potential first-year sales volume (in units) of a new product currently under consideration. The cumulative probability ogive was obtained by a procedure similar to that outlined in Problems 6-14 and 6.15.

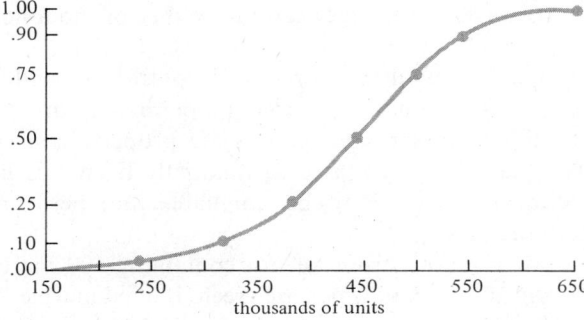

1. Divide the horizontal axis into classes characterized by a common class interval. Select the number of classes in such a fashion that the underlying pattern in the data will be reflected in the resulting distribution. Raise perpendiculars at the class limits. Referring to the cumulative probability

ogive, determine the probability that the volume of demand will fall within the various class intervals. Present this information in the format of a percentage frequency distribution.

2. Determine the mean and standard deviation of the distribution, basing your computations on the percentage frequency distribution developed in part 1. Use the formulas suggested in Problem 6.15. Interpret the results.

Problem 6.17

You are contemplating constructing an addition to your home and will require that a foundation be dug. You randomly select the name of an earth-removal contractor, Dom Iaria, from the Yellow Pages and make an appointment to discuss your project. Your deep-seated feeling is that about 20 percent of the earth-removal firms are unreliable and do sloppy work.

1. Prior to your meeting with Dom, what probabilities should be assigned to the following two states of the universe? Assume that the two states indicated are mutually exclusive and collectively exhaustive.

> State a: Unreliable and sloppy work, i.e., $P(U)$?
> State b: Reliable and competent work, i.e., $P(R)$?

2. Upon arrival at Dom's residence, you are informed that Dom has been unexpectedly called away and could not keep the appointment. His wife indicates that he had tried to reach you but apparently you had already left. As you are leaving, you notice that Dom's property and equipment have been kept in immaculate condition. You have now acquired some additional information. The new information, K, can be described as follows:

> K = Dom failed to keep the appointment for what *may* have been legitimate reasons. Furthermore, his property and equipment appear to be maintained in excellent condition.

Required: Provide probability assignments to the following conditional events:

> $P(K \mid U) = P(K \mid$ Dom is unreliable and performs sloppy work)
> $P(K \mid R) = P(K \mid$ Dom is reliable and performs competent work)

Approach: Consider yourself faced with the necessity of committing yourself to a $1000 wager. The wager can take either of the following two forms:

a. You are given a list of all earth-removal contractors with whom a similar experience could have occurred (appointment broken for an apparently legitimate reason and immaculate property and equipment observed). One name is to be selected randomly from this list. If the contractor selected turns out to be unreliable and performs sloppy work, you lose the $1000.

b. You are to draw a marble from an urn containing 100 marbles, K of which are red and $100 - K$ of which are green. If a red marble is selected, you lose the $1000.

Suppose the urn contained 10 red marbles ($K = 10$). Which gamble would you prefer? If you prefer a, you believe the probability that Dom is unreliable and performs sloppy work is less than 10 percent. If you prefer b, you believe the probability is greater than 10 percent. If you are indifferent between a and b, you believe the probability is approximately 10 percent. Proceed to vary the composition of the marbles in the urn (in your mind) until you reach a point of indifference between the two gambles. Suppose this turns out to be the case where $K = 25$. The probability you should assign to the conditional event "Dom is unreliable and performs sloppy work" is 25 percent. This technique is referred to in the literature as the use of an equivalent urn. The idea is to vary the contents of the urn until the two gambles are considered equivalent.

Using the equivalent-urn approach, generate the probabilities *you* would assign to the two conditional events referred to above under required.

c. How would a classicist react to the request in part b?
d. Using the probabilities generated in part b, revise the probability assignments to the states of the universe.
  i. the contractor selected will be unreliable and performs sloppy work, i.e., $P(U \mid K)$
  ii. the contractor selected will be reliable and performs competent work, i.e., $P(R \mid K)$

Problem 6.18

The Miller Analogies Test is occasionally required for consideration for admittance to a graduate program. The testee is not informed of his test results. The results are only sent to graduate schools upon request. You recently took the test and felt at the time that the probability that you obtained a score of ≥ 600 was 0.4 and the probability that you achieved a score less than 600 was 0.6. Today, you received a letter indicating that you had been accepted into the graduate program. You believe the probability that you would have been accepted, if your score was below 600, is 0.30 and the probability that you would have been accepted if your score was equal to or greater than 600 is 0.70. Given the new information that you have been accepted into the graduate program, revise the probabilities assigned to the states of the universe:

State 1: Your score on the Miller Analogies Test was $\geq$ 600.
State 2: Your score on the Miller Analogies Test was $<$ 600.

THE CONCEPT OF A DISCRETE PROBABILITY DISTRIBUTION; THE BINOMIAL PROBABILITY DISTRIBUTION

7

Consider the following experiment: A coin is to be flipped twice, and the total number of heads observed is to be recorded. The possible outcomes to the experiment are depicted in Figure 7.1. The outcomes "no heads" and "two heads" each can occur only one way, whereas the outcome "one head" can be observed two different ways (i.e., H, T or T, H). Each flip of the coin constitutes an independent trial. The probability of occurrence for each sequence of observations is generated by assigning the appropriate conditional probabilities to each branch and then multiplying the probabilities encountered in moving from the origin to the associated endpoint for each sequence. Given that the coin is a fair coin, the probability of a head being observed is 0.5 on any one trial and remains constant from trial to trial regardless of the outcome on previous trials (Figure 7.2).

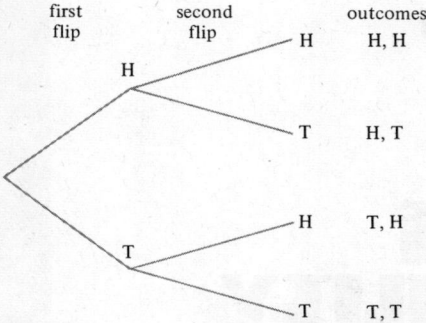

Figure 7.1

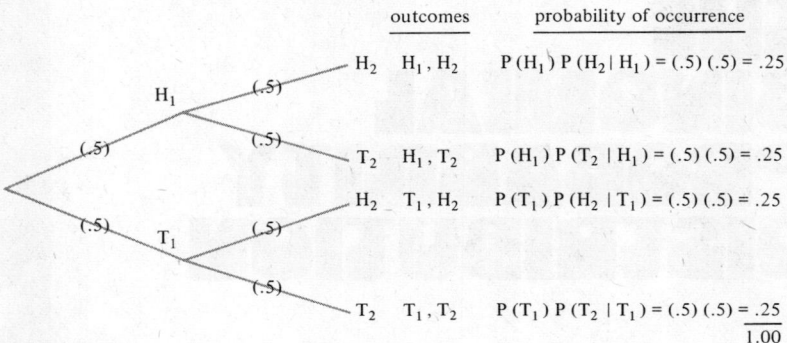

Figure 7.2

A list of categories encompassing all possible outcomes to an experiment together with their associated probability of occurrence is called a *probability distribution*. The list must be collectively exhaustive and mutually exclusive. Collectively exhaustive signifies that every conceivable outcome to the experiment must be assignable to one of the categories. The mutually exclusive constraint stipulates that the categories must be defined in such a way that none of the outcomes can be construed as falling within the definition of more than one of the categories. In summation, each outcome must be associated with one and only one of the categories listed. A probability distribution describing the pattern of outcomes for the experiment in Figure 7.2 is provided below:

List of categories encompassing all possible outcomes	Probability of occurrence
0 heads	0.25
1 head	0.50
2 heads	0.25
	1.00

There are alternative acceptable descriptions of the same experiment, provided that the list of categories satisfies the dual constraints—mutually exclusive and collectively exhaustive. To illustrate, the outcomes to the experiment (flipping a fair coin twice) could alternatively have been described in the following fashion:

List of categories encompassing all possible outcomes	Probability of occurrence
Head on first flip, head on second	0.25
Head on first flip, tail on second	0.25
Tail on first flip, head on second	0.25
Tail on first flip, tail on second	0.25
	1.00

Consider the following experiment: three flips of an unfair coin weighted such that heads will appear 80 percent of the time. The outcomes of this experiment in association with their relative likelihood of occurrence are summarized by the probability distribution shown in Figure 7.3. The procedure for generating the probabilities for the first two branches is illustrated below.

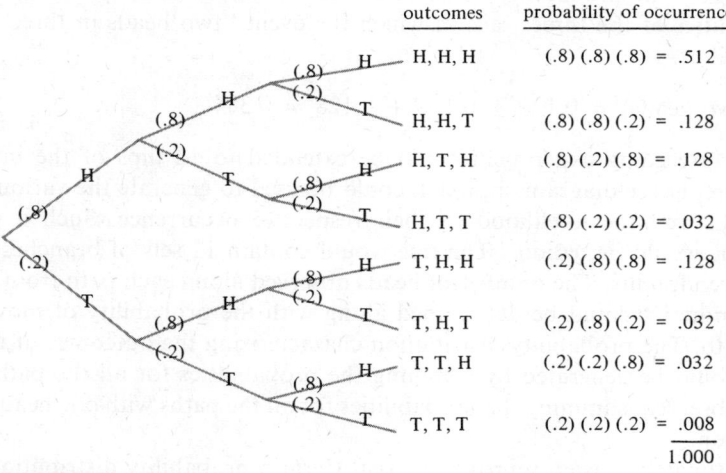

outcomes	probability of occurrence
H, H, H	(.8) (.8) (.8) = .512
H, H, T	(.8) (.8) (.2) = .128
H, T, H	(.8) (.2) (.8) = .128
H, T, T	(.8) (.2) (.2) = .032
T, H, H	(.2) (.8) (.8) = .128
T, H, T	(.2) (.8) (.2) = .032
T, T, H	(.2) (.2) (.8) = .032
T, T, T	(.2) (.2) (.2) = .008
	1.000

Figure 7.3

First branch: head on first flip; head on second flip; head on third flip.

$$P(H_1 \cap H_2 \cap H_3) = P(H_1) \cdot P(H_2 \mid H_1) \cdot P(H_3 \mid H_1 \cap H_2)$$

However, the events are independent. Therefore,

$P(H_1) \cdot P(H_2 \mid H_1) \cdot P(H_3 \mid H_1 \cap H_2)$ simplifies to $P(H_1) \cdot P(H_2) \cdot P(H_3)$.

$$P(H_1 \cap H_2 \cap H_3) = P(H_1) \cdot P(H_2) \cdot P(H_3) = (0.8)(0.8)(0.8) = 0.512$$

Similarly, for the second branch,

$$P(H_1 \cap H_2 \cap T_3) = P(H_1) \cdot P(H_2 \mid H_1) \cdot P(T_3 \mid H_1 \cap H_2)$$
$$= P(H_1) \cdot P(H_2) \cdot P(T_3)$$

Substituting,

$$P(H_1 \cap H_2 \cap T_3) = (0.8)(0.8)(0.2) = 0.128$$

and so on.

Combining similar outcomes, we obtain the following probability distribution description of the pattern of outcomes to the experiment:

List of all possible outcomes	Probability of occurrence	
0 heads	0.008	= 0.008
1 head	0.032 + 0.032 + 0.032	= 0.096
2 heads	0.128 + 0.128 + 0.128	= 0.384
3 heads	0.512	= 0.512
		1.000

Note that the outcome "two heads" can occur three different ways (HHT, HTH, and THH). The probability of observing two heads is obtained by summing the probabilities of the three ways in which the event "two heads in three flips" can occur. Thus,

$$P(\text{two heads}) = 0.128 + 0.128 + 0.128 = 0.384$$

The preceding experiment could be extended to 12 flips of the unfair coin. As before, a tree-diagram approach could be used to generate the various possible outcomes and the likelihood of their respective occurrence. Such a procedure however, would be tedious. The tree would contain 12 sets of branches and $(2)^{12}$ or 4096 endpoints. The number of heads observed along each path from the origin to an endpoint could be determined along with the probability of moving along each path. The probability distribution characterizing the outcomes of the experiment would be generated by summing the probabilities for all the paths with no heads observed, summing the probabilities for all the paths with one head observed, and so on.

Fortunately, easier approaches exist. Certain probability distribution patterns appear when certain types of experiments are performed. By analyzing the nature of an experiment, one can often determine in advance which probability distribution pattern is appropriate. Formulas can then be used to compute the probabilities of the various outcomes directly. Even better, when the underlying computations have already been performed, the probabilities can be read directly from a table.

Some of the more important categories of probability distributions that tend to occur with regularity are

1. The binomial distribution
2. The hypergeometric distribution

3. The Poisson distribution
4. The normal distribution
5. The t or student's distribution
6. The χ^2 (chi-square) distribution
7. The F distribution

It is possible to describe the necessary and sufficient conditions under which a given distribution will provide an appropriate description of the probability pattern of the outcomes to an experiment. In this chapter, concern will be limited to the binomial and the hypergeometric probability distributions. The Poisson distribution is investigated in Chapter 10; the normal distribution is examined in Chapter 11. Chapters 17 and 19 deal with the t distribution and the chi-square distribution, respectively.

The binomial distribution provides the appropriate description of the probabilities of occurrence of the possible outcomes to an experiment whenever the outcomes to the experiment are generated by a Bernoulli process. Two characteristics must be satisfied if a process is to be considered a Bernoulli process.

1. The outcome on any given trial must be capable of being classified into one of two mutually exclusive and collectively exhaustive categories. To illustrate:

defective or nondefective
acceptable or nonacceptable
correct or incorrect
works or does not work
head or tail

An experiment that has many possible outcomes, such as rolling a die (which has six possible outcomes) or drawing a card from a deck of cards (which has 52 possible outcomes), can still satisfy the first constraint of a Bernoulli process by appropriate adjustment of the interpretation of observations of the experiment. For example, the outcomes can be reclassified into the following categories:

rolling a one; not rolling a one
drawing an ace; not drawing an ace

2. The second requirement is that the outcome on any given trial of the experiment must be independent of the outcomes on other trials.

outcome I = head $P(\text{outcome I}) = p$

outcome II = tail $P(\text{outcome II}) = 1 - p = q$

Recalling the definition of independent events, the second requirement implies that the probability of success (i.e., p) must remain constant from one trial to another in order for the binomial probability distribution to be appropriate.

The above prerequisites are met by the illustrative examples employed previously, (1) two flips of a fair coin and (2) three flips of an unfair coin. Therefore, a binomial probability distribution can be used to generate the probabilities directly.

Actually, there is no such thing as *the* binomial probability distribution, but rather there is a family of binomial distributions linked together by certain mathematical relationships possessed in common—sort of a family trait.[1] The common

---

[1] In fact, when we examine the other common probability distributions, it will be discovered that they are also families with distinctive mathematical traits.

factor or family trait characterizing the binomial distribution family is that they can all be generated by the mathematical expression known as the binomial expansion, that is,

$$(p + q)^n$$

p = probability of outcome I on any given trial

q = probability of outcome II on any given trial or $1 - p$

n = number of trials in the experiment

Consider $(p + q)^2$. Algebraically, this is interpreted as an instruction to square the binomial (two-term) expression contained in parentheses; that is, multiply the expression $(p + q)$ by itself. To multiply the two expressions, simply multiply each term in the second expression by each term in the first expression and collect similar terms.

$$p + q$$
$$p + q$$
$$\overline{}$$
$$p^2 + pq$$
$$pq + q^2$$
$$\overline{}$$
$$p^2 + 2pq + q^2$$

Thus,

$$(p + q)^2 = p^2 + 2pq + q^2$$

Referring to the first illustration—two flips of a fair coin—the terms n, p, and q take on the following interpretation:

$n = 2$ (two flips or two trials)

$p = \frac{1}{2}$ (probability of a head on any one flip of the coin)

$q = \frac{1}{2}$ (probability of a tail on any one flip of the coin)

Interpreting the three terms in the expansion in the context of the above specifications,

p^2 corresponds to the observance of two heads. The implicit coefficient of one indicates that there is only one way this outcome can occur.

$2pq$ corresponds to one head and one tail. The coefficient 2 indicates that there are two ways this outcome can occur (i.e., H, T, or T, H).

q^2 corresponds to two tails. Again, the implicit coefficient of one indicates that there is only one way this outcome can occur.

Each possible outcome to the experiment (where outcome refers to the number of "successes" observed) is represented by a term in the expansion. To obtain the probability associated with each of the outcomes, simply substitute the values for p and q into the corresponding algebraic expression and solve.

Note that the sum of the probabilities of the terms in the expansion sum to one. This follows from the fact that p and q sum to one, and one raised to any power equals one.

$$(p + q)^2 = (\tfrac{1}{2} + \tfrac{1}{2})^2 = (1)^2 = 1$$

List of categories encompassing all possible outcomes	*Probability of occurrence*
0 heads	$q^2 = (\frac{1}{2})^2 \quad = \frac{1}{4} = 0.25$
1 head	$2pq = 2(\frac{1}{2})(\frac{1}{2}) = \frac{2}{4} = 0.50$
2 heads	$p^2 = (\frac{1}{2})^2 \quad = \frac{1}{4} = 0.25$
	$\overline{}$
	1.00

Referring to the experiment, three flips of an unfair coin weighted such that heads will occur 80 percent of the time, the binomial expansion is as follows:

$$(p + q)^3 = (p + q)(p + q)(p + q)$$

$$p + q$$
$$p + q$$
$$\overline{}$$
$$p^2 + pq$$
$$\qquad pq + q^2$$
$$\overline{}$$
$$p^2 + 2pq + q^2$$

$$(p + q)^3 = (p^2 + 2pq + q^2)(p + q)$$

$$p^2 + 2pq + q^2$$
$$p + q$$
$$\overline{}$$
$$p^3 + 2p^2q + pq^2$$
$$\qquad p^2q + 2pq^2 + q^3$$
$$\overline{}$$
$$p^3 + 3p^2q + 3pq^2 + q^3$$

$$(p + q)^3 = \quad p^3 \quad + 3p^2q \quad + 3pq^2 \quad + \quad q^3$$

| three heads | two heads | one head | no heads |

Note that the coefficient 3 in the second term indicates that there are three ways in which two heads can be obtained. Similarly, the coefficient 3 in the third term indicates that there are three ways in which only one head can be observed. (Check the tree diagram constructed previously to verify this.)

Given: p = probability of a head = (0.8)

q = probability of a tail = (0.2)

The probability distribution describing the pattern of the outcomes becomes:

List of all possible outcomes	Probability of occurrence		
0 heads	q^3	$(0.2)^3$	$= 0.008$
1 head	$3pq^2$	$3(0.8)(0.2)^2$	$= 0.096$
2 heads	$3p^2q$	$3(0.8)^2(0.2)$	$= 0.384$
3 heads	p^3	$(0.8)^3$	$= 0.512$
			$\overline{1.000}$

The binomial expansion eliminates the necessity of constructing a tree diagram to generate a description of the pattern of outcomes to an experiment such as 12 flips of a fair coin, but expanding $(p + q)^{12}$ is no picnic either:

$$(p + q)^{12} = (p + q)(p + q)(p + q)(p + q)(p + q)(p + q)$$
$$\times (p + q)(p + q)(p + q)(p + q)(p + q)(p + q)$$

Fortunately, the following characteristics enable one to write binomial expansions directly.

The expansion of $(p + q)^n$ possesses the following properties:

1. It contains $n + 1$ terms. For example, $(p + q)^2$ contains 3 terms, $(p + q)^3$ contains 4 terms, and $(p + q)^{12}$ contains 13 terms.
2. The first term is p^n. In each succeeding term, the exponent on p declines by one until the $(n + 1)$th term, for which the exponent on p is zero. $p^0 = 1$ and therefore is usually not expressed explicitly.
3. The exponent on q is zero in the first term of the expansion and, since $q^0 = 1$, is normally omitted. In each succeeding term the exponent on q increases by one until the $(n + 1)$th or last term, which is q^n. Note that the sum of the exponents on p and q sum to n in every term.
4. The coefficients: The coefficient of the first term is always equal to 1. The coefficient of the second term is always equal to n. The coefficient of all terms after the first term can be determined by the following relationship: Multiply the exponent on p in the *preceding* term by the coefficient of the *preceding* term and divide the resulting product by the numerical position of the *preceding* term in the expansion.

The expansion of $(p + q)^3$ is illustrated below:

	Without coefficients	Coefficients	With coefficients
First term	p^3q^0 or p^3	1	p^3
Second term	p^2q^1	$\dfrac{(3)(1)}{1} = 3$	$3p^2q$
Third term	p^1q^2	$\dfrac{(2)(3)}{2} = 3$	$3pq^2$

Fourth term $\qquad p^0 q^3$ or $q^3 \qquad \dfrac{(1)(3)}{3} = 1 \qquad q^3$

or

$$(p + q)^3 = p^3 + 3p^2q + 3pq^2 + q^3$$

Similarly,

$$(p + q)^4 = p^4 + 4p^3q + 6p^2q^2 + 4pq^3 + q^4$$

1	$\dfrac{(4)(1)}{1}$	$\dfrac{(3)(4)}{2}$	$\dfrac{(2)(6)}{3}$	$\dfrac{(1)(4)}{4}$

$$(p + q)^{12} = p^{12} + 12p^{11}q + 66p^{10}q^2 + 220p^9q^3 + 495p^8q^4 + 792p^7q^5$$
$$+ 924p^6q^6 + 792p^5q^7 + 495p^4q^8 + 220p^3q^9 + 66p^2q^{10}$$
$$+ 12pq^{11} + q^{12}$$

Note that if the coefficients of the 13 terms in the expansion are summed (i.e., $1 + 12 + 66 + 220 + 495 + 792 + 924 + 792 + 495 + 220 + 66 + 12 + 1$), they total 4096, the number of endpoints in the corresponding tree diagram. The coefficient of 792 on the sixth term of the expansion indicates that of the 4096 paths through the tree diagram, 792 contain 7 successes and 5 failures out of 12 trials. Note that the probability of 7 successes out of 12 trials will equal $\frac{792}{4096}$ only in the instance where the two possible outcomes on each trial are equally likely (i.e., where $p = q = \frac{1}{2}$). In all other instances, the values for p and q must be substituted into the expressions to generate the probability associated with each outcome.

The probability distribution characterizing the outcomes to an experiment consisting of flipping an unfair coin six times (weighted so as to turn up heads 80 percent of the time) is described by the following expansion:

$$(p + q)^6 = p^6 + 6p^5q + 15p^4q^2 + 20p^3q^3 + 15p^2q^4 + 6pq^5 + q^6$$

Substituting, we obtain:

List of all possible outcomes	Probability of occurrence			
0 heads	q^6	$= (0.2)^6$	$= 0.000064$	$= 0.000064$
1 head	$6pq^5$	$= 6(0.8)(0.2)^5$	$= 6(0.8)(0.00032)$	$= 0.001536$
2 heads	$15p^2q^4$	$= 15(0.8)^2(0.2)^4$	$= 15(0.64)(0.0016)$	$= 0.015360$
3 heads	$20p^3q^3$	$= 20(0.8)^3(0.2)^3$	$= 20(0.512)(0.008)$	$= 0.081920$
4 heads	$15p^4q^2$	$= 15(0.8)^4(0.2)^2$	$= 15(0.4096)(0.04)$	$= 0.245760$
5 heads	$6p^5q$	$= 6(0.8)^5(0.2)$	$= 6(0.32768)(0.2)$	$= 0.393216$
6 heads	p^6	$= (0.8)^6$	$= 0.262144$	$= 0.262144$
				1.000000

7.1 GENERATION OF AN INDIVIDUAL TERM IN A BINOMIAL EXPANSION

Frequently, rather than requiring the entire probability distribution of outcomes to a binomial experiment, all that is desired is the probability of a particular outcome. A formula is available for computing this information directly:

$$C\binom{n}{r} p^r q^{n-r}$$

where

n = number of trials in the experiment
p = probability of success on any one trial
q = probability of failure on any one trial
r = number of successes out of n trials in the outcome of concern

$$C\binom{n}{r} = \frac{n!}{r!\,(n-r)!}$$

$C\binom{n}{r}$ refers to the number of permutations of n objects that can be obtained where r of the objects are of one kind (success) and $(n-r)$ objects are of the other (failures). Thus, $C\binom{n}{r}$ can be used to generate the number of ways in which one can obtain r successes and $(n-r)$ failures in n trials.

If the formula

$$C\binom{n}{r} = \frac{n!}{r!\,(n-r)!}$$

is not familiar, refer to Technical Note 6 at the end of the chapter, where the concepts and formulas for permutations and combinations are reviewed. Particularly note extension 3, which discusses the interpretation of $C\binom{n}{r}$ as it applies to the binomial distribution.

EXAMPLE 7.1

The probability of obtaining two heads in three flips of a fair coin:

$$n = 3 = \text{number of trials}$$
$$p = P(\text{head}) = 0.5$$
$$q = P(\text{tail}) = 0.5$$
$$r = 2 = \text{number of successes}$$

$$C\binom{n}{r} p^r q^{n-r} = \frac{n!}{r!\,(n-r)!} \, p^r q^{n-r}$$

$$= \frac{3!}{2!\,1!} \, (0.5)^2 (0.5)^1$$

$$= \frac{3 \cdot 2!}{2! \cdot 1!} \, (0.25)(0.5)$$

$$= 3(0.125)$$

$$= 0.375$$

EXAMPLE 7.2

The probability of obtaining two heads in three flips of an unfair coin weighted so as to turn up heads 80 percent of the time:

$$n = 3$$

$$p = P(\text{head}) = 0.8$$

$$q = P(\text{tail}) = 0.2$$

$$r = 2$$

$$C \binom{n}{r} p^r q^{n-r} = \frac{n!}{r!\,(n-r)!} p^r q^{n-r}$$

$$= \frac{3!}{2!\,1!}\,(0.8)^2 (0.2)^1$$

$$= 3(0.64)(0.2) = 3(0.128)$$

$$= 0.384$$

EXAMPLE 7.3

The probability of obtaining three heads in six flips of an unfair coin weighted so as to turn up heads 80 percent of the time:

$$n = 6$$

$$p = P(\text{head}) = 0.8$$

$$q = P(\text{tail}) = 0.2$$

$$r = 3$$

$$C \binom{n}{r} p^r q^{n-r} = \frac{n!}{r!\,(n-r)!} p^r q^{n-r}$$

$$= \frac{6!}{3!\,3!}\,(0.8)^3 (0.2)^3$$

$$= \frac{6 \cdot 5 \cdot 4 \cdot 3!}{3!\,3 \cdot 2 \cdot 1}\,(0.512)(0.008)$$

$$= 20(0.512)(0.008)$$

$$= 0.081920$$

Although the computations pursued up to this point are useful as an aid to understanding the rationale behind the probability assignments, they are seldom necessary in practice. To illustrate: The expansion of $(p + q)^3$ is identical for any binomial experiment consisting of three trials.

$$(p + q)^3 = p^3 + 3p^2q + 3pq^2 + q^3$$

Moreover, if $p = 0.04$, then $p^3 + 3p^2q + 3pq^2 + q^3$ becomes $(0.04)^3 + 3(0.04)^2 \times (0.96) + 3(0.04)(0.96)^2 + (0.96)^3$ or

$$0.000064 + 0.004608 + 0.110592 + 0.884736$$

List of outcomes	*Probability of occurrence*
0 successes	0.884736
1 success	0.110592
2 successes	0.004608
3 successes	0.000064
	1.000000

Clearly, whenever a binomial experiment consisting of three trials is performed and the probability of success on any one trial is 0.04, the probability distribution of outcomes is identical to that computed above. The above distribution could thus be stored and referred to for any binomial experiment consisting of three trials with the probability of success on any one trial equal to 0.04 (i.e., $n = 3$, $p = 0.04$), thereby eliminating the necessity of repeating the calculations. Note that the two parameters n and p are sufficient for identification of the relevant binomial probability distribution. (In fact, the above table can provide double service, since the same probabilities can be used to describe a binomial experiment of three trials where the probability of success on any one trial is 0.96. Due to the symmetrical nature of the binomial distribution when expressed in p's and q's, what previously would have been interpreted as success would now be interpreted as failure—thus, zero successes with $p = 0.04$ is identical in likelihood to three successes with $p = 0.96$.)

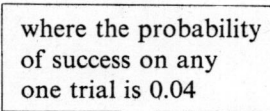

where the probability of success on any one trial is 0.04

Outcomes	*Probability of occurrence*	*Outcomes*
0 successes	0.884736	3 successes
1 success	0.110592	2 successes
2 successes	0.004608	1 success
3 successes	0.000064	0 successes

where the probability of success on any one trial is 0.96

More efficiently:

	$p = 0.04$	
$r = 0$	0.884736	$3 = r$
1	0.110592	2
2	0.004608	1
3	0.000064	0
	$p = 0.96$	

The r's at the left refer to the number of successes when the value of p at the top of the column is used. The r's at the right refer to the number of successes when using the value of p at the bottom of the column.

In a similar fashion, probability distributions can be generated for $p = 0.01$, $p = 0.02, \ldots, p = 0.50$ for binomial experiments consisting of three trials. By assigning $(1 - p)$ to the base of each column, the probability distributions associated with $p = 0.99, p = 0.98, \ldots$, and so on are also provided. The resulting distributions appear in Table 7.1.

Table 7.1

$$N = 3$$

R P	01	02	03	04	05	06	07	08	09	10	
0	9703	9412	9127	8847	8574	8306	8044	7787	7536	7290	3
1	0294	0576	0847	1106	1354	1590	1816	2031	2236	2430	2
2	0003	0012	0026	0046	0071	0102	0137	0177	0221	0270	1
3	0000	0000	0000	0001	0001	0002	0003	0005	0007	0010	0
	99	98	97	96	95	94	93	92	91	90	P R

R P	11	12	13	14	15	16	17	18	19	20	
0	7050	6815	6585	6361	6141	5927	5718	5514	5314	5120	3
1	2614	2788	2952	3106	3251	3387	3513	3631	3740	3840	2
2	0323	0380	0441	0506	0574	0645	0720	0797	0877	0960	1
3	0013	0017	0022	0027	0034	0041	0049	0058	0069	0080	0
	89	88	87	86	85	84	83	82	81	80	P R

R P	21	22	23	24	25	26	27	28	29	30	
0	4930	4746	4565	4390	4219	4052	3890	3732	3579	3430	3
1	3932	4015	4091	4159	4219	4271	4316	4355	4386	4410	2
2	1045	1133	1222	1313	1406	1501	1597	1693	1791	1890	1
3	0093	0106	0122	0138	0156	0176	0197	0220	0244	0270	0
	79	78	77	76	75	74	73	72	71	70	P R

R P	31	32	33	34	35	36	37	38	39	40	
0	3285	3144	3008	2875	2746	2621	2500	2383	2270	2160	3
1	4428	4439	4444	4443	4436	4424	4406	4382	4354	4320	2
2	1989	2089	2189	2289	2389	2488	2587	2686	2783	2880	1
3	0298	0328	0359	0393	0429	0467	0507	0549	0593	0640	0
	69	68	67	66	65	64	63	62	61	60	P R

R P	41	42	43	44	45	46	47	48	49	50	
0	2054	1951	1852	1756	1664	1575	1489	1406	1327	1250	3
1	4282	4239	4191	4140	4084	4024	3961	3894	3823	3750	2
2	2975	3069	3162	3252	3341	3428	3512	3594	3674	3750	1
3	0689	0741	0795	0852	0911	0973	1038	1106	1176	1250	0
	59	58	57	56	55	54	53	52	51	50	P R

The concept can be extended to encompass binomial experiments of any number of trials. Such a collection of tables has been constructed and is reproduced in the Appendix as Table C.

Using Table C, verify the following probability distributions previously calculated:

$n = 2, p = 0.50$

$n = 3, p = 0.04$

$n = 3, p = 0.80$

$n = 6, p = 0.80$

Often, the issue of interest is not "what is the probability of *exactly three* heads in six flips of an unfair coin weighted so as to turn up heads 20 percent of the time" but rather "what is the probability of *three or more* heads in six flips of an unfair coin weighted so as to turn up heads 20 percent of the time." Such tables can easily be generated from tables of binomial probability distributions (individual terms) — Appendix Table C. See Tables 7.2(a) and 7.2(b). Tables of decumulative binomial probabilities can be found in texts of standard statistical tables.

Table 7.2(a) Individual probabilities, 0.20

r	
0	0.2621
1	0.3932
2	0.2458
3	0.0819
4	0.0154
5	0.0015
6	0.0001

Table 7.2(b) Decumulative[a] probabilities, 0.20

r		
0	1.0000	0.0001 + 0.0015 + 0.0154 + 0.0819 + 0.2458 + 0.3932 + 0.2621
1	0.7379	0.0001 + 0.0015 + 0.0154 + 0.0819 + 0.2458 + 0.3932
2	0.3447	0.0001 + 0.0015 + 0.0154 + 0.0819 + 0.2458
3	0.0989	0.0001 + 0.0015 + 0.0154 + 0.0819
4	0.0170	0.0001 + 0.0015 + 0.0154
5	0.0016	0.0001 + 0.0015
6	0.0001	0.0001

[a] In a decumulative binomial probability distribution table, the r value refers to r or more successes.

7.2 BINOMIAL APPROXIMATION

Technically, the use of the binomial distribution to generate the probabilities of the various outcomes to an experiment is justified only if the experiment is characterized as a Bernoulli process (i.e., a series of Bernoulli trials). Recall that two conditions must be met for a Bernoulli process:

1. The outcome on any one trial must fall into one of two mutually exclusive and collectively exhaustive categories.
2. The outcome on any given trial of the experiment must be independent of the outcomes on the other trials. When trials are independent, it means that the probability of success remains constant from trial to trial.

In many instances, the experiment performed involves sampling from a finite universe, for example, a collection of 1000 billings. Suppose a sample of size 3 were to be drawn without replacement from this universe and each observation in the sample classified as to either "calculation correctly performed" or "calculation incorrectly performed." The experiment described meets the first condition of a Bernoulli process but not the second. The probability of having selected a correctly computed billing on a given trial will vary depending on what occurred on the other two trials, thereby violating the independence requirement. Suppose that the true percentage of incorrectly calculated billings in the universe is 5 percent (i.e., 50 of the 1000 billings). The probability distribution of sample outcomes will be generated (1) by direct calculation, and (2) by a binomial approximation under the assumption that p remains constant at 0.05 from trial to trial (see Figure 7.4).

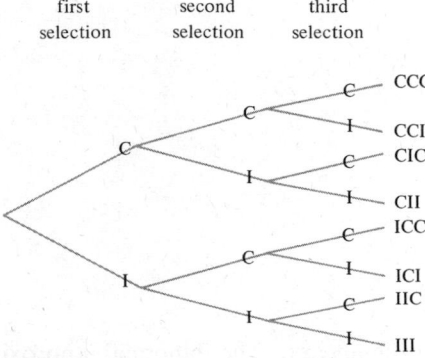

| first selection | second selection | third selection |

Figure 7.4

1. By direct calculation:

$$\text{CCC} \quad \frac{950}{1000} \cdot \frac{949}{999} \cdot \frac{948}{998} = \frac{854,669,400}{997,002,000} = 0.85724$$

$$\text{CCI} \quad \frac{950}{1000} \cdot \frac{949}{999} \cdot \frac{50}{998} = \frac{45,077,500}{997,002,000} = 0.04521$$

$$\text{CIC} \quad \frac{950}{1000} \cdot \frac{50}{999} \cdot \frac{949}{998} = \frac{45,077,500}{997,002,000} = 0.04521$$

$$\text{CII} \quad \frac{950}{1000} \cdot \frac{50}{999} \cdot \frac{49}{998} = \frac{2,327,500}{997,002,000} = 0.00233$$

$$\text{ICC} \quad \frac{50}{1000} \cdot \frac{950}{999} \cdot \frac{949}{998} = \frac{45,077,500}{997,002,000} = 0.04521$$

$$\text{ICI} \quad \frac{50}{1000} \cdot \frac{950}{999} \cdot \frac{49}{998} = \frac{2,327,500}{997,002,000} = 0.00233$$

$$\text{IIC} \quad \frac{50}{1000} \cdot \frac{49}{999} \cdot \frac{950}{998} = \frac{2,327,500}{997,002,000} = 0.00233$$

$$\text{III} \quad \frac{50}{1000} \cdot \frac{49}{999} \cdot \frac{48}{998} = \frac{117,600}{997,002,000} = 0.00012$$

0.99998   (round off error)

List of outcomes	Probability of occurrence
0 correct	0.0001
1 correct	0.0070
2 correct	0.1356
3 correct	0.8572
	0.9999

2. Using the binomial approximation:
$n = 3, p = 0.95$

List of outcomes	Probability of occurrence
0 correct	0.0001
1 correct	0.0071
2 correct	0.1354
3 correct	0.8574
	1.0000

The difference in results is insignificant. The binomial approximation is therefore preferred because of the comparative simplicity of computation. The binomial approximation generates a more than satisfactory approximation when the universe is large in comparison with the sample size. A rough rule of thumb is that the approximation will suffice provided that the size of the sample is less than 20 percent of the size of the universe. If the sample size is large relative to the size of the universe and drawings are made without replacement, the hypergeometric probability distribution must be used to generate the probability distribution describing the pattern of sample outcomes. (See Technical Note 7.) The hypergeometric distribution is more cumbersome to work with than the binomial. Fortunately, in practice the binomial approximation is often sufficiently accurate for decision making, although the hypergeometric might be the distribution technically appropriate in the given situation.

7.3 COMPUTATION OF THE MEAN AND STANDARD DEVIATION OF A BINOMIAL DISTRIBUTION

Binomial experiment
$n = 3, p = 0.80$

List of outcomes (number of successes)	Probability of occurrence
x	$P(x)$
0	0.0080
1	0.0960
2	0.3840
3	0.5120
	1.0000

Formulas:

$$\mu_x = \frac{\Sigma\,(WX)}{\Sigma\,(W)}$$

$$\sigma_x = \sqrt{\frac{\Sigma\,[W(X - \mu_x)^2]}{\Sigma\,(W)}}$$

where W, the weights, refer to the probability of occurrence. Thus, the formulas become

$$\mu_x = \frac{\Sigma\,[P(X) \cdot X]}{\Sigma\,[P(X)]}$$

$$\sigma_x = \sqrt{\frac{\Sigma\,[P(X) \cdot (X - \mu_x)^2]}{\Sigma\,[P(X)]}}$$

x	$P(X)$	$P(X) \cdot X$	$X - \mu_x$	$(X - \mu_x)^2$	$P(X) \cdot (X - \mu_x)^2$
0	0.0080	0	-2.4	5.76	0.04608
1	0.0960	0.0960	-1.4	1.96	0.18816
2	0.3840	0.7680	-0.4	0.16	0.06144
3	0.5120	1.5360	$+0.6$	0.36	0.18432
	1.0000	2.4000			0.48000

$$\mu_x = \frac{\Sigma\,[P(X) \cdot X]}{\Sigma\,[P(X)]} = \frac{2.4}{1} = 2.4$$

$$\sigma_x = \sqrt{\frac{\Sigma\,[P(X) \cdot (X - \mu_x)^2]}{\Sigma\,[P(X)]}} = \sqrt{\frac{0.48}{1}} = \sqrt{0.48} = 0.6928$$

Although the mean and standard deviation of a binomial distribution can be computed as above by the standard formulas, a more efficient computation exists— one that does not even require the generation of the probability distribution of outcomes. All that is needed are the parameters of the distribution, n and p. The formulas are

$$\mu_x = np = 3(0.8) = 2.4$$

$$\sigma_x = \sqrt{npq} = \sqrt{3(0.8)(0.2)} = \sqrt{3(0.16)} = \sqrt{0.48} = 0.6928$$

A proof of the short-cut formulas is provided in Technical Note 8.

TECHNICAL NOTE 6

Combinations and permutations

A *combination* is a collection of r objects where the ordering or sequence in which the objects occur is of no significance. To illustrate: Imagine an elective that is restricted to an enrollment of three students. Any group of three students would be considered a combination if concern is limited to which three students attend class (i.e., where the students sit is of no significance—just who are the members of the group).

A *permutation* is a collection of r objects where ordering is of significance. Suppose the room where the class meets contains only three seats. Each possible seating plan is a different permutation. Given the same group of three students, if the seating arrangement were to be changed, it would constitute a different permutation—it would still reflect the same combination even though the arrangement was changed, however, since the composition of the group is unchanged.

A set of formulas can readily be derived for determining the number of possible combinations or the number of possible different permutations of a given size that can be generated out of a given pool of objects.

Permutations

The formula for determination of the number of different permutations of size r that can be formed out of a pool of n objects is derived as follows:

Suppose the pool consists of five students (labeled a, b, c, d, and e). Three students are to be selected out of the pool. The question: How many different possible seating arrangements of size 3 can be generated out of the pool of five students?

There are five students, any one of whom could be assigned to the first seat. Once the first seat has been assigned to a student, there will be four students left in the pool, any one of whom could be assigned to the second seat. Given that the first and second seats have been assigned, three students would remain in the pool, any one of whom could be assigned to the third seat. The tree diagram describing the available choices is comprised of five branches at the first stage, four branches at the second stage, and three branches at the third stage.

The total number of endpoints in the tree is equal to the number of branches at the first stage multiplied by the number of branches at the second stage multiplied by the number of branches at the third stage. Thus, the tree contains (5) × (4) × (3) or 60 endpoints. Each endpoint corresponds to a different path through the tree. Each path through the tree corresponds to a different permutation. Thus, the relationship used to determine the number of endpoints in the tree can be used to generate the number of different possible permutations of size 3. The symbol $P\binom{n}{r}$ is interpreted as the number of different permutations that can be formed out of a pool of n objects taking r objects at a time.

$$P\binom{n}{r} = n(n-1)(n-2)\cdots(n-r+1)$$

where

$n =$ number of choices for the first position or the total number of items in the pool

$n - 1 =$ number of choices remaining for the second position

$n - 2 =$ number of choices remaining for the third position

$n - r + 1 =$ number of choices remaining for the last position

Thus,

$$P\binom{5}{3} = 5 \cdot 4 \cdot 3 = 60$$

where

$$(n - r + 1) = (5 - 3 + 1) = 3$$

The first factor is always equal to n, each succeeding factor is one less than the factor preceding it, and the total number of factors always equals r. Thus, $P\binom{9}{4}$ indicates that the first factor will be 9 and that the total number of factors will be 4. Therefore,

$$P\binom{9}{4} = 9 \cdot 8 \cdot 7 \cdot 6 = 3024 \text{ permutations}$$

In the special case where the number of items selected is equal to the number of items in the pool (i.e., where $n = r$), the first factor will be n, there will be n factors, and the last factor will be 1. Note that $(n - r + 1) = 1$ when n is equal to r. Thus,

$$P\binom{7}{7} = 7 \cdot 6 \cdot 5 \cdot 4 \cdot 3 \cdot 2 \cdot 1$$
$$= 7!$$

7! is read as 7 *factorial* and is a shorthand notation for writing the product of the successive integers starting with 7 and going to 1. 7! is considerably more efficient than writing out $7 \cdot 6 \cdot 5 \cdot 4 \cdot 3 \cdot 2 \cdot 1$. Note: 0! is defined as being equal to 1. In general, where $n = r$, $P\binom{n}{r} = n!$

Combinations

Suppose that there are three students in the pool and three students are to be selected. How many combinations are there? Clearly, only one. How many permutations could be created out of this one combination taking all three at a time? Recall that when $n = r$, $P\binom{n}{r} = n! = r! = 3! = 3 \cdot 2 \cdot 1 = 6$. Thus, six permutations could be generated.

Suppose that there are five students in the pool and five students to be selected. Again, there is only one combination, and from that one combination of size 5 there can be created $P\binom{n}{r} = r! = 5!$ permutations of size 5. The relationship between permutations and combinations should now be evident.

In general, for every combination there will be $r!$ permutations. If the symbol $C\binom{n}{r}$ is used to represent the number of possible combinations of n objects taking r at a time, then the relationship between combinations and permutations can be expressed as follows:

$$P\binom{n}{r} = r! \, C\binom{n}{r}$$

where

$$P\binom{n}{r} = \text{the number of permutations of } n \text{ objects taking } r \text{ at a time}$$

$$C\binom{n}{r} = \text{the number of combinations of } n \text{ objects taking } r \text{ at a time}$$

Solving for $C\binom{n}{r}$, we obtain:

$$C\binom{n}{r} = \frac{P\binom{n}{r}}{r!}$$

Substituting $n(n-1)\cdots(n-r+1)$ for $P\binom{n}{r}$, we obtain

$$C\binom{n}{r} = \frac{n(n-1)\cdots(n-r+1)}{r!}$$

Multiplying numerator and denominator of a fraction by the same value (other than zero) leaves the value of the fraction unchanged; therefore,

$$C\binom{n}{r} = \frac{n(n-1)\cdots(n-r+1)}{r!} = \frac{(n)(n-1)\cdots(n-r+1)}{r!} \cdot \frac{(n-r)!}{(n-r)!}$$

Since

$$n(n-1)\cdots(n-r+1)\cdot(n-r)! = n!$$

then

$$C\binom{n}{r} = \frac{n!}{r!\,(n-r)!}$$

The above constitutes a general formula for determining the number of combinations of n objects taking r at a time.

To illustrate, the number of combinations of 3 students that could be generated out of a pool of 15 students is determined as follows:

$$C\binom{n}{r} = \frac{n!}{r!\,(n-r)!}$$

$$C\binom{15}{3} = \frac{15!}{3!\,12!} = \frac{\overset{5}{\cancel{15}} \cdot \overset{7}{\cancel{14}} \cdot 13 \cdot \cancel{12!}}{\cancel{3} \cdot \cancel{2} \cdot 1 \cdot \cancel{12!}} = 455$$

From the same pool, 2730 permutations of size 3 could have been created. There will be $r!$ permutations for each combination. Therefore,

$$P\binom{n}{r} = r!\,C\binom{n}{r}$$

$$= 3!\,(455)$$

$$= 3 \cdot 2 \cdot 1(455)$$

$$= 6(455)$$

$$= 2730$$

Computed directly:

$$P\binom{n}{r} = P\binom{15}{3} = 15 \cdot 14 \cdot 13 = 2730$$

In the special case where the number selected equals the number in the pool from which selections are to be made, the formula for the number of combinations simplifies to $C\binom{n}{r} = 1$.

Where $r = n$:

$$C\binom{n}{r} = \frac{n!}{r!\,(n-r)!} \quad \text{becomes} \quad \frac{n!}{n!\,(n-n)!} = \frac{n!}{n!\,0!} = \frac{n!}{n!} = 1$$

In summary, we have the following relationships:
For permutations:

In general: $\quad P\binom{n}{r} = n(n-1)\cdots(n-r+1)$

In the special case where $n = r$: $\quad P\binom{n}{r} = n!$

For combinations:

In general: $\quad C\binom{n}{r} = \frac{n!}{r!\,(n-r)!}$

In the special case where $n = r$: $\quad C\binom{n}{r} = 1$

Remember that the above formulas are based on the assumption that objects once removed from the pool are not replaced.

Extensions

1. To find the total number of possible permutations of n items taking *no more* than r at a time (i.e., taking less than r at a time would be acceptable), simply obtain the sum of the following terms:

$$P\binom{n}{r} + P\binom{n}{r-1} + P\binom{n}{r-2} + \cdots + P\binom{n}{1}$$

2. Consider the following: A menu provides a choice of six different appetizers, nine different entrees, and seven different desserts. How many different ways could an individual order an appetizer, entree, and dessert?

This problem differs from those discussed previously in that each of the selections is made out of a separate pool. Thus, the correct solution is the product of three different permutations:

$$P\binom{6}{1} \cdot P\binom{9}{1} \cdot P\binom{7}{1} = 6 \cdot 9 \cdot 7 = 378$$

3. Finally, let us examine the situation where the objects to be selected are to be drawn from a pool consisting of n_1 indistinguishable elements of one kind (say,

green marbles) and n_2 indistinguishable elements of another kind (say, blue marbles). Suppose five items are to be drawn. How many different ways can the five drawings be made? A tree diagram depicting the various alternatives is presented in Figure 7.5. If the five items in the observation GGGBB were unique, there would be 5! permutations possible. If the three green marbles were distinguishable, there would be $3 \cdot 2 \cdot 1$ or 3! ways they could be assigned to their three slots. If the green marbles are indistinguishable, all 3! ways are considered the same result and are counted only once. Similarly, the 2! ways the two blue marbles could be arranged are also counted just once. There are 3! 2! ways GGGBB could occur, but they all look the same and thus are counted only once. Thus, 5! would be an overstatement for indistinguishable elements and should be adjusted to 5!/(3! 2!) or, in general, $n!/[r! \, (n - r)!]$, where n is the total number of observations of one kind and $n - r$ is the number of observations of the other kind.

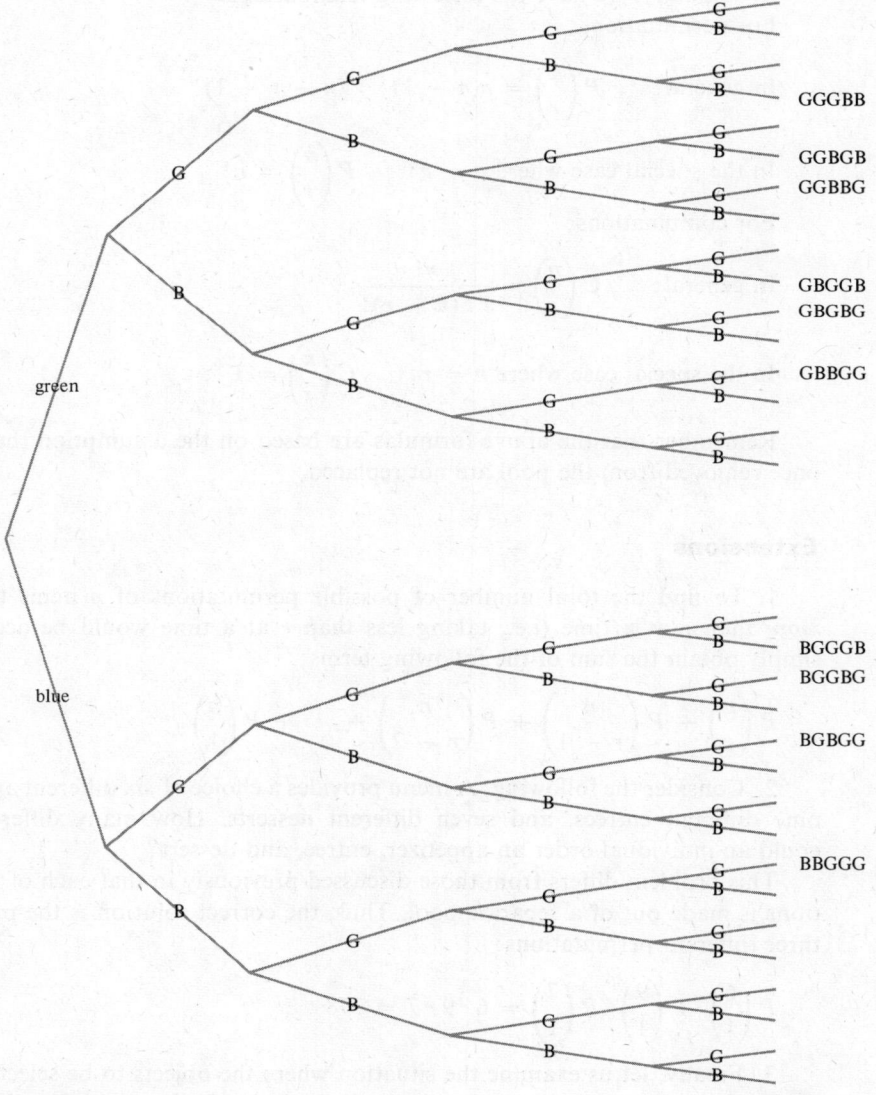

Figure 7.5

Referring to the tree in Figure 7.5, note that there are 32 endpoints. By observation, it can be observed that three green, two blue are observed at ten of the endpoints. This same information is generated by the following formula:

$$C\binom{n}{r} = \frac{n!}{r!\,(n-r)!}$$

$$C\binom{5}{3} = \frac{5!}{3!\,2!} = \frac{5 \cdot 4 \cdot 3!}{3!\,2 \cdot 1} = 10$$

Thus, the formula previously used for determining the number of combinations of n objects taking r at a time will also generate the number of permutations of size n that consist of r items of one kind and $n-r$ items of another kind. It follows that the expression $C\binom{n}{r}$ can be used to generate directly the coefficients in a binomial expansion where n is the number of trials and r is the number of successes [it is implicitly understood that $(n-r)$ is the number of failures].

TECHNICAL NOTE 7

The hypergeometric distribution

Consider an urn containing eight red marbles and two green marbles. A sample of size 5 is to be selected.

Case A: Determine the probability of observing the sample outcome three red out of five when drawings are made with replacement. *Note:* The probability distribution of sample outcomes that is theoretically appropriate when sampling with replacement is the binomial distribution.

Case B: Determine the probability of observing the sample outcome three red out of five when drawings are made without replacement. *Note:* The probability distribution of sample outcomes that is theoretically appropriate when sampling without replacement from a finite population is the hypergeometric distribution.

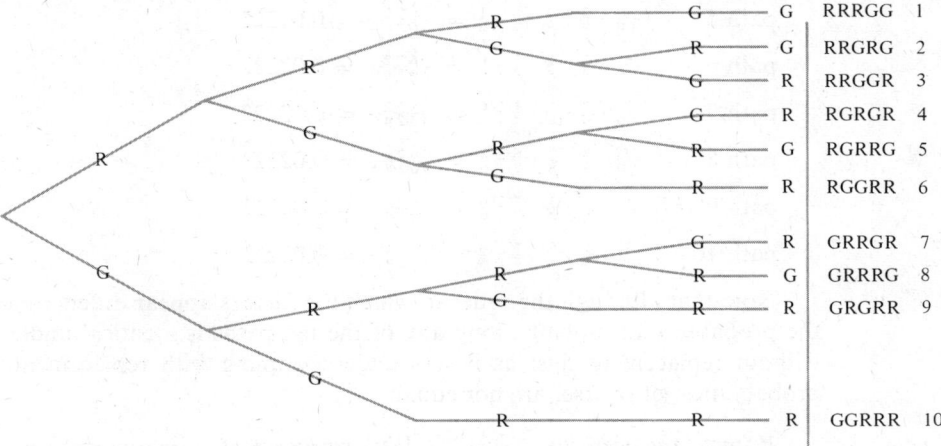

Figure 7.6

The tree diagram in Figure 7.6 is only a partial description of the outcomes to the sampling experiment, because it depicts only those branches that pertain to the particular sample event three red observed out of five. Since discussion will be restricted to this event, such a simplification is appropriate.

The tree in Figure 7.6 depicts the various ways in which three red marbles could be observed out of five and applies to drawings both with and without replacement. The difference in treatment occurs when assigning probabilities to the branches. Note that the total number of ways three reds and two greens could have been observed can be computed directly by the formula:

$$C\binom{n}{r} = \frac{n!}{r!\,(n-r)!} = \frac{5!}{3!\,2!} = \frac{5 \cdot 4 \cdot 3!}{3!\,2 \cdot 1} = 10$$

Case A: Sampling with replacement. The probability of moving along any one path is $p^3 q^2 = (0.8)^3 (0.2)^2$.

path 1 $\frac{8}{10} \cdot \frac{8}{10} \cdot \frac{8}{10} \cdot \frac{2}{10} \cdot \frac{2}{10} = (0.8)^3 (0.2)^2$

path 2 $\frac{8}{10} \cdot \frac{8}{10} \cdot \frac{2}{10} \cdot \frac{8}{10} \cdot \frac{2}{10} = (0.8)^3 (0.2)^2$

$\vdots$

path 10 $\frac{2}{10} \cdot \frac{2}{10} \cdot \frac{8}{10} \cdot \frac{8}{10} \cdot \frac{8}{10} = (0.8)^3 (0.2)^2$

Thus, the probability of observing three red out of five when sampling with replacement is $10(0.8)^3 (0.2)^2$ and can be generated directly by the binomial expression $C\binom{n}{r} p^r q^{n-r}$.

Case B: Sampling without replacement from a finite universe. The computation of the probability of moving along each path in the diagram is presented below:

path 1 $\frac{8}{10} \cdot \frac{7}{9} \cdot \frac{6}{8} \cdot \frac{2}{7} \cdot \frac{1}{6} = \frac{672}{30240} = 0.02222$

path 2 $\frac{8}{10} \cdot \frac{7}{9} \cdot \frac{2}{8} \cdot \frac{6}{7} \cdot \frac{1}{6} = \frac{672}{30240} = 0.02222$

path 3 $\frac{8}{10} \cdot \frac{7}{9} \cdot \frac{2}{8} \cdot \frac{1}{7} \cdot \frac{6}{6} = \frac{672}{30240} = 0.02222$

path 4 $\frac{8}{10} \cdot \frac{2}{9} \cdot \frac{7}{8} \cdot \frac{1}{7} \cdot \frac{6}{6} = \frac{672}{30240} = 0.02222$

path 5 $\frac{8}{10} \cdot \frac{2}{9} \cdot \frac{7}{8} \cdot \frac{6}{7} \cdot \frac{1}{6} = \frac{672}{30240} = 0.02222$

path 6 $\frac{8}{10} \cdot \frac{2}{9} \cdot \frac{1}{8} \cdot \frac{7}{7} \cdot \frac{6}{6} = \frac{672}{30240} = 0.02222$

path 7 $\frac{2}{10} \cdot \frac{8}{9} \cdot \frac{7}{8} \cdot \frac{1}{7} \cdot \frac{6}{6} = \frac{672}{30240} = 0.02222$

path 8 $\frac{2}{10} \cdot \frac{8}{9} \cdot \frac{7}{8} \cdot \frac{6}{7} \cdot \frac{1}{6} = \frac{672}{30240} = 0.02222$

path 9 $\frac{2}{10} \cdot \frac{8}{9} \cdot \frac{1}{8} \cdot \frac{7}{7} \cdot \frac{6}{6} = \frac{672}{30240} = 0.02222$

path 10 $\frac{2}{10} \cdot \frac{1}{9} \cdot \frac{8}{8} \cdot \frac{7}{7} \cdot \frac{6}{6} = \frac{672}{30240} = 0.02222$

Note that although the order in which the factors appear differs on each path, the probability of moving along any of the ten paths is identical under sampling without replacement, just as it was under sampling with replacement. The two probabilities, of course, are not equal.

Without replacement	*With replacement*
$\frac{2}{10} \cdot \frac{1}{9} \cdot \frac{8}{8} \cdot \frac{7}{7} \cdot \frac{6}{6} = 0.02222$	$(0.8)^3 (0.2)^2 = 0.02048$

The explanation is that the probability of observing a red marble (or a green for that matter) does not remain constant from trial to trial when drawings are made without replacement.

Without replacement, the probability of moving along any one path is

$$\frac{(8 \cdot 7 \cdot 6) \cdot (2 \cdot 1)}{10 \cdot 9 \cdot 8 \cdot 7 \cdot 6}$$

Thus, the probability of observing three red out of five when sampling without replacement is

$$10 \frac{(8 \cdot 7 \cdot 6) \cdot (2 \cdot 1)}{10 \cdot 9 \cdot 8 \cdot 7 \cdot 6} = \frac{2}{9}$$

This term could also have been computed directly using the formula for an individual term in the hypergeometric distribution.

$$\text{Probability of } r \text{ successes out of } n \text{ observations} = \frac{C\binom{N_1}{r} \cdot C\binom{N_2}{n-r}}{C\binom{N}{n}}$$

where

N = total number of elements in the universe

N_1 = number of elements of one kind (successes)

p = proportion of successes in the universe

$N_1 = pN$

N_2 = number of elements of the other kind (failure)

q = proportion of failures in the universe

$N_2 = qN$

n = size of sample

r = number of successes in the sample

$n - r$ = number of failures in the sample

$N_1 + N_2 = N$

$r + (n - r) = n$

$$\text{Probability of three successes out of five observations} = \frac{C\binom{8}{3} \cdot C\binom{2}{2}}{C\binom{10}{5}}$$

$$= \frac{(8!/3! \, 5!) \cdot (2!/2! \, 0!)}{(10!/5! \, 5!)}$$

$$= \frac{8!}{3! \, 5!} \cdot \frac{5! \, 5!}{10!}$$

$$= \frac{8! \, 5!}{3! \, 10!}$$

$$= \frac{8! \, 5 \cdot 4 \cdot 3!}{3! \, 10 \cdot 9 \cdot 8!}$$

$$= \tfrac{20}{90}$$

$$= \tfrac{2}{9}$$

General tables have been constructed for the binomial distribution, because any distribution with parameters n and p can be served by the same table. Unfortunately, this is not true for the hypergeometric distribution. Different tables would be required for a constant n and p for each possible universe size or N. As a result, although a few hypergeometric tables have been computed, in most instances it would be necessary to resort to calculation by the formulas. Fortunately, as previously mentioned, the binomial approximation will usually suffice, and the probabilities can be read from a table.

TECHNICAL NOTE 8

Proof that the mean and variance of a binomial distribution are np and npq, respectively

Proof: $\mu_x = np$. The outcome on any trial of a binomial experiment is either a success or a failure. Let X represent the number of successes on a given trial. X can take a value of zero or one on any given trial. X_i is defined as the number of successes on the ith trial.

The expected number of successes on the first trial can be computed as follows, where p equals the probability of success on any given trial (and, of course, q, the probability of failure).

X_1	$P(X_1)$	$P(X_1) \cdot X_1$
1	p	p
0	q	0
	1.0	p

Thus,

$$\sum [P(X_1) \cdot X_1] = p$$

$$E(X_1) = \frac{\sum [P(X_1) \cdot X_1]}{\sum [P(X_1)]}$$

$$= \frac{p}{1}$$

$$= p$$

Similarly, the expected number of successes on the second trial would be

X_2	$P(X_2)$	$P(X_2) \cdot X_2$
1	p	p
0	q	0
	1.0	p

$$E(X_2) = \frac{\sum [P(X_2) \cdot X_2]}{\sum [P(X_2)]}$$

$$= \frac{p}{1}$$

$$= p$$

Since the expected value of a sum is equal to the sum of the expected values, it follows that

$$E(X_1 + X_2) = E(X_1) + E(X_2) = p + p = 2p$$

It is a simple extension to the general case of n trials:

$$E(X_1 + X_2 + X_3 + \cdots + X_n) = np$$

Proof: $\sigma_x^2 = npq$. On p. 93 it was demonstrated that

$$\text{variance } (X) = \frac{\Sigma (X_2)}{N} - \left(\frac{\Sigma X}{N} \right)^2$$

This can alternatively be expressed as

$$\text{Var } (X) = E(X^2) - [E(X)]^2$$

X	$P(X)$	$P(X) \cdot X$	X^2	$P(X) \cdot X^2$
1	p	p	1	p
0	q	0	0	0
	1	p		p

$$E(X) = \frac{\Sigma [P(X) \cdot X]}{\Sigma [P(X)]} = \frac{p}{1} = p$$

$$E(X^2) = \frac{\Sigma [P(X) \cdot X^2]}{\Sigma [P(X)]} = \frac{p}{1} = p$$

Substituting,

$$\begin{aligned}
\text{Var } (X) &= E(X^2) - [E(X)]^2 \\
&= p - (p)^2 \\
&= p - p^2 \\
&= p(1 - p) \\
&= pq
\end{aligned}$$

The variance of a sum is equal to the sum of the variances. Therefore,

$$\begin{aligned}
\text{Var } (X_1 + X_2) &= \text{Var } (X_1) + \text{Var } (X_2) \\
&= pq + pq \\
&= 2pq
\end{aligned}$$

$$\begin{aligned}
\text{Var } (X_1 + X_2 + X_3) &= \text{Var } (X_1) + \text{Var } (X_2) + \text{Var } (X_3) \\
&= pq + pq + pq \\
&= 3pq
\end{aligned}$$

In general,

$$\begin{aligned}
\text{Var } (X_1 + X_2 + X_3 \cdots X_n) &= \text{Var } (X_1) + \text{Var } (X_2) + \cdots + \text{Var } (X_n) \\
&= npq
\end{aligned}$$

PROBLEMS

Problem 7.1

In a file of 5000 freight bills, 5 percent contain errors. An auditor randomly selects 13 bills from this file. Using the binomial approximation—

1. What is the probability that exactly 3 of the 13 contain errors?
2. What is the probability that none of the 13 contain errors?
3. What is the probability that 3 or more of the 13 freight bills contain errors?
4. What is the probability that no more than 2 of the 13 contain errors?
5. What is the probability that exactly 1 of the 13 contains an error?

Problem 7.2

A firm estimates that 3 percent of its accounts receivable cannot be collected. Of 100 accounts selected at random, what is the probability that 8 or more will subsequently prove to be uncollectable?

Problem 7.3

The binomial probability distribution is related to the special rule of multiplication in the same fashion that the hypergeometric distribution is related to the general rule of multiplication. Discuss.

Problem 7.4

When the manufacturing process for producing wheel bearings is set up correctly, 95 percent of the bearings produced will be of acceptable quality. If a sample of size 50 is selected from a process that has been set up correctly, what is the probability of:
1. Observing no defectives?
2. Observing one defective?
3. Observing two defectives?
4. Observing two or less defectives?
5. Observing five or more defectives?

Problem 7.5

Problem 5.1 was solved using the addition and multiplication rules. Generate the probabilities asked for in Problem 5.1 using the binomial probability distribution. Which of the problems cannot be solved by the binomial? Why?

Problem 7.6

Problem 5.6 was solved using the addition and multiplication rules. Generate the probabilities asked for in Problem 5.6 using the binomial. Which of the problems cannot be solved by the binomial? Why?

Problem 7.7

Problem 5.7 was solved using the addition and multiplication rules. Generate the probabilities asked for in Problem 5.7 using the binomial.

Problem 7.8

The probability of an individual being neurotic is 22 percent, and this characteristic is randomly distributed throughout the population.
1. If you have been assigned to a dormitory with two individuals to a room, what is the probability that your roommate will be neurotic?

2. If you have two roommates, what is the probability that they will both be neurotic?
3. If you have two roommates, what is the probability that all three of you will be neurotic?
4. If you have two roommates, what is the probability that two of the three of you will be neurotic?
5. What is the probability that more than 22 of the 100 students in the dorm will be neurotic?

Problem 7.9

The firm that supplies a transistor component to your firm can only guarantee that 90 percent of the components produced will be acceptable to your manufacturing standards.

1. If 8 of the components are purchased, what is the probability that all 8 will be acceptable?
2. If 9 of the components are purchased, what is the probability that 8 or more will be acceptable?
3. If 10 of the components are purchased, what is the probability that 8 or more will be acceptable?
4. If 11 of the components are purchased, what is the probability that 8 or more will be acceptable?
5. If 12 of the components are purchased, what is the probability that 8 or more will be acceptable?
6. What is the minimum number of components that should be purchased to ensure the receipt of at least 8 acceptable components 90 percent of the time?

Problem 7.10

A budding young actress by the name of Holly Meyer has a part consisting of five lines in the current Silver Masque Production. Unfortunately, Holly has a great deal of difficulty remembering her lines. She is just as likely to forget one line as another, and the likelihood that she will forget any one line is totally unaffected by her degree of success or failure up to that point. As a result of extensive rehearsal, the probability that Holly will forget any one line has been reduced to 10 percent.

1. What is the probability that Holly will forget her first line but get the remaining four lines correct?
2. What is the probability that Holly will remember her first three lines and forget her last two?
3. What is the probability that Holly will forget two of her five lines?
4. What is the probability that Holly will forget all five lines?
5. It is a well-known theatre adage that to wish a performer good luck is to bring bad luck. Thus, the reverse psychology is used, and a performer wishes another good luck by saying "break a leg." Suppose an unknowing back-stage visitor wishes Holly "good luck," and Holly as a result gets so nervous that the probability that she will forget any one line increases to 20 percent. What is the probability now that she will forget all five of her lines?

Problem 7.11

Mark Jellison, a hoopster on the college varsity basketball team, historically makes 75 percent of his foul-shot opportunities. If Mark has 15 foul-shot opportunities in the next contest, what is the probability:
1. That he will make all 15?
2. That he will convert on at least 12?
3. That he will convert on no more than 5?

Problem 7.12

1. What conditions must be satisfied if a binomial probability distribution is to provide an accurate description of the probability of occurrence of the possible outcomes to a sampling experiment?
2. What conditions must be satisfied if a hypergeometric probability distribution is to provide an accurate description of the probability of occurrence of the possible outcomes to a sampling experiment?
3. Under what conditions can the binomial distribution legitimately be used to provide an approximation to the hypergeometric distribution? What is the advantage of such an approximation?

Problem 7.13

As manager of the Shady Dealings Motel, you have established that there is a 40 percent probability that an individual who has made a reservation but has not arrived by 11 p.m. will not show at all.
1. Suppose that it is just past 11 p.m. and you have reservations pending but not met. A person who has not made a reservation comes to the desk and wants a room. If you have decided that you will accept a 20 percent chance of getting "burned" (not having a room when one was reserved), should you rent a room to this person? Show supporting calculations.
2. Suppose that two people who do not have reservations show up after 11 p.m. (with four reservations not met). Given the decision rule specified in part 1, should you rent the rooms to these individuals?

Problem 7.14

Past experience has indicated that about 40 percent of customers' accounts are more than 30 days old. If that is so:
1. What is the probability of getting *exactly* three accounts that are more than 30 days old in a random sample of nine accounts?
2. What is the probability of getting less than three such accounts in a random sample of nine?

Problem 7.15

Given 5000 parts, 4500 of which are acceptable and 500 of which are defective, what is the probability of obtaining three acceptables out of three selected? Selections are made *without* replacement. Justify the approach utilized to answer this question.

Problem 7.16

A common decision procedure utilized to decide whether or not to reject lots is the following: "Inspect a sample of five articles selected at random from a lot of 1000 parts. If none of the five articles are defective, accept the lot; if one or more articles are defective, reject the lot."

1. What is the probability that a lot containing 2 percent defectives will be accepted?
2. What is the probability that a lot containing 6 percent defectives will be accepted?
3. What is the probability that a lot containing 10 percent defectives will be accepted?
4. On the basis of these calculations, do you think that the decision rule offers protection against accepting lots of inferior quality?

Permutations and combinations

Problem 7.17

A parking attendant has a row that contains seven parking spaces. He has seven cars to park. How many different ways can he park the seven cars in the seven spaces?

Problem 7.18

The Brand X Cereal Company manufactures ten different brands of cereal. It has decided to offer these cereals in a six-pack option.

1. Assuming that duplications are not permitted in the six pack, how many different options (different six packs) are available to Brand X?
2. Assuming that one duplication is permitted, how many different options (different six packs) are available to Brand X?

Problem 7.19

Given that there are 40 students in the class and that the instructor has decided to award six A's by drawing names randomly out of a hat, compute the number of different groups of six students to which the A's could be awarded.

Problem 7.20

1. Given 51 entrants in the Miss America Contest, how many different ways could the five finalists be constituted?
2. Given the five finalists, how many different ways could the rankings of fourth runner up, third runner up, second runner up, first runner up, and winner be assigned?

BAYES' RULE REVISITED

8

An illustrative case involving revision of a prior probability distribution on the basis of new information obtained through binomial sampling.

In Chapter 6, the underlying procedure involving the use of Bayes' rule to revise probabilities in the light of new information was developed. In Chapter 7, a familiarity was obtained with the concept of a probability distribution in general and with the binomial probability distribution in particular. In this chapter, the above concepts are combined, and their use in a decision-making context is illustrated.

Consider the following information for illustrative purposes: Brenner Die and Foundry, Inc., a job shop-type operation, produces a wide range of machined parts. A representative example is a worm gear (No. 674-B) for which Brenner, Inc., receives orders periodically over the year. It is Brenner's policy to produce these parts in lots of 1000. Typically, 50 such lots are run per year. The setup required for producing worm gears is quite elaborate. When set up correctly, 97 percent of the gears produced fall within the tolerance limits allowable; the remaining 3 percent are defective. The company ships all parts and accepts returns on defective parts, crediting the customer's account. The company must absorb the freight costs on defective parts. Historical records indicate that this has amounted to an average of $1 per defective part returned. In addition, the company must bear the burden of unrecovered costs (manufacturing, shipping, etc.). Cost records indicate that this amounts to approximately $4 for each part returned.

Two major setup errors can occur:

Setup error type A: When this type of setup error occurs, approximately 8 percent of the parts produced will lie outside the tolerance limits (i.e., the proportion defective of the gears produced will run approximately 8 percent). Setup error type B: When this type of setup error is made, 25 percent of the gears produced will be defective (i.e., a proportion defective of 25 percent).

In the past, Brenner, Inc., has used its own personnel to set up the process for producing worm gears (No. 674-B) and has had moderate success. Historically, 60 percent of the time the process has been set up correctly; 30 percent of the time a type A setup error has been made; and 10 percent of the time a type B setup error has been made.

Brenner, Inc., has been offered an alternative. V. P. Cronin Associates, a firm specializing in performing complicated setups, has offered to send a team of expert mechanics to check the setups prior to any production being run. The team, using sophisticated techniques Brenner cannot duplicate, would determine if any setup errors had been made and rectify them. Should the proportion defective exceed 3 percent after such a check, V. P. Cronin has indicated that Brenner would be reimbursed for any incremental costs incurred. The charge for the service is a fixed $400 per call, whether the team has to reconstruct the setup or not.

Brenner is thus faced with the following decision:

Alternative I: Continue as in the past.
Alternative II: Call in the team of expert mechanics and have the setups checked before any production is run.

Brenner conducted the following analysis:

UNDER ALTERNATIVE I

The freight reimbursement and nonrecoverable out-of-pocket costs the firm incurs due to defective parts varies depending on the percentage of defective parts produced. The proportion defective could take on any of the following three values, depending on the circumstances:

Circumstance	Proportion defective
Correct setup	$p_1 = 0.03$
Setup error type A	$p_2 = 0.08$
Setup error type B	$p_3 = 0.25$

When p_1 is the proportion defective, 3 percent of the parts produced will be defective, or 30 parts $[(0.03)(1000) = 30]$. At $5 per part, the expected freight reimbursement and nonrecoverable out-of-pocket costs would be $150.

When p_2 is the proportion defective, 8 percent of the parts produced will be defective, or 80 parts $[(0.08)(1000) = 80]$. At $5 per part, the expected freight reimbursement and nonrecoverable out-of-pocket costs would be $400.

When p_3 is the proportion defective, 25 percent of the parts produced will be defective, or 250 parts $[(0.25)(1000) = 250]$. At $5 per part, expected freight reimbursement and nonrecoverable out-of-pocket costs would be $1250.

Over a period of a year, 50 such setups would be made. Of the 50 setups, it is expected that 60 percent of the time or on 30 occasions a freight reimbursement and out-of-pocket nonrecoverable cost of $150 will be incurred; 30 percent of the time or on 15 occasions a freight reimbursement and nonrecoverable out-of-pocket cost of $400 will be incurred; and 10 percent of the time or on 5 occasions a freight reimbursement and nonrecoverable out-of-pocket cost of $1250 will be incurred.

Over the 50 setups, the total expected freight reimbursement and nonrecoverable out-of-pocket costs would total to $16,750 and average out to $335 per setup (see Table 8.1).

Table 8.1 Computation of total and average freight reimbursement and nonrecoverable out-of-pocket costs

Freight reimbursement and nonrecoverable out-of-pocket costs ($)	Expected frequency per year	Computation of total expected per year
150	30	4500
400	15	6000
1250	5	6250
	50	16750

Total freight reimbursement and nonrecoverable out-of-pocket costs	$= \$16,750$
Average freight reimbursement and nonrecoverable out-of-pocket costs per setup	$= \dfrac{\$16,750}{50} = \$335$

UNDER ALTERNATIVE II

Bringing in the team of expert mechanics guarantees a correct setup, thereby ensuring a proportion defective of 3 percent and a freight reimbursement and nonrecoverable out-of-pocket cost on defective items of $150 per setup. The reduction in expected freight reimbursement and nonrecoverable out-of-pocket costs of $185 (from $335 to $150) is not enough, however, to offset the $400 fee incurred by calling in V. P. Cronin Associates. Clearly, as supported by the computations in Table 8.2, Brenner should select alternative I: Continue as is.

Brenner rejected V. P. Cronin Associates' proposal.

V. P. Cronin Associates then presented the following alternative proposition. Instead of calling in the team of expert mechanics prior to any production being run, under the new proposal Brenner, Inc., would produce a sample of ten items, evaluate these, and on the basis of the sample information decide whether or not to let the process run or to call in the team from V. P. Cronin Associates. Intuitively, the new proposal sounded better to Brenner than the original proposal, but he was confused as to how to use the new information once the sample had been selected and evaluated. V. P. Cronin said she would return the next day with a diagram

Table 8.2 Comparative cost of pursuing Alternative I and Alternative II

Alternative I: Continue as is		Alternative II: Call in V. P. Cronin Assoc.	
Average freight reimbursement and nonrecoverable out-of-pocket costs	$335	Average freight reimbursement and nonrecoverable out-of-pocket costs	$150
		Cost incurred by calling in V. P. Cronin Associates	$400
	$335		$550

that would clearly explain the way the new information could be used in deciding whether or not to call in V. P. Cronin Associates.

V. P. Cronin was aware of how to revise probabilities in the light of new information (a management consultant acquaintance had introduced her to the technique in a similar situation some months earlier). As a starting point, she developed Table 8.3 for the situation where a sample of size 10 drawn from the process contained 4 defectives.

Table 8.3

K = New information = 6 acceptable observed out of a sample of size 10

States of the universe	Priors	Conditionals	Joint probabilities	Revised or posterior probabilities
$p_1 = 0.03$	0.60	0.0001		
$p_2 = 0.08$	0.30	0.0052		
$p_3 = 0.25$	0.10	0.1460		
	1.00			

Realizing that she would have to explain the underlying reasoning to Brenner, V. P. Cronin stopped at this point and reflected on her entries in the conditional column.

States of the universe	Conditionals
$p_1 = 0.03$	0.0001
$p_2 = 0.08$	0.0052
$p_3 = 0.25$	0.1460

"First of all, I know that the sampling procedure is a Bernoulli process—10 trials, on each trial the part observed is either defective or acceptable, and the probability of selecting an acceptable part remains constant from trial to trial (the sample of size 10 can be considered as having been selected out of the universe of infinite parts that the process *could* produce once set up.) This means that I can find the probability of observing the particular sample outcome 6 acceptable out of 10 observed by going to a binomial table and finding the probability that corresponds to $n = 10$, r (the number of successes) $= 6$, and p (the probability of a success on any one trial) equal to 0.97, 0.92, or 0.75 depending on the state of the universe. The probabilities 0.97, 0.92, and 0.75 presume that success is defined as the part observed being acceptable."

Ironically, V. P. Cronin thought to herself, if she were to look at the problem from her own point of view, defective parts observed would be successes, for that would increase the likelihood that Brenner, Inc., would have to call in the team of expert mechanics. Looked at from this viewpoint, r (the number of successes) would be 4 instead of 6, and p (the probability of success on any one trial) would become 0.03, 0.08, and 0.25, respectively. Looking these up in the table, V. P. Cronin arrived at exactly the same set of probabilities. Thus, it is apparent that it makes no difference which outcome on a trial is labeled as a success. The probability of 4 defectives out of 10 and the probability of 6 acceptable out of 10 are alternative ways of describing the same outcome (Figure 8.1).

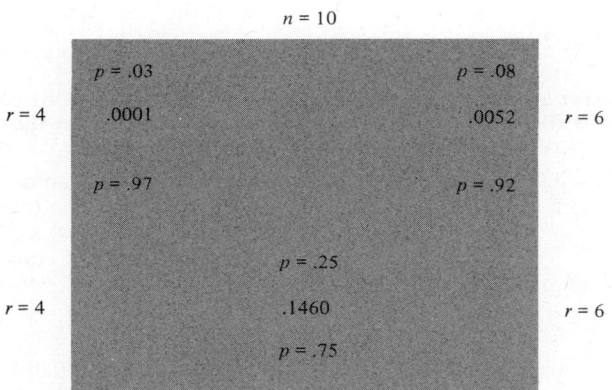

Figure 8.1 Reproduction of a segment of the binomial table associated with 10 trials.

As for interpretation of the entries in the conditional probability column, the first conditional (0.0001), associated with $p = 0.03$, can be interpreted as follows: If 10,000 samples of size 10 were to be drawn out of a universe containing 3 percent defectives, the result "6 acceptable out of 10" would be expected to be observed only once. Similarly, the second conditional (0.0052) states that if 10,000 samples of size 10 were to be selected out of a universe containing 8 percent defectives, the sample result "6 acceptable out of 10" would be expected to be observed on only 52 occasions, and the third conditional (0.1460) infers that of every 10,000 samples of size 10 drawn out of a universe with 25 percent defective, 1,460 of the samples would be expected to contain "6 acceptable out of 10."

Returning to the table, V. P. Cronin then computed the joint probability column (Table 8.4) and paused to reflect again.

Interpreting the joint probability column, V. P. Cronin recalled that the sum of the joint probabilities equaled the probability of event "k" or, in this instance, out of every 100,000 times a setup was made and subsequently a sample of size 10 drawn, the sample result "6 acceptable out of 10" would be expected to be observed

Table 8.4

K = new information = 6 acceptable observed out of a sample of size 10

States of the universe	Priors	Conditionals	Joint probabilities	Revised or posterior probabilities
$p_1 = 0.03$	0.60	0.0001	0.00006	
$p_2 = 0.08$	0.30	0.0052	0.00156	
$p_3 = 0.25$	0.10	0.1460	0.01460	
	1.00		0.01622	

1,622 times. Of the 1,622 times, 6 would have been samples selected out of the 60,000 times the process was operating under p_1, 156 would have been selected out of the 30,000 times the process was operating under p_2, and 1460 would have been selected out of the 10,000 times p_3 was the relevant state of the universe (i.e., when 25 percent of the universe was defective).

V. P. Cronin then generated the posterior probabilities by dividing each of the entries in the joint probability column by the sum of the joint probabilities. The result is reproduced as Table 8.5(a).

Table 8.5(a) Computation of the revised or posterior probabilities of the various states of the universe given the additional information that a sample of size 10 revealed 6 acceptable and 4 defective parts

K = new information = 6 acceptable observed out of a sample of size 10.

States of the universe	Priors	Conditionals	Joint probabilities	Revised or posterior probabilities
$p_1 = 0.03$	0.60	0.0001	0.00006	$\dfrac{0.00006}{0.01622} = 0.004$
$p_2 = 0.08$	0.30	0.0052	0.00156	$\dfrac{0.00156}{0.01622} = 0.096$
$p_3 = 0.25$	0.10	0.1460	0.01460	$\dfrac{0.01460}{0.01622} = 0.900$
	1.00		0.01622	1.000

Interpreting the posterior probabilities[1]: If every time the sample result "6 acceptable out of 10" were observed, the process was allowed to continue to run, 4 out of every 1000 times it would be operating under proportion defective one ($p_1 = $ 3 percent defective), 96 out of every 1000 times it would be operating under proportion defective two ($p_2 = 8$ percent defective), and 900 out of every 1000 times it would be operating under proportion defective three ($p_3 = 25$ percent defective).

If Brenner continued to operate with the process as is over many such decisions (after observing 6 acceptable out of 10), the expected value of freight reimbursement

---

[1] The interpretation presented above presumes, of course, that the prior probabilities are steady-state probabilities of 0.60, 0.30, and 0.10.

plus nonrecoverable out-of-pocket costs would be equal to $1164 per set up. The supporting computations are presented in Table 8.5(b). The *expected value* of a variable is obtained by multiplying each value of the random variable by the probability of occurrence of that value. The expected value of the variable is equal to the sum of the products. The expected value is similar in computation to a weighted arithmetic mean. The weights are always the probabilities of occurrence of the values of the variable and sum to one. If the experiment were to be performed many times, the mean value of the variable observed would be expected to approach the expected value of the variable.

Table 8.5(b) Computation of expected freight reimbursement plus nonrecoverable out-of-pocket costs associated with allowing the process to continue to run after observing 6 acceptable, 4 defectives out of a sample of size 10

State of the universe	Conditional out-of-pocket costs	Probability of occurrence	Expected value
$p_1 = 0.03$	150	0.004	$(150)(0.004) = \quad 0.60$
$p_2 = 0.08$	400	0.096	$(400)(0.096) = \quad 38.40$
$p_3 = 0.25$	1250	0.900	$(1250)(0.900) = 1125$
		1.000	1164

On the other hand, if instead of letting the process run, Brenner, Inc., called in V. P. Cronin Associates (at a cost of $400), a freight reimbursement and non-recoverable out-of-pocket cost of $150 would be guaranteed. The comparative costs of pursuing alternative I and alternative II given the new information are summarized in Table 8.5(c). Clearly, whenever faced with the information "6 acceptable out of 10," Brenner, Inc., should call in V. P. Cronin Associates. Such an action over time would result in expected savings of $614 per setup.

Table 8.5(c) Comparative cost of pursuing alternative I and alternative II given the new information 6 acceptable, 4 defective observed out of a sample of size 10

After observing 6 acceptable out of 10

Alternative I: Continue to run		Alternative II: Call in V. P. Cronin Assoc.	
Average freight reimbursement and nonrecoverable out-of-pocket costs	$1164	Average freight reimbursement and nonrecoverable out-of-pocket costs	$150
		Cost incurred by calling in V. P. Cronin Assoc.	$400
	$1164		$550

V. P. Cronin developed tables for other sample outcomes as well. Tables 8.6(a), 8.6(b), and 8.6(c) contain the underlying computations associated with

Table 8.6(a) Computation of the revised or posterior probabilities of the various states of the universe given the additional information that a sample of size 10 revealed 7 acceptable and 3 defectives

K = New information = 7 acceptable observed out of a sample of size 10

States of the universe	Priors	Conditionals	Joint probabilities	Posterior or revised probabilities
$p_1 = 0.03$	0.60	0.0026	0.00156	$\dfrac{0.00156}{0.03688} = 0.042$
$p_2 = 0.08$	0.30	0.0343	0.01029	$\dfrac{0.01029}{0.03688} = 0.279$
$p_3 = 0.25$	0.10	0.2503	0.02503	$\dfrac{0.02503}{0.03688} = 0.679$
	1.00		0.03688	1.000

Table 8.6(b) Computation of expected freight reimbursement plus nonrecoverable out-of-pocket costs associated with allowing the process to continue to run after observing 7 acceptable, 3 defective out of a sample of size 10

States of the universe	Conditional freight reimbursement and non recoverable out-of-pocket costs	Probability of occurrence	Expected value	
$p_1 = 0.03$	150	0.042	(0.042)(150) =	6.30
$p_2 = 0.08$	400	0.279	(0.279)(400) =	111.60
$p_3 = 0.25$	1250	0.679	(0.679)(1250) =	848.75
		1.000		966.65

Table 8.6(c) Comparative cost of pursuing alternative I and alternative II given the new information 7 acceptable, 3 defective observed out of a sample of size 10

After observing 7 acceptable out of 10

Alternative I: Continue to run		Alternative II: Call in V. P. Cronin Assoc.	
Average freight reimbursement and nonrecoverable out-of-pocket costs	$966.65	Average freight reimbursement and nonrecoverable out-of-pocket costs	$150
		Cost incurred by calling in V. P. Cronin Assoc.	$400
	$966.65		$550

evaluating the appropriate action to take when 7 acceptable parts are observed out of a sample of size 10. Clearly, whenever faced with the information "7 acceptable out of 10," Brenner, Inc., should call in V. P. Cronin Associates. Such an action over time would result in expected savings of $416.65 per setup.

Note that although calling in V. P. Cronin Associates is still the better alternative, the margin of benefit is not as great for "7 acceptable out of 10 observed" as it was for "6 acceptable out of 10" observed.

Tables 8.7(a), 8.7(b), and 8.7(c) contain the underlying computations associated with determining the appropriate action to take when 8 acceptable parts are observed out of a sample of size 10. Once again, the optimal action is to call in

Table 8.7(a) Computation of the revised or posterior probabilities of the various states of the universe given the additional information that a sample of size 10 revealed 8 acceptable and 2 defectives

K = New information = 8 acceptable out of a sample of size 10

States of the universe	Priors	Conditionals	Joint probabilities	Posterior or revised probabilities
$p_1 = 0.03$	0.60	0.0317	0.01902	$\dfrac{0.01902}{0.09152} = 0.208$
$p_2 = 0.08$	0.30	0.1478	0.04434	$\dfrac{0.04434}{0.09152} = 0.484$
$p_3 = 0.25$	0.10	0.2816	0.02816	$\dfrac{0.02816}{0.09152} = 0.308$
	1.00		0.09152	1.000

Table 8.7(b) Computation of expected freight reimbursement plus nonrecoverable out-of-pocket costs associated with allowing the process to continue to run after observing 8 acceptable, 2 defective out of a sample of size 10

States of the universe	Conditional freight reimbursement and nonrecoverable out-of-pocket costs	Probability of occurrence	Expected value
$p_1 = 0.03$	150	0.208	$(0.208)(150) = 31.20$
$p_2 = 0.08$	400	0.484	$(0.484)(400) = 193.60$
$p_3 = 0.25$	1250	0.308	$(0.308)(1250) = 385.00$
		1.000	609.80

Table 8.7(c) Comparative cost of pursuing alternative I and alternative II given the new information 8 acceptable, 2 defective observed out of a sample of size 10

After observing 8 acceptable out of 10

Alternative I: Continue to run		Alternative II: Call in V. P. Cronin Assoc.	
Average freight reimbursement and nonrecoverable out-of-pocket costs	$609.80	Average freight reimbursement and nonrecoverable out-of-pocket costs	$150
		Cost incurred by calling in V. P. Cronin Assoc.	$400
	$609.80		$550

V. P. Cronin Associates. Such an action whenever the sample result "8 acceptable out of 10 observed" is obtained will over time result in expected savings of $59.80. Note the further shrinkage in the margin of benefit associated with the optimal decision.

Tables 8.8(a), 8.8(b), and 8.8(c) contain the underlying computations associated

Table 8.8(a) Computation of the revised or posterior probabilities of the various states of the universe given the additional information that a sample of size 10 revealed 9 acceptable, 1 defective

States of the universe	Priors	Conditionals	Joint probabilities	Posterior or revised probabilities
$p_1 = 0.03$	0.60	0.2281	0.13686	$\dfrac{0.13686}{0.26894} = 0.509$
$p_2 = 0.08$	0.30	0.3777	0.11331	$\dfrac{0.11331}{0.26894} = 0.421$
$p_3 = 0.25$	0.10	0.1877	0.01877	$\dfrac{0.01877}{0.26894} = 0.070$
	1.00		0.26894	1.000

Table 8.8(b) Computation of expected freight reimbursement plus nonrecoverable out-of-pocket costs associated with allowing the process to continue to run after observing 9 acceptable, 1 defective out of a sample of size 10

State of the universe	Conditional freight reimbursement and nonrecoverable out-of-pocket costs	Probability of occurrence	Expected value	
$p_1 = 0.03$	150	0.509	(0.509)(150) =	76.35
$p_2 = 0.08$	400	0.421	(0.421)(400) =	168.40
$p_3 = 0.25$	1250	0.070	(0.070)(1250) =	87.50
		1.000		332.25

Table 8.8(c) Comparative cost of pursuing alternative I and alternative II given the new information 9 acceptable, 1 defective observed out of a sample of size 10

After observing 9 acceptable out of 10

Alternative I: Continue to run		Alternative II: Call in V. P. Cronin Assoc.	
Average freight reimbursement and nonrecoverable out-of-pocket costs	$332.25	Average freight reimbursement and nonrecoverable out-of-pocket costs	$150
		Cost incurred in calling in V. P. Cronin Assoc.	$400
	$332.25		$550

with determining the appropriate action to take when 9 acceptable parts are observed out of a sample of size 10. Note that the optimal action has switched to alternative I. Whenever faced with the information "9 acceptable out of 10," Brenner, Inc., should *not* call in V. P. Cronin Associates but rather should run the process as is.

The results of the preceding computations are summarized in Table 8.9.

Table 8.9 Optimal action alternatives associated with varying number of acceptable parts observed out of a sample of size 10

| Number of acceptable parts observed | Optimal action I. Continue as is II: Call in V. P. Cronin Assoc. | |
	Computed directly	Inferred from the turning point
0		II
1		II
2		II
3		II
4		II
5		II
6	II	II
7	II	II
8	II	II
9	I	I
10		I

The pattern is clear: The point at which the optimal action changes from II (call in V. P. Cronin Associates) to I (continue to operate) is when the number of acceptable parts observed is 9 or more. Once the turning point is determined, the other entries can be made directly without bothering with the underlying computations.

After taking a sample of size 10 and determining how many of the 10 were acceptable, a clerk could simply consult the table and institute the correct procedure:

   I Continue as is and run the process.
   II Call in the team of expert mechanics from V. P. Cronin Associates.

A more efficient way of summarizing the results of the above analysis is to use a decision rule of the form (n, c), where n refers to the sample size and c is the critical number or the number of successes at which the optimal action changes from one alternative to the other. The optimal decision rule for sample size 10 for Brenner, Inc., is $(n, c) = (10, 9)$.

The next day, V. P. Cronin arrived at Brenner, Inc., armed with the diagram reproduced as Figure 8.2, which she used to support her proposal in the following manner: Suppose you started out at the origin (before sampling) 100,000 different times—this would be equivalent to 100,000 different setups. On the basis of past experience (as reflected by your priors), you would expect to proceed down the path labeled p_1 60,000 times, down the path labeled p_2 30,000 times, and down the path designated p_3 on 10,000 occasions.

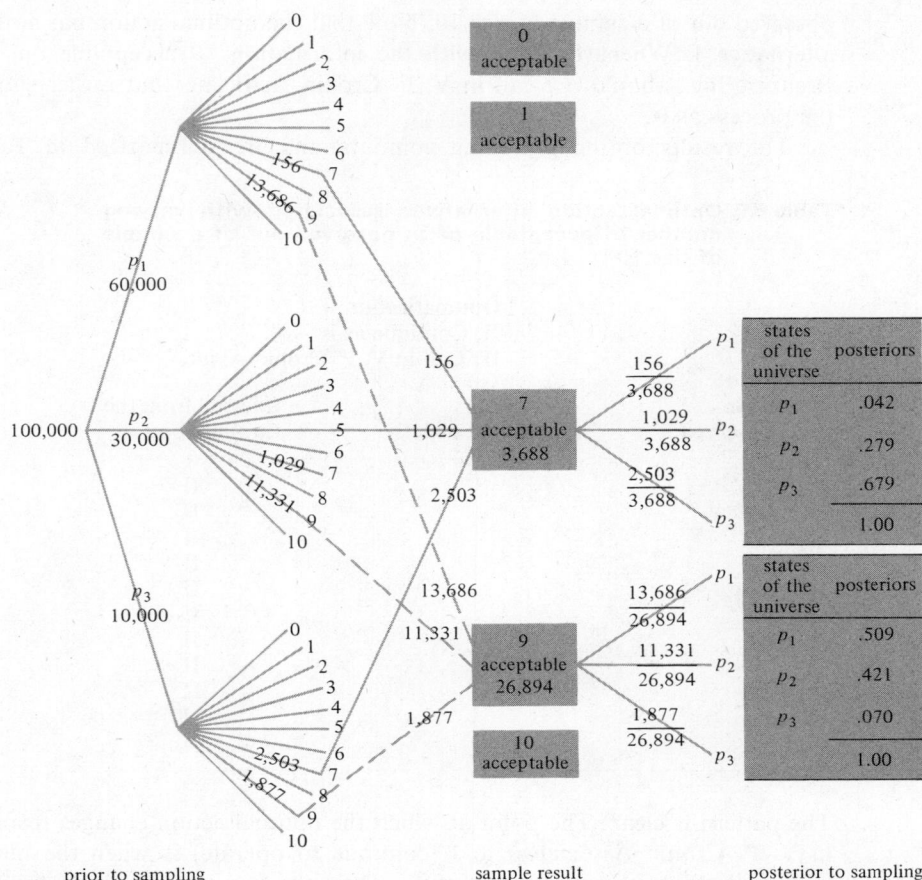

Figure 8.2

Now let us suppose that a sample of size 10 is drawn from each of the 100,000 setups. There are $n + 1$ possible sample outcomes or 11 possible sample outcomes —none acceptable, one acceptable, two acceptable,..., ten acceptable. The sampling procedure is a Bernoulli process; therefore, the binomial probability distribution is used to calculate the probability of occurrence of each of the 11 outcomes.

To illustrate: Of the 60,000 occasions on which the sample would be expected to be selected out of a universe consisting of 3 percent defectives and 97 percent acceptables, the sample result "7 acceptable out of 10" would be expected to occur 0.26 percent of the time or on 156 occasions.

	Conditional probability of the given sample result	Number of occasions
$n = 10$ $r = 7$ $p = 0.97$	Prob $\begin{pmatrix} 7 \text{ acceptable} \\ \text{out of } 10 \end{pmatrix} = 0.0026$	$(0.0026)\,(60000) = 156$

A similar line of reasoning generates the statement that the sample result "9 acceptable out of 10" would be expected to be observed on 13,686 of the 60,000 occasions the sample would be expected to be drawn from a universe consisting of 3 percent defective parts.

	Conditional probability of the given sample result	Number of occasions
$n = 10$ $r = 9$ $p = 0.97$	Prob $\left(\begin{array}{c}\text{9 acceptable}\\\text{out of 10}\end{array}\right) = 0.2281$	$(0.2281)(60000) = 13,686$

Similarly, of the 30,000 occasions on which the sample would be expected to be selected out of a universe containing 8 percent defectives and 92 percent acceptables, the sample result "7 acceptable out of 10" would be expected to be observed 1,029 times; the sample result "9 acceptable out of 10" would be expected to be observed 11,331 times; and so on.

Of the 10,000 occasions on which the sample would be expected to be selected out of a universe containing 25 percent defectives and 75 percent acceptables, the sample result "7 acceptable out of 10" would be expected to be observed 2503 times; the sample result "9 acceptable out of 10" would be expected to be observed 1877 times; and so on.

Referring to Figure 8.2, out of 100,000 times Brenner starts out at the origin, it would be expected that he would end up in the box representing the sample result "7 acceptable out of 10" 3688 times. 156 of these would be expected to have been drawn from a process containing 3 percent defectives, 1029 from a process containing 8 percent defectives, and 2503 from a process containing 25 percent defectives. Thus, if Brenner were to continue to operate after observing "7 defectives out of 10," it would be expected that out of every 3688 occasions the decision to continue to operate was made, on 156 occasions (or 4.2 percent of the time), 3 percent of the parts produced would be defective; on 1029 occasions (or 27.9 percent of the time), 8 percent of the parts produced would be defective; and on 2503 occasions (or 67.9 percent of the time), 25 percent of the parts produced would be defective.

States of the universe	Posterior probabilities
$p_1 = 0.03$	0.042
$p_2 = 0.08$	0.279
$p_3 = 0.25$	0.679
	1.000

Using the revised probabilities and the conditional cost associated with each state of the universe, Brenner could then reach a determination as to whether or not to call in V. P. Cronin Associates.

Brenner completely grasped the rationale and accepted V. P. Cronin's second proposition, but with one reservation. Why a sample of size 10 instead of, say, a sample of size 15 or 20? V. P. Cronin admitted she had just made an arbitrary choice—sort of woman's intuition. She promised to check with the consultant from whom she had picked up the technique of revising probabilities in the light of new information to determine if a technique was available that could be used to determine the optimal sample size to select. The consultant replied that indeed there was and agreed to meet with them at a later date to explain the relevant techniques. The ensuing discussion, concerned with the determination of optimal sample size, is presented in Chapter 9.

PROBLEMS

Problem 8.1

A Gordon machine used by your firm is set up correctly 80 percent of the time. If the machine is set up correctly 90 percent of the parts produced will be acceptable. If it is set up incorrectly, 65 percent of the parts produced will be acceptable. You have just evaluated a sample from a pilot production run and found 15 out of the 20 items sampled to be acceptable. Your decision rule is to stop production and check the settings on the machine if the probability of the machine being set up incorrectly exceeds 40 percent.

1. Given the sample result, what is the probability that the machine is *set up correctly*?
2. Given the decision rule, what action should the firm take?

Problem 8.2

When a particular machine is set up correctly, 3 percent of the parts produced will be defective. When the machine is set up incorrectly, 9 percent of the parts produced will be defective. It is known that correct setups occur 80 percent of the time. A sample of size 20 has just been drawn, and 5 of the 20 were defective. What is the probability that the process is set up correctly?

Problem 8.3

Ig Knight, quality control manager of Asbestos Gloves, Inc., examined the results of the latest test. It disclosed 38 acceptable of 50 items tested. The product is produced by two different machines. The tag indicating whether it was produced by machine A or machine B has become lost in transit. Knight feels that the batch was just as likely to have come from machine A, which produces 82 percent acceptable, as from machine B, which produces 73 percent acceptable.

1. What is the probability of observing "38 acceptable of 50 items tested" if the batch tested was produced by machine A?
2. What is the probability of observing "38 acceptable of 50 items tested" if the batch tested was produced by machine B?
3. What probability (given the sample result) should Knight assign to "the batch was produced by machine A"?

Problem 8.4

Consider the following two urns. Urn I contains two red marbles and eight green marbles. Urn II contains seven red marbles and three green marbles. Initially, there is an 80 percent chance that urn I will be selected. Once an urn is selected, all subsequent draws will be from the same urn, and after each draw the marble removed will be replaced.

1. Without any information as to the results of draws from the selected urn, what is the probability that urn II was selected?
2. Given the information that six red marbles were obtained in ten draws, revise your estimates of the probabilities of urn I and urn II being selected.

Problem 8.5

Consider the following three urns:

Urn I	Urn II	Urn III
8 red	4 red	2 red
2 green	6 green	8 green
10 marbles	10 marbles	10 marbles

A die is to be rolled:

If a 1 turns up, select marbles from urn I
If a 2 or a 3 turns up, select marbles from urn II
If a 4, 5, or 6 turns up, select marbles from urn III

1. What are $P(\text{I})$, $P(\text{II})$, and $P(\text{III})$?
2. Given the information that three red marbles were selected in five draws (with replacement), what is the probability that the urn selected was urn I? Urn II? Urn III?
3. What would the answers to part 2 have been if the marbles were selected without replacement? (Note: the hypergeometric distribution must be used to generate the conditionals.)

Problem 8.6

Recall Problem 6.2. A buddy of yours has recently left for Australia, leaving in your care two black books (A and B). He swears to you that in book A, 90 percent of the girls are fabulous, while book B contains only 30 percent fabulous girls. The books are identical and not marked A or B but strictly contain a list of names. (He also mentioned that the names are randomly good and bad). You select one book and start to date.

1. What prior probability would you assign to the selected book being book A?
2. What revised probability would you assign to the selected book being book A given the information that three of the first five girls dated turned out to be fabulous?
3. Revise the probability assigned in part 2 on the basis of the additional information that two of the next three girls dated are terrible.
4. Instead of the two-stage revision reflected in parts 2 and 3, make a one-stage revision of the probabilities established in part 1 on the basis of the information "four fabulous of eight girls dated."
5. Justify intuitively why parts 3 and 4 generate the same revised probabilities.

Problem 8.7

When Johnson, Inc.'s manufacturing process is set up correctly, 5 percent of the parts produced will be defective. When the process is set up incorrectly, 12 percent of the parts produced will be defective. Correct setups occur 90 percent of the time. A sample of size 5 has just been drawn, and three of the five were defective.

1. Given the above information, what is the probability that the process is set up correctly?
2. Instead of treating the new information as a combination, that is, "3 out of 5," treat it as a permutation, namely, A_1, D_2, D_3, A_4, D_5.
    a. What is the probability of observing the particular permutation referred to above: first part acceptable, second defective, third defective, fourth acceptable, and fifth defective
        (i) Given a correct setup
        (ii) Given an incorrect setup
    b. Revise the original prior probabilities, treating the new information as a permutation instead of a combination.
3. What is the relationship between the *formulas* for the probability of the combination "3 defective out of 5" and the probability of observing the particular permutation "A_1, D_2, D_3, A_4, D_5"?
4. The posterior probabilities generated in part 1 and in part 2(b) are identical, Drawing on the relationship in part 3, explain why this is so. In general. would the revised probabilities always be the same whether the new information is treated as a permutation or a combination? Justify your answer.

Problem 8.8

The marketing manager has assessed an 80 percent probability of a new product being successful and a 20 percent probability of its being a failure. A market test is set up. Seventy percent of the individuals tested typically respond positively when the product later turns out to be successful. Twenty-five percent typically respond positively when the product later turns out to be a failure. The test responses for 100 individuals disclosed 68 positive responses.

1. What is the probability of observing the test result "68 positive responses out of 100" if the product is ultimately destined to be a failure?
2. What is the probability of observing the test result "68 positive responses out of 100" if the product is ultimately destined to be a success?
3. What probability should the marketing manager assign to the product ultimately being a success given the test result that "68 respondents of the 100 reacted positively"?

Problem 8.9

Given:

$p_1 = 0.03$ is to be interpreted as "the state of the universe is that 3 percent of all parts produced will be defective"

$p_2 = 0.08$ is to be interpreted as "the state of the universe is that 8 percent of all parts produced will be defective"

$p_3 = 0.25$ is to be interpreted as "the state of the universe is that 25 percent of all parts produced will be defective"

and the computations reproduced in Table 8.10.

Table 8.10 Computation of the revised or posterior probabilities of the various states of the universe given the additional information that a sample of size 10 revealed 6 acceptable and 4 defective parts

K = New information = 6 acceptable observed out of a sample of size 10

States of the universe	Priors	Conditionals	Joint	Posteriors
$p_1 = 0.03$	0.60	0.0001	0.00006	$\dfrac{0.00006}{0.01622} = 0.004$
$p_2 = 0.08$	0.30	0.0052	0.00156	$\dfrac{0.00156}{0.01622} = 0.096$
$p_3 = 0.25$	0.10	0.1460	0.01460	$\dfrac{0.01460}{0.01622} = 0.900$
	1.00		0.01622	1.000

1. Interpret or explain the meaning of the value 0.0001 entered in the conditional probability column.
2. Interpret or explain the meaning of the value 0.00006 entered in the joint probability column.
3. Interpret or explain the meaning of the value 0.01622 entered as the sum of the joint probability column.
4. Interpret or explain the meaning of the value 0.004 entered in the posterior probability column.

Problem 8.10

The following questions refer to the illustrative example in Chapter 8 (Brenner Die and Foundary, Inc.).

1. Given the decision rule (10, 9), under what conditions would Brenner call in V. P. Cronin Associates?
2. a. Given that the true state of the universe is a 3 percent defective rate, what is the probability that application of the decision rule (10, 9) will result in calling in V. P. Cronin Associates?
   b. Given that the true state of the universe is an 8 percent defective rate, what is the probability that application of the decision rule (10, 9) will result in calling in V. P. Cronin Associates?
   c. Given that the true state of the universe is a 25 percent defective rate, what is the probability that application of the decision rule (10, 9) will result in calling in V. P. Cronin Associates?
3. a. Determine the probability of the joint event "a 3 percent defective rate *and* V. P. Cronin Associates is called in."
   b. Determine the probability of the joint event "an 8 percent defective rate *and* V. P. Cronin Associates is called in."
   c. Determine the probability of the joint event "a 25 percent defective rate *and* V. P. Cronin Associates is called in."
4. Determine the percentage of all setups on which V. P. Cronin Associates will subsequently be called in.

5. Given that V. P. Cronin Associates is called in:
   a. What is the probability that the setup is correct?
   b. What is the probability that a type A setup error has been committed?
   c. What is the probability that a type B setup error has been committed?
6. V. P. Cronin estimates her cost structure as follows:
   a. Cost to check out a correct setup, $50.
   b. Cost to detect a type A setup error, break down the process, and set it up correctly, $325.
   c. Cost to detect a type B setup error, break down the process, and set it up correctly, $450.

   What will it cost V. P. Cronin on the average per service call?
7. V. P. Cronin is unwilling to consider the arrangement unless she can average $80 over and above cost per setup. What implication does this have with respect to the $400 fee associated with the original offer?

DETERMINATION OF AN OPTIMAL DECISION RULE— BINOMIAL SAMPLING: A BAYESIAN APPROACH

9

Returning to the scene of Brenner Die and Foundry, Inc. (Chapter 8), we find V. P. Cronin and Steve Brenner awaiting the arrival of Ken Moore, V. P. Cronin's consultant acquaintance, and his presentation of a technique for determination of optimal sample size. A few days previous, Ken had dropped by to gather data for the presentation. Ken's first order of business consisted of summarizing this information in the conditional payoff table presented as Table 9.1. A column is assigned to each action alternative and a row established for each of the three possible states of the universe. In each of the six cells comprising the body of the payoff table, Ken entered the conditional dollar loss that would be incurred by Brenner Die and Foundry given the selection of the action alternative noted at the head of the column and the subsequent materialization of the state of nature indicated by the row.

Table 9.1

States of the universe	Action 1: Continue as is	Action 2: Call in V. P. Cronin Assoc.
$p_1 = 0.03$ $p_2 = 0.08$ $p_3 = 0.25$		

To illustrate: Given the selection of action alternative 1 (continue operating), the resultant dollar loss is described by the following relationship:

$$A_{i1} = (\text{XLOTSZ})(\text{PRODEF}_i)(\text{XOPCST})$$

where

A_{i1} refers to the dollar loss associated with the selection of action 1 and the occurrence of the ith state of nature. A_{i1} has a double subscript. In conformance with custom, the first subscript identifies the row association in the payoff table and the second subscript identifies the column.

XLOTSZ refers to the lot size. Worm gears (No. 674-B) are produced in lots of size 1000.

PRODEF$_i$ refers to the proportion of defective parts expected when production takes place under the conditions associated with the ith state of nature. From previously gathered data, the following conditions are known to exist:

State of nature	Proportion defective
1. Correct setup	0.03
2. Type A setup error	0.08
3. Type B setup error	0.25

XOPCST refers to the expected freight reimbursement and nonrecoverable out-of-pocket costs incurred with the production of a defective part. Records generated a figure of $5.

Substituting, Ken obtained:

$$A_{11} = (1000)(0.03)(\$5) = \$150$$

$$A_{21} = (1000)(0.08)(\$5) = \$400$$

$$A_{31} = (1000)(0.25)(\$5) = \$1250$$

By comparison, if Brenner Die and Foundry elected to pursue action 2 (Call in V. P. Cronin Associates), the defective rate would be guaranteed at 3 percent, with the associated dollar loss of $150. To this figure must be added V. P. Cronin's charge. Letting CNSTCST represent the consultant's cost, the dollar loss associated with selection of action 2 is described by the following relationship:

$$A_{i2} = (\text{XLOTSZ})(\text{PRODEF}_1)(\text{XOPCST}) + \text{CNSTCST}$$

Substituting:

$$A_{i2} = (1000)(0.03)(\$5) + \$400$$

$$= \$550$$

Thus, the total dollar loss associated with the selection of action 2 is equal to $550 regardless of the original state of the universe.

Entering the conditional dollar losses, the payoff table appears as shown in Table 9.2.

Table 9.2 Conditional dollar loss payoff table

States of nature	Action 1: Continue as is	Action 2: Call in V. P. Cronin Assoc.
$p_1 = 0.03$	150	550
$p_2 = 0.08$	400	550
$p_3 = 0.25$	1250	550

Given the existence of a particular state of nature, an optimum action can be identified. In this case, where the entries represent dollar losses, the optimum action associated with each state of nature is the action associated with the minimum dollar entry in the associated row of the payoff table. Opportunity loss refers to the profit forgone (or additional cost incurred) as a result of having selected other than the optimal action given a particular state of nature. The conditional opportunity loss associated with each cell of the payoff table can be generated by determining the difference between each of the payoffs in a row and the payoff associated with the optimum action in that row. Performing this operation, Ken developed the conditional opportunity loss payoff table presented as Table 9.3.

Table 9.3 Conditional opportunity loss payoff table

States of nature	Action 1: Continue as is	Action 2: Call in V. P. Cronin Assoc.
$p_1 = 0.03$	0	400
$p_2 = 0.08$	0	150
$p_3 = 0.25$	700	0

Ken then pointed out that the *expected* opportunity loss associated with each action alternative can be generated by the following relationship:

$$\text{EXOL}_j = \sum_{i=1}^{n} [(\text{PRIOR}_i)(\text{OL}_{ij})]$$

where

EXOL_j refers to the expected opportunity loss associated with the jth action alternative

PRIOR_i refers to the probability of occurrence of the ith state of nature

OL_{ij} refers to the conditional opportunity loss associated with the selection of the jth action alternative and the occurrence of the ith state of nature

Historically, a defective rate of 0.03 has been observed 60 percent of the time; a defective rate of 0.08, 30 percent of the time; and a defective rate of 0.25, 10 percent of the time. Substituting, Ken obtained the expected opportunity loss associated with each of the action alternatives as follows:

$$EXOL_1 = (PRIOR_1)(OL_{11}) + (PRIOR_2)(OL_{21}) + (PRIOR_3)(OL_{31})$$
$$= (0.6)(0) + (0.3)(0) + (0.1)(700)$$
$$= \$70$$

$$EXOL_2 = (PRIOR_1)(OL_{12}) + (PRIOR_2)(OL_{22}) + (PRIOR_3)(OL_{32})$$
$$= (0.6)(400) + (0.3)(150) + (0.1)(0)$$
$$= \$285$$

The optimum action to select as a consistent course of action is the act associated with the *minimum* expected opportunity loss. Thus, action 1 would be the optimal action alternative under conditions of uncertainty. The minimum expected opportunity loss of $70 represents the average profit forgone as a result of consistently selecting action 1 under conditions of uncertainty. Ken pointed out that if Steve was in a position where he could determine which state of nature would materialize prior to selection of his course of action, Steve would always be able to select the optimal act and thus would incur an average opportunity loss of zero. The maximum potential improvement given this information is $70. It follows that the maximum value of sample information is $70. Any sample plan costing more than $70 per sample could not possibly pay its way. Thus, only sampling plans involving an expenditure of $70 or less need be considered. Sampling plans typically involve a fixed preparation cost and a variable cost per item sampled. Assuming a linear relationship, the cost of sampling is described in general by the relationsihp

$$SMPCST = FXCST + VARCST(n)$$

where

$$SMPCST = \text{the cost of obtaining the sample}$$
$$FXCST = \text{the fixed cost of sampling}$$
$$VARCST = \text{the variable cost of sampling}$$
$$n = \text{the sample size}$$

For illustrative purposes, Ken assumed that the fixed cost involved in setting up the sampling procedure would amount to $30 and that the variable cost per item sampled would be approximately $0.25. Ken indicated that Steve would have to get together with his cost accountant to thrash out the actual figures. Substituting, Ken generated the following description of the cost of sampling:

$$SMPCST = \$30.00 + 0.25(n)$$

To determine the maximum sample size Steve should be willing to consider, Ken substituted $70.00 for SMPCST and solved for n.

$$\$70.00 = \$30.00 + 0.25(n)$$
$$0.25n = \$40.00$$
$$n = 160$$

The optimal decision rule for each sample size from 1 to 160 can then be determined by following the procedure utilized by V. P. Cronin for determination of the optimal decision rule for sample size 10.[1] Steve interjected the comment that the theory was fine, but the number pushing was getting out of hand. Stating that he couldn't agree more, Ken then provided Steve and V. P. Cronin with a computer program designed to perform this task together with the printout reproduced as Table 9.4.

---

[1] See Chapter 8.

Table 9.4 Printout-program reproduced in Technical Note 1

```
                    CONDITIONAL DOLLAR LOSS TABLE
                    ACTION 1      ACTION 2

    PRODEF(1)= .03        150.          550.

    PRODEF(2)= .08        400.          550.

    PRODEF(3)= .25       1250.          550.

                    CONDITIONAL OPPORTUNITY LOSS TABLE
                    ACTION 1      ACTION 2

    PRODEF(1)= .03        0.00        400.00

    PRODEF(2)= .08        0.00        150.00

    PRODEF(3)= .25      700.00          0.00

                    EXPECTED OPPORTUNITY LOSS
                    ACTION 1      70.00
                    ACTION 2     285.00

    THE MINIMUM EXPECTED OPPORTUNITY LOSS IS     70.00,
    WHICH IS ASSOCIATED WITH ACTION    1..
    THIS FIGURE    70.00IS ALSO THE EXPECTED VALUE OF PERFECT INFORMATION.

    MAXIMUM SAMPLE SIZE  160
```

```
    CRITICAL NUMBER FOR SAMPLE SIZE    1    IS    1
    CRITICAL NUMBER FOR SAMPLE SIZE    2    IS    2
    CRITICAL NUMBER FOR SAMPLE SIZE    3    IS    3
    CRITICAL NUMBER FOR SAMPLE SIZE    4    IS    3
    CRITICAL NUMBER FOR SAMPLE SIZE    5    IS    4
    CRITICAL NUMBER FOR SAMPLE SIZE    6    IS    5
    CRITICAL NUMBER FOR SAMPLE SIZE    7    IS    6
    CRITICAL NUMBER FOR SAMPLE SIZE    8    IS    7
    CRITICAL NUMBER FOR SAMPLE SIZE    9    IS    8
    CRITICAL NUMBER FOR SAMPLE SIZE   10    IS    9
    CRITICAL NUMBER FOR SAMPLE SIZE   11    IS   10
    CRITICAL NUMBER FOR SAMPLE SIZE   12    IS   10
    CRITICAL NUMBER FOR SAMPLE SIZE   13    IS   11
    CRITICAL NUMBER FOR SAMPLE SIZE   14    IS   12
    CRITICAL                              15   IS
    CRITI
    CRIT        UMBER FOR SAMP
    CR          L NUMBER FOR SAMPLE SIZE          34
    C           CAL NUMBER FOR SAMPLE SIZE   40   IS   35
                ICAL NUMBER FOR SAMPLE SIZE   41   IS   36
                RITICAL NUMBER FOR SAMPLE SIZE   42   IS   36
    CRITICAL NUMBER FOR SAMPLE SIZE   43   IS   37
    CRITICAL NUMBER FOR SAMPLE SIZE   44   IS   38
    CRITICAL NUMBER FOR SAMPLE SIZE   45   IS   39
    CRITICAL NUMBER FOR SAMPLE SIZE   46   IS   40
    CRITICAL NUMBER FOR SAMPLE SIZE   47   IS   41
    CRITICAL NUMBER FOR SAMPLE SIZE   48   IS   41
    CRITICAL NUMBER FOR SAMPLE SIZE   49   IS   42
    CRITICAL NUMBER FOR SAMPLE SIZE   50   IS   43
    CRITICAL NUMBER FOR SAMPLE SIZE   51   IS   44
    CRITICAL NUMBER              SIZE   52   IS   45
    CRITICAL                             53   IS
    CRITIC
    CRITIC
    CRIT        BER FOR SAMPLE SI
    CR          NUMBER FOR SAMPLE SIZE             224
                AL NUMBER FOR SAMPLE SIZE  146   IS  125
                ICAL NUMBER FOR SAMPLE SIZE  147   IS  125
                ITICAL NUMBER FOR SAMPLE SIZE  148   IS  126
    RITICAL NUMBER FOR SAMPLE SIZE  149   IS  127
    CRITICAL NUMBER FOR SAMPLE SIZE  150   IS  128
    CRITICAL NUMBER FOR SAMPLE SIZE  151   IS  129
    CRITICAL NUMBER FOR SAMPLE SIZE  152   IS  130
    CRITICAL NUMBER FOR SAMPLE SIZE  153   IS  131
    CRITICAL NUMBER FOR SAMPLE SIZE  154   IS  131
    CRITICAL NUMBER FOR SAMPLE SIZE  155   IS  132
    CRITICAL NUMBER FOR SAMPLE SIZE  156   IS  133
    CRITICAL NUMBER FOR SAMPLE SIZE  157   IS  134
    CRITICAL NUMBER FOR SAMPLE SIZE  158   IS  135
    CRITICAL NUMBER FOR SAMPLE SIZE  159   IS  136
    CRITICAL NUMBER FOR SAMPLE SIZE  160   IS  137
```

A copy of the program appears in both the Instructor's Manual and in the Student Solution Guide. Ken stated that the repetitive nature of the computational process involved in generating the optimal decision rules for various sample sizes was made to order for computer programming.

Table 9.5 Data requirements

ACTS = the number of action alternatives
SU   = the number of states of the universe
XLOTSZ
XOPCST
CNSTCST
$PRODEF_i$, one for each state of nature $\Big\}$ Previously defined
$PRIOR_i$, one for each state of nature
FXCST
VARCST

Table 9.6

```
    SAMPLE SIZE 11, 0 ACCEPTABLES

             PRIOR        CONJ        JOINT         POST
   SU(1)      .600     .0000000     .0000000     .0000000
   SU(2)      .300     .0000000     .0000000     .0000108
   SU(3)      .100     .0000002     .0000000     .9999892

 ELSACT1=, 1249.99              ELSACT2=,  550.00
    SAMPLE SIZE 11, 1 ACCEPTABLES

             PRIOR        CONJ        JOINT         POST
   SU(1)      .600     .0000000     .0000000     .0000000
   SU(2)      .300     .0000000     .0000000     .0000414
   SU(3)      .100     .0000079     .0000008     .9999586

 ELSACT1=, 1249.96              ELSACT2=,  550.00
    SAMPLE SIZE 11, 2 ACCEPTABLES

             PRIOR        CONJ        JOINT         POST
   SU(1)      .600     .0000000     .0000000     .0000001
   SU(2)      .300     .0000000     .0000000     .0001588
   SU(3)      .100     .0001180     .0000118     .9998411

 ELSACT1=, 1249.86              ELSACT2=,  550.00
    SAMPLE SIZE 11, 3 ACCEPTABLES

             PRIOR        CONJ        JOINT         POST
   SU(1)      .600     .0000000     .0000000     .0000006
   SU(2)      .300     .0000002     .0000001     .0006085
   SU(3)      .100     .0010622     .0001062     .9993910

 ELSACT1=, 1249.48              ELSACT2=,  550.00
    SAMPLE SIZE 11, 4 ACCEPTABLES

             PRIOR        CONJ        JOINT         POST
   SU(1)      .600     .0000000     .0000000     .0000060
   SU(2)      .300     .0000050     .0000015     .0023294
   SU(3)      .100     .0063729     .0006373     .9976656

 ELSACT1=, 1248.01              ELSACT2=,  550.00
    SAMPLE SIZE 11, 5 ACCEPTABLES

             PRIOR        CONJ        JOINT         POST
   SU(1)      .600     .0000003     .0000002     .0000643
   SU(2)      .300     .0000798     .0000239     .0088666
   SU(3)      .100     .0267663     .0026766     .9910691

 ELSACT1=, 1242.39              ELSACT2=,  550.00
    SAMPLE SIZE 11, 6 ACCEPTABLES

             PRIOR        CONJ        JOINT         POST
   SU(1)      .600     .0000094     .0000056     .0006751
   SU(2)      .300     .0009179     .0002754     .0331354
   SU(3)      .100     .0802989     .0080239     .9661894

 ELSACT1=, 1221.09              ELSACT2=,  550.00
```

Table 9.6 (*Continued*)

```
    SAMPLE SIZE 11, 7 ACCEPTABLES

              PRIOR        COND       JOINT        POST
    SU(1)     .600       .0002160    .0001296    .0066119
    SU(2)     .300       .0075403    .0022621    .1154211
    SU(3)     .100       .1720691    .0172069    .8779670

  ELSACT1=,  1144.52              ELSACT2=,   550.00
        SAMPLE SIZE 11, 8 ACCEPTABLES

              PRIOR        COND       JOINT        POST
    SU(1)     .600       .0034916    .0020949    .0512057
    SU(2)     .300       .0433567    .0130070    .3179242
    SU(3)     .100       .2581036    .0258104    .6308700

  ELSACT1=,   923.44              ELSACT2=,   550.00
        SAMPLE SIZE 11, 9 ACCEPTABLES

              PRIOR        COND       JOINT        POST
    SU(1)     .600       .0376314    .0225789    .2298116
    SU(2)     .300       .1662008    .0498602    .5074861
    SU(3)     .100       .2581036    .0258104    .2627023

  ELSACT1=,   565.84              ELSACT2=,   550.00
        SAMPLE SIZE 11,10 ACCEPTABLES

              PRIOR        COND       JOINT        POST
    SU(1)     .600       .2433500    .1460100    .5286869
    SU(2)     .300       .3822618    .1146786    .4152391
    SU(3)     .100       .1548622    .0154862    .0560740

  ELSACT1=,   315.49              ELSACT2=,   550.00
  CRITICAL NUMBER FOR SAMPLE SIZE   11   IS   10
```

Providing Steve and V. P. Cronin with the data list in Table 9.5, Ken stated that all the user of the program need do is supply the values for the inputs in the list and the computer would perform all the underlying computations. The printout would be generated in approximately one minute. Furthermore, Ken pointed out that the program had been written in such a fashion that a printout describing the underlying computations involved in generating the revised or posterior probabilities for selected sample sizes could be easily obtained. To illustrate, Ken provided the underlying computations supporting the determination of the optimal decision rule for sample size 11. The printout is reproduced as Table 9.6. ELSACT 1 refers to the expected *dollar* loss associated with the selection of action 1 after observing a particular sample result. ELSACT 2 refers to the expected *dollar* loss associated with the selection of action 2 after observing a particular sample result.

At this point, Steve remarked that although he fully grasped all that had been said, he was unable to perceive how he was any better off. True, he knew the optimal decision rule for all sample sizes from 1 to 160, but exactly how was he to decide which one to use? Ken replied that the global optimal decision rule (best of all the local optimal decision rules) would be the one that would lead to the greatest reduction in expected opportunity loss after taking into consideration the cost of sampling. To illustrate the procedure, Ken utilized the local optimum decision rule developed for sample size 10 by V. P. Cronin Associates.

Referring to the payoff table (Table 9.7), Ken indicated for each cell whether

Table 9.7 Action alternatives

States of the universe	Action 1: Continue as is	Action 2: Call in V. P. Cronin Assoc.
$p_1 = 0.03$	Correct action	Error
$p_2 = 0.08$	Correct action	Error
$p_3 = 0.25$	Error	Correct action

selection of the associated act would lead to a correct action or to an error. As previously demonstrated, there is an opportunity loss associated with each cell. Each cell represents the occurrence of a particular joint event. When the decision rule leads to a correct action, a zero opportunity loss is incurred. When the decision rule leads to an incorrect action, the opportunity loss associated with the corresponding cell will be incurred (see Table 9.8).

Table 9.8 Conditional opportunity loss table

States of nature	Action 1: Continue as is	Action 2: Call in V. P. Cronin Assoc.
$p_1 = 0.03$	$OL_{11} = 0$	$OL_{12} = 400$
$p_2 = 0.08$	$OL_{21} = 0$	$OL_{22} = 150$
$p_3 = 0.25$	$OL_{31} = 700$	$OL_{32} = 0$

To obtain the expected opportunity loss associated with consistent application of a given decision rule, simply multiply the conditional opportunity loss associated with each of the cells by the probability that the decision rule will lead to the joint event represented by that cell.

Ken proceeded to illustrate using the decision rule (10, 9).

Given $p_1 = 0.03$: The probability that action 1 will be selected is equal to the probability that 9 or more acceptables will be observed in a sample of size 10. Turning to the table of the binomial, Ken obtained the following probabilities:

$$n = 10, r = 9, \quad (1 - p_1) = 0.97 \quad 0.2281$$

$$n = 10, r = 10, (1 - p_1) = 0.97 \quad 0.7374$$

$$\overline{ 0.9655}$$

Thus, when the true state of the universe is $p_1 = 0.03$, the decision rule (10, 9) will lead to a correct decision and a zero opportunity loss 96.55 percent of the time. The decision rule would lead to an error whenever less than 9 acceptables are observed. It follows that 3.45 percent of the time when $p_1 = 0.03$ is the true state of the universe, the decision rule (10, 9) will lead to an error and an opportunity loss of \$400. Of course, $p_1 = 0.03$ is not always the true state of the universe. This occurs only 60 percent of the time. Thus, the opportunity loss OL_{11} will be incurred only $(0.6)(0.9655)$ or 57.93 percent of the time the decision rule (10, 9) is employed. Similarly, the opportunity loss OL_{12} will be incurred only $(0.6)(0.0345)$ or 2.07 percent of the time the decision rule (10, 9) is employed. Note that 57.93 plus 2.07 equals 60 percent.

In a similar fashion, the probability of incurring the opportunity loss associated with the other four cells in the table can be determined. Note that the decision rule leads to an error when *less than 9* acceptables are observed when $p = 0.08$ and an error when *9 or more* acceptables are observed when $p_3 = 0.25$ is the true state of the universe. This is because when $p = 0.25$, the correct action switches to "call in V. P. Cronin Associates." Once the probabilities have been generated, they are multiplied by the associated conditional opportunity losses to obtain the expected opportunity loss associated with consistent application of the decision rule (10, 9). To depict this process in a formal framework, Ken developed Table 9.9.

Table 9.9 Determination of the expected opportunity loss associated with the decision rule (10, 9)

X = Number of acceptables observed
C = Critical number or 9

Cell$_{ij}$	State of the universe	Prior	$P(X \geq C)^a$	$P(X < C)$	Joint probability	OL$_{ij}$	(Joint) (OL$_{ij}$)
1, 1	$p_1 = 0.03$	0.6	0.9655		0.57930	0	—
1, 2	$p_1 = 0.03$	0.6		0.0345	0.02070	400	8.2800
2, 1	$p_2 = 0.08$	0.3	0.8121		0.24363	0	—
2, 2	$p_2 = 0.08$	0.3		0.1879	0.05637	150	8.4555
3, 1	$p_3 = 0.25$	0.1	0.2440		0.02440	700	17.0800
3, 2	$p_3 = 0.25$	0.1		0.7560	0.07560	0	—
							33.8155

^{a} Obtained by summing the appropriate terms in the binomial distribution table in Appendix C. These values could have been looked up directly in a decumulative binomial distribution table.

If the decision rule (10, 9) is consistently applied to determine whether or not to call in V. P. Cronin Associates, the expected opportunity loss (profit forgone on the average per setup) would amount to $33.82. Contrasting this figure with the expected opportunity loss of $70 associated with the optimum action without sampling, it is clear that the value of the sample is $36.18. This value is referred to as the expected value of sample information or EVSI. Of course, sample information is obtained at a cost. To obtain the *expected net gain from sampling* (ENGS), it is necessary to subtract the cost of sampling.

$$\text{ENGS} = \text{EVSI} - \text{SMPCST}$$

$$\text{SMPCST} = \text{FXCST} + \text{VARCST}(n)$$

Substituting:

$$\text{SMPCST} = \$30 + \$0.25(10)$$
$$= \$32.50$$

Therefore,

$$\text{ENGS}(10, 9) = \text{EVSI}(10, 9) - \text{SMPCST}(10)$$
$$= \$36.18 - \$32.50$$
$$= \$3.68$$

Ken pointed out that a sample size of 10 and subsequent application of the decision rule (10, 9) is an economically preferable alternative to not sampling at all, but not necessarily the optimal sample size to select. To illustrate, Ken supplied Steve and V. P. Cronin with the printout reproduced as Table 9.10. As you probably anticipated, Ken had computerized this phase of the analysis as well.

Referring to the printout, Steve observed that the EOL associated with the decision rule (20, 17) is $19.58. The expected value of sample information, EVSI, is the difference between $70.00 and this figure, or $50.42. SMPCST(20) = $30 + 0.25(20) or $35. Subtracting SMPCST(20) from EVSI(20, 17) generates an ENGS(20, 17) of $15.42. It is apparent that the decision rule (20, 17) is economically preferable to the decision rule (10, 9). Although Ken felt that it was a useful

exercise to see the computations underlying the determination of the EOL for each
of the local optimal decision rules, he indicated that he usually suppressed this
portion of the printout (along with the others he had used in the process of explain-
ing the underlying computational procedures) in favor of the summary printout
reproduced as Table 9.11.

Surveying the printout, Steve observed that the maximum ENGS was associated
with the decision rule (44, 38). Actually, he thought, any of the local optimal
decision rules in that neighborhood generated about the same expected payoff.
Steve felt more comfortable knowing that a margin for error existed. Steve was
impressed. "Let me be sure I've got it straight. I select a sample of size 44. If 38
or more of the parts examined are acceptable, I continue to operate. If less than 38
are acceptable, I call in V. P. Cronin Associates." "Not quite," Ken responded.
Steve reacted with one of those "I-knew-it-was-too-good-to-be-true" expressions.
Ken pointed out that the analysis they had performed is what is referred to as
pre-posterior analysis. The function of pre-posterior analysis is to determine the

Table 9.10

STATE OF THE UNIVERSE	PRIOR	SAMPLE SIZE 20 CRITICAL VALUE, 17 P(X GR. OR EQ. C)	P(X LT. C)	JOINT	OPPORT LOSS	EXPECTED OPPORT LOSS
1	.600	.9973312		.5983987	0.00	0.00
1	.600		.0026688	.0016013	400.00	.64
2	.300	.9293848		.2788154	0.00	0.00
2	.300		.0706152	.0211846	150.00	3.18
3	.100	.2251560		.0225156	700.00	15.76
3	.100		.7748440	.0774844	0.00	0.00
						19.58

Table 9.11

SAMPLE SIZE	CRITICAL NUMBER	EXPECTED OPPORTUNITY LOSS EOL	EXPECTED VALUE OF SAMPLE INFORMATION EVSI	COST OF SAMPLING	EXPECTED NET GAIN FROM SAMPLING ENGS
1	1	63.30	6.70	30.25	-23.55
2	2	60.47	9.53	30.50	-20.97
3	3	60.45	9.55	30.75	-21.20
4	3	54.47	15.53	31.00	-15.47
5	4	48.78	21.22	31.25	-10.03
6	5	43.84	26.16	31.50	-5.34
7	6	39.87	30.13	31.75	-1.62
8	7	36.90	33.10	32.00	1.10
9	8	34.91	35.09	32.25	2.84
10	9	33.82	36.18	32.50	3.68
11			36.46	32.75	3.71
12			38.56	33.00	5
1	13	23.55	.65	33.25	
	14	22.41		33.50	.29
	15	21.44	48.	33.75	
18	16	20.90	49.10		14.31
19	17	20.76	49.24		14.60
20	17	19.58	50.42	35.00	14.49
21	18	17.88	52.12	35.25	15.42
22	19	16.52	53.48	35.50	16.87
23	20	15.49	54.51	35.75	17.98
24	21	14.76	55.24	36.00	18.76
25	22	14.30	55.70	36.25	19.24
26	23	14.08	55.92	36.50	19.45
27	23	14.06	55.94	36.75	19.42
28	24	12.89	57.11	37.00	19.19
29		.95	58.05	37.25	20
			58.77	37.50	
			59.30	37.75	
			.64	38.00	

Table 9.11 (*Continued*)

	25	11.?			?1
?0	26	11.23			?0.80
31	27	10.70			21.27
?2	28	10.36	59.?	?0	21.55
??	29	10.19	59.82	?8.25	21.64
34	30	10.15	59.85	38.50	21.57
?5	30	9.54	60.46	38.75	21.35
36	31	8.88	61.12	39.00	21.71
37	?2	8.38	61.62	39.25	22.12
?8	33	8.0?	61.98	39.50	22.37
?9	34	7.78	62.22	39.75	22.48
40	35	7.66	62.34	40.00	22.47
41	36	7.64	62.36	40.25	22.34
42	36	7.19	62.81	40.50	22.11
43	37	6.73	63.27	40.75	22.31
44	38	6.38	63.62	41.00	22.52
45	39	6.14	63.86	41.25	22.62
4?	40	5.98	64.02	41.50	22.61
47	41	5.91	64.09	41.75	22.52
48	41	5.89	64.11	42.00	22.34
49	42	5.49	64.51	42.25	22.11
5?	43	5.17	64.83	42.50	22.26
51	44	4.93	65.07	42.75	22.33
52	45	4.76	65.24	43.00	22.32
53	46	?.66	65.34	43.25	22.24
54			65.33	43.50	22.09
55			?5.48	43.75	21.?
			?	44.00	
	86	.9?		44.25	?8
	87	.91			?3.59
??3	88	.86	69.?		13.39
104	89	.83	69.17		13.17
105	90	.?1	69.19	??.25	12.94
106	91	.79	69.21	56.50	12.71
107	?-	?79	69.21	56.75	12.4?
108		79	69.24	57.00	1?
1??			?9.28	57.25	
				57.50	
	131	.18		57.75	.32
	132	.17	6?		1.08
?5?	133	.16	69.84		.84
157	134	.15	69.85		.60
15?	135	.15	69.85	69.50	.35
15?	136	.15	69.85	69.75	.10
1?0	137	.15	69.85	70.00	-.15

optimal action prior to generating any actual sample information. Sometimes there will not be any economically justifiable sampling plan. Had that been the case with the worm gears, Steve's optimal action would have been to continue operating, as that would have been the best terminal decision given his *prior* probabilities. Often, however, the optimal course of action is to generate more information (select a sample) before making a terminal decision. In the illustrative example, the optimal act was to select a sample of size 44 and condition subsequent action on the sample evidence observed. After a particular sample has been selected and the result observed, the new information obtained is used to revise the prior probabilities assigned to the various states of the universe. The revised priors are then utilized in the same fashion as the original priors as inputs in evaluating which of the following action alternatives would be economically preferable:

Action 1 Continue as is
Action 2 Call in V. P. Cronin Associates
Action 3 Postpone arriving at a terminal decision and gather additional
  evidence

This phase of the analysis has come to be referred to as *posterior analysis*. The function of posterior analysis is to determine the optimal action to take *posterior* to selecting a sample and observing a particular sample result. The procedure is identical to pre-posterior analysis. The only difference is the use of posterior probabilities in place of the priors used in the pre-posterior analysis.

EPILOGUE[2]

To illustrate posterior analysis, Ken adjusted the program so that it would generate the posterior probabilities associated with the various possible outcomes that could occur with a sample of size 44. A portion of the resulting printout is reproduced as Table 9.12.

Table 9.12

```
        SAMPLE SIZE 44,34 ACCEPTABLES

             PRIOR        COND        JOINT         POST
    SU(1)     .600     .0000005     .0000003     .0000225
    SU(2)     .300     .0015644     .0004693     .0339103
    SU(3)     .100     .1337072     .0133707     .9660672

ELSACT1=, 1221.15                   ELSACT2=,  550.00
        SAMPLE SIZE 44,35 ACCEPTABLES

             PRIOR        COND        JOINT         POST
    SU(1)     .600     .0000048     .0000029     .0002217
    SU(2)     .300     .0051403     .0015421     .1185711
    SU(3)     .100     .1146062     .0114606     .8812072

ELSACT1=, 1148.97                   ELSACT2=,  550.00
        SAMPLE SIZE 44,36 ACCEPTABLES

             PRIOR        COND        JOINT         POST
    SU(1)     .600     .0000388     .0000233     .0017855
    SU(2)     .300     .0147783     .0044335     .3396727
    SU(3)     .100     .0859546     .0085955     .6585418

ELSACT1=,  959.31                   ELSACT2=,  550.00
        SAMPLE SIZE 44,37 ACCEPTABLES

             PRIOR        COND        JOINT         POST
    SU(1)     .600     .0002715     .0001629     .0097198
    SU(2)     .300     .0367461     .0110238     .6576606
    SU(3)     .100     .0557543     .0055754     .3326196

ELSACT1=,  680.30                   ELSACT2=,  550.00
        SAMPLE SIZE 44,38 ACCEPTABLES

             PRIOR        COND        JOINT         POST
    SU(1)     .600     .0016173     .0009704     .0354100
    SU(2)     .300     .0778437     .0233531     .8521581
    SU(3)     .100     .0308116     .0030812     .1124319

ELSACT1=,  486.71                   ELSACT2=,  550.00
        SAMPLE SIZE 44,39 ACCEPTABLES
             PRIOR        COND        JOINT         POST
    SU(1)     .600     .0080452     .0048271     .1014820
    SU(2)     .300     .1377236     .0413171     .8686213
    SU(3)     .100     .0142207     .0014221     .0298967

ELSACT1=,  400.04                   ELSACT2=,  550.00
        SAMPLE SIZE 44,40 ACCEPTABLES

             PPIOR        COND        JOINT         POST
    SU(1)     .600     .0325160     .0195096     .2456011
    SU(2)     .300     .1979776     .0593933     .7476856
    SU(3)     .100     .0053328     .0005333     .0067133

ELSACT1=,  344.31                   ELSACT2=,  550.00
        SAMPLE SIZE 44,41 ACCEPTABLES

             PPIOR        COND        JOINT         POST
    SU(1)                          .0615425     .4795460
    S                               .364       .5192378
                                               .0012162
```

[2] The epilogue discusses *posterior analysis* (i.e., how to determine the optimal action *after* having selected a sample and having obtained some sample evidence). This section may be omitted without loss of continuity.

Consider the action Steve should take after having observed the specific sample outcome "34 acceptables out of 44." Referring to Table 9.12, the revised or posterior probabilities are determined to be

$P(p_1 = 0.03) = 0.0000225$

$P(p_2 = 0.08) = 0.0339103$

$P(p_3 = 0.25) = 0.9660672$

The original priors $P(p_1 = 0.03) = 0.60$, $P(p_2 = 0.08) = 0.30$, and $P(p_3 = 0.25) = 0.10$ are replaced by the revised priors reflecting the sample evidence, 34 acceptable of 44 observed. The program for determination of the global optimum decision rule would then be rerun. Ken assumed that the cost of a second sample would involve a reduction from \$30 to \$5 for the fixed-cost portion and that the variable cost per item sampled would remain at \$0.25. The printout reproduced in Table 9.13 discloses that the minimum expected opportunity loss associated with the optimum action without sampling further is \$5.10. The cost of a sample of size 1 would be \$5.25. There is no economic justification for further sampling. The decision should be to call in V. P. Cronin Associates.

Table 9.13 Determination of optimum decision rule after observing 34 acceptable out of 44

```
                    CONDITIONAL OPPORTUNITY LOSS TABLE
                    ACTION 1        ACTION 2

   PRODEF(1)= .03        0.00           400.00

   PRODEF(2)= .08        0.00           150.00

   PRODEF(3)= .25      700.00             0.00

                    EXPECTED OPPORTUNITY LOSS
                    ACTION 1        676.25
                    ACTION 2          5.10
THE MINIMUM EXPECTED OPPORTUNITY LOSS IS       5.10,
WHICH IS ASSOCIATED WITH ACTION     2..
THIS FIGURE     5.10IS ALSO THE EXPECTED VALUE OF PERFECT INFORMATION.

MAXIMUM SAMPLE SIZE     0
```

Portions of the printouts associated with the incorporation of the revised priors generated after observing 35 acceptable, 36 acceptable, 37 acceptable, 38 acceptable, 39 acceptable, and 40 acceptable are reproduced in Tables 9.14 through 9.19. A summary is presented in Table 9.20.

Table 9.14 Determination of optimum decision rule after observing 35 acceptable out of 44

```
                      CONDITIONAL OPPORTUNITY LOSS TABLE
                         ACTION 1      ACTION 2

       PRODEF(1)= .03       0.00         400.00

       PRODEF(2)= .08       0.00         150.00

       PRODEF(3)= .25     700.00           0.00

                      EXPECTED OPPORTUNITY LOSS
                      ACTION 1       616.85
                      ACTION 2        17.87

   THE MINIMUM EXPECTED OPPORTUNITY LOSS IS     17.87,
   WHICH IS ASSOCIATED WITH ACTION     2..
   THIS FIGURE    17.87IS ALSO THE EXPECTED VALUE OF PERFECT INFORMATION.

   MAXIMUM SAMPLE SIZE   51
```

SAMPLE SIZE	CRITICAL NUMBER	EXPECTED OPPORTUNITY LOSS EOL	EXPECTED VALUE OF SAMPLE INFORMATION EVSI	COST OF SAMPLING	EXPECTED NET GAIN FROM SAMPLING ENGS
1	1	464.06	-446.18	5.25	-451.43
2	2	349.71	-331.84	5.50	-337.34
3	3	264.18	-246.30	5.75	-252.05
4	4	200.23	-182.35	6.00	-188.35
5	5	152.46	-134.58	6.25	-140.83
6	6	116.80	-98.93	6.50	-105.43
7	7	90.22	-72.35	6.75	-79.10
8	8	70.43	-52.56	7.00	-59.56
9	9	55.72	-37.85	7.25	-45.10
10	10	44.82	-26.95	7.50	-34.45
11	11	36.76	-18.88	7.75	-26.63
12	12	30.81	-12.94	8.00	-20.9?
13	13	26.45	-8.58	8.25	-1?
14		??	-5.40	8.50	
15			-3.10	8.75	
					.26
	40	9.??			-7.46
	40	9.37			-7.50
45	41	8.8?	9.05	16.25	-7.20
46	42	8.44	9.44	16.50	-7.06
47	43	8.17	9.71	16.75	-7.04
48	44	8.00	9.87	17.00	-7.13
49	45	7.92	9.96	17.25	-7.29
50	46	7.90	9.97	17.50	-7.53
51	46	7.52	10.36	17.75	-7.39

Table 9.15 Determination of optimum decision rule after observing 36 acceptable out of 44

```
                        CONDITIONAL OPPORTUNITY LOSS TABLE
                          ACTION 1        ACTION 2

    PRODEF(1)= .03          0.00            400.00

    PRODEF(2)= .08          0.00            150.00

    PRODEF(3)= .25        700.00              0.00

                     EXPECTED OPPORTUNITY LOSS
                       ACTION 1        460.98
                       ACTION 2         51.67

THE MINIMUM EXPECTED OPPORTUNITY LOSS IS      51.67,
WHICH IS ASSOCIATED WITH ACTION        2..
THIS FIGURE     51.67IS ALSO THE EXPECTED VALUE OF PERFECT INFORMATION.

MAXIMUM SAMPLE SIZE  186
```

SAMPLE SIZE	CRITICAL NUMBER	EXPECTED OPPORTUNITY LOSS EOL	EXPECTED VALUE OF SAMPLE INFORMATION EVSI	COST OF SAMPLING	EXPECTED NET GAIN FROM SAMPLING ENGS
50	45	13.86	37.81	17.50	20.31
51	45	13.14	38.52	17.75	20.77
52	46	12.38	39.28	18.00	21.28
53	47	11.83	39.84	18.25	21.59
54	48	11.45	40.22	18.50	21.72
55	49	11.22	40.44	18.75	21.69
56	50	11.13	40.54	19.00	21.54
57	50	10.99	40.68	19.25	21.43
58	51	10.28	41.38	19.50	21.88
59	52	9.75	41.91	19.75	22.16
60	53	9.37	42.30	20.00	22.30
61	54	9.12	42.55	20.25	22.30
62	55	8.98	42.68	20.50	22.18
63	56	8.94	42.72	20.75	21.97
64	56	8.58	43.08	21.00	22.08
65	57	8.08	43.58	21.25	22.33
66	58	7.71	43.95	21.50	22.45
67	59	7.45	44.22	21.75	22.47 ✶
68	60	7.29	44.38	22.00	22.38
69	61	7.21	44.45	22.25	22.20
70	61	7.20	44.46	22.50	21.96
71	62	6.74	44.93	22.75	22.18
72	63	6.38	45.28	23.00	22.28
73	64	6.12	45.54	23.25	22.29
74	65	5.95	45.72	23.50	22.22
75	66	5.85	45.82	23.75	22.07
76	67	5.81	45.85	24.00	21.85
77	67	5.64	46.02	24.25	21.77
78	68	5.31	46.36	24.50	21.86
79	69	5.06	46.61	24.75	21.86
80	70	4.88	46.79	25.00	21.79
81	71	.76	46.90	25.25	21
82			46.96	25.50	
83				25.75	

Table 9.16 Determination of optimal decision rule after observing 37 acceptable out of 44

```
                    CONDITIONAL OPPORTUNITY LOSS TABLE
                        ACTION 1      ACTION 2

      PRODEF(1)= .03        0.00          400.00

      PRODEF(2)= .08        0.00          150.00

      PRODEF(3)= .25      700.00            0.00

                    EXPECTED OPPORTUNITY LOSS
                    ACTION 1      232.83
                    ACTION 2      102.54

 THE MINIMUM EXPECTED OPPORTUNITY LOSS IS    102.54,
 WHICH IS ASSOCIATED WITH ACTION      2..
 THIS FIGURE    102.54IS ALSO THE EXPECTED VALUE OF PERFECT INFORMATION.

 MAXIMUM SAMPLE SIZE  390
```

SAMPLE SIZE	CRITICAL NUMBER	EXPECTED OPPORTUNITY LOSS EOL	EXPECTED VALUE OF SAMPLE INFORMATION EVSI	COST OF SAMPLING	EXPECTED NET GAIN FROM SAMPLING ENGS
50	44	14.57	87.97	17.50	70.47
51	44	13.81	88.73	17.75	70.98
52	45	12.99	89.55	18.00	71.55
53	46	12.38	90.16	18.25	71.91
54	47	11.95	90.59	18.50	72.09
55	48	11.68	90.85	18.75	72.10
56	49	11.57	90.96	19.00	71.96
57	49	11.43	91.11	19.25	71.86
58	50	10.69	91.85	19.50	72.35
59	51	10.12	92.42	19.75	72.67
60	52	9.70	92.84	20.00	72.84
61	53	9.42	93.12	20.25	72.87
62	54	9.26	93.27	20.50	72.77
63	55	9.22	93.32	20.75	72.57
64	55	8.85	93.69	21.00	72.69
65	56	8.33	94.21	21.25	72.96
66	57	7.93	94.61	21.50	73.11
67	58	7.65	94.89	21.75	73.14 *
68	59	7.47	95.07	22.00	73.07
69	60	7.38	95.15	22.25	72.90
70	60	7.37	95.17	22.50	72.67
71	61	6.90	95.64	22.75	72.89
72	62	6.52	96.01	23.00	73.01
73	63	6.25	96.29	23.25	73.04
74	64	6.06	96.48	23.50	72.98
75	65	5.95	96.59	23.75	72.84
76	66	5.91	96.62	24.00	72.62
77	66	5.74	96.80	24.25	72.55
78	67	5.40	97.14	24.50	72.64
79	68	5.14	97.40	24.75	72.5
80			97.59	25.00	
81				25.25	

Table 9.17 Determination of optimal decision rule after observing 38 acceptable out of 44

```
                        CONDITIONAL OPPORTUNITY LOSS TABLE
                        ACTION 1        ACTION 2

        PRODEF(1)= .03        0.00          400.00

        PRODEF(2)= .08        0.00          150.00

        PRODEF(3)= .25      700.00            0.00

                        EXPECTED OPPORTUNITY LOSS
                        ACTION 1        78.70
                        ACTION 2       141.99

    THE MINIMUM EXPECTED OPPORTUNITY LOSS IS    78.70,
    WHICH IS ASSOCIATED WITH ACTION    1..
    THIS FIGURE    78.70 IS ALSO THE EXPECTED VALUE OF PERFECT INFORMATION.

    MAXIMUM SAMPLE SIZE  294
```

SAMPLE SIZE	CRITICAL NUMBER	EXPECTED OPPORTUNITY LOSS EOL	EXPECTED VALUE OF SAMPLE INFORMATION EVSI	COST OF SAMPLING	EXPECTED NET GAIN FROM SAMPLING ENGS
50	43	9.16	69.54	17.50	52.04
51	43	8.70	70.00	17.75	52.25
52	44	8.20	70.51	18.00	52.51
53	45	7.81	70.90	18.25	52.65
54	46	7.53	71.17	18.50	52.67 *
55	47	7.35	71.35	18.75	52.60
56	48	7.28	71.42	19.00	52.42
57	48	7.19	71.51	19.25	52.26
58	49	6.74	71.96	19.50	52.46
59	50	6.38	72.32	19.75	52.57
60	51	6.11	72.59	20.00	52.59
61	52	5.93	72.77	20.25	52.52
62	53	5.83	72.87	20.50	52.37
63	54	5.80	72.90	20.75	52.15
64	54	5.57	73.13	21.00	52.13
65	55	5.25	73.45	21.25	52.20
66	56	5.00	73.70	21.50	52.20
67	57	4.82	73.89	21.75	52.14
68	58	4.70	74.00	22.00	52.00
69	59	4.64	74.06	22.25	51.81
70	59	4.64	74.07	22.50	51.57
71	60	4.34	74.36	22.75	51.61
72	61	4.11	74.59	23.00	51.59
73	62	3.94	74.77	23.25	51.52
74	63	3.81	74.89	23.50	51.39
75	64	3.74	74.96	23.75	51.21
76	65	3.72	74.98	24.00	50.98
77	65	3.61	75.09	24.25	50.84
78	66	3.40	75.30	24.50	50.80
79			75.47	24.75	5.
80			75.59	25.00	

Table 9.18 Determination of optimal decision rule after observing 39 acceptable out of 44

```
                    CONDITIONAL OPPORTUNITY LOSS TABLE
                      ACTION 1      ACTION 2

    PRODEF(1)= .03        0.00          400.00

    PRODEF(2)= .08        0.00          150.00

    PRODEF(3)= .25      700.00            0.00

                    EXPECTED OPPORTUNITY LOSS
                      ACTION 1      20.93
                      ACTION 2     170.89

THE MINIMUM EXPECTED OPPORTUNITY LOSS IS     20.93,
WHICH IS ASSOCIATED WITH ACTION     1..
THIS FIGURE     20.93IS ALSO THE EXPECTED VALUE OF PERFECT INFORMATION.

MAXIMUM SAMPLE SIZE    63
```

SAMPLE SIZE	CRITICAL NUMBER	EXPECTED OPPORTUNITY LOSS EOL	EXPECTED VALUE OF SAMPLE INFORMATION EVSI	COST OF SAMPLING	EXPECTED NET GAIN FROM SAMPLING ENGS
1	0	20.93	0.00	5.25	-5.25
2	1	20.49	.44	5.50	-5.06
3	2	20.13	.79	5.75	-4.96
4	~	20.12	.81	6.00	-5.19
~~	21	9.52			
	21	9.00	11.92		
	22	8.60	12.33		
28	23	8.30	12.63	12.00	
29	24	8.12	12.80	12.25	.55
30	25	8.07	12.86	12.50	.36
31	25	7.82	13.11	12.75	.36
32	26	7.36	13.57	13.00	.57
33	27	6.99	13.94	13.25	.69
34	28	6.71	14.22	13.50	.72 ✱
35	29	6.52	14.41	13.75	.66
36	30	6.42	14.50	14.00	.50
37	31	6.42	14.51	14.25	.26
38	31	6.05	14.87	14.50	.37
39	32	5.72		14.75	.46
40	33			.00	.47
41	34				.40
42	35				.27
43	~				.07
44					
45					

Table 9.19 Determination of optimal decision rule after observing 40 acceptable out of 44

```
                    CONDITIONAL OPPORTUNITY LOSS TABLE
                      ACTION 1      ACTION 2

    PRODEF(1)= .03        0.00          400.00

    PRODEF(2)= .08        0.00          150.00

    PRODEF(3)= .25      700.00            0.00

                    EXPECTED OPPORTUNITY LOSS
                      ACTION 1       4.70
                      ACTION 2     210.39

THE MINIMUM EXPECTED OPPORTUNITY LOSS IS      4.70,
WHICH IS ASSOCIATED WITH ACTION     1..
THIS FIGURE      4.70IS ALSO THE EXPECTED VALUE OF PERFECT INFORMATION.

MAXIMUM SAMPLE SIZE    -1
```

Table 9.20 Summary of results after revising priors on the basis of the observed result in a sample of size 44

Observed result (number acceptable)	Maximum sample size that could be considered	Optimal action (action 1, action 2, or global decision rule)	ENGS	
34 or less	0	Action 2	Negative	⎛None of the 51
35	51	Action 2	Negative	possibilities would pay their way⎞
36	186	(67, 59)	$22.47	
37	390	(67, 58)	$73.14	
38	294	(57, 49)	$52.16	
39	63	(34, 28)	$0.72	
40 or more	0	Action 1	Negative	

At this point, a question was raised by V. P. Cronin. She pointed out that prior to taking a sample of size 44, the maximum potential improvement from sampling was $70. Table 9.16, on the other hand, indicated that after having taken a sample of size 44 and subsequently observing the specific outcome 37 acceptables, the expected value of perfect information increases to $102.54. V. P. Cronin said that intuitively this bothered her. After generating sample information, she would have expected to observe a decrease in the potential gain from *additional* information. Ken explained that her expectation correctly described the pattern of behavior when the sample information clearly supported one of the postulated states of the universe. On some occasions, however, the sample evidence obtained is such that it increases the uncertainty of the decision maker with respect to the true state of the universe. Observing 37 acceptables out of 44 falls into this category. As a result, the potential reduction in uncertainty from further sampling is increased as reflected by an increased EVPI.

The concepts employed by Ken Moore in his presentation to Steve Brenner and V. P. Cronin can be extended to analyses involving continuous priors and nonbinomial sampling plans. The required extensions are beyond the scope of the present text. The interested reader is referred to the following sources.

References

1. Pratt, J. W., Raiffa, H., and Schlaifer, R., *Introduction To Statistical Decision Theory*, New York: McGraw-Hill, 1965.
2. Schlaifer, R., *Analysis of Decisions Under Uncertainty*, New York: McGraw-Hill, 1969.
3. Degroot, M. H., *Optimal Statistical Decisions*, New York: McGraw-Hill, 1970.
4. Jedamus, P., and Frame, R., *Business Decision Theory*, New York: McGraw-Hill 1969.

PROBLEMS

Problem 9.1

Intuitively justify the statement: "Given that the optimal decision rule associated with sample size 46 is (46, 40), the critical number associated with the optimal decision rule for sample size 47 must be at least 40."

Problem 9.2

Determine the optimal decision rule associated with sample size 17 for the text example, utilizing the procedure illustrated in Table 9.6. The desired result is the line of output in Table 9.4 reading

Critical number for sample size 17 is _____

Problem 9.3

Given the optimal decision rule for sample size 50 recorded in Table 9.4, determine the expected opportunity loss associated with that decision rule. Utilize the format illustrated in Table 9.9. Using the data provided in the text illustration, compute the expected net gain from sampling associated with use of this decision rule. Compare your result with the printout for sample size 50 in Table 9.11.

Problem 9.4

Determine the optimal decision rule associated with sample size 100 for the text example utilizing the procedure illustrated in Table 9.6. The desired result is the line of output in Table 9.4 reading

Critical number for sample size 100 is _____

Hint: Observe the critical number associated with sample size 102 as indicated in Table 9.11.

Problem 9.5

Referring to the optimal decision rule associated with sample size 100 as determined in Problem 9.4, determine the expected opportunity loss associated with that decision rule. Use the format illustrated in Table 9.9. Utilizing the data provided in the text illustration, compute the expected net gain from sampling associated with use of this decision rule.

Problem 9.6

What does it indicate if the optimal (n, c) rule has a negative ENGS?

Problem 9.7

Suppose the optimal (n, c) rule in a given situation (not the text example) has a sampling cost of $19,000. You select a sample, apply the rule, and it indicates that the product in question should not be developed. The cost of sampling shows up on the books as a $19,000 expense. Your superior calls you on the carpet to justify this expenditure. "How," he inquires, "can you justify this expenditure for $19,000 when we have nothing to show for it?" Justify it, if you can.

Problem 9.8

Clearly define the term "opportunity loss."

Problem 9.9

Why is an (n, c) decision rule of the form $(n, 0)$ a meaningless decision rule?

Problem 9.10

Discuss fully the interpretation of EOL (expected opportunity loss).

Problem 9.11

Refer to the joint probability column in Table 9.9:

Cell_{ij}	Joint probability	OL_{ij}
1, 1	0.57930	0
1, 2	0.02070	400
2, 1	0.24363	0
2, 2	0.05637	150
3, 1	0.02440	700
3, 2	0.07560	0

The probability that the decision rule (10, 9) will lead to an incorrect decision is $0.02070 + 0.05637 + 0.02440 = 0.10147$. The probability of making an incorrect decision is not the appropriate parameter for evaluating a decision rule, however, because some wrong decisions are more serious than others. Explain the significance of this statement, referring to Table 9.9 in formulating your answer.

Problem 9.12

Discuss the process by which one would determine if $(n, c) = (10, 3)$ is the optimal (n, c) rule associated with sample size 10.

Problem 9.13

Discuss the process by which one would determine the optimal sample size to select given that the local optimal decision rule $[(n, c)$ rule] has been determined for all possible sample sizes worthy of consideration.

Problem 9.14

Suppose that the optimal decision rule associated with sample size 20 is (20, 4) and that the EVPI associated with the use of this rule is $32,684. A sample of size 20 is then selected and zero defectives observed. The probabilities assigned to the possible states of the universe are then revised. The EVPI is computed in association with the new optimal decision rule associated with sample size 20. Assume that the new optimal decision rule is also (20, 4). The EVPI after incorporating the new information is $56,321. This result puzzles some of your friends. Explain how the acquisition of additional information can sometimes *increase*, rather than decrease, the value of additional information.

Problem 9.15

A monthly book club is considering offering its membership the opportunity to purchase a book containing reproductions of famous lithographs. The setup cost is quite expensive ($100,000). The price to members is $15, and the variable cost associated with the proposal is $10. The break-even point is a volume of 20,000 [$100,000/($15 − $10)]. The book club has a membership of 100,000. Thus, 20

percent of the membership would have to participate to make the venture worthwhile. The probability of various levels of participation have been assessed as
follows:

Proportion participating	Probability
0.10	0.10
0.15	0.30
0.20	0.30
0.25	0.20
0.30	0.10
	1.00

1. Determine the conditional dollar payoffs associated with the cells in the
following conditional profit matrix.

Proportion participating	Action 1: Offer the book	Action 2: Do not offer the book
0.10		
0.15		
0.20		
0.25		
0.30		

2. Determine the conditional opportunity loss associated with each cell in the
payoff matrix referred to in part 1.
3. Determine the optimal action under conditions of uncertainty.
4. Determine EVPI (expected value of perfect information).
5. If the fixed cost of sampling is $3000 and the variable cost per elementary
sampling unit is $25, what is the largest sample size you would consider
under the assumption that this largest size would provide perfect
information)?
6. Assume a sample of size 20 has already been selected. Eight of the respondents indicate that they would participate in such a program. Revise the
probabilities assigned to the proportion participating distribution in the
light of this new information. What decision would you now make—offer
the book or not offer the book?
7. Assuming a sample has not yet been selected, determine the optimal decision
rule associated with sample size 20. The program provided in the Student
Guide can be used to perform this task.
8. Given the optimal decision rule associated with sample size 20 as determined
in part 7. Determine:
   a. The expected opportunity loss associated with the application of the
   optimal decision rule associated with sample size 20.
   b. The net expected gain from sampling.

9. Determine the global optimal decision rule, that is, the optimal sample size and its associated optimal decision rule. Note: The program provided in the Student Guide *must* be used to perform this task. The magnitude of the computational task is too great to approach this task in any other way.

Problem 9.16

A major oil company has been approached with the following proposal. A magnificent stereo system is to be offered to its credit card holders for $400. The oil company will be charged set-up costs amounting to $250,000 and a variable charge of $300 per stereo produced. The oil company has 100,000 credit card holders. To break even, the firm would have to generate enough contribution to recover the $250,000 set up charge. A contribution of $100 is realized on each set sold. Thus, 2500 sets would have to be sold to recover the $250,000 set-up charge. To break even, 2.5 percent of the credit card holders would have to participate in the offer. The market research group has assessed the following probabilities for the proportion of the credit holders participating in the offer:

Proportion participating	Probability
0.01	0.20
0.02	0.40
0.03	0.25
0.04	0.10
0.05	0.05
	1.00

1. Determine the conditional dollar payoffs associated with the cells in the following conditional profit matrix.

Proportion participating	Action 1: Accept the proposal	Action 2: Reject the proposal
0.01		
0.02		
0.03		
0.04		
0.05		

2. Determine the conditional opportunity loss associated with each cell in the payoff matrix referred to in part 1.
3. Determine the optimal action under conditions of uncertainty.
4. Determine EVPI (expected value of perfect information).
5. If the fixed cost of sampling is $2000 and the variable cost per elementary sampling unit is $50, what is the largest sample size you would consider (under the assumption that this largest size would provide perfect information)?

6. Assume that a sample of size 100 has been selected. Exactly five of the respondents indicate that they would participate in such a program. Revise the probabilities assigned to the proportion participating distribution in the light of this new information. What decision would you now make—accept the proposal or reject it?

7. Assuming that a sample has not yet been selected, determine the optimal decision rule associated with sample size 100. (The program provided in the Student Guide can be used to perform this task.)

8. Given the optimal decision rule associated with sample size 100 as determined in part 7. Determine:
   a. The expected opportunity loss associated with the application of the optimal decision rule associated with sample size 100.
   b. The net expected gain from sampling.

9. Determine the global optimal decision rule, that is, the optimal sample size and its associated optimal decision rule. Note: The program provided in the Student Guide *must* be used to perform this task. The magnitude of the computational task is too great to approach this task in any other way.

Problem 9.17

The techniques described in Chapter 9 may be applied to the case *Robinson Abrasives Company*, ICH C40R 79R1. Copies of this case may be obtained by writing to: Intercollegiate Case Clearing House, Harvard Business School, Soldiers Field Road, Boston, Massachusetts 02163.

Approach this case under the following assumption: If the decision is made to run a batch with the old bonding material, a full batch of 1000 grinding wheels is to be run. *Any manufacturing cost incurred in producing the test batch is to be considered as part of the cost of sampling.*

1. Determine the conditional dollar payoffs associated with the cells in the following conditional payoff matrix:

Events	Action 1: Run a batch with the old bonding material	Action 2: Discard the old bonding material
Proportion defective = 0.00		
Proportion defective = 0.05		
Proportion defective = 0.20		

2. Determine the conditional opportunity loss associated with each cell in the payoff matrix referred to in part 1.

3. Determine the optimal action under conditions of uncertainty *without sampling.*

4. Determine EVPI (expected value of perfect information).

5. Determine the fixed and variable costs of sampling from the case. Given this cost information, what is the largest sample size you would consider (under the assumption that this largest size would provide perfect information)?

6. Determine the global optimal decision rule, that is, the optimal sample size and its associated optimal decision rule. The program provided in the Student Guide can be used to perform this task.

7. Repeat steps 5 and 6 with the following changes in the case parameters:
   a. Change the fixed cost of sampling from $50 to $30.
   b. Change the variable cost of sampling from $25 (i.e., 15 + 10) to $16 (i.e., 15 + 1).

8. Assume that a sample of size 4 has been selected and two defectives observed. Revise the probabilities assigned to the proportion defective distribution in the light of new information. What decision would you now make—accept the proposal, reject the proposal, or sample further?

9. Discuss the validity of the technique as it has been applied here to Robinson Abrasives. In particular, discuss the implication of the fact that this is a one-time, nonrepeatable decision.

THE POISSON AND EXPONENTIAL DISTRIBUTIONS

10

The Poisson probability distribution is a discrete probability distribution bearing a close similarity to the binomial probability distribution discussed in Chapter 7.[1] The Poisson distribution is generally employed to describe the behavior pattern of a process in which some event of interest occurs at varying random intervals over a continuum of time, length, or space. Such a process is referred to as a Poisson process. Some examples follow.

(A) Consider a horizontal line as representing the continuous passage of time and dots recorded on that line as representing the occurrence of some given event such as incoming calls received at a switchboard.

---

[1] The Poisson distribution takes its name from Siméon Poisson, a French mathematician (1781–1840).

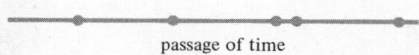

passage of time

(B) Consider a horizontal line as representing a coil of electrical wire produced by a machine process. Dots on the line indicate the location of defects in the wire.

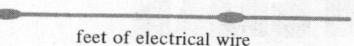

feet of electrical wire

(C) Consider the following rectangular configuration as representing a roll of linoleum produced by a manufacturing process and dots within the configuration as representing the location of defects.

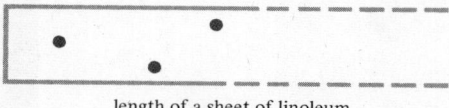

length of a sheet of linoleum

Theoretically, each continuum represented above could be subdivided into subsections such that at most one event (as depicted by a dot) could occur in each subdivision.

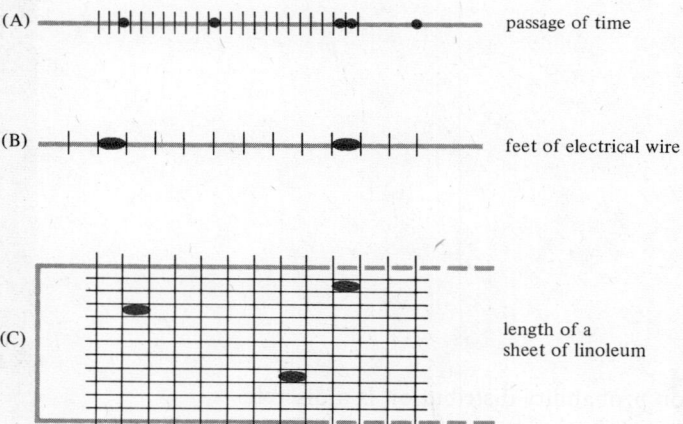

Provided that the underlying "experiment" meets the requirements of a Bernoulli process, the subdivisions can be treated as a sequence of Bernoulli trials. The Poisson distribution can then be used to generate the probability distribution indicating the relative frequency with which one could expect to observe the occurrence of the event in question over repeated observations. Each segment of the process observed will ordinarily be comprised of a large number of these small subdivisions.

(A) In the case of incoming telephone calls, one might be concerned with the distribution of calls per minute. The subdivisions might be 1 second apiece. Thus, each segment of interest would contain 60 subdivisions.

(B) In the case of the coil of electrical wire, one might be concerned with the number of defects per 1000 yards. Each subdivision might be 1 in. in length. Thus, each segment of interest would contain 36,000 subdivisions.

(C) In the case of sheets of linoleum, one might be concerned with the distribution per 100 square-yard segment. Each subdivision might consist of 1 square in. Each segment of interest would thus be comprised of 12,960,000 subdivisions.

In order for the subdivisions to qualify as a series of Bernoulli trials, the following constraints must be met:

1. The subdivisions must be sufficiently small that the probability of two or more events occurring in any one subdivision is so unlikely that it may be considered to be zero. This is tantamount to saying that there is only one of two possible outcomes for each subdivision. There will either be a single occurrence of the event in question or the event will not occur at all.
2. The occurrence or lack of occurrence of the event in question in one subdivision must have no effect on the probability of occurrence of the event in any other subdivision. Moreover, the probability of observing the event must be the same for all subdivisions. This is tantamount to saying that the subdivisions consist of a series of independent trials.

Up to this point, the above could be construed as a description of a binomial experiment. There is a major fundamental difference, however. In a binomial experiment, one is concerned with the number of successes out of a specified number of determinable trials. Each trial and its outcome is capable of being observed. In the case of the Poisson, the number of successes is again the subject of concern, but it becomes virtually impossible to determine the total number of trials. For a Poisson distribution, the total number of trials is the total number of opportunities for the event in question to occur. One can readily observe the occasions when the event did occur. The difficulty is in observing occasions when the event did not occur.

How does one determine the total number of occasions when incoming calls could have occurred but did not?
How does one determine the total number of chances for defects to occur in some specified length of electrical wire?
How does one determine the total number of opportunities for defects to occur in some specified area of linoleum?

This, then, is the fundamental difference between the Poisson and binomial distributions. Failure, in the binomial sense, is observable. Failure, in the Poisson sense, refers to an occasion when success could have occurred but did not. Failure in the Poisson sense, usually is not observable. Fortunately, it is not necessary to know the number of trials (or the number of opportunities for success to have occurred) in order to utilize the Poisson distribution. The only information required is the expected or mean number of successes for the segment of concern for the process under investigation. This information can usually be derived from past observations and is designated by the symbol λt.

λ (pronounced lambda) refers to the expected number of successes over some specified segment of the process.
t refers to the multiple of the segment on which λ is based. For example, if λ is based on a 5-second interval and the pattern of outcomes for 30-second intervals is desired, then $t = 6$.

EXAMPLE 10.1

If λ refers to the expected number of calls per 10-second interval and you are interested in the distribution of calls per minute, then $\lambda t = 6\lambda$.

EXAMPLE 10.2

If λ refers to the expected number of defects per 100 yards of electrical wire and you are interested in the distribution of the number of defects per 1000-yd interval, then $\lambda t = 10\lambda$.

EXAMPLE 10.3

In the special case where λ has been determined for exactly the segment of interest to the analyst, $t = 1$ and λt simplifies to λ.

In general, the probability of observing exactly x Poisson successes over the interval t is equal to

$$P(X) = \frac{(\lambda t)^x e^{-\lambda t}}{x!}$$

where

x is a random variable representing the number of occurrences (or successes) and is restricted to a nonnegative integer

λt is the expected mean number of occurrences of the event in question over the interval t

e is the base of the natural logarithm or 2.71828

Both the mean and the variance of the Poisson distribution always equal the same value, namely λt. Thus, the Poisson distribution is totally characterized by the single parameter λt, or its mean. Tables of the Poisson probabilities have been calculated for varying values of λt and are reproduced in the Appendix as Table D.

10.1 ESTIMATION OF λt

To describe a Poisson probability distribution fully, it is sufficient to determine the parameter λt. Note that it is not necessary to determine the size of the subdivision within which at most one event could occur. Moreover, it is not necessary to ascertain how many such subdivisions there are. All that is necessary is to verify that it would be *theoretically* possible to do so. If one can then determine λt, the expected number of occurrences of the event per some interval t, the Poisson probability distribution can be used.

Consider the historical record with respect to the frequency of incoming telephone calls shown in Table 10.1. The mean rate of calls per 2-min interval, or λ, is computed by the following ratio

$$\lambda = \frac{T}{N} = \frac{423}{110} = 3.846$$

where

$N = \sum f$, or the total number of 2-min intervals observed

$T = \sum fk$, or the total number of incoming calls over the intervals observed

Table 10.1

k Number of calls observed in a 2-min interval	f Frequency of occurrence	fk Total number of calls
0	3	0
1	7	7
2	12	24
3	23	69
4	31	124
5	17	85
6	9	54
7	5	35
8	2	16
9	1	9
	110	423

In this case t refers to a 2-min interval; thus:

$t = 1$

$\lambda = 3.846$

and

$\lambda t = 3.846$

If t were to refer to a 1-min interval, then

$t = \frac{1}{2}$

$\lambda = 3.846$

and

$\lambda t = 1.923$

If t were to refer to a 4-min interval, then

$t = 2$

$\lambda = 3.846$

and

$\lambda t = 7.692$

As illustrated above, once the mean number of occurrences has been obtained for some convenient interval (2 min in the example above), it is a simple matter to adjust λ by t in order to obtain the estimated mean number of occurrences for some other interval of interest. As the length of t changes, the value of λt changes proportionately.

10.2 THE NATURE OF THE POISSON DISTRIBUTION

The Poisson probability distribution for certain selected values of λt is reproduced as Table 10.2 from Table D in the Appendix. The distributions are also depicted graphically in Figure 10.1.

Table 10.2 Representative Poisson distributions for selected values of λt

Number of successes observed	$\lambda t = 0.1$	$\lambda t = 0.2$	$\lambda t = 0.6$	$\lambda t = 1.0$	$\lambda t = 4.0$	$\lambda t = 8.0$
0	0.9048	0.8187	0.5488	0.3679	0.0183	0.0003
1	0.0905	0.1637	0.3293	0.3679	0.0733	0.0027
2	0.0045	0.0164	0.0988	0.1839	0.1465	0.0107
3	0.0002	0.0011	0.0198	0.0613	0.1954	0.0286
4	0.0000	0.0001	0.0030	0.0153	0.1954	0.0573
5		0.0000	0.0004	0.0031	0.1563	0.0916
6			0.0000	0.0005	0.1042	0.1221
7				0.0001	0.0595	0.1396
8					0.0298	0.1396
9					0.0132	0.1241
10					0.0053	0.0993
11					0.0019	0.0722
12					0.0006	0.0481
13					0.0002	0.0296
14					0.0001	0.0169
15						0.0090
16						0.0045
17						0.0021
18						0.0009
19						0.0004
20						0.0002
21						0.0001

The Poisson distribution is positively skewed. When λt is small, the distribution has a reverse J-shaped pattern. Thus, for small values of λt, the probability of observing x events in the interval t continuously declines and rapidly approaches zero as x increases. As the value of λt increases, the degree of skewness declines, approaching the symmetrical pattern of the normal distribution as the limiting case.[2] The above behavior patterns are clearly reflected in the Poisson distributions depicted in Figure 10.1.

EXAMPLE 10.4

Given that λ, or the mean rate of calls per *2-min interval*, is equal to 3.846 calls; generate the probability distribution describing the pattern of outcomes likely to be observed for *4-min intervals*.

$$\lambda = 3.846$$

$$t = 2$$

$$\lambda t = 7.692$$

$$P(X) = \frac{(\lambda t)^x e^{-\lambda t}}{x!} = \frac{(7.70)^x e^{-7.70}}{x!}$$

Referring to Table D in the Appendix (using 7.7 as the closest value for λt), we obtain the values in Table 10.3.

---

[2] The normal probability distribution is discussed in Chapter 11.

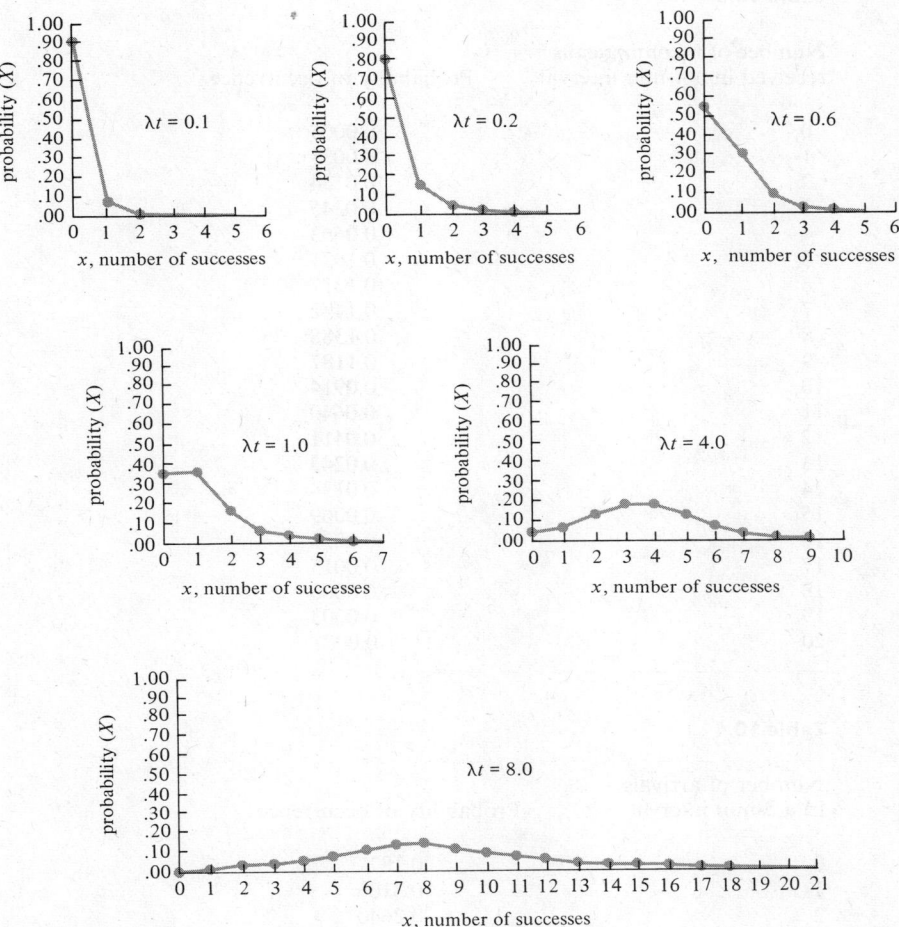

Figure 10.1

EXAMPLE 10.5

1. Given that the mean rate of arrival of license applicants at the Registry of Motor Vehicles is 1.7 arrivals per *5-min interval* between the hours 9:00 a.m. and 11:00 a.m., generate the probability distribution description of the pattern of arrivals expected to be observed for *5-min intervals*.

$$\lambda = 1.7$$

$$t = 1$$

$$\lambda t = 1.7$$

$$P(X) = \frac{(\lambda t)^x e^{-\lambda t}}{x!} = \frac{(1.7)^x e^{-1.7}}{x!}$$

Referring to Table D in the Appendix, the probability distribution given in Table 10.4 is obtained.

Table 10.3

Number of incoming calls received in a 4-min interval	Probability of occurrence
0	0.0005
1	0.0035
2	0.0134
3	0.0345
4	0.0663
5	0.1021
6	0.1311
7	0.1442
8	0.1388
9	0.1187
10	0.0914
11	0.0640
12	0.0411
13	0.0243
14	0.0134
15	0.0069
16	0.0033
17	0.0015
18	0.0006
19	0.0003
20	0.0001

Table 10.4

Number of arrivals in a 5-min interval	Probability of occurrence
0	0.1827
1	0.3106
2	0.2640
3	0.1496
4	0.0636
5	0.0216
6	0.0061
7	0.0015
8	0.0003
9	0.0001

2. Given the information in part 1, generate the probability distribution describing the expected pattern of arrivals for *10-min intervals*.

$$\lambda = 1.7$$

$$t = 2$$

$$\lambda t = 3.4$$

$$P(X) = \frac{(\lambda t)^x e^{-\lambda t}}{x!} = \frac{(3.4)^x e^{-3.4}}{x!}$$

Table 10. 5

Number of arrivals in a 10-min interval	Probability of occurrence
0	0.0334
1	0.1135
2	0.1929
3	0.2186
4	0.1858
5	0.1264
6	0.0716
7	0.0348
8	0.0148
9	0.0056
10	0.0019
11	0.0006
12	0.0002

EXAMPLE 10.6

Local law enforcement authorities have been advised that weekday robberies for suburban towns of similar income and population characteristics are Poisson distributed with a mean rate of occurrence of 0.8 robberies per day. Generate the probability distribution describing the relative frequency with which they could expect days with zero robberies, one robbery, two robberies, and so on.

$$\lambda = 0.8$$

$$t = 1$$

$$\lambda t = 0.8$$

$$P(X) = \frac{(\lambda t)^x e^{-\lambda t}}{x!} = \frac{(0.8)^x e^{-0.8}}{x!}$$

Table 10.6

Number of robberies observed (in one day)	Probability of occurrence
0	0.4493
1	0.3595
2	0.1438
3	0.0383
4	0.0077
5	0.0012
6	0.0002
7	0.0000

10.3 THE POISSON APPROXIMATION TO THE BINOMIAL

The Poisson distribution provides a useful approximation of the probabilities generated by a binomial probability distribution under the special conditions where

n is large and p is quite small.[3] A rule of thumb, customarily quoted, is that a Poisson approximation is legitimate provided $p \leq .05$ and $n \geq 20$. The fact that the Poisson provides a good approximation under these conditions is quite useful since computations involving the binomial become unwieldy when n is large.

EXAMPLE 10.7

Suppose that there are 1000 transistors in a space capsule and the probability that any one transistor will fail is 0.0001. What is the probability that all 1000 transistors will function properly?

Approaching the problem as a binomial, the desired probability would be

$$C \binom{1000}{1000} (0.0001)^0 (0.9999)^{1000}$$

which simplifies to

$$(0.9999)^{1000}$$

It is apparent that manipulation of $(0.9999)^{1000}$ is unwieldy.

A Poisson approximation of the desired probability is obtained as follows: The mean of a binomial probability distribution is equal to np. Therefore,

$$\lambda = np$$

or the expected mean number of occurrences over the interval in question, namely, the 1000 transistors.

Substituting:

$$\lambda = 1000(0.0001)$$

$$\lambda = .1$$

$$t = 1$$

Therefore,

$$P(X) = \frac{(\lambda t)^x e^{-\lambda t}}{x!} = \frac{(1)^x e^{-.1}}{x!}$$

$$P(X = 0) = \frac{(1)^0 e^{-.1}}{0!} = e^{-.1}$$

From Table E in the Appendix, $e^{-.1}$ equals 0.90484. The probability could have been obtained directly, of course, from the table of Poisson probabilities, Table D in the Appendix. From the Table, when $\lambda = .1$, $P(X = 0) = 0.9048$.

The above computations disclose that all 1000 transistors would be expected to function correctly on approximately 90 of every 100 missions.

EXAMPLE 10.8

When $n = 100$ and $p = 0.01$, the binomial probability distribution table provides the probability distribution for number of successes given in Table 10.7.

---

[3] If np is held constant, while n increases without limit and p approaches zero, the binomial probability distribution approaches the Poisson as a limiting case. A proof is provided in Technical Note 9 at the end of the chapter.

Table 10.7

Number of successes	Probability of occurrence
0	0.3660
1	0.3697
2	0.1849
3	0.0610
4	0.0149
5	0.0029
6	0.0005
7	0.0001
8	0.0000

The Poisson approximation is as follows:

$$\lambda = np = 100(0.01) = 1$$

$$t = 1$$

Thus,

$$\lambda t = 1$$

$$P(X) = \frac{(\lambda t)^x e^{-\lambda t}}{x!} = \frac{(1)^x e^{-1}}{x!}$$

From Table E in the Appendix, e^{-1} equals 0.36788. Therefore,

$$P(X) = \frac{1^x (0.36788)}{x!}$$

It is not necessary to compute the individual terms, since this information is recorded in the table of Poisson probabilities. From Table D in the Appendix, we obtain the values in Table 10.8.

Table 10.8

Number of successes	Probability of occurrence Poisson approximation with $\lambda = 1.0$
0	0.3679
1	0.3679
2	0.1839
3	0.0613
4	0.0153
5	0.0031
6	0.0005
7	0.0001
8	0.0000

EXAMPLE 10.9

When operating as designed, a manufacturing process produces 2 percent defectives. Given that a sample of size 300 is to be selected:

1. What is the probability of observing exactly ten defectives?
2. What is the probability of observing ten or more defectives?
3. Determine the number of defectives that you would not expect to exceed more than 5 percent of the time if the process is operating correctly.

Using the Poisson approximation,

$$\lambda = np = 300(0.02) = 6$$

$$t = 1$$

Therefore,

$$\lambda t = 6$$

$$P(X) = \frac{(\lambda t)^x e^{-\lambda t}}{x!} = \frac{6^x e^{-6}}{x!}$$

$$e^{-6} = 0.00248$$

Therefore,

$$P(X) = \frac{6^x (0.00248)}{x!}$$

The above formula could be used to compute the desired probabilities directly. Turning instead to the table of Poisson probabilities, the following values are obtained:

1. $P(X = 10)$ when $\lambda = 6$ is 0.0413.
2. $P(X \geq 10)$ when $\lambda = 6$ is 0.0838.

$$P(X = 10) = 0.0413$$
$$P(X = 11) = 0.0225$$
$$P(X = 12) = 0.0113$$
$$P(X = 13) = 0.0052$$
$$P(X = 14) = 0.0022$$
$$P(X = 15) = 0.0009$$
$$P(X = 16) = 0.0003$$
$$P(X = 17) = 0.0001$$
$$\overline{\qquad\qquad 0.0838}$$

3. $P(X \geq 10) = 0.0838$

$$P(X \geq 11) = 0.0838 - 0.0413 = 0.0425$$

Thus, the number of defectives that you would not expect to exceed more than 5 percent of the time when the process is operating correctly is 11. In fact, you would not expect to observe 11 or more defectives more than 4.25 percent of the time.

10.4 THE EXPONENTIAL DISTRIBUTION

The Poisson distribution generates the probability that there will be x occurrences of the event in question during some interval t. A related question is: What is the pattern of time (or distance, etc.) that elapses between successive occurrences of the event in question? The time (or distance, etc.) that elapses between successive occurrences is referred to as the *inter-arrival time*. The exponential distribution generates the probability distribution of inter-arrival times characterizing the activity of a Poisson process.

The Poisson distribution is a discrete distribution. Outcomes are restricted to nonnegative integers such as 0, 1, 2, 3, and so on. The exponential distribution is a continuous distribution, and the variable of interest, inter-arrival time, can take on any value between zero and plus infinity.

The exponential distribution is characterized by the function

$$P(t) = \frac{1}{\mu} e^{-t/\mu}$$

where

t represents the amount of time elapsed before the occurrence of the next event
e is the base of the natural logarithm
$\mu = 1/\lambda$, where λ is the parameter of the associated Poisson process. Recall that λ refers to the mean number of occurrences in some stipulated interval. μ, or $1/\lambda$, refers to the mean time elapsed between successive Poisson occurrences.

To illustrate: If $\lambda = 5$ per hour, then $\mu = 1/\lambda = 1/5$ or an average time elapse between occurrences of 20 min.

The mean of the exponential distribution is $1/\lambda$, and the *variance* equals $1/\lambda^2$. Thus, the mean and the *standard deviation* of the exponential distribution are equal. Recall that the mean and the *variance* of the Poisson distribution are equal and have a value of λ.

The exponential distribution is skewed to the right and takes a reverse J shape, as depicted in Figure 10.2.

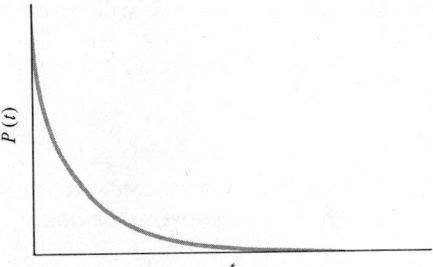

Figure 10.2

The following relationship provides a convenient method for deriving the probabilities of the exponential distribution.

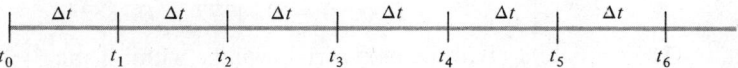

The horizontal line represents continuous time subdivided into segments of interval Δt. Given that the occurrence of events in question are Poisson distributed, the probability that no event will occur in any given time interval, Δt, is, in general,

$$P(X) = \frac{(\lambda \Delta t)^x e^{-\lambda \Delta t}}{x!}$$

where

Δt is defined as an elapse of time sufficiently small that at most one event can occur

λ is the mean number of successes in Δt; thus $\lambda \Delta t = \lambda$

$x = 0$

Substituting:

$$P(X = 0) = \frac{(\lambda)^0 e^{-\lambda}}{0!} = e^{-\lambda}$$

It follows that the probability that *exactly one event* will occur in any given time interval, Δt, as defined above, is in general,

$$P(X = 1) = 1 - e^{-\lambda}$$

The probability that five such time periods, Δt, will elapse before the occurrence of the next event is

$$P\begin{pmatrix}\text{no event}\\\text{in period}\\1\end{pmatrix} \cdot P\begin{pmatrix}\text{no event}\\\text{in period}\\2\end{pmatrix} \cdot P\begin{pmatrix}\text{no event}\\\text{in period}\\3\end{pmatrix} \cdot P\begin{pmatrix}\text{no event}\\\text{in period}\\4\end{pmatrix} \cdot P\begin{pmatrix}\text{no event}\\\text{in period}\\5\end{pmatrix}$$

$$\cdot P\begin{pmatrix}\text{an event}\\\text{in period}\\6\end{pmatrix}$$

$$= (e^{-\lambda \Delta t})(e^{-\lambda \Delta t})(e^{-\lambda \Delta t})(e^{-\lambda \Delta t})(e^{-\lambda \Delta t})(1 - e^{-\lambda \Delta t})$$
$$= (e^{-5\lambda \Delta t})(1 - e^{-\lambda \Delta t})$$
$$= (e^{-\lambda(5 \Delta t)})(1 - e^{-\lambda \Delta t})$$
$$= e^{-\lambda(5 \Delta t)} - e^{-\lambda(6 \Delta t)}$$

In general, the probability that elapsed inter-arrival time will be between t_n and t_{n+1} is $e^{-\lambda t_n} - e^{-\lambda t_{n+1}}$. More formally:

$$P(t_n < t < t_{n+1}) = e^{-\lambda t_n} - e^{-\lambda t_{n+1}}$$

EXAMPLE 10.10

Given that the mean rate of arrival of license applicants at the Registry of Motor Vehicles is 1.7 arrivals per 5-min interval between the hours of 9:00 a.m. and 11:00 a.m.

1. What is the probability that the next arrival will arrive within 1 min of the last arrival?

$$\lambda = 1.7$$

$$e^{-(1.7)(0)} - e^{-(1.7)(1/5)} = e^{-0.00} - e^{-0.34}$$

From Table E in the Appendix,

$$e^{-0.00} - e^{-0.34} = 1.00000 - 0.71177 = 0.28823$$

The probability that the next arrival will be within 1 min is 28.8 percent.

2. What is the probability that the next arrival will arrive within 2 min of the last arrival?

$$\lambda = 1.7$$

$$e^{-(1.7)(0)} - e^{-(1.7)(2/5)} = e^{-0.00} - e^{-0.68}$$

From Table E in the Appendix,

$$e^{-0.00} - e^{-0.68} = 1.00000 - 0.50662 = 0.49338$$

The probability that the next arrival will be within 2 min is 49.338 percent.

3. What is the probability that between 1 and 2 min will elapse before the next arrival?

$$\lambda = 1.7$$

$$e^{-(1.7)(1/5)} - e^{-(1.7)(2/5)} = e^{-0.34} - e^{-0.68}$$

From Table E in the Appendix,

$$e^{-0.34} - e^{-0.68} = 0.71177 - 0.50662 = 0.20515$$

The probability that between 1 and 2 min will elapse before the next arrival is 20.515 percent.

Table 10.9(a) Probability distribution of inter-arrival times

Inter-arrival time	Probability of occurrence
0–1 min	0.82796
1–2 min	0.14184
2–3 min	0.02531
3–4 min	0.00408

Table 10.9(b) Supporting computations for Table 10.9(a)

Inter-arrival time (1-min intervals)	$e^{-\lambda t_n} - e^{-\lambda t_{n+1}}$
0–1 min	$= e^{-(3.525)(0)} - e^{-(3.525)(0.5)}$ $= e^{-0.00} - e^{-1.76}$ $= 1.00000 - 0.17204$ $= 0.82796$
1–2 min	$= e^{-3.525(0.5)} - e^{-3.525(1)}$ $= e^{-1.76} - e^{-3.50}$ $= 0.17204 - 0.03020$ $= 0.14184$
2–3 min	$= e^{-3.525(1)} - e^{-3.525(1.5)}$ $= e^{-3.525} - e^{-5.288}$ $= 0.03020 - 0.00499$ $= 0.02531$
3–4 min	$= e^{-3.525(1.5)} - e^{-3.525(2)}$ $= e^{-5.288} - e^{-7.050}$ $= 0.00499 - 0.00091$ $= 0.00408$

EXAMPLE 10.11

Given that λ, or the mean rate of incoming calls per 2-min interval, is equal to 3.525 calls; generate a probability distribution of the pattern of inter-arrival times.

The probability distribution of the inter-arrival times is given in Table 10.9(a) and the supporting computations in Table 10.9(b).

TECHNICAL NOTE 9

This Technical Note gives a derivation of the Poisson distribution as a limiting case of the binomial distribution when np is held constant, while n increases without limit and p approaches zero.

The formula for an individual term in the binomial probability distribution is

$$\frac{n!}{x!\,(n-x)!}\,p^x q^{n-x}$$

Expanding the factorial in the numerator, we obtain

$$\frac{n(n-1)(n-2)\cdots(n-x+1)(n-x)!}{x!\,(n-x)!}\,p^x q^{n-x}$$

Cancelling generates the following equivalent relationship:

$$\frac{n(n-1)(n-2)\cdots(n-x+1)}{x!}\,p^x q^{n-x}$$

The expected number of occurrences for the binomial is equal to np.

Let $\lambda = np$. Solving for p,

$$p = \frac{\lambda}{n}$$

Substituting,

$$\frac{n(n-1)(n-2)\cdots(n-x+1)}{x!}\left(\frac{\lambda}{n}\right)^x\left(1-\frac{\lambda}{n}\right)^{n-x}$$

$$\left(\frac{\lambda}{n}\right)^x = \frac{\lambda^x}{n^x}$$

Multiplying $(n)(n-1)(n-2)\cdots(n-x+1)$ by $1/n^x$ is equivalent to

$$\frac{n(n-1)(n-2)\cdots(n-x+1)}{n\cdot n\cdot n\cdot\;\cdots\;\cdot n}$$

Letting $x = 4$,

$$\frac{(n)(n-1)(n-2)(n-3)}{n\cdot n\cdot n\cdot n}$$

is equivalent to

$$\frac{n}{n}\cdot\frac{(n-1)}{n}\cdot\frac{(n-2)}{n}\cdot\frac{(n-3)}{n}$$

or

$$(1)\left(1 - \frac{1}{n}\right) \cdot \left(1 - \frac{2}{n}\right) \cdot \left(1 - \frac{3}{n}\right)$$

In general,

$$(1)\left(1 - \frac{1}{n}\right)\left(1 - \frac{2}{n}\right) \cdots \left(1 - \frac{x-1}{n}\right)$$

Substituting,

$$\frac{(1)(1 - 1/n)(1 - 2/n) \cdots [1 - (x - 1)/n]}{x!} \lambda^x \left(1 - \frac{\lambda}{n}\right)^{n-x}$$

Writing $(1 - \lambda/n)^{n-x}$ as

$$\left[\left(1 - \frac{\lambda}{n}\right)^{-n/\lambda}\right]^{-\lambda} \left(1 - \frac{\lambda}{n}\right)^{-x}$$

and substituting, we obtain

$$\frac{(1)(1 - 1/n)(1 - 2/n) \cdots [1 - (x - 1)/n]}{x!} \lambda^x \left[\left(1 - \frac{\lambda}{n}\right)^{-n/\lambda}\right]^{-\lambda} \left(1 - \frac{\lambda}{n}\right)^{-x}$$

Letting n go to infinity, while x and λ remain fixed,

$$(1)\left(1 - \frac{1}{n}\right)\left(1 - \frac{2}{n}\right) \cdots \left(1 - \frac{x-1}{n}\right) \text{ approaches } (1)(1)(1) \cdots (1) = 1$$

$$\left(1 - \frac{\lambda}{n}\right)^{-x} \text{ approaches } (1)^{-x} = \frac{1}{(1)^x} = \frac{1}{1} = 1$$

$$\left(1 - \frac{\lambda}{n}\right)^{-n/\lambda} \text{ approaches } e = 2.71818, \text{ the base of the natural logarithm}$$

Thus,

$$\frac{(1)(1 - 1/n)(1 - 2/n) \cdots [1 - (x - 1)/n]}{x!} \lambda^x \left[\left(1 - \frac{\lambda}{n}\right)^{-n/\lambda}\right]^{-\lambda} \left(1 - \frac{\lambda}{n}\right)^{-x}$$

approaches

$$\frac{(1)}{x!} \lambda^x e^{-\lambda}$$

This, of course, is the formula for finding the probability of an individual term in the Poisson distribution:

$$P(x) = \frac{(\lambda)^x e^{-\lambda}}{x!}$$

PROBLEMS

Problem 10.1

In a large metropolitan center, the ALA patrols the expressway during peak travel hours, offering towing and road-call service free of charge as a public service. The number of cars serviced *per hour* is Poisson distributed with a mean of 1.5.

1. What is the probability on a given day that there will be no cars serviced during a 2-hr period?
2. Generate the probability distribution of the number of cars serviced daily if the service is offered 4 hr per day.

Problem 10.2

The mean number of defects in bolts of cloth produced by Quality Mills has been determined to be five defects per bolt of cloth.
1. What is the probability that a bolt of cloth will have no defects?
2. What is the probability that a bolt of cloth will have seven or more defects?

Problem 10.3

A production process has been determined to operate at a 2 percent defective rate.
1. Using the binomial probability distribution, generate the probability distribution describing the number of defects observed per lot of size 100.
2. Repeat part 1, using the Poisson probability distribution approximation. Evaluate the quality of the approximation.
3. Using the Poisson probability distribution approximation, generate a description of the probability distribution describing the number of defects observed per lot of size 500.

Problem 10.4

The number of arrivals at the university computer facility quick-batch service counter between 12 noon and 2 p.m. is Poisson distributed with a mean of 1.2 per minute.
1. What is the probability of no arrivals during a given *1-min* interval?
2. What is the probability of no arrivals during a given *2-min* interval?
3. What is the probability of no arrivals during a given *5-min* interval?

Problem 10.5

The distributor of Ida Hoe Potatoes claims that the number of spoiled potatoes per 5-lb bag is Poisson distributed with a mean of 0.2 potatoes per bag. If the distributor's claim is correct:
1. What is the probability of a customer purchasing a bag containing no spoiled potatoes?
2. What is the probability of a customer purchasing a bag containing one spoiled potato?
3. What is the probability of a customer purchasing a bag containing two spoiled potatoes? Three spoiled potatoes? Four spoiled potatoes?
4. What is the probability of a customer purchasing a bag containing five or more spoiled potatoes?

Problem 10.6

A switchboard for a telephone answering service receives on the average 3.8 calls per minute. Assuming that the distribution of incoming calls is Poisson distributed, determine the following.

1. The probability that during any given minute there will be no incoming calls.
2. The probability there will be exactly three incoming calls during any given minute.
3. The probability of four or more calls during any given minute.

Problem 10.7

If the probability that a single item selected from a manufacturing process will be defective is 0.012, what is the probability that a sample of size 200 will contain exactly zero defectives? Exactly one defective? Exactly two defectives? Exactly three defectives? Exactly four defectives? Five or more defectives?

Problem 10.8

A particular manufacturing operation is subject to occasional stoppages due to the failure of an intricate switching mechanism. The pattern of failures has been characterized by a Poisson distribution with a mean time elapsed between failures of 10 hr. The manufacturing operation is conducted four times a year, each time running for 100 hr of continuous production. Failure of the switching mechanism involves a slight delay, provided switching mechanisms are in inventory. If no switching mechanisms are available, a special emergency order must be placed and an expensive shutdown is incurred. The cost of a shutdown is estimated at $4000. The cost of carrying switching mechanisms is inventory is estimated at $150 per switch carried. Given the objective of minimizing expected cost, what is the optimal number of switching mechanisms to carry in inventory?

Problem 10.9

A bartender in a large restaurant numbers among his responsibilities the fulfilling of cocktail orders for waitresses. During the hours of 6 to 9 p.m., the arrival of waitresses is Poisson distributed with a mean arrival rate of 40 per hour. The exponential distribution describes the distribution of inter-arrival times. Generate the probabilities associated with the list of outcomes provided below:

Time elapsed
between successive
arrivals
Inter-arrival time

0–1 min

1–2 min

2–3 min

3–4 min

4–5 min

5–6 min

6–7 min

7–8 min

Problem 10.10

Reliable Electronics, Inc., claims that the failures of a particular transistorized component are Poisson distributed, with an average of only three failures per 100,000 hr of operation.

1. What is the probability that at least 25,000 hr will elapse before the first failure is observed?
2. What is the probability that at least 50,000 hr will elapse before the first failure is observed?
3. What is the probability that at least 75,000 hr will elapse before the first failure is observed?
4. What is the probability that at least 100,000 hr will elapse before the first failure is observed?
5. What is the probability that at least 150,000 hr will elapse before the first failure is observed?
6. What is the probability that at least 200,000 hr will elapse before the first failure is observed?

Problem 10.11

Referring to Problem 10.1: The number of cars serviced per hour was Poisson distributed with a mean of 1.5. The exponential distribution describes the distribution of inter-arrival times. Generate the probabilities associated with the list of outcomes below:

*Time elapsed
between successive
arrivals*

0–20 min

20–40 min

40–60 min

60–80 min

80–100 min

100–120 min

120–140 min

140–160 min

160–180 min

180–200 min

Problem 10.12

Incoming calls at a switchboard were observed to be Poisson distributed with a mean of 3.6 calls per minute. The exponential distribution describes the distribution of inter-arrival times. Using intervals of 10-second duration, generate a probability distribution of the inter-arrival times.

Problem 10.13

Customer arrivals at a pizza parlor during the rush hour are Poisson distributed with a mean arrival rate of 8 per hour (i.e., $a = 8$). Pizza preparation time is 6 min, assuming no waiting time. Thus, the mean service rate is 10 per hour ($s = 10$). Given that arrivals are Poisson distributed and service time is constant, the following formulas are applicable provided a, the arrival rate, is less than s, the service rate.

$$\text{expected length of the waiting line} = \frac{a^2}{2s(s - a)}$$

$$\text{expected percentage of the time the facility will be idle} = 1 - \frac{a}{s}$$

$$\text{expected waiting time of a unit before entering the system} = \frac{a}{2s(s - a)}$$

1. Determine the expected length of the waiting line at the pizza parlor during the rush hour.
2. Determine the expected percentage of the time the facility will be idle during the rush-hour period.
3. Determine the expected waiting time of a unit before entering the system.
4. If the time to prepare a pizza could be reduced to 4 min, how would the characteristics of the system as computed in parts 1, 2, and 3 be altered?

Problem 10.14

Nan Tucket runs a bicycle repair shop. Customer requests for service are Poisson distributed with a mean rate of 2 requests per hour ($a = 2$). Service times are distributed exponentially. The average service time is 20 min. Thus, the mean service rate is 3 per hour ($s = 3$). The following formulas can be used to determine the characteristics of the system provided that arrivals are Poisson distributed, service times are exponentially distributed, and the service rate is greater than the arrival rate:

$$\text{expected length of the waiting line} = \frac{a^2}{s(s - a)}$$

$$\text{expected percentage of the time the facility will be idle} = 1 - \frac{a}{s}$$

$$\text{expected waiting time of a unit before entering the system} = \frac{a}{s(s - a)}$$

Using the above formulas, determine the characteristics of the system described above.

Problem 10.15

Arrivals of employees at a tool crib counter serviced by a single attendant are Poisson distributed with a mean arrival rate of 6 per hour. Service times are exponentially distributed with a mean service time of 6 min (or a mean service time of 10 per hour). The following formulas can be used to determine the characteristics of the system provided that arrivals are Poisson distributed, service times are exponentially distributed, and the service rate is greater than the arrival rate:

$$\begin{array}{l} \text{expected length of} \\ \text{the waiting line} \end{array} = \frac{a^2}{s(s-a)}$$

$$\begin{array}{l} \text{expected percentage} \\ \text{of the time the} \\ \text{facility will be idle} \end{array} = 1 - \frac{a}{s}$$

$$\begin{array}{l} \text{expected waiting} \\ \text{time of a unit} \\ \text{before entering the system} \end{array} = \frac{a}{s(s-a)}$$

Using the above formulas, determine the characteristics of the system described above.

Problem 10.16

Referring to Problem 10.15: Assume that the hourly wage rate for the tool crib attendant is $2.00 and that the time of employees serviced by the tool crib is valued at $5 per hour. Determine the expected cost of operating the system for an 8-hr period.

$$\begin{array}{l} \text{expected time} \\ \text{an employee spends in} \\ \text{the system} \end{array} = \begin{array}{l} \text{expected waiting time} \\ \text{before entering} \\ \text{the system} \end{array} + \begin{array}{l} \text{expected} \\ \text{service} \\ \text{time} \end{array}$$

$$= \frac{a}{s(s-a)} + \frac{1}{s}$$

$$= \frac{1}{s-a}$$

Problem 10.17

Referring to Problem 10.16: If a more efficient attendant could service employees with an average service time of 4 min, what is the maximum hourly wage you could pay the attendant without increasing the expected cost of operating the system?

THE NORMAL PROBABILITY DISTRIBUTION: A CONTINUOUS PROBABILITY DISTRIBUTION

11

11.1 DISCRETE VERSUS CONTINUOUS DISTRIBUTIONS

In Chapter 7, a probability distribution was defined as a list of all possible outcomes to an experiment along with the associated probability of occurrence of each of the outcomes. It was pointed out that certain probability distribution patterns tend to occur with regularity. Moreover, it was observed that by analyzing underlying characteristics of the "experiment" involved, it is often possible to determine in advance which probability distribution pattern will emerge. In Chapter 7, discussion centered on two discrete probability distributions—the binomial and the hypergeometric distributions. In discrete probability distributions, outcomes are restricted to certain values among the range of values on the number line spanned by the experiment. Continuous probability distributions deal with experiments in which outcomes can take on any value over the range of possible values. Discrete probability distributions are often treated as if they were continuous in situations where there are many outcomes and the gaps between permissible values on the number line are insignificant. In this chapter, a probability distribution of the continuous type will be examined—the normal probability distribution.

11.2 CHARACTERISTICS OF THE NORMAL DISTRIBUTION

The normal probability distribution is one of the most useful and important distributions in statistical analysis. Carl Gauss, in the process of obtaining repeated measurements of the orbit of a planet, observed that measurements tended to be distributed around the mean value in the pattern of a normal distribution. Gauss believed the explanation for this phenomenon to be the fact that each observation consisted of the true magnitude of the object being measured plus a measurement error. The magnitude and direction of the error associated with any one observation was attributed to chance influences operating at that moment. Note the symmetrical pattern reflected in Figure 11.1. A symmetrical pattern emerges because errors in measurement in one direction are just as likely to occur as errors in measurement in the other direction. The bell-shaped pattern is also in accordance with expectations. The larger the magnitude of the error, the less the likelihood of its occurrence. The net influence of these factors cause the observations to be distributed in a bell-shaped, unimodal pattern around the true magnitude of the object being measured. This distribution has come to be known as the *normal curve of error* because of the tendency for measurement errors in repeated observations of the same object to reflect this pattern. The normal probability distribution pattern, however, is not restricted to repeated measurements of some true magnitude. Whenever values of a variable differ from their mean as a result of a myriad of independent chance influences operating on the phenomenon being observed, the resulting probability distribution depicting values of the variable and their likelihood of occurrence will tend to be a normal probability distribution. Many variables, such as human characteristics (height, weight, I.Q.), dimensions and attributes of machine processes (diameters of parts, tensile strength, weight), patterns of human behavior (demand for a product), and so on, are subject to just such a battery of influences and tend to be described by normal probability distributions.

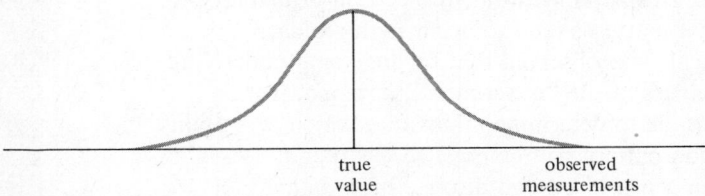

Figure 11.1 The normal probability distribution.

Perhaps the most important application of the normal distribution is with respect to sampling distributions. It can be demonstrated that if all possible samples of a given size were to be drawn from a universe and the value of certain statistics of those samples computed (for example, the arithmetic mean), that the values of the sample statistics tend to be distributed around the value of the corresponding universe parameter in the pattern of a normal probability distribution. The above relationship holds regardless of the shape of the universe, provided that sample size is sufficiently large. Although not all sample statistics behave in this fashion, enough of them do to place the normal probability distribution in a predominant position in the area of statistical inference. The role of the normal distribution in this context is discussed in Chapters 12–15.

Although the normal probability distribution is sometimes referred to as the Gaussian distribution, it was actually discovered not by Gauss, but by Alexander Demoivre in 1733. Demoivre was investigating the behavior of the binomial distribution as n (the number of trials) was allowed to increase to infinity. He discovered that it approached the pattern of a normal probability distribution regardless of the value of p (the probability of success on any one given trial). A number of other probability distributions approach the normal probability distribution as a limiting form, which further emphasizes the strategic position of the normal distribution in statistical analysis.

The binomial probability distribution, the hypergeometric probability distribution, and the normal probability distribution can each be thought of as representing a family of distributions. All members of the same family are described by the same underlying mathematical relationship. In the case of discrete distributions, the mathematical relationship is referred to as the probability mass function, and it generates the probability of occurrence associated with a given discrete outcome.

To illustrate, the probability mass function for the binomial probability distribution is

$$P(n, r) = \frac{n!}{r!\,(n-r)!}\, p^r q^{n-r}$$

where

$P(n, r)$ = probability of r successes in n trials
n = the number of trials
r = the number of successes
p = probability of success on any one trial
q = probability of failure on any one trial

Substitution of particular values for n and p identify a particular member of the binomial probability distribution family. The value of r serves to designate a particular outcome in the distribution of that family member.

The probability mass function for the hypergeometric probability distribution is

$$P(n, r) = \frac{C\binom{N_1}{r} \cdot C\binom{N_2}{n-r}}{C\binom{N}{n}}$$

where

N = total number of elements in the universe
N_1 = total number of elements of one kind in the universe (successes)
N_2 = total number of elements of the other kind in the universe (failures)
n = size of sample
r = number of successes in the sample

$n - r$ = number of failures in the sample

$C\begin{pmatrix} N_1 \\ r \end{pmatrix}$ = number of combinations of N_1 items taking r at a time

$C\begin{pmatrix} N_2 \\ n - r \end{pmatrix}$ = number of different combinations of $n - r$ things that can be generated out of a pool of N_2 items

$C\begin{pmatrix} N \\ n \end{pmatrix}$ = number of different combinations of n things that can be generated out of a pool of N items

$P(n, r)$ = probability of r successes out of n trials (drawings made without replacement)

In the case of a continuous probability distribution, the mathematical relationship is referred to as the probability density function. The probability density function indicates the relative likelihood of occurrence of a particular value on the number line by the magnitude of the function associated with that outcome. Total area under a density function is equal to one. The probability of any specific outcome or group of outcomes is equal to the ratio of the area under the curve associated with those outcomes to the total area under the curve. It is often stated that the probability of occurrence of any given specific outcome under a continuous probability density function is zero, because although the density function has a value (height of the area under the curve associated with that outcome), the corresponding span on the number line is infinitely close to zero (base of the area under the curve associated with that outcome). Any value multiplied by a value infinitely close to zero generates a product infinitely close to zero. The distinctive mathematical family trait of the normal probability distribution is its density function:

$$y = \frac{1}{\sqrt{2\pi}\, \sigma} e^{-(1/2)[(X-\mu)/\sigma]^2}$$

where π and e are symbols that the literature has assigned to mathematical values that have such useful properties and are so frequently employed that they deserve an identity of their own.

π is the ratio of the circumference of a circle to the diameter of a circle and is equal to 3.14159

e is the base of the natural log and is approximately equal to 2.7182[1]

μ is the arithmetic mean of the distribution

σ is the standard deviation of the distribution

X refers to the value of an outcome to the experiment—a location on the number line

y is the value of the density function or the height of the function above a given value of X being referenced on the horizontal axis or number line

---

[1] When the exponent to e is fairly complicated, there is a convenient mathematical shorthand notation often used to make equations more readable. In this case, if we let

$$x = -\frac{1}{2}\left(\frac{X - \mu}{\sigma}\right)^2$$

then e^x may be written as exp (x), and the equation above becomes

$$y = \frac{1}{\sqrt{2\pi\sigma}} \exp\left[-\frac{1}{2}\left(\frac{X - \mu}{\sigma}\right)^2\right]$$

We shall use this shorthand form whenever the exponent to e is complicated.

The parameters of the normal probability distribution are μ and σ. Given the value of μ and σ, the value of y (magnitude of the density function) can be determined for any given value of X. By inserting all possible values of X into the function, the entire distribution can be derived mathematically. Thus, knowledge of μ and σ and the fact that a distribution is normally distributed enables one to describe completely a normal probability distribution. Given a different mean and/or standard deviation, a different normal curve is derived. All such curves derived, however, possess the common family characteristic

$$y = \frac{1}{\sqrt{2\pi}\,\sigma} \exp\left[-\frac{1}{2}\left(\frac{X-\mu}{\sigma}\right)^2\right]$$

If the standard deviation is held constant and the mean varied, the effect is to maintain the same shape but to vary the location of the curve, as shown in Figure 11.2(a). If the mean is held constant and the standard deviation increased, the effect is a flattening of the curve, as shown in Figure 11.2(b); the reverse effect occurs with a decrease in the standard deviation, as shown in Figure 11.2(c).

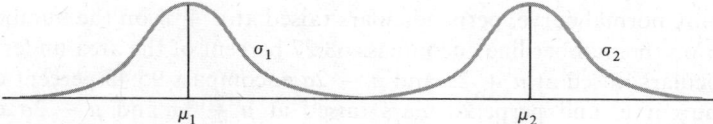

Figure 11.2(a) $\mu_1 > \mu_2$ but $\sigma_1 = \sigma_2$.

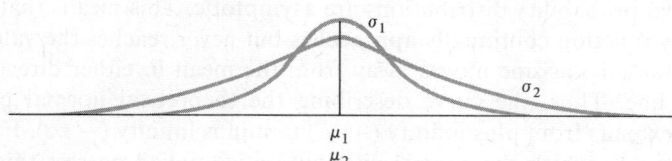

Figure 11.2(b) $\mu_2 = \mu_1$ but $\sigma_2 > \sigma_1$.

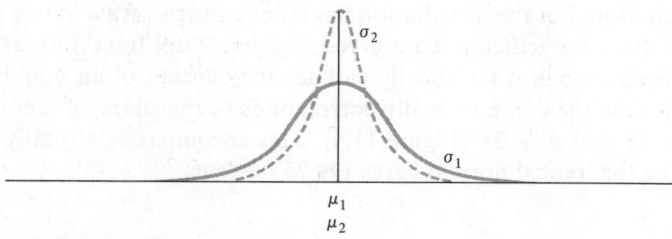

Figure 11.2(c) $\mu_2 = \mu_1$ but $\sigma_2 < \sigma_1$.

Although the physical appearance of the curves varies depending on the value of the standard deviation, certain characteristics are common to all normal curves. They are all symmetrical, unimodal, and bell shaped. Every normal probability distribution has inflection points at exactly one standard deviation away from the mean. At $\mu - \sigma$, the density function changes from increasing at an increasing rate to increasing at a decreasing rate. At $\mu + \sigma$, the function changes from decreasing at an increasing rate to decreasing at a decreasing rate. Such changeover points are referred to as points of inflection (Figure 11.3).

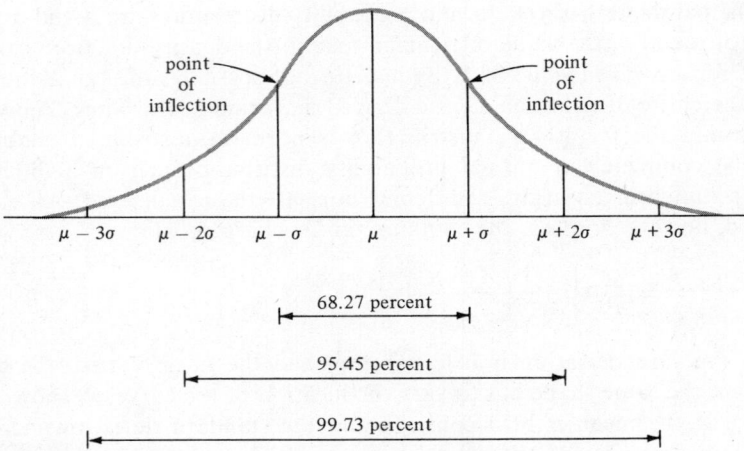

Figure 11.3

For any normal curve, perpendiculars raised at $\mu + \sigma$ on the number line and at $\mu - \sigma$ on the number line encompass 68.27 percent of the area under the curve. Perpendiculars raised at $\mu + 2\sigma$ and $\mu - 2\sigma$ encompass 95.45 percent of the area under the curve, and perpendiculars raised at $\mu + 3\sigma$ and $\mu - 3\sigma$ encompass 99.73 percent of the area under the curve. Once one has proceeded a distance of three standard deviations from the mean in both directions along the number line, virtually all of the area under the curve is encompassed.

All normal probability distributions are asymptotic. This means that the value of the density function continually approaches but never reaches the value of zero (the horizontal axis) as one moves away from the mean in either direction along the number line. Thus, the curve describing the theoretical normal probability distribution extends from plus infinity ($+ \infty$) to minus infinity ($- \infty$). Few, if any, statistical series to which the normal distribution is applied possess this attribute with respect to the range of possible outcomes. For example, numerical grades on an exam given to a large number of students would be expected to approximate a normal distribution, but the distribution has a finite range with a low of zero and a high of 100. It is nevertheless considered legitimate to treat this as normally distributed, even though it technically violates the concept of an infinite range of outcomes, because the curve basically corresponds to the shape of a normal curve between $\mu + 3\sigma$ and $\mu - 3\sigma$ (Figure 11.4). This encompasses virtually the entire area under the theoretical normal curve (99.73 percent).

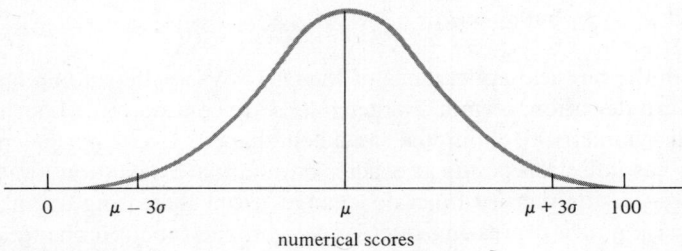

Figure 11.4

11.3 THE STANDARDIZED NORMAL DISTRIBUTION

Not only is it possible to compute the height of a normal curve for any given point on the number line by means of the density function

$$y = \frac{1}{\sqrt{2\pi}\sigma} \exp\left[-\frac{1}{2}\left(\frac{X-\mu}{\sigma}\right)^2\right]$$

but it is also possible to compute the proportion of the total area under the curve that lies between any two given points along the number line. This is accomplished by drawing on the tools of integral calculus and integrating the function

$$y = \frac{1}{\sqrt{2\pi}\sigma} \exp\left[-\frac{1}{2}\left(\frac{X-\mu}{\sigma}\right)^2\right]$$

over the range of values on the number line between the two points of concern.

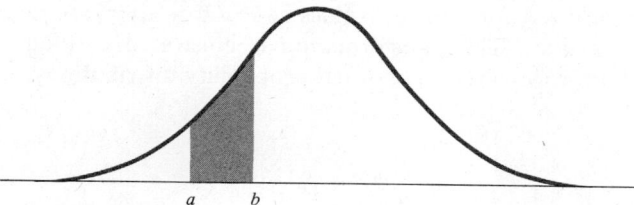

Figure 11.5

In Figure 11.5, integrating

$$y = \frac{1}{\sqrt{2\pi}\sigma} \exp\left[-\frac{1}{2}\left(\frac{X-\mu}{\sigma}\right)^2\right]$$

over the range a to b provides a value equal to the ratio of the shaded area to the total area under the curve. Recalling that relative area is equivalent to relative frequency of occurrence, the value obtained can be interpreted as the probability that the variable will take on a value between a and b. The direct computation of the relative area under the curve lying between any two given points on the number line by this approach can be quite unwieldy. A more desirable approach would be to have this information already calculated and recorded in tables. Then one could simply consult the appropriate table to obtain the proportion of total area under the curve lying between any two points on the number line. The fact that different curves are obtained with different values of μ and σ presents a problem. A separate table would be needed for each μ and σ combination. Such an arrangement is just not feasible. Fortunately, every normal probability distribution can be converted into a common-denominator normal probability distribution called the *standardized normal distribution*. The standardized normal probability distribution has a mean of zero and a standard deviation of one. One table showing the areas under the curve for the standardized normal distribution is all that need be constructed. Any given normal probability distribution is simply converted into common denominator form prior to consulting the table.

The nature of the conversion process is illustrated in Figure 11.6. Figure 11.6 is a generalized representation of a normal probability distribution. The objective of the transformation is to convert the original distribution, with a mean of μ and a standard deviation of σ, into a distribution with a mean of zero and a standard deviation of one. The first step is a change in location—the distribution is shifted to the zero point on the number line by subtracting the value of μ from each of the values in the distribution. This constitutes a change from the original unit scale (X scale) to a scale that expresses each value on the number line in terms of the amount and direction by which it differs from the mean as measured in original units ($X - \mu$ scale). The final step in the conversion is to divide each of the $X - \mu$ values by the value of the standard deviation. This expresses each of the values on the number line as a deviation from the mean in standard deviation units. The effect of this conversion is to create a distribution with a mean of zero and a standard deviation of one. Thus, any normal probability distribution can be converted to standardized form by applying the transformation $(X - \mu)/\sigma$ to the values on the number line, where X represents the values as expressed in the original number scale. $z = (X - \mu)/\sigma$ represents the new number scale when the distribution is converted to standardized form. The z values $[z = (X - \mu)/\sigma]$ are referred to as standard normal deviates. The transformation is illustrated in Figures 11.7(a), 11.7(b), and 11.7(c) for three specific normal probability distributions.

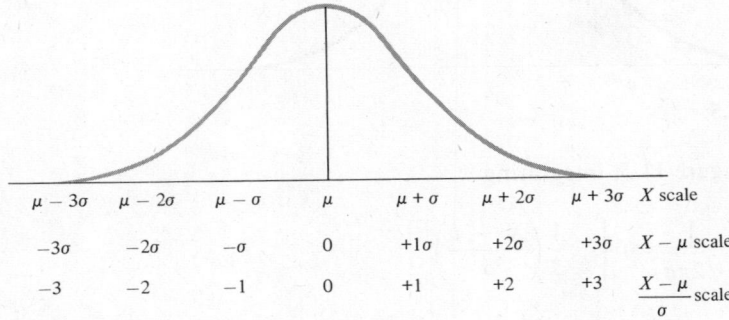

$\mu - 3\sigma$	$\mu - 2\sigma$	$\mu - \sigma$	μ	$\mu + \sigma$	$\mu + 2\sigma$	$\mu + 3\sigma$	X scale
-3σ	-2σ	$-\sigma$	0	$+1\sigma$	$+2\sigma$	$+3\sigma$	$X - \mu$ scale
-3	-2	-1	0	$+1$	$+2$	$+3$	$\dfrac{X-\mu}{\sigma}$ scale

Figure 11.6

A table has been constructed for the standardized normal distribution in which the proportion of the area under the curve that lies between a perpendicular raised at the maximum ordinate (the mean) and a perpendicular raised at varying standard deviation units away from the mean is recorded. The table entitled "Areas Under the Normal Curve" is reproduced in the Appendix as Table F. Obviously, the greater the number of standard deviation units between the mean and the value on the number line where the perpendicular is raised, the greater the proportion of the area under the curve included. Note that the table only refers to half the distribution. Due to the symmetrical nature of the distribution, a given number of standard deviation units away from the mean will encompass the same proportion of total area under the curve regardless of whether the point is located to the left or right of the mean. Thus, the table serves either half of the distribution equally well.

z values or standard normal deviates[2] are listed in the perimeter of the table.

[2] Standard normal deviates are deviations from the mean expressed in standard deviation units.

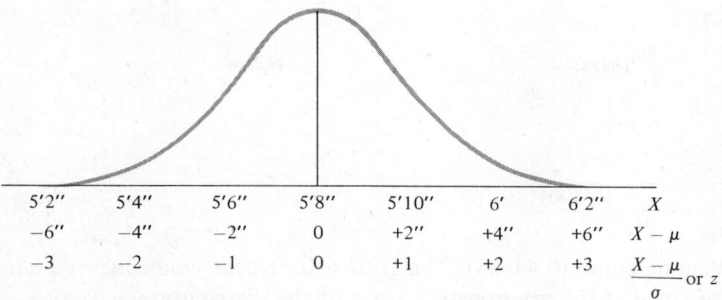

Figure 11.7(a) Height of Massachusetts residents: $\mu = 5'\, 8''$, $\sigma = 2''$.

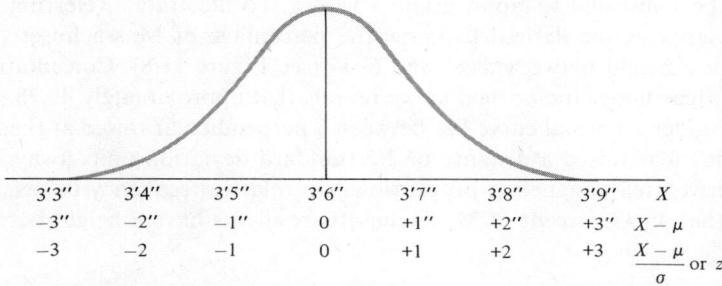

Figure 11.7(b) Height of African pygmies: $\mu = 3'\, 6''$, $\sigma = 1''$.

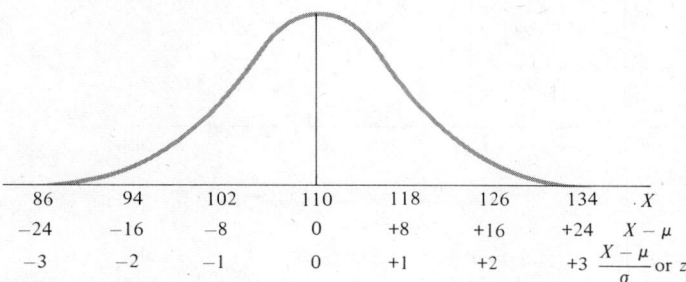

Figure 11.7(c) I.Q. in the freshman class at average university: $\mu = 110$ points, $\sigma = 8$ points.

The corresponding percentage of total area under the curve that lies between the mean and a point the given number of standard deviates away is recorded in the body of the table. To illustrate: Suppose that one wished to determine the percentage of the area under the curve that lies between the mean and a point 2.33 standard deviation units away. One would look for the z value 2.33 in the outer perimeter of the table. The units and the tenths digits determine the row, and the hundredths digit determines the column. The corresponding percentage of the area under the curve is the value located at the intersection of the 2.3 row and 0.03 column or 0.4901. Thus, 49.01 percent of the area under the curve is contained between the mean and a point on the number line 2.33 standard deviation units away (Table 11.1).

Note that a movement along the number line a distance of 3 standard deviation units away from the mean for all intents and purposes includes 50 percent of the

Table 11.1

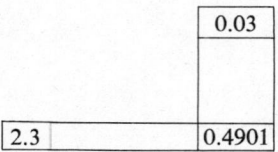

	0.03
2.3	0.4901

area under the curve (0.49865). The entire 50 percent would never be included, of course, because of the asymptotic nature of the distribution.

To utilize the table, it is not necessary to convert the entire distribution into standardized form—in fact, all that is required is that the particular X values of concern be converted to appropriate z values. To illustrate: Referring to Figure 11.7(a), suppose one desired to know the percentage of Massachusetts residents who have a height between 5' 8" and 6' 1" (see Figure 11.8). Consultation of the table of areas under the normal curve reveals that approximately 49.38 percent of the area under a normal curve lies between a perpendicular raised at the mean and a perpendicular raised a distance of 2.5 standard deviation units away. Recalling that relative area assigned is proportional to relative frequency of occurrence, it follows that 49.38 percent of Massachusetts residents have a height between 5' 8" and 6' 1".

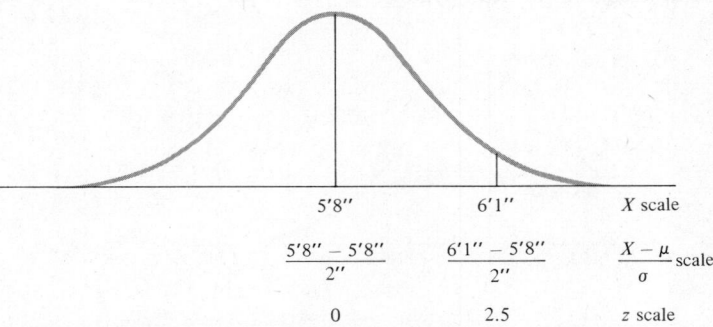

Figure 11.8 Height of Massachusetts residents: $\mu = 5'\,8''$, $\sigma = 2''$.

Note that to use the table of areas, one of the perpendiculars must always be raised at the mean. Frequently, the interval of interest on the number line does not possess this characteristic. For example, one might wish to determine the percentage of Massachusetts residents with a height between 5' 10" and 6' (Figure 11.9). Neither 5' 10" nor 6' correspond to the mean of the distribution. The area of interest (shaded area in Figure 11.9) can nevertheless be determined by finding the area under the curve between 5' 8" and 6' and then subtracting from it the area lying between 5' 8" and 5' 10". In this manner, the area under the curve between 5' 10" and 6' can be isolated. Although some manipulation may be required to obtain the area between the two points of interest on the number line, the question can be resolved for any two points under the curve. In the following section, a series of illustrative examples will be utilized to demonstrate the type of manipulation necessary to answer questions raised about relationships described by a normal probability distribution.

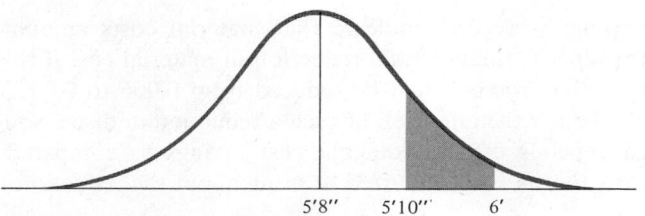

Figure 11.9 Height of Massachusetts residents: $\mu = 5'\,8''$, $\sigma = 2''$.

11.4 ILLUSTRATIVE EXAMPLES

One of the products manufactured by the Vantage Manufacturing Company is a candy bar called the Flaky Coconut Bar. The manufacturing process has been designed in such a fashion that the mean weight of all bars produced when the process is operating correctly is 1.5 oz. The weights of individual bars produced vary around this value, of course, due to the myriad of random influences operating on the process at any one point in time. The extent of variation is controllable and is inherent in the design of the manufacturing process. Extent of variability and the cost of the associated production process vary inversely. Thus, in designing the production process, the manufacturer must make a trade-off between the amount of variability and the costliness of the process. Vantage Manufacturing Company compromised on a production process characterized by a standard deviation of 0.006 oz. Weights of individual candy bars will thus be distributed in the pattern of a normal distribution with a mean equal to 1.5 oz and a standard deviation equal to 0.006 oz. (Recall that whenever values of a variable differ from their mean as a result of a myriad of independent chance influences operating on the phenomenon being observed, the resulting probability distribution depicting values of the variable and their likelihood of occurrence will tend to be a normal probability distribution.) The normal curve depicting the distribution of weights of Flaky Coconut Bars produced by this process will be used to illustrate the manner in which the table of areas is used to generate information about a process or variable described by a normal probability distribution.

In particular, answers will be sought to the following questions:

1. a. What percentage of the candy bars produced will have a weight between 1.5 oz and 1.512 oz?
   b. What percentage of the candy bars produced will have a weight between 1.488 oz and 1.5 oz?
   c. What percentage of the candy bars produced will have a weight between 1.491 oz and 1.509 oz?
   d. What percentage will have a weight between 1.512 oz and 1.518 oz?
   e. What percentage will have a weight between 1.488 oz and 1.491 oz?
   f. What percentage of the candy bars produced will have a weight greater than 1.515 oz?
   g. What percentage will have a weight less than 1.482 oz?
   h. Ten percent of the candy bars will weigh more than what value?
   i. The middle third of the candy bars produced will have a weight between what two values?
2. If company policy is to have no more than 3 percent of the candy bars produced contain a weight of less than 1.48 oz, what is the minimum level at which the mean weight could be established and still meet this requirement?

3. Suppose company records indicate that material costs amount to $0.04 per ounce. What is the potential reduction in material cost if the standard deviation of the process could be reduced from 0.006 to 0.002? (Keep in mind that whether the initiation of such a reduction in dispersion would be warranted depends on the potential cost savings as compared to incremental costs that must be incurred to bring about the reduction.)

1.a. What percentage of the candy bars produced will have a weight between 1.5 oz and 1.512 oz?

A graphic representation (Figure 11.10) of the situation is a useful aid in perceiving the approach to take to generate the desired information. In Problem 1a, the information desired is the proportion of the total area under the curve that lies between 1.5 oz and 1.512 oz. The question to be resolved can be restated as: What proportion of the total area under the curve lies between the mean and a point 0.012 oz away (1.512 − 1.5)? This, in turn, must be expressed in standard deviation units before the table of areas can be consulted. Given that the value of a standard deviation is 0.006 oz, a distance of 0.012 oz is equivalent to 0.012/0.006 or 2 standard deviation units.

$$z = \frac{X - \mu}{\sigma} = \frac{1.512 - 1.500}{0.006} = \frac{0.012}{0.006} = 2$$

Turning to the table of areas, the information desired is the proportion of total area under the curve that lies between a perpendicular raised at the mean and a perpendicular raised at a distance of 2 standard deviation units away along the number line. The table of areas indicates this to be 0.4772. Thus, 47.72 percent of the candy bars produced will have a weight between 1.5 oz and 1.512 oz.

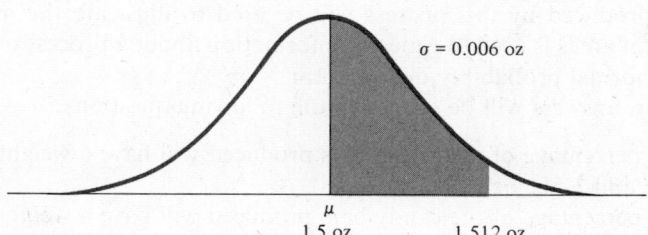

Figure 11.10

1.b. What percentage of the candy bars produced will have a weight between 1.488 oz and 1.5 oz?

The graphic presentation is shown in Figure 11.11. Note that Problem 1b is a mirror image of Problem 1a. The question to be resolved, once again, is: What proportion of the total area under the curve lies between the mean and a point 0.012 oz away (1.488 − 1.500)? In this case, however, the z value is negative, indicating that the region of concern lies to the left of the mean rather than to the right

$$z = \frac{X - \mu}{\sigma} = \frac{1.488 - 1.500}{0.006} = \frac{-0.012}{0.006} = -2$$

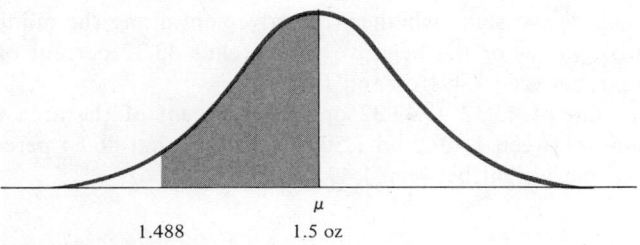

Figure 11.11

Recalling that the table of areas refers to either half of the curve because of the symmetrical nature of the normal distribution, one finds that 47.72 percent of the area under the curve lies between a perpendicular raised at the mean and a perpendicular raised at a distance of − 2 standard deviation units away along the number line. Given that relative area under the curve is proportional to relative frequency of occurrence, one can conclude that 47.72 percent of the candy bars produced will have a weight between 1.488 oz and 1.5 oz.

1.c. What percentage of the candy bars produced will have a weight between 1.491 oz and 1.509 oz?

The graphic presentation is shown in Figure 11.12. In Problem 1c, the area of interest is not bounded by the mean and therefore cannot be determined directly from the table of areas. The area can, however, be segmented into two areas, each of which is bounded by the mean on one side. One simply finds the proportion of total area under the curve for each of these segments and then combines them to find the proportion of total area under the curve lying in the region of interest.

The proportion of total area lying between 1.500 oz and 1.509 oz is determined as follows:

$$z = \frac{X - \mu}{\sigma} = \frac{1.509 - 1.500}{0.006} = \frac{0.009}{0.006} = 1.5$$

The table of areas reveals that 43.32 percent of the area under the curve lies between a perpendicular raised at the mean and a perpendicular raised at a point 1.5 standard deviation units away along the number line.

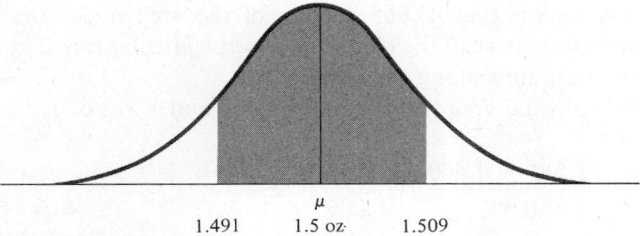

Figure 11.12

The proportion of total area lying between 1.491 oz and 1.500 oz is determined similarly:

$$z = \frac{X - \mu}{\sigma} = \frac{1.491 - 1.500}{0.006} = \frac{-0.009}{0.006} = -1.5$$

The percentage will be the same whether the movement along the number line is toward the positive values or the negative values, thus 43.32 percent of the area under the curve lies between 1.491 oz and 1.500 oz.

Therefore, a total of 43.32 + 43.32 or 86.64 percent of the area under the curve will be found between 1.491 and 1.509. It follows that 86.64 percent of the candy bars will have a weight between 1.491 oz and 1.509 oz.

1.d. What percentage of the candy bars produced will have a weight between 1.512 oz and 1.518 oz?

The graphic presentation is shown in Figure 11.13. In Problem 1c, the boundaries of the region of interest were on opposite sides of the mean and the proportion of total area under the curve in the region of interest was determined by combining two segments. In Problem 1d, the boundaries both lie on the same side of the mean. The appropriate procedure in this situation is to find the difference between the area encompassed by two segments under the curve. First determine the area between 1.500 oz and 1.518 oz. Then subtract from this value the area contained between 1.500 oz and 1.512 oz. The result will be the area under the curve lying between 1.512 and 1.518 oz.

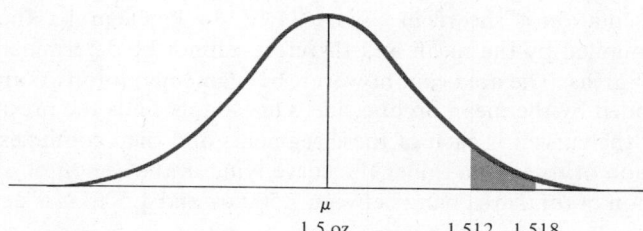

$$\mu$$
1.5 oz   1.512   1.518

Figure 11.13

The proportion of total area lying between 1.500 and 1.518 oz is

$$z = \frac{X - \mu}{\sigma} = \frac{1.518 - 1.500}{0.006} = \frac{0.018}{0.006} = 3$$

The table of areas reveals that 49.865 percent of the area under the curve lies between a perpendicular raised at the mean and a perpendicular raised at a distance of 3 standard deviation units along the number line.

The proportion of total area lying between 1.500 and 1.512 oz is

$$z = \frac{X - \mu}{\sigma} = \frac{1.512 - 1.500}{0.006} = \frac{0.012}{0.006} = 2$$

The table of areas reveals that 47.72 percent of the area under the curve lies between a perpendicular raised at the mean and a perpendicular raised at a point 2 standard deviation units away along the number line.

Therefore, (49.865 − 47.72) or 2.145 percent of the area under the curve must lie between 1.512 oz and 1.518 oz. Accordingly, 2.145 percent of the candy bars will have a weight between 1.512 oz and 1.518 oz.

1.e. What percentage of the candy bars produced will have a weight between 1.488 oz and 1.491 oz?

The graphic presentation is shown in Figure 11.14. The proportion of total area lying between 1.488 oz and 1.500 oz is

$$z = \frac{X - \mu}{\sigma} = \frac{1.488 - 1.500}{0.006} = \frac{-0.012}{0.006} = -2$$

According to the table of areas, 47.72 percent of the area under the curve lies between the mean of the distribution and a point 2 standard deviation units away along the number line.

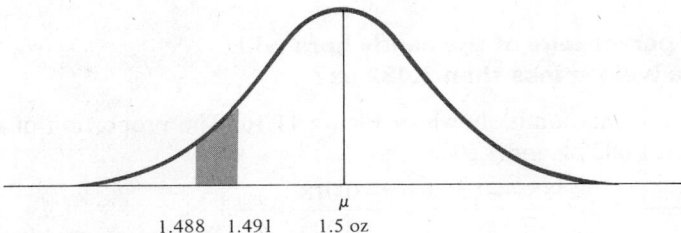

1.488   1.491   1.5 oz

Figure 11.14

The proportion of total area lying between 1.491 oz and 1.500 oz is

$$z = \frac{X - \mu}{\sigma} = \frac{1.491 - 1.500}{0.006} = \frac{-0.009}{0.006} = -1.5$$

The table of areas indicates that 43.32 percent of the area under the curve will be found between the mean of the distribution and a point 1.5 standard deviation units away along the number line.

The difference between the two (47.72 − 43.32) or 4.4 percent is the percentage of the area under the curve lying between 1.488 oz and 1.491 oz. Therefore, 4.4 percent of the candy bars have a weight between 1.488 oz and 1.491 oz.

1.f. What percentage of the candy bars produced will have a weight greater than 1.515 oz?

The graphic presentation is shown in Figure 11.15. The region of interest is in the tail of the distribution. It is known that 50 percent of the area under the curve lies to the right of the mean value of the distribution. One need simply find the percentage of the area under the curve lying between the mean and the value on the number line where the region of interest starts and subtract this from 50 percent to obtain the percentage of the area under the curve lying in the region of interest.

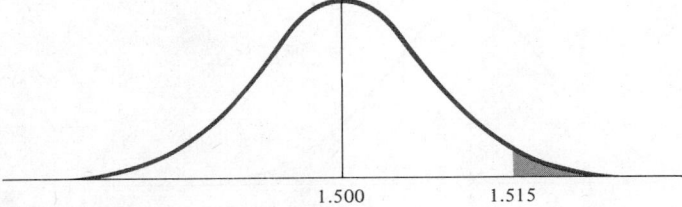

1.500   1.515

Figure 11.15

The proportion of total area lying between 1.500 oz and 1.515 oz is

$$z = \frac{X - \mu}{\sigma} = \frac{1.515 - 1.500}{0.006} = \frac{0.015}{0.006} = 2.5$$

The table of areas indicates that 49.38 percent of the area under the curve will be found between the mean of the distribution and a point 2.5 standard deviation units away along the number line. Thus, (50.00 − 49.38) or 0.62 percent of the area under the curve lies more than 2.5 standard deviation units away from the mean in a positive direction along the number line (the same statement would hold true with respect to a negative direction along the number line due to the symmetrical nature of the curve). Accordingly, 0.62 percent of the candy bars have a weight greater than 1.515 oz.

1.g. What percentage of the candy bars will have a weight less than 1.482 oz?

The graphic presentation is shown in Figure 11.16. The proportion of total area lying between 1.482 oz and 1.500 oz is

$$z = \frac{X - \mu}{\sigma} = \frac{1.482 - 1.500}{0.006} = \frac{-0.018}{0.006} = -3$$

The table of areas indicates that 49.865 percent of the area under the curve will be found between the mean of the distribution and a point 3 standard deviation units away along the number line. Thus, (50.000 − 49.865) or 0.135 percent of the area under the curve lies more than 3 standard deviations away from the mean in a negative direction along the number line. It follows that 0.135 percent of the candy bars have a weight less than 1.482 oz.

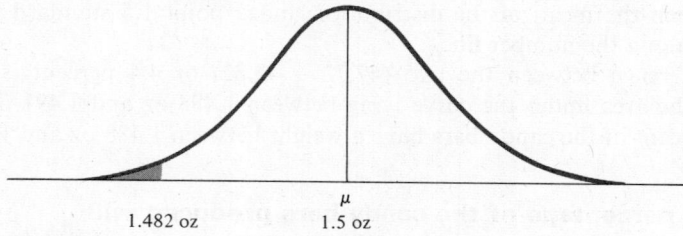

1.482 oz μ
                 1.5 oz

Figure 11.16

1.h. Ten percent of the candy bars will weigh more than what value?

The graphic presentation is shown in Figure 11.17. In this instance the probability is known and the unknown is the associated value of the variable on the number line.

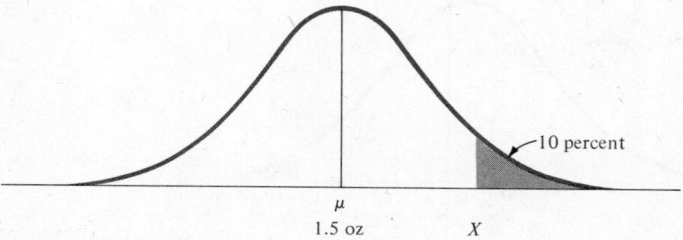

10 percent

μ
1.5 oz X

Figure 11.17

It is known that 50 percent of the area under the curve lies to the right of the mean. Ten percent of the area under the curve lies to the right of the unknown value. Therefore, it follows that 40 percent of the area under the curve lies between the mean and the unknown. The next step is to consult the table of areas to determine how many standard deviation units one must proceed along the number line in order to reach a point such that 40 percent of the area under the curve will be contained between a perpendicular raised at that point and a perpendicular raised at the mean.

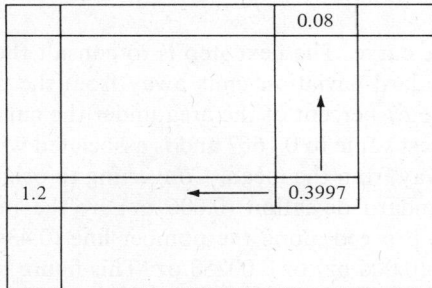

		0.08	
		↑	
1.2		← 0.3997	

One scans the body of the table of areas and locates the percentage closest to 40 percent. The associated number of standard deviation units is then read from the perimeter of the table. The table indicates that 39.97 percent of the area under the curve will be found between the mean and a point 1.28 standard deviation units away. To determine the corresponding value of the variable on the number line, it is necessary to convert from standard deviation units into original units (i.e., ounces). One standard deviation unit is equivalent to 0.006 oz, therefore a distance of 1.28 standard deviation units is equivalent to a distance of (1.28) multiplied by (0.006 oz) or 0.00768 oz. Thus, the value of the variable beyond which 10 percent of the area under the curve lies is $\mu + 1.28\sigma$ or $1.50000 + 0.00768$ or 1.50768 oz.

The 10 percent of the candy bars with the greatest weight will have values greater than or equal to 1.50768 oz (see Figure 11.18).

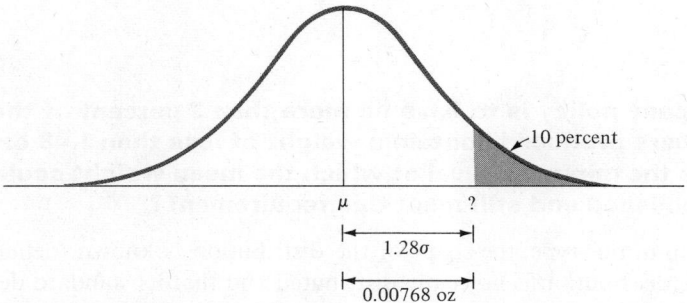

Figure 11.18

1.i. The middle third of the candy bars produced will have a weight between what two values?

The graphic presentation is shown in Figure 11.19. The middle third can be segmented into the portion lying to the right of the mean along the number line and the portion lying to the left of the mean. Each of these segments contains

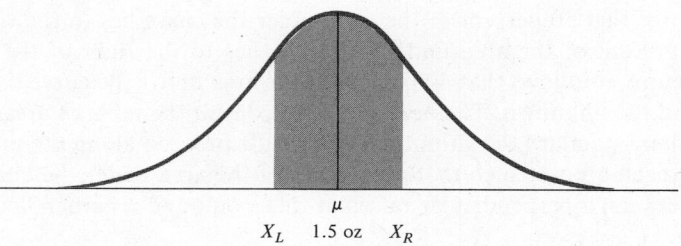

Figure 11.19

16.67 percent of the area under the curve. The next step is to consult the table of areas to determine how many standard deviation units away from the mean one must proceed in order to include 16.67 percent of the area under the curve. In the body of the table, 0.1664 is the closest value to 0.1667 and is associated with a span of 0.43 standard deviation units away from the mean. Converting to original units by multiplying the value of a standard deviation (0.006 oz) by the number of standard deviation units one must proceed along the number line (0.43 standard deviation units), one obtains (0.43)(0.006 oz) or 0.00258 oz. This figure represents the amount that must be added to and subtracted from the distribution mean in order to obtain the values of the variable that enclose the middle third of the area under the curve. The two values are 1.49742 oz and 1.50258 oz. Thus, the middle third of the candy bars produced will have a weight between 1.49742 and 1.50258 oz (see Figure 11.20).

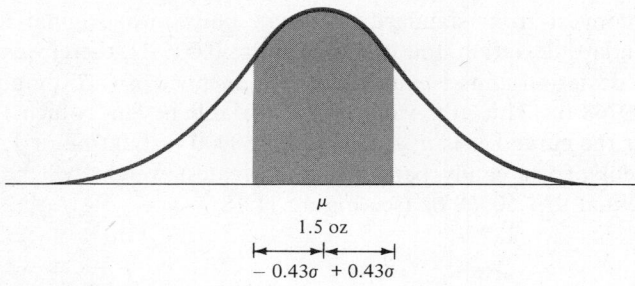

Figure 11.20

2. If company policy is to have no more than 3 percent of the candy bars produced contain a weight of less than 1.48 oz, what is the minimum level at which the mean weight could be established and still meet this requirement?

In a situation of this type, the shape of the distribution is known (defined by the fact that the distribution is normally distributed and that its standard deviation is 0.006 oz). The objective is to determine at what value of the variable on the number line the distribution should be located in order to satisfy the constraint "no more than 3 percent with a weight of less than 1.48 oz" in an optimal manner. In this case, one builds from the known value on the number line, 1.48 oz, to the desired unknown value, the value of the mean, μ.

It is clear from Figure 11.21 that if 3 percent of the area under the curve is to lie below the level of 1.48 oz, then 47 percent of the area under the curve lies between 1.48 oz and the value of the mean. Reference to the table of areas discloses that one must proceed a distance of 1.88 standard deviation units along the number line away from the mean in order to enclose 46.99 percent of the area under the curve. Thus, the value 1.48 oz should be located 1.88 standard deviation units away from the mean. Converting to original units, this is a distance of (1.88)(0.006 oz) or 0.01128 oz. By adding 0.01128 oz to the value 1.48 oz, one can determine the desired value for the mean of the distribution or 1.49128 oz. In conclusion, to satisfy the constraint "no more than 3 percent of the candy bars produced should contain a weight of less than 1.48 oz" in an optimal manner, the process should be set up so that the mean weight of candy bars produced is 1.49128 oz. A mean weight greater than 1.49128 oz would result in less than 3 percent with a weight of less than 1.48 oz, but would increase the cost of production unnecessarily (average material cost would be increased).

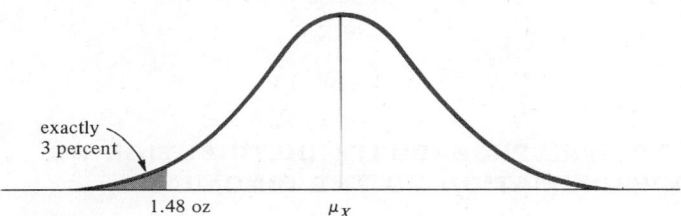

exactly 3 percent

1.48 oz μ_X

Figure 11.21

3. **Company records indicate that material costs amount to $0.04 per ounce. What is the potential reduction in material cost if the standard deviation of the process could be reduced from 0.006 oz to 0.002 oz?**

Once again, the shape of the distribution is known (defined by the fact that the distribution is normally distributed and that its standard deviation is 0.002 oz). Given the constraint of "no more than 3 percent of production with a weight of less than 1.48 oz," the objective is to determine the new optimal location on the number line for the mean of the distribution.

Figure 11.22 reflects the fact that 47 percent of the area under the curve lies between a perpendicular raised at the mean and a perpendicular raised at 1.48 oz on the number line. Reference to the table of areas discloses that one must proceed a distance of 1.88 standard deviation units along the number line away from the mean in order to encompass 46.99 percent of the area under the curve. Thus, the value 1.48 oz should be located 1.88 standard deviation units away from the mean. Converting to original units, this is a distance of 1.88(0.002) oz or 0.00376 oz. Note that the value of a standard deviation unit is 0.002 oz now rather than 0.006 oz. By adding 0.00376 oz to 1.48 oz, one can determine the new optimal value for the mean of the distribution, namely, 1.48376 oz. The reduction in dispersion (the value of a standard deviation unit was reduced from 0.006 oz to 0.002 oz) enabled the firm to lower the average amount of material per candy bar from 1.49128 oz to 1.48376 oz, an average saving of 0.00752 oz per bar produced. Since material costs amount to $0.04 per bar, this constitutes a dollar saving of (0.00752)($0.04) or $0.0003008 per bar produced. To determine the total dollar saving over a given period, multiply the saving per bar by the volume of production associated with that period. To illustrate: If production per month amounted to a million candy bars, monthly savings generated by the reduction in dispersion from 0.006 oz to 0.002 oz would equal (1,000,000)($0.0003008) or $300.80. Whether or not the initiation of the above reduction in dispersion is warranted depends on the magnitude of any incremental costs that are associated with such an undertaking as compared to the $300.80 monthly saving.

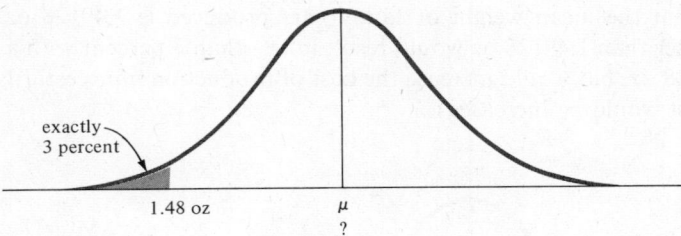

Figure 11.22

11.5 THE NORMAL PROBABILITY DISTRIBUTION AS AN APPROXIMATION TO THE BINOMIAL

Previously, it was mentioned that discrete probability distributions are often treated as if they were continuous in situations where there are many outcomes and the gaps between permissible values on the number line are insignificant. Recall that Demoivre discovered that the binomial probability distribution (a discrete probability distribution) approaches the normal probability distribution (a continuous probability distribution) as its limiting case when n, the number of trials, is allowed to approach infinity and p is held constant. In practice, a normal approximation to the binomial is considered legitimate if n is greater than or equal to 30 and the products np and $n(1 - p)$ have a value of at least 5. Where use of the binomial probability distribution would involve the successive computation of a number of discrete probabilities in order to generate the information of concern, the use of the normal approximation can often save considerable computational effort and still yield satisfactory results.

Consider a sample of size 100 to be drawn (with replacement) from a universe consisting of 500,000 billings. If the true percentage of correctly calculated billings in the universe is 80 percent, what is the probability that the sample will contain 85 or more correctly calculated billings?

Each performance of the experiment would consist of 100 Bernoulli trials each time it was conducted. The probability distribution describing all possible outcomes to the experiment as well as the corresponding probability of their occurrence would be a binomial probability distribution with $n = 100$ and $p = 0.80$.

The probability that the sample will contain 85 or more correctly calculated billings could be determined directly from the binomial probability distribution by successively performing the computation

$$\frac{n!}{(n - r)!\, r!}\, p^r q^{n-r} \qquad \text{for } r = 85, 86, 87, 88, \ldots, 100$$

When $r = 85$,

$$\frac{n!}{(n-r)!\,r!}\ p^r q^{n-r} = \frac{100!}{15!\,85!}\ (0.8)^{85}(0.2)^{15} = 0.0481$$

Thus, the probability of observing *exactly* 85 successes out of 100 trials is 0.0481. Similarly, one could generate the probability of observing *exactly* 86 successes out of 100 trials, and so on. Summing these terms would generate the probability of observing *85 or more* successes out of 100 trials.

Number of successes out of 100 Trials	Probability of occurrence
85	0.0481
86	0.0335
87	0.0216
88	0.0128
89	0.0069
90	0.0034
91	0.0015
92	0.0006
93	0.0002
94	0.0001
95	0.0000
96	0.0000
97	0.0000
98	0.0000
99	0.0000
100	0.0000[3]
	0.1287

Thus, direct calculation from the binomial distribution generates (through a somewhat tedious process) a probability of 12.87 percent that 85 or more successes will be observed in the sample of 100 billings to be drawn, given that 80 percent of the billings in the universe have been computed correctly.

Alternatively, an approximation to this value could have been generated by using the normal approximation to the binomial as the basis for computation.

$$n = 100$$

$$np = 100(0.80) = 80$$

$$n(1 - p) = 100(0.20) = 20$$

---

[3] For $r = 95$ through $r = 100$, the probability when rounded to four places becomes zero. For example, the probability associated with $r = 100$ is not zero but in reality is $(0.80)^{100}$.

The necessary conditions for employing the normal approximation are met:

1. $n \geq 30$
2. np and $n(1 - p) \geq 5$

Utilizing the formulas derived in Chapter 7, the mean and standard deviation of the binomial distribution are obtained as follows:

$$\mu = np = 100(0.80) = 80$$

$$\sigma = \sqrt{npq} = \sqrt{100(0.80)(0.20)} = \sqrt{16} = 4$$

It is necessary to employ a continuity correction when utilizing a continuous probability distribution to represent a discrete probability distribution. The discrete outcome "85 correct billings" is represented by the span on the number line encompassing 84.5 through 85.5. In general, any outcome X is represented by the span $X - 0.5$ through $X + 0.5$. Thus, the group of outcomes associated with *85 or more* successes is associated with the area under the curve lying to the right of the value 84.5 on the number line (Figure 11.23.)

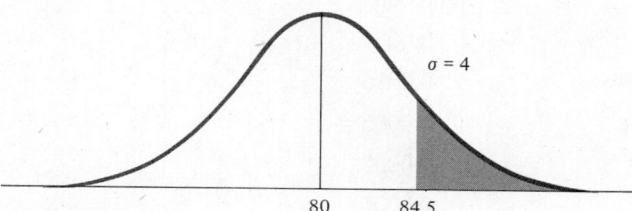

Figure 11.23

The approach is to determine the percentage of the area under the curve lying between the value of the mean, 80, and the value 84.5 on the number line. Subtracting this value from 0.5000 generates the probability of observing 85 or more correct calculations in the sample of size 100.

From 80 to 84.5 is a distance of 4.5 correct calculations. Dividing this by the value of a standard deviation (4 correct calculations) one finds that 84.5 lies a distance of 1.125 standard deviation units away from the mean. The table of areas indicates that 37.08 percent of the area under the curve lies between the mean and a point 1.13 standard deviation units away. It follows that 12.92 percent of the area under the curve lies more than 1.13 standard deviation units away from the mean in a positive direction along the number line. Thus, the normal approximation provides an estimate that 12.92 percent of the time a sample of size 100 would disclose 85 or more correct calculations if the true proportion of correct calculations in the universe is 80 percent. As is to be expected, this approximation compares quite favorably with the value of 12.87 percent obtained by direct computation from the binomial probability distribution.

PROBLEMS

Problem 11.1

The process designed to fill 4-oz jars of Minim Freeze Dried Coffee is set to operate with a mean fill of 4.005 oz and a standard deviation of 0.003 oz. The distribution of the weight of contents for individual jars is normally distributed.

1. What percentage of the jars will have a content weight between 4 oz and 4.005 oz?
2. What percentage of the jars will have a content weight between 3.999 oz and 4.005 oz?
3. What percentage of the jars will have a content weight between 4.002 oz and 4.009 oz?
4. What percentage of the jars will have a content weight between 4.007 oz and 4.009 oz?
5. What percentage of the jars will have a content weight between 3.998 oz and 4.000 oz?
6. What percentage of the jars produced will have a content weight greater than 4.005 oz?
7. What percentage of the jars produced will have a content weight greater than 4.007 oz?
8. What percentage of the jars produced will have a content weight less than 4 oz?
9. The 5 percent of the jars with the heaviest content weight will contain at least what amount of freeze dried coffee?
10. The middle 50 percent of the jars in terms of weight of contents will contain between what two amounts of freeze dried coffee?

Problem 11.2

If the makers of Minim Freeze Dried Coffee (described in Problem 11.1) wish to produce no more than 4 percent of their production with a content weight of less than 4 oz, at what value should the mean content weight of the filling process be established? (Assume that the standard deviation of the filling process is fixed at 0.003 oz.)

Problem 11.3

The lives of Eastinghouse extra-life lightbulbs are normally distributed with a mean equal to 1350 hr and a standard deviation equal to 18 hr.
1. What percentage of the bulbs will have a life between 1350 and 1377 hr?
2. What percentage of the bulbs will have a life between 1341 and 1350 hr?
3. What percentage of the bulbs will have a life between 1338 and 1365 hr?
4. What percentage of the bulbs will have a life between 1365 and 1377 hr?
5. What percentage of the bulbs will have a life between 1338 and 1344 hr?
6. What percentage of the bulbs will last longer than 1386 hr?
7. What percentage of the bulbs will last less than 1323 hr?
8. The 10 percent of the bulbs with the longest life will last longer than how many hours?
9. The 20 percent of the bulbs with the shortest life will last no longer than how many hours?
10. What percentage of the bulbs will wear out during the 1350th hour? (Be careful: Where does the 1350th hour fall on the number line?)

Problem 11.4

Daily demand for Tony's Super Subversive Subs is normally distributed with a mean equal to 120 and a standard deviation equal to 15. Super Subversive Subs must be made up in advance due to time constraints during the period when Tony's shop is open for business.

1. Ten percent of the time, demand for Super Subversive Subs will exceed what value?
2. Fifteen percent of the time, Tony will expect demand to be less than what quantity?
3. What percentage of the time will daily demand be for at least 140 Super Subversive Subs?
4. How many Super Subversive Subs should Tony regularly prepare if he wishes to ensure no unsatisfied customers 90 percent of the time?
5. Alfredo is considering opening a competing shop across the street. If he does, Tony expects that mean demand for his subs will decline to 95 and the standard deviation will fall to 12. Given that Alfredo opens the competing shop, how many Super Subversive Subs should Tony regularly prepare if he wishes to ensure no unsatisfied customers 90 percent of the time?

Problem 11.5

Rodent Bill, notorious small game hunter, estimates that the number of rounds required for him to meet his daily quota of rat pelts is normally distributed with a mean of 15 and a standard deviation of 1.3.

1. What percentage of the time will fewer than 13 rounds be required?
2. If Rodent Bill carries only 18 rounds, what percentage of the time will he be expected to fall short of his quota?
3. If Rodent Bill's objective is to carry the minimum number of rounds necessary to ensure meeting the quota of ten rat pelts 99 percent of the time, how many rounds should he carry?
4. If Rodent Bill were to increase the amount of time he spends at the practice range, how would the mean and standard deviation of the distribution referred to above be affected?
5. Given the increased practice time, would you expect the distribution of number of rounds required to meet the daily quota to continue to be normally distributed? Why or why not?

Problem 11.6

The diameter of piston rings produced by Tilden Automotive Parts is normally distributed with a mean of 2 in. and a standard deviation of 0.003 in.

1. If the diameter of a piston ring is less than 1.9925 in., the part will be too tight to work effectively. What percentage of production would be expected to fall into this category?
2. If the diameter of a piston ring exceeds 2.0075 in., the part will be too loose to operate effectively. What percentage of production would be expected to fall into this category?
3. Suppose Tilden Automotive Parts desired 99.73 percent of the parts produced when the process is operating correctly to fall within the tolerance limits indicated (i.e., between 1.9925 and 2.0075 in. in diameter). In designing the process (assuming a universe mean of 2 in.), at what value should the standard deviation be established?
4. The middle 50 percent of the piston rings produced (in terms of diameter) will have a diameter between what two values?

5. If the process were to slip a setting such that the mean diameter of parts produced increased to 2.001 in. (assume the standard deviation continued to be 0.003 in.):
   a. What percentage of the parts produced would now fall outside the tolerance limits?
   b. What percentage would be too tight?
   c. What percentage would be too loose?

Problem 11.7

The fuel pump of an automobile has been set so as to pump 0.005 oz of gasoline into each cylinder per cycle. However, there are times when a bit more or less than that amount is actually pumped in. Generally, the distribution of the actual amount pumped may be approximated by a normal distribution with a standard deviation of 0.001 oz.

1. If the engine of this automobile becomes flooded whenever more than 0.008 oz is pumped into any one cylinder, what will be the proportion of times in which the fuel is pumped that the engine will become flooded?
2. What is the likelihood that one squirt pumped will be between 0.004 and 0.005 oz?

Problem 11.8

Past experience has demonstrated that the amount of "rope" needed in order for students to "hang themselves" in the statistics course is normally distributed with a mean of 10 yards and a standard deviation equal to 3.3 yards. What length of "rope" should the instructor extend (on a per student basis) in order to provide 90 percent of his students with sufficient "rope" to avail themselves of this opportunity? (Draw a diagram to aid you in perceiving the relationship desired.)

Problem 11.9

The distribution of diameters of ball bearings produced by Company A can be approximated closely by a normal distribution with a mean of 0.380 and a standard deviation of 0.004 in. The percentage of total production characterized by a diameter between 0.375 and 0.384 is designated by the shaded area in the diagram below:

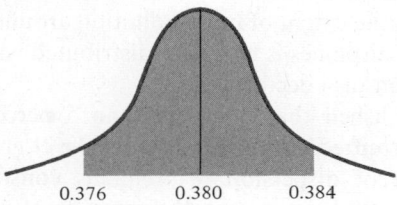

| 0.376 | 0.380 | 0.384 |

The distance between 0.384 and 0.376 is equivalent to 2 standard deviation units $[z = (0.384 - 0.376)/0.004 = 0.008/0.004 = 2]$. Explain carefully why it would be incorrect to look up 2.0 standard deviation units in the table of areas to determine the percentage of the area under the curve contained between 0.376 and 0.384 in.

Problem 11.10

Early Riser has determined that the length of time it takes him to commute from his residence to his place of work is normally distributed with a mean of 55 min and a standard deviation of 5 min.

1. How much time must Early allow if he wishes to arrive at his place of work on time 95 percent of the time?

2. Early is currently considering the purchase of a home located much closer to his place of work. He estimates that commuting time from this new location would be normally distributed with a mean of 25 min and a standard deviation of 3 min. Early still intends to abide by the constraint of allowing enough commuting time to be on time for work 95 percent of the time. If he purchases the new residence, how much later can Early rise and still meet the constraint?

Problem 11.11

Assume that the lives of Deuce flashlight batteries are normally distributed with a mean of 756 min and a standard deviation of 35 min. The proportion of the batteries with a life between 721 min and 791 min is depicted by the shaded area in the diagram below:

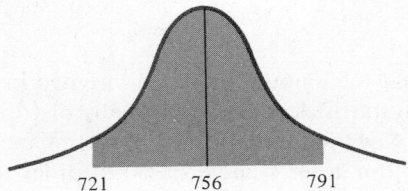

The procedure for determining the proportion of total area between 721 and 791 is to find the proportion of total area between 721 and 756 and the proportion between 756 and 791 and then add these two proportions together. Respond fully to the following question: Is the proportion of total area between 721 and 791 overstated by this procedure because the probability of the specific outcome 756 is counted twice?

Problem 11.12

Recently, the Federal Aviation Administration has been considering the establishment of more stringent limitations on the extent of noise pollution around airports. At the present time, the noise level of airplanes is normally distributed with a mean of 100 decibels and standard deviation of 6 decibels.

1. Suppose a regulation is established that no more than 5 percent of the airplanes at an airport are permitted to generate a noise level greater than 105 decibels. If the measure of dispersion (σ) remains constant, what changes if any, must be made with respect to the mean level of noise in order to comply with the regulation?

2. If the mean level of noise does not change (i.e., it remains at 100 decibels), what adjustment must be made to the standard deviation in order to meet this regulation?

Problem 11.13

Explain why the probability of the event $(X \leq 7)$ is for all intents and purposes identical to the probability of the event $(X < 7)$ given that X is described by a continuous distribution.

Problem 11.14

A decibel scale printed by the State Audobon Society indicates that acceptable noise levels should not exceed 45 decibels.
1. If the noise level in an urban area is normally distributed with a mean of 52 and a standard deviation of 7 decibels, what percentage of the locations in the urban area have a noise level exceeding the acceptable level?
2. If the noise level in a suburban community is normally distributed with a mean of 36 and a standard deviation of 6 decibels, what percentage of the locations in the suburban area have a noise level exceeding the acceptable level?

Problem 11.15

Would you expect the content of cans of 12 oz net weight Niblet Corn filled by a machine process to be normally distributed? Why or why not?

Problem 11.16

1. Referring to Problem 11.15, it is company policy that no more than 10 percent of production should contain less than the advertised net weight of 12 oz. Given that the variability in the production process is characterized by a standard deviation of 0.16 oz, at what value should the process mean be established in order to meet this constraint?
2. Given that corn costs $0.004 per ounce, what is the potential cost saving per can if the variability in the filling process as represented by the standard deviation could be reduced to 0.02 oz?
3. In words, how would you decide whether or not the necessary changes to bring about the reduction in variability should be instituted?

SAMPLING: THE CONCEPT AND THE DESIGN

12

Knowledge of the nature of the environment in which a
decision is to be implemented is of great importance
in determining the action to pursue. A crucial factor in
deciding whether or not to locate a store in a particular
location could be the proportion of family units
residing in that geographical area with a family income
over $10,000. Whether or not to continue a particular
advertising campaign may be dependent on the
percentage of individuals in the market segment one is
attempting to reach who are recognizing and
responding to the message. A production control
statistician will periodically assess the production
process to determine if the tensile strength of bars being
produced has departed from the norm stipulated by
product quality requirements. An auditor would be
concerned with ascertaining the number of billings that
have been totaled and posted correctly in the previous

accounting period. A credit manager, having instituted a more liberal credit policy, would be interested in determining whether average dollar amount outstanding over 30 days per account has increased as compared to the level that existed prior to the institution of the new policy. In each of the above situations, one approach to obtaining the desired information is to take a census. A census consists of an examination of each and every unit in the population or universe in question. The auditor, for example, would examine each and every one of the billings prepared during the last accounting period to ascertain whether or not the totaling and posting had been correctly done. An alternative approach is for the auditor to select a sample out of the universe or totality of items under consideration, determine the number of incorrect billings contained in the sample, and project from this an estimate of the total number of incorrect billings in the universe as a whole. Methods have been designed to increase the likelihood of selecting elements to be incorporated in the sample in such a way that the likelihood of obtaining a representative sample will be increased. A number of these techniques are examined later in the chapter.

12.1 ADVANTAGES OF SAMPLING AS COMPARED TO A CENSUS

There are a number of reasons why an analyst might prefer a sample to a complete census when attempting to determine characteristics of the population of interest.

1. *An approximation may be sufficient.* In many instances, the approximation of the universe parameter by the value of the sample statistic is more than adequate to enable the analyst to make a decision. The cost of obtaining an approximation from a sample is, of course, significantly less than the expenditure required for a complete census. For example, in the store location illustration, it is legitimate to presume that sufficient accuracy for arriving at a decision would be obtained if the sample estimate were to be within 5 percent of the true proportion for the universe. It would certainly not be necessary to know the exact percentage of family units with an income over $10,000. The objective is to reach a correct decision, and any accuracy obtained above and beyond the accuracy required for this purpose has no economic value. As the increased accuracy comes at a cost and has no value, it should be forgone. Interestingly, it is conceivable that a sample result could be more accurate than the result obtained by a complete census due to the data-gathering and computational errors that can occur in the handling and processing of large masses of census data. Controls designed to obtain greater accuracy are considerably more feasible when dealing with small amounts of data.

2. *Immediacy.* In many situations, information is needed now. The time required to take a complete census frequently exceeds the time available for gathering information prior to having to make a decision. Approximate information when needed is much to be preferred over census information too late to be of any use.

3. *Provision of more detailed information.* A sample enables the investigator to obtain more detailed information. The more detailed and complex the question-naire, the more demanding the requirements for interviewer skill, supervision, and training. One can afford a detailed examination with a few items, but with a census the cost would be prohibitive.

4. *Inaccessibility of the universe.* A census may be impossible in some situations due to the inaccessibility of some elements in a finite population or where the universe in question is an infinite one. For example, the production control

statistician measuring the tensile strength of bars is dealing with the theoretical distribution of all bars that could conceivably be produced by the process if it continued to operate indefinitely into the future in exactly the same fashion as it is operating when the sampling study was undertaken. Clearly, such a universe is inaccessible.

5. *The sampling process may be of a destructive nature.* Finally, the process involved in investigating the characteristics of the elements in the universe may be destructive in nature. Tensile strength of a bar is measured by subjecting it to a pulling process until it breaks. Where the sampling process is of a destructive nature, a census is out of the question.

12.2 THE FRAME

In conducting a study, the analyst is concerned with estimating the characteristics of the elements comprising a particular universe or population. The population of interest should be carefully defined and preliminary steps taken to ensure that the universe actually sampled is the one intended to be sampled. To accomplish this, it is important to develop an adequate frame. A frame is a listing or designation of the elements in the population to be sampled. The sampling unit is referred to as the elementary unit. In the store location illustration, where the concern is family income, the elementary unit is the family. The universe the analyst intends to sample is referred to as the target population. The universe that is actually sampled is the working population. Any estimates obtained are estimates of characteristics possessed by the working population. Whether or not these estimates can also be projected to the target population depends on the degree of correspondence between the target population and the working population. Perfect correspondence is seldom attainable. Usually, a close enough approximation can be obtained to ensure workable results. If an acceptable working population cannot be obtained, the study should be terminated.

One of the most quoted examples of an inadequate working population relates to the presidential election poll conducted in 1936 under the auspices of the Literary Digest. The frame or working population consisted of names obtained from telephone directories and automobile registration lists. When this frame was sampled, the resulting prediction was that Landon, the Republican candidate, would win in a landslide. It is a matter of history, of course, that Roosevelt won, receiving over 60 percent of the vote. The explanation, much to the Digest's chagrin, was the lack of correspondence between the working population and the target population. The target population was all eligible voters, while the working population consisted primarily of members of the middle and upper classes, a group with a predominantly Republican orientation. Many times, ready-made frames exist: telephone directories, member lists of associations, city directories, and so on. As the Literary Digest poll results illustrate, however, these ready-made frames must correspond closely to the target population if the study is to yield valid results. The importance of an adequate frame or working population cannot be overemphasized.

12.3 RANDOM VERSUS NONRANDOM SAMPLING

Sampling techniques can be classified on the basis of whether they generate a random or a nonrandom sample.

Random Sampling

The distinguishing characteristic of random sampling (or probability sampling) is that every elementary unit in the universe has *some known* chance (not necessarily equal for each unit) of being included in the sample selected. It must be possible to determine in advance the probability of each elementary unit's inclusion in the sample to be selected. With probability sampling, the elementary units selected are determined by *chance*. Different samples would be obtained if the analyst were to repeat the procedure a number of times.

Nonrandom Sampling

The distinguishing characteristic of nonrandom sampling (or nonprobability sampling) is that the elementary units included in the sample are not determined by chance, but rather are the result of conscious selection by the investigator. The investigator selects those sampling units that in his opinion would serve most effectively as representative values for the population under investigation. Every elementary unit does not have a chance of being included in the sample to be selected. The elements included are the result of a judgment decision by the investigator. If the procedure were to be repeated a number of times, the investigator would repeatedly choose the same sample, since in his judgment that sample is *the* representative sample for the universe under investigation.

Nonrandom sampling leads to a single-valued estimate of the universe parameter under investigation. A point estimate states that the value of the universe parameter is a particular value on the number line.

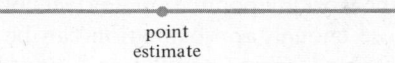

point
estimate

The major advantage of random sampling over nonrandom sampling is that it enables the analyst not only to generate an estimate of the universe parameter but also to assign a probability that the estimate is correct. With random sampling, the estimate takes the form of an interval estimate known as a confidence interval. A confidence-interval estimate states that the universe parameter has a value that lies somewhere inside a stipulated interval on the number line.

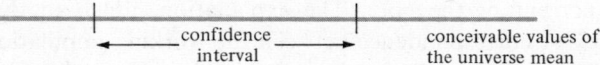

confidence   conceivable values of
interval   the universe mean

All other things being equal, the wider the interval estimate, the higher the probability that the estimate is correct. The procedure and underlying theory that allows such intervals to be constructed is presented in Chapter 13. A brief word is in order here, however, as to interpretation. A "99 percent confidence interval" refers to an estimate that has been obtained by a procedure that leads to the generation of correct interval estimates 99 percent of the time. To clarify: Suppose that the analyst repeated the sampling procedure 1000 times. He would generate 1000 different samples and 1000 sample means. The letters recorded on the number line in Figure 12.1 represent the values of 5 of the 1000 sample means.

In association with each sample mean, the analyst could develop a "99 percent confidence interval." The 99 percent confidence intervals associated with the 5 means depicted in Figure 12.1 are reproduced in Figure 12.2. To illustrate, the

"99 percent confidence interval" associated with the sample mean A runs from A_L to A_U on the number line. (A_L refers to the lower limit of the confidence interval and A_U to the upper limit.)

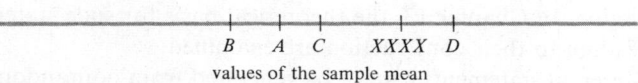

values of the sample mean

Figure 12.1

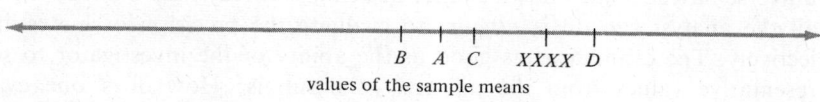

values of the sample means

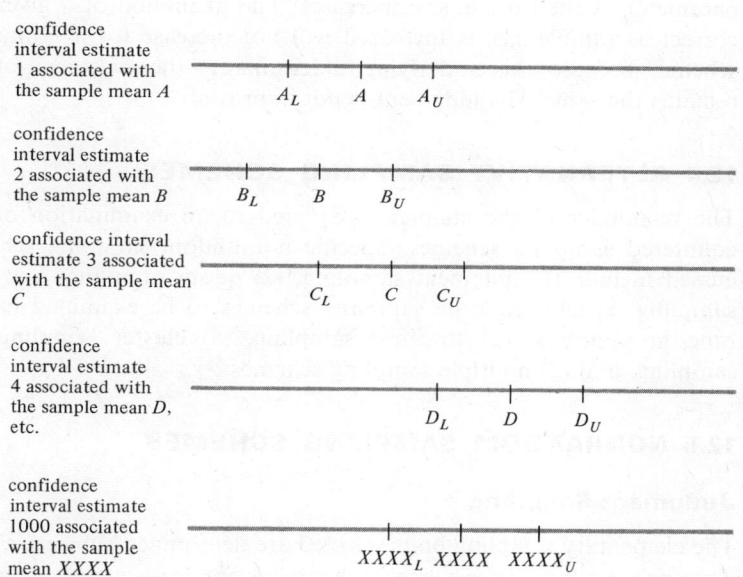

confidence interval estimate 1 associated with the sample mean A

confidence interval estimate 2 associated with the sample mean B

confidence interval estimate 3 associated with the sample mean C

confidence interval estimate 4 associated with the sample mean D, etc.

confidence interval estimate 1000 associated with the sample mean XXXX

Figure 12.2

On the basis of the first sample, the analyst would estimate that the universe parameter has a value between A_L and A_U on the number line.

On the basis of the second sample, the analyst would estimate that the universe parameter has a value between B_L and B_U on the number line.

On the basis of the thousandth sample, the analyst would estimate that the universe parameter has a value between XXXX_L and XXXX_U.

All of the above statements would not be correct. The procedure that leads to the generation of such statements can, however, be expected to produce correct statements 99 percent of the time (i.e., it would be expected that in the long run, the ratio of correct statements to the total number of statements generated would approximate 990/1000 or 99 percent). Unfortunately, the analyst has no way of telling which statements are correct and which are not. One can, however, assign a probability that any one statement will be correct. Recall that the probability of success is equal to the ratio of the number of ways an experiment can generate a success to the total number of ways an outcome can occur. Thus, the probability that the procedure will lead to the generation of a correct statement is equal to 99 chances out of a 100 or 99 percent.[1] Moreover, as the sample size is enlarged, the interval on the number line to which such statements refer becomes narrower and narrower, thereby enabling the investigator to generate more precise estimates of the universe value. In Chapter 13, the theoretical basis for such statements and the procedure leading to their construction are examined.

Confidence-interval statements cannot be generated from nonrandom samples. The ability to make such statements is predicated on each elementary unit in the universe having some known chance of being selected. With a nonrandom sample, all the analyst can do is *attempt* to evaluate the precision of his estimate subjectively. The estimate is as good as the ability of the investigator to select representative values from the universe—the rub is: How does one evaluate an investigator's ability to do this? A second advantage of random sampling is that random sampling schemes result in more and more precise estimates of the universe parameter as the sample size increases. The likelihood of a given estimate being correct as sample size is increased will not increase for a nonrandom sampling scheme, because the underlying determinant—the *judgment* of the selector—remains the same. His judgment is not improved.

12.4 ALTERNATIVE SAMPLING SCHEMES

The remainder of the chapter is devoted to an examination of commonly encountered sampling schemes. Specific nonrandom sampling schemes to be discussed include (1) judgment sampling, (2) quota sampling, and (3) convenience sampling. Specific random sampling schemes to be examined include (1) simple random sampling, (2) stratified sampling, (3) cluster sampling, (4) systematic sampling, and (5) multiple sampling schemes.

12.5 NONRANDOM SAMPLING SCHEMES

Judgment Sampling

The elementary sampling units selected are determined subjectively by the analyst. The analyst selects those elements that in his opinion are most clearly representative of the working population under investigation. Recall that the primary advantage of random sampling schemes is the ability to state that the value of the universe parameter falls within a very narrow range of values on the number line and has a known probability, say 99 percent, of being correct. The larger the sample size, the narrower the interval associated with a 99 percent probability of being correct becomes. Unfortunately, for small sample sizes the interval is quite wide and the advantage of being able to make an interval estimate, even though with a high probability of being correct, becomes quite meaningless. For small sample sizes,

---

[1] Confidence intervals are interpreted differently by Bayesians and classicists. A Bayesian would say that there is a 99 percent probability that any one statement is correct. The 99 percent represents the strength of the Bayesian's conviction that the true state of nature is that the statement is correct. The strict classicist disagrees. The strict classicist states that any one statement is either correct (in which case the probability that it is correct is 1.00) or incorrect (in which case the probability that it is correct is 0.00). The strict classicist attaches the 99 percent probability to the likelihood that the *procedure* will generate correct statements but not to the probability that any one statement is correct. The classicist position is totally objective, while the Bayesian introduces subjective considerations. Take your pick.

therefore, judgment samples often present a more desirable alternative. They are frequently less costly and, if the investigator is knowledgeable about the nature of the working population, a small judgment sample may be more representative of the working population than a small random sample of the same size.

Quota Sampling

The analyst identifies various strata in the working population and then selects as representatives of each stratum those elementary sampling units that in his judgment best represent the stratum in question. As with any nonrandom method of sample selection, the precision of the sample estimate is only as good as the judgmental ability of the investigator conducting the study. Unfortunately, there is no objective way of measuring this. It simply becomes a matter of strength of conviction.

Convenience Sampling or the Chunk

The investigator selects a group of elementary sampling units from the working population solely on the grounds of convenience. Such an approach presumes that there is no reason for any one chunk from the population to be different than any other chunk with respect to the characteristic under investigation. For example, if one wished to obtain an estimate of the mean quality-point average for the sophomore class, one could simply extract a "chunk" of record cards out of the registrar's file and generate an average quality-point average for the chunk to project to the universe. Convenience samples are often used when rough estimates will suffice.

12.6 RANDOM SAMPLING SCHEMES

Simple Random Sampling

A simple random sampling scheme is one in which every item in the working population has an equal opportunity of being included in the sample to be selected. Moreover, if a sample of size n is to be selected from the universe, each possible sample of size n that could be drawn has an equal opportunity of being selected. For example, if the universe consists of 60 items and a sample of size 5 is to be selected (without replacement), the number of different samples that could be selected (with no two samples consisting of the same five elementary sampling units) is equal to

$$C\binom{60}{5} = \frac{60!}{55!\ 5!} = 5,461,512$$

Each of the 5,461,512 samples would have an equal opportunity of being selected. The probability of any one particular sample being selected is $1/5,461,512$. To obtain a simple random sample, it is only necessary to ensure that the method of sampling selection guarantees that on each draw from the population, those elements in the working population that have not been previously selected all have an equal opportunity of being selected on the remaining draws necessary to complete a given size sample. A procedure designed to accomplish this is first to identify each elementary sampling unit in the working population and assign to it a unique identification number. If a sample of size 5 is to be drawn, five numbers are selected from a pool consisting of all the identification numbers. Selection must take place in such a fashion that those identification numbers not already selected have an equal opportunity of being selected on any remaining draws to be made. The five elementary sampling units whose identification numbers correspond to the numbers drawn constitute the simple random sample. A table of random numbers is the device typically used to select the identification numbers of the elementary sampling units that will constitute the sample.

A table of random numbers consists of row upon row of the digits 0, 1, 2, 3, 4, 5, 6, 7, 8, and 9 arranged in such a fashion that the digit 0 is just as likely to be followed by itself as it is by any of the other digits 1 through 9. In fact, each of the digits 0 through 9, when encountered in a table of random digits, is just as likely to be followed by any one of the one-digit numbers 0 through 9 as by any other. For that matter, any two-digit number, when encountered in the table, is just as likely to be followed by any one of the two-digit numbers 00–99 as by any other; any three-digit number is just as likely to be followed by one of the three-digit numbers 000–999 as by any other, and so on. A table of random numbers consists of page after page of digits arranged in this fashion. Some such tables contain over a million digits. A segment of such a table is reproduced in the Appendix as Table G. A small section of Table G is reproduced in Table 12.1.

Table 12.1

3395	3381	1862	3250	8614	5683	6757	5628	2551	6971
6081	6526	3028	2338	5702	8819	3679	4829	9909	4712
3470	9879	2935	1141	6398	6387	5634	9589	3212	7963
0432	8641	5020	6612	1038	1547	0948	4278	0020	6509
4995	5596	8286	8377	8567	8237	3520	8244	5694	3326
8246	6718	3851	5870	1216	2107	1387	1621	5509	5772
7825	8727	2849	3501	3551	1001	0123	7873	5926	6078
6258	2450	2962	1183	3666	4156	4454	8239	4551	2920
3235	5783	2701	2378	7460	3398	1223	4688	3674	7872
2525	9008	5997	0885	1053	2340	7066	5328	6412	5054
5852	9739	1457	8999	2789	9068	9829	1336	3148	7875
0440	3769	7864	4029	4494	9829	1339	4910	1303	9161
0820	4641	2375	2542	4093	5364	1145	2848	2792	0431
7114	2842	8554	6881	6377	9427	8216	1193	8042	8449
6558	9301	9096	0577	8520	5923	4717	0188	8545	8745
0345	9937	5569	0279	8951	6183	7787	7808	5149	2185
7430	2074	9427	8422	4082	5629	2971	9456	0649	7981
8030	7345	3389	4739	5911	1022	9189	2565	1982	8577
6272	6718	3849	4715	3156	2823	4174	8733	5600	7702
4894	9847	5611	4763	8755	3388	5114	3274	6681	3657

The use of a table of random numbers will be illustrated for the situation where a simple random sample of size 5 is to be drawn without replacement from a universe of size 60. First, every elementary sampling unit in the working population is assigned a unique identification number. It will be presumed that the consecutive two-digit numbers from 01 to 60 are assigned to the 60 elementary sampling units in the working population. Next, a page in the table of random numbers and a starting point on that page are selected arbitrarily. Let us suppose that the arbitrarily selected starting point turns out to be the circled two-digit number 74 in Table 12.1. There is no elementary sampling unit with the identification number 74, therefore this value is disregarded. The direction in which movement takes place within the table is an arbitrary decision. If the analyst decided to continue moving across the row and to the right, the five identification numbers selected would be 60, 33, 12, 23, and 46. The number 98 is ignored because there is no elementary sampling unit identified with that two-digit number. If by chance a two-digit number is repeated, it is ignored on its second appearance. If the analyst had elected to move down the column to obtain his sample of five items, the sample would have consisted of the five elementary sampling units associated with the identification numbers 10, 27, 44, 40, and 59. (Note that in this sequence the number 40 appears twice and therefore is ignored on its second appearance.) If the process were to be repeated an infinite number of times (always initiated by a random start), each of the possible samples of size 5 would be expected to appear with a relative frequency of occurrence of 1/5,461,512.

The procedure described above is quite satisfactory with a small working population. With a large working population, the process of assigning identification numbers to each elementary sampling unit becomes prohibitive with respect to both time and expenditure. Moreover, the working population is often spread out geographically or composed of clearly identified strata possessing unique characteristics with respect to the trait under study. Whenever any of the above situations arise, alternative sampling schemes that are sophisticated mutants of simple random sampling yield significantly better results for the same expenditure of time or money. As a result, simple random sampling is seldom used in practice. The theory of simple random sampling warrants discussion, however, since it is a basic ingredient of all random sampling schemes, regardless of the degree of sophistication possessed.

In the following sections, attention will be directed toward sampling schemes designed to:

1. provide guaranteed representation in the sample for each stratum in the population identified as being substantially unique with respect to the trait under investigation;
2. overcome the problem of having to assign a unique identification number to each elementary sampling unit in the working population; and
3. avoid selection of samples composed of elementary sampling units that are dispersed geographically (thereby increasing the cost of sampling) in situations where geographical disbursement is not necessary to ensure representative sampling.

Stratified Random Sampling

Suppose the analyst is concerned with estimating the average age of the students in a statistics class and that the class composition is as follows:

Number of veterans	24
Number of nonveterans	16
Total class size	40

Further presume that the average age of the veterans is 25 years and the average age of the nonveterans is 21 years. If a simple random sampling scheme were to be employed whereby each possible combination for the given sample size has an equal chance of being selected, it is conceivable that the sample selected might contain all veterans, in which case the estimated average age projected to the universe would be on the high side. Similarly, the sample selected could contain all nonveterans, thereby causing the estimated average age of the universe to be significantly in error on the low side. In situations where the analyst can clearly identify strata within the working population that differ significantly with respect to the trait under investigation, nonrepresentative samples (such as all veterans or all nonveterans) can be prevented by selecting a stratified random sample. Where stratification is meaningful, a stratified random sample provides a more precise interval estimate of a universe parameter than a simple random sample of the same size.

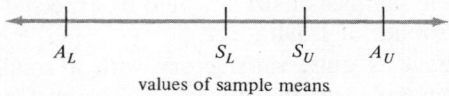

values of sample means

For a sample of size n, a simple random sample would place the estimated value of the universe parameter within the interval from, say, A_L to A_U. For a meaningful stratified random sample of the same size, the corresponding interval estimate would be significantly smaller, say, between S_L and S_U.

The first step in stratified random sampling is to partition the universe into discernable strata that differ significantly with respect to the characteristic of interest from one stratum to another but have a great deal of homogeneity with respect to the characteristic within each stratum.[2] Successful stratification is characterized by the dispersion as measured within each stratum (individual observations as dispersed around the stratum mean) being significantly smaller than the overall dispersion (individual observations for the entire working population as distributed around the overall working population mean). This is often expressed as follows: The within-stratum variance should be significantly smaller than the overall variance. In the illustration, the working population would be partitioned into two strata, the veteran group and the nonveteran group. A sample would then be drawn from each stratum and a representative value for each stratum obtained. To obtain a projected estimate of the universe mean, a weighted mean of the representative values for each stratum is computed. The weight assigned to each stratum mean is equivalent to the proportion its stratum size is to the size of the working population.

Suppose that the mean age for the sample drawn from the veteran stratum is 24.6 years and the mean age for the sample drawn from the nonveteran group is 20.3 years.

$$\overline{X}_V = 24.6 \text{ years}$$

$$\overline{X}_{NV} = 20.3 \text{ years}$$

Weights:

Veterans $\frac{24}{40}$ or $0.6 = W_V$

Nonveterans $\frac{16}{40}$ or $0.4 = W_{NV}$

---

[2] Stratified random sampling is to random sampling what quota sampling is to the judgment sample.

The estimate of the mean for the overall working population is computed as follows:

$$\overline{X}_{\text{overall}} = (W_{\text{V}})(\overline{X}_{\text{V}}) + (W_{\text{NV}})(\overline{X}_{\text{NV}})$$
$$= (0.6)(24.6) + (0.4)(20.3) = 14.76 + 8.12 = 22.88 \text{ years}$$

It is extremely important that the representative value for each stratum be weighted by the proportion that that stratum is to the size of the universe. The weights must be assigned in this fashion if the estimator is to provide an unbiased estimate of the universe parameter. That is,

$$\text{Ave}(\overline{X}_{\text{overall}}) = \mu_{\text{overall}}$$

PROPORTIONAL VERSUS NONPROPORTIONAL

Stratified samples can be either proportional or nonproportional. Proportional stratified sampling refers to the fact that the percentage of items in the sample to be selected from each stratum is proportional to the size of that stratum to the size of the working population. If a sample of size 10 is to be selected and veterans comprise 60 percent of the working population, then the number of elementary sampling units to be selected from the veteran stratum would be 60 percent of 10 or 6. Proportional stratified sampling yields satisfactory results if the dispersion in the various strata is of approximately the same magnitude. If there is a significant difference in dispersion from stratum to stratum, sample estimates will be much more efficient if nonproportional stratified random sampling is used. One sampling scheme is said to be more efficient than another when the sample estimates generated by that scheme tend to cluster more closely around the universe parameter being estimated. An estimator of a universe parameter should possess the following characteristics.

It should be unbiased. An estimator is unbiased when the average value of the sample statistic is equal to the universe parameter being estimated. Two unbiased estimators previously discussed (see pp. 82–87) are

$$\text{Ave}\,(\overline{X}) = \mu_x$$
$$\text{Ave}\left[\frac{\sum (X - \overline{X})^2}{n - 1}\right] = \sigma_x{}^2$$

Unbiasedness is obviously a desirable characteristic. Clearly, the sample estimates should cluster around the value the analyst wishes to estimate.

It should be efficient. Efficiency with respect to sample size means that the sample estimates should be clustered as closely as possible to the universe parameter being estimated for a given sample size. To illustrate: Both the sample mean and the sample median are unbiased estimators of the universe mean when the universe is normally distributed. However, for any given sample size, the sample means cluster more closely around the universe mean than do the sample medians. Thus, although both are unbiased estimators of the universe mean, the sample mean is the unbiased *efficient* estimator of the universe mean. The more efficient estimator is the one with the smaller variance.

Another example involves stratified sampling. Where stratification is meaningful, a stratified random sample will be more efficient than a simple random sample of the same size.

A sampling design is considered efficient with respect to cost if the sample estimates cluster more closely around the universe parameter being estimated than they would for any alternative sampling scheme involving an equivalent dollar expenditure.

It should be consistent. An estimator is considered to be consistent if the sample estimates cluster more and more closely around the universe parameter being estimated as sample size increases.

It can be demonstrated mathematically that if it is desired that the aggregate sample mean be as efficient an estimator of the overall mean of the working population as possible for a given sample size, then the proportion of the overall sample to be selected from each stratum should be assigned on the basis of the following relationship[3]:

$$n_{\text{stratum}_i} = \frac{N_{\text{stratum}_i} \sigma_{\text{stratum}_i}}{\sum (N_{\text{stratum}_i} \sigma_{\text{stratum}_i})} n$$

where

n_{stratum_i} = number of elementary sampling units to be selected from the ith stratum

N_{stratum_i} = number of elementary sampling units in the ith stratum

$\sigma_{\text{stratum}_i}$ = standard deviation for the ith stratum

n = aggregate sample size

An examination of the formula reveals that the larger the relative size of the measure of dispersion for a given stratum, the greater the proportion of the aggregate sample that should be selected from that particular stratum. Intuitively, this makes sense—if a stratum is characterized by very little dispersion, the value of the items in that stratum are quite similar. Just a few items drawn from a stratum of that type provide a very efficient estimate of the stratum mean. If another stratum is characterized by wide dispersion, additional items assigned to that stratum would provide a much more substantial payoff in terms of increased efficiency than if they were assigned to the stratum characterized by little dispersion.

Returning to the veteran–nonveteran example: If the standard deviation of the veteran stratum is known to be 5 years and the standard deviation of the nonveteran stratum is 1.875 years, the optimal allocation of a sample of size 10 to the two strata would be determined as follows:

$$\sigma_V = 5 \text{ years} \qquad N_V = 24$$

$$\sigma_{NV} = 1.875 \text{ years} \qquad N_{NV} = 16$$

$$
\begin{aligned}
n_V &= \frac{N_V \sigma_V}{N_V \sigma_V + N_{NV} \sigma_{NV}} (n) \\[2mm]
&= \frac{(24)(5)}{(24)(5) + (16)(1.875)} (10) \\[2mm]
&= \frac{120}{120 + 30} (10) \\[2mm]
&= \frac{120}{150} (10) \\[2mm]
&= (0.8)(10) \\[2mm]
&= 8
\end{aligned}
$$

[3] Taro Yamane, *Elementary Sampling Theory*, Englewood Cliffs, N.J.: Prentice-Hall, 1967, p. 136.

$$n_{NV} = \frac{N_{NV}\sigma_{NV}}{N_V\sigma_V + N_{NV}\sigma_{NV}}\,(n)$$

$$= \frac{(16)(1.875)}{(24)(5) + (16)(1.875)}\,(10)$$

$$= \frac{30}{120 + 30}\,(10)$$

$$= \frac{30}{150}\,(10)$$

$$= (0.2)(10)$$

$$= 2$$

The optimal sample allocation is 2 to the nonveteran stratum and 8 to the veteran stratum. This signifies that the means of all stratified samples of size 10 consisting of 8 veterans and 2 nonveterans will be more closely clustered around the value of the overall universe mean than would the sample means associated with any other stratified sampling plan of size 10 (say, for example, all stratified samples consisting of 7 veterans and 3 nonveterans).

By reassigning 4 of the elements to be drawn from stratum NV to stratum V, a slight reduction in efficiency in the estimate of the mean for stratum NV is incurred, but this is more than offset by the increased efficiency in the estimate of the mean for stratum V. As mentioned earlier, it can be demonstrated mathematically that the optimum allocation of the aggregate sample to the various strata is described by the following relationship:

$$n_{\text{stratum}_i} = \frac{N_{\text{stratum}_i}\sigma_{\text{stratum}_i}}{\sum (N_{\text{stratum}_i}\sigma_{\text{stratum}_i})}\,(n)$$

This will provide maximum efficiency for a sample of a given size. If cost of sampling varies from stratum to stratum, the proportion selected from each stratum will depart from this relationship if the analyst's objective is to obtain maximum efficiency for a given dollar expenditure as opposed to a given sample size.[4]

It is important to note that even though the stratified sample is selected on a nonproportional basis, the weights assigned to each stratum should still be in proportion to the size of the respective stratum to the size of the working population. The preceding analysis demonstrated that 8 of the 10 elementary sampling units should be selected from the veteran stratum and 2 from the nonveteran stratum. The respective stratum means should then be weighted by the proportion of elements in the given stratum to the proportion of elements in the overall working population.

If the 2 selected from the nonveteran stratum are 20.7 and 21.3 years, then

$$\bar{X}_{NV} = \frac{\sum X_{NV}}{n_{NV}} = \frac{20.7 + 21.3}{2} = \frac{42}{2} = 21 \text{ years}$$

---

[4] The extension for determination of optimal stratum allocation given the objective of obtaining maximum efficiency for a given dollar expenditure will be left to more advanced texts. See Taro Yamane, *Elementary Sampling Theory*, Englewood Cliffs, N.J.: Prentice-Hall, 1967.

If the 8 selected from the veteran stratum are 24.7, 23.0, 28.2, 31.3, 27.2, 25.2, 22.0, and 23.1, then

$$\bar{X}_V = \frac{\sum X_V}{n_V} = \frac{204.7}{8} = 25.6 \text{ years}$$

The two sample means are then weighted by the proportion their respective stratum size is to the size of the overall working population to obtain the estimate of the overall universe mean.

$$
\begin{aligned}
\bar{X}_{\text{overall}} &= (W_{nV})(\bar{X}_{nV}) + (W_V)(\bar{X}_V) \\
&= (\tfrac{16}{40})(21) + (\tfrac{24}{40})(25.6) \\
&= (0.4)(21) + (0.6)(25.6) \\
&= 8.4 + 15.36 \\
&= 23.76 \text{ years}
\end{aligned}
$$

Cluster or Area Sampling

In cluster sampling, the working population is once again subdivided into subsets. Unlike stratified sampling, however, where each subset or stratum is designed to be distinctly different from the others with respect to the trait under investigation, the subsets in cluster sampling are designed to be as similar to each other as possible. Each cluster should be designed in such a fashion that the diversity of the working population is reflected within the cluster. The clusters are referred to as primary sampling units, and each is assigned a unique identification number. Random sampling techniques are then used to select primary sampling units (or clusters) from the working population. Then, within each of the selected subsets, either a complete census is taken (referred to as single-stage sampling) or further sampling takes place (referred to as multistage sampling or subsampling). The decision on whether to census or sample at the second stage depends primarily on the cost differential and whether or not a time restriction exists. If there is little cost differential and no time restrictions, the analyst might as well take a census of the selected primary sampling units. Although a cluster sampling plan is not as efficient as a simple random sampling plan of the same size, it may be more efficient than a simple random sampling plan involving the same dollar expenditure. (A much larger cluster sample can be obtained for a given expenditure.) The considerable cost saving is associated primarily with the fact that the analyst does not have to assign an identification number to each elementary sampling unit in the working population. In cluster sampling, identification numbers must be assigned only to the elementary sampling units in the selected clusters and not even then if the analyst elects to take a complete census of the selected clusters. *Area sampling* is a form of cluster sampling in which the clusters are defined on the basis of geographical location. Perhaps the biggest obstacle to designing a successful area sampling scheme is the tendency for similar elementary sampling units to be located in close proximity to each other, thereby making it difficult to subdivide the working population into subsets possessing similar characteristics.[5]

---

[5] Of course, all sorts of combinations of these various sampling schemes are possible, for example, stratified cluster sampling.

Systematic Sampling

A systematic sampling scheme involves moving through a listing of the working population and selecting every kth item for inclusion in the sample, where $k = N/n$. For example, if the sample size is 50 and the size of the working population is 1000, then $k = 1000/50 = 20$. Thus, every twentieth item would be selected for inclusion in the sample. Systematic sampling involves a random start in that every one of the first k items is provided an equal opportunity of being selected. All subsequent elementary sampling units selected are located at intervals of k within the working population. Given k and the same starting point, the same sample would always be selected. Systematic sampling is more efficient than simple random sampling if similar elementary sampling units tend to be located together. When this is the case, systematic sampling tends to generate results quite similar to those obtained from stratified sampling. When the working population can be easily arranged in the form of an array, systematic sampling schemes can be quite efficient. The main disadvantage of systematic sampling schemes is that they can generate very unrepresentative samples if certain segments of the working population that are similar to each other but dissimilar to the remainder of the population tend to be located at intervals throughout the population that correspond to k, the sampling interval. For example, if a sample involved examining every tenth household and the working population was arranged so that every tenth household fell on a street corner, the sample would yield biased results if the characteristic of interest is significantly different for people living on corner lots as compared to those who do not.

Multiple Sampling Schemes

Decision making takes place in an environment of uncertainty. Sampling is a device that enables the decision maker to reduce the uncertainty about the environment in which decisions must be made. Faced with a go–no go decision, sample results may clearly indicate which action alternative should be taken. In other instances it is advisable to obtain additional information before reaching a decision (i.e., take another sample). The advantage of multiple sampling schemes is that in those instances where a small sample provides sufficient evidence on which to base a decision, the cost of the larger sample is avoided. Multiple sampling schemes are particularly advantageous in situations where the cost of sampling is quite expensive. Sequential sampling, an extreme form of multiple sampling, involves examining only one elementary sampling unit at each sampling stage. With the observation of each successive elementary sampling unit, a decision to go, not to go, or to sample further is made.

12.7 SAMPLING VERSUS NONSAMPLING ERROR

In selecting among alternative random sampling schemes, the analyst must ascertain the degree of efficiency the sampling scheme must possess in order for meaningful decision making to ensue. An examination of the various sampling schemes would reveal which provides the desired degree of efficiency at minimum cost given the nature of the working population under investigation. It is, of course, not to be expected that the value of the sample statistic will be equal to the universe parameter it is being used to estimate. The difference observed between the value of the true universe parameter and the corresponding value of a sample statistic can be attributed to the combined influence of two types of error—sampling error and nonsampling error. Sampling error refers to that portion of the difference that can be attributed to the fact that the computation was based on a sample rather than on a complete census of the working population. For any given sample size, the magnitude of expected sampling error differs from one sampling scheme to another. For example, the expected sampling error associated with a simple random sample will be greater than that expected in association with a stratified random sample of the same size (provided that stratification is meaningful). Moreover, given any particular random sampling scheme (provided it is consistent), expected sampling error will decline as the sample size is increased. This reduction in sampling error is obtained at a cost. Increases in sample size reflect increased expenditure. In addition, as the size of a sampling study increases, the likelihood of the occurrence of nonsampling error increases. Nonsampling error refers to that portion of the difference between the true value of the universe parameter and the actual observed value of the sample statistic that can be attributed to carelessness in defining the frame, carelessness in data collection, carelessness in data tabulation and manipulation, and so on. When the size of a sampling study is small, controls designed to maintain high interviewer skill and accuracy in data tabulation and manipulation are quite feasible. As the size of the study increases, however, controls break down and less highly skilled technicians must be used. As a result, a greater degree of nonsampling error is liable to be introduced. In designing a sample study, the objective of the analyst is to constrain the degree of error in his estimate to the levels stipulated by the situation in which the estimate is to be used. This means selecting a sampling scheme–sample size combination such that the combined sum of expected sampling error and expected nonsampling error (very difficult to measure objectively in advance) will fall within the specified limits. If a number of sampling schemes will accomplish this, it goes without saying that the scheme selected is the one that accomplishes this result at minimum cost.

In Chapter 13, the theoretical basis for measuring and controlling sampling error is discussed. The procedure for generating interval estimates of universe parameters from random samples will also be treated. In the remainder of the text, discussion will be couched in the framework of simple random sampling. The concepts introduced can be extended to the other forms of random sampling. The treatment of these extensions will be left to more advanced texts.

PROBLEMS

Problem 12.1

What is the primary advantage of a probability sampling plan as compared to a nonprobability sampling plan?

Problem 12.2

Although probability sampling plans are usually considered preferable to nonprobability sampling plans, an exception arises when sampling size is quite small. Explain.

Problem 12.3

The characteristics of a good estimator are that it be (1) unbiased, (2) efficient, and (3) consistent. Explain what is meant by each of these characteristics.

Problem 12.4

If a sampling plan is consistent, it becomes more and more efficient as sample size is increased. Explain.

Problem 12.5

How can a stratified random sampling plan be more efficient than a cluster random sampling plan in one sense and less efficient than that same cluster random sampling plan in another sense?

Problem 12.6

What is the difference between a nonrandom sampling plan and a nonprobability sampling plan?

Problem 12.7

Nonrandom sampling techniques generate point estimates, whereas random sampling techniques generate interval estimates.
1. Why do nonrandom sampling techniques not generate interval estimates?
2. What advantage is associated with the interval estimate of the random sampling plan as compared to the point estimate of the nonrandom sampling plan?

Problem 12.8

1. Which nonprobability sampling plan is similar to stratified random sampling?
2. How are they similar?
3. How are they dissimilar?

Problem 12.9

1. What is "the target population"?
2. What is "the working population"?
3. What problem arises if there is not a close correspondence between the target and working populations?

Problem 12.10

Why is a simple random sampling plan usually an inferior alternative to the more sophisticated random sampling plans?

Problem 12.11

What is sampling error? Discuss fully.

Problem 12.12

If a sampling plan is efficient, what statement can you make about the sampling error associated with that sampling plan?

Problem 12.13

Northeastern University operates a 5-year undergraduate cooperative program in which students alternate quarters between the classroom and work assignments. You are interested in estimating the average weekly earnings of undergraduate business majors on their cooperative work assignments. Majors include accounting, management, finance, and marketing. Would a simple random sample selected from the pool of all undergraduate business majors be the appropriate sampling technique, or should you stratify? What information would you want to know in answering this question?

Problem 12·14

1. What is a simple random sample?
2. What is a stratified random sample?
3. Based on the statistical use of the term, which of the above would be more efficient? Explain.

Problem 12.15

When computing the overall sample mean,
1. What desirable property is achieved by weighting the representative value for each stratum by the percentage the number of elements in that stratum is to the number of elements in the total universe under consideration?
2. Would the answer to part 1 be the same for both proportional and non-proportional stratified random sampling plans? Explain.

Problem 12.16

You are interested in estimating the average commuting time for an undergraduate student at the University. Eighty percent of the student body reside at the dormitory complex and the other 20 percent commute from neighboring communities. A small sample drawn from each of these groups has generated an estimated standard deviation of 3 min for the commuting times of dormitory residents and 12 min for the standard deviation of the nondormitory residents.
1. Given that a sample of 180 is to be selected, what allocation among the two strata would be associated with the most efficient stratified random sampling plan?
2. Assume that the sample has been selected and that it has generated a mean commuting time of 15 min for the dormitory residents and a mean commuting time of 45 min for the nondormitory residents. Compute the overall sample mean in such a fashion that the sample mean will be an unbiased estimator of the universe mean.

Problem 12.17

A major department store, Phileans, wishes to determine the average amount spent over the October–November–December period by customers using one of the two following charge formats:
  a. Revolving credit (installment repayment plan)
  b. 30-day charge (lump sum repayment plan)
Seventy percent of Philean's charge account customers have revolving credit and 30 percent have 30-day accounts.

1. Given that a stratified random sampling plan is to be used, what proportion of the sample should be drawn from each stratum if:
   a. the standard deviation is the same for both categories?
   b. the standard deviation for the revolving credit accounts is twice that of the 30-day charge accounts?
   c. the standard deviation for the revolving credit accounts is four times that of the 30-day charge accounts?
2. Using the formula

$$n_{(stratum_i)} = \frac{N_{(stratum_i)}\sigma_{(stratum_i)}}{\sum [N_{(stratum_i)}\sigma_{(stratum_i)}]} (n)$$

and the answer to part 1a, identify in general the situations where proportional stratified sampling would be appropriate.

Problem 12.18

What is the difference between proportional stratified random sampling plans and nonproportional stratified random sampling plans?

Problem 12.19

You are interested in estimating the average amount spent on entertainment in a geographical location where you are considering opening another in your chain of bowling alley–movie theater complexes. Four basic income strata have been identified, and the proportion each is to the total population in the area has been identified:

Under $6,000	20 percent
$6,000–$10,000	50 percent
$10,000–$14,000	20 percent
$14,000 and over	10 percent

Similar studies in other geographical areas have indicated that the magnitude of dispersion describing the pattern of entertainment expenditures for each of these sectors has consistently demonstrated the following pattern:

	Magnitude of dispersion
Under $6,000 ·	k
$6,000–$10,000	$1.4k$
$10,000–$14,000	$2k$
$14,000 and over	$3k$

1. If a stratified random sample of size 160 is to be selected, what proportion of the sample should be selected from each of the strata?
2. Stratified random sampling generates not only an overall sample mean, but also a representative value for each stratum. In terms of the example above, how would this additional information prove useful to the decision maker?

ESTIMATION OF A UNIVERSE PARAMETER: THE UNIVERSE MEAN

13

The theory of sampling is based on the interrelationships that exist between three different but related probability distributions: (1) the distribution of the universe, (2) the distribution of the sample, and (3) the sampling distribution of the statistic.

13.1 THE DISTRIBUTION OF THE UNIVERSE

The distribution of the universe refers to the totality of items under consideration. It is a probability distribution that describes the manner in which the individual observations comprising the universe are distributed along a number line measuring magnitudes of the trait under investigation. The distribution can be characterized by any of a number of shapes, depending on the nature of the trait under investigation.

To illustrate: Suppose that the population of

interest is the sophomore class consisting of 1200 students and that for each student the following three characteristics are known:

X = quality point average
Y = parent's income
Z = performance on a recent exam where the students fell into two basic subgroups—those who studied and did very well and those who did not study and performed poorly.

When the characteristic of interest is the quality point average of the students, the distribution of the universe would tend to be normally distributed. Individual X values (quality point averages) would be recorded on the number line, the mean would be designated as μ_x, and the standard deviation as σ_x. The x subscript serves to indicate that μ_x is the mean of the individual X values and that σ_x is the standard deviation of the individual X values.

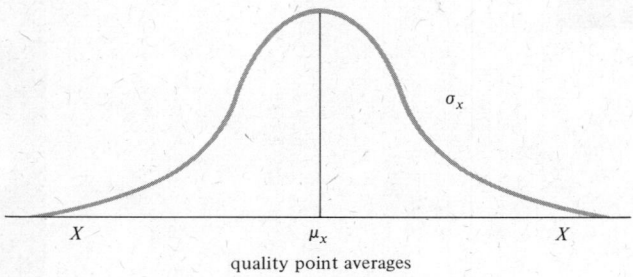

quality point averages

When the characteristic of interest is parent's income, the distribution would tend to be skewed to the right. Individual Y values (parent's income) would be recorded on the number line, the mean would be designated as μ_y, and the standard deviation as σ_y.

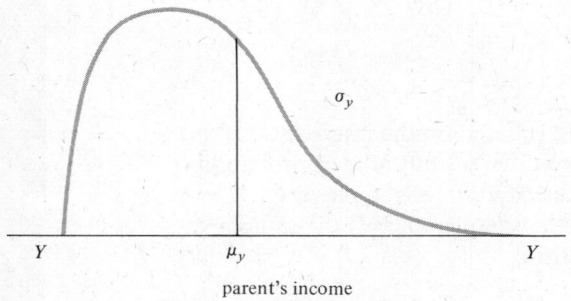

parent's income

The distribution of exam scores would be U shaped. Individual Z values would be recorded on the number line, the mean would be designated as μ_z, and the standard deviation as σ_z.

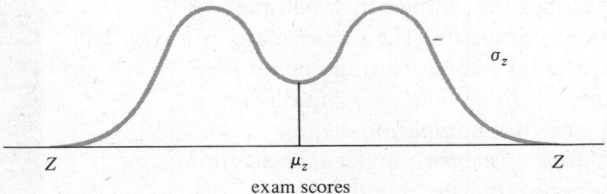

exam scores

Note that in a bimodal distribution such as the exam scores, the mean would not serve very effectively as a representative value.

13.2 THE DISTRIBUTION OF THE SAMPLE

If a sample of size 50 were to be drawn from the aforementioned population of 1200 students, the distribution of the *individual* observations of the 50 items in the sample along the number line would be referred to as the distribution of the sample. The shape of the distribution of the sample tends to take on a pattern somewhat similar to that of the universe from which it has been selected. The larger the sample selected, the closer the correspondence to the shape of the universe. Thus, one would expect the distribution of the sample to be somewhat normal when considering quality point averages, skewed to the right when considering parent's incomes, and U shaped when score on the test is the trait under examination. The mean of the sample would be expected to differ from the mean of the universe, of course, due to sampling error. If the trait of interest is quality point average, the symbols $\bar{X}$ and $\hat{\sigma}_x$ would be used to designate the mean and standard deviation of the sample, respectively. Different symbols are used for the sample and for the universe in order to distinguish clearly those occasions when one is referring to descriptive measures of the universe (parameters) as contrasted with situations where one refers to descriptive measures of a sample drawn from that universe (statistics). The custom in symbol representation is summarized in Table 13.1.

Table 13.1

Variable of interest	Symbol	Universe parameters		Sample statistics	
		Mean	Standard deviation	Mean	Standard deviation
Quality point average	X	μ_x	σ_x	$\bar{X}$	$\hat{\sigma}_x$
Parent's income	Y	μ_y	σ_y	$\bar{Y}$	$\hat{\sigma}_y$
Test score	Z	μ_z	σ_z	$\bar{Z}$	$\hat{\sigma}_z$

13.3 THE SAMPLING DISTRIBUTION OF THE STATISTIC

Descriptive measures of a universe such as the universe mean, median, or variance are referred to as parameters. When a sample is drawn from a universe and its mean, median, or variance is computed, the resulting value is referred to as a statistic. In sampling, the sample statistic is used as an estimate of the corresponding universe parameter. Two cases previously mentioned in Chapters 3 and 4 were $\bar{X}$ used as an estimate of μ_x and $(\hat{\sigma}_x)^2$ used as an estimate of σ_x^2.

The term sampling distribution of the statistic is a general term. In this chapter, attention is restricted to a particular statistic, namely, the sample mean. Thus, the specific distribution of concern will be the sampling distribution of the mean. (Other possibilities include the sampling distribution of the median, the sampling distribution of the variance, etc.) The sampling distribution of the mean refers to the theoretical distribution that would be obtained if all possible samples of some specified size that could be drawn from the universe were to be determined, the value of the arithmetic mean for each of these samples computed, and the resulting values summarized in the form of a probability distribution.

If a universe consists of 40 elementary sampling units and the designated sample size is 5, the number of possible samples that could be selected with no two samples containing the same 5 elementary sampling units would be equal to the number of combinations of 40 items taking 5 at a time.

$$C\binom{40}{5} = \frac{40!}{5!\,35!} = 658,008$$

Computing the mean for each of the 658,008 samples and grouping them into a probability distribution generates the sampling distribution of the mean for all possible samples of size 5 drawn from the universe of 40 items.

13.4 THREE THEOREMS

The sampling distribution of the mean possesses certain distinguishing characteristics that enable it to serve as the foundation on which the theory of sampling has been constructed. The characteristics are summarized in three important statistical theorems.

1. The *central limit theorem* states that for almost all populations (virtually without regard to the shape of the original population), the sampling distribution of the mean derived from that universe will be approximately normally distributed provided that the associated sample size is sufficiently large. When the original distribution is itself normally distributed, the sampling distribution of the mean derived from that universe will be normally distributed regardless of the associated sample size. When the original distribution is not normally distributed, the sample size required before the associated sampling distribution of the mean is normally distributed varies directly with the degree of departure of the original universe from normality. It has been demonstrated empirically that it is reasonable to presume that a sampling distribution of the mean associated with a sample size greater than or equal to 30 will be normally distributed regardless of the shape of the original universe. It is important to recognize that for any given original universe, there are associated with it a number of different sampling distributions of the mean—in fact, there is a different sampling distribution associated with each feasible sample size.

Any particular sampling distribution of the mean is a function of two things (assuming simple random sampling without replacement): (1) the original universe from which the samples are to be selected (its size, shape, and standard deviation) and (2) the simple random sample size.

The main significance of the central limit theorem is that it enables the analyst to operate with the knowledge that the sampling distribution of the mean will be normally distributed provided that the associated sample size is 30 or more.[1] Thus, for sample sizes equal to 30 or more, the analyst can draw on the tenets of statistical theory and eliminate the tedious task of actually constructing the distribution physically before any conclusions as to its shape and mathematical characteristics can be drawn. The central limit theorem is restricted to sample sizes of 30 or more. The statistician, however, is not at a total loss with respect to describing the behavior pattern for sampling distributions of the mean associated with sample sizes less than 30. Two additional situations in which the analyst can predict the behavior pattern of the sampling distribution of the mean are discussed below.

---

[1] Note that the lower limit of 30 refers to the sample size, not the number of samples.

Note that both require the universe from which the samples are to be selected to be normally distributed.

SITUATION 1
Provided that the original distribution is normally distributed and the standard deviation of the universe is known, the sampling distributions of the mean derived from that universe will be normally distributed for all sample sizes (including sample size 1).

SITUATION 2
When the original population is normally distributed and *the standard deviation of the universe is unknown*, the sampling distributions of the mean derived from that universe will be described by a distribution pattern referred to as the *t* or student's distribution. The *t* distributions associated with sample sizes greater than 30 are virtually indistinguishable from normal probability distributions (central limit theorem). As a result, the normal distribution is used in association with sample sizes of 30 or more, and the *t* distribution is used in association with sample sizes less than 30. The body of theory that has evolved around the *t* or student's distribution is referred to as small sample theory and will be deferred to Chapter 17 (you may breathe a temporary sigh of relief). The basic theory discussed in this chapter (large sample theory) does not differ from that associated with small sample theory in concept, but only with respect to the distribution within which the theoretical constructions are to be developed.

2. The second theorem relates to the fact that the expected value of the mean of the sampling distribution of the mean is equal in value to the original universe mean from which the theoretical sampling distribution is derived. This theorem holds regardless of the sample size to which the sampling distribution of the mean refers. Algebraically, the second theorem is expressed as

$$\bar{\bar{X}} = \mu_x$$

Intuitively, this is an easy theorem to justify. With simple random sampling, each item has an equal opportunity of being selected each time an element is selected from the universe (with replacement). If each item in the universe is selected an equal number of times, the average value of the sample means has to equal the mean of the universe.

To illustrate: Consider all possible samples of size 3 drawn with replacement that could be selected from a universe comprised of the four equiprobable values 1, 2, 3, and 4. Figure 13.1 is a tree diagram depicting the various ways a sample of size 3 could be selected (with replacement). Each of these possibilities is also listed in Table 13.2. Note that although each of the ways a sample of size 3 can be selected is equally likely, the likelihood of particular sample compositions are not. (The composition 1, 1, 1 can occur only one way, whereas the sample composition comprised of 1, 2, 3 can occur six ways). All samples characterized by the same composition have been assigned a common identifying letter (for example, each of the six possibilities associated with the sample composition 1, 2, 3 have been assigned the letter E).

There are 64 possible samples, each containing 3 elementary sampling units for a total of 192 elementary sampling units. Each of the elements in the original universe is represented an equal number of times, 48. Therefore, the arithmetic mean of all the sample means is equivalent to a weighted mean of the original

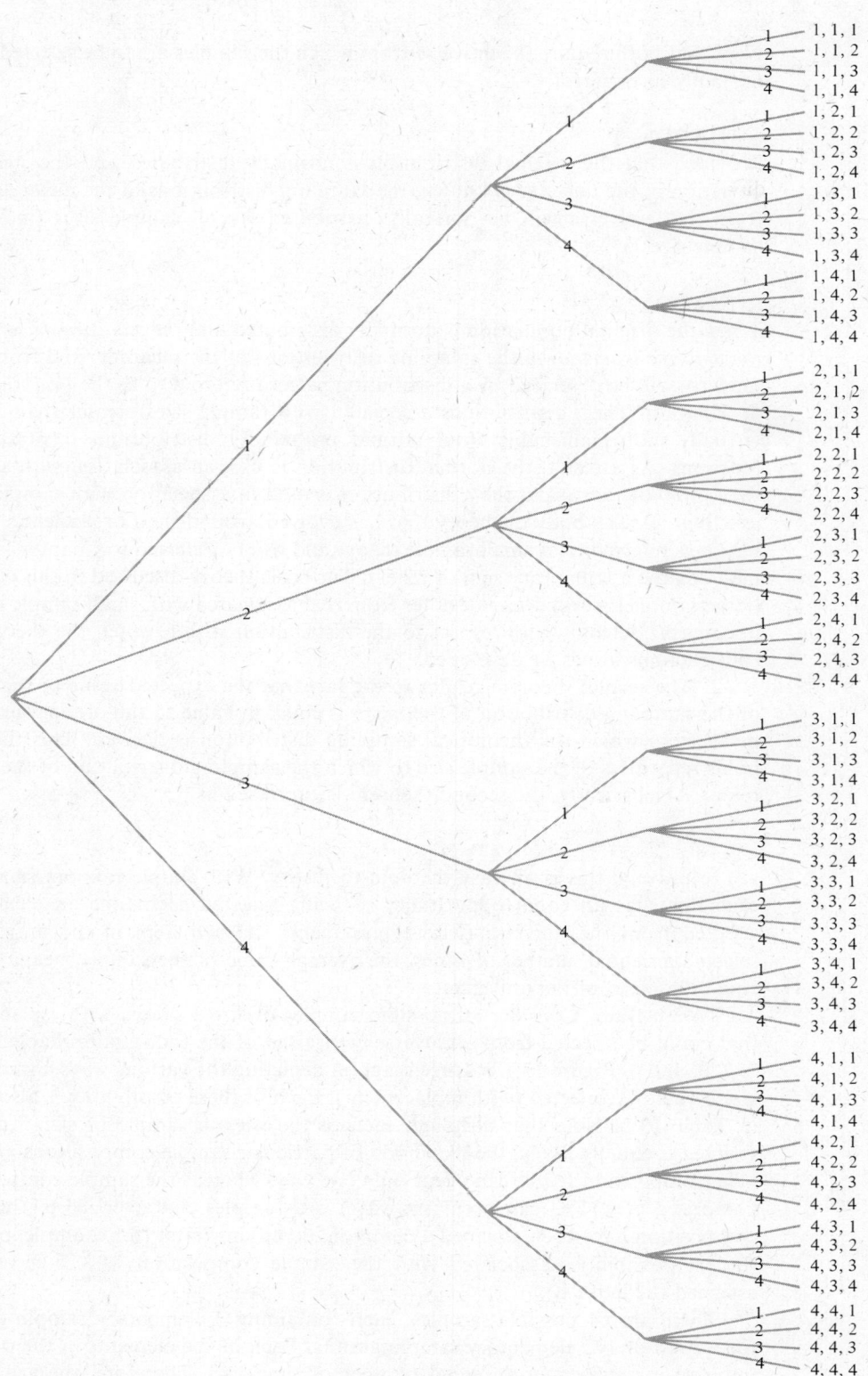

Figure 13.1

Table 13.2 List of possible sample permutations: Samples of size 3 drawn with replacement from the universe consisting of the equiprobable values 1, 2, 3, and 4

111	224 L	344 P
112 A	231 E	411 C
113 B	232 J	412 F
114 C	233 K	413 H
121 A	234 M	414 I
122 D	241 F	421 F
123 E	242 L	422 L
124 F	243 M	423 M
131 B	244 N	424 N
132 E	311 B	431 H
133 G	312 E	432 M
134 H	313 G	433 O
141 C	314 H	434 P
142 F	321 E	441 I
143 H	322 J	442 N
144 I	323 K	443 P
211 A	324 M	444
212 D	331 G	
213 E	332 K	
214 F	333	
221 D	334 O	
222	341 H	
223 J	342 M	
	343 O	

Table 13.3 Distribution of possible sample compositions and their means: Samples of size 3 drawn with replacement from a universe consisting of the equiprobable values 1, 2, 3, and 4

List of all possible sample compositions	Frequency of occurrence	Associated sample mean, $\bar{X}$
111	1	1
222	1	2
333	1	3
444	1	4
A 112	3	1.33
B 113	3	1.67
C 114	3	2.00
D 122	3	1.67
E 123	6	2.00
F 124	6	2.33
G 133	3	2.33
H 134	6	2.67
I 144	3	3.00
J 223	3	2.33
K 233	3	2.67
L 242	3	2.67
M 243	6	3.00
N 244	3	3.33
O 334	3	3.33
P 344	3	3.67

values in the universe where the assigned weights are all equal. It follows that $\bar{\bar{X}} = \mu_x$.

Grouping together all sample compositions with the same arithmetic mean generates the sampling distribution of the mean for all possible samples of size 3 drawn with replacement from the universe consisting of the equiprobable values 1, 2, 3, and 4 (Table 13.4).

Table 13.4 Sampling distribution of the mean: All possible samples of size 3 drawn with replacement from a universe consisting of the equiprobable values 1, 2, 3, and 4

Values of the sample mean $\bar{X}$	Frequency with which a given sample mean would be observed f	$f\bar{X}$
1	1	1
1.33	3	3.99
1.67	6	10.02
2.00	10	20.00
2.33	12	27.96
2.67	12	32.04
3.00	10	30.00
3.33	6	19.98
3.67	3	11.01
4.00	1	4.00
	64	160.00

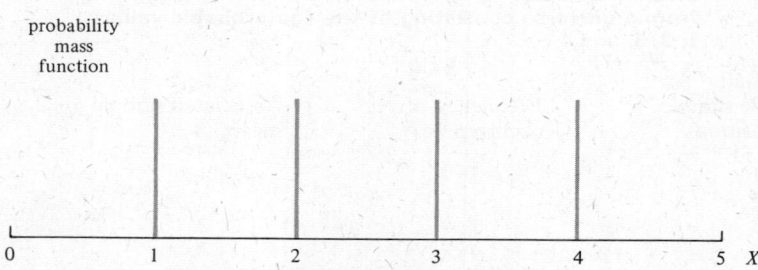

probability mass function

0   1   2   3   4   5 X

values of the individual items in the distribution

Figure 13.2(a) Distribution of the universe consisting of the four equiprobable values 1, 2, 3, and 4.

Observe that the bell-shaped pattern predicted for sample sizes greater than 30 has already begun to emerge, even though the original universe is a uniform distribution (all values equally likely) and the sample size is only 3 [Figures 13.2(a) and 13.2(b)]. Note that the arithmetic mean of the distribution $[\bar{\bar{X}} = \sum (f\bar{X})/\sum (f) = 160/64 = 2.5]$ is equal in value to the mean of the universe from which the theoretical sampling distribution was derived ($\mu_x = 2.5$). $\bar{\bar{X}} = \mu_x$ holds whether sampling takes place with or without replacement. Table 13.5 describes the sampling distribution of the mean associated with all possible samples of size 2 drawn *without* replacement from a universe consisting of the four equiprobable values 1, 2, 3, and 4. Table 13.6 describes the sampling distribution of the mean associated

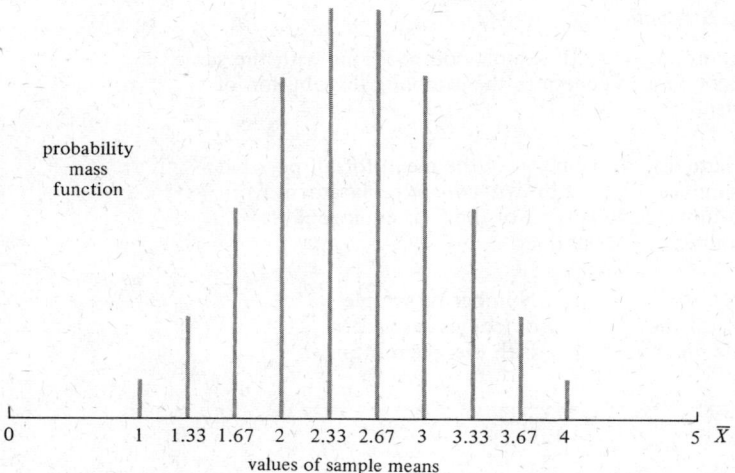

probability
mass
function

values of sample means

Figure 13.2(b) Sampling distribution of the mean: all possible samples of size 3 drawn with replacement from a universe consisting of the equiprobable values 1, 2, 3, and 4.

Table 13.5 All possible samples of size 2 drawn without replacement from a universe consisting of the four equiprobable values 1, 2, 3, and 4

a. A tree diagram depicting all possible ways a sample of size 2 could be drawn *without* replacement.

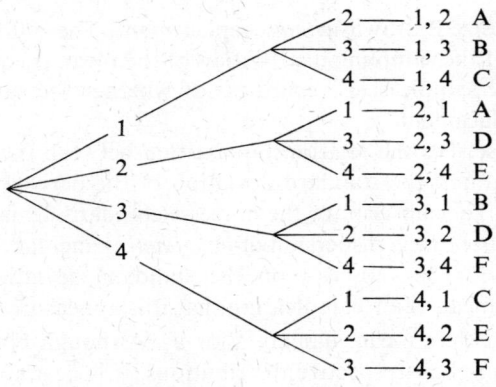

2 —— 1, 2	A	
3 —— 1, 3	B	
4 —— 1, 4	C	
1 —— 2, 1	A	
3 —— 2, 3	D	
4 —— 2, 4	E	
1 —— 3, 1	B	
2 —— 3, 2	D	
4 —— 3, 4	F	
1 —— 4, 1	C	
2 —— 4, 2	E	
3 —— 4, 3	F	

Note: There are 12 possible ways, each containing 2 elementary sampling units for a total of 24 elementary sampling units. A survey of the 24 elementary sampling units discloses that each member of the original universe occurs 6 times.

b. Distribution of all possible sample compositions of size 2 drawn *without* replacement and their associated means.

List of all possible sample compositions	Frequency of occurrence	Associated sample means
A 1, 2	2	1.5
B 1, 3	2	2.0
C 1, 4	2	2.5
D 2, 3	2	2.5
E 2, 4	2	3.0
F 3, 4	2	3.5
	——	
	12	

Table 13.5 *(Continued)*

c. Grouping together all sample compositions with the same arithmetic mean generates the sampling distribution of the mean.

Sampling distribution of the mean for all possible samples of size 2 drawn *without* replacement from a universe consisting of the four equiprobable values 1, 2, 3, and 4

Values of the sample mean $\bar{X}$	Number of sample outcomes associated with the given mean f	$f\bar{X}$
1.5	2	3.0
2.0	2	4.0
2.5	4	10.0
3.0	2	6.0
3.5	2	7.0
	12	30.0

---

d. As expected, $\bar{\bar{X}} = \sum (f\bar{X})/\sum (f) = 30/12 = 2.5$ is equal in value to the mean of the original universe from which the above theoretical sampling distribution of the mean was derived.

---

with all possible samples of size 3 drawn *without* replacement. These illustrations demonstrate that the mean of the sampling distribution of the mean is equal to the mean of the universe for all sample sizes regardless of whether the samples are selected with or without replacement.

3. The third theorem describes the relationship existing between the standard deviation of the universe, σ_x, and the standard deviation of the derived sampling distribution of the mean, $\sigma_{\bar{x}}$. The subscript of the universe standard deviation is an x, indicating that it is a measure of the dispersion of *individual* elementary sampling units around their mean, μ_x. The subscript on the standard deviation of the sampling distribution of the mean is an $\bar{x}$, indicating that it is a measure of dispersion of *sample means* around their mean, namely $\bar{\bar{X}}$ or μ_x. Although both σ_x and $\sigma_{\bar{x}}$ are standard deviations for their respective distributions, it is desirable to give them different names. One can think of the universe as the parent distribution and the sampling distribution as the derived distribution or offspring. Referring to both by the same name would be a source of confusion. Everyone is familiar with the confusion that arises when father and son are both called Bob. To avoid this, the standard deviation of the sampling distribution of the mean is called the *standard error of the mean*. (The dispersion in the sampling distribution is attributable to sampling error—hence the term the standard error of the mean.)

The relationship between the standard deviation of the universe and the standard error of the mean (the standard deviation of the derived sampling distribution of the mean) is described by the following algebraic relationship:

$$\sigma_{\bar{x}} = \sigma_x \cdot \frac{1}{\sqrt{n}} \cdot \sqrt{\frac{N-n}{N-1}}$$

Table 13.6 All possible samples of size 3 drawn without replacement from a universe consisting of the four equiprobable values 1, 2, 3, and 4

a. A tree diagram depicting all possible ways a sample of size 3 could be drawn *without* replacement.

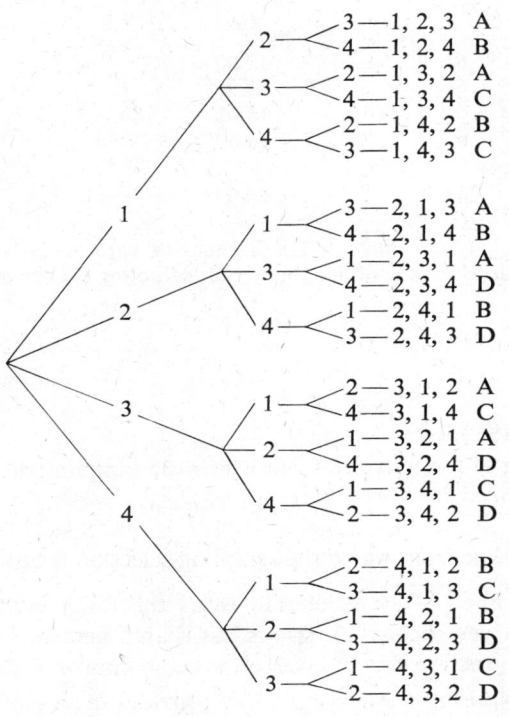

Note: There are 24 possible ways, each containing 3 elementary sampling units for a total of 72 elementary sampling units. A survey of the 72 elementary sampling units discloses that each member of the original universe occurs 18 times.

b. Distribution of all possible sample compositions of size 3 drawn *without* replacement and their associated means.

List of all possible sample compositions	Frequency of occurrence	Associated sample means
A 1, 2, 3	6	2
B 1, 2, 4	6	2.33
C 1, 3, 4	6	2.67
D 2, 3, 4	6	3.00
	24	

c. Grouping together all sample compositions with the same arithmetic mean generates the sampling distribution of the mean.

Sampling distribution of the mean for all possible samples of size 3 drawn *without* replacement from a universe consisting of the four equiprobable values 1, 2, 3, and 4

Table 13.6 *(Continued)*

Values of the sample mean $\overline{X}$	Number of sample outcomes associated with the given mean f	$f\overline{X}$
2.00	6	12.00
2.33	6	13.98
2.67	6	16.02
3.00	6	18.00
	24	60.00

d. As expected, $\bar{\overline{X}} = \sum (f\overline{X})/\sum (f) = 60/24 = 2.5$ is equal in value to μ_x, the mean of the original universe from which the above theoretical sampling distribution of the mean was derived.

where

$\sigma_{\bar{x}}$ = standard error of the mean

σ_x = standard deviation of the universe from which the sampling distribution of the mean is derived

n = the sample size

N = the size of the universe from which the sampling selection is to take place

The expression $\sqrt{(N - n)/(N - 1)}$ is referred to as the finite multiplier or finite correction factor. Provided the sample size is less than 5 percent of the size of the universe, the finite correction factor takes on a value approximately equal to one, and the simpler expression $\sigma_{\bar{x}} = \sigma_x \cdot 1/\sqrt{n}$ provides a reasonable description of the relationship between the standard deviation of the universe and the standard error of the mean. The relationship $\sigma_{\bar{x}} = \sigma_x \cdot (1/\sqrt{n}) \cdot \sqrt{(N - n)/(N - 1)}$ (which can be approximated by $\sigma_{\bar{x}} = \sigma_x \cdot 1/\sqrt{n}$ when $n/N <$ 5 percent) can be demonstrated by mathematical proof. The proof will be left to more advanced books in statistical analysis. An intuitive justification of the relationship, however, aids in understanding the role of the components of the formula. Referring to $\sigma_{\bar{x}} = \sigma_x \cdot 1/\sqrt{n}$, note that $\sigma_{\bar{x}}$ varies directly with σ_x and inversely with the square root of sample size.

INFLUENCE OF σ_x

For any given sample size, the greater the dispersion existing among the elementary sampling units comprising the universe, the greater the dispersion one would expect to observe among the sample means derived from that universe.

INFLUENCE OF n

When sample size is small, extreme values exert significant influence on the means of samples in which they are included. As the sample size is increased, the potential impact of extreme values becomes less and less significant. Accordingly, the sample means cluster more and more closely together. This is accompanied by a decrease in the value of $\sigma_{\bar{x}}$, reflecting the decreased distance along the number line required to include any given percentage of the sample means. Figure 13.3 describes the way in which the clustering increases as the sample size increases from a low of size 1 to a maximum of size N.

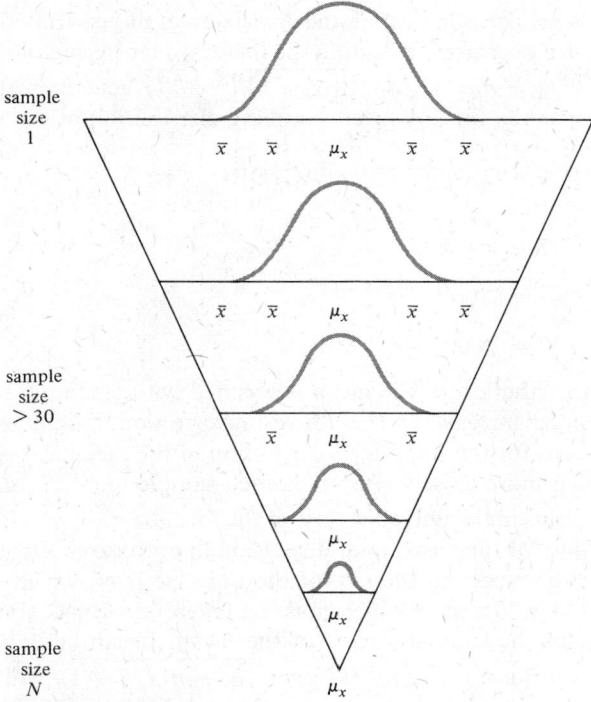

sample size 1

sample size > 30

sample size N

Figure 13.3

Note that with samples of size 1, the shape of the sampling distribution of the mean is identical to that of the universe. As sample size increases, the sampling distribution of the mean becomes less and less skewed. By the time a sample size of 30 is reached, the sampling distribution of the mean is for all intents and purposes normally distributed.

Sample size of size 1: With a sample of size 1, every elementary sampling unit is a sample, and the sampling distribution of the mean is identical to the distribution of the universe. As expected, σ_x equals $\sigma_{\bar{x}}$.

$$\sigma_{\bar{x}} = \sigma_x \cdot \frac{1}{\sqrt{n}}$$

$$= \sigma_x \cdot \frac{1}{\sqrt{1}}$$

$$= \sigma_x$$

Sample size of size N: The sampling distribution of the mean associated with a sample of size N consists of just one sample (a sample the size of the universe is a complete census). The mean of the universe is equal to the mean of the sample (there is no sampling error), and $\sigma_{\bar{x}}$ is equal to zero. With $n = N$, the finite multiplier would have to be used. With $n = N$, $\sqrt{(N - n)/(N - 1)}$ is equal to zero. Therefore,

$$\sigma_{\bar{x}} = \sigma_x \cdot \frac{1}{\sqrt{n}} \cdot \sqrt{\frac{N - n}{N - 1}}$$

$$= \sigma_x \cdot \frac{1}{\sqrt{n}} \cdot (0)$$

$$= 0$$

Clearly, as the sample size is increased, the dispersion in the associated sampling distribution of the mean decreases, reflecting the increased clustering of the sample means. The formula $\sigma_{\bar{x}} = \sigma_x \cdot (1/\sqrt{n}) \cdot \sqrt{(N - n)/(N - 1)}$ generates the appropriate measure of dispersion for any given sampling distribution of the mean.

ROLE OF THE FINITE CORRECTION FACTOR
Given the following:

Universe A	*Universe B*
$\sigma_x = \$10$	$\sigma_x = \$10$
$N = 17$	$N = 10{,}001$

If the sampling distribution of the mean associated with a sample size of size 16 were to be developed for each of the above universe populations, one would expect the sample means for the sampling distribution of the mean associated with universe A to be much more closely clustered (each sample includes 16 of the 17 elementary sampling units in the universe). Use of the formula $\sigma_{\bar{x}} = \sigma_x \cdot 1/\sqrt{n}$ would generate the same value for the measure of dispersion in both cases. Obviously, the size of the sample with respect to the size of the universe from which it is to be selected is a factor, the influence of which must be taken into consideration when computing the value of the standard error of the mean. It can be demonstrated mathematically that inclusion of the factor $\sqrt{(N - n)/(N - 1)}$ will correctly reflect the influence of universe size. Again, the mathematical proof is left to more advanced texts. An empirical demonstration of the effect of the finite multiplier is provided below:

For universe A:

$$\sigma_{\bar{x}} = \sigma_x \cdot \frac{1}{\sqrt{n}} \sqrt{\frac{N - n}{N - 1}}$$

$$= (\$10) \left(\frac{1}{\sqrt{16}}\right) \sqrt{\frac{17 - 16}{17 - 1}}$$

$$= (\$10) \left(\frac{1}{4}\right) \sqrt{\frac{1}{16}}$$

$$= \$2.50\sqrt{0.0625}$$

$$= \$2.50(0.25)$$

$$= \$0.625$$

With respect to universe A, taking into consideration the size of the universe reduces the value of the standard error of the mean from \$2.50 (before adjustment) to \$0.625 (after adjustment).

For universe B:

$$\sigma_{\bar{x}} = \sigma_x \cdot \frac{1}{\sqrt{n}} \sqrt{\frac{N - n}{N - 1}}$$

$$= \$10 \cdot \left(\frac{1}{\sqrt{16}}\right) \sqrt{\frac{10{,}001 - 16}{10{,}001 - 1}}$$

$$= \$2.50 \sqrt{\frac{9985}{10,000}}$$

$$= \$2.50\sqrt{0.9985}$$

$$= \$2.50(0.999)$$

$$= \$2.4975$$

With respect to universe B, taking into consideration the size of the universe has virtually no effect on the value of the standard error of the mean. This is always the case when the universe size is large relative to the size of the sample to be selected. For this reason, it is customary to ignore the finite correction factor whenever the sample size amounts to less than 5 percent of the size of the universe. The correction factor is referred to as the finite correction factor because it is only used in association with finite universes (an infinite universe is infinitely large and no finite sample size can exceed 5 percent of infinity). Even with finite universes, the correction factor is only utilized when $n/N > 5$ percent.

In Tables 13.8 and 13.9, the relationship is demonstrated empirically for sample sizes of 2 and 3, respectively, drawn *with replacement* from the universe consisting of the four equiprobable values 1, 2, 3, and 4. Note that when drawing with replacement that there is no limit to sample size—no matter how large the sample selected, the universe can never be exhausted. Sampling with replacement is thus equivalent to sampling from an infinite universe, and the relationship between the universe standard deviation and the standard error of the mean is described by

$$\sigma_{\bar{x}} = \sigma_x \cdot \frac{1}{\sqrt{n}}$$

The computation of the universe standard deviation is presented in Table 13.7.

Table 13.7 Computation of the universe standard deviation for the universe consisting of the four equiprobable values 1, 2, 3, and 4

X	$X - \mu_x$	$(X - \mu_x)^2$
1	-1.5	2.25
2	-0.5	0.25
3	$+0.5$	0.25
4	$+1.5$	2.25
10	-2.0	5.00
	$+2.0$	
	0	

Universe mean:

$$\mu_x = \frac{\sum X}{N} = \frac{10}{4} = 2.5$$

Universe standard deviation:

$$\sigma_x = \sqrt{\frac{\sum (X - \mu_x)^2}{N}} = \sqrt{\frac{5.00}{4}}$$

$$= \sqrt{1.25} \qquad = 1.118$$

Table 13.8

a. Direct computation of the standard error of the mean (standard deviation of the sampling distribution of the mean) for the sampling distribution of the mean associated with all possible samples of size 2 drawn *with replacement* from a universe consisting of the four equiprobable values 1, 2, 3, and 4
This table is based on Table 4.4, p. 85.

Values of the sample mean X	Number of sample permutations associated with the given mean f	fX	$X - \bar{\bar{X}}$	$(X - \bar{\bar{X}})^2$	$f(X - \bar{\bar{X}})^2$
1.0	1	1	-1.5	2.25	2.25
1.5	2	3	-1.0	1.00	2.00
2.0	3	6	-0.5	0.25	0.75
2.5	4	10	0	0	0.00
3.0	3	9	$+0.5$	0.25	0.75
3.5	2	7	$+1.0$	1.00	2.00
4.0	1	4	$+1.5$	2.25	2.25
	16	40			10.00

$$\bar{\bar{X}} = \frac{\sum (f\,X)}{\sum (f)} = \frac{40}{16} = 2.5$$

$$\sigma_{\bar{x}} = \sqrt{\frac{\sum f(X - \bar{\bar{X}})^2}{\sum (f)}}$$

$$= \sqrt{\frac{10}{16}}$$

$$= \sqrt{0.625} = 0.79$$

The value of the standard error of the mean computed directly is 0.79.

b. Computation of the standard error of the mean by the relationship $\sigma_{\bar{x}} = \sigma_x \cdot 1/\sqrt{n}$

$$\sigma_{\bar{x}} = \sigma_x \cdot \frac{1}{\sqrt{n}}$$

$$= 1.118 \cdot \frac{1}{\sqrt{n}}$$

$$= 1.118 \cdot \frac{1}{\sqrt{2}}$$

$$= \frac{1.118}{1.41421}$$

$$\doteq 0.79$$

Note the correspondence between the value of the standard error of the mean as computed directly in part a. and the standard error of the mean as computed by the relationship

$$\sigma_{\bar{x}} = \sigma_x \cdot \frac{1}{\sqrt{n}}$$

Table 13.9

a. Direct computation of the standard error of the mean (standard deviation of the sampling distribution of the mean) for the sampling distribution of the mean for all possible samples of size 3 drawn with replacement from a universe consisting of the four equiprobable values 1, 2, 3, and 4 (refer to Table 13.4)

Values of the sample mean $\bar{X}$	Number of sample outcomes associated with the given mean f	$f\bar{X}$	$\bar{X} - \bar{\bar{X}}$	$(\bar{X} - \bar{\bar{X}})^2$	$f(\bar{X} - \bar{\bar{X}})^2$
1.00	1	1.00	-1.50	2.2500	2.2500
1.33	3	3.99	-1.17	1.3689	4.1067
1.67	6	10.02	-0.83	0.6889	4.1334
2.00	10	20.00	-0.50	0.2500	2.5000
2.33	12	27.96	-0.17	0.0289	0.3468
2.67	12	32.04	$+0.17$	0.0289	0.3468
3.00	10	30.00	$+0.50$	0.2500	2.5000
3.33	6	19.98	$+0.83$	0.6889	4.1334
3.67	3	11.01	$+1.17$	1.3689	4.1067
4.00	1	4.00	$+1.50$	2.2500	2.2500
	64	160.00			26.6738

$$\bar{\bar{X}} = \frac{\sum (f\bar{X})}{\sum (f)} = \frac{160}{64} = 2.5$$

$$\sigma_{\bar{x}} = \sqrt{\frac{\sum f(\bar{X} - \bar{\bar{X}})^2}{\sum (f)}}$$

$$= \sqrt{\frac{26.6738}{64}}$$

$$= \sqrt{0.4167}$$

$$= 0.646$$

The value of the standard error of the mean computed directly is 0.645. (Note that the means of samples of size 3 are more closely clustered than the means of samples of size 2.)

b. Computation of the standard error of the mean by the relationship $\sigma_{\bar{x}} = \sigma_x \cdot 1/\sqrt{n}$

$$\sigma_{\bar{x}} = \sigma_x \cdot \frac{1}{\sqrt{n}}$$

$$= 1.118 \left(\frac{1}{\sqrt{3}}\right)$$

$$= 1.118 \left(\frac{1}{1.73205}\right)$$

$$= \frac{1.118}{1.73205}$$

$$= 0.646$$

Note the correspondence between the value of the standard error of the mean as computed directly and the standard error of the mean as computed by the relationship

$$\sigma_{\bar{x}} = \sigma_x \cdot \frac{1}{\sqrt{n}}$$

Table 13.10

---

a. Direct computation of the standard error of the mean (standard deviation of the sampling distribution of the mean) for the sampling distribution of the mean associated with all possible samples of size 2 drawn *without replacement* from a universe consisting of the four equiprobable values 1, 2, 3, and 4 (refer to Table 13.5)

Values of the sample mean $\bar{X}$	Number of sample outcomes associated with the given mean f	$f\bar{X}$	$(\bar{X} - \bar{\bar{X}})$	$(\bar{X} - \bar{\bar{X}})^2$	$f(\bar{X} - \bar{\bar{X}})^2$
1.5	2	3.0	-1.0	1.00	2.00
2.0	2	4.0	-0.5	0.25	0.50
2.5	4	10.0	0	0	0
3.0	2	6.0	$+0.5$	0.25	0.50
3.5	2	7.0	$+1.0$	1.00	2.00
	12	30.0			5.00

$$\bar{\bar{X}} = \frac{\sum(f\bar{X})}{\sum(f)} = \frac{30}{12} = 2.5$$

$$\sigma_{\bar{x}} = \sqrt{\frac{\sum f(\bar{X} - \bar{\bar{X}})^2}{\sum(f)}}$$

$$= \sqrt{\frac{5.00}{12}} = \sqrt{0.4167} = 0.646$$

The value of the standard error of the mean computed directly is 0.646

b. Computation of the standard error of the mean by the relationship

$$\sigma_{\bar{x}} = \sigma_x \cdot \frac{1}{\sqrt{n}} \sqrt{\frac{N-n}{N-1}}$$

$$= 1.118 \left(\frac{1}{\sqrt{2}}\right) \sqrt{\frac{4-2}{4-1}}$$

$$= 1.118 \left(\frac{1}{1.41421}\right) \sqrt{\frac{2}{3}}$$

$$= \frac{1.118}{1.41421} (\sqrt{0.67})$$

$$= (0.79)(0.818)$$

$$= 0.64622$$

Note the correspondence between the value of the standard error of the mean as computed directly and the standard error of the mean as computed by the relationship

$$\sigma_{\bar{x}} = \sigma_x \cdot \frac{1}{\sqrt{n}} \sqrt{\frac{N-n}{N-1}}$$

To achieve this correspondence, it was necessary to include the finite multiplier.

In Tables 13.10 and 13.11, the relationship is demonstrated for sample size 2 and sample size 3, respectively, drawn *without replacement*, from the universe consisting of the four equiprobable values 1, 2, 3, and 4. When drawing without replacement, the universe would be exhausted with a sample of size 4. Samples of size 2 or 3 drawn from a finite universe of size 4 are significantly greater than 5 percent of the universe. Clearly, the finite correction factor is required, and the appropriate description of the relationship between the standard deviation of the universe and the standard error of the mean is

$$\sigma_{\bar{x}} = \sigma_x \cdot \frac{1}{\sqrt{n}} \sqrt{\frac{N-n}{N-1}}$$

Table 13.11

a. Direct computation of the standard error of the mean (standard deviation of the sampling distribution of the mean) for the sampling distribution of the mean associated with all possible samples of size 3 drawn *without replacement* from a universe consisting of the four equiprobable values 1, 2, 3, and 4 (refer to Table 13.6)

Values of the sample mean $\bar{X}$	Number of sample outcomes associated with the given mean f	$f\bar{X}$	$(\bar{X} - \bar{\bar{X}})$	$(\bar{X} - \bar{\bar{X}})^2$	$f(\bar{X} - \bar{\bar{X}})^2$
2.00	6	12.00	−0.50	0.2500	1.5000
2.33	6	13.98	−0.17	0.0289	0.1734
2.67	6	16.02	+0.17	0.0289	0.1734
3.00	6	18.00	+0.50	0.2500	1.5000
	24	60.00			3.3468

$$\bar{\bar{X}} = \frac{\sum (f\bar{X})}{\sum (f)} = \frac{60.00}{24} = 2.5$$

$$\sigma_{\bar{x}} = \sqrt{\frac{\sum f(\bar{X} - \bar{\bar{X}})^2}{\sum (f)}}$$

$$= \sqrt{\frac{3.3468}{24}} = \sqrt{0.1394} = 0.373$$

The value of the standard error of the mean computed directly is 0.373

b. Computation of the standard error of the mean by the relationship

$$\sigma_{\bar{x}} = \sigma_x \frac{1}{\sqrt{n}} \sqrt{\frac{N-n}{N-1}}$$

$$= 1.118 \left(\frac{1}{\sqrt{3}}\right) \sqrt{\frac{4-3}{4-1}} = 1.118 \left(\frac{1}{1.73205}\right) \sqrt{\frac{1}{3}}$$

$$= \frac{1.118}{1.73205} \sqrt{0.3333} = (0.646) \sqrt{0.3333} = (0.646)(0.577)$$

$$= 0.373$$

Incorporation of the finite correction factor generates a correspondence between the value of the standard error of the mean as computed directly and the standard error of the mean as computed by the relationship

$$\sigma_{\bar{x}} = \sigma_x \cdot \frac{1}{\sqrt{n}} \cdot \sqrt{\frac{N-n}{N-1}}$$

13.5 SIGNIFICANCE OF THE THREE THEOREMS

The three theorems discussed in preceding sections constitute the foundation on which a significant portion of statistical inference has been built. The central limit theorem indicates that the sampling distribution of the mean associated with samples of a size greater than 30 will be normally distributed regardless of the shape of the original universe. Thus, the shape of the distribution is known without having to undergo the mechanics of actual construction. Recall that a normal distribution is totally described if its mean and standard deviation are known. The second theorem indicates that $\bar{X} = \mu_x$ and the third theorem that $\sigma_{\bar{x}} = \sigma_x(1/\sqrt{n})(\sqrt{N - n}/(N - 1))$. Thus, if one knows the mean and the standard deviation of the universe, the generation of the mean and standard error of the mean for any derived sampling distribution of the mean is merely a matter of algebraic manipulation. Unfortunately, the mean and standard deviation of the universe are seldom known. As a matter of fact, the objective of the sampling process usually is to provide estimates of these values. The beauty of sampling, however, is that a single sample selected from the universe provides the basis for generating an estimate of both the universe mean and the universe standard deviation.

Recall that $\bar{X}$ is an unbiased estimator of μ_x and $\hat{\sigma}_x{}^2$ is an unbiased estimator of $\sigma_x{}^2$ when computed by the formula

$$\hat{\sigma}_x{}^2 = \frac{\sum (X - \bar{X})^2}{n - 1}$$

$\sqrt{\hat{\sigma}_x{}^2}$ or $\hat{\sigma}_x$ is used as an estimate of σ_x, and $\bar{X}$ is used as an estimate of μ_x. With these two values and the knowledge that the sampling distribution of the mean is normally distributed, an estimate of the sampling distribution of the mean can be obtained. Thus, a single sample provides all the information needed to generate an estimate of the sampling distribution of the mean associated with that particular sample size (or any other sample size for that matter).

13.6 CONFIDENCE INTERVAL ESTIMATE OF THE UNIVERSE MEAN: THE CONCEPT

Consider once again the universe consisting of the 1200 students comprising the sophomore class. Presume that the variable of interest is parent's income. Suppose that the universe mean is $10,000 and that the universe standard deviation is $240. Consider the sampling distribution of the mean associated with all possible samples of size 36 drawn from that universe. The mean of the sampling distribution of the mean is equal to $10,000, and the standard error of the mean is equal to $40.

$$\bar{Y} = \mu_y = \$10{,}000$$

$$\sigma_{\bar{y}} = \sigma_y \cdot \frac{1}{\sqrt{n}} = \$240 \left(\frac{1}{\sqrt{36}}\right) = \$240(\tfrac{1}{6}) = \$40$$

The central limit theorem indicates that the sampling distribution of the mean associated with sample size 36 will be normally distributed (Figure 13.4).

Perpendiculars raised at a point on the number line 2 standard errors of the mean greater than μ_y and at a point 2 standard errors of the mean less than μ_y will encompass approximately 95.44 percent of the total area under the curve (Figure 13.5).

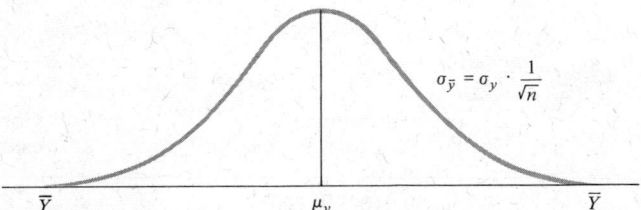

Figure 13.4 Sampling distribution of the mean associated
with all possible samples of size 36: $\overline{\overline{Y}} = \mu_y = \$10,000$;
$\sigma_y = \$240$.

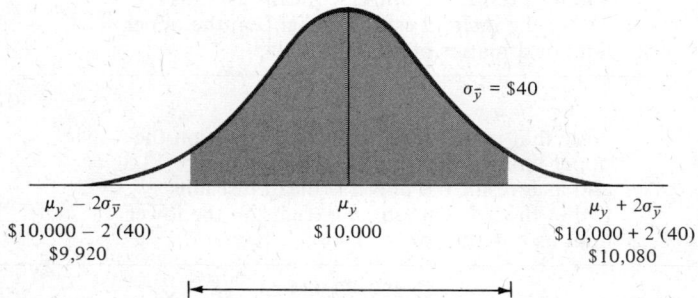

Figure 13.5 Sampling distribution of the mean associated with all
possible samples of size 36: $\overline{\overline{Y}} = \mu_y = \$10,000$; $\sigma_y = \$240$.

Recalling that area under a curve is proportional to relative frequency of occurrence, if 95.44 percent of the area lies between \$9,920 and \$10,080, then 95.44 percent of the means of all possible samples of size 36 have a value between \$9,920 and \$10,080. Another way of stating this is to say that 95.44 percent of the values of the means of all possible samples of size 36 lie within 2 standard errors of the universe mean.

Suppose that a sample of size 36 is selected out of the universe, and the value of the sample mean is \$9,970. If 2 standard errors of the mean are added and subtracted to the value of that sample mean, an interval estimate of the universe mean will be generated. The interval estimate takes the form of a statement, namely, that the universe mean has a value somewhere in the interval from \$9,890 to \$10,050. Given our state of omniscience, it is clear that this statement is correct. An interval estimate of this type could be developed in association with each and every sample mean. Some of these are illustrated in Table 13.12.

In general, if a mean selected at random from the universe falls within the shaded region, the resulting interval estimate of the universe mean is correct. In Figure 13.5, 95.44 percent of all possible sample means associated with samples of size 36 fall within the shaded region. Therefore, 95.44 percent of all possible interval estimates would be correct. With simple random sampling, each sample has an equal opportunity of being selected. Given that only one sample is to be drawn, the probability that the one sample selected will lead to a correct interval estimate is equal to the ratio of the number of ways a correct interval estimate could be generated to the total number of interval estimates possible or 95.44 percent.

If the analyst desires to increase the probability of a correct estimate (given a particular sample size), the interval associated with the estimate must be widened. For example, if one obtained interval estimates by adding and subtracting 2.5

Table 13.12

$$\mu_y = \$10{,}000, \; \sigma_{\bar{y}} = \$40$$

Sample mean $\bar{Y}$	Interval estimate $\bar{Y} - 2\sigma_{\bar{y}}$ to $\bar{Y} + 2\sigma_{\bar{y}}$	Correctness of estimate
a. $10,020	$9,940–$10,100	Correct
b. $9,920	$9,840–$10,000	Correct
	Note that in this instance, $\bar{Y}$ falls right on the lower limit of the shaded region in Figure 13.5 and as a result the universe mean just falls within the interval estimate (right on the upper limit as a matter of fact).	
c. $10,080	$10,000–$10,160	Correct
	Note that in this instance, $\bar{Y}$ falls right on the upper limit of the shaded region in Figure 13.5 and as a result the universe mean just falls within the interval estimate (right on the lower limit as a matter of fact).	
d. $9,900	$9,820–$9,980	Incorrect
	The sample mean $9,900 falls in the unshaded region in the left-hand tail of the distribution and therefore not within 2 standard errors of the universe mean.	
e. $10,090	$10,010–$10,170	Incorrect
	The sample mean $10,090 falls in the unshaded region in the right-hand tail of the distribution and therefore not within 2 standard errors of the universe mean.	

standard errors of the mean to the value of the sample mean, 98.76 percent of all such interval estimates associated with samples of size 36 would be correct.

If one wished to maintain the same level of confidence in the correctness of the statement (say, a 90 percent probability of the statement being correct) but wished to narrow the interval within which the estimate is to be made, a larger sample size must be used. With a larger sample size, the sample means cluster more closely around the universe mean. In the associated sampling distribution, this is reflected by a reduction in the value of the standard error of the mean. The reduction in the value of the standard error of the mean is significant in the sense that one need proceed a shorter distance along the number line on either side of the universe mean in order to include any given percentage of all sample means than is the case for smaller sample sizes. This behavior pattern is illustrated in Figure 13.6.

Referring to Figure 13.6, one could conclude that:

1. Sixty-eight percent of all sample means of size 36 lie within $40 of the universe mean; therefore, 68 percent of all confidence interval statements of the form $\bar{Y} \pm \$40$ (sample mean plus or minus $40) would be correct, where $\bar{Y}$ is the mean of a sample of size 36.

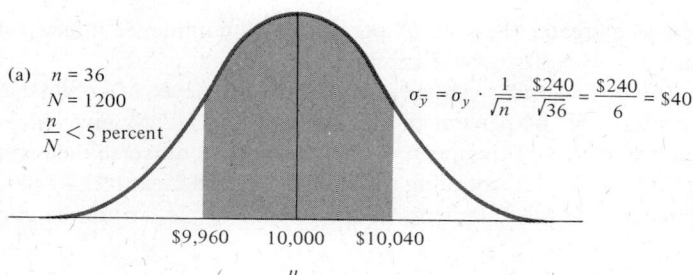

(a) $n = 36$
$N = 1200$
$\dfrac{n}{N} < 5$ percent

$$\sigma_{\bar{y}} = \sigma_y \cdot \frac{1}{\sqrt{n}} = \frac{\$240}{\sqrt{36}} = \frac{\$240}{6} = \$40$$

$9,960   10,000 $10,040

μ_y

It would be necessary to proceed plus and minus \$40 along the number line
from the value of the universe mean to include 68 percent of all sample
means associated with a sample size of size 36.

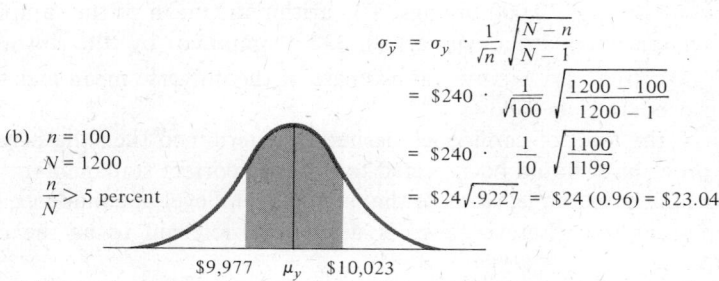

(b) $n = 100$
$N = 1200$
$\dfrac{n}{N} > 5$ percent

$$\sigma_{\bar{y}} = \sigma_y \cdot \frac{1}{\sqrt{n}} \sqrt{\frac{N-n}{N-1}}$$

$$= \$240 \cdot \frac{1}{\sqrt{100}} \sqrt{\frac{1200-100}{1200-1}}$$

$$= \$240 \cdot \frac{1}{10} \sqrt{\frac{1100}{1199}}$$

$$= \$24\sqrt{.9227} = \$24(0.96) = \$23.04$$

$9,977 μ_y $10,023

It would be necessary to proceed plus and minus \$23.04 along the number
line from the value of the universe mean to include 68 percent of all sample
means associated with a sample size of size 100.

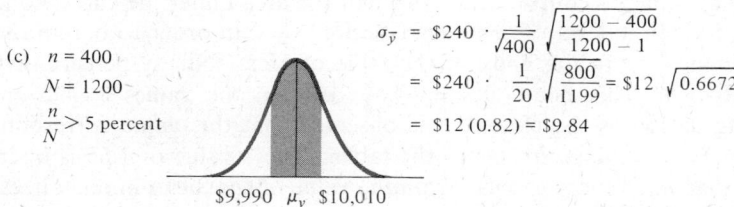

(c) $n = 400$
$N = 1200$
$\dfrac{n}{N} > 5$ percent

$$\sigma_{\bar{y}} = \$240 \cdot \frac{1}{\sqrt{400}} \sqrt{\frac{1200-400}{1200-1}}$$

$$= \$240 \cdot \frac{1}{20} \sqrt{\frac{800}{1199}} = \$12 \sqrt{0.6672}$$

$$= \$12(0.82) = \$9.84$$

$9,990 μ_y $10,010

It would be necessary to proceed plus and minus \$9.84 along the number line
from the value of the universe mean to include 68 percent of all sample
means associated with a sample size of size 400.

Figure 13.6 Sampling distribution of the mean associated with sample
sizes 36, 100, and 400: $\mu_y = \$10,000$; $\sigma_y = \$240$.

2. Sixty-eight percent of all sample means of size 100 lie within \$23.04 of the
universe mean; therefore, 68 percent of all confidence interval statements of
the form $\bar{Y} \pm \$23.04$ would be correct, where $\bar{Y}$ is the mean of a sample of
size 100.

3. Sixty-eight percent of all sample means of size 400 lie within \$9.84 of the
universe mean; therefore, 68 percent of all confidence interval statements of
the form $\bar{Y} \pm \$9.84$ would be correct, where $\bar{Y}$ is the mean of a sample of
size 400.

4. In general, 68 percent of all sample means of any given size lie within 1
standard error of the mean of their respective universe means. Therefore,
given any particular sampling distribution of the mean associated with a

sample size greater than 30, 68 percent of all confidence interval statements of the form $\overline{Y} \pm 1\sigma_{\bar{y}}$ would be correct.

5. This line of reasoning can be extended to any level of confidence desired. For example, 95.44 percent of all sample means of any given size lie within 2 standard errors of the mean of their respective universe means; therefore, given any particular sampling distribution of the mean, 95.44 percent of all confidence interval statements of the form $\overline{Y} \pm 2\sigma_{\bar{y}}$ would be correct.

13.7 CONFIDENCE INTERVAL ESTIMATE OF THE UNIVERSE MEAN: THE PROCEDURE

The situation: A simple random sample of 64 billings has been selected from a universe consisting of 20,000 billings. The arithmetic mean of the sample is $223, and the sample standard deviation is $32 (computed by the formula $\hat{\sigma}_x = \sqrt{\sum (X - \overline{X})^2 / (n - 1)}$. An interval estimate of the universe mean is desired.

The procedure is as follows.

1. First, the level of confidence desired is determined (i.e., the percentage of time the procedure should be expected to produce correct statements).

In the billings example, suppose that a 90 percent level of confidence is desired. The 90 percent (or whatever level is desired) is referred to as the confidence coefficient.

2. Determine how many standard errors of the mean one must add and subtract to the value of the universe mean in order to obtain two points on the number line equidistant from the universe mean between which "90 percent" of the area under the curve is contained.

Questions of this type are answered by the table of areas under the normal curve. The table, of course, deals with half the area under the curve, so it is necessary to divide the level of confidence desired by 2 in order to obtain the relevant value to look up in the body of the table of areas. Ninety percent divided by 2 equals 0.4500. This figure lies midway between the values 0.4495 and 0.4505. Selecting 0.4505 as the closest, it is observed that this value corresponds to a z value of 1.65 in the perimeter of the table.[2] The z value of 1.65 is interpreted as stating that when moving away from the value of the mean in one direction along the number line, it is necessary to go a distance of 1.65 standard error units in order to include approximately 45 percent of the area under the curve.

3. The third step is the determination of the value of the standard error of the mean. The standard error of the mean is related to the standard deviation of the universe by the formula

$$\sigma_{\bar{x}} = \frac{\sigma_x}{\sqrt{n}} \sqrt{\frac{N - n}{N - 1}}$$

which simplifies to

$$\sigma_{\bar{x}} = \frac{\sigma_x}{\sqrt{n}} \quad \text{when} \quad \frac{n}{N} < 5 \text{ percent}$$

---

[2] By selecting 0.4505 instead of 0.4495, the procedure will generate correct statements slightly more than 90 percent of the time instead of slightly less than 90 percent of the time.

Use of this formula requires knowledge of the universe standard deviation. If the universe standard deviation is unknown, its value is estimated from the selected sample by the formula

$$\hat{\sigma}_x = \sqrt{\frac{\sum (X - \bar{X})^2}{n - 1}}$$

When the value obtained for $\hat{\sigma}_x$ is substituted into the formula for the standard error of the mean, an estimated value for the standard error of the mean is obtained:

$$\hat{\sigma}_{\bar{x}} = \frac{\hat{\sigma}_x}{\sqrt{n}}$$

where $\hat{\sigma}_x$ is an estimate of σ_x and $\hat{\sigma}_{\bar{x}}$ is an estimate of $\sigma_{\bar{x}}$. Specifically in the case of the billings illustration,

$$\hat{\sigma}_{\bar{x}} = \frac{\hat{\sigma}_x}{\sqrt{n}}$$

$$= \frac{\$32}{\sqrt{64}}$$

$$= \frac{\$32}{8}$$

$$= \$4$$

4. The fourth step is to multiply the size of the confidence interval estimate in standard error units (determined in step 2) by the value of the standard error of the mean (determined in step 3) to obtain the amount to be added and subtracted to and from the sample mean.

$$\begin{pmatrix} \text{number of} \\ \text{standard} \\ \text{errors} \end{pmatrix} \begin{pmatrix} \text{value of} \\ \text{the standard} \\ \text{error} \end{pmatrix} = z\hat{\sigma}_{\bar{x}} = (1.65)(\$4) = \$6.60$$

5. Add and subtract the value obtained in step 4 to the sample mean to obtain the upper and lower confidence limits of the confidence interval.

upper confidence level limit

$$\bar{X} + z\hat{\sigma}_{\bar{x}} = \$223 + 1.65(\$4) = \$223 + \$6.60 = \$229.60$$

lower confidence level limit

$$\bar{X} - z\hat{\sigma}_{\bar{x}} = \$223 - 1.65(\$4) = \$223 - \$6.60 = \$216.40$$

6. The last step is to make the confidence interval statement, "I am 90 percent confident that the statement is correct when I state that the value of the universe mean lies somewhere in the interval $216.40–$229.60." The justification for such a statement is that the procedure used will generate correct statements "90 out of every 100 times."

13.8 CONFIDENCE STATEMENTS ABOUT THE MAGNITUDE OF EXPECTED ERROR: THE CONCEPT

Expected error as used in this context refers to the amount of sampling error likely to be observed in association with a sample estimate of the universe parameter. Sampling error refers to the difference observed between the sample statistic and the corresponding value of the universe parameter. The level of confidence refers to the percentage of time the procedure leads to correct statements of the form "the sampling error does not exceed E where $E = \varkappa\sigma_{\bar{x}}$."

To illustrate, it is known that 95.44 percent of all sample means associated with a given sample size will fall within 2 standard errors of the universe mean. Therefore, one could be 95.44 percent confident that one's statement would be correct when one stated that "the sampling error associated with a particular sample mean will not exceed $2\sigma_{\bar{x}}$."

$$\text{level of confidence} = 95.44 \text{ percent}$$
$$E = 2\sigma_{\bar{x}}$$

Similarly, it is known that 80 percent of all sample means associated with a given sample size will fall within 1.28 standard errors of the universe mean. Therefore, one could be 80 percent confident that one's statement would be correct when one stated that "the sampling error associated with a particular sample mean will not exceed $1.28\sigma_{\bar{x}}$."

$$\text{level of confidence} = 80 \text{ percent}$$
$$E = 1.28\sigma_{\bar{x}}$$

13.9 CONFIDENCE STATEMENTS ABOUT THE MAGNITUDE OF EXPECTED ERROR: THE PROCEDURE

The situation: A simple random sample of 64 billings has been selected from a universe consisting of 20,000 billings. The arithmetic mean of the sample is $223, and the sample standard deviation computed by the formula $\hat{\sigma}_x = \sqrt{\sum (X - \bar{X})^2}/\sqrt{(n - 1)}$ is $32. What statement can be made with a 90 percent probability of being correct about the magnitude of sampling error associated with the sample estimate of $223?

The procedure: The formula for expected sampling error is

$$E = \varkappa\sigma_{\bar{x}}$$

when σ_x is known and

$$E = \varkappa\hat{\sigma}_{\bar{x}}$$

when σ_x is unknown and its value is estimated from a sample, where

$E =$ Magnitude of sampling error.

$\varkappa =$ The number of standard error units associated with the desired level of confidence. The desired level of confidence refers to the percentage of the time that the procedure will lead to correct statements about the magnitude of sampling error.

$\sigma_{\bar{x}} =$ The standard error of the mean,

$$\sigma_{\bar{x}} = \sigma_x \frac{1}{\sqrt{n}} \sqrt{\frac{N - n}{N - 1}}$$

$\hat{\sigma}_{\bar{x}}$ = An estimate of the value of the standard error of the mean when σ_x is
unknown; the value of σ_x is estimated from the sample by the formula
$\hat{\sigma}_x = \sqrt{\sum (X - \bar{X})^2/(n - 1)}$. $\hat{\sigma}_x$ is then used as an input to compute an
estimate of the standard error of the mean by the formula

$$\hat{\sigma}_{\bar{x}} = \hat{\sigma}_x \cdot \frac{1}{\sqrt{n}} \sqrt{\frac{N - n}{N - 1}}$$

Procedural steps:

1. Determine the level of confidence desired (i.e., the percentage of the time it
is desired that the procedure should generate correct statements about the magnitude of sampling error). In the billings illustration, 90 percent confidence is
desired.

2. Determine the number of standard errors of the mean one must add and
subtract to the value of the universe mean in order to encompass the given percentage of the total area under the curve. The table of areas indicates that one
must proceed a distance of 1.65 standard error units along the number line in
one direction from the mean in order to include approximately 45 percent of the
area under the curve. Thus, the z value associated with a 90 percent level of
confidence is 1.65.

3. Determine the value of the standard error of the mean. σ_x is unknown in
this case, therefore, $\hat{\sigma}_{\bar{x}}$ is used as an estimate of $\sigma_{\bar{x}}$.

$$\hat{\sigma}_{\bar{x}} = \frac{\hat{\sigma}_x}{\sqrt{n}} \quad \text{because} \quad \frac{n}{N} < 5 \text{ percent}$$

$$\hat{\sigma}_{\bar{x}} = \frac{\$32}{\sqrt{64}} = \frac{\$32}{8} = \$4$$

4. Multiply the z value determined in step 2 by the value of the standard error
of the mean computed in step 3 to determine the likely magnitude of sampling
error:

$$z\hat{\sigma}_{\bar{x}} = 1.65(\$4) = \$6.60$$

5. The final step is to generate the statement about expected sampling error:
"I am 90 percent confident that the statement is correct when I state that the
magnitude of sampling error will not exceed \$6.60." The justification for such a
statement is that the procedure used to generate such statements will, on the
average, generate correct statements 90 out of every 100 times.

There is a close family relationship between confidence interval statements and
statements about the magnitude of expected sampling error.

90 percent confidence interval	90 percent confidence statement on expected sampling error
$\bar{X} \pm 1.65\sigma_{\bar{x}}$	$1.65\sigma_{\bar{x}}$
or	or
$\bar{X} \pm 1.65\hat{\sigma}_{\bar{x}}$	$1.65\hat{\sigma}_{\bar{x}}$

80 percent confidence interval	80 percent confidence statement on expected sampling error
$\overline{X} \pm 1.28\sigma_{\overline{x}}$	$1.28\sigma_{\overline{x}}$
or	or
$\overline{X} \pm 1.28\hat{\sigma}_{\overline{x}}$	$1.28\hat{\sigma}_{\overline{x}}$

It should be clear from the preceding examples that confidence intervals can be written as $\overline{X} \pm E$, where $E = z\sigma_{\overline{x}}$ or $E = z\hat{\sigma}_{\overline{x}}$, depending on whether or not the universe standard deviation is known. The two approaches constitute different ways of looking at the same relationship.

13.10 DETERMINATION OF SAMPLE SIZE

Up to this point, the sample has been treated as if it had been selected arbitrarily and then subsequently the associated confidence interval or confidence statement about the magnitude of expected sampling error generated. In actuality, sample size is not an arbitrary determination, but rather is the result of a careful reflection on the part of the analyst with respect to three things:

1. The maximum allowable sampling error: The reason for sampling is to provide an estimate of a universe characteristic to aid the manager in arriving at a decision. For example, one informational input in attempting to decide whether or not to locate a store in a particular geographical location is average family income in that geographical area. For purposes of reaching a decision, it is not necessary to know average family income to the nearest penny. In fact, the decision maker may feel that an estimate that is within $500 of the universe mean family income for that geographical area is sufficiently accurate. If such were the case, maximum allowable sampling error would be set at $500.

2. The investigator must also reach a decision on the level of confidence he requires that the allowable sampling error will not be exceeded. It must be determined what percentage of the time the sampling procedure employed will be allowed to generate estimates that differ from the universe mean by more than the maximum allowable sampling error.

Suppose the decision maker in the store location illustration is willing to accept a procedure that will generate sample estimates that are more than $500 away from the value of the universe mean 2 percent of the time. This is equivalent to requiring 98 percent confidence.

3. The investigator must have an estimate of the degree of dispersion in the universe from which the sample is selected.

    a. In some instances the value of the universe standard deviation is known. For example, a machine process designed to fill 8-oz cans of beans does not fill each can to exactly 8 oz. Variation around the mean value of 8 oz is inherent in the filling process and, in fact, the extent to which this variability occurs is built into the operation at the time the process is originally set up.

    b. Occasionally, studies similar to the one currently being undertaken have occurred elsewhere, and the standard deviation observed in such studies may provide a good approximation of the measure of dispersion in the current study.

c. If no information exists, the analyst usually selects a small preliminary sample from the universe, observes the standard deviation for the small pilot sample, and projects this value as a rough estimate of the standard deviation of the universe.

13.11 DERIVATION OF THE FORMULA FOR THE DETERMINATION OF SAMPLE SIZE

Recalling the formula for maximum allowable sampling error:

$$E = z\sigma_{\bar{x}}$$

or

$$E = z\frac{\sigma_x}{\sqrt{n}}$$

Multiplying both sides by $\sqrt{n}$:

$$\sqrt{n}\,E = z\sigma_x$$

Dividing both sides by E:

$$\sqrt{n} = \frac{z\sigma_x}{E}$$

Squaring both sides:

$$n = \left(\frac{z\sigma_x}{E}\right)^2$$

Examination of the above formula demonstrates that sample size is a function of the three elements discussed above, namely, E, the maximum allowable sampling error, z, the number of standard error units associated with the level of confidence specified, and σ_x, the standard deviation of the universe.

Referring once again to the store location illustration, suppose the decision maker decides as follows:

1. Maximum allowable sampling error is $500.
2. The sample estimate should not differ from the universe mean by more than the maximum allowable sampling error of $500 more than 2 percent of the time.

The approach leading to the determination of appropriate sample size is reflected diagrammatically in Figures 13.7(a), 13.7(b), and 13.7(c). As the sample size increases, the distance one has to go along the number line in order to include 98 percent of the sample means decreases. In effect, the analyst is searching for the sampling distribution within which one would have to proceed exactly $500 on either side of the universe mean to accomplish this objective (i.e., to encompass 98 percent of the sample means). The sample size associated with this sampling distribution is the desired sample size. According to the diagrams in Figure 13.7, the desired sample size is 136. In actuality, this value is generated algebraically by solving the formula $n = (z\sigma_x/E)^2$ for n.

(a) Sampling distribution of the mean associated with samples of size 34

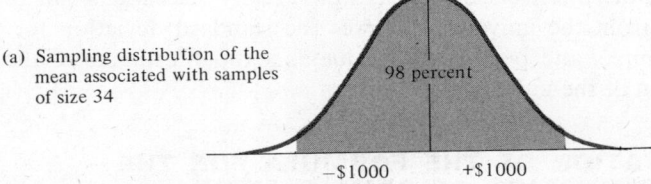

Figure 13.7(a) Sampling distribution of the mean associated with samples of size 34.

(b) Sampling distribution of the mean associated with samples of size 60.

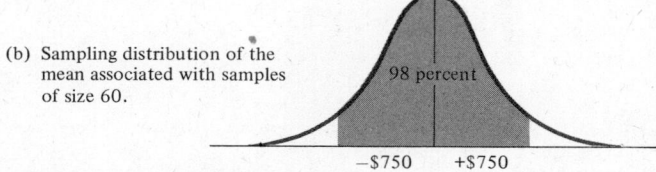

Figure 13.7(b) Sampling distribution of the mean associated with samples of size 60.

(c) Sampling distribution of the mean associated with samples of size 136

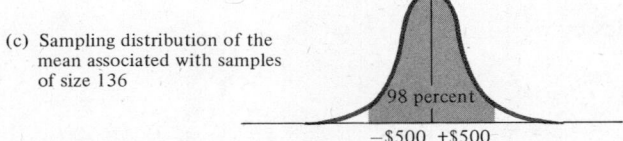

Figure 13.7(c) Sampling distribution of the mean associated with samples of size 136.

Computation of Appropriate Sample Size for the Store Location Illustration

z refers to the distance along the number line one must go (proceeding in both a positive and negative direction from the mean) in order to include 98 percent of the area under the curve. Between the mean and a point 2.33 standard errors away, 49.01 percent of the area under the curve would be included. Proceeding 2.33 standard errors in both directions includes 98 percent. Therefore, the z value associated with 98 percent confidence is 2.33.

E refers to the maximum allowable sampling error or $500. σ_x, the standard deviation of the universe is unknown. A small pilot sample would be selected from the universe, and on the basis of the dispersion in the sample, an estimate of the universe standard deviation generated, using the formula $\hat{\sigma}_x = \sqrt{\sum (X - \overline{X})^2} / \sqrt{(n-1)}$. Presume that a sample of size 50 has been selected and that σ_x, when computed by the formula $\hat{\sigma}_x = \sqrt{\sum (X - \overline{X})^2 / (n-1)}$, equalled $2500.

Substituting into the formula for n, we obtain

$$n = \left(\frac{z\sigma_x}{E}\right)^2 = \left[\frac{(2.33)(2500)}{500}\right]^2 = [(2.33)(5)]^2 = (11.65)^2 = 135.72$$

Thus, if a sample of size 136 is selected from the universe, one can be 98 percent certain that one's statement is correct when one states that the sampling error (difference between the sample mean and the universe mean) does not exceed $500. The justification for such a statement is that the procedure leads, on the average, to sample estimates that differ from the universe mean by less than $500, 98 out of every 100 times.

The above analysis presumes that the sample size will be greater than 30 and that as a result the sampling distribution of the mean is normally distributed. It also presumes that sample size will be less than 5 percent of the universe and therefore that the finite multiplier can be ignored. If the latter assumption does not correspond to reality, the formula $E = z(\sigma_x/\sqrt{n})\sqrt{(N - n)/(N - 1)}$ must be solved for n to generate a formula for appropriate sample size given E, z, an estimate of σ_x, and N.

PROBLEMS

Problem 13.1

1. What is the central limit theorem?
2. What is the significance of the central limit theorem for sampling theory?

Problem 13.2

1. What is a "sampling distribution of a statistic"?
2. What information is provided by the "standard error"?

Problem 13.3

When constructing a confidence interval, the analyst usually concludes with a statement such as "I am 80 percent confident that the statement that the true mean of the universe lies between $63.4 and $70.7 is correct." Explain thoroughly the underlying relationships that allow an analyst to believe that 80 percent of the statements made would be correct. Use a diagram and refer to it in answering this question.

Problem 13.4

$\sqrt{(N - n)/(N - 1)}$ is called the finite correction factor.
1. When should it be used?
2. What effect does it have when used?
3. On the basis of logical reasoning and common sense, justify the inclusion of the finite multiplier in those situations where its use is appropriate.

Problem 13.5

Why is confidence greater at the 95 percent level than at the 80 percent level?

Problem 13.6

In what sense are the standard deviation of the universe and the standard error of the mean similar? In what sense do they differ?

Problem 13.7

A random sample of size 50 is selected from daily production records of personnel employed by a large manufacturing concern. The arithmetic mean of the sample is 150 units and the standard deviation is 30 units. Calculate a 98 percent confidence interval for the mean output of all employees.

Problem 13.8

Northeastern University is interested in establishing the average distance in miles that evening students commute to its Huntington Avenue campus. A sample of size 400 is selected at random from the evening enrollment. The mean of the sample is 18.53 miles and the sample standard deviation $[\hat{\sigma}_x = \sqrt{\sum (X - \bar{X})^2/(n - 1)}]$ is 5.06 miles. Construct a 95 percent confidence interval estimate of the value of the universe mean.

Problem 13.9

Given a universe of 1000 items, the number of different samples of size 400 that could be selected is equal to

$$C\binom{1000}{400} \quad \text{or} \quad \frac{1000!}{(400!)(600!)}$$

The number of different samples of size 600 that could be selected is equal to

$$C\binom{1000}{600} \quad \text{or} \quad \frac{1000!}{(400!)(600!)}$$

It follows that the same number of sample means would exist for the sampling distribution of the mean associated with all possible samples of size 400 as would exist for the sampling distribution associated with all possible samples of size 600. Which of the two sampling distributions would have the smaller measure of dispersion? Justify your answer.

Problem 13.10

On a recent instructor evaluation, one of the questions asked was "What is your overall evaluation of the course?" Two different instructors using different approaches received the following scores (where score refers to the mean quality point as assigned by the students in the instructor's class).

Instructor 1 *(lecture approach)*	*Instructor 2* *(programmed learning approach)*
100 student responses	36 student responses
mean score 2.2	mean score 1.6
standard deviation 0.1	standard deviation 0.1

1. If the mean of 2.2 for instructor 1 is to be used as a point estimate of the overall attitude of the students (past, present, and future) he would have in the course, what statement can be made with 90 percent confidence about the magnitude of expected sampling error?
2. If the mean of 1.6 for instructor 2 is to be used as a point estimate of the overall attitude of the students (past, present, and future) he would have in the course, what statement can be made with 90 percent confidence about the magnitude of expected sampling error?
3. The computations in parts 1 and 2 should disclose that the expected magnitude of sampling error is larger for the instructor 2 estimate than for the instructor 1 estimate. Very carefully and explicitly, explain the reason for the differences in magnitude of the expected sampling error. (Draw on sampling theory in answering this question.)

Problem 13.11

You wish to take a sample to determine the average amount spent by a college freshman in the United States in a college of business administration for textbooks during his freshman year. It is imperative that your answer fall within a confidence interval of $\pm$ $0.50 of the universe mean. (Imperative is interpreted as 99.73 percent confidence in the correctness of your estimate.) A small preliminary pilot study yielded a standard deviation of $6 $[\hat{\sigma}_x = \sqrt{\sum (X - \overline{X})^2/(n - 1)}]$. How large a sample will be required to meet the stipulated constraints?

Problem 13.12

Nixon and Halderman form a partnership to distribute fertilizer. They wish to estimate the average amount of fertilizer used on a farm per planted acre in order to determine the extent of the potential market. The estimate should be within 1.2 bags. They desire 99.73 percent confidence in the correctness of the estimate. A small preliminary study has been undertaken which yielded a standard deviation of 3.2 bags $[\hat{\sigma}_x = \sqrt{\sum (X - \overline{X})^2/(n - 1)}]$. Using the above information, determine the number of elementary sampling units that would have to be included in the sample in order to meet the precision and reliability specifications.

Problem 13.13

A random sample of 81 games is drawn from a universe of high school football games known to be distributed in an approximately normal fashion. The average number of plays per game for the sample is 67 with a standard deviation of 9. Between what two values can you state that the average number of plays per game for the universe will fall and be 99 percent confident of being correct?

Problem 13.14

An estimate of average family income in a particular geographical locality is desired. Knowledge of this neighborhood characteristic will be a useful input in deciding whether or not to open a branch at a new shopping center currently under construction. From past studies, it is reasonably certain that the standard deviation of the universe will not exceed $1800. For the sample estimate of the universe mean to be useful, it should be within $500 of the true value of the universe mean. The location committee has indicated a willingness to accept a sampling procedure that generates estimates lying outside this $500 range 5 percent of the time. Given the above constraints, how large a sample should the analyst select?

Problem 13.15

A major department store desires to measure the impact of Christmas shopping on the average dollar balance of its revolving charge accounts. To measure this, the department store records the account balance as of October 31 and the new balance as of December 31. The number of payment periods required by the customer to return the balance to the October 31 level is then determined. A sample of 64 accounts is selected at random. The mean of the sample is computed to be 7.3 payment periods and the standard deviation to be 1.6 payment periods. Using the above information, generate a 90 percent confidence interval estimate of the mean number of payment periods required to return the balance to the October 31 level for the universe.

Problem 13.16

One of the responsibilities of a wool agent is to negotiate with sheep ranchers for the purchase of wool. The rancher is paid so much per pound for dirty wool. A typical sheep yields 10 lb of dirty wool when sheared. After being subjected to cleaning, the 10 lb of dirty wool can shrink to as little as 5 lb of clean wool. The amount of clean wool yielded per pound of dirty wool varies from sheep to sheep. The agent will typically select a random sample of sheared wool and send it to a laboratory for cleaning. The results of the test influence the price per pound of dirty wool at which the rancher and the agent agree to do business. Suppose the sample consists of 49 cores (1 lb apiece) selected at random from the total "harvest." The mean amount of clean wool yielded is 0.72 lb of clean wool per pound of dirty wool. The standard deviation of the sample is 0.14 lb. If the agent uses the sample mean of 0.72 lb as an estimate of the universe mean, what statement can he make with respect to the magnitude of sampling error associated with the estimate with a 90 percent probability of being correct?

Problem 13.17

County medical examiners are frequently asked to testify in court as to the time of death of a victim. Their statement is usually of the form "The time of death was between time a and time b." In cross examining the medical examiner, what sort of questions might you ask in order to bring into clear focus the underlying statistical basis that enables him to make such statements? How could you determine the degree of confidence that should be placed in the accuracy of his estimate?

Problem 13.18

The Accounts Receivable Manager for I. M. Overdoo, Inc., has decided to utilize sampling techniques to estimate the average age of the 1000 outstanding accounts. A simple random sample of 36 accounts generates a mean age of 24 days and a standard deviation of 4 days [i.e., $\sqrt{\sum (X - \bar{X})^2/(n - 1)} = 4$ days].

1. What statement can be made with respect to the magnitude of the size of the error in the estimate with a level of confidence of 90 percent that the statement will be correct?
2. What is the probability that the estimate will be in error by an amount exceeding 1 day?

Problem 13.19

The sampling distribution of the mean. What is it? What are its characteristics? How does it behave with changes in the associated sample size?

Problem 13.20

Consider yourself employed by the Brand K Cereal Corporation. You desire to estimate μ_x, the average number of raisins that a competitor puts in its 15-oz boxes of Raisin Bran. A random sample of 100 15-oz boxes of the competitor's Raisin Bran has been selected from various supermarkets and the number of raisins in each box counted. The following distribution was obtained.

Number of Raisins	Number of boxes
25–35	10
36–44	30
45–55	40
56–64	15
65–75	5
	100

1. Determine the value of the sample mean.
2. Determine the standard deviation of the sample. This value is to be employed as an estimate of the unknown standard deviation of the universe.
3. Using the results in parts 1 and 2, generate a 95 percent confidence interval estimate of μ_x.
4. Choose from among the following statements the one that gives the correct interpretation of the interval obtained in part 3:
   a. The probability is 0.95 that the population mean lies within this interval.
   b. Ninety-five percent of future sample means will fall within this interval.
   c. Ninety-five percent of the confidence intervals obtained in this manner, assuming that repeated samples are drawn and the limits obtained as in part 3, will contain the true mean within them.
5. Suppose it is desired that the estimate, $\overline{X}$, should be no further from μ_x than 1 raisin (i.e., The half-width of the confidence interval should be no greater than 1). Moreover, 99 percent confidence that this occurs is desired. Determine the sample size required in order to meet the constraints specified above.

ESTIMATION: THE BAYESIAN VERSUS THE CLASSICAL POSITION

14

14.1 CONCEPTUAL COMPARISON OF THE BAYESIAN AND THE CLASSICAL APPROACH

The Classical Approach

The main theme of Chapter 13 was the investigation of the theory and procedure underlying the classical approach to the generation of confidence interval estimates. The distinctive characteristic of the classical approach is that the information considered in forming the confidence interval estimate is restricted to observed sample evidence. The classical estimation procedure is based on the known manner in which sample means are distributed around the unknown universe mean (recall the three theorems, pp. 314–329, and see Figure 14.1).

To a classicist, the horizontal axis of the estimating distribution (the theoretical sampling distribution of the

347

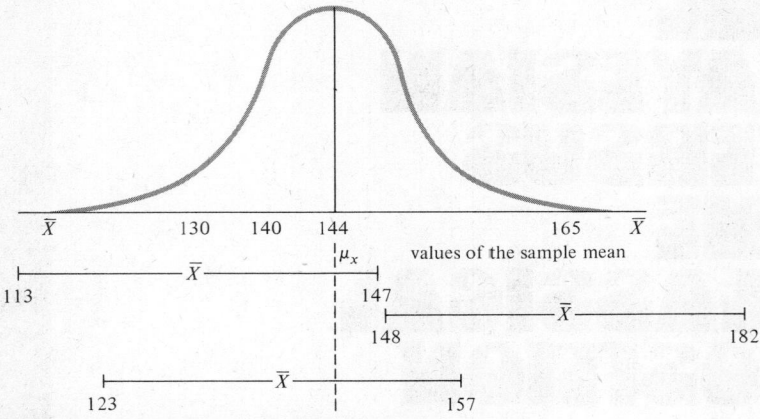

Figure 14.1 Sampling distribution of the mean.

mean) reflects values of $\overline{X} \mid \mu_x$ and the vertical axis records the relative likelihood of these values, $P(\overline{X} \mid \mu_x)$. When a classicist constructs a 90 percent confidence interval, the 90 percent is interpreted as referring to the probability that the procedure will lead to the generation of a correct statement. The classicist is absolutely opposed to the interpretation that the 90 percent refers to the probability that the true universe mean lies within the specified interval. In the eyes of a classicist, a unique true value exists for the universe mean, and therefore the value of the universe mean cannot be treated as a random variable. To a classicist, any given confidence interval statement is either correct (in which case the probability that it is correct is 1.0) or incorrect (in which case the probability that it is correct is 0.0). A classicist is willing to generate statements with respect to the probability that *the procedure* will lead to the generation of correct statements and is willing to agree that the probability that μ_x lies within a stipulated interval is either 1.0 or 0.0 (which of the two he doesn't know). This is as much as the classicist is willing to say.

The Bayesian Approach

The Bayesian approach contrasts with the classical approach in two important respects:

1. Bayesians believe that the unknown value of the universe parameter, μ_x, can and should be treated as a random variable.
2. Bayesians take the position that any estimate generated should reflect all the information at the decision maker's disposal. This is reflected by the assignment of a prior estimating distribution, which is used in conjunction with observed sample evidence to form a posterior estimating distribution.

Invariably, the analyst or decision maker has certain previously held beliefs prior to sampling with respect to the likelihood that certain values will actually turn out to be the true value of the universe mean. These convictions are a reflection of past experience, common sense, and intuition. A Bayesian believes that a prior distribution can and should be developed to reflect such convictions held by the decision maker (Figure 14.2).

The horizontal axis of the prior probability distribution records values that the decision maker feels could conceivably be the true value of the universe mean. The

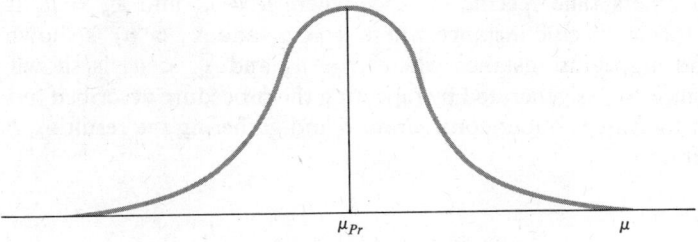

possible values of the universe mean

Figure 14.2 Prior estimating distribution.

vertical axis indicates the decision maker's relative strength of conviction with
respect to the given values of μ_x actually being the true mean. The degree of uncertainty of the decision maker with respect to the true value of μ is measured by
the dispersion in his prior estimating distribution. Using a $_{pr}$ subscript to designate
prior, the standard deviation of the prior estimating distribution will be represented
by the symbol σ_{pr}.

Suppose the Bayesian received new information consisting of a sample of size
100 with a mean of $\overline{X}_a$. Utilizing Bayes' rule, the prior estimating distribution would
be revised to incorporate the new information. A confidence interval statement
would then be developed from the revised estimating distribution.

The first step in the revision process is to develop a conditional probability
distribution that reflects the relative likelihood of observing $\overline{X}_a$ given each of the
possible values of the universe mean, just as we have done in Chapters 6, 8, and 9
with discrete priors. This distribution is reproduced as Figure 14.3.

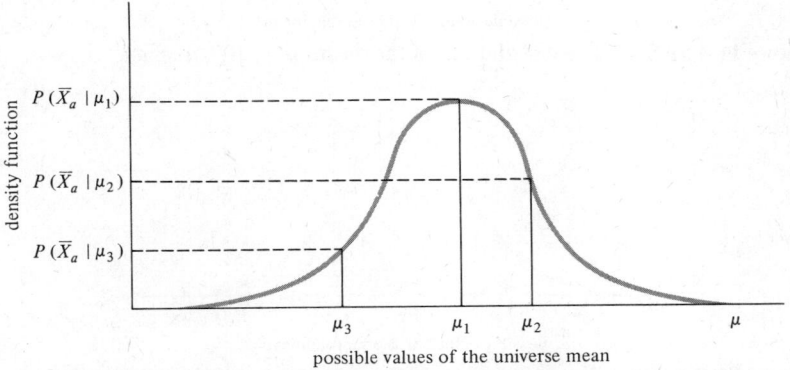

possible values of the universe mean

Figure 14.3 Conditional probability distribution of $\overline{X}_a \mid \mu$: $n = 100$.

Conceivable values of the universe mean are recorded on the horizontal axis.
The relative likelihood of observing the specific sample result, $\overline{X}_a$, in a sample of
size 100 selected from a universe with a mean equal to the associated value recorded
on the horizontal axis, is indicated on the vertical axis. Each likelihood in Figure
14.3 is generated from a separate sampling distribution of the mean. To illustrate,
the supporting sampling distributions of the mean associated with the three values
μ_1, μ_2, and μ_3 are reproduced in Figure 14.4. Note in each case that $\overline{X}_a$ is only one
of the possible values of the sample mean that could have been observed. Also note
that $\overline{X}_a$ is more likely to be observed in association with some values of the universe

mean than others. The specific instance where $\mu = \mu_1$ and $\overline{X}_a = \mu_1$ is shown in
Figure 14.4(a), a specific instance where $\mu = \mu_2$ and $\overline{X}_a > \mu_2$ is shown in Figure
14.4(b), and a specific instance where $\mu = \mu_3$ and $\overline{X}_a < \mu_3$ is shown in Figure
14.4(c). Figure 14.3 is generated by repeating the procedure described in Figure 14.4
for each of the values under consideration and gathering the resulting $P(\overline{X}_a \mid \mu)$ in
one distribution.

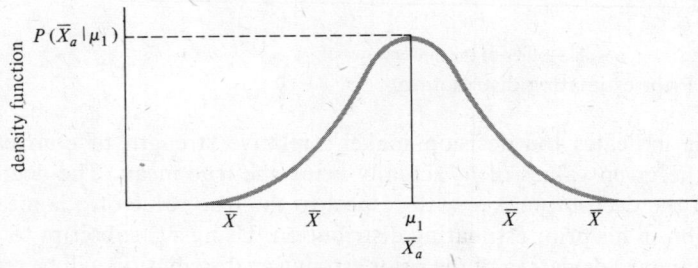

possible values of the sample mean

Figure 14.4(a) Sampling distribution of the mean: $n = 100$, $\mu = \mu_1$.

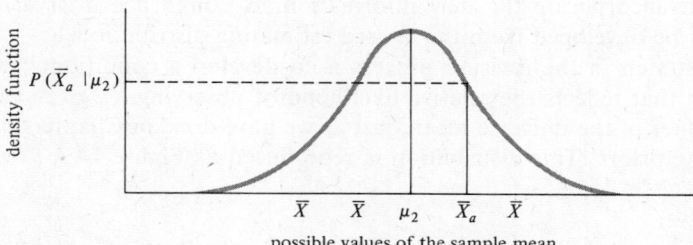

possible values of the sample mean

Figure 14.4(b) Sampling distribution of the mean: $n = 100$, $\mu = \mu_2$.

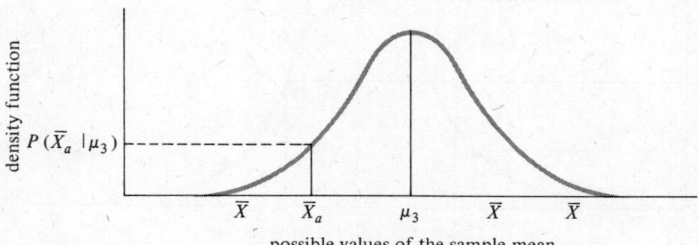

possible values of the sample mean

Figure 14.4(c) Sampling distribution of the mean: $n = 100$, $\mu = \mu_3$.

Figure 14.3 reflects the way a Bayesian would summarize the information
contributed by the sample evidence. By simply changing the labels on the horizontal
and vertical axis, the same distribution becomes the sampling distribution of the
mean associated with $n = 100$ and $\mu_x = \overline{X}_a$. In its latter form, the distribution
represents the information contributed by the sample as viewed by a classicist.
The two distributions are alternative ways of looking at the same information, the
difference in viewpoint being reflected by the labeling of the axes. In either case, the
measure of dispersion is the standard error of the mean, $\sigma_{\overline{x}}$.

To generate the posterior or revised estimating distribution, the Bayesian would multiply $P(\mu)$ by $P(\overline{X}_a \mid \mu)$ for each conceivable value of μ. The resulting products form a revised probability distribution reflecting the analyst's strength of conviction with respect to the given value of μ being the true mean *after* incorporation of the new sample evidence.

From Bayes' rule:

$$P(\mu \mid \overline{X}_a) = \frac{P(\mu \cap \overline{X}_a)}{P(\overline{X}_a)} = \frac{P(\mu) \cdot P(\overline{X}_a \mid \mu)}{P(\overline{X}_a)}$$

The area under a normal curve is equal to one. Referring to Figure 14.3, $P(\overline{X}_a)$ is equal to the total area under the curve. Substituting:

$$P(\mu \mid \overline{X}_a) = \frac{P(\mu) \cdot P(\overline{X}_a \mid \mu)}{P(\overline{X}_a)} = \frac{P(\mu) \cdot P(\overline{X}_a \mid \mu)}{1}$$

Thus, the revised likelihood of μ given $\overline{X}_a$ simply becomes $P(\mu) \cdot P(\overline{X}_a \mid \mu)$. The revised estimating distribution is reproduced as Figure 14.5.

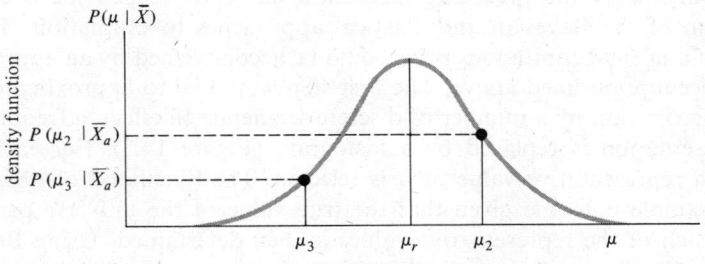

Figure 14.5 Revised estimating distribution.

The revised distribution is then utilized to generate statements with respect to the likelihood that the true value of μ lies between two specified conceivable values of the mean. It is stated without proof that if the prior estimating distribution is normal and the sample size is sufficiently large to ensure a normal sampling distribution of the mean, the revised estimating distribution will also be normal.[1] The standard deviation of the revised estimating distribution is designated by the symbol σ_r. Once the revised distribution is obtained, the confidence interval is constructed by the same procedure outlined in the classical approach.

There is one situation where the Bayesian and the classicist approaches lead to the generation of identical confidence interval estimates. This occurs when the Bayesian has no preconceived beliefs to cause him to consider certain conceivable values of the true mean more likely than others. In such cases, the Bayesian assigns a uniform or rectangular prior characterized by an infinite variance. A uniform distribution (Figure 14.6) is one in which each conceivable outcome is assigned an equal likelihood.[2]

Recall that the revised distribution is obtained by multiplying $P(\mu)$ by $P(\overline{X}_a \mid \mu)$ for each conceivable value of μ. If $P(\mu)$ is identical for each value of μ, the revised

---

[1] The procedure for dealing with nonnormality is left to more advanced texts.

[2] The term diffuse prior is often used in the literature to refer to the special case where a uniform prior with an infinite variance is assigned.

$P(\mu)$

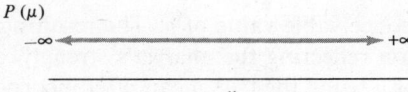

μ

Figure 14.6 A uniform prior.

distribution will be identical to the conditional probability distribution describing $P(\overline{X}_a \mid \mu)$. It follows that the classicist and the Bayesian with a uniform prior construct their confidence interval estimates from distributions with identical properties. They still differ, of course, in interpretation. The classicist is 90 percent confident that his statement that "μ has a value between CL_L and CL_U" is correct. The Bayesian, on the other hand, is 90 percent confident that μ has a value between CL_L and CL_U.

In conclusion, if one accepts the Bayesian arguments, it would appear that the classical approach will generate correct confidence interval estimates only on those occasions where the decision maker has no knowledge with respect to the parameter he is attempting to estimate. Such situations seldom, if ever, occur.

The purpose of the preceding discussion has been to provide a conceptual comparison of the Bayesian and classical approaches to estimation. The actual revision of a normal continuous prior could be accomplished by an approximation of the procedure outlined above. The first step would be to approximate the continuous distribution by a number of discrete segments. In effect, a frequency polygon representation is replaced by a histogram (Figure 14.7). For each discrete segment, a representative value of μ is selected. The likelihood of observing the specified sample outcome given that the true value of the universe parameter is equal to each of the representative values is then determined. Using Bayes' rule, the relative likelihood of the value of the true universe mean being equal to each of the representative values, given the incorporation of the new information, is then generated. Provided that the segments are sufficiently narrow, the approximation to the continuous distribution will be quite good. Fortunately, the typical procedure employed for revision of a continuous normal prior reflects the incorporation of a number of shortcuts and is considerably simpler than the procedure outlined in the conceptual comparison. In fact, in practice the continuous revision procedure is often used to generate an approximation of the revision of a discrete probability distribution. Provided that the number of discrete outcomes is sufficiently large, the continuous revision procedure provides a reasonable approximation at a considerable saving in effort. The following section describes the short-cut procedure employed for revision of a normal continuous prior.

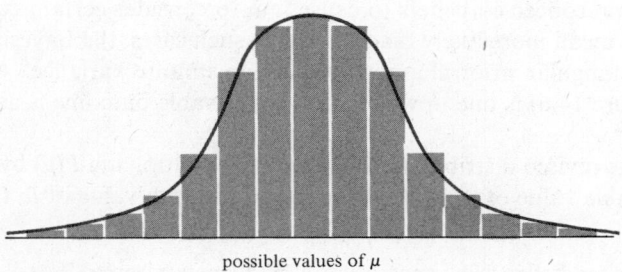

possible values of μ

Figure 14.7

14.2 REVISION OF A NORMAL CONTINUOUS
PRIOR: THE PROCEDURE

The procedure discussed below assumes that the new information is generated by a sampling plan associated with a sample size sufficiently large to ensure that the theoretical sampling distribution of the mean will be normally distributed. Consider the following illustrative example.

A large gasoline distributor desires to estimate the average number of miles between oil changes for customers patronizing service stations in its chain. Appropriate car maintenance requires that the oil be changed roughly every 2500 miles. If it appears that, on the average, car owners change the oil considerably less frequently than every 2500 miles, the distributor intends to institute a promotional program to emphasize the importance of timely oil changes (and, in the process, sell more oil). When pressed for an answer by his statistical analyst (a Bayesian), the distributor indicated that his best guess was that the true average number of miles between oil changes for customers patronizing service stations in his chain was 3500 miles. The analyst then asked the distributor if he would be willing to accept an even bet that his estimate is within 500 miles of the true value. After a little thought, the distributor replied that he believed he would probably be as likely to win as lose if he were to bet that his estimate is within 300 miles of the true universe mean. Thus, he would accept an even bet that his estimate is within 300 miles of the true value. The analyst then retired to his office with the objective of capturing the distributor's prior beliefs about the true value of the universe mean in the form of a continuous distribution. The analyst felt he was reasonably justified in assuming a continuous normal prior in that the distributor believed his estimate was as likely to be in error in one direction as the other and that small estimating errors were more likely than large ones. The analyst therefore concluded that the distributor's prior estimating distribution, reflecting his strength of conviction as to the likelihood that the true value of the universe mean was equal to various values, was normally distributed around the value of the distributor's best estimate (namely, 3500 miles). From the information provided by the distributor, the analyst was also able to generate an estimate of the standard deviation of the distributor's prior estimating distribution (Figure 14.8).

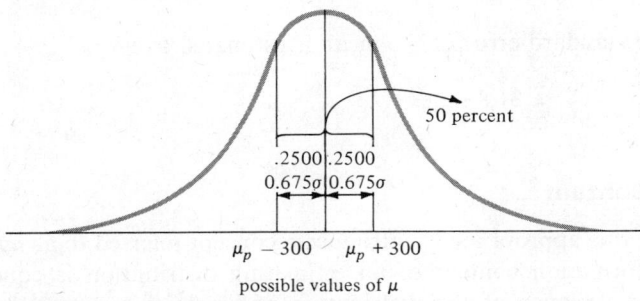

Figure 14.8 Prior estimating distribution.

By assigning a 50 percent probability that the estimate was within 300 miles, the distributor in effect stated that there was a 50 percent probability that his esti-

mate was within 0.675 standard deviation units of the true universe mean. It
follows that

$$0.675\sigma_{pr} = 300 \text{ miles}$$

Solving for σ_{pr}:

$$\sigma_{pr} = \frac{300}{0.675}$$

$$= 444 \text{ miles}$$

Thus, the analyst concluded that the distributor's prior estimating distribution
was normally distributed with a mean equal to 3500 miles and a standard deviation
equal to 444 miles (Figure 14.9).

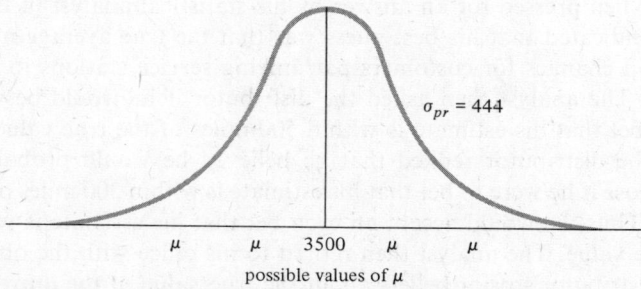

Figure 14.9 Distributor's prior estimating distribution.

Service station operators in the distributor's chain were asked to check the
sticker insider the car door to determine the mileage differential between oil
changes. Company policy required that such stickers be placed inside the door of
all cars serviced. Appropriate safeguards were built into the sampling plan to
ensure that the observations collected would be representative of the target
population.

400 observations were obtained.

The sample mean was 3800 miles.

The sample standard deviation, $\hat{\sigma}_x = \sqrt{\sum (X - \bar{X})^2/(n - 1)}$, was equal to
636 miles.

Substituting, the standard error of the mean is estimated to be

$$\hat{\sigma}_{\bar{x}} = \frac{\hat{\sigma}_x}{\sqrt{400}} = \frac{636}{20} = 31.8 \text{ miles}$$

Information Content

At this juncture, it is appropriate to introduce a concept referred to as *information
content*. The information content of an estimating distribution is equal to the
reciprocal of the variance of the distribution and reflects the precision of the
estimates generated by that distribution. The greater the dispersion in the estimating distribution, the less the precision associated with the estimates. It follows
that the greater the dispersion in the estimating distribution, the less the information content. Thus, there is an inverse relationship between the dispersion of a
distribution and its information content.

The information content of the Bayesian's prior estimating distribution is measured by the following relationship:

$$IC_{pr} = \frac{1}{(\sigma_{pr})^2}$$

where

IC_{pr} = information content of the prior
$(\sigma_{pr})^2$ = variance of the prior estimating distribution

The information content of the sampling distribution of the mean is determined as follows:

$$IC_{\bar{x}} = \frac{1}{(\sigma_{\bar{x}})^2}$$

where

$IC_{\bar{x}}$ = information content of the sampling distribution of the mean
$(\sigma_{\bar{x}})^2$ = variance of the sampling distribution of the mean

Recall that $(\sigma_{\bar{x}})^2 = \sigma_x^2/n$. Substituting:

$$IC_{\bar{x}} = \frac{1}{\sigma_x^2/n} = \frac{n}{\sigma_x^2}$$

It follows that the information content of the sampling distribution of the mean varies directly with sample size and inversely with the magnitude of dispersion in the universe sampled. Although this distribution has the same properties as the sampling distribution of the mean (normal, same mean, same variance), it should be noted that it really is not the sampling distribution of the mean but rather the conditional likelihood distribution, $\bar{X} \mid \mu_x$.

The Bayesian desires to combine the information reflected in the prior estimating distribution with the information obtained from sampling (as reflected by the distribution of $\bar{X} \mid \mu_x$) to obtain a posterior or revised estimating distribution. Given that the prior estimating distribution is normally distributed and that sample size is sufficiently large to ensure a normally distributed sampling distribution of the mean, the revised estimating distribution will also be normal with a mean and information content determined as follows:

$$\mu_r = \frac{IC_{pr} M_{pr} + IC_{\bar{x}} \bar{X}}{IC_{pr} + IC_{\bar{x}}}$$

$$IC_r = IC_{pr} + IC_{\bar{x}}$$

Examination of the formulas for μ_r and IC_r reveals the following:

1. The mean of the revised estimating distribution is a weighted average of the mean of the prior and the mean of the likelihood distribution. The weights indicate the relative importance of the components in the computation of the weighted average. The relative importance assigned to each of the estimates is the information content of its associated distribution.
2. The information content of the revised distribution is a reflection of the precision that can be attached to the estimates generated from that distribution. The information content of the revised distribution is the sum of the information content of the prior and the likelihood distributions.

Substituting, the analyst obtained the following results:

$$IC_{pr} = \frac{1}{(\sigma_{pr})^2} = \frac{1}{(444)^2} = \frac{1}{197136} = 0.0000051$$

$$IC_{\bar{x}} = \frac{n}{(\sigma_x)^2} = \frac{400}{(636)^2} = \frac{400}{404496} = 0.0009889$$

$$\mu_r = \frac{IC_{pr}\,\mu_{pr} + IC_{\bar{x}}\,\overline{X}}{IC_{pr} + IC_{\bar{x}}}$$

$$= \frac{(0.0000051)(3500) + (0.0009889)(3800)}{0.0000051 + 0.0009889}$$

$$= \frac{0.01785 + 3.75782}{0.0009940}$$

$$= \frac{3.77567}{0.0009940}$$

$$= 3798 \text{ miles}$$

$$IC_r = IC_{pr} + IC_{\bar{x}}$$

$$= 0.0000051 + 0.0009889$$

$$= 0.0009940$$

IC_r is, of course, the reciprocal of the variance of the revised estimating distribution.

$$IC_r = \frac{1}{\sigma_r^2}$$

Solving for σ_r:

$$\sigma_r^2 = \frac{1}{IC_r}$$

$$\sigma_r = \sqrt{\frac{1}{IC_r}} = \frac{1}{\sqrt{IC_r}}$$

Substituting:

$$\sigma_r = \frac{1}{\sqrt{IC_r}} = \frac{1}{\sqrt{0.0009940}} = \frac{1}{0.0315} \cong 31.4 \text{ miles}$$

The posterior or revised estimating distribution is thus determined to be normally distributed with a mean equal to 3798 miles and a standard deviation equal to 31.4 miles (Figure 14.10).

The Bayesian then generated the following 95 percent confidence interval:

$\mu_r \pm z\sigma_r$
$3798 \pm 1.96(31.4)$
3798 ± 61.544
Lower confidence limit = 3736.5 miles
Upper confidence limit = 3859.5 miles

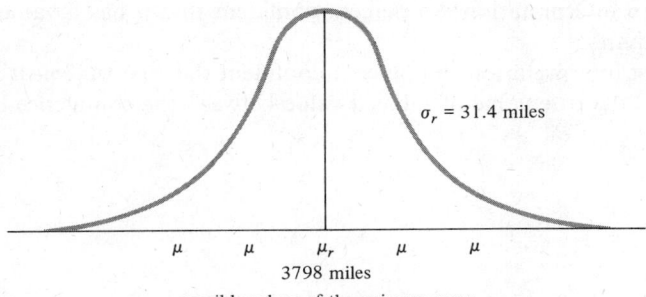

$\sigma_r = 31.4$ miles

μ μ μ_r μ μ
3798 miles
possible values of the universe mean

Figure 14.10 Distributor's revised estimating distribution.

Interpreting the result, the analyst concluded that the gasoline distributor could be 95 percent confident that the true universe mean had a value between 3736.5 miles and 3859.5 miles. Any of the values in this range are sufficiently greater than 2500 miles to justify instituting the promotional campaign.

The Bayesian analyst noted that in this instance there was very little difference in the estimating interval he obtained and that which would have been generated by some of his confirmed classicist colleagues. The confidence interval of the classicist would have been generated from the theoretical sampling distribution of the mean and taken the following form:

$$\bar{X} \pm z\hat{\sigma}_{\bar{x}}$$

For a 95 percent confidence interval:

$3800 \pm 1.96(31.8)$
3800 ± 62.3
Lower confidence limit $= 3737.7$
Upper confidence limit $= 3862.3$

The intervals are virtually the same. The explanation, of course, is that in this instance the information on which the Bayesian estimate is based is dominated by the sample information. Approximately 99.5 percent of the information content of the revised distribution is contributed by the sample information.[3] Clearly, the larger the sample size, the smaller the dispersion in the likelihood distribution $(\mu \mid \bar{X})$, and the greater the information content of the likelihood distribution.[4]

Furthermore, it should be noted that the greater the uncertainty of the decision maker as to the true value of the universe parameter, the greater the dispersion in the estimating prior, and the smaller the information content of the estimating prior. The extreme case is the situation where the decision maker has absolutely no prior feelings about the value of the universe parameter and assigns a diffuse prior. The variance of a diffuse prior is for all intents and purposes equal to infinity. The information content of the diffuse prior is therefore $1/\infty$ or zero. In such a situation $\mu_r = \bar{X}$ and $IC_r = IC_{\bar{x}}$. In this situation, the Bayesian and the classicist generate identical confidence intervals. They would, of course, still differ the following in interpretation.

[3] $IC_{\bar{x}}/(IC_{\bar{x}} + IC_{pr}) = 0.0009889/0.0009940 = 0.995$.
[4] Conversely, the smaller the sample size, the more influence the prior beliefs of the decision maker will have on the parameters of the revised estimating distribution.

The Bayesian interpretation: 95 percent confident that μ has a value between the confidence limits.

The classicist interpretation: 95 percent confident that the *statement* is correct when he states, "the true value of μ has a value between the confidence limits."

PROBLEMS

Problem 14.1

Compare and contrast the following measures of dispersion: σ_{pr}, $\sigma_{\bar{x}}$, and σ_r.

Problem 14.2

Given:
  a. a normal prior estimating distribution with $\mu_{pr} = 50$ and $\sigma_{pr} = 2$
  b. a sample of size 100 with $\bar{x} = 46$ and $\hat{\sigma}_x = 15$
Required:
  1. Referring to the prior estimating distribution, determine the probability that $\mu_x \leq 47$.
  2. Determine the parameters of the revised estimating distribution, namely, μ_r and σ_r.
  3. Referring to the revised estimating distribution, determine the probability that $\mu_x \leq 47$.

Problem 14.3

Referring to Problem 14.2:
  1. Calculate a numerical measure of the information content of the prior estimating distribution.
  2. Calculate a numerical measure of the information content of the sampling distribution of the mean.
  3. Calculate a numerical measure of the information content of the revised estimating distribution. How is the value obtained in part 3 related to the values computed in parts 1 and 2?
  4. Suppose the sample size had been 225 (with $\bar{x}$ and $\hat{\sigma}_x$ retaining the same values). How would the information content of the three distributions change? Explain why this is in accordance with expectations.

Problem 14.4

Given:
  a. a normal prior estimating distribution with $\mu_{pr} = 3000$ and $\sigma_{pr} = 200$.
  b. a sample of size 100 with $\bar{x} = 2700$ and $\hat{\sigma}_x = 150$
Required:
  1. Referring to the prior estimating distribution, determine the probability that $\mu_x \geq 3100$.
  2. Determine the parameters of the revised estimating distribution, namely, μ_r and σ_r.
  3. Referring to the revised estimating distribution, determine the probability that $\mu_x \geq 3100$.

Problem 14.5

Referring to Problem 14.4:

1. Calculate a numerical measure of the information content of the prior estimating distribution.
2. Calculate a numerical measure of the information content of the sampling distribution of the mean.
3. Calculate a numerical measure of the information content of the revised estimating distribution. How is the value obtained in part 3 related to the values computed in parts 1 and 2?
4. Suppose the sample size had been 900 (with $\bar{x}$ and $\hat{\sigma}_x$ retaining the same values). How would the information content of the three distributions change? Explain why this is in accordance with expectations.

Problem 14.6

Given:
   a. a normal prior estimating distribution with $\mu_{pr} = 3000$ and $\sigma_{pr} = 10$.
   b. a sample of size 100 with $\bar{x} = 2950$ and $\hat{\sigma}_x = 25$.
Required:

1. Referring to the prior estimating distribution, determine the probability that $\mu_x \leq 2975$.
2. Determine the parameters of the revised estimating distribution, namely, μ_r and σ_r.
3. Referring to the revised estimating distribution, determine the probability that $\mu_x \leq 2975$.

Problem 14.7

1. Referring to Problem 14.6, perform steps 1–4 as described in Problem 14.to
2. Compare the results when the questions are applied to Problem 14.6 5. those obtained when they were applied to Problem 14.4. Discuss fully the implications of these observations.

Problem 14.8

Discuss the concept referred to as "information content."

Problem 14.9

1. How is the information content of a posterior or revised estimating distribution related to the information content of the prior estimating distribution and the information content of the sampling distribution of the mean?
2. As the sample size increases, what is the impact on the information content of the three distributions referred to in part 1?

Problem 14.10

The controller has been asked to submit an estimate of the average dollar value per order processed. The billing clerk estimated that the average dollar value of orders processed was $25. Furthermore, he indicated that he was about 50 percent certain that the average was between $20 and $30. To supplement these feelings and to generate additional information, a random sample of 100 billings is selected. The sample mean is $18.70 and the standard deviation ($\hat{\sigma}_x$) is $6.20.

1. Utilizing a classical approach, generate a 95 percent confidence interval estimate of μ_x, the average dollar value per order processed.
2. Construct a prior estimating distribution of μ_x based on the billing clerk's observations and the assumption of normality.
3. Generate a revised estimating distribution of μ_x by incorporating the sample evidence with the prior estimating distribution developed in part 2.
4. What statement would a Bayesian make with 95 percent certainty based on the distribution developed in part 3? Contrast this with the classical statement generated in part 1.

Problem 14.11

Referring to Problem 14.10, perform steps 1–4 as described in Problem 14.3.

Problem 14.12

Territorial Imperative, Inc., is considering adding a new product to their line. The sales manager has evaluated the potential impact of the product as follows:

I believe that it is as likely that monthly sales per salesman will increase by $500 or more as it is that they will increase by less than $500. Furthermore, I am 50 percent confident that the sales increase will be between $200 and $800.

Before making a final commitment, the product is to be test marketed by 36 salesmen. After a reasonable shake-down period, average sales for the test group were found to have increased by $320 with a standard deviation ($\hat{\sigma}_x$) of $50.

1. Utilizing a classical approach, generate a 95 percent confidence interval estimate of μ_x, average monthly dollar sales increase per salesman.
2. Construct a prior estimating distribution of μ_x based on the sales manager's beliefs and the assumption of normality.
3. Generate a revised estimating distribution of μ_x by incorporating the sample evidence with the prior estimating distribution developed in part 2.
4. What statement would a Bayesian make with 95 percent certainty based on the distribution developed in part 3? Contrast this with the classical statement generated in part 1.

Problem 14.13

Referring to Problem 14.12, perform steps 1–4 as described in Problem 14.3.

Problem 14.14

The president of Liberal Bank and Trust has become increasingly concerned that depositors with checking accounts are not being sufficiently discouraged from overdrawing their accounts by the current penalty schedule. Current policy is to pay all checks presented against overdrawn accounts and to charge $1 per check. Bank managers have the discretion of waiving the penalty fee, and the president feels that this action is being "liberally" applied by the bank managers, in effect creating a "no charge for overdrafts" system. After careful reflection, the president has come up with the following *assessments*:

For those with overdrawn accounts:
a. The average number of checks paid on overdrawn accounts for the year is about 25 checks. He is 50 percent certain that the average figure is between 20 and 30 checks.

b. The average number of dollars overdrawn per day over the period that the accounts are overdrawn is about $60. He is 50 percent certain that the average figure is between $40 and $80.

Before issuing an ultimatum, the president decides to gather some supporting evidence. A sample of size 100 is drawn from the universe of accounts that have been overdrawn at least once over the last year. The sample disclosed that the average number of "rubber" checks was 12 with a standard deviation of 3 and that the average amount overdrawn per day (during the overdrawn period) was $20 with a standard deviation of $4.

1. Utilizing a classical approach, generate a 95 percent confidence interval estimate for each of the universe parameters referred to above.

   μ_x = average number of checks cashed on an overdrawn account per year
   μ_y = average number of dollars overdrawn per day over the period that the accounts are overdrawn.

2. Construct prior estimating distributions for μ_x and μ_y based on the president's beliefs and the assumption of normality.
3. Generate revised estimating distributions of μ_x and μ_y by incorporating the sample evidence with the prior estimating distributions developed in part 2.
4. What statements would a Bayesian make with respect to μ_x and μ_y with 95 percent certainty based on the distributions developed in part 3? Contrast these with the classical statements generated in part 1.

Problem 14.15

Referring to Problem 14.14, perform steps 1–4 as described in Problem 14.3 for both μ_x and μ_y.

Problem 14.16

Compare and contrast the classical and the Bayesian procedures for generating confidence interval estimates.

Problem 14.17

1. Under what conditions would the limiting values (i.e., lower and upper confidence limits) for the Bayesian and the classical confidence interval statements be identical?
2. Referring to the situation posited in part 1, what would be the implications for the information content of the prior estimating distribution, the sampling distribution of the mean, and the revised estimating distribution?
3. Referring to the situation posited in part 1, would the interpretation of the confidence interval statements be the same? If not, how would they differ?

TESTING HYPOTHESES CONCERNING THE VALUE OF A UNIVERSE PARAMETER: THE UNIVERSE MEAN

15

Statistical inference encompasses two major types of inferential procedure whereby inferences with respect to universe parameters are made on the basis of sample evidence.

1. Statistical estimation. Statistical estimation is concerned with generating statements about the value of universe parameters. The statements generally take one of two forms:

   a. A confidence interval statement that asserts that the value of the universe parameter falls within a particular interval along the number line. The probability of the correctness of the assertion is also provided.

   b. A point estimate of the value of the universe parameter and a statement about the maximum magnitude of sampling error one should expect in association with that point estimate. In addition, a statement is provided indicating the probability that the assertion with respect to the maximum magnitude of sampling error is correct.

In each of the above cases, the value of the universe parameter is unknown and the objective is to provide an estimate of that value with some known probability of being correct. The theoretical basis and the procedure for generating statements of this type was the focus of Chapter 13.

2. Test of hypotheses. The second category deals with situations where a pre-conceived belief or claim as to the value of the universe parameter or characteristic exists. It is the objective of the statistical analyst in such situations to draw on the tools of statistical inference to determine whether evidence collected through sampling can be considered as supporting the claim or offering conflicting evidence. If the evidence conflicts with the hypothesis, the originally held belief is rejected.

Consider a parent who starts out with the hypothesis that his son, Jeff, is basically honest. Monday, Jeff comes home with a new Yo-Yo. The parent natur-ally inquires where the Yo-Yo was obtained. "I found it in the field," Jeff responds. The father weighs the evidence, matches it up with his hypothesis about his son's moral character, and believes the boy. On Tuesday, Jeff is the proud possessor of a new jackknife. Subsequent parental inquiry brings forth the response, "I found it in the field." It is possible that this could have happened, but the likelihood of such a pair of events occurring due to chance is pretty slim. Clearly, parental suspicion is aroused, but the boy is given the benefit of the doubt. Thursday, Jeff arrives home with a second new Yo-Yo to add to his already burgeoning pile of recent acquisi-tions. "You won't believe this," Jeff starts to say—He is right; nobody believes him. The probability that such a string of events could have occurred solely due to chance is so small that chance is rejected as an explanation and another explanation is sought (known in the vernacular as "the third degree").

The preceding example, with which only too many of us can identify, captures the basic elements of hypothesis testing:

1. Establish the hypothesis.
2. Collect some evidence.
3. Inquire as to whether or not the evidence observed is consistent with the hypothesis.
4. If it is consistent with the hypothesis, accept the hypothesis; if it is not, reject it.

The body of statistical analysis that has been developed to fulfill this function is referred to under the general heading of testing of hypotheses. Hypotheses can be generated and tested with respect to any one of a number of universe para-meters. The basic underlying theory and approach is quite similar for all para-meters. A number of these will be discussed in subsequent chapters. In this chapter, attention is restricted to the universe mean. Once the concepts and approach have been mastered with respect to the case of the universe mean, it will be a simple task to extend our understanding to include other universe parameters.

15.1 ESTABLISHMENT OF THE HYPOTHESIS

Letting μ_x represent the true value of the universe mean and K a value on the number line, the hypothesis to be tested will take on one of the following three forms.

1. $\mu_x = K$, that is, the true value of the universe mean is equal to some specified value, K.
2. $\mu_x \geq K$, that is, the true value of the universe mean is equal to or greater than some specified value, K.
3. $\mu_x \leq K$, that is, the true value of the universe mean is equal to or less than some specified value, K.

In each of the above situations, the hypothesis being tested is referred to as the *null hypothesis*. The null hypothesis states that there is *no difference* between the actual state of the universe and the state of the universe as it has been postulated in the hypothesis. If the null hypothesis is rejected, it is implied that some alternative description of the state of the universe is accepted. There will always be an *alternate hypothesis* associated with any given *null hypothesis*. Acceptance (or rejection) of one implicitly implies rejection (or acceptance) of the other. Table 15.1 presents the alternate hypotheses that constitute the counterparts to the three types of null hypothesis previously introduced. Illustrations of each category are presented below.

Table 15.1 Null hypotheses and their corresponding alternate hypotheses

	Null hypothesis	Alternate hypothesis
1.	$\mu_x = K$	$\mu_x \neq K$ i.e., $\mu_x > K$ or $\mu_x < K$
2.	$\mu_x \geq K$	$\mu_x < K$
3.	$\mu_x \leq K$	$\mu_x > K$

Illustration 1

Uniform Gear Works has developed a process for producing gears with an average diameter of 2.5 in. Any gears with a diameter greater than 2.503 in. are too loose to perform properly, and any gears with a diameter less than 2.497 in. are too tight to fit the shaft. The process has been designed in such a way that 99.73 percent of the gears produced when the process is operating correctly fall within the tolerances specified for usable gears (the value of a standard deviation is equal to 0.001 in.).

If one of the settings in the process shifts so that the average diameter of gears produced is greater than 2.5 in., an excessive number of gears will be produced with a diameter greater than 2.503 in. Similarly, if the process slips a setting so that average diameter becomes less than 2.5 in., an excessive number of gears will be produced with a diameter less than 2.497 in. The firm periodically selects samples of process output in an effort to detect shifts in machine settings as early as possible. Upon evaluation of each sample, the question is asked, "Is this sample result consistent with the hypothesis that $\mu_x = 2.5$ in.?"

H_{null}: $\mu_x = 2.5$ in.

$H_{\text{alternate}}$: $\mu_x \neq 2.5$ in.

that is, $\mu_x > 2.5$ in.

or

$\mu_x < 2.5$ in.

In situations of this type, the analyst is concerned with detecting departures from the hypothesized value of the mean in either direction along the number line. Assume that a sample of size 9 is selected. For the null hypothesis to be accepted, the sample mean observed must be consistent with those values of the sample mean that could reasonably be expected to be associated with the sampling distribution of the mean depicted in Figure 15.1.

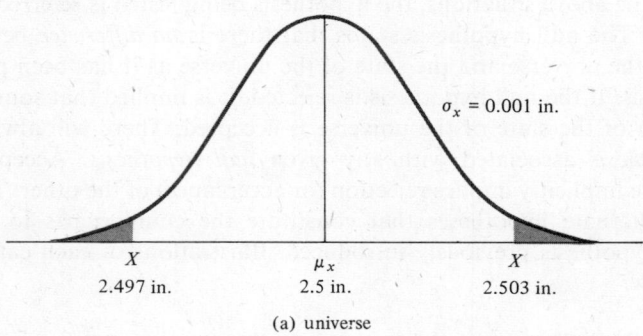

$\sigma_x = 0.001$ in.

| X | μ_x | X |
| 2.497 in. | 2.5 in. | 2.503 in. |

(a) universe

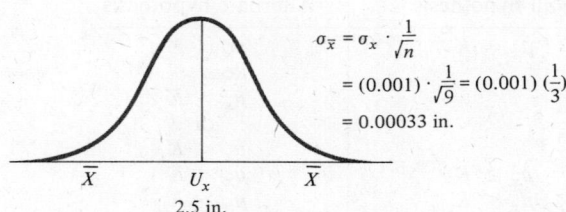

$\sigma_{\bar{x}} = \sigma_x \cdot \dfrac{1}{\sqrt{n}}$

$= (0.001) \cdot \dfrac{1}{\sqrt{9}} = (0.001)\left(\dfrac{1}{3}\right)$

$= 0.00033$ in.

| $\overline{X}$ | U_x | $\overline{X}$ |
| | 2.5 in. | |

(b) sampling distribution of the mean; n = 9

Figure 15.1 (a) Universe. **(b)** Sampling distribution of the mean: $n = 9$.

Note that the sampling distribution of the mean as depicted in Figure 15.1 is normally distributed even though it is associated with a sample size of only size 9. This is in accordance with expectations if the universe from which the elementary sampling units are selected is normally distributed and the standard deviation of that universe is known. All sampling distributions of the mean derived from such universes will be normally distributed regardless of sample size.

Certain values of the sample mean as recorded on the number line are consistent with the hypothesis that $\mu_x = 2.5$ in. and others are not. In effect, the number line is segmented into those regions that contain values of the sample mean that are consistent with the hypothesized mean (known as *the acceptance region*) and those regions that contain values of the sample mean that are inconsistent with the hypothesized mean (known as *the rejection region*).

To illustrate: Referring to Figure 15.2, those values of the sample mean that fall between A and B might be considered as reasonably likely to have been selected from a universe with a mean of 2.5 in. and a standard deviation of 0.001 in. Although a sample of size 9 could have a mean greater than B, the probability of this occurring due to chance selection from a universe matching up to the hypothesized universe is so small that chance would be rejected as an explanation. If the difference between the sample mean and the postulated value of the universe mean (i.e., sampling error) is so large that it cannot reasonably be attributed to chance, it is referred to as a *significant difference*. Just when a difference is large enough to be considered significant varies depending on the situation and the attitude of the analyst. A rough rule of thumb is, if the probability of a difference of the observed magnitude occurring due to chance is less than 0.05 (5 percent), reject chance selection in sampling as a reasonable explanation and conclude that the difference is significant. Clearly, the less the probability of an observed difference occurring due to chance selection in sampling, the more significant the difference. Thus, if the probability of an observed difference occurring due to chance is 0.001 (0.1 percent), the analyst should not only conclude that the difference is significant but in fact should be virtually *certain* that something other than chance selection in sampling is the explanation. Whenever the observed difference is considered significant, another explanation is sought. In this instance, the analyst would conclude that a setting has probably slipped and the mean of the universe from which the gears in the sample were actually selected is greater than 2.5 in. Similarly, although a sample of size 9 could have a mean less than *A*, the probability of such a value occurring due to chance selection from the hypothesized universe is again so small that this explanation is rejected. In this case, the alternative hypothesis that the mean of the universe from which the samples were drawn is less than 2.5 (due to slippage in a setting) would be accepted intuitively. Necessary steps to rectify the situation would then be undertaken.

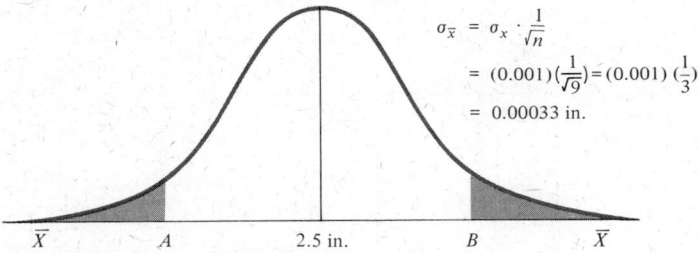

$$\sigma_{\bar{x}} = \sigma_x \cdot \frac{1}{\sqrt{n}}$$
$$= (0.001)(\frac{1}{\sqrt{9}}) = (0.001)(\frac{1}{3})$$
$$= 0.00033 \text{ in.}$$

Figure 15.2 Sampling distribution of the mean: $n = 9$.

The number line is thus divided into an acceptance region between the values *A* and *B* and two rejection regions—one encompassing all values of the sample mean less than *A* and the other encompassing all values of the sample mean greater than *B* (Figure 15.3). Whenever the analyst is concerned with detecting values of the sample mean that are either too large or too small to be consistent with the hypothesis being tested, rejection regions are located in both tails of the distribution. Such situations are referred to as *two-tailed tests*.

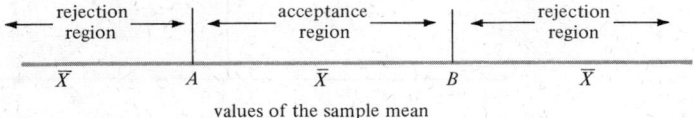

values of the sample mean

Figure 15.3

Illustration 2

Consumer's Advice, a publication dedicated to calling outstanding products to the attention of its readers, is considering recommending Goliath's new Triple-Ply Tires. Before committing itself to such an endorsement, however, Consumer's

Advice wishes to verify the manufacturer's claim that the average life of the tires
will be at least 35,000 miles. A sample of 100 tires has been sent to a laboratory that
will simulate road experience and provide Consumer's Advice with the actual life
observed for each of the 50 tires. The staff of Consumer's Advice will then ask the
question, "Is this sample result consistent with the hypothesis that $\mu_x \geq 35,000$
miles?"

H_{null}: $\mu_x \geq 35,000$ miles

$H_{\text{alternate}}$: $\mu_x < 35,000$ miles

In this case, the analyst is concerned only with detecting if the true universe mean is
less than 35,000 miles. (A universe mean greater than 35,000 miles is an extra bonus
to the consumer, but it is not relevant to the purpose of the study.) For the null
hypothesis to be accepted, the value of the sample mean observed must be consistent with values of the sample mean that could reasonably be expected to be
associated with one of the sampling distributions of the mean associated with
universe means of 35,000 or more. Figure 15.4 displays a few of these distributions.

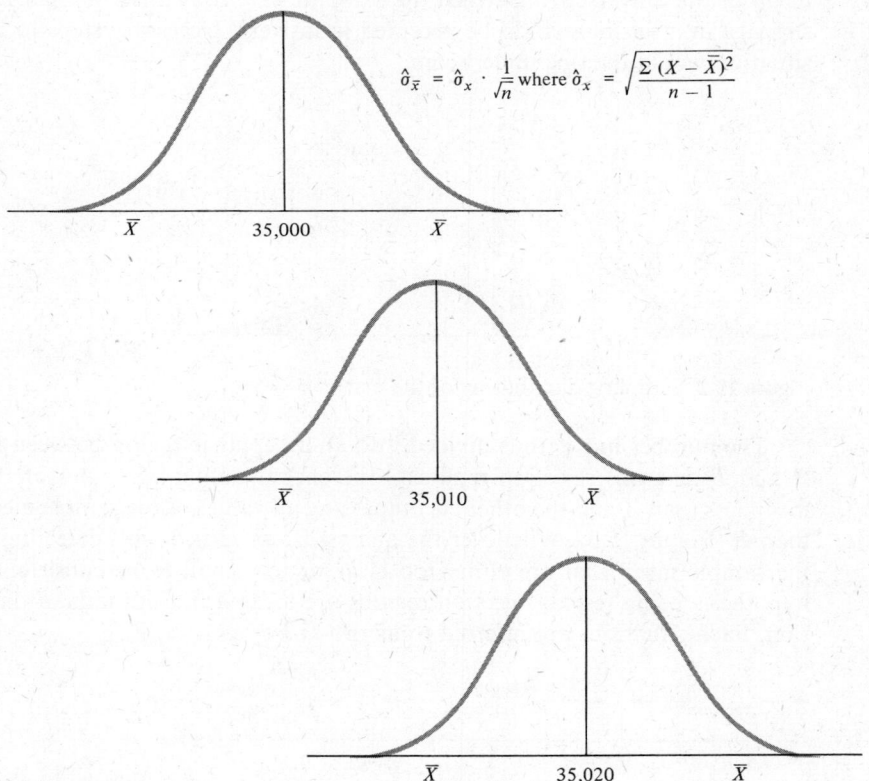

$$\hat{\sigma}_{\bar{x}} = \hat{\sigma}_x \cdot \frac{1}{\sqrt{n}} \text{ where } \hat{\sigma}_x = \sqrt{\frac{\Sigma(X-\bar{X})^2}{n-1}}$$

Figure 15.4

If the observed sample mean is consistent with any of the sampling distributions
of the mean depicted in Figure 15.4, the null hypothesis is accepted. (Although
they are not all reflected in Figure 15.4, this statement holds for all sampling
distributions associated with a universe mean greater than or equal to 35,000 miles.)
If the hypothesis is accepted for the minimally desirable of the possible sampling

distributions of the mean (namely, the one associated with $\mu_x = 35.000$), it will automatically be acceptable for all the other desirable sampling distributions of the mean (namely, the ones associated with $\mu_x > 35,000$). Thus, if the test is performed in association with the sampling distribution of the mean associated with the mean, 35,000 miles, a decision rule can be formulated that will lead to the acceptance of the null hypothesis for all values of the sample mean consistent with universe means $\geq 35,000$ miles. The analyst is concerned only with detecting if the universe mean is less than 35,000 miles, therefore the entire rejection region is located in the left-hand tail of the distribution (Figure 15.5).

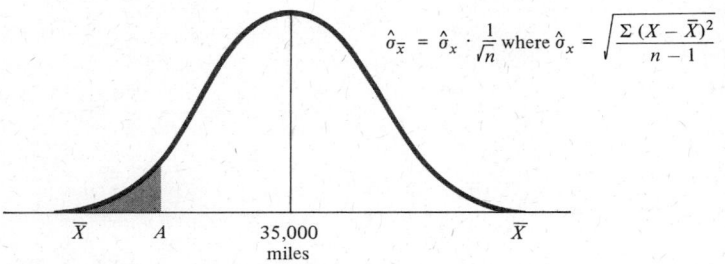

$$\hat{\sigma}_{\bar{x}} = \hat{\sigma}_x \cdot \frac{1}{\sqrt{n}} \text{ where } \hat{\sigma}_x = \sqrt{\frac{\Sigma (X - \bar{X})^2}{n - 1}}$$

Figure 15.5(a) Sampling distribution of the mean.

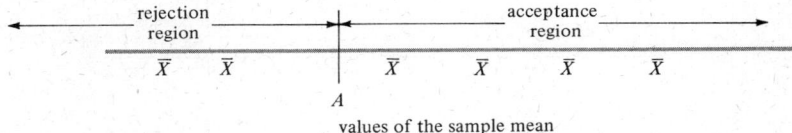

Figure 15.5(b)

Whenever the analyst is concerned solely with detecting values of the sample mean that are too small to be consistent with the hypothesis under investigation, the rejection region is located in the left-hand tail of the distribution. Such situations are referred to as *one-tailed tests*.

Illustration 3

The Federal Trade Commission is concerned with evaluating the advertising claims of Gaseous Bottlers, Inc. Gaseous Bottlers, Inc., claims that their latest sugar-free concoction, Epidermis,[1] contains an average of less than 1 oz of cyclamates per quart. The FTC has selected a sample of 200 bottles of Epidermis. Upon obtaining the average cyclamate content per quart in the sample, the statistician for the FTC will pose the question, "Is this sample result consistent with the hypothesis that $\mu_x \leq 1$ oz?"

H_{null}: $\mu_x \leq 1$ oz
$H_{alternate}$: $\mu_x > 1$ oz

In a situation of this type, it is the objective of the analyst to detect when the universe mean is larger than the maximum hypothesized value. If cyclamates per quart average to less than 1 oz, that's great, but it is not important with respect to the test at hand. What the FTC is attempting to detect is whether the average

---

[1] The name "Epidermis" was suggested by an advertising agency that plans to build a promotional campaign around the slogan, "Give me some skin."

cyclamate content is greater than 1 oz per quart. For the null hypothesis to be
accepted, the value of the sample mean observed must be consistent with those
values of the sample mean that could reasonably be expected to occur in association with sampling distributions of the mean associated with universe means of
1 oz or less. Figure 15.6 displays a few of these distributions.

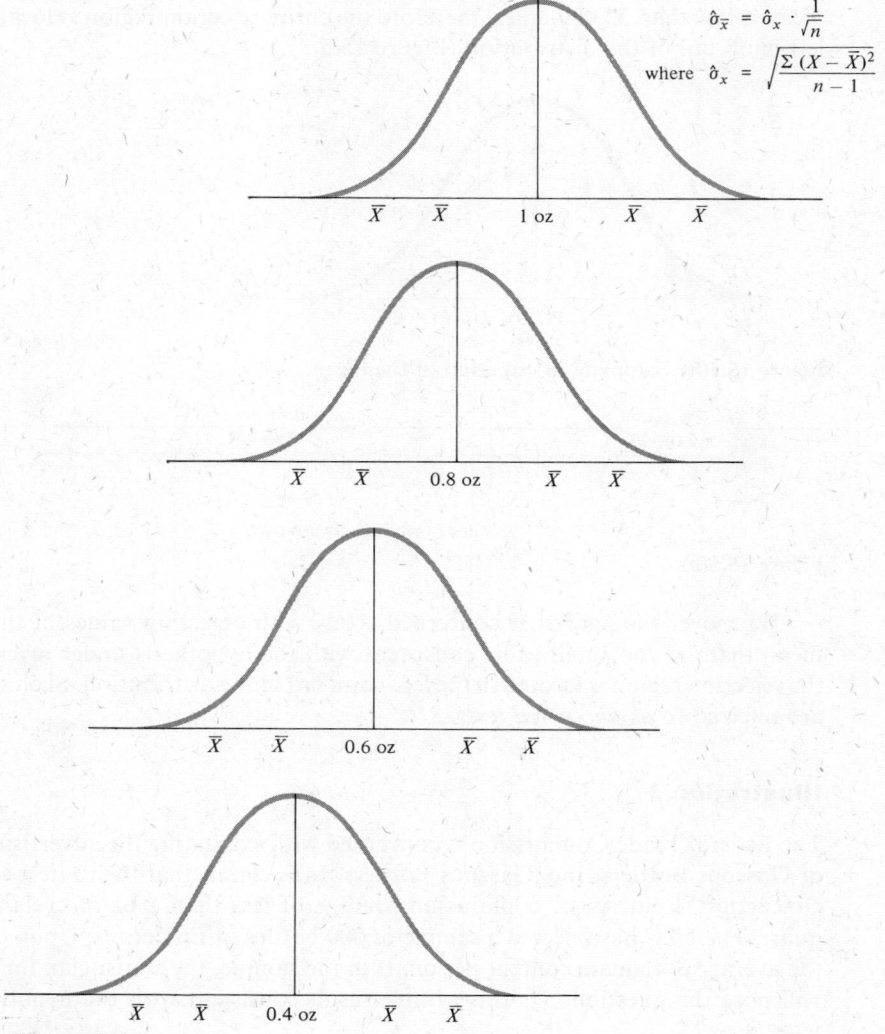

$$\hat{\sigma}_{\bar{x}} = \hat{\sigma}_x \cdot \frac{1}{\sqrt{n}}$$

$$\text{where } \hat{\sigma}_x = \sqrt{\frac{\Sigma (X - \bar{X})^2}{n - 1}}$$

Figure 15.6

The null hypothesis is accepted if the sample mean selected is consistent with
any of the sampling distributions of the mean depicted in Figure 15.6 (in fact, with
any sampling distribution of the mean associated with a universe mean from 0 to
1.0 oz). If the null hypothesis is accepted for the least desirable of the sampling
distributions of the mean (namely, the one associated with a mean of 1 oz), it will
automatically be acceptable for all the other desirable sampling distributions of the
mean (namely, the ones associated with $\mu_x < 1$ oz). Thus, if the test is performed in
association with the sampling distribution of the mean associated with the mean,
1 oz, a decision rule can be formulated that will lead to the acceptance of the null
hypothesis for all values of the sample mean consistent with universe means ≤ 1 oz.

The analyst is concerned solely with detecting if the universe mean is greater than 1 oz, therefore the entire rejection region is located in the right-hand tail of the distribution (Figure 15.7).

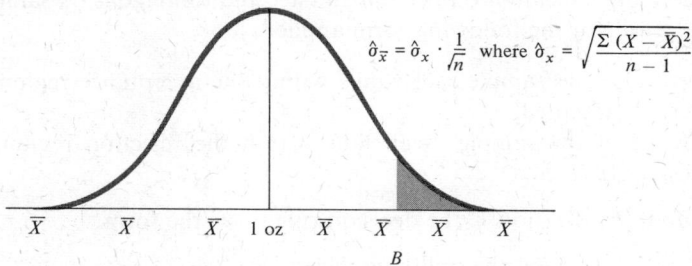

$$\hat{\sigma}_{\bar{x}} = \hat{\sigma}_x \cdot \frac{1}{\sqrt{n}} \quad \text{where } \hat{\sigma}_x = \sqrt{\frac{\Sigma (X - \bar{X})^2}{n-1}}$$

Figure 15.7(a) Sampling distribution of the mean.

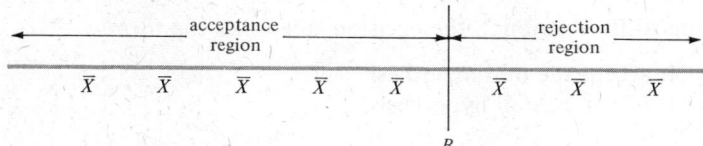

Figure 15.7(b)

Whenever the analyst restricts his concern to detecting values of the sample mean that are too large to be consistent with the hypothesis, the rejection region is located in the right-hand tail of the distribution and therefore involves a *one-tailed test*. The preceding relationships are summarized in Table 15.2.

Table 15.2

Analyst's objective	Null hypothesis	Alternative hypothesis	Type of test	Location of rejection region
1. To avoid accepting the null hypothesis when the true mean is *greater* or *less* than the postulated value	$\mu_x = K$	$\mu_x \neq K$ i.e., $\mu_x > K$ or $\mu_x < K$	Two-tailed	The rejection region is located in both the right- and left-hand tails of the sampling distribution associated with the mean, K
2. To avoid accepting the null hypothesis when the true mean is less than the postulated value	$\mu_x \geq K$	$\mu_x < K$	One-tailed	The rejection region is located in the left-hand tail of the sampling distribution associated with the mean, K
3. To avoid accepting the null hypothesis when the true mean is greater than the postulated value	$\mu_x \leq K$	$\mu_x > K$	One-tailed	The rejection region is located in the right-hand tail of the sampling distribution associated with the mean, K

15.2 DETERMINATION OF THE CRITICAL VALUE(S)

Testing hypotheses involves partitioning the number line containing all possible values of the sample mean into rejection and acceptance regions. A sample is then selected and a rule of the following form applied:

If the value of the sample mean falls within the acceptance region, the null hypothesis is accepted.

If the value of the sample mean falls within the rejection region, the null hypothesis is rejected.

Referring to Illustration 1, the decision rule takes the form

If $A \leq \overline{X} \leq B$, accept the null hypothesis.
If $\overline{X} < A$ or $\overline{X} > B$, reject the null hypothesis.

Referring to Illustration 2, the decision rule takes the form

If $\overline{X} \geq A$, accept the null hypothesis.
If $\overline{X} < A$, reject the null hypothesis.

Referring to Illustration 3, the decision rule takes the form

If $X \leq B$, accept the null hypothesis.
If $X > B$, reject the null hypothesis.

The matrix in Table 15.3 depicts the possible outcomes associated with the application of any given decision rule.

Table 15.3

Action alternatives / States of nature	Action alternative I: Accept the hypothesis on the basis of sample evidence	Action alternative II: Reject the hypothesis on the basis of sample evidence
State of nature I: Hypothesis is true	Correct decision	Type I error
State of nature II: Hypothesis is false	Type II error	Correct decision

There are two action alternatives (represented by the columns in the matrix):

Action alternative I   Accept the hypothesis on the basis of sample evidence
Action alternative II   Reject the hypothesis on the basis of sample evidence

There are two possible states of nature (represented by the rows in the matrix):

State of nature I   The hypothesis is true
State of nature II   The hypothesis is false

Application of the decision rule results in one of the four possible outcomes represented by the four cells in the body of the matrix.

1. The hypothesis is accepted on the basis of sample evidence *and* the hypothesis is true. In this situation, the decision rule leads to a correct action.
2. The hypothesis is rejected on the basis of sample evidence *and* the hypothesis is false. Again, the decision rule leads to a correct action.
3. The hypothesis is rejected on the basis of sample evidence although the hypothesis is actually true. In this case, the decision rule leads to an error (an incorrect conclusion is drawn). Rejecting a true hypothesis is known as a Type I error.
4. The hypothesis is accepted on the basis of sample evidence although the hypothesis is actually false. Again, the decision rule leads to an error. Accepting a false hypothesis is known as a Type II error.

Thus, two types of error can occur in association with any given decision rule.[2]

1. Type I error: rejection of a true hypothesis
2. Type II error: acceptance of a false hypothesis

The placement of the critical value separating a rejection and acceptance region exerts a significant influence on the probability that Type I and Type II errors will be committed. In the next section, the factors influencing the probability of committing errors of each of these types will be examined. In the analysis that precedes the establishment of the rejection and acceptance regions, the decision maker must weigh the seriousness of the consequences associated with committing Type I and Type II errors in conjunction with the probability of each of these types of errors being committed. These probabilities vary depending on the decision rule selected (a decision rule specifies the acceptance–rejection regions). The decision rule selected should be one that limits the probability of committing Type I and Type II errors to those levels considered desirable by the decision maker.

15.3 TYPE I ERRORS

A Type I error refers to the rejection of a true hypothesis. The probability of committing a Type I error is customarily referred to as the *critical probability* and designated by the Greek letter alpha (α). A critical probability of 10 percent indicates that the maximum probability that a Type I error will be committed is 10 percent. The probability of a Type I error is equivalent to the probability of a sample mean falling within the rejection region when the hypothesis is actually true.

15.4 TYPE I ERRORS AND TWO-TAILED TESTS

The sampling distribution of the mean associated with all possible samples of size 9 in the Uniform Gear Works example (Illustration 1) is reproduced as Figure 15.8.

---

[2] A convenient way to distinguish Type I from Type II errors is to associate the single horizontal slash in the letter T with Type I errors (T for rejecting a *true*) and the two horizontal slashes in the letter F with Type II errors (F for accepting a *false*).

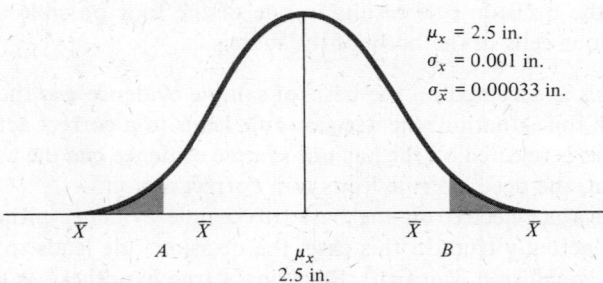

Figure 15.8 Uniform Gear Works: sampling distribution
of the mean associated with all possible samples of size 9.
$\mu_x = 2.5$ in.; $\sigma_x = .001$ in.; $\sigma_{\bar{x}} = .00033$ in.

The decision rule employed takes the form

If $A \leq \bar{X} \leq B$, accept the null hypothesis.
If $\bar{X} < A$ or $\bar{X} > B$, reject the null hypothesis.

The probability of committing a Type I error is equal to the probability that
the value of a simple random sample of size 9 will have a sample mean that falls
in the shaded region (rejection region) of Figure 15.8. If the decision maker is
willing to take a chance of committing a Type I error only 10 percent of the time,
the values A and B should be located in such a fashion that 5 percent of the area
under the curve lies to the left of point A and 5 percent to the right of point B.[3] It
follows that if 5 percent lies to the right of point B, then 45 percent of the area
under the curve must lie between point B and the mean. The table of areas indicates
that 45 percent of the area under the curve will be contained between the mean
and a point 1.65 standard errors away. Therefore, point B will be equal to $\mu_x +
1.65\sigma_{\bar{x}}$. Substituting:

$$\begin{aligned}
\text{point } B &= \mu_x + 1.65\sigma_{\bar{x}} \\
&= 2.5 + 1.65(0.00033) \\
&= 2.5 + 0.0005445 \\
&= 2.5005445 \text{ in.}
\end{aligned}$$

Due to the symmetrical nature of the curve and the fact that 45 percent of the area
under the curve also lies between the mean and point A, point A will equal
$\mu_x - 1.65\sigma_{\bar{x}}$. Substituting:

$$\begin{aligned}
\text{point } A &= \mu_x - 1.65\sigma_{\bar{x}} \\
&= 2.5 - 1.65(0.00033) \\
&= 2.5 - 0.0005445 \\
&= 2.4994555 \text{ in.}
\end{aligned}$$

Thus, the decision rule takes the form

Accept the null hypothesis ($\mu_x = 2.5$) if the value of the sample mean falls
within the range 2.4994555–2.5005445 in.
Reject the null hypothesis if the value of the sample mean is below 2.4994555 in.
or above 2.5005445 in.

---

[3] Note: Two-tailed tests do not always consist of the critical probability being divided equally
between the two tails of the distribution. Equal division presumes that errors in either direction
are equally undesirable. An illustration of a situation where this is not the case is introduced
in Chapter 16.

If the decision maker believes that a 10 percent probability of committing a Type I error is too high and desires to lower this probability to 5 percent, the dividing lines separating the rejection regions from the acceptance region must be pushed farther out into the tails of the curve. A probability of committing a Type I error equal to 5 percent is equivalent to a probability of $\overline{X} > B$ equal to 2.5 percent and a probability of $\overline{X} < A$ equal to 2.5 percent. This, in turn, is equivalent to 47.5 percent of the area under the curve being contained between μ_x and point A and 47.5 percent between μ_x and point B. The table of areas under the normal curve indicates that 47.5 percent of the area under a normal curve is contained between the mean and a point 1.96 standard errors away. Therefore,

$$\text{point } A = \mu_x - 1.96\sigma_{\overline{x}} = 2.5 - 1.96(0.00033) = 2.5 - 0.0006468$$
$$= 2.4993532$$
$$\text{point } B = \mu_x + 1.96\sigma_{\overline{x}} = 2.5 + 1.96(0.00033) = 2.5 + 0.0006468$$
$$= 2.5006468$$

The decision rule associated with a 5 percent probability of committing a Type I error is

Accept the null hypothesis ($\mu_x = 2.5$) if the value of the sample mean falls within the range 2.4993532–2.5006468 in.
Reject the null hypothesis if the value of the sample mean is less than 2.4993532 in. or exceeds 2.5006468 in.

15.5 TYPE I ERRORS AND ONE-TAILED TESTS

When Type I errors are examined in conjunction with one-tailed tests, two additional considerations are introduced:

1. The entire critical probability is assigned to one tail of the distribution.
2. In the case of a one-tailed test, there is more than one value of the universe mean that is consistent with the null hypothesis. As a result, the critical probability now refers to the maximum probability of committing a Type I error rather than to *the* probability of a Type I error as is the case with a two-tailed test.[4]

Illustration 2, the situation where *Consumer's Advice* is concerned with testing the validity of the claims made on behalf of Goliath's new Triple-Ply Tires, will be used to investigate the role of Type I errors in conjunction with one-tailed tests. The sampling distribution of the mean associated with all possible samples of size 100 is reproduced as Figure 15.9.

In the case of Illustration 2, the universe standard deviation is unknown. Therefore, $\hat{\sigma}_x = \sqrt{\sum (X - \overline{X})^2/(n - 1)}$ is used as an estimate of σ_x and inserted in the formula $\hat{\sigma}_{\overline{x}} = \hat{\sigma}_x \cdot 1/\sqrt{n}$ to generate an estimate of $\sigma_{\overline{x}}$. Assuming a sample size of 100 and $\hat{\sigma}_x$ equal to 320 miles (where $\hat{\sigma}_x = \sqrt{\sum (X - \overline{X})^2/(n - 1)}$), the estimated sampling distribution of the mean would have a standard error of the

---

[4] With a two-tailed test, acceptance of the null hypothesis means acceptance of *a* value for the universe mean. With a one-tailed test, acceptance of the null hypothesis means accepting that the universe mean is one of a number of acceptable values.

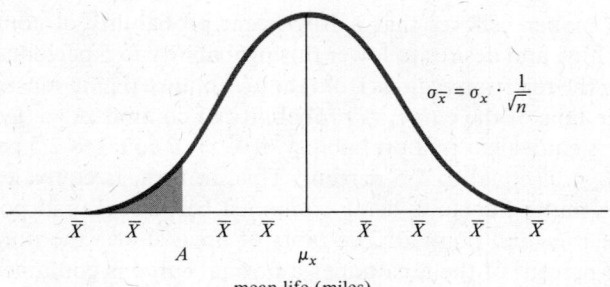

Figure 15.9 Goliath's Triple-Ply Tires: sampling distribution of the mean life associated with all possible samples of size 100.

mean equal to 32 miles. Assuming the minimum acceptable value for the universe mean (35,000 miles), the associated sampling distribution of the mean is reproduced as Figure 15.10.

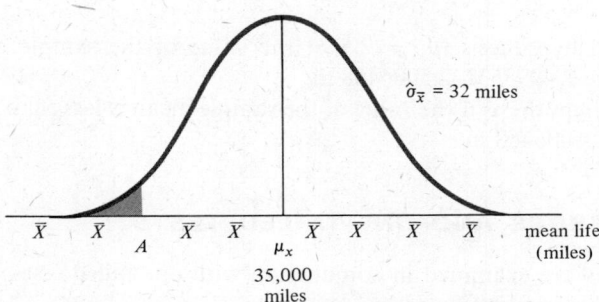

Figure 15.10 Goliath's Triple-Ply Tires: sampling distribution of the mean life associated with all possible samples of size 100. $\mu_x = 35,000$ miles; $\hat{\sigma}_x = 320$ miles; $\hat{\sigma}_{\bar{x}} = 32$ miles.

The probability of committing a Type I error is equal to the probability that a simple random sample of size 100 will have a sample mean that falls within the shaded region of Figure 15.10. Given that the decision maker is willing to accept a 10 percent chance of committing a Type I error, this is equivalent to 10 percent of the area under the curve falling to the left of point A on the number line (i.e., $P(\bar{X} < A) = 0.10$). This, in turn, is equivalent to stating that 40 percent of the area under the curve must lie between the mean and point A. The table of areas under the normal curve indicates that 39.97 percent of the area under the normal curve lies between the mean and a point 1.28 standard errors away. Therefore, point A will be equal to $\mu_x - 1.28\hat{\sigma}_{\bar{x}}$. Substituting:

$$
\begin{aligned}
\text{point } A &= \mu_x - 1.28\hat{\sigma}_{\bar{x}} \\
&= 35,000 - 1.28(32) \\
&= 35,000 - 40.96 \\
&= 34,959.04 \text{ miles}
\end{aligned}
$$

The decision rule:

Accept the null hypothesis if $\bar{X} \geq 34,959.04$ miles.
Reject the null hypothesis if $\bar{X} < 34,959.04$ miles.

The above decision rule is associated with an α of 10 percent (i.e., the *maximum* probability of committing a Type I error is 10 percent). The explanation for the emphasis on the word maximum is that the test is being performed in conjunction with the least acceptable situation (namely, with the sampling distribution associated with a universe mean of 35,000). If the true universe mean is greater than 35,000, the probability of committing a Type I error will be less than the critical probability of 10 percent. The greater the amount by which the true value of the universe mean exceeds the minimally acceptable value of 35,000 miles, the smaller the probability becomes of committing a Type I error. To illustrate: The probability of committing a Type I error using the above decision rule will be investigated for the case where the true value of the universe mean actually is 35,010 and for the case where the true value of the universe mean actually is 35,020 miles.

WHEN $\mu_x = 35,010$ MILES. Referring to Figure 15.11, the probability of committing a Type I error when the true value of the universe mean is 35,010 miles is equal to the proportion of the total area under the curve lying to the left of the critical value 34959.04 on the number line. The distance between 35,010 and 34,959.04 in miles is equal to approximately 51 miles. Dividing by the value of a standard error, we obtain $z = (\overline{X} - \mu_x)/\hat{\sigma}_{\overline{x}} = -51/32 = -1.59$; this is equivalent to a distance of 1.59 standard error units. Between the mean and a point 1.59 standard errors away will lie 44.41 percent of the area under the curve. If 44.41 percent of the area under the curve lies between point A and the mean, it necessarily follows that 5.59 percent lies below point A. Therefore, the probability that a Type I error will be committed if the true value of the universe mean is 35,010 miles (given the preceding decision rule) is 5.59 percent.

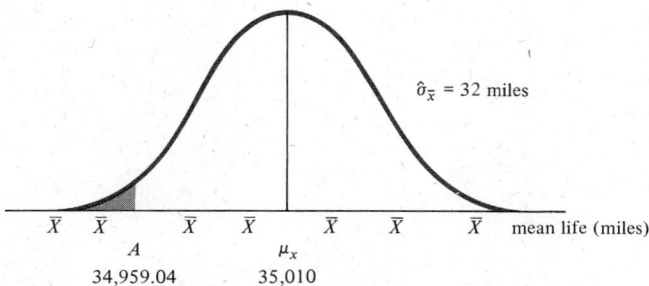

Figure 15.11 Goliath's Triple-Ply Tires: sampling distribution of the mean life associated with all possible samples of size 100. $\mu_x = 35,010$ miles; $\hat{\sigma}_x = 320$ miles; $\hat{\sigma}_{\overline{x}} = 32$ miles.

WHEN $\mu_x = 35,020$ MILES. Referring to Figure 15.12, the probability of committing a Type I error, when the true value of the universe mean is 35,020 miles, is equal to the proportion of the total area under the curve lying to the left of the critical value 34,959.04 miles on the number line.

The distance between 35,020 and 34,959.04 is approximately 61 miles or, expressed in standard error units, a distance of 1.91 standard errors

$$\left[z = \frac{\overline{X} - \mu_x}{\hat{\sigma}_{\overline{x}}} = \frac{-61}{32} = -1.91 \right]$$

Between the mean and a point 1.91 standard errors away, 47.19 percent of the area under the normal curve lies. It follows that 2.81 percent of the area under the curve

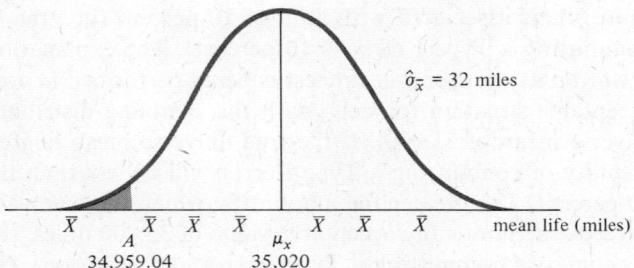

Figure 15.12 Goliath's Triple-Ply Tires: sampling distribution of the mean life associated with all possible samples of size 100. $\mu_x = 35,020$ miles; $\hat{\sigma}_x = 320$ miles; $\hat{\sigma}_{\bar{x}} = 32$ miles.

must lie within the shaded region. Thus, the probability that a Type I error will be committed given that the true universe mean is 35,020 is 2.81 percent. Figure 15.13 displays the probability of committing a Type I error (given the decision rule: accept the null hypothesis if $\bar{X} \geq 34,959.04$ miles; reject if $\bar{X} < 34,959.04$ miles) for each of the three cases examined, namely:

$$\mu_x = 35,000 \text{ miles}$$
$$\mu_x = 35,010 \text{ miles}$$
$$\mu_x = 35,020 \text{ miles}$$

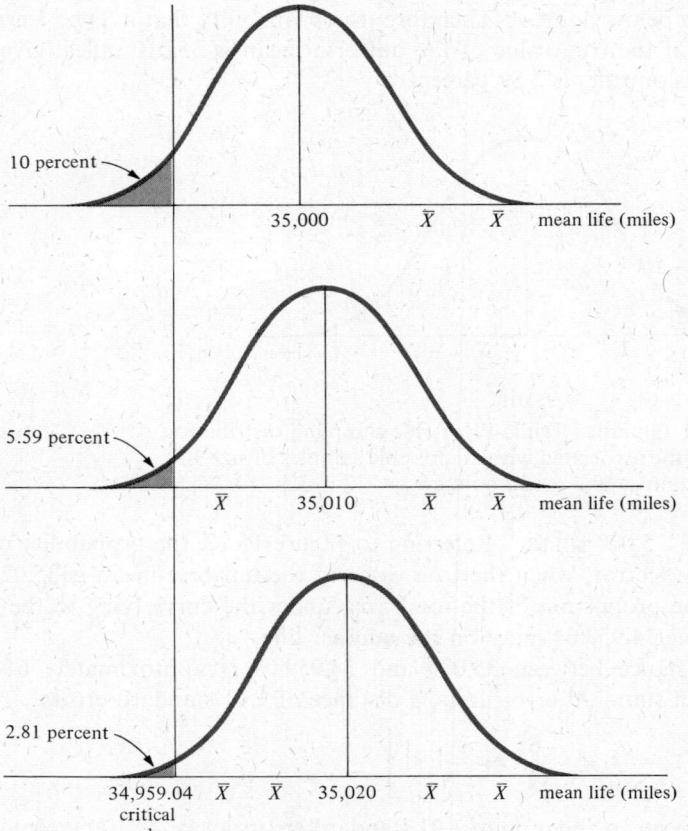

Figure 15.13 Probability of committing a Type I error for each of the three cases: $\mu_x = 35,000$; $\mu_x = 35,010$; $\mu_x = 35,020$.

As clearly depicted in Figure 15.13, the greater the amount by which the value of the true universe mean exceeds the minimally acceptable value of 35,000 miles, the smaller the probability of committing a Type I error. Following a similar procedure, the probability of committing a Type I error (given a particular decision rule) can be ascertained for all conceivable values of the universe mean. Figure 15.14 is an error curve depicting the probability of committing a Type I error over the range of conceivable values for the universe mean. Note that the probability of a Type I error is maximized when the true value of the universe mean is equal to the minimally acceptable value (namely, 35,000 miles). The probability of a Type I error at this maximum point is equal to the critical probability. As the true value of the universe mean exceeds the minimally acceptable value by greater and greater amounts, the probability of committing a Type I error declines sharply. Observe that the probability of committing a Type I error is zero for all values lying to the left of the minimally acceptable value. This, of course, is due to the fact that if $\mu_x < 35,000$ miles, the null hypothesis is false. Type I errors can occur only when the null hypothesis is true.

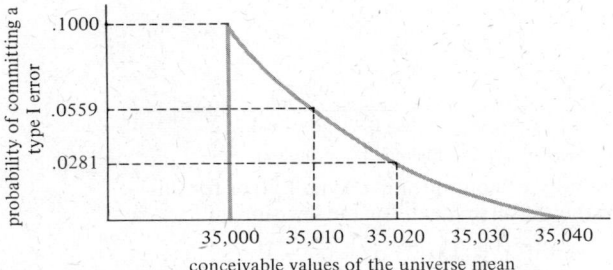

Figure 15.14 Probability of committing a Type I error for all conceivable values of the universe mean for the decision rule:
accept if $\bar{X} \geq 34,959.04$
reject if $\bar{X} < 34,959.04$

If the rejection region is located in the right-hand tail, as it is for Illustration 3, the error curve describing the probability of committing a Type I error appears with the tail extending to the left. Figure 15.15(a) develops the decision rule for Illustration 3 based on a critical probability of 8 percent, a sample size of 25, and a sample standard deviation of 0.15 oz $[\hat{\sigma}_x = \sqrt{\sum (X - \bar{X})^2/(n - 1)} = 0.15 \text{ oz}]$. Figure 15.15(b) depicts the probability of committing a Type I error for all conceivable values of the universe mean given the decision rule developed in Figure 15.15(a). (Normality is assumed even though the sample size is less than 30.)

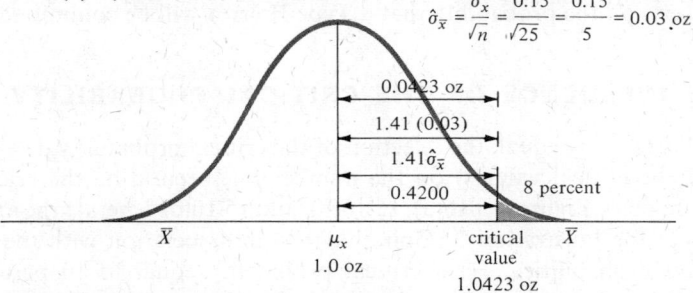

Figure 15.15(a) Determination of the decision rule associated with the null hypothesis $\mu_x \leq 1.0$ oz and a critical probability of 8 percent.

Given an 8 percent critical probability, 42 percent of the area under the curve is contained between the mean and the critical value. This is equivalent to a distance of 1.41 standard errors. A single standard error equals 0.03 oz. Therefore, the value of the critical value is described by the following relationship:

$$\text{critical value} = 1.0 + 1.41(0.03 \text{ oz})$$
$$= 1.0 + 0.0423$$
$$= 1.0423 \text{ oz}$$

The decision rule is:

Accept the null hypothesis if $\overline{X} \leq 1.0423$ oz.
Reject the null hypothesis if $\overline{X} > 1.0423$ oz.

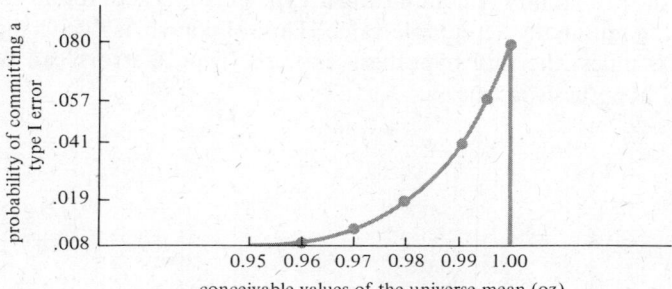

Figure 15.15(b) Probability of committing a Type I error for all conceivable values of the universe mean for the decision rule:

accept if $\overline{X} \leq 1.0423$ oz
reject if $\overline{X} > 1.0423$ oz

15.6 TYPE II ERRORS

Up to this point, Type I errors have been examined in isolation. In actuality, in selecting an appropriate decision rule, the implications of committing a Type II error must also be considered. A Type II error refers to the acceptance of a false hypothesis and is designated by the Greek letter beta, β. The probability that a Type II error will be committed is a function of three factors: (1) the level of critical probability selected, (2) the degree of falseness of the hypothesis (a Type II error refers to the acceptance of a false hypothesis—therefore a Type II error can occur only when the null hypothesis is false), and (3) the sample size.

Let us now turn to an examination of the nature of the influence each of these elements exerts on the probability that a Type II error will be committed.

15.7 THE INFLUENCE OF THE CRITICAL PROBABILITY

All other things being equal, the selection of the critical probability determines the location of the critical value(s) on the number line separating the rejection and acceptance regions. Figures 15.16(a), 15.16(b), and 15.16(c) generate the appropriate decision rules for Illustration 2 (Goliath Tires) in association with the following three critical probabilities: (1) a critical probability equal to 10 percent; (2) a critical probability equal to 5 percent; and (3) a critical probability equal to 1 percent.

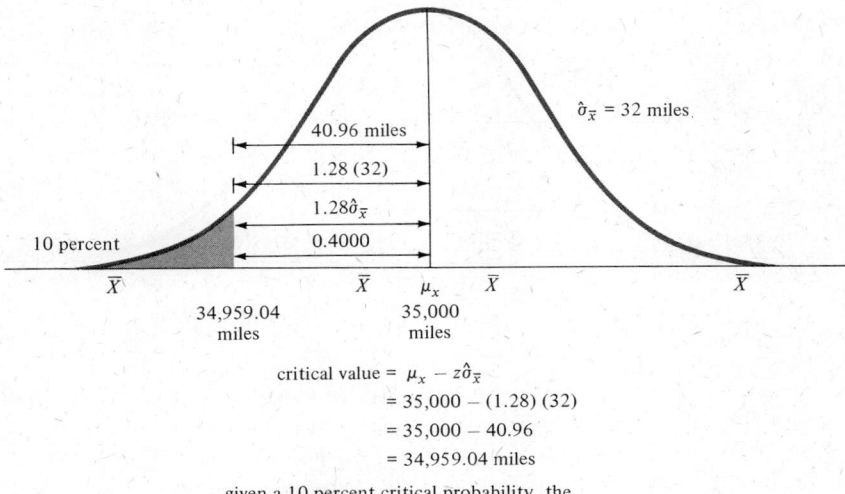

$$\text{critical value} = \mu_x - z\hat{\sigma}_{\bar{x}}$$
$$= 35,000 - (1.28)\,(32)$$
$$= 35,000 - 40.96$$
$$= 34,959.04 \text{ miles}$$

given a 10 percent critical probability, the appropriate decision rule is

accept the null hypothesis if $\bar{X} \geq 34,959.04$ miles

reject the null hypothesis if $\bar{X} < 34,959.04$ miles

Figure 15.16(a) Illustration 2—Goliath Tires: development of the decision rule associated with a 10 percent critical probability. $\mu_x = 35,000$ miles; $\hat{\sigma}_x = 320$ miles; $n = 100$; $\hat{\sigma}_{\bar{x}} = 32$ miles. Null hypothesis: $\mu_x \geq 35,000$ miles.

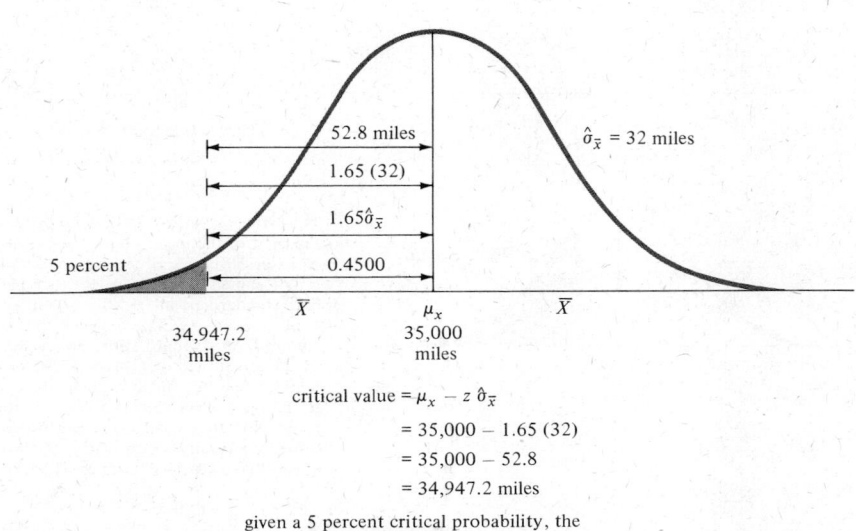

$$\text{critical value} = \mu_x - z\,\hat{\sigma}_{\bar{x}}$$
$$= 35,000 - 1.65\,(32)$$
$$= 35,000 - 52.8$$
$$= 34,947.2 \text{ miles}$$

given a 5 percent critical probability, the appropriate decision rule is

accept the null hypothesis if $\bar{X} \geq 34,947.2$ miles

reject the null hypothesis if $\bar{X} < 34,947.2$ miles

Figure 15.16(b) Illustration 2—Goliath Tires: development of the decision rule associated with a 5 percent critical probability. $\mu_x = 35,000$ miles; $\hat{\sigma}_x = 320$ miles; $n = 100$; $\hat{\sigma}_{\bar{x}} = 32$ miles. Null hypothesis: $\mu_x \geq 35,000$ miles.

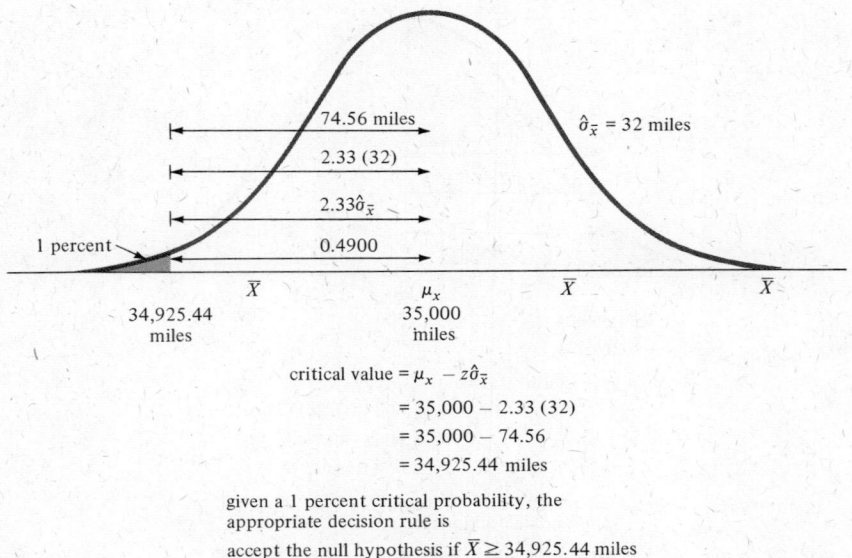

$$\text{critical value} = \mu_x - z\hat{\sigma}_{\bar{x}}$$
$$= 35,000 - 2.33\,(32)$$
$$= 35,000 - 74.56$$
$$= 34,925.44 \text{ miles}$$

given a 1 percent critical probability, the appropriate decision rule is

accept the null hypothesis if $\bar{X} \geq 34,925.44$ miles

reject the null hypothesis if $\bar{X} < 34,925.44$ miles

Figure 15.16(c) Illustration 2—Goliath Tires: development of the decision rule associated with a 1 percent critical probability. $\mu_x = 35,000$ miles; $\hat{\sigma}_x = 320$ miles; $n = 100$; $\hat{\sigma}_{\bar{x}} = 32$ miles. Null hypothesis: $\mu_x \geq 35,000$ miles.

Figures 15.17(a), 15.17(b), and 15.17(c) illustrate the impact of varying the critical probability on the probability of committing a Type II error. The true value of the universe mean will be assumed to equal 34,950 miles. (Given $\mu_x = 34,950$, the null hypothesis $\mu_x \geq 35,000$ miles would be false.) In Figure 15.17, the probability of committing a Type II error (accepting $\mu_x \geq 35,000$ when μ_x actually

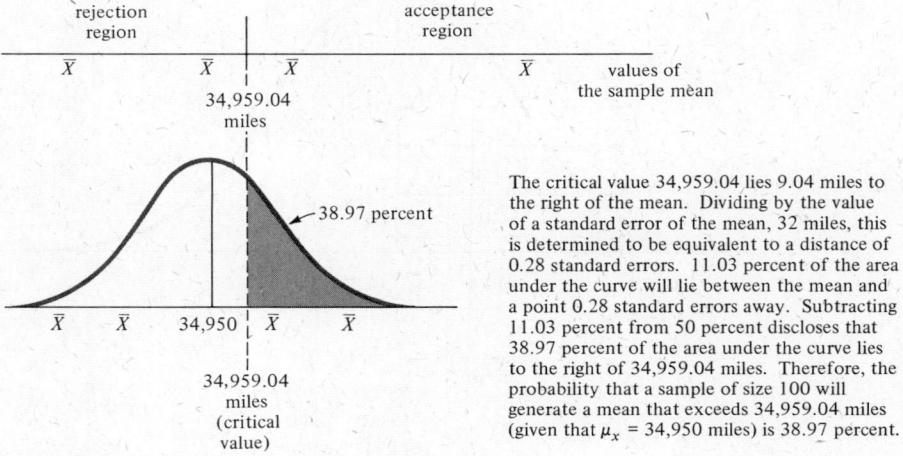

The critical value 34,959.04 lies 9.04 miles to the right of the mean. Dividing by the value of a standard error of the mean, 32 miles, this is determined to be equivalent to a distance of 0.28 standard errors. 11.03 percent of the area under the curve will lie between the mean and a point 0.28 standard errors away. Subtracting 11.03 percent from 50 percent discloses that 38.97 percent of the area under the curve lies to the right of 34,959.04 miles. Therefore, the probability that a sample of size 100 will generate a mean that exceeds 34,959.04 miles (given that $\mu_x = 34,950$ miles) is 38.97 percent.

probability (type II error) = 38.97 percent

Figure 15.17(a) Illustration 2—Goliath Tires: probability of committing a Type II error given the null hypothesis $\mu_x \geq 35,000$ miles, given the true value of the universe mean = 34,950 miles. Decision rule based on a 10 percent critical probability:

accept null hypothesis if $\bar{X} \geq 34,959.04$ miles
reject null hypothesis if $\bar{X} < 34,959.04$ miles

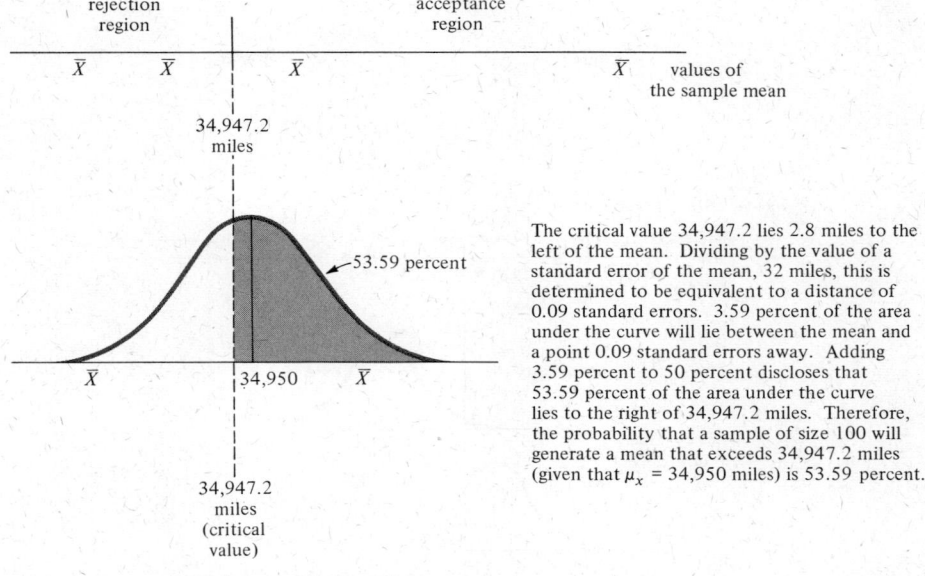

The critical value 34,947.2 lies 2.8 miles to the left of the mean. Dividing by the value of a standard error of the mean, 32 miles, this is determined to be equivalent to a distance of 0.09 standard errors. 3.59 percent of the area under the curve will lie between the mean and a point 0.09 standard errors away. Adding 3.59 percent to 50 percent discloses that 53.59 percent of the area under the curve lies to the right of 34,947.2 miles. Therefore, the probability that a sample of size 100 will generate a mean that exceeds 34,947.2 miles (given that μ_x = 34,950 miles) is 53.59 percent.

probability (type II error) = 53.59 percent

Figure 15.17(b) Illustration 2—Goliath Tires: probability of committing a Type II error given the null hypothesis $\mu_x \geq 35,000$ miles, given the true value of the universe mean = 34,950 miles. Decision rule based on a 5 percent critical probability:

accept null hypothesis if $\bar{X} \geq 34,947.2$ miles
reject null hypothesis if $\bar{X} < 34,947.2$ miles

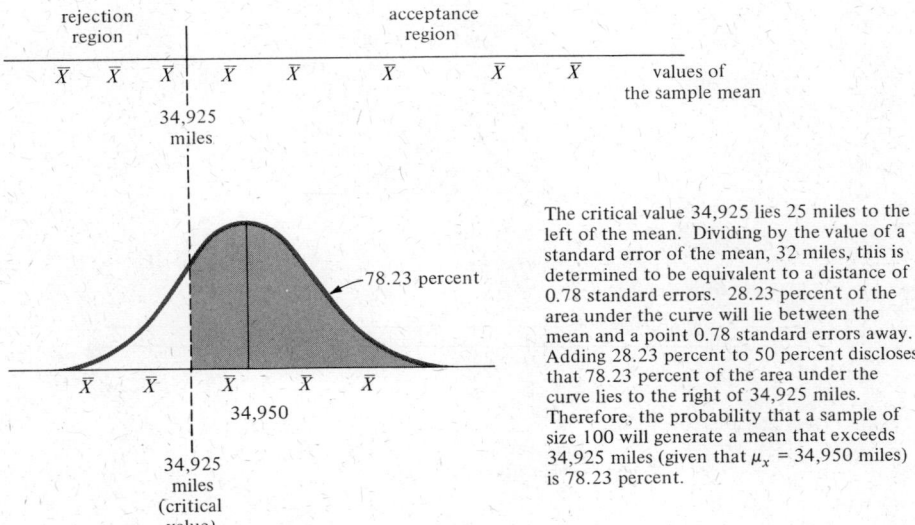

The critical value 34,925 lies 25 miles to the left of the mean. Dividing by the value of a standard error of the mean, 32 miles, this is determined to be equivalent to a distance of 0.78 standard errors. 28.23 percent of the area under the curve will lie between the mean and a point 0.78 standard errors away. Adding 28.23 percent to 50 percent discloses that 78.23 percent of the area under the curve lies to the right of 34,925 miles. Therefore, the probability that a sample of size 100 will generate a mean that exceeds 34,925 miles (given that μ_x = 34,950 miles) is 78.23 percent.

probability (type II error) = 78.23 percent

Figure 15.17(c) Illustration 2—Goliath Tires: probability of comitting a Type II error given the null hypothesis $\mu_x \geq 35,000$ miles, given the true value of the universe mean = 34,950 miles. Decision rule based on a 1 percent critical probability:

accept null hypothesis if $\bar{X} \geq 34,925$ miles
reject null hypothesis if $\bar{X} < 34,925$ miles

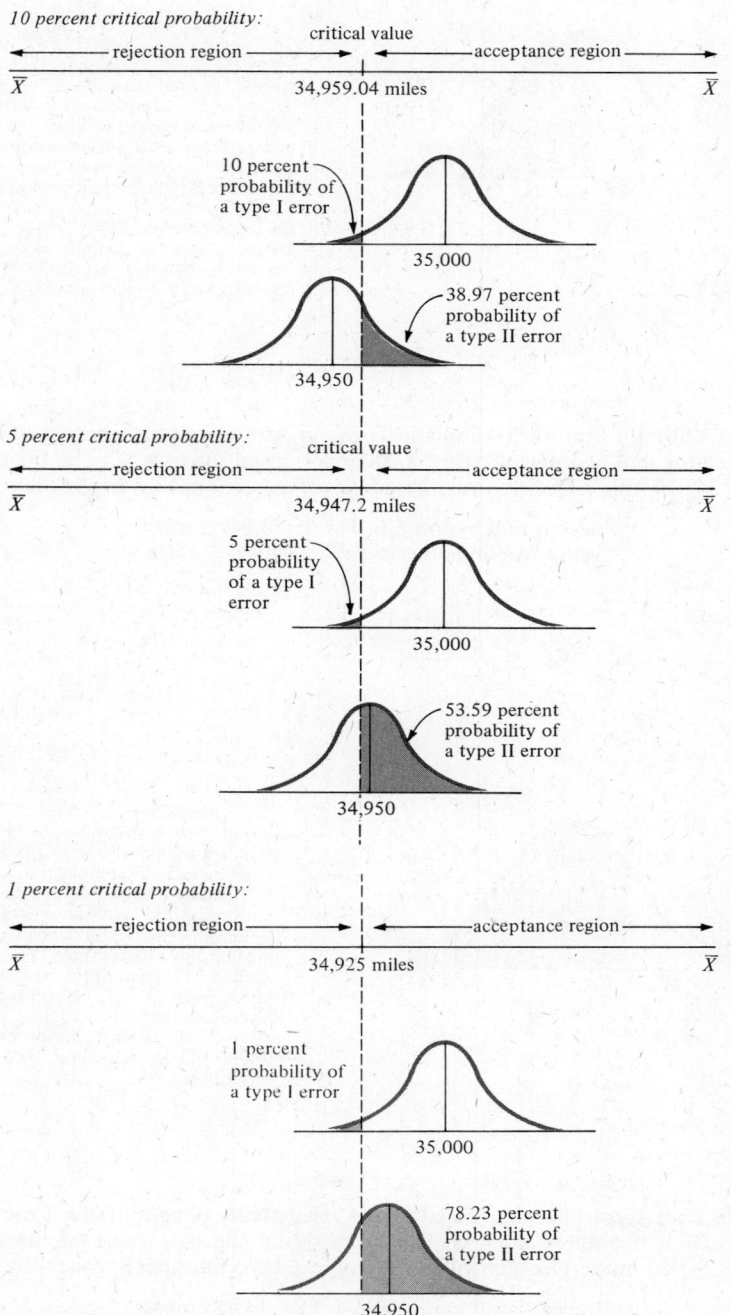

Figure 15.18

equals 34,950) is computed for each of the three decision rules developed in Figure 15.16. Note that the probability of committing a Type II error is equal to the probability that a sample mean will fall within the acceptance region given that the null hypothesis is actually false. In each of the diagrams in Figure 15.17, the probability of a Type II error is indicated by the shaded section under the appropriate sampling distribution of the mean.

Figure 15.18 depicts the probability of committing a Type I error (when $\mu_x = 35,000$) and the probability of committing a Type II error (when $\mu_x = 34,950$) for the three decision rules associated with critical probabilities of 10 percent, 5 percent, and 1 percent, respectively.

An examination of Figure 15.18 discloses that as the critical probability is decreased, the critical value separating the rejection and acceptance regions is pushed farther out into the tail of the distribution. This results in a decrease in the proportion of sample means falling within the rejection region (given that the null hypothesis is true). Unfortunately, as the rejection region is pushed farther to the left along the number line, a larger proportion of the sample means associated with unacceptable sampling distributions of the mean fall within the acceptance region. Thus, a decrease in the critical probability, while reducing the probability that a Type I error will be committed, simultaneously results in an increase in the probability of committing a Type II error. This is true in association with each conceivable value of μ_x less than the minimally acceptable value of 35,000 miles.

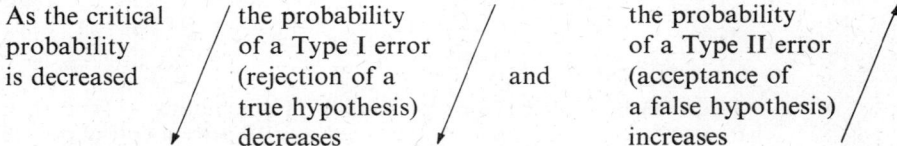

As the critical probability is decreased / the probability of a Type I error (rejection of a true hypothesis) decreases / and the probability of a Type II error (acceptance of a false hypothesis) increases /

Changing the value of the critical probability has an opposite effect on the two types of errors. (The probability of a Type II error varies inversely and the probability of a Type I error varies directly with a change in the critical probability.)

15.8 THE INFLUENCE OF THE DEGREE OF FALSENESS OF THE HYPOTHESIS

All other things being equal, the probability of committing a Type II error decreases as the true value of the universe mean departs to a greater and greater extent from the acceptable values of the universe mean as specified by the null hypothesis. Figure 15.19 depicts the probability of committing a Type II error given values for the true universe mean of 34,950, 34,940, 34,930, and 34,920 miles.

As Figure 15.19 clearly demonstrates, the greater the extent to which the true value of the universe mean departs from the range of acceptable values of the universe mean as specified by the null hypothesis, the less the likelihood that a sample mean will have a value falling within the acceptance region. The less the probability of a sample mean falling within the acceptance region, the less the likelihood that a Type II error will be committed. Figure 15.15(b) presented a Type I error curve. A Type I error curve shows the probability of committing a Type I error in association with each conceivable value of the universe mean. There is, of course, a different Type I error curve for each decision rule. In a similar fashion, a Type II error curve can be developed that depicts the probability of committing

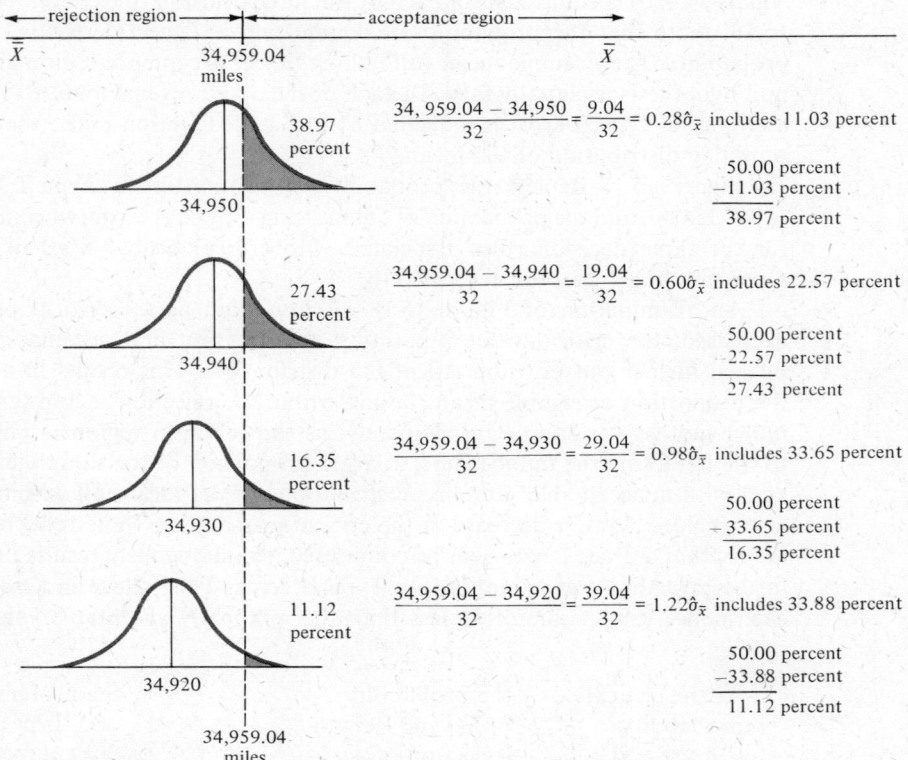

Figure 15.19 Goliath Tires: probability of committing a Type II error. $n = 100$;
$\hat{\sigma}_x = 320$; $\hat{\sigma}_{\bar{x}} = 32$. Decision rule based on a 10 percent critical probability:

accept if $\bar{X} \geq 34,959.04$ miles
reject if $\bar{X} < 34,959.04$ miles

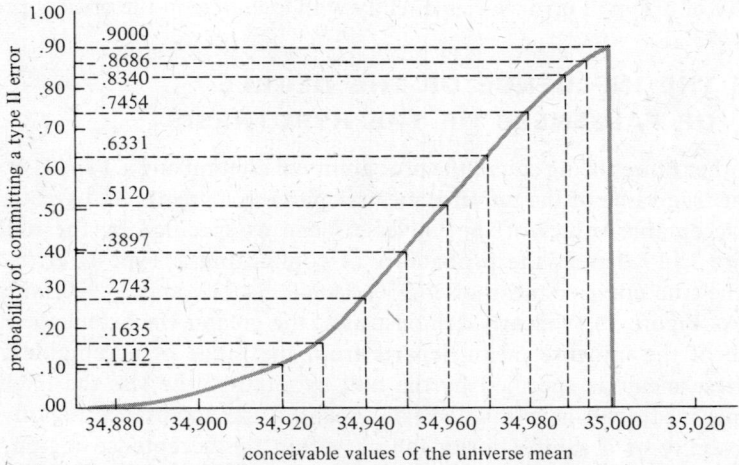

Figure 15.20 Goliath Tires: probability of committing a Type II error
for all conceivable values of the universe mean. Sample size = 100;
critical probability = 10 percent. For the decision rule:

accept if $\bar{X} \geq 34,959.04$ miles
reject if $\bar{X} < 34,959.04$ miles

a Type II error in association with each conceivable value of the universe mean.
A Type II error curve based on computations similar to those presented in Figure
15.19 is presented in Figure 15.20.

An examination of Figure 15.20 discloses that the maximum probability of
committing a Type II error is associated with the value lying just outside the range
of acceptable values of the universe mean (i.e., $\mu_x = 34,999.9999$). The maximum
probability of committing a Type II error is always equal to 1.0 − critical probability). In Figure 15.20, this would be equal to 1.0 − 0.10 or 90 percent. As the
true value of the universe mean departs to a greater and greater extent from the
range of acceptable values of the universe mean, the probability of committing a
Type II error declines. The probability of committing a Type II error is zero, of
course, for all values of the universe mean consistent with the null hypothesis
being true. It is impossible to commit a Type II error when the null hypothesis is
true.

The Type I error curve and the Type II error curve associated with any given
decision rule can be combined in a single diagram. Such a diagram is referred to as
an *error curve* and depicts the probability that a given decision rule will lead to an
error in association with each conceivable value of the universe mean. Whether a
given error is a Type I or Type II depends on whether the value of the universe
mean corresponds to one of the values required to satisfy the null hypothesis
(Type I) or one of the values corresponding to the alternative hypothesis (Type II).
The solid lines in Figure 15.21 display the error curve composed of the Type I
error curve of Figure 15.14 and the Type II error curve of Figure 15.20.

A set of diagrams summarizing the error characteristics of alternative decision

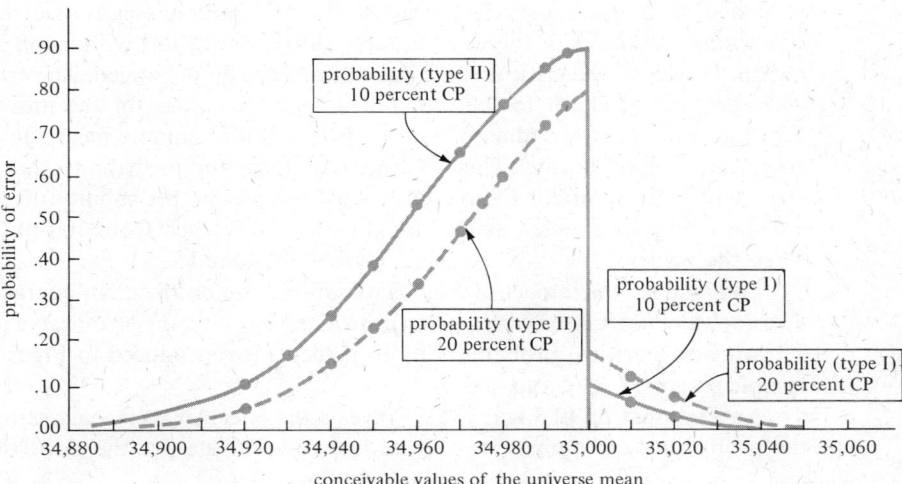

Figure 15.21 Goliath Tires: probability of committing an error (Type I or Type II) for
all conceivable values of the universe mean.

rules can provide a useful input in deciding which decision rule is most appropriate in a given situation. The dotted line in Figure 15.21 displays the error characteristics of the decision rule associated with a critical probability of 20 percent.[5] Notice how an increase in the critical probability is accompanied by an increase in the likelihood of a Type I error in association with each of the values of the universe mean in the acceptable range and lowers the probability of committing a Type II error in association with each of the values of the universe mean in the unacceptable range. This trade-off (a decrease in the probability of committing a Type II error obtained at the cost of an increase in the probability of committing a Type I error) is one of the factors the analyst must consider in selecting the critical probability.

15.9 THE INFLUENCE OF SAMPLE SIZE

Another parameter of the decision rule that exerts considerable influence over the probability of committing a Type II error is the sample size. By increasing the sample size, the decision maker can reduce the probability of committing a Type II error in association with each possible value of the universe mean to whatever level is desired. The increase in sample size accomplishes this through a twofold effect.

1. One impact of an increase in sample size is to cause the critical value on the number line separating the acceptance and rejection regions to be located closer to the range of hypothesized values of the universe mean. This reduction in the distance one must proceed in order to reach the critical value(s) is a reflection of the closer clustering of sample means around the mean of the universe from which they have been selected. To illustrate this effect, the critical value separating the rejection and acceptance regions in Illustration 2 (Goliath Tires) is computed in Table 15.4 for sample size 100, sample size 256, and sample size 400. The critical probability is held constant at 10 percent to isolate the effect of increasing sample size.

2. A second effect of increased sample size is to cause closer clustering of the sample means associated with universe means in the unacceptable range. As a result, given that the critical value separating the acceptance and rejection regions remained unchanged, a smaller proportion of the sample means associated with a given universe mean in the unacceptable range would fall within the acceptance region. However, the critical value has not remained unchanged. The critical value will have moved closer to the range of acceptable values for the universe mean. Thus, an even greater reduction in the proportion of sample means falling in the acceptance region occurs. This, of course, reduces the probability that a Type II error will be committed. Figure 15.22 depicts how the probability of committing a Type II error decreases as the sample size is increased from 100 to 256 to 400 when the value of the universe mean equals 34,940 miles.

To illustrate the impact of increasing sample size on the error characteristics of decision rules associated with a given critical probability, the error curve for Goliath Tires (given a critical probability of 10 percent) is reproduced in Figure 15.23 for sample sizes 100, 256, and 400.

An examination of Figure 15.23 reveals the effect of increasing sample size on the probability of committing a Type II error. The steeper the slope of the error

---

[5] The shape of the error curve for two-tailed tests differs with respect to pattern but is similar in interpretation. For an illustration, see Problem 15.7.

Table 15.4 Goliath Tires: computation of the critical value associated with a 10 percent critical probability

Sample standard deviation = 320 miles
Null hypothesis: $\mu_x \geq 35,000$ miles

Sample size = 100

$$\text{Critical value} = \mu_x - 1.28\hat{\sigma}_{\bar{x}} \qquad \hat{\sigma}_{\bar{x}} = \frac{\hat{\sigma}_x}{\sqrt{n}}$$

$$= 35,000 - 1.28(32) \qquad = \frac{320}{\sqrt{100}}$$

$$= 35,000 - 40.96 \qquad = \frac{320}{10}$$

$$= 34,959.04 \text{ miles} \qquad = 32 \text{ miles}$$

Sample size = 256

$$\text{Critical value} = \mu_x - 1.28\hat{\sigma}_{\bar{x}} \qquad \hat{\sigma}_{\bar{x}} = \frac{\hat{\sigma}_x}{\sqrt{n}}$$

$$= 35,000 - 1.28(20) \qquad = \frac{320}{\sqrt{256}}$$

$$= 35,000 - 25.6 \qquad = \frac{320}{16}$$

$$= 34,974.4 \text{ miles} \qquad = 20 \text{ miles}$$

Sample size = 400

$$\text{Critical value} = \mu_x = 1.28\hat{\sigma}_{\bar{x}} \qquad \hat{\sigma}_{\bar{x}} = \frac{\hat{\sigma}_x}{\sqrt{n}}$$

$$= 35,000 - 1.28(16) \qquad = \frac{320}{\sqrt{400}}$$

$$= 35,000 - 20.48 \qquad = \frac{320}{20}$$

$$= 34,979.52 \text{ miles} \qquad = 16 \text{ miles}$$

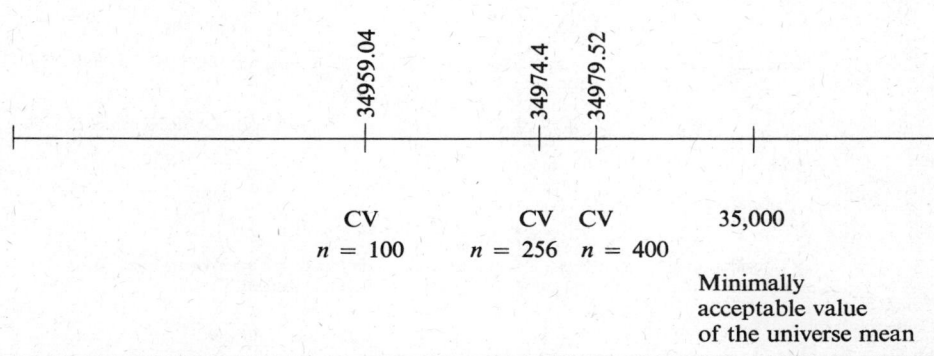

curve, the more rapidly the probability of committing a Type II error declines as the true value of the universe mean departs farther and farther from the range of acceptable values of the universe mean as specified by the null hypothesis. The greater the extent to which the value of the true universe mean departs from the acceptable region, the more serious the consequences associated with committing a Type II error. An inspection of Figure 15.23 reveals that it is in the range where Type II errors are most serious that the greatest impact of increasing sample size is felt. In addition, for one-tailed tests, note that an increase in sample size also causes the probability of committing a Type I error to decline more rapidly as the true value of the universe mean penetrates the acceptable range to a greater degree. The probability of committing a Type I error in a two-tailed test is unaffected by a change in sample size, of course, because there is only one value of the universe mean that satisfies the null hypothesis in a two-tailed test.

Every decision rule has a different set of error characteristics. The decision maker must evaluate the probabilities of committing Type I and Type II errors together with the seriousness of the consequences associated with committing

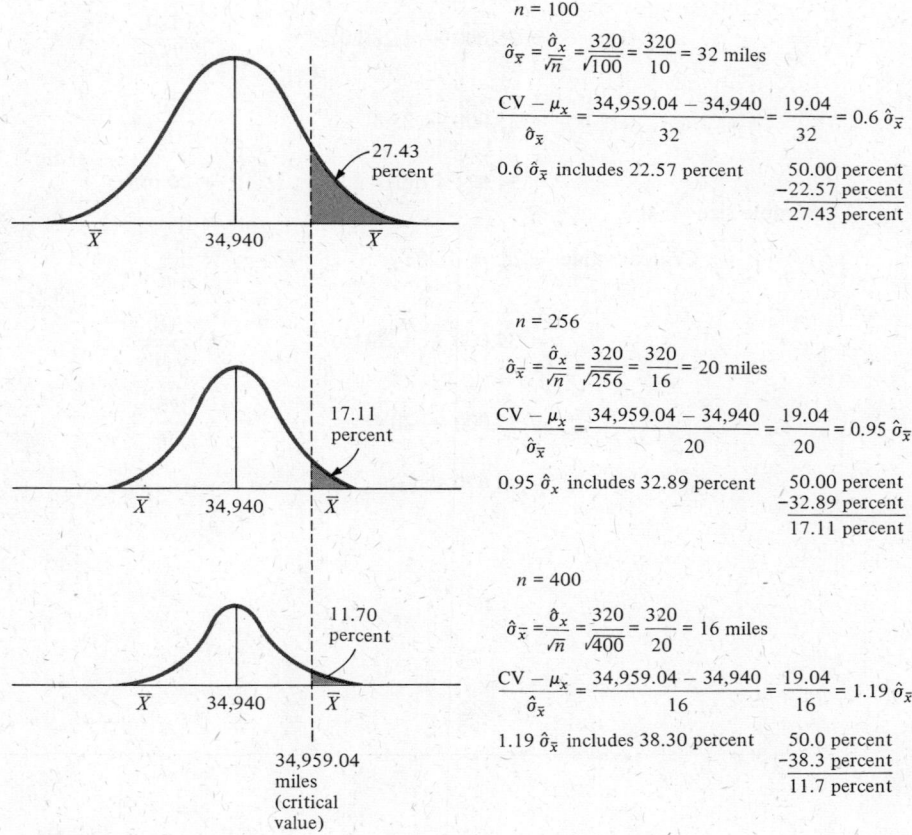

$n = 100$

$$\hat{\sigma}_{\bar{x}} = \frac{\hat{\sigma}_x}{\sqrt{n}} = \frac{320}{\sqrt{100}} = \frac{320}{10} = 32 \text{ miles}$$

$$\frac{CV - \mu_x}{\hat{\sigma}_{\bar{x}}} = \frac{34{,}959.04 - 34{,}940}{32} = \frac{19.04}{32} = 0.6\ \hat{\sigma}_{\bar{x}}$$

$0.6\ \hat{\sigma}_{\bar{x}}$ includes 22.57 percent   50.00 percent
                                                  -22.57 percent
                                                  27.43 percent

$n = 256$

$$\hat{\sigma}_{\bar{x}} = \frac{\hat{\sigma}_x}{\sqrt{n}} = \frac{320}{\sqrt{256}} = \frac{320}{16} = 20 \text{ miles}$$

$$\frac{CV - \mu_x}{\hat{\sigma}_{\bar{x}}} = \frac{34{,}959.04 - 34{,}940}{20} = \frac{19.04}{20} = 0.95\ \hat{\sigma}_{\bar{x}}$$

$0.95\ \hat{\sigma}_x$ includes 32.89 percent   50.00 percent
                                                  -32.89 percent
                                                  17.11 percent

$n = 400$

$$\hat{\sigma}_{\bar{x}} = \frac{\hat{\sigma}_x}{\sqrt{n}} = \frac{320}{\sqrt{400}} = \frac{320}{20} = 16 \text{ miles}$$

$$\frac{CV - \mu_x}{\hat{\sigma}_{\bar{x}}} = \frac{34{,}959.04 - 34{,}940}{16} = \frac{19.04}{16} = 1.19\ \hat{\sigma}_{\bar{x}}$$

$1.19\ \hat{\sigma}_{\bar{x}}$ includes 38.30 percent   50.0 percent
                                                    -38.3 percent
                                                    11.7 percent

(a) Considering only the increased clustering in the sampling distribution of the mean associated with $\mu_x = 34{,}940$ miles and neglecting the change in the critical value (CV).

(a)

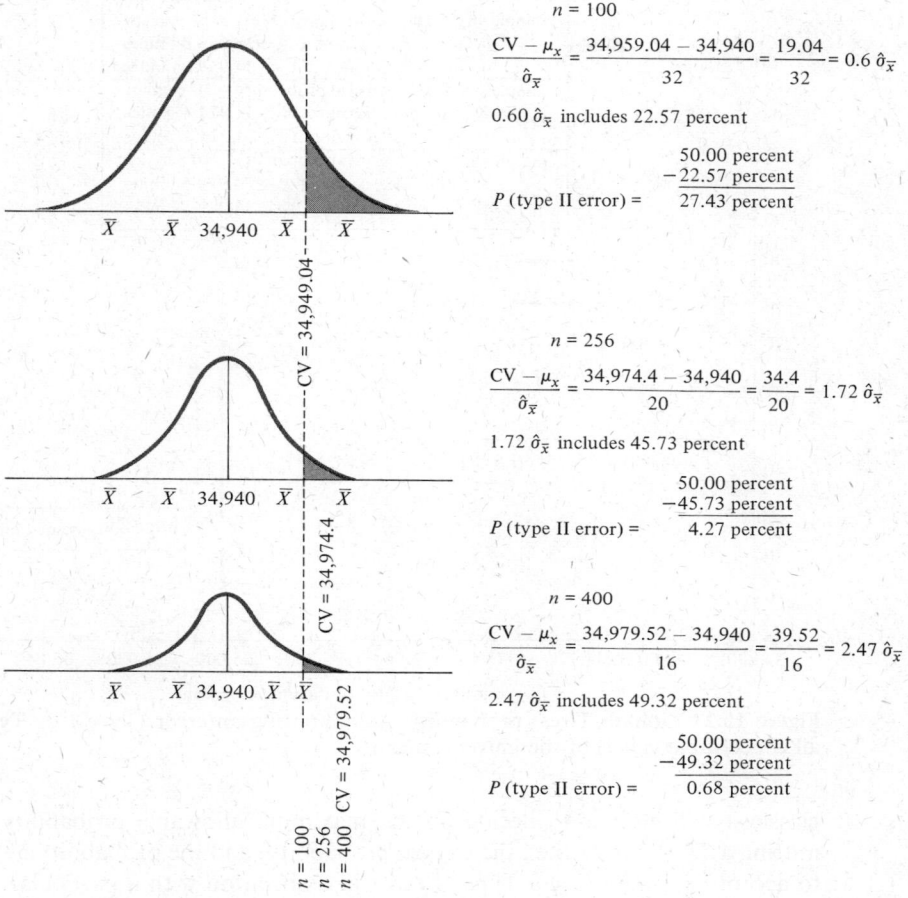

$$n = 100$$

$$\frac{CV - \mu_x}{\hat{\sigma}_{\bar{x}}} = \frac{34,959.04 - 34,940}{32} = \frac{19.04}{32} = 0.6\,\hat{\sigma}_{\bar{x}}$$

$0.60\,\hat{\sigma}_{\bar{x}}$ includes 22.57 percent

	50.00 percent
	−22.57 percent
P (type II error) =	27.43 percent

$$n = 256$$

$$\frac{CV - \mu_x}{\hat{\sigma}_{\bar{x}}} = \frac{34,974.4 - 34,940}{20} = \frac{34.4}{20} = 1.72\,\hat{\sigma}_{\bar{x}}$$

$1.72\,\hat{\sigma}_{\bar{x}}$ includes 45.73 percent

	50.00 percent
	−45.73 percent
P (type II error) =	4.27 percent

$$n = 400$$

$$\frac{CV - \mu_x}{\hat{\sigma}_{\bar{x}}} = \frac{34,979.52 - 34,940}{16} = \frac{39.52}{16} = 2.47\,\hat{\sigma}_{\bar{x}}$$

$2.47\,\hat{\sigma}_{\bar{x}}$ includes 49.32 percent

	50.00 percent
	−49.32 percent
P (type II error) =	0.68 percent

(b) Incorporating the effect of the change in the critical value (see Table 15-4 for the computation of the critical values).

(b)

Figure 15.22 Goliath Tires: probability of committing a Type II error associated with a 10 percent critical probability for the universe mean value = 34,940 miles for sample sizes 100, 256, and 400. (a) Considering only the increased clustering in the sampling distribution of the mean associated with μ_x = 34,940 miles and neglecting the change in the critical value (CV). (b) Incorporating the effect of the change in the critical value (see Table 15.4 for the computation of the critical values).

errors of each of these types.[6] A computer program can be developed that evaluates the above characteristics for various critical probabilities and sample sizes. Given an appropriate decision criterion, the computer program can select from among these alternatives the optimal decision rule. More generally, the approach the

---

[6] Some analysts prefer to summarize the characteristics of a decision rule in an operating characteristic curve and a power curve rather than with an error curve. The information conveyed is approximately the same. Operating characteristic curves and power curves are discussed briefly in Technical Note 10 at the end of the chapter.

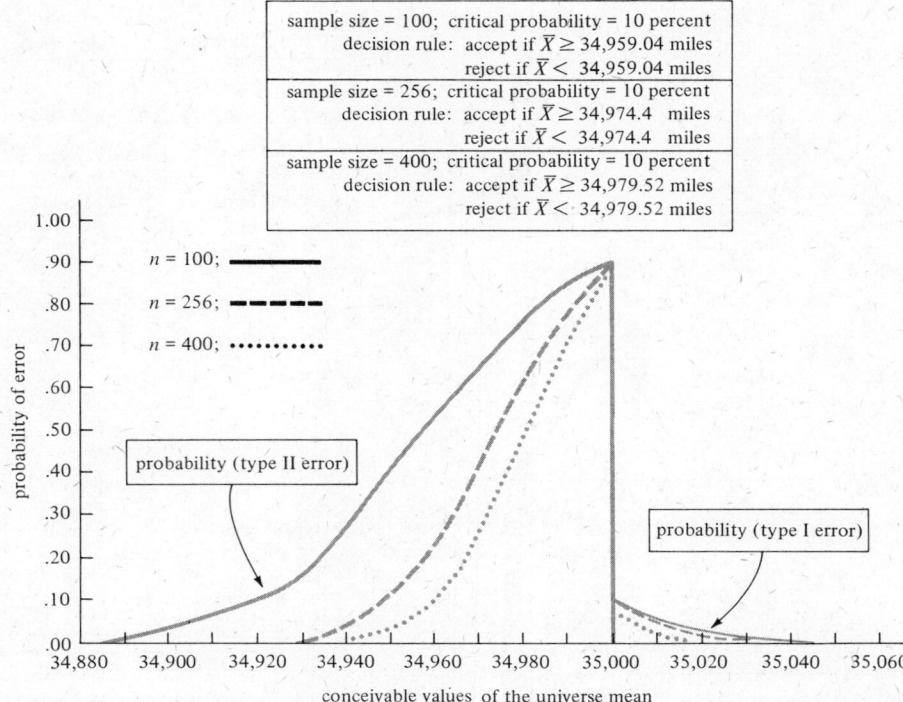

Figure 15.23 Goliath Tires: probability of committing an error (Type I or Type II) for all conceivable values of the universe mean.

classicist will elect is to decide on the maximum allowable probability of committing a Type I error (i.e., the critical probability) and the probability he is willing to accept of committing a Type II error in association with a particular value (or values) of the universe mean associated with the alternate hypothesis. With this information (and an estimate of the universe standard deviation), the analyst can determine the appropriate sample size to select and the optimal decision rule to employ. The procedure for determination of the optimal sample size to employ when testing a hypothesis (classical approach) is the subject of Chapter 16. The Bayesian approach to the determination of optimal sample size was presented in Chapter 9. Both the classicist and the Bayesian utilize the error-curve concept. Two major differences in approach should be noted:

1. The Bayesian explicitly introduces a prior estimating distribution reflecting the relative likelihood that each of the possible values of the universe parameter (and the associated probability of committing an error as reflected by the error curve) will be observed.
2. In addition, an explicit loss function is introduced that reflects the loss that would be incurred given the occurrence of a particular value of the universe parameter.

The optimum decision rule is the rule that minimizes the net expected loss. After the classical approach is examined in Chapter 16, the two approaches will be contrasted and compared in greater depth.

Table 15.5 Summary of factors influencing the probability of committing Type I and Type II errors

Factor	Nature of the influence on the probability of committing a Type I error	Nature of the influence on the probability of committing a Type II error
An increase in the critical probability (under control of the analyst)	Increases the probability of committing a Type I error	Decreases the probability of committing a Type II error
A decrease in the critical probability (under control of the analyst)	Decreases the probability of committing a Type I error	Increases the probability of committing a Type II error
An increase in sample size (under control of the analyst)	The maximum probability of committing a Type I error remains unchanged (equal to the critical probability)	The maximum probability of committing a Type II error remains unchanged; it will equal 1.0 − critical probability
	Given a one-tailed test, the probability of committing a Type I error will be decreased for all values of the universe mean lying within the acceptable range with the exception of the minimally acceptable value of the universe mean .	The probability of committing a Type II error will be decreased for all values of the universe mean that are considered unacceptable according to the null hypothesis when conducting a one-tailed test
Location of the true value of the universe mean (this factor is not under control of the analyst)	A Type I error can occur only when the universe mean falls within the range of values specified by the *null* hypothesis	A Type II error can occur only when the universe mean falls within the range of values specified by the *alternate* hypothesis
	The probability of committing a Type I error declines as the true value of the universe mean penetrates farther into the range of values of the universe mean that are considered as acceptable by the null hypothesis	The probability of committing a Type II error declines as the true value of the universe mean departs from the range of acceptable values to a greater and greater degree

TECHNICAL NOTE 10

Operating characteristic curves and power curves

The Operating Characteristic Curve

The operating characteristic curve (or OC curve) is a curve displaying the probability that the null hypothesis will be accepted in association with each conceivable value of the universe mean. For those values of the universe mean that are in the acceptance range (as specified by the null hypothesis), the OC curve shows the probability that the decision rule will lead to a correct decision (acceptance of the null hypothesis). For those values of the universe mean falling in the unacceptable range (as implied by the null and as specified by the alternate hypothesis), the OC curve shows the probability that the decision rule will lead to an incorrect decision. An incorrect decision in this range is to accept the null hypothesis and is referred to as a Type II error. There is, of course, a different OC curve associated with each conceivable decision rule. A different decision rule is generated whenever there is a change in sample size, a change in the critical probability, or a change in the division of the critical probability between the two tails of the distribution. The OC curve for Goliath Tires associated with a 10 percent critical probability and a sample size of 100 is reproduced as Figure 15.24. Note that for the range of values of the universe mean specified as unacceptable by the null hypothesis, the OC curve in Figure 15.24 is identical to the error curve as displayed in Figure 15.23.

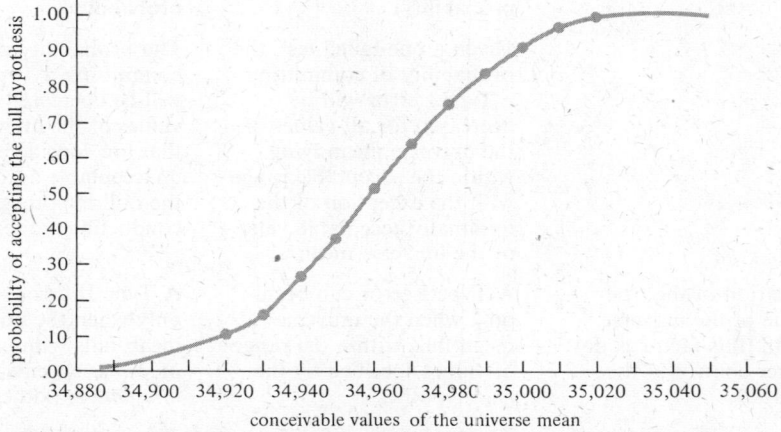

Figure 15.24 Goliath Tires: operating characteristic curve. Sample size = 100; critical probability = 10 percent.

The Power Curve

The power curve is a curve displaying the probability that the null hypothesis will be *rejected* in association with each conceivable value of the universe mean. For those values of the universe mean that are in the unacceptable range, the power curve shows the probability that the decision rule will lead to a correct decision (rejection of the null hypothesis). For those values of the universe mean that are in the acceptable range, the power curve shows the probability that the decision rule will lead to an incorrect decision. An incorrect decision in this range is to reject the null hypothesis and is referred to as a Type I error. The power curve is the mirror image of the operating characteristic curve associated with the same decision rule. The power curve for Goliath Tires associated with a 10 percent critical probability and a sample size of 100 is reproduced as Figure 15.25. Note that for the range of values of the universe mean specified as acceptable by the null hypothesis, the power curve in Figure 15.25 is identical to the error curve as displayed in Figure 15.23. Thus, the same sort of information about the error characteristics of a decision rule is displayed by an error curve such as Figure 15.23 or by the combination of an operating characteristic curve and a power curve. The approach used is up to the discretion of the decision maker. The position of the author is a preference for the error curve.

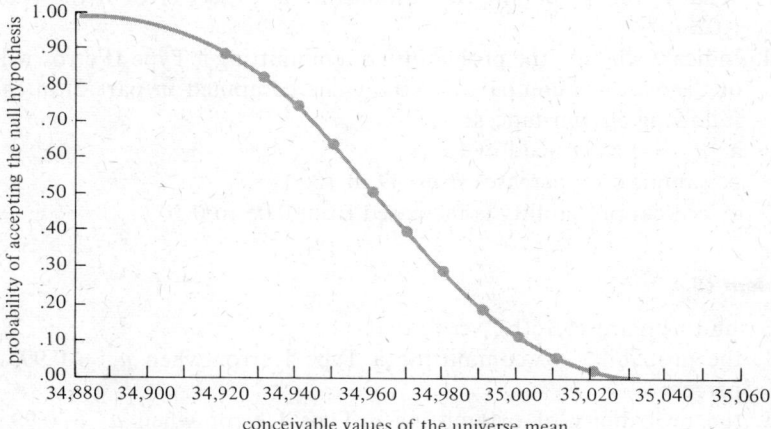

Figure 15.25 Goliath Tires: power curve. Sample size = 100; critical probability = 10 percent.

PROBLEMS

Problem 15.1

1. Distinguish between Type I and Type II errors.
2. The probability of committing a Type II error is a function of three main factors. What are they? Illustrate the nature of the influence each exerts on the probability that a Type II error will be committed.

Problem 15.2

When in correct adjustment, a machine produces parts that have a mean diameter of 0.058 in. with a standard deviation of 0.010 in. A part will not be serviceable if its diameter exceeds 0.088 in. or falls below 0.028 in. The manufacturer is thus equally concerned with detecting shifts in the mean of the distribution in either direction.

1. Devise the appropriate *null hypothesis* and *decision rule* for verifying if the machine is in perfect adjustment for a sample size of 100 and a critical probability of 0.05.
2. Suppose a sample of size 100 has been selected and its mean is found to be 0.05732 in. Should the null hypothesis be accepted or rejected? Why?
3. Can the decision rule formulated in part 1 be used in association with samples of size 50? Explain fully and carefully why or why not.

Problem 15.3

Healthy Cola has claimed in its advertising that 16-oz bottles of their beverage contain, on the average, less than 1 oz of cyclamates and that the production process is so designed that 99.73 percent of the bottles produced will contain an amount of cyclamates no more than 1.3 oz nor less than 0.7 oz. ($\sigma_x = 0.1$ oz). The Federal Trade Commission takes a sample of size 49 to test the validity of Healthy Cola's claim.

1. Design the appropriate decision rule using a critical probability of 5 percent.
2. What is the probability of committing a Type II error if the true value of the universe mean is 1.028 oz.?
3. What is the probability of committing a Type I error if the true mean is 1.028 oz?
4. Indicate whether the probability of committing a Type II error will increase or decrease (as compared to the value computed in part 2) in each of the following circumstances:
   a. $\mu_x = 1.01$ instead of 1.028
   b. sample size increases from 49 to 100
   c. critical probability is increased from 0.05 to 0.10

Problem 15.4

Referring to Figure 15.15(b), verify that:
1. the probability of committing a Type I error when $\mu_x = 0.995$ oz is 5.7 percent
2. the probability of committing a Type I error when $\mu_x = 0.99$ oz is 4.1 percent
3. the probability of committing a Type I error when $\mu_x = 0.98$ oz is 1.9 percent
4. the probability of committing a Type I error when $\mu_x = 0.97$ oz is 0.8 percent

Problem 15.5

Referring to Figure 15.20 in the text, verify that the probability of committing a Type II error is
1. 86.86 percent when $\mu_x = 34,995$ miles
2. 83.40 percent when $\mu_x = 34,990$ miles
3. 74.54 percent when $\mu_x = 34,980$ miles
4. 63.31 percent when $\mu_x = 34,970$ miles
5. 51.20 percent when $\mu_x = 34,960$ miles

Problem 15.6

Is it possible to commit a Type I error and a Type II error at the same time? Discuss.

Problem 15.7

1. Construct the error curve (reflecting the probability of committing both a Type I and a Type II error) that describes the error pattern associated with the decision rule generated on p. 374. The decision rule is reproduced below.
   Take a sample of size 9.
   If $2.4994555 \le \overline{X} \le 2.5005445$ in., accept the null hypothesis that $\mu_x = 2.5$ in.
   If $\overline{X} < 2.4994555$ or > 2.5005445 in., reject the null hypothesis.
   Let the vertical axis run from 0.00 to 1.00 in increments of 0.10. Let the horizontal axis extend from 2.4980 in. to 2.5020 in. in increments of 0.0002 in. To draw the curve, plot the probability of committing an error (either

Type I or Type II, whichever is appropriate) in association with each of the following possible values for the universe mean.

2.4985	2.50001
2.4990	2.5001
2.4991	2.5003
2.4992	2.5004
2.4994	2.5006
2.4996	2.5008
2.4997	2.5009
2.4999	2.5010
2.49999	2.5015
2.50000	

2. Discuss the differences and similarities of error curves describing one-tailed tests and error curves describing two-tailed tests.

Problem 15.8

The FTC desires to conduct a test in an attempt to ascertain the validity of the advertising claim of Crunchy Chocolate Bars that the average weight in ounces of their chocolate bars is 8 oz. A sample of size 64 generated a sample standard deviation $[\hat{\sigma}_x = \sqrt{\sum (X - \overline{X})^2/(n - 1)}]$ of 0.3 oz. A 5 percent critical probability is to be employed.

1. State the null hypothesis.
2. Generate the appropriate decision rule to test the null hypothesis stated in part 1.

Problem 15.9

1. Referring to Problem 15.8, construct a Type II error curve. Let the vertical axis run from 0.00 to 1.00 in increments of 0.10. Let the horizontal axis extend from 7.85 to 8.00 in increments of 0.01. Problem 15.8 provides the point associated with $\mu_x = 7.999$ oz. To draw the curve, plot the points corresponding to $\mu_x = 7.89$, $\mu_x = 7.999$, $\mu_x = 7.93$, and $\mu_x = 7.96$.
2. Roughly sketch in the effect on the Type II error curve of an increase in sample size. (No computations are necessary—simply indicate the rough impact with a dotted line.)

Problem 15.10

An organizational design consultant recently conducted an experiment at a large factory in the Southwest. He randomly selected a group of 25 workers and allowed them to work *without formal supervision* for 3 months. The mean productivity per man for that period was 5000 units with a standard deviation of 500 units. (Assume that the universe is normally distributed.) The mean productivity per man for a typical 3-month period for the entire factory under the *formal supervision* setup is 4700 units. Using a critical probability of 2.5 percent, test the null hypothesis that the mean productivity of the factory under the proposed new setup (no formal supervision) would be no higher than under the formal supervision setup (i.e., null hypothesis: $\mu_x \leq 4700$).

Problem 15.11

Given:
    a. sample size = 100
    b. critical probability = 10 percent
    c. null hypothesis: $\mu_x \geq 35{,}000$ miles
    d. decision rule:
        Accept if $\overline{X} \geq 34{,}959.04$ miles
        Reject if $\overline{X} < 34{,}959.04$ miles
    e. the probability of committing a Type II error when $\mu_x = 34{,}940$ miles is 0.2743

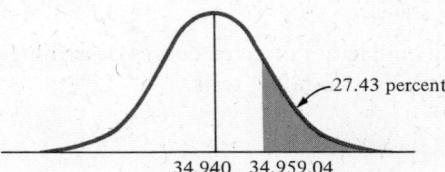

27.43 percent

34,940 34,959.04

Required:
    1. If the critical probability were to be changed to 5 percent, would the probability of committing a Type II error when $\mu_x = 34{,}940$ be increased, decreased, or remain the same? Justify your answer (intuitively, not mathematically).
    2. If the sample size were to be increased from 100 to 400, would the probability that a Type II error would be committed when $\mu_x = 34{,}940$ be increased, decreased, or remain the same? Justify (intuitively, not mathematically).

Problem 15.12

Given the decision rule:
    Take a sample of size 100.
    If $\overline{X} \geq 82$, accept the null hypothesis that $\mu_x \geq 85$
    If $\overline{X} < 82$, reject the null hypothesis
and

$$\hat{\sigma}_{\bar{x}} = \frac{\hat{\sigma}_x}{\sqrt{n}} = \frac{8}{\sqrt{100}} = \frac{8}{10} = 0.8$$

    1. What is the probability that a Type II error will be committed when the true value of $\mu_x = 81.5$?
    2. What is the probability that a Type I error will be committed when the true value of $\mu_x = 83$?

Problem 15.13

When in correct adjustment, a filling mechanism is set to fill containers with a mean fill of 14 oz. Variability in the fill per container is characterized by a standard deviation of 0.3 oz. The manufacturer is equally concerned with determining if the process mean has shifted in either direction.
    1. Why would the manufacturer be concerned if the process mean shifted to a value less than 14 oz.? Why would he be concerned if the process mean shifted to a value greater than 14 oz?

2. Devise the appropriate *null hypothesis* and *decision rule* for verifying if the filling mechanism is in perfect adjustment for a sample size of 36 and a critical probability of 0.02.
3. What is the probability of committing a Type II error when $\mu_x = 13.9$?
4. What is the probability of committing a Type II error when $\mu_x = 13.8$?
5. What is the probability of committing a Type II error when $\mu_x = 14.2$?
6. The possibility of committing a Type I error is associated with what values of the universe mean?
7. Suppose that a sample of size 36 is selected and a sample mean equal to 13.97 observed. Should the null hypothesis be accepted or rejected? Justify your answer.

Problem 15.14

A cooperative work program has claimed that the mean weekly wage on job placements is $110. You desire to test this claim. A critical probability of 5 percent is to be employed. You are concerned only with detecting if the true wage is less than $110.
1. State the null hypothesis.
2. A sample of 49 students has been selected. The sample mean is $108.40 and the sample standard deviation is equal to $14. Should the null hypothesis be accepted or rejected? Justify your answer.

Problem 15.15

An appliance store offers a service contract on appliances that it sells. The service contract costs $20 for a 2-year period. The store manager claims that average repairs on the appliance you have purchased are at least $32 for the 2-year period (*for those appliances requiring a service call*). The manager allows you to select the records for 36 accounts from a group of accounts that have just experienced their second birthday and that have required service of some kind. The sample mean is $30.10 and the standard deviation is equal to $12.
1. State the null hypothesis.
2. Employing a critical probability of 10 percent, determine if the null hypothesis should be accepted or rejected. Justify your answer.
3. The store manager has indicated that 40 percent of the appliances sold require no service over the 2-year period. What is the probability that your 2-year repair bill will be less than $25, given that you accept the hypothesis that repair expenditures for the other 60 percent of the appliances will be normally distributed with a mean equal to $32 and a standard deviation equal to $12?

Problem 15.16

A manufacturer of printing presses employs an air-driven tool for tightening bolts. The tool is designed to apply 180 ft-lb of pressure. If too much pressure is applied, the bolt may break. If too little pressure is applied, vibration will cause excessive wear. The manufacturer is equally concerned with avoiding both situations. The tool is to be monitored periodically to check that it is operating according to specification. A critical probability of 1 percent is to be employed.
1. State the null hypothesis.
2. State the alternate hypothesis.

3. A sample of size 36 is selected. The sample mean is determined to be 178.6 ft-lb and the sample standard deviation is equal to 6 ft-lb of pressure. Should the null hypothesis be accepted or rejected? Justify your answer.

Problem 15.17

The U.S. Post Office is in the process of replacing a number of older vehicles in their fleet. In order to be considered, new vehicles must average at least 14 miles per gallon.
   1. State the null hypothesis.
   2. A sample of 49 vehicles has been selected of a model currently under investigation. The sample mean is 13.82 mpg and the standard deviation is equal to 0.21 mpg. A 10 percent critical probability is to be employed. Should the null hypothesis be accepted or rejected? Justify your answer.

Problem 15.18

One of the major gasoline companies has become increasingly concerned with the magnitude of the average dollar balance in delinquent accounts. One executive claims that the figure is over $180. You desire to test this claim. A critical probability of 10 percent is to be employed.
   1. State the null hypothesis.
   2. A sample of 121 delinquent accounts is selected. The sample mean is $172 and the sample standard deviation ($\hat{\sigma}_x$) is $34.10. Should the null hypothesis be accepted or rejected? Justify your answer.

TESTS OF HYPOTHESES: DETERMINATION OF OPTIMAL SAMPLING SIZE— A CLASSICAL APPROACH

16

16.1 DETERMINATION OF APPROPRIATE SAMPLE SIZE

The Uniform Gear Works illustration introduced in Chapter 15 will be used to illustrate the procedure employed to determine the optimal sample size when testing hypotheses. Recall that the diameters of gears produced, when the process is operating as designed, are normally distributed with a mean equal to 2.5 in. and a standard deviation of 0.001 in. If the process slips a setting such that the average diameter of gears produced is either greater or less than 2.5 in., the proportion of unacceptable gears produced will exceed allowable proportions. The firm plans to select samples of process output periodically in an effort to detect shifts in machine settings as early as possible.

The null hypothesis is $\mu_x = 2.5$ in. The error characteristics of the decision rule to be employed are described as follows:

1. No more than a 10 percent probability of committing a Type I error.
2. No more than a 10 percent probability of committing a Type II error when $\mu_x = 2.499$ in.
3. No more than a 10 percent probability of committing a Type II error when $\mu_x = 2.501$ in.

The sample size required to meet the above error restrictions and the specific decision rule to be employed are determined as follows:

The error constraints are depicted in Figure 16.1. Diagramming the error constraints in this fashion facilitates the generation of the inputs needed to determine optimal sample size.

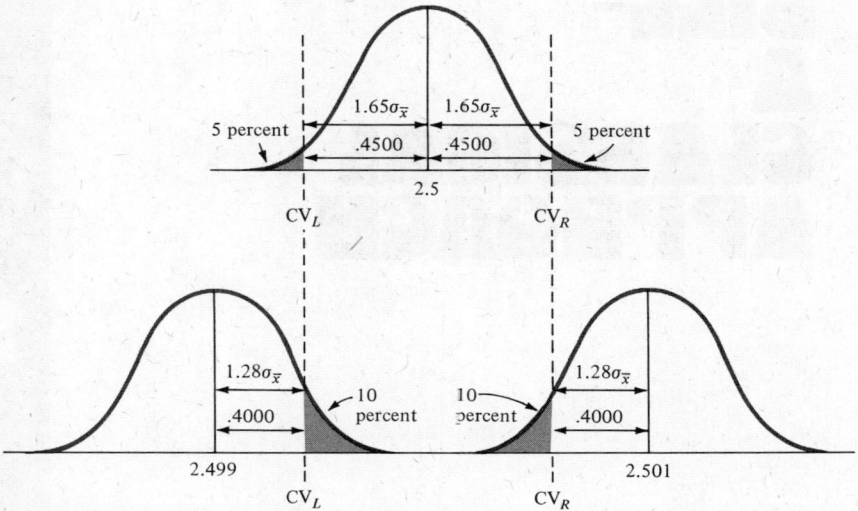

Figure 16.1

Procedure

1. Determine the formula describing the critical value that satisfies the constraint placed on the probability of committing a Type I error. Referring to Figure 16.1, the critical value separating the rejection region in the left-hand segment of the number line from the acceptance region is determined as follows:

$$CV_L = 2.5 - 1.65\sigma_{\bar{x}}$$

$$= 2.5 - 1.65\frac{\sigma_x}{\sqrt{n}}$$

$$= 2.5 - 1.65\left(\frac{0.001}{\sqrt{n}}\right)$$

2. Determine the formula describing the critical value that satisfies the constraint placed on the probability of committing a Type II error. Referring to Figure

16.1, the critical value separating the rejection region in the left-hand segment of the number line from the acceptance region is determined as follows:

$$CV_L = 2.499 + 1.28\sigma_{\bar{x}}$$

$$= 2.499 + 1.28 \frac{\sigma_x}{\sqrt{n}}$$

$$= 2.499 + 1.28 \left(\frac{0.001}{\sqrt{n}}\right)$$

The nature of the constraints placed on the decision rule are such that the same pattern describes the critical value separating the rejection region and acceptance region in the right-hand segment of the number line and therefore need not be repeated here.

3. Set the two alternative descriptions of the critical value in the left-hand segment of the number line equal to each other and solve for n. It is not necessary to do this for the critical value in the right-hand region in this problem because of the symmetry of the constraints—if the constraints place nonsymmetrical demands on the two critical values in a two-tailed test, the procedure must be performed for both critical values and the larger of the two sample sizes used. A problem illustrating this procedure is presented in the next example.

Equating the two formulas describing the constraints placed on the critical value in the left-hand segment of the number line and solving for n:

$$2.5 - 1.65\left(\frac{0.001}{\sqrt{n}}\right) = 2.499 + 1.28\left(\frac{0.001}{\sqrt{n}}\right)$$

$$0.001 - 1.65\left(\frac{0.001}{\sqrt{n}}\right) = 1.28\left(\frac{0.001}{\sqrt{n}}\right)$$

Multiplying both sides by $\sqrt{n}$:

$$0.001\sqrt{n} - (1.65)(0.001) = (1.28)(0.001)$$
$$0.001\sqrt{n} - 0.00165 = 0.00128$$
$$0.001\sqrt{n} = 0.00293$$
$$\sqrt{n} = 2.93$$
$$n = 8.58$$

or a sample size of 9.

4. Generate the decision rule to be employed in association with a sample of size 9.

Critical value in the left-hand segment of the number line

$$CV_L = 2.5 - 1.65\left(\frac{0.001}{\sqrt{9}}\right)$$

$$= 2.5 - 1.65(0.00033)$$
$$= 2.5 - 0.0005445$$
$$= 2.4994555 \text{ in.}$$

Critical value in the right-hand segment of the number line

$$CV_R = 2.5 + 1.65\left(\frac{0.001}{\sqrt{9}}\right)$$

$$= 2.5 + 1.65(0.00033)$$
$$= 2.5 + 0.0005445$$
$$= 2.5005445 \text{ in.}$$

The decision rule associated with the null hypothesis $\mu_x = 2.5$ is:

Take a sample of size 9 and compute its mean:
If $2.4994555 \leq \overline{X} \leq 2.5005445$, accept the null hypothesis.
If $\overline{X} < 2.4994555$ or $\overline{X} > 2.5005445$, reject the null hypothesis.

Two alternative formulations for the above decision rule are occasionally used.

FIRST ALTERNATIVE FORMULATION

The decision rule is expressed in the form of a comparison of two factors: (1) the number of standard errors (or z values) the critical value lies away from the hypothesized value of the universe mean, and (2) the number of standard errors (or z values) the observed value of the sample mean departs from the hypothesized value of the universe mean.

For the above illustration, it is necessary to proceed 1.65 standard errors from the hypothesized value of the mean to reach the rejection regions. Thus, if the sample mean lies within 1.65 standard errors of the hypothesized mean, the null hypothesis should be accepted.

If the sample mean lies more than 1.65 standard errors away from the population mean, the null hypothesis should be rejected.

The first alternative formulation of the decision rule is

Accept the null hypothesis if $z \leq |1.65|$[1]
Reject the null hypothesis if $z > |1.65|$

where z refers to the number of standard errors the sample mean departs from the postulated value of the universe mean.

SECOND ALTERNATIVE FORMULATION

The decision rule is based on a comparison of two factors: (1) the critical probability, that is, the maximum probability of a sampling error occurring due to chance selection in sampling that the decision maker is willing to accept as consistent with the null hypothesis, and (2) the probability that a sample mean could differ from the value of the hypothesized mean by as much as the observed sample mean solely due to sampling error, that is, chance selection in sampling.

For the illustration referred to above, the critical probability was established at 10 percent. Thus, the second alternative formulation for the above decision rule is as follows: If the probability of the observed sample mean deviating from the postulated value of the universe mean due to sampling error is less than 10 percent, reject chance selection in sampling as a reasonable explanation and reject the null hypothesis. If the probability of observing a sampling error that great or more is greater than 10 percent, accept the null hypothesis.

The above three formulations are equivalent representations of the same relationship. Any one of the three lead to the same conclusion. The three alternative formulations expressed in their most efficient form are summarized below:

    I. If $2.4994555 \leq \overline{X} \leq 2.5005445$, accept.
       If $\overline{X} < 2.4994555$ or $\overline{X} > 2.5005445$, reject.
   II. If $z \leq |1.65|$, accept.
      If $z > |1.65|$, reject.
  III. If Prob. (observed sampling error or more) $\geq$ 10 percent, accept.
      If Prob. (observed sampling error or more) $<$ 10 percent, reject.

---

[1] $|1.65|$ is interpreted as the absolute value 1.65.

16.2 DETERMINATION OF APPROPRIATE
SAMPLE SIZE—A SECOND EXAMPLE

A pharmaceutical firm has recently developed a new headache remedy called Detach. The primary ingredient in the drug is a compound referred to as X-L32. If the drug contains too little X-L32, it will be ineffectual. If it contains too much, use of the drug will bring on extreme nausea. The drug is manufactured by a process that supplies an average of 3 cc of X-L32 to a unit dosage. The actual amount of X-L32 in a unit dosage varies around 3 cc in the pattern of a normal curve with a standard deviation of 0.15 cc. The dispersion in the distribution is relatively constant, but the mean has been known to shift due to a slippage in process settings.

Given the null hypothesis $\mu_x = 3$ cc and the following constraints placed on the decision rule to be employed, determine the appropriate sample size and decision rule.

Constraints (see Figure 16.2):

1. No more than a 10 percent critical probability of committing a Type I error. In this case, the manufacturer is much more concerned with avoiding accepting $\mu_x = 3$ cc when the distribution has shifted to the right than when it has shifted to the left. To reflect this concern, the critical probability will be divided up so that 8 percent is allocated to the right tail and 2 percent to the left tail.
2. No more than a 5 percent probability of committing a Type II error when $\mu_x = 2.8$ cc.
3. No more than a 1 percent probability of committing a Type II error when $\mu_x = 3.1$ cc.

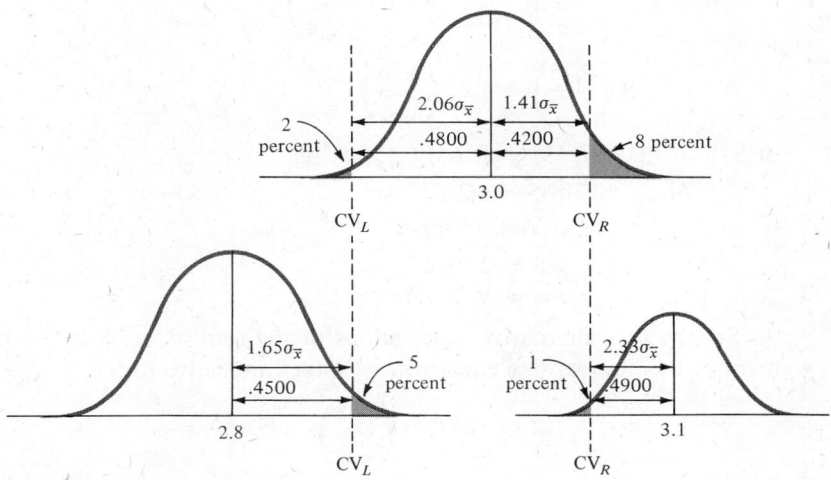

Figure 16.2

Procedure

1. Determine the formulas describing the critical values that satisfy the constraints placed on the probability of committing a Type I error. Refer to Figure 16.2.

$$CV_L = 3.0 - 2.06\sigma_{\bar{x}} \qquad\qquad CV_R = 3.0 + 1.41\sigma_{\bar{x}}$$

$$= 3.0 - 2.06\,\frac{\sigma_x}{\sqrt{n}} \qquad\qquad = 3.0 + 1.41\,\frac{\sigma_x}{\sqrt{n}}$$

$$= 3.0 - 2.06\left(\frac{0.15}{\sqrt{n}}\right) \qquad\qquad = 3.0 + 1.41\left(\frac{0.15}{\sqrt{n}}\right)$$

2. Determine the formulas describing the critical values that satisfy the constraints placed on the probability of committing a Type II error. Referring to Figure 16.2:

$$CV_L = 2.8 + 1.65\sigma_{\bar{x}} \qquad\qquad CV_R = 3.1 - 2.33\sigma_{\bar{x}}$$

$$= 2.8 + 1.65\,\frac{\sigma_x}{\sqrt{n}} \qquad\qquad = 3.1 - 2.33\,\frac{\sigma_x}{\sqrt{n}}$$

$$= 2.8 + 1.65\left(\frac{0.15}{\sqrt{n}}\right) \qquad\qquad = 3.1 - 2.33\left(\frac{0.15}{\sqrt{n}}\right)$$

3. Set the two alternative descriptions of the critical value in the left-hand segment of the number line equal to each other and solve for n.

$$CV_L = 3.0 - 2.06\left(\frac{0.15}{\sqrt{n}}\right)$$

$$CV_L = 2.8 + 1.65\left(\frac{0.15}{\sqrt{n}}\right)$$

$$3.0 - 2.06\left(\frac{0.15}{\sqrt{n}}\right) = 2.8 + 1.65\left(\frac{0.15}{\sqrt{n}}\right)$$

$$0.2 - 2.06\left(\frac{0.15}{\sqrt{n}}\right) = 1.65\left(\frac{0.15}{\sqrt{n}}\right)$$

$$0.2\sqrt{n} - (2.06)(0.15) = (1.65)(0.15)$$

$$0.2\sqrt{n} - 0.3090 = 0.2475$$

$$0.2\sqrt{n} = 0.5565$$

$$\sqrt{n} = 2.7825$$

$$n = 7.74 \quad\text{or}\quad 8$$

4. Set the two alternative descriptions of the critical value in the right-hand segment of the number line equal to each other and solve for n.

$$CV_R = 3.0 + 1.41\left(\frac{0.15}{\sqrt{n}}\right)$$

$$CV_R = 3.1 - 2.33\left(\frac{0.15}{\sqrt{n}}\right)$$

$$3.0 + 1.41\left(\frac{0.15}{\sqrt{n}}\right) = 3.1 - 2.33\left(\frac{0.15}{\sqrt{n}}\right)$$

$$1.41\left(\frac{0.15}{\sqrt{n}}\right) = 0.1 - 2.33\left(\frac{0.15}{\sqrt{n}}\right)$$

$$(1.41)(0.15) = 0.1\sqrt{n} - (2.33)(0.15)$$
$$0.2115 = 0.1\sqrt{n} - 0.3495$$
$$0.5610 = 0.1\sqrt{n}$$
$$5.61 = \sqrt{n}$$

$$n = 31.47 \quad \text{or} \quad 32$$

5. Choose the larger of the sample sizes computed in steps 3 and 4 and use this value (i.e., 32) to compute the values for the critical values.

$$CV_L = 3.0 - 2.06 \left(\frac{0.15}{\sqrt{32}}\right)$$

$$= 3.0 - 2.06 \left(\frac{0.15}{5.66}\right)$$

$$= 3.0 - 2.06(0.0265)$$
$$= 3.0 - 0.0546$$
$$= 2.9454$$

$$CV_R = 3.0 + 1.41 \left(\frac{0.15}{\sqrt{32}}\right)$$

$$= 3.0 + 1.41 \left(\frac{0.15}{5.66}\right)$$

$$= 3.0 + 1.41(0.0265)$$
$$= 3.0 + 0.0374$$
$$= 3.0374$$

6. Formulate the decision rule (the three alternative formulations are presented below, but any one of the three would be sufficient).

Take a sample of size 32.

  I. If $2.9454 \leq \bar{X} \leq 3.0374$, accept the null hypothesis.
    If $\bar{X} < 2.9454$ or $\bar{X} > 3.0374$, reject the null hypothesis.
 II. If $-2.06 \leq z \leq 1.41$, accept.
    If $z < -2.06$ or $z > 1.41$, reject.
III. If Prob. $\begin{pmatrix} \text{observing a sampling error in} \\ \text{a positive direction of the} \\ \text{observed amount or more} \end{pmatrix} \geq 8$ percent, accept.

    If Prob. $\begin{pmatrix} \text{observing a sampling error in} \\ \text{a negative direction of the} \\ \text{observed amount or more} \end{pmatrix} \geq 2$ percent, accept.

    If Prob. $\begin{pmatrix} \text{observing a sampling error in} \\ \text{a positive direction of the} \\ \text{observed amount or more} \end{pmatrix} < 8$ percent, reject.

    If Prob. $\begin{pmatrix} \text{observing a sampling error in} \\ \text{a negative direction of the} \\ \text{observed amount or more} \end{pmatrix} < 2$ percent, reject.

16.3 DETERMINATION OF APPROPRIATE SAMPLE SIZE FOR A ONE-TAILED TEST

The illustration referring to the magazine, *Consumer's Advice*, and its attempt to verify the advertising claim of Goliath Tires that the average life of their new Triple-Ply Tires exceeds 35,000 miles will be used as the "vehicle" to illustrate the approach for one-tailed tests.

Given the null hypothesis $\mu_x \geq 35,000$ miles and the following constraints on the decision rule to be employed, determine the appropriate sample size and decision rule.

Constraints (see Figure 16.3):

1. No more than a 5 percent probability of committing a Type I error.
2. No more than a 10 percent probability of committing a Type II error when $\mu_x = 34,960$ miles.

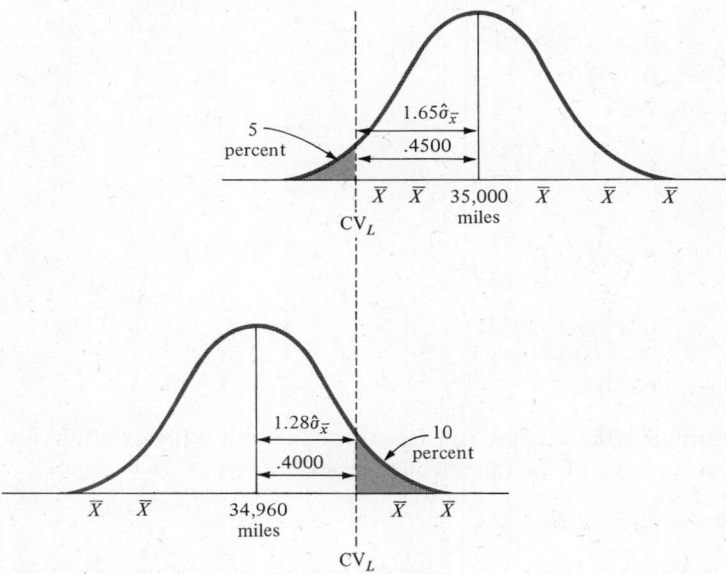

Figure 16.3

A preliminary sample of size 100 has already been taken and generated a sample standard deviation equal to 320 miles $[\hat{\sigma}_x = \sqrt{\sum (X - \overline{X})^2/(n - 1)}]$. (If the standard deviation of the universe is unknown, an estimate of its value must be generated from similar studies, judgment, or a preliminary sample.)

Procedure

1. Determine the formula describing the critical value that satisfies the constraint placed on the probability of committing a Type I error. Referring to Figure 16.3:

$$CV_L = 35,000 - 1.65\hat{\sigma}_{\overline{x}}$$

$$= 35,000 - 1.65 \frac{\hat{\sigma}_x}{\sqrt{n}}$$

$$= 35,000 - 1.65 \left(\frac{320}{\sqrt{n}}\right)$$

2. Determine the formula describing the critical value that satisfies the constraint placed on the probability of committing a Type II error. Referring to Figure 16.3:

$$CV_L = 34{,}960 + 1.28\hat{\sigma}_{\bar{x}}$$

$$= 34{,}960 + 1.28\,\frac{\hat{\sigma}_x}{\sqrt{n}}$$

$$= 34{,}960 + 1.28\left(\frac{320}{\sqrt{n}}\right)$$

3. Set the two alternative descriptions of the critical value equal to each other and solve for n.

$$CV_L = 35{,}000 - 1.65\left(\frac{320}{\sqrt{n}}\right)$$

$$CV_L = 34{,}960 + 1.28\left(\frac{320}{\sqrt{n}}\right)$$

$$35{,}000 - 1.65\left(\frac{320}{\sqrt{n}}\right) = 34{,}960 + 1.28\left(\frac{320}{\sqrt{n}}\right)$$

$$40 - 1.65\left(\frac{320}{\sqrt{n}}\right) = 1.28\left(\frac{320}{\sqrt{n}}\right)$$

$$40\sqrt{n} - (1.65)(320) = (1.28)(320)$$
$$40\sqrt{n} - 528 = 409.60$$
$$40\sqrt{n} = 937.60$$
$$\sqrt{n} = 23.44$$

$$n = 549.43 \quad \text{or} \quad 550$$

4. Formulate the decision rule.

$$CV_L = 35{,}000 - 1.65\left(\frac{320}{\sqrt{550}}\right)$$

$$= 35{,}000 - 1.65\left(\frac{320}{23.45}\right)$$

$$= 35{,}000 - 1.65(13.65)$$
$$= 35{,}000 - 22.52$$
$$= 34{,}977.48 \text{ miles}$$

Any of the following three formats would be appropriate:

Take a sample of size 550.

  I. If $\bar{X} \geq 34{,}977.48$ miles, accept the null hypothesis.
    If $\bar{X} < 34{,}977.48$ miles, reject the null hypothesis.
 II. If $z \geq -1.65$, accept.
    If $z < -1.65$, reject.

III. If the probability of a sampling error in a negative direction of the observed amount or more is equal to or more than 5 percent, accept the null hypothesis; if it is less than 5 percent, reject the null hypothesis.

5. Application of the decision rule: Suppose a sample of size 550 is selected and generates a mean equal to 34,970 miles.

I. 34,970 < 34,977.48, therefore reject.

II.

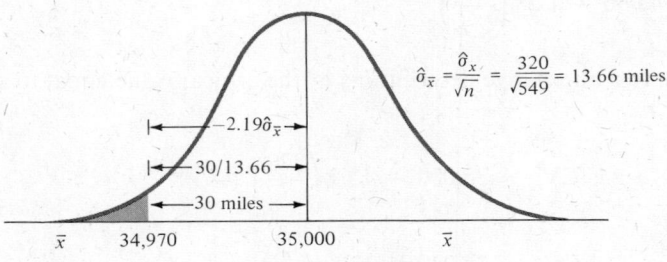

$$\hat{\sigma}_{\bar{x}} = \frac{\hat{\sigma}_x}{\sqrt{n}} = \frac{320}{\sqrt{549}} = 13.66 \text{ miles}$$

−2.19 < 1.65; therefore, reject

III. Between the universe mean and a point 2.19 standard errors away, 48.57 percent of the area under the curve will fall. Subtracting from 50 percent, we find that 1.43 percent of the sample means would differ from the universe mean in a negative direction by as much or more than the observed sample mean. The decision maker is willing to accept chance selection in sampling as the explanation of observed sampling error as long as the probability of such a sampling error being observed exceeds 5 percent. In this case, it does not. The difference is considered to be too significant to attribute reasonably to chance, and the null hypothesis is rejected.

16.4 COMPARISON: THE BAYESIAN VERSUS THE CLASSICAL APPROACH

Any given decision rule selected by a classicist possesses a set of error characteristics typically summarized in an error curve (see Figure 15.23). For each conceivable value of the universe parameter, the error curve depicts the probability that implementation of the decision rule will lead to an error (selection of other than the optimal action). Errors are of two types. If application of the decision rule leads to the acceptance of a false hypothesis, it is referred to as a Type II error. If application of the decision rule leads to the rejection of a true hypothesis, it is referred to as a Type I error. The probability of committing an error (either Type I or Type II) in association with each possible state of the universe can be determined for any given decision rule. Selection of a given decision rule therefore implies a willingness to accept the conditional error risks reflected in the associated error curve. The classicist contends that the relative seriousness of the consequences of committing an error in association with each conceivable value of the unknown universe parameter is implicitly taken into consideration in choosing an acceptable error curve (and hence a decision rule). The critical probability selected determines the risk of committing a Type I error. In theory, the more serious the consequences associated with commission of a Type I error, the lower the critical probability selected. In practice, the critical probability customarily assigned is either 0.01 or

0.05. The consequences associated with accepting the null hypothesis when it is actually false are then evaluated for strategic values of the universe parameter associated with the alternate hypothesis. The classicist then specifies the risk he is willing to take of incurring those consequences. In effect, he specifies the probability of committing a Type II error for the strategic values of the universe parameter. The classicist contends that he implicitly considers the likelihood that the given strategic value of the universe parameter could occur when he specifies the associated conditional probability of committing a Type II error that he will accept. Using the technique described in this chapter, the analyst then determines the sample size that satisfies the given constraints. Given the sample size, the critical value(s) associated with the decision rule possessing the desired error characteristics can be determined.

In contrast, the Bayesian considers explicitly what the classicist attempts to incorporate implicitly. The Bayesian *explicitly* introduces a prior estimating distribution reflecting his attitude with respect to the likelihood that each of the possible values of the universe parameter is the actual value of the universe parameter. For a given specified sample size, the probability of committing an error is then determined for each conceivable value of the universe parameter. The resulting conditional distribution is, of course, an error curve (see Figure 15.23). The prior estimating distribution and the error curve are then combined (using the general rule of multiplication) to generate a distribution depicting the Bayesian's assessment of the probability of the joint event—the universe parameter is equal to μ_x *and* an error (Type I or Type II) is committed—for each conceivable value of μ_x. An *explicit* loss function is then introduced which reflects the loss that would be incurred in association with each conceivable value of the universe parameter. The expected loss associated with any given decision rule is obtained by multiplying the conditional losses as reflected by the loss function by the probability that they will be incurred. The net expected loss is obtained by adding the cost of using the decision rule (cost of sampling) to the expected loss associated with that decision rule. The optimal decision rule is the decision rule that minimizes the net expected loss. (This is equivalent to maximizing the net gain from sampling.) The Bayesian approach was illustrated in Chapter 9 for a discrete probability distribution. The techniques can be extended to continuous distributions. The extensions will be left to more advanced texts.

Given that one can determine the appropriate loss function and generate a prior estimating distribution for μ_x, it would appear that the Bayesian approach has much to offer. The interested reader is referred to the previously mentioned sources for a detailed treatment of the nuances of the Bayesian approach.[2]

In those situations where the decision maker has difficulty generating an explicit loss function and explicitly generating a prior estimating distribution, he must of necessity fall back on the classical approach. At least his attitude on both of the above factors will be implicitly considered in generating a decision rule.

PROBLEMS

Problem 16.1

I.Q. required to perform a given job is estimated at 110. An I.Q. considerably less than this will not be sufficient to ensure satisfactory performance. An individual

---

[2] See Chapter 9, p. 233.

with an I.Q. considerably greater than this would be bored with the job. To evaluate
applicants, a battery of I.Q. tests is to be given. Performance on any one segment
is an estimate of the individual's true I.Q. Dispersion of individual performances
around the true I.Q. is claimed to be normally distributed with a standard deviation
of 15 I.Q. points.

Constraints:

a. No more than a 10 percent chance of rejecting an individual when his I.Q.
actually is 110 (two-tailed test).
b. No more than a 10 percent chance of accepting an individual when his I.Q.
actually is 95.
c. No more than a 20 percent chance of accepting an individual when his I.Q.
actually is 120.

Given the above constraints, determine the sample size required (i.e., the
number of segments an applicant should be required to complete) and the associated decision rule (i.e., critical values).

Problem 16.2

Lids on jars of fruit cocktail are attached by an automatic machine process set to
apply the lids in such a fashion that the average pressure required for removal is
32 in.-lb. Although the mean of the process has been known to shift, the variability
in the process is considered stable at $\sigma_x = 2$ in.-lb. Lids attached too tightly are
undesirable because they are hard to remove, and lids that are attached too loosely
can cause spoilage. Management desires a decision rule to test the hypothesis that
$\mu_x = 32$ in.-lb subject to the following constraints:

a. No more than a 10 percent probability of committing a Type I error (evenly
distributed between the two tails).
b. No more than a 5 percent chance of committing a Type II error when
$\mu_x = 33$ in.-lb.
c. No more than a 1 percent chance of committing a Type II error when
$\mu_x = 30$ in.-lb.

Required:

1. Determine the minimum necessary sample size required in order to meet all
three constraints specified above.
2. State the associated decision rule.
3. Apply the decision rule, assuming a sample of the size stipulated in part 1
is taken and its mean is determined to be 31.7 in.-lb.

Problem 16.3

Your firm purchases steel rods from a number of suppliers. The average breaking
strength of rods shipped must meet a specified level of 2000 lb pressure per square
inch (psi), and dispersion in the lot should not exceed 25 psi. Whenever a shipment
arrives, it is to be subjected to a test to determine whether or not it meets standards.
It is assumed that the standard deviation of the universe is 25 psi.

The test must provide for:

a. No more than a 1 percent probability of committing a Type I error when
$\mu_x = 2000$ psi.
b. No more than a 10-percent probability of committing a Type II error when
$\mu_x = 1990$ psi.

Required:
1. State the null hypothesis.
2. Determine the appropriate sample size.
3. Given the sample size determined in part 2, generate the appropriate decision rule to be used to test the null hypothesis stated in part 1.

Problem 16.4

Sweetetuth Manufacturers desires to establish a test to verify whether or not the production process for producing their Neverenuf candy bar is operating in accordance with specifications. When operating correctly, the average weight is 8 oz and the universe standard deviation is 0.3 oz. The mean of the process is likely to shift due to slippage in machine settings, but the standard deviation is considered quite stable. Sweetetuth Manufacturers is concerned with detecting when the process is operating with a mean less than 8 oz, since an FTC ruling against them for false advertising could damage their public image. They are also concerned with detecting shifts of the process mean to values greater than 8 oz, because this causes costs of production to increase.
1. State the null hypothesis.
2. Given the following constraints:
    a. No more than a 5 percent probability of committing a Type I error (divided up so that 2 percent is assigned to the left tail and 3 percent to the right tail).
    b. No more than a 10 percent probability of committing a Type II error when $\mu_x = 7.9$ oz.
    c. No more than a 1 percent probability of committing a Type II error when $\mu_x = 8.1$ oz.
    Determine the appropriate sample size.
3. Given the sample size determined in part 2, generate the appropriate decision rule to be employed to test the null hypothesis stated in part 1.

Problem 16.5

Consumer's Advice magazine desires to conduct a laboratory test in an attempt to ascertain the validity of the advertising claim of Reliable Products, Inc., that the average life of their Ace battery is at least 250 hr. A small preliminary sample has already been tested. It generated an estimate of the standard deviation of the universe equal to 10 hr.
1. State the null hypothesis.
2. Given the following constraints:
    a. No more than a 10 percent probability of committing a Type I error.
    b. No more than a 5 percent probability of committing a Type II error when $\mu_x = 245$ hr.
    Determine the appropriate sample size.
3. Given the sample size determined in part 2, generate the appropriate decision rule to be used to test the null hypothesis stated in part 1.

Problem 16.6

1. If the constraint "no more than a 40 percent chance of committing a Type II error when μ_x equals 247 hr" were to be added to the constraints specified in Problem 16.5, what change, if any, would occur? Support your answer.

2. If the constraint "no more than a 20 percent chance of committing a Type II error when μ_x equals 247 hr" were to be added to the constraints specified in Problem 16.5, what change, if any, would occur? Support your answer.

Problem 16.7

It is claimed that the average life of a particular transistor is at least 175 hr. A small preliminary sample provided an estimate of the standard deviation of the universe equal to 20 hr.

1. State the null hypothesis.
2. Given the following constraints:
   a. No more than a 10 percent chance of committing a Type I error when $\mu_x = 175$ hr.
   b. No more than a 5 percent chance of committing a Type II error when $\mu_x = 170$ hr.
   Determine the appropriate sample size.
3. Given the sample size determined in part 2, generate the appropriate decision rule to be used to test the null hypothesis stated in part 1.

Problem 16.8

If the constraint "no more than a 1 percent chance of committing a Type I error when μ_x equals 180 hr" were to be added to the constraints specified in Problem 16.7, what change, if any, would occur?

Problem 16.9

A mail order firm specializing in servicing rural communities is considering expanding its clientele to include suburban communities with certain specified income characteristics at certain minimum distances from large shopping centers. To justify the printing and mailing of catalogs to this market, the firm feels that it must average $30 in sales per catalog mailed. A sample is to be selected from a representative suburban community to test the hypothesis that average dollar sales per catalog sent will be at least $30. The standard deviation of dollar sales for catalogs sent to rural communities is $5. This figure is to be employed as the best estimate of the standard deviation for the suburban communities. The decision rule to be employed must reflect the following error characteristics:

a. No more than a 5 percent chance of committing a Type I error when $\mu_x = \$30$.
b. No more than a 10 percent chance of committing a Type II error when $\mu_x = \$28$.
c. No more than a 1 percent chance of committing a Type II error when $\mu_x = \$26$.

Required:

1. State the null hypothesis.
2. Determine the appropriate sample size.
3. Given the sample size determined in part 2, generate the appropriate decision rule to be employed to test the null hypothesis stated in part 1.

Problem 16.10

A sample equal to the size determined to be appropriate in part 2 of Problem 16.9 is selected. The standard deviation of this sample is $10, reflecting a wider variation in dollar sales for the suburban communities. The firm decides to recompute the sample size required to meet the error characteristics specified in Problem 16.9, using the new $10 figure as the best estimate of the universe standard deviation.

1. Determine the new appropriate sample size.
2. Given the sample size determined in part 1, determine the new decision rule to be employed.

Problem 16.11

Discuss the relationship between the error constraints specified in determining the decision rules in Problems 16.1 through 16.5 and the concept of the error curve as developed in Chapter 15.

Problem 16.12

Some error constraints are dominated by others. Discuss.

THE *t* DISTRIBUTION [SMALL SAMPLE THEORY]

17

The techniques examined in Chapters 13–16 under the general headings of statistical estimation and hypothesis testing were based on a knowledge of the underlying sampling distribution of the statistic. In Chapter 13, discussion focused on two situations where the analyst could conclude that the underlying sampling distribution of the mean would be normally distributed:

1. If the original population is *normally distributed* and the *standard deviation of the universe is known*, all sampling distributions of the mean derived from that universe are normally distributed[1] regardless of the sample size.

2. Regardless of the shape of the original universe and regardless of whether the standard deviation of the universe is known, all sampling distributions of the mean derived from the universe are

[1] Exactly normally distributed.

417

normally distributed[2] provided that they are associated with a sample size of 30 or more (central limit theorem).

Still another situation exists in which the analyst can draw on the tenets of statistical theory to arrive at a specification of the nature of an underlying sampling distribution of the mean, namely: *If the original population is normally distributed and the standard deviation of the universe is unknown (and therefore must be estimated from a sample), the sampling distribution of the mean derived from that universe will be described by a t distribution with n − 1 degrees of freedom (where n refers to sample size).* When the sample size exceeds 30, the description provided by the appropriate *t* distribution is virtually identical to that provided by the normal distribution. Accordingly, the more general normal curve description is used in situations where the sample size exceeds 30. Estimation and testing of hypotheses in situations where the underlying sampling distribution of the mean is normally distributed were developed in Chapters 13 and 15. This chapter extends the concepts of estimation and hypothesis testing to situations where the underlying sampling distribution of the mean is *t* distributed.[3] Such situations are characterized by (1) a normally distributed universe, (2) an unknown universe standard deviation, and (3) a sample size less than 30.

Because the *t* distribution is applied only when the sample size is less than 30, the body of statistical analysis that has evolved around the *t* distribution is referred to as *small sample theory*. The basic theory discussed in this chapter (small sample theory) does not differ from that associated with *large sample theory* (Chapters 13–16) in concept. The major difference lies in the distribution within which the theoretical constructions are developed.

17.1 CHARACTERISTICS OF THE *t* DISTRIBUTION

1. The *t* distribution is similar to the normal distribution in that it is symmetrical, bell shaped, and is asymptotic extending from minus infinity to plus infinity along the number line.

2. The *standardized* normal curve has a mean equal to zero and a standard deviation equal to one.

$$N(0, 1)$$

The *standardized t* distribution also has a mean equal to zero, but its standard deviation is described by the relationship $n/(n - 2)$, where *n* refers to sample size.

$$t\left(0, \frac{n}{n - 2}\right)$$

The first implication of a standard deviation equal to $n/(n - 2)$ is that the nature of dispersion in the standardized distribution is a function of sample size. Thus, there is a different standardized *t* distribution associated with each sample size.

---

[2] Not exactly normally distributed, but sufficiently close to be reasonably approximated by a normal distribution. The less the departure of the universe from a normal distribution, the greater the correspondence for any particular sample size. In addition, the larger the sample size, regardless of the shape of the universe, the closer the correspondence of the sampling distribution to the shape of a normal distribution.

[3] As in Chapter 13, discussion will be restricted to a particular statistic, the sample mean. In Chapter 18, three other universe parameters will be examined: the universe proportion, the difference between two universe means, and the difference between two universe proportions.

The second implication is that as the sample size increases, the value $n/(n - 2)$ approaches the limiting value of one. It is this characteristic that allows the normal approximation of the *t* distribution to be used once the associated sample size exceeds 30.

    3. Graphically, the standardized *t* distribution is flatter than the normal distribution, with a larger proportion of the area under the curve located in the tails of the distribution. This implies that one must proceed a greater distance along the number line away from the mean under a standardized *t* distribution to include any given percentage of the area under the curve than would be the case for the standardized normal distribution. As sample size is increased, the associated *t* distribution corresponds more and more closely to the pattern of the standardized normal curve. One table of areas services the standardized normal curve. A separate table of areas must be prepared for each standardized *t* distribution. Each such table is identified with a particular number of degrees of freedom.[4]

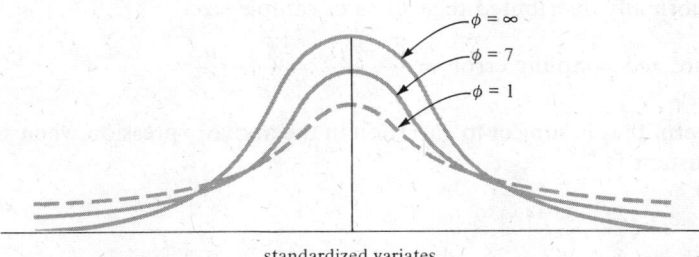

standardized variates

Figure 17.1 Graphic depictions of standardized *t* distributions associated with $\phi = 1$, $\phi = 7$, and $\phi = \infty$ where ϕ refers to the number of degrees of freedom.

    To enhance understanding of the *t* distribution, it is useful to investigate the impact of not knowing the standard deviation of the universe on the ability of the analyst to generate estimates of universe parameters.

    Suppose the analyst desires to construct a 90 percent confidence interval estimate of the universe mean. The procedure used should be such that 90 percent of the statements constructed by that procedure will be correct. A correct statement in this sense is interpreted as the value of the universe mean lying within the confidence limits specified by the statement.

Situation A: Normally Distributed Universe; Universe Standard Deviation Known

Ninety percent confidence interval statements take the form

$$\overline{X} \pm \varkappa_{0.05}\sigma_{\bar{x}}$$

---

[4] To compute the sample standard deviation, the sample mean must first be determined. Therefore, one degree of freedom is lost (see Chapter 4, p. 82). In other circumstances, more than one degree of freedom may be lost. In general, one degree of freedom is lost for each constant that must be estimated from the sample data before the deviations can be computed. Thus, a particular sample size can be associated with varying numbers of degrees of freedom, depending on what is being estimated from the sample. In recognition of this, *t* tables are referenced by the number of degrees of freedom (rather than sample size) in order to generate one set of tables that is generally applicable to all situations described by *t* distributions.

where $\varkappa_{0.05}$ refers to the number of standard errors one must proceed in one direction from the mean in order to encompass 45 percent of the area under the normal curve

$$\varkappa_{0.05} = 1.65$$

Therefore,

$$\overline{X} \pm 1.65 \frac{\sigma_x}{\sqrt{n}}$$

Although the value of the sample mean varies from sample to sample, the width of the interval constructed around each sample mean remains constant. The term $1.65(\sigma_x/\sqrt{n})$ is a constant and does not vary from sample to sample (assuming that sample size is not changed). See Figure 17.2. The sampling error reflected in the estimates is attributable solely to variation in the sample means. When sampling from a normal universe, sampling error attributable solely to variation in sample means is normally distributed regardless of sample size.

$$\text{standardized sampling error} = \frac{\overline{X} - \mu_x}{\sigma_{\overline{x}}} = \frac{\overline{X} - \mu_x}{\sigma_x/\sqrt{n}}$$

The only term that is subject to variation in the above expression when sample size is held constant is $\overline{X}$.

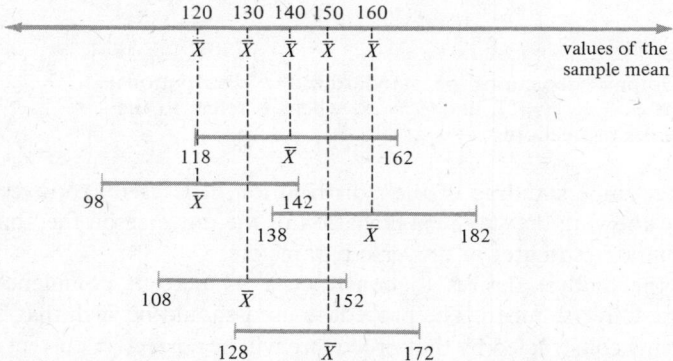

Figure 17.2 Representative 90 percent confidence interval statements based on a specified fixed sample size drawn from a normally distributed universe when the universe standard deviation is known.

Situation B: Normally Distributed Universe; Universe Standard Deviation Unknown

Ninety percent confidence interval statements take the form

$$\overline{X} \pm t_{0.05}\hat{\sigma}_{\overline{x}}$$

where $t_{0.05}$ refers to the number of standard errors one must proceed in one direction from the mean in order to encompass 45 percent of the area under the appropriate t distribution. $t_{0.05}$ varies depending on the number of degrees of freedom. Thus, we have

$$\overline{X} \pm t_{0.05} \frac{\hat{\sigma}_x}{\sqrt{n}}$$

In this situation the sample provides not only a sample mean but also an estimate of the standard deviation of the universe [i.e., $\hat{\sigma}_x = \sqrt{\sum (X - \overline{X})^2/(n - 1)}$]. Thus, when the confidence interval statement is constructed, the analyst must contend not only with the variation among sample means but also with the variation among sample standard deviations. The width of the confidence interval as well as the sample mean varies from sample to sample. See Figure 17.3.

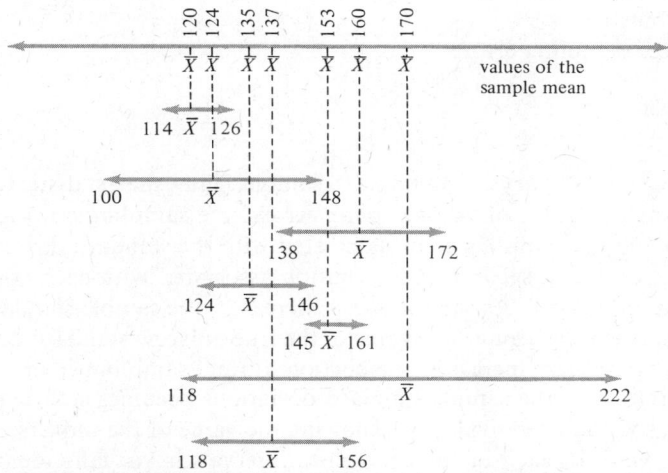

Figure 17.3 Representative confidence interval statements based on a specified fixed sample size (less than 30) drawn from a normally distributed universe when the universe standard deviation is unknown.

In order for the procedure to generate a 90 percent confidence interval, the interval estimates must be constructed in such a way that the amount added to and subtracted from the sample mean will be large enough to include 90 percent of the sampling error to be observed. Each interval is of equivalent width in standard error units but, because of variation in the value of the sample standard deviation, varies from sample to sample when measured in original units. In this situation, the standardized sampling error [i.e., $(\overline{X} - \mu_x)/(\hat{\sigma}_x/\sqrt{n})$] associated with any estimate is comprised of the combined net influence of the sampling error associated with the variation among the sample means *plus* the sampling error associated with the variation among the sample standard deviations. Given that the universe from which the samples are selected is normally distributed and the sample size is known (more precisely, the number of degrees of freedom is known), the probability distribution of this combined net sampling error can be described.[5]

If the sampling error is attributable solely to the variation in sample means

[5] W. M. Gosset, a statistician for a brewery that disapproved of his publishing activities (hence he wrote under the pen name "student.") investigated the properties of the standardized sampling error under the conditions described above. ("The Probable Error of the Mean," *Biometrika*, 1908). Gosset arrived at a general mathematical description of the manner in which the sampling errors behaved given various degrees of freedom. The results have been captured and described in a series of distributions that have subsequently come to be referred to as *t* or student's distributions.

and the universe from which the samples are selected is normally distributed, then the standardized sampling error will be normally distributed or z distributed.

$$z = \frac{\overline{X} - \mu_x}{\sigma_{\bar{x}}} = \frac{\overline{X} - \mu_x}{\sigma_x/\sqrt{n}}$$

If the standardized sampling error is attributable to the combined net variation in sample means and in sample standard deviations and the universe from which the samples are selected is normally distributed, the standardized sampling error will be t distributed.

Standardized sampling error:

$$t = \frac{\overline{X} - \mu_x}{\hat{\sigma}_{\bar{x}}} = \frac{\overline{X} - \mu_x}{\hat{\sigma}_x/\sqrt{n}}$$

Note that both $\overline{X}$ and $\hat{\sigma}_x$ vary from sample to sample. Thus, the t statistic represents the combined net influence of two sampling errors: the sampling error associated with the mean and the sampling error associated with the standard deviation.

There is, of course, a separate t distribution associated with each conceivable level of degrees of freedom. As sample size is increased, the sample standard deviations cluster more closely around the value of the true universe standard deviation.[6] Eventually, as sample size increases, the portion of total sampling error associated solely with variation in the sample standard deviations becomes sufficiently small that the effect is virtually equivalent to knowing the value of the universe standard deviation (the width of each of the confidence intervals is virtually identical). As sample size increases and the portion of sampling error attributable to variation in the sample standard deviations decreases, the t distribution approaches the pattern of the normal distribution as a limiting case. Technically, there is a difference between the t distribution description of the sampling distribution of the mean associated with sample size 100 when the universe standard deviation is unknown and the normal distribution description of the sampling distribution of the mean associated with sample size 100 when the universe standard deviation is known. For sampling distributions associated with a sample size of 30 or more, however, virtually all the variation observed is attributable to variation in the sample means, and the normal approximation is appropriate. Technically, a separate table of areas exists for each level of degrees of freedom. These tables are similar to the table of areas under the normal curve (Table F in the Appendix). The information contained in these separate tables is usually compressed into one table by recording the number of standard errors one must proceed under the appropriate t distribution for only those areas under the curve most frequently referenced. Figure 17.4 depicts the relationship between the individual tables and the overall summary table usually employed.

Referring to Table H in the Appendix (a portion of which is reproduced in Figure 17.4), the columnar headings refer to the percentage of the area under the appropriate t distribution that lies more than t standard errors to the right of the mean (or, for that matter, to the left of the mean given the symmetrical nature of the t distribution). Each row in the table refers to a particular number of degrees of freedom. Each row contains those t values most frequently referenced. To provide the information for all t values, each row would have to be expanded into a

---

[6] The sampling distribution of the sample variance is chi-square distributed. The χ^2 (chi-square) distribution is discussed in Chapter 19.

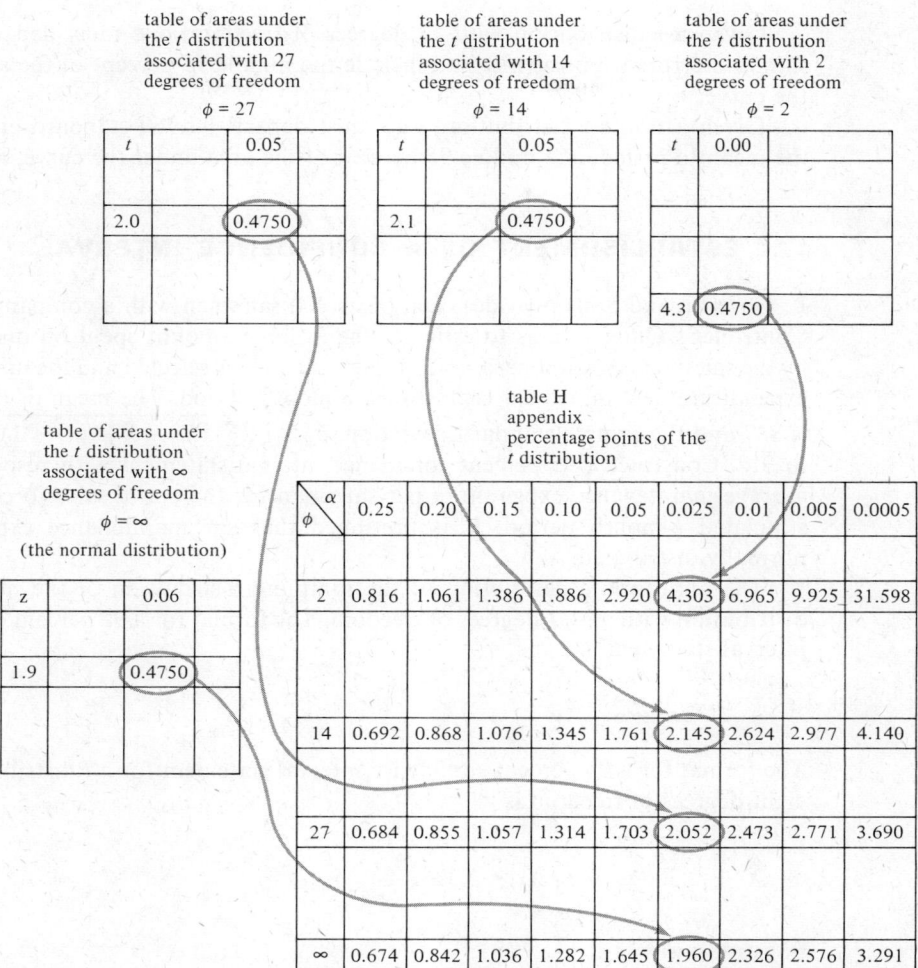

table of areas under the *t* distribution associated with 27 degrees of freedom

$\phi = 27$

t		0.05	
2.0		0.4750	

table of areas under the *t* distribution associated with 14 degrees of freedom

$\phi = 14$

t		0.05	
2.1		0.4750	

table of areas under the *t* distribution associated with 2 degrees of freedom

$\phi = 2$

t	0.00		
4.3	0.4750		

table of areas under the *t* distribution associated with ∞ degrees of freedom

$\phi = \infty$

(the normal distribution)

z		0.06	
1.9		0.4750	

table H appendix percentage points of the *t* distribution

α / ϕ	0.25	0.20	0.15	0.10	0.05	0.025	0.01	0.005	0.0005
2	0.816	1.061	1.386	1.886	2.920	4.303	6.965	9.925	31.598
14	0.692	0.868	1.076	1.345	1.761	2.145	2.624	2.977	4.140
27	0.684	0.855	1.057	1.314	1.703	2.052	2.473	2.771	3.690
∞	0.674	0.842	1.036	1.282	1.645	1.960	2.326	2.576	3.291

Figure 17.4

full-blown table of areas. The entries in the body of the table refer to the number of standard errors one must depart from the mean to either the right or the left to obtain the value on the number line beyond which the percentage of the area indicated at the top of the column would lie. This distance varies depending on the number of degrees of freedom. As the number of degrees of freedom increases, the distances one must depart (*t* values) approach the corresponding distances under the normal curve (*z* values).

Given a *t* distribution with 11 degrees of freedom, 45 percent of the area under the curve lies between the mean and a point 1.796 standard errors away. It follows that 5 percent of the values exceed the value of the mean by more than 1.796 standard errors.

Given a *t* distribution with 2 degrees of freedom, one must depart ± 4.303 standard errors from the mean to include the middle 95 percent of the area under the curve.

Given a *t* distribution with 14 degrees of freedom, one must depart ± 2.145 standard errors from the mean to include the middle 95 percent of the area under the curve.

Given a t distribution with 27 degrees of freedom, one must depart ± 2.052 standard errors from the mean to include the middle 95 percent of the area under the curve.

Given a normal distribution, one must depart ± 1.960 standard errors from the mean to include the middle 95 percent of the area under the curve, and so on.

17.2 ESTABLISHMENT OF A CONFIDENCE INTERVAL

L. Gozman and Sons provides each of its 400 salesmen with a company car. The Controller's Office desires to estimate the average amount spent on maintenance by the salesmen. A random sample of nine salesmen is selected and the maintenance expenditure determined for each over a 6-month period. The mean of the sample is \$87 and the sample standard deviation, $\hat{\sigma}_x = \sqrt{\sum (X - \bar{X})^2/(n - 1)}$, amounts to \$12. Construct a 95 percent confidence interval statement with respect to the average maintenance expenditure per salesman for the entire sales force over the stipulated 6-month period. It is presumed that car maintenance expenses are normally distributed.

The appropriate description of the sampling distribution of the mean is a t distribution with $n - 1$ degrees of freedom. The format for a 95 percent confidence interval statement is

$$\bar{X} \pm t_{0.025}\hat{\sigma}_{\bar{x}}$$

The format for a 95 percent confidence interval statement for a t distribution with eight degrees of freedom is

$$\bar{X} \pm 2.306\sigma_{\bar{x}} \quad \text{or} \quad \bar{X} \pm 2.306 \frac{\hat{\sigma}_x}{\sqrt{n}}$$

Substituting:

$$87 \pm 2.306 \frac{12}{\sqrt{9}}$$

$$87 \pm 2.306(\tfrac{12}{3})$$
$$87 \pm 2.306(4)$$
$$87 \pm 9.224$$
$$77.776 \leftrightarrow 96.224$$

The analyst would be 95 percent confident that his statement is correct when he states that "the mean amount spent on car maintenance over the stipulated 6-month period is somewhere between \$77.78 and \$96.22." The justification for such a statement is that the procedure used will generate correct statements 95 percent of the time. An increase in sample size would lead to a decrease in the width of the confidence interval. The increase in sample size has a twofold effect: (1) It decreases the value of the standard error of the mean, and (2) it decreases the number of standard errors required to encompass the stipulated percentage of the area under the curve. This second effect becomes negligible once sample size exceeds 30.

17.3 CONFIDENCE STATEMENT ON EXPECTED SAMPLING ERROR

Diners Club charge slips provide a location for a recording of the tip. As a service to their cardholders, Diners Club desires to estimate the mean tip as a percentage of the cost of the meal as recorded on the charge slips by its cardholders. (Only charge slips involving expenditures at restaurants and only charge slips with tips recorded on the slip are considered. It is presumed that cash tips were left in those instances where no tip was recorded on the charge slip.) A sample of size 25 is selected at random. The mean tip expressed as a percentage of the cost of the meal on the 25 charge slips is 18 percent and the standard deviation of the sample is 3 percent $[\hat{\sigma}_x = \sqrt{\sum (X - \overline{X})^2/(n - 1)}]$. If the Diners Club were to use 18 percent as a point estimate of the universe mean, what statement could be made with a 90 percent probability of being correct about the magnitude of sampling error associated with the estimate?

The appropriate description of the sampling distribution of the mean is a t distribution with $n - 1$ degrees of freedom. The format for a 90 percent confidence statement with respect to the magnitude of sampling error is

$$E = t_{0.05}\hat{\sigma}_{\overline{x}}$$

The format for a 90 percent confidence statement with respect to the magnitude of sampling error associated with a t distribution with 24 degrees of freedom is

$$E = 1.711\hat{\sigma}_{\overline{x}} \quad \text{or} \quad E = 1.711 \frac{\hat{\sigma}_x}{\sqrt{n}}$$

Substituting:

$$E = 1.711 \frac{0.03}{\sqrt{25}}$$

$$= 1.711 \left(\frac{0.03}{5}\right)$$

$$= 1.711(0.006)$$
$$= 0.010266$$

The Diners Club would be 90 percent confident that their statement is correct when they state that "the estimate is not off by more than 1 percentage point." The justification for this statement is that 90 percent of the sample means generated by this procedure would be within 1.711 estimated[7] standard errors of the mean of the true mean value for the universe.

17.4 TEST OF A HYPOTHESIS: ONE-TAILED TEST

You are currently engaged in negotiations with a builder with respect to the heating system to be installed in a home currently under construction. The builder claims that the heating system he recommends will heat your home for no more than

[7] Ninety percent of the sample means generated by this procedure would be within 1.65 *actual* standard errors of the mean of the universe mean. The reason for the difference between 1.711 and 1.65 is that when the universe standard deviation is unknown, one must also take into account the variability in the sample standard deviations.

$310. The builder has installed 450 of these heating systems in homes similar to the one you are constructing. The builder allows you to select a random sample of size 4 out of this group. You contact the four individuals designated in the sample and ascertain that their average heating bill was $325 with a standard deviation of $30 $[\hat\sigma_x = \sqrt{\sum (X - \overline{X})^2/(n - 1)}]$. Using a critical probability of 5 percent, test the validity of the builder's claim.

The appropriate description of the sampling distribution of the mean is a t distribution with $n - 1$ degrees of freedom (it is reasonable to conclude that heating expenditures would be normally distributed). In this case, it is a t distribution with three degrees of freedom (Figure 17.5).

$$\begin{aligned}
\text{Null hypothesis:} &\quad \mu_x \le \$310 \\
\text{Alternate hypothesis:} &\quad \mu_x > \$310
\end{aligned}$$

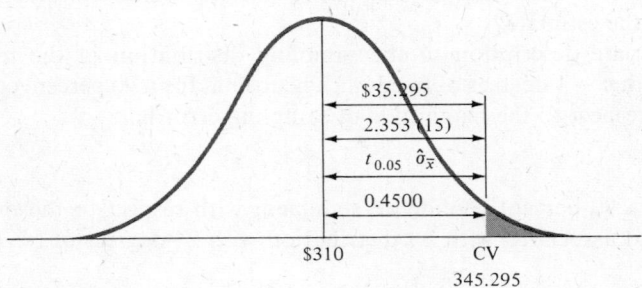

Figure 17.5

The critical value should be located such that 5 percent of the area under the t distribution associated with three degrees of freedom lies to the right of the critical value. Thus, 45 percent of the area under the curve lies between the mean and the critical value. Table H, *Percentage Points of the t Distribution*, indicates that 45 percent of the area under a t distribution with three degrees of freedom is contained between the mean and a point 2.353 *estimated* standard errors away. Thus, the critical value is equal to $\mu_x + 2.353\hat\sigma_{\overline{x}}$.

Substituting:

$$\text{CV} = \mu_x + 2.353\hat\sigma_{\overline{x}} \qquad \hat\sigma_{\overline{x}} = \frac{\hat\sigma_x}{\sqrt{n}}$$

$$= 310 + 2.353(15) \qquad = \frac{30}{\sqrt{4}}$$

$$= \$310 + \$35.295 \qquad = \frac{30}{2}$$

$$= \$345.295 \qquad = \$15$$

Three equivalent alternative formulations of the decision rule are given below:

I. Accept the null hypothesis ($\mu_x \le \$310$) if $\overline{X} \le \$345.295$. Reject if $\overline{X} > \$345.295$.

II. Accept the null hypothesis if $t \le +2.353$.
Reject the null hypothesis if $t > +2.353$.

III. Accept the null hypothesis if the probability of selecting a sample mean that *exceeds* $310 by at least as much as the one observed ≥ 5 percent. Reject if the probability of obtaining such a mean solely by chance selection in sampling is less than 5 percent.

Application of the decision rules:

I. $325 < 345.295, therefore accept.

II. $t = \dfrac{\overline{X} - \mu_x}{\hat{\sigma}_{\bar{x}}} = \dfrac{325 - 310}{15} = \dfrac{+15}{15} = +1$

$+1 < +2.353$, therefore accept.

III. In Table H, *Percentage Points of the t Distribution*, read across the row associated with three degrees of freedom. The observed mean exceeds the hypothesized value of the universe mean by one standard error. The table shows that 20 percent of the sample means will exceed the universe mean by at least 0.978 standard errors and 15 percent will exceed the universe mean by at least 1.25 standard errors. Thus, the probability of selecting a sample mean that exceeds the universe mean by at least as much as the one observed is at least 15 percent and probably closer to 20 percent. Fifteen percent[+] > 5 percent, therefore the null hypothesis should be accepted.

17.5 TEST OF A HYPOTHESIS: TWO-TAILED TEST

An industrial engineer employed by American Obstacle, a well-known manufacturer of eyeglass components, is in the process of establishing an incentive wage system for piecework operations involved in the assembly of eyeglass frames. Management believes that the average number of correct assemblies that could reasonably be performed per hour is 153. The industrial engineer wishes to verify this figure before developing the incentive wage scheme. He is concerned with detecting if the true mean number of assemblies is less than 153 per hour. If he designs an incentive scheme around a mean of 153 when the true mean is less than 153, incentive levels would be too difficult to obtain and lead to worker frustration. He is equally concerned with detecting if the true mean is greater than 153. If he designs an incentive scheme around 153 when the true mean is greater than 153, incentive levels will be too easy to obtain and defeat the objective of stimulating worker productivity. The industrial engineer selected 11 workers at random and observed their output levels. The mean output for the 11 workers is 158 correct assemblies per hour with a standard deviation of 8 correct assemblies $[\hat{\sigma}_x = \sqrt{\sum (X - \overline{X})^2/(n - 1)}]$. Assuming that worker performances are normally distributed, test the hypothesis that the true mean number of correct assemblies per hour is 153. Use a critical probability of 5 percent.

The appropriate description of the sampling distribution of the mean is a *t* distribution with $n - 1$ degrees of freedom. In this case it is a *t* distribution with ten degrees of freedom (Figure 17.6).

Null hypothesis: $\mu_x = 153$
Alternate hypothesis: $\mu_x \neq 153$

that is,

$$\mu_x > 153$$

or

$$\mu_x < 153$$

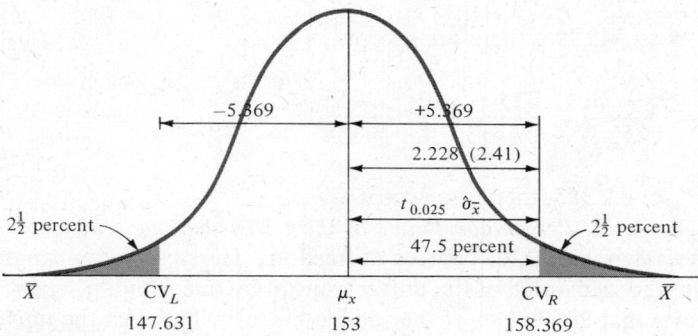

Figure 17.6

The critical values should be located such that 2.5 percent of the area under the t distribution associated with 10 degrees of freedom lies to the right of CV_R and 2.5 percent lies to the left of CV_L. Thus, 47.5 percent of the area under the curve lies between the mean and each of the critical values. Table H, *Percentage Points of the t Distribution*, indicates that 47.5 percent of the area under a t distribution with ten degrees of freedom is contained between the mean and a point 2.228 *estimated* standard errors away. Thus, the critical values are equal to $\mu_x - 2.228\hat{\sigma}_{\bar{x}}$ and $\mu_x + 2.228\hat{\sigma}_{\bar{x}}$. Substituting:

$$CV_L = \mu_x - 2.228\hat{\sigma}_{\bar{x}} \qquad \hat{\sigma}_{\bar{x}} = \frac{\hat{\sigma}_x}{\sqrt{n}}$$

$$= 153 - 2.228(2.41) \qquad = \frac{8}{\sqrt{11}}$$

$$= 153 - 5.369 \qquad = \frac{8}{3.317}$$

$$= 147.631 \qquad = 2.41$$

$$CV_R = \mu_x + 2.228\hat{\sigma}_{\bar{x}}$$

$$= 153 + 2.228(2.41)$$

$$= 153 + 5.369$$

$$= 158.369$$

The decision rule can be equivalently expressed in any one of the following formats:

I. Accept the null hypothesis ($\mu_x = 153$) if $147.631 \leq \bar{X} \leq 158.369$.
Reject if $\bar{X} < 147.631$ or $\bar{X} > 158.369$.

II. Accept the null hypothesis if $-2.228 \leq t \leq +2.228$.
Reject if $t < -2.228$ or $t > +2.228$.
This alternative could be also expressed as

Accept if $|t| \leq 2.228$
Reject if $|t| > 2.228$

III. Accept the null hypothesis if the probability of selecting a sample mean that departs from the hypothesized mean by at least as much as the sample mean observed (regardless of the direction of the deviation) is at least 5 percent. Reject if the probability of observing such a difference is less than 5 percent.

Application of the decision rules:

I. $147.631 \leq 158 \leq 158.369$; therefore accept (but sleep a little uneasy).

II. $|t| = \dfrac{|\overline{X} - \mu_x|}{\hat{\sigma}_{\overline{x}}} = \dfrac{|158 - 153|}{2.41} = \dfrac{5}{2.41} = 2.07$

$2.07 < 2.228$, therefore accept, or $-2.228 \leq +2.07 \leq +2.228$; therefore accept.

III. In Table H, *Percentage Points of the t Distribution*, read across the row associated with 10 degrees of freedom. The observed mean lies $|2.07|$ standard errors away from the hypothesized value of the universe mean. The table discloses that 2.5 percent of the sample means will exceed the universe mean by 2.228 estimated standard errors (and 2.5 percent will have a value 2.228 estimated standard errors or more less than the universe mean). Therefore, 5 percent lie $|2.228|$ standard errors or more away from the universe mean. Obviously, something greater than 5 percent must lie $|2.07|$ standard errors away or more. Five percent$^+$ $>$ 5 percent, therefore accept.

PROBLEMS

Problem 17.1

A large mail order house is considering the institution of a policy of absorbing postal charges on merchandise. A random sample of 16 orders has been selected and the postal charges associated with the shipment of each order determined. The mean postal charge for the 16 orders comprising the sample is $1.35 and the sample standard deviation is $0.18 [i.e., $\hat{\sigma}_x = \sqrt{\sum (X - \overline{X})^2/(n - 1)} = \$0.18$]. Construct a 95 percent confidence interval estimate of the true average postal charge should the firm decide to institute the policy on all orders. Assume that the postal charges are normally distributed. (A statistical technique, chi-square analysis, can be utilized to test the validity of such an assumption. Chi-square tests are discussed in Chapter 19.)

Problem 17.2

The general formula for a 95 percent confidence interval statement for a sampling distribution of the mean that is *t* distributed is

$$\overline{X} \pm t_{0.025}\hat{\sigma}_{\overline{x}}$$

As the sample size is increased, the width of the confidence interval declines. There are two forces operative that account for this behavior pattern.
1. What are the two forces referred to above?
2. Explain why one force continues to be operative when sample size is greater than 30, whereas the other is considered insignificant for sample sizes greater than 30.

Problem 17.3

The proprietor of a restaurant is conducting a cost analysis with respect to an entree he is considering adding to his menu. One of the basic ingredients is subject to considerable variation in price. Rather than vary the price of the entree to reflect the variation in the price of the basic ingredient, he intends to base the menu price on the average cost of the ingredient. A random sample of size 9 is selected from historical records. The sample mean is $0.98 per pound and the sample standard deviation is equal to $0.12 $[\hat{\sigma}_x = \sqrt{\sum (X - \bar{X})^2/(n - 1)} = \$0.12]$.

If the sample mean of $0.98 is to be used as an estimate of the universe mean, what statement can be made with respect to the expected magnitude of sampling error associated with the estimate? Assume that a 90 percent probability of generating a correct statement is desired. Also assume that the daily price of the ingredient is normally distributed.

Problem 17.4

You have recently taken a part-time job waiting on tables. The manager claimed that your tips would average at least $25 per night. Tips for the first five evenings were as follows: $21, $18, $27, $20, and $24. At a critical probability of 0.05, test the hypothesis that $\mu_x \geq \$25$.

Problem 17.5

When operating as designed, a machine produces parts with a mean diameter of 0.714 in. with a standard deviation of 0.008 in. Slippage in a setting can cause the universe mean to shift, although dispersion in the distribution will remain constant. A random sample of 16 parts is selected. The sample mean is 0.709 in.

1. Using a critical probability of 0.05, test the hypothesis that the machine is still in perfect adjustment.
2. Why does the normal distribution, rather than the t distribution, provide the appropriate description of the sampling distribution of the mean in the situation posited above?

Problem 17.6

Given that the sample size is less than 30.

1. Under what conditions does the normal distribution provide an appropriate description of the sampling distribution of the mean?
2. Under what conditions does the t distribution provide an appropriate description of the sampling distribution of the mean?
3. Under what conditions would neither the normal nor the t distribution provide an appropriate description of the sampling distribution of the mean?

Problem 17.7

A prospective supplier of a transistorized component has claimed that the average life of the components will be at least 200 hr. When tested, a sample of 16 components generated a mean life of 192 hr with a standard deviation of 6.3 hr [i.e., $\hat{\sigma}_x = \sqrt{\sum (X - \bar{X})^2/(n - 1)} = 6.3$]. Using a critical probability of 0.05, test the hypothesis that the supplier's claim is true.

Problem 17.8

When operating correctly, the filling mechanism on cans of stewed tomatoes has a mean fill of 14.5 oz with a standard deviation of 0.01 oz. A weighing device monitors the can as it is being filled. At a given reading, a response mechanism triggers a shut-off valve. The mechanism is quite sensitive, and both the mean and the standard deviation of the filling process are subject to variation. A recent sample of nine cans was characterized by a mean fill of 14.503 oz and a standard deviation of 0.012 oz.

1. Does the normal distribution or the t distribution provide the appropriate description of the sampling distribution of the mean? Discuss.
2. Using a critical probability of 10 percent, test the hypothesis that the true process mean is 14.5 oz.

Problem 17.9

Max M. Izeoutput, recently appointed business manager of the Southeast Medical Clinic, desires to ascertain the average amount of time doctors actually spend with patients in order to schedule appointments on as tight a schedule as possible. The clinic has numerous examination rooms, so that a doctor can simultaneously handle a number of patients. A sample of size 9 disclosed a mean of 8 min with a standard deviation of 1.2 min [i.e., $\sqrt{\sum (X - \bar{X})^2/(n - 1)} = 1.2$]. Assuming that the time doctors spend with patients is normally distributed, generate a 95 percent confidence interval estimate of the mean time doctors spend with patients.

Problem 17.10

H. R. Botch, Inc., a firm specializing in personalized income tax returns, has some part-time opportunities available. They offer $5 per completed return. To be worth your while, you must make at least $20 per hour. You spent an evening at the local Botch branch and observed the proceedings. You randomly selected nine customers and observed the time required to complete their returns. The mean time was 13 min with a standard deviation of 3.5 min [i.e., $\sqrt{\sum (X - \bar{X})^2/(n - 1)} = 3.5$]. Moreover, you overheard one of the employees remarking that there was always someone waiting to be serviced. You are willing to take a 10 percent chance of turning down the job when it actually would have been worth your while. Test the hypothesis that the mean time to complete a customer's income tax return is less than or equal to 15 min.

THREE OTHER PARAMETERS

In this chapter, attention is directed to the following universe parameters: (1) the universe proportion, (2) the difference between two universe means, and (3) the difference between two universe proportions.

18.1 THE UNIVERSE PROPORTION

Often the parameter of interest to the decision maker is the proportion of elements in the universe possessing a particular trait of interest. To illustrate: The decision maker may be concerned with estimating the percentage of total production that is defective or the proportion of individuals in a market survey expressing a preference for one packaging display over another. In such situations, the relevant statistical distribution is the sampling distribution of the proportion.

It is convenient, when visualizing the universe

from which the sample is to be drawn, to think in terms of two strata: the stratum comprised of those who possess the trait of interest and the stratum comprised of those who do not. If an element in the universe possesses the trait of interest, it is assigned the discrete value one. If an element in the universe belongs to the stratum that does not possess the trait of interest, the discrete value zero is assigned.

A geometric representation of such a binary distribution is depicted in Figure 18.1. The height of the probability mass function associated with each stratum reflects the proportional size of that strata to the size of the total universe.

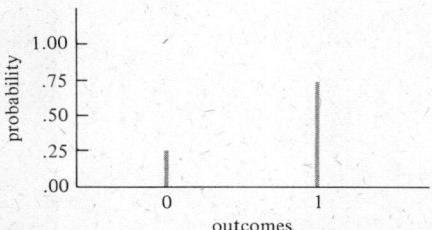

Figure 18.1 Probability distribution description of a universe in which 75 percent of the elements possess the trait of interest.

If one were to draw all possible samples of, say, size 100 from this universe and describe each sample outcome in terms of the number of successes observed, the resulting sampling distribution would be binomally distributed. Recall from Chapter 7 that the mean of a binomial distribution is equal to np and its standard deviation is equal to $\sqrt{npq}$. From Chapter 11, recall that the binomial probability distribution can be approximated by a normal curve with mean equal to np and standard deviation equal to $\sqrt{npq}$ provided the products np and $n(1-p)$ both exceed 5. Referring to Figure 18.2, where r is defined as the number of successes observed, it follows that

$$\bar{r} = np$$
$$\sigma_r = \sqrt{npq}$$

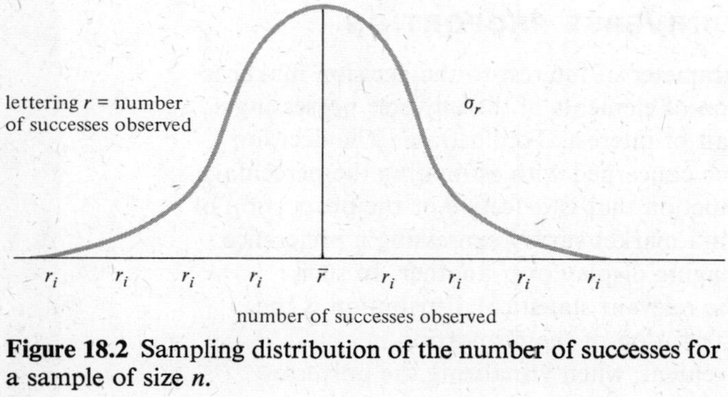

lettering r = number of successes observed σ_r

number of successes observed

Figure 18.2 Sampling distribution of the number of successes for a sample of size n.

If each sample is evaluated on the basis of the proportion of successes rather than the number of successes, the appropriate sampling distribution is obtained by simply dividing each of the observed r values by the sample size. The probability of occurrence is, of course, unchanged. The probability of observing r_i successes in a sample of size n is equivalent to observing a proportion of successes equal to r_i/n in a sample of size n (Figure 18.3).

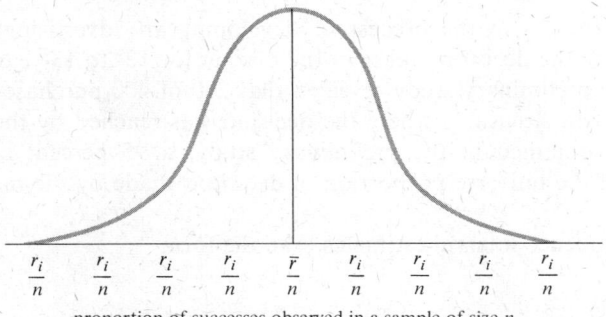

$$\frac{r_i}{n} \quad \frac{r_i}{n} \quad \frac{r_i}{n} \quad \frac{r_i}{n} \quad \frac{\bar{r}}{n} \quad \frac{r_i}{n} \quad \frac{r_i}{n} \quad \frac{r_i}{n} \quad \frac{r_i}{n}$$

proportion of successes observed in a sample of size n

Figure 18.3 Sampling distribution of the proportion of successes for a sample of size n.

Equivalent transformations of the mean and the standard deviation generate the following relationships:

$$\text{mean} = \frac{\bar{r}}{n} = \frac{np}{n} = p$$

$$\text{standard deviation} = \frac{\sigma_r}{n} = \frac{\sqrt{npq}}{n} = \sqrt{\frac{npq}{n^2}} = \sqrt{\frac{pq}{n}}$$

Consistent with the terminology previously developed in Chapter 13, the measure of dispersion for the sampling distribution of the proportion is designated as the standard error of the proportion and is represented by the symbol σ_p. Thus,

$$\sigma_p = \sqrt{\frac{pq}{n}}$$

It should be noted that the standard error of the proportion requires the incorporation of the finite correction factor when sample size is greater than 5 percent of the universe.

$$\sigma_p = \sqrt{\frac{pq}{n}} \cdot \sqrt{\frac{N-n}{N-1}} \quad \text{when} \quad \frac{n}{N} > 5 \text{ percent}$$

Thus, provided the products np and $n(1-p)$ both exceed 5, the sampling distribution of the proportion can be approximated by a normal curve with mean equal to p and standard deviation (i.e., standard error of the proportion) equal to $\sqrt{pq/n}$, where p equals the proportion of elements in the original universe possessing the trait of interest.[1] *If the constraints $np > 5$ and $n(1-p) > 5$ are not satisfied, the*

---

[1] The appropriate adjustment should be made when $n/N > 5$ percent.

underlying statistical estimates and tests must be performed within the framework of the binomial probability distribution.[2]

18.2 APPLICATIONS

Determination of a Confidence Interval

Market Oriented, Inc., is in the process of developing an advertising program designed to influence the decision maker in the market for 12- to 13-year-old girls wearing apparel. A preliminary study revealed that 280 of 400 purchase decisions investigated dealt with situations where the decision was reached by the mother. Utilizing the data obtained in the preliminary study, a 95 percent confidence interval estimate of the universe proportion of decisions made by the mother will be computed.

In general, confidence interval estimates take the form

$$\hat{p} \pm x\hat{\sigma}_p$$

where

$\hat{p}$ = the proportion of successes observed in the sample

x = the number of standard errors associated with the level of confidence specified

$\hat{\sigma}_p = \sqrt{\dfrac{\hat{p}\hat{q}}{n}}$ = the estimated value of the standard error of the proportion where $\hat{p}$, the sample proportion, is used as an estimate of the unknown universe parameter, p.

Referring to the example above,

$$\hat{p} = \frac{280}{400} = .70$$

$$x = 1.96$$
$$n = 400$$

Substituting:

$$\hat{p} \pm x\sqrt{\frac{\hat{p}\hat{q}}{n}} = 0.7 \pm 1.96\sqrt{\frac{(0.7)(0.3)}{400}}$$

$$= 0.7 \pm 1.96\sqrt{\frac{0.21}{400}}$$

$$= 0.7 \pm 1.96\sqrt{0.000525}$$

$$= 0.7 \pm 1.96(0.023)$$
$$= 0.7 \pm 0.04508$$
$$= 0.65492 \text{ to } 0.74508$$

Market Oriented, Inc., is 95 percent confident that their statement is correct when they state that the true proportion of decisions reached by the mother for the

---

[2] Recall from Chapter 10 that the Poisson distribution can be used to generate approximations of binomial probabilities at a considerable saving in computational effort when n is large and p is small.

market category investigated is between 65.5 percent and 74.5 percent. The justification for such a statement is that the procedure used to generate such statements will, on the average, yield correct statements 95 out of every 100 times.

The above analysis could also have generated a statement with respect to the magnitude of expected sampling error. In general, such statements take the form

$$E = z \sqrt{\frac{\hat{p}\hat{q}}{n}}$$

Substituting:

$$E = 1.96 \sqrt{\frac{(0.7)(0.3)}{400}}$$

$$= 0.04508$$

Market Oriented, Inc., could then state with 95 percent confidence that its statement is correct when they state that the point estimate of 0.7 does not differ from the true universe proportion by more than 0.04508.

Determination of Sample Size (Estimation)

In general, a statement on the magnitude of expected sampling error takes the form

$$E = z \sqrt{\frac{pq}{n}}$$

assuming that $n/N \leq 5$ percent.
Solving for n:

$$\frac{E}{z} = \sqrt{\frac{pq}{n}}$$

$$\left[\frac{E}{z}\right]^2 = \frac{pq}{n}$$

$$n\left[\frac{E}{z}\right]^2 = pq$$

$$n = pq\left[\frac{1}{(E/z)^2}\right]$$

$$n = pq\left[\frac{z}{E}\right]^2$$

Examination of the formula reveals that sample size is a function of three factors, namely, (1) E, the maximum allowable sampling error, (2) z, the number of standard error units associated with the level of confidence specified, and (3) p, the proportion of elements in the universe possessing the trait of interest.

Invariably, the universe proportion is unknown, and an estimate of its value must be provided. It is interesting to note the behavior of the factor pq as the value of the universe parameter p changes.

As depicted in Figure 18.4, pq is maximized when p is equal to 0.50. It follows, all other things being equal, that the required sample size will be maximized when p equals 0.50. Given the lack of a preconceived attitude as to the value of p, the analyst solves for n when p equals 0.50. This ensures a sample size large enough to satisfy the precision and confidence constraints. If the analyst has strong cause to believe that p differs from 0.50, substitution of his best estimate can result in a significant reduction in required sample size.

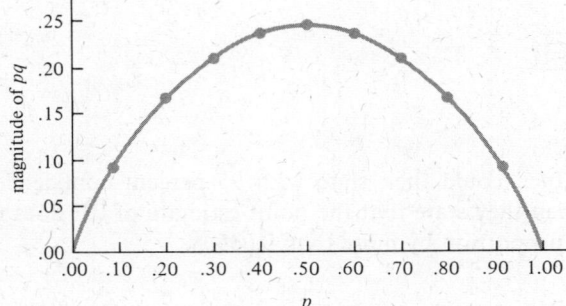

Figure 18.4

The procedure for determining appropriate sample size is illustrated in the context of the following situation: Offering potential subscribers the opportunity to participate in a sweepstakes with a large monetary prize is a device often used to encourage magazine subscriptions. Recently, the courts have ruled that eligibility in such sweepstakes must be accorded to all responding to the proposal, whether they subscribe to the magazine or not. Moreover, publishers are required to state this fact in promotion materials. Although publishers state that the chances of winning will in no way be affected by whether the respondent subscribes or not, return envelopes are constructed in such a fashion that a window reveals whether or not the respondent has elected to subscribe. The advertising agency representing one such publisher wishes to ascertain the proportion of individuals responding who believe that their chances of winning will be improved if they subscribe to the magazine. Assume that the advertising agency stipulates a maximum allowable sampling error of 0.02 (2 percentage points) and a 98 percent level of confidence that the sample estimate will not differ from the universe mean by more than the allowable sampling error.

1. Using $p = 0.50$, required sample size is determined as follows:

$$n = pq \left(\frac{z}{E}\right)^2$$

$$= (0.5)(0.5)\left(\frac{2.33}{0.02}\right)^2$$

$$= (0.25)(116.5)^2$$
$$= (0.25)(13,572.25)$$
$$= 3393.06$$

A sample of size 3393 is required to meet the precision and confidence constraints.

2. If the advertising agency believed that at most 20 percent of the respondents believed subscribing would improve their chances of winning the sweepstakes, the required sample size reduces to 2172.

$$n = (0.2)(0.8)\left(\frac{2.33}{0.02}\right)^2$$

$$= (0.16)(13{,}572.25)$$
$$= 2171.56$$

Test of a Hypothesis

A public accounting firm wishes to test the hypothesis that no more than 3 percent of a large set of freight billings contain clerical errors. Given the following constraints, determine the appropriate sample size and decision rule.

Constraints (Figure 18.5):

1. No more than a 5 percent probability of committing a Type I error.
2. No more than a 10 percent probability of committing a Type II error when $p = 0.05$.

The null hypothesis:

$p \leq 0.03$

The alternate hypothesis:

$p > 0.03$

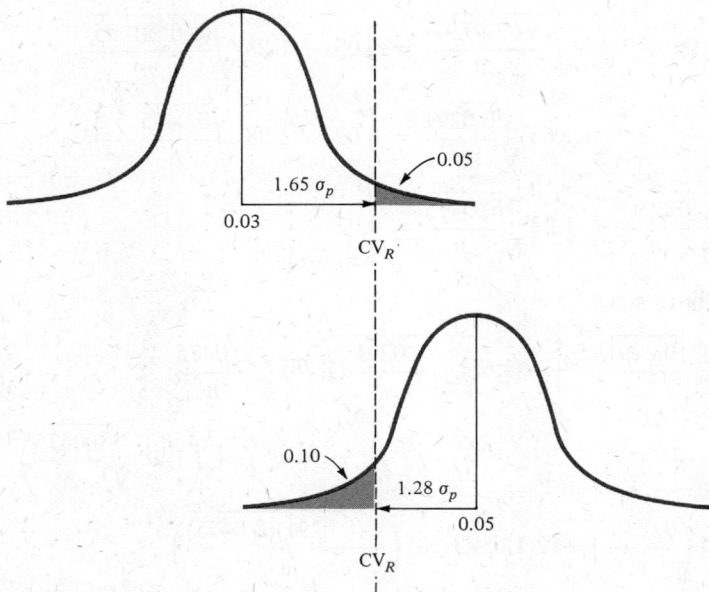

Figure 18.5 Graphic representation of the constraints.

Procedure:

1. Determine the formula describing the critical value that satisfies the constraint placed on the probability of committing a Type I error. As depicted in Figure 18.5:

$$CV_R = 0.03 + 1.65\sigma_p$$

$$= 0.03 + 1.65\sqrt{\frac{pq}{n}}$$

$$= 0.03 + 1.65\sqrt{\frac{(0.03)(0.97)}{n}}$$

2. Determine the formula describing the critical value that satisfies the constraint placed on the probability of committing a Type II error. As depicted in Figure 18.5:

$$CV_R = 0.05 - 1.28\sigma_p$$

$$= 0.05 - 1.28\sqrt{\frac{pq}{n}}$$

$$= 0.05 - 1.28\sqrt{\frac{(0.05)(0.95)}{n}}$$

Note that the value of σ_p differs from that utilized in step 1. This, of course, is a reflection of the fact that the magnitude of σ_p is a function of the universe parameter, p.

3. Set the two alternative descriptions of the critical value equal to each other and solve for n.

$$0.03 + 1.65\sqrt{\frac{(0.03)(0.97)}{n}} = 0.05 - 1.28\sqrt{\frac{(0.05)(0.95)}{n}}$$

$$1.65\sqrt{\frac{0.0291}{n}} = 0.02 - 1.28\sqrt{\frac{0.0475}{n}}$$

$$1.65\sqrt{\frac{0.0291}{n}} + 1.28\sqrt{\frac{0.0475}{n}} = 0.02$$

Squaring both sides:

$$\left(1.65\sqrt{\frac{0.0291}{n}}\right)^2 + 2(1.65)\sqrt{\frac{0.0291}{n}}(1.28)\sqrt{\frac{0.0475}{n}}$$

$$+ \left(1.28\sqrt{\frac{0.0475}{n}}\right)^2 = 0.0004$$

$$2.7225\left(\frac{0.0291}{n}\right) + 2(1.65)(1.28)\left(\frac{\sqrt{(0.0291)(0.0475)}}{n}\right)$$

$$+ 1.6384\left(\frac{0.0475}{n}\right) = 0.0004$$

Multiplying both sides by n, we obtain

$$0.0004n = 2.7225(0.0291) + 2(2.112)(\sqrt{0.00138225}) + 1.6384(0.0475)$$
$$0.0004n = 0.07922475 + 4.224(0.0372) + 0.077824$$
$$0.0004n = 0.07922475 + 0.1571328 + 0.07782400$$
$$0.0004n = 0.31418155$$
$$4n = 3141.8155$$
$$n = 785.454$$

4. Formulate the decision rule:

$$CV_R = 0.03 + 1.65 \sqrt{\frac{0.0291}{786}}$$

$$= 0.03 + 1.65\sqrt{0.00003702}$$
$$= 0.03 + 1.65(0.0061)$$
$$= 0.03 + 0.010065$$
$$= 0.040$$

The decision rule:
Select a sample of size 786.

If $p \leq 0.040$, accept the null hypothesis.
If $p > 0.040$, reject the null hypothesis.

18.3 THE DIFFERENCE BETWEEN TWO UNIVERSE MEANS

The underlying theory will be developed within the framework of the following situation. An instructor utilized two different pedagogical techniques to teach different sections of a course. One section was taught utilizing a conventional lecture format and the other utilizing a programmed learning approach. At the conclusion of the course, each student was asked to complete an evaluation sheet. One of the questions asked was, "What is your overall evaluation of the course?" The score assigned was the mean quality point assigned by the students in the class measured on a scale from 0 to 4. The standard deviations were computed using $\sqrt{\sum (X - \bar{X})^2/(n - 1)}$ in order to obtain an unbiased estimate of the universe standard deviation. The different approaches received the following scores:

Section 1 (Conventional lecture approach)		Section 2 (Programmed learning approach)	
$\bar{X}_1$ mean score	2.22	$\bar{X}_2$ mean score	2.16
$\hat{\sigma}_{x_1}$ standard deviation	0.11	$\hat{\sigma}_{x_2}$ standard deviation	0.09
n_1 sample size	64	n_2 sample size	36

THE CONVENTIONAL LECTURE METHOD. Theoretically, the universe from which the sample has been drawn consists of a probability distribution describing the pattern of individual responses to the student evaluation for all students (past, present, and future) likely to participate in such a course at this school. This distribution is depicted in Figure 18.6(a) with a mean equal to μ_{x_1} and a standard deviation

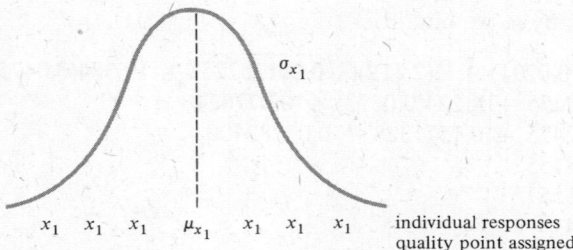

σ_{x_1}

$x_1 \quad x_1 \quad x_1 \quad \mu_{x_1} \quad x_1 \quad x_1 \quad x_1$ individual responses
quality point assigned

Figure 18.6(a) Theoretical probability distribution: individual responses—conventional lecture method.

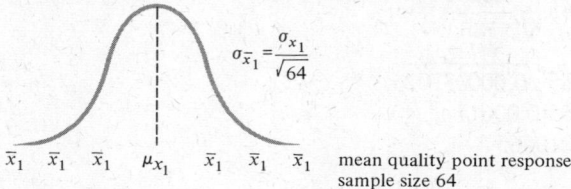

$\sigma_{\bar{x}_1} = \dfrac{\sigma_{x_1}}{\sqrt{64}}$

$\bar{x}_1 \quad \bar{x}_1 \quad \bar{x}_1 \quad \mu_{x_1} \quad \bar{x}_1 \quad \bar{x}_1 \quad \bar{x}_1$ mean quality point response
sample size 64

Figure 18.6(b) Theoretical probability distribution: sampling distribution of the mean—sample size 64.

equal to σ_{x_1}. The sample of 64 students can be considered as having been randomly selected from this universe. The theoretical sampling distribution of the mean for all possible samples of size 64 that could be drawn from such a universe is depicted in Figure 18.6(b). The sampling distribution of the mean has a mean equal to μ_{x_1} and a standard error, $\sigma_{\bar{x}_1}$, equal to $\sigma_{x_1}/\sqrt{n_1}$.

THE PROGRAMMED LEARNING APPROACH. A similar line of reasoning generates the distributions depicted in Figures 18.7(a) and 18.7(b) for the programmed learning approach.

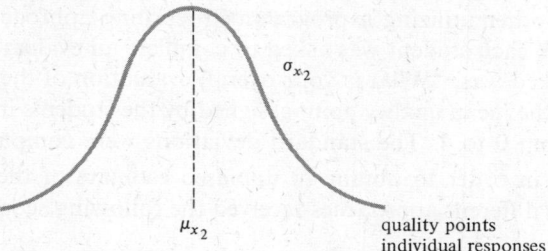

σ_{x_2}

μ_{x_2} quality points
individual responses

Figure 18.7(a) Theoretical probability distribution: individual responses—programmed learning approach.

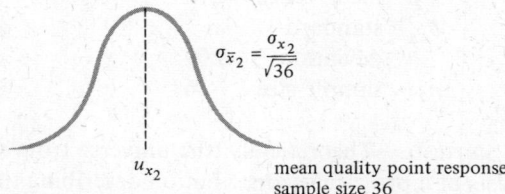

$\sigma_{\bar{x}_2} = \dfrac{\sigma_{x_2}}{\sqrt{36}}$

u_{x_2} mean quality point response
sample size 36

Figure 18.7(b) Theoretical probability distribution: sampling distribution of the mean—sample size 36.

A new probability distribution can be generated from the distributions depicted in Figures 18.6(b) and 18.7(b) by comparing the mean for each and every sample in the distribution described by Figure 18.6(b) with the mean for each and every sample in the distribution described by Figure 18.7(b). The new probability distribution is referred to as the sampling distribution of the difference between two sample means and is depicted in Figure 18.8.

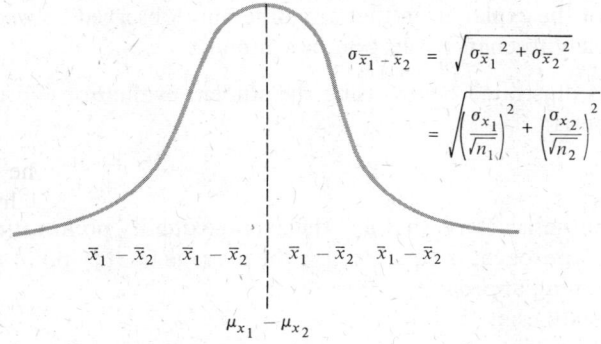

$$\sigma_{\bar{x}_1 - \bar{x}_2} = \sqrt{\sigma_{\bar{x}_1}^2 + \sigma_{\bar{x}_2}^2}$$

$$= \sqrt{\left(\frac{\sigma_{x_1}}{\sqrt{n_1}}\right)^2 + \left(\frac{\sigma_{x_2}}{\sqrt{n_2}}\right)^2}$$

$$\bar{x}_1 - \bar{x}_2 \quad \bar{x}_1 - \bar{x}_2 \quad \bar{x}_1 - \bar{x}_2 \quad \bar{x}_1 - \bar{x}_2$$

$$\mu_{x_1} - \mu_{x_2}$$

difference between sample means

Figure 18.8 Sampling distribution of the difference between two sample means.

It is stated without proof that if $\bar{x}_1$ and $\bar{x}_2$ are independent and normally distributed, then the statistic $\bar{x}_1 - \bar{x}_2$ is also normally distributed. Moreover, the mean of the sampling distribution of the difference between two sample means is equal to $\mu_{x_1} - \mu_{x_2}$. The standard deviation (known as the standard error of the difference between two sample means) is related to the dispersion in the original universes from which the samples were drawn and to sample size. Again, without proof, this relationship is as follows:

$$\sigma_{\bar{x}_1 - \bar{x}_2} = \sqrt{(\sigma_{\bar{x}_1})^2 - (\sigma_{\bar{x}_2})^2} = \sqrt{\left(\frac{\sigma_{x_1}}{\sqrt{n_1}}\right)^2 + \left(\frac{\sigma_{x_2}}{\sqrt{n_2}}\right)^2} = \sqrt{\frac{(\sigma_{x_1})^2}{n_1} + \frac{(\sigma_{x_2})^2}{n_2}}$$

In summary, provided that $\bar{x}_1$ and $\bar{x}_2$ are independent and normally distributed, the sampling distribution of the difference between two sample means is normally distributed with mean equal to $\mu_{x_1} - \mu_{x_2}$ and standard deviation equal to

$$\sqrt{\frac{(\sigma_{x_1})^2}{n_1} + \frac{(\sigma_{x_2})^2}{n_2}} \ .$$

The sampling distribution of the difference can be used to generate confidence interval statements and statements on the magnitude of expected sampling error. Generally, such statements take the following form:

Confidence interval:

$$(\bar{x}_1 - \bar{x}_2) \pm z\sigma_{\bar{x}_1 - \bar{x}_2}$$

$$(\bar{x}_1 - \bar{x}_2) \pm z\sqrt{\frac{(\sigma_{x_1})^2}{n_1} + \frac{(\sigma_{x_2})^2}{n_2}}$$

Expected sampling error:

$$E = z\sigma_{\bar{x}_1 - \bar{x}_2} \quad \text{or} \quad z\sqrt{\frac{(\sigma_{x_1})^2}{n_1} + \frac{(\sigma_{x_2})^2}{n_2}}$$

Usually, however, the sampling distribution of the difference is used to test the null hypothesis that there is no difference between the universe means for the universes from which the samples have been drawn. Acceptance of the null hypothesis implies acceptance of the explanation that any difference observed between $\bar{x}_1$ and $\bar{x}_2$ is attributable solely to chance selection in sampling.

The technique is illustrated below using the student evaluation illustration. The null hypothesis:

$$\mu_{x_1} - \mu_{x_2} = 0$$

That is, the null hypothesis asserts that μ_{x_1}, the average quality point rating for the conventional lecture approach, is equal to μ_{x_2}, the average quality point rating for the programmed learning approach.

The alternate hypothesis:

$$\mu_{x_1} - \mu_{x_2} \neq 0$$

That is, $\mu_{x_1} - \mu_{x_2} > 0$, implying that the conventional lecture approach has a higher rating, or $\mu_{x_1} - \mu_{x_2} < 0$, implying that the programmed learning approach has a higher rating.

A decision rule allowing a 5 percent probability of committing a Type I error is to be used.

The procedure:

Compute the probability of observing an $\bar{x}_1 - \bar{x}_2$ statistic that deviates from a $\mu_{x_1} - \mu_{x_2}$ of zero by at least as much as the $(\bar{x}_1 - \bar{x}_2)$ observed.

If the probability $\geq$ 5 percent, accept the null hypothesis.
If the probability $<$ 5 percent, reject the null hypothesis.

Computation of the standard error of the difference between two sample means:

$$\sigma_{\bar{x}_1 - \bar{x}_2} = \sqrt{\frac{(\sigma_{x_1})^2}{n_1} + \frac{(\sigma_{x_2})^2}{n_2}}$$

$$= \sqrt{\frac{(0.11)^2}{64} + \frac{(0.09)^2}{36}}$$

$$= \sqrt{\frac{0.0121}{64} + \frac{0.0081}{36}}$$

$$= \sqrt{0.000189 + 0.000225}$$
$$= \sqrt{0.000414}$$
$$= 0.02$$

The postulated sampling distribution of the difference between two sample means is normally distributed with mean equal to zero and standard deviation (standard error of the difference) equal to 0.02.

Application of the procedure:

1. Determine the amount the observed sample statistic of the difference between two sample means deviates from the postulated universe mean difference of zero.

$$(\bar{x}_1 - \bar{x}_2) - (\mu_{x_1} - \mu_{x_2}) = (2.22 - 2.16) - 0.00 = 0.06 - 0.00 = 0.06$$

2. Express the deviation in standard error units.

$$\frac{(\bar{x}_1 - \bar{x}_2) - (\mu_{x_1} - \mu_{x_2})}{\sigma_{\bar{x}_1 - \bar{x}_2}} = \frac{0.06}{0.02} = 3\sigma_{\bar{x}_1 - \bar{x}_2}$$

3. Using the table of areas, determine the percentage of sample statistics that are expected to deviate from the universe mean difference by at least $3\sigma_{\bar{x}_1 - \bar{x}_2}$.

$3\sigma_{\bar{x}_1 - \bar{x}_2}$ includes 0.49865

Multiply this value by 2 in order to recognize deviations on both sides of the mean difference.

$$2(0.49865) = 0.9973$$

Therefore 99.73 percent of the sample statistics differs by $3\sigma_{\bar{x}_1 - \bar{x}_2}$ or less. It follows that 0.27 percent differ by 3 standard errors or more.

4. Compare the probability determined in step 3 with the critical probability to determine whether or not to accept the null hypothesis; 0.27 percent < 5 percent. Therefore the null hypothesis would be rejected.

The analyst would conclude that there is a difference in student attitude toward the two approaches. The observed difference of $+0.06$ would lead the analyst to conclude that μ_{x_1} or the conventional lecture approach would receive a higher mean score.

Two alternative formats in which the decision rule can be expressed are:

    I. If $\varkappa > |1.96|$, reject.
      If $\varkappa \leq |1.96|$, accept.
    II. If $CV_L \leq (\bar{x}_1 - \bar{x}_2) \leq CV_R$, accept.
      If $(\bar{x}_1 - \bar{x}_2) < CV_L$ or $(\bar{x}_1 - \bar{x}_2) > CV_R$, reject.

where

$$\begin{aligned} CV_R &= (\mu_{x_1} - \mu_{x_2}) + \varkappa\sigma_{\bar{x}_1 - \bar{x}_2} \\ &= 0 + 1.96(0.02) \\ &= +0.0392 \end{aligned}$$

and

$$\begin{aligned} CV_L &= (\mu_{x_1} - \mu_{x_2}) - \varkappa\sigma_{\bar{x}_1 - \bar{x}_2} \\ &= 0 - 1.96(0.02) \\ &= -0.0392 \end{aligned}$$

Any one of the decision rules is interchangeable with the others. They generate identical conclusions—in this case, reject the null hypothesis.

18.4 THE DIFFERENCE BETWEEN TWO UNIVERSE PROPORTIONS

The concept of the sampling distribution of the difference between two sample means can be extended to the sampling distribution of the difference between two sample proportions.

The following symbols will be used:

p_1 = the proportion possessing the attribute of interest in the first universe

p_2 = the proportion possessing the attribute of interest in the second universe

$\hat{p}_1$ = the proportion possessing the attribute of interest in the sample selected from universe 1

$\hat{p}_2$ = the proportion possessing the attribute of interest in the sample selected from universe 2

n_1 = the sample size selected from universe 1

n_2 = the sample size selected from universe 2

Provided that n_1 and n_2 are sufficiently large, the sampling distribution of the difference between two proportions is normally distributed with mean equal to $p_1 - p_2$ and standard deviation (standard error of the difference between two sample proportions) equal to

$$\sigma_{p_1 - p_2} = \sqrt{(\sigma_{p_1})^2 + (\sigma_{p_2})^2} = \sqrt{\frac{p_1(1 - p_1)}{n_1} + \frac{p_2(1 - p_2)}{n_2}}$$

The sampling distribution of the difference can be used to generate confidence interval statements and statements on the magnitude of expected sampling error. Generally, such statements take the following form:

Confidence interval:

$$(\hat{p}_1 - \hat{p}_2) \pm z\hat{\sigma}_{p_1 - p_2}$$

$$(\hat{p}_1 - \hat{p}_2) \pm z\sqrt{\frac{\hat{p}_1(1 - \hat{p}_1)}{n_1} + \frac{\hat{p}_2(1 - \hat{p}_2)}{n_2}}$$

where $\hat{p}_1$ and $\hat{p}_2$ are used as estimates of the unknown universe parameters p_1 and p_2.

Expected sampling error:

$$E = z\hat{\sigma}_{p_1 - p_2} \quad \text{or} \quad z\sqrt{\frac{\hat{p}_1(1 - \hat{p}_1)}{n_1} + \frac{\hat{p}_2(1 - \hat{p}_2)}{n_2}}$$

Test of the Hypothesis That There Is No Difference Between Two Universe Proportions

It has been postulated that the cleats on football shoes are largely responsible for some 50,000 serious knee injuries every season. Some medical experts contend that use of a soccer-type shoe will reduce the incidence of knee injuries. Conventional football shoes have only seven cleats, which means that more weight is carried on each cleat than is the case for soccer shoes, which have 14 cleats with flatter points. Players are injured when cleats fix themselves so deeply into the ground that the player is unable to move his foot. The rubber soccer shoe has a molded rubber sole with blunt points, which does not allow the foot to sink deeply into the turf. A large metropolitan area introduced the use of a soccer-type shoe. The reported incidence of knee injuries was 36 out of 400 participating players. The year before, using a conventional football shoe, 92 knee injuries were reported among 400 participating players. Using the above data, test the null hypothesis that there would be no difference between the proportion of knee injuries incurred with the use of a soccer-type shoe and the proportion of knee injuries incurred with the use of a conventional football cleat for the United States as a whole.

Conventional football shoe	Soccer-type shoe
Sample size $n_1 = 400$	Sample size $n_2 = 400$
Number of knee injuries $x_1 = 92$	Number of knee injuries $x_2 = 36$

The null hypothesis:

$$p_1 - p_2 = 0$$

The alternate hypothesis:

$$p_1 - p_2 \neq 0$$

that is, either $p_1 - p_2 > 0$ or $p_1 - p_2 < 0$.

A decision rule allowing a 10 percent probability of committing a Type I error is to be used.

THE DECISION RULE

Compute the probability of observing a $\hat{p}_1 - \hat{p}_2$ statistic that deviates from the postulated $p_1 - p_2$ of zero by at least as much as the $\hat{p}_1 - \hat{p}_2$ observed.

If the probability $\geq$ 10 percent, accept the null hypothesis.
If the probability $<$ 10 percent, reject the null hypothesis.

COMPUTATION OF THE STANDARD ERROR OF THE DIFFERENCE BETWEEN TWO SAMPLE PROPORTIONS

The hypothesis being tested is that $p_1 = p_2$. The best estimate of this postulated common proportion of knee injuries is a weighted average of the observed sample proportions.

Letting $\hat{p}$ represent the postulated common universe proportion,

$$\hat{p} = \frac{n_1 \hat{p}_1 + n_2 \hat{p}_2}{n_1 + n_2}$$

$$= \frac{400(0.23) + 400(0.09)}{800}$$

$$= \frac{92 + 36}{800}$$

$$= \frac{128}{800}$$

$$= 0.16$$

Given that $p_1 = p_2 = p$, the formula

$$\hat{\sigma}_{p_1 - p_2} = \sqrt{\frac{\hat{p}_1(1 - \hat{p}_1)}{n_1} + \frac{\hat{p}_2(1 - \hat{p}_2)}{n_2}}$$

can be simplified by substituting the estimated common proportion

$$\hat{p} = \frac{n_1 \hat{p}_1 + n_2 \hat{p}_2}{n_1 + n_2}$$

for $\hat{p}_1$ and $\hat{p}_2$. Thus,

$$\hat{\sigma}_{p_1-p_2} = \sqrt{\frac{\hat{p}(1-\hat{p})}{n_1} + \frac{\hat{p}(1-\hat{p})}{n_2}}$$

Factoring out the common expression $\hat{p}(1-\hat{p})$ and setting $(1-\hat{p})$ equal to q, we obtain

$$\hat{\sigma}_{p_1-p_2} = \sqrt{\hat{p}\hat{q}\left(\frac{1}{n_1} + \frac{1}{n_2}\right)}$$

Substituting:

$$\hat{\sigma}_{p_1-p_2} = \sqrt{(0.16)(0.84)(\tfrac{1}{400} + \tfrac{1}{400})}$$

$$= \sqrt{(0.1344)(\tfrac{2}{400})}$$

$$= \sqrt{\frac{0.1344}{200}}$$

$$= \sqrt{0.000672}$$

$$= 0.026$$

The postulated sampling distribution of the difference between two sample proportions is normally distributed with mean equal to zero and standard deviation (standard error of the difference between two sample proportions) equal to 0.026.

APPLICATION OF THE DECISION RULE

1. Determine the amount the observed sample statistic deviates from the postulated universe mean difference of zero.

$$(\hat{p}_1 - \hat{p}_2) - (p_1 - p_2) = (\tfrac{92}{400} - \tfrac{36}{400}) - 0.00 = 0.23 - 0.09 = 0.14$$

2. Express the deviation in standard error units.

$$\frac{(\hat{p}_1 - \hat{p}_2) - (p_1 - p_2)}{\hat{\sigma}_{p_1-p_2}} = \frac{0.14}{0.026} = 5.4\hat{\sigma}_{p_1-p_2}$$

3. Using the table of areas, determine the percentage of sample statistics that is expected to deviate from the universe mean difference by at least $5.4\hat{\sigma}_{p_1-p_2}$. The z value associated with 5.4 standard errors is not indicated in Table F; however, $3.8\sigma_{p_1-p_2}$ includes at least 0.4999. Multiply this value by 2 in order to recognize deviations on both sides of the mean difference

$$2(0.4999) = 0.9998$$

99.98 percent of the sample statistics would differ by $3.8\sigma_{p_1-p_2}$ or less. It follows that 0.02 percent would differ by 3.8 standard errors or more. Obviously an even smaller percentage would differ by 5.4 standard errors or more.

4. Compare the probability determined in step 3 with the critical probability to determine whether or not to accept the null hypothesis. 0.2 percent < 10 percent. Therefore the null hypothesis would be rejected. The analyst would conclude that there is a difference in the proportion of knee injuries incurred with the two types of shoe. Further the analyst would conclude that the observed evidence, a positive value for $\hat{p}_1 - \hat{p}_2$, implies that $p_1 > p_2$. Thus, the statistical evidence supports

the claim that use of the soccer type shoe will decrease the proportion of knee injuries incurred among participating players.

PROBLEMS

Problem 18.1

A Boston newspaper conducts a telephone subscription campaign in which individuals are asked to subscribe for 10 weeks, with all proceeds going to a well-known local charity. All subsequent revenues generated from subscriptions maintained beyond the 10-week period accrue to the newspaper. The managing editor wishes to estimate the percentage of the individuals who subscribe under this program who continue to maintain their subscription after the expiration of the 10-week period. A sample of size 100 contained 42 individuals who had continued to maintain their subscription after the expiration of the 10-week period. Construct a 95 percent confidence interval estimate of the percentage of all subscribers enrolled under this program who will maintain their subscription after the expiration of the 10-week period.

Problem 18.2

Brady Enterprises is considering offering a stock option to its employees (100,000 employees), but first wishes to estimate the percentage of the employees who will take advantage of such a program. How large a sample would have to be selected in order to be 95 percent confident that the error in the estimate will not be more than 5 percent? Proceed under the assumption that the analysts have absolutely no idea as to the true percentage.

Problem 18.3

The Rook of the Month Club is considering offering the book *How to Cheat at Monopoly* by I. T. Tea as a special purchase to its membership. The club wishes to estimate the percentage of the membership who will take advantage of this opportunity, since there is to be only one printing. Thirty-five of 100 subscribers selected at random from the mailing list indicated that they would purchase the book.

1. Construct a 95 percent confidence interval estimate of the proportion of the membership who will participate in the special offer.
2. Top management indicates that the sampling error associated with the estimate in part 1 is much too large. In fact, they desire to be 95 percent confident that the estimate generated by the sampling study will not differ from the true universe proportion by more than 1 percent. Determine the sample size required to meet this request. Use the 0.35 provided by the preliminary sample as the estimate of p, the universe parameter.
3. For what range of values of p would the actual sample size required to meet the precision constraints be less than that specified in part 2?
4. For what range of values of p would the actual sample size required to meet the precision constraints be greater than that specified in part 2?
5. For what range of values of p, the universe parameter, would the sample size determined in part 2 be neither too large nor too small?

Problem 18.4

A behaviorist, in a discussion with some restaurant owner friends, took issue with the comment that most people were very pleased with their dining experience at the restaurant. He pointed out that many people are reluctant to make statements that will cause unpleasantness and as a result will say that everything was fine when in reality they are dissatisfied. To demonstrate his point, he set up the following experiment. The owner was to inquire of each party whether or not they had enjoyed their dining experience. Outside, in the parking lot, the behaviorist and his wife, playing the role of prospective diners, would ask each departing party about the merits of the restaurant. Of 100 parties who responded favorably to the owner, 10 indicated some dissatisfaction to the behaviorist and his wife. What statement can the behaviorist make to the restaurant owner about the percentage of customers who have been indicating they are satisfied when in fact they are somewhat disgruntled? (Presume that the behaviorist has specified a 95 percent confidence level that his statement will be correct.)

Problem 18.5

Consider a decision rule to be used for testing a null hypothesis concerning the value of the universe mean. In constructing the Type II error curve, the value of the standard error of the mean is presumed to be the same regardless of the value of the universe mean. When constructing a Type II error curve in association with a null hypothesis concerning the value of the universe proportion, the appropriate sampling distribution is the sampling distribution of the proportion. Explain why it is not legitimate to assume that the standard error of the proportion is the same regardless of the value of the universe proportion.

Problem 18.6

At a town meeting, 60 percent of the eligible voters present voted in favor of a particular proposal (60 for; 40 against). Assume that less than 5 percent of the eligible voters attended and that those who did attend can be considered as a random sample from the eligible voter list (an heroic assumption).

1. If the true proportion of eligible voters who would have voted favorably is 50 percent, what is the probability of selecting a random sample of size 100 in which 60 percent or more would have voted favorably for this issue?

2. Assuming that the objective is to reflect correctly the posture of *all* eligible voters, what is the probability that the proposal will be passed by a random sample of 100 eligible voters when the true universe proportion in favor of the proposal is exactly 0.50? (More than 50 percent must vote favorably if the legislation is to be passed.)

3. If the true proportion of eligible voters who would have voted favorably is equal to 0.40, what is the probability of obtaining a random sample of 100 of the eligible voters that will pass the legislation?

Problem 18.7

The binomial distribution always provides the true description of the sampling distribution of the proportion. Under what conditions is it considered legitimate to approximate the sampling distribution of the proportion with a normal distribution?

Problem 18.8

A distillery evaluates the quality of each batch by selecting a random sample of eight employees from a pool of 1500 employees to conduct the following taste test: Each participant is provided a taste of the standard blend and a taste of the latest batch. He is then asked to indicate which is the standard blend. The ideal situation would be when none of the potential participants (entire pool of all employees) can differentiate between the two. When this occurs, one would expect 50 percent of the employees to identify the standard blend correctly and 50 percent to identify the latest batch as the standard blend (i.e., $p = 0.50$).

1. Which distribution provides the appropriate description of the sampling distribution of the proportion in the situation described above? Why?
2. What is the appropriate decision rule for testing the null hypothesis $p = 0.50$ for a sample of size 8 and a willingness to take a 7.04 percent chance of committing a Type I error?
3. What is the probability of committing a Type II error given the decision rule specified in part 2 for a sample of size 8 when $p = 0.20$?
4. Suppose that 50 employees are selected at random to perform the test.
   a. Specify the appropriate decision rule for testing the null hypothesis $p = 0.50$ for $n = 50$, $p = 0.50$ and an 11.86 percent chance of committing a Type I error. *Use the binomial distribution.*
   b. *Using the normal curve approximation* of the sampling distribution of the proportion for $n = 50$ and $p = 0.50$, determine the appropriate decision rule for testing the null hypothesis $p = 0.50$ in association with an 11.86 percent chance of committing a Type I error.
5. a. Using the normal curve approximation, determine the probability of committing a Type II error in association with the decision rule generated in part 4b when $p = 0.40$.
   b. Using the binomial distribution, determine the probability of committing a Type II error in association with the decision rule generated in part 4a when $p = 0.40$.

Problem 18.9

A public accounting firm wishes to test the hypothesis that no more than 2 percent of a large set of transactions have been recorded incorrectly. The following risk specifications have been stipulated:
   a. No more than a 5 percent probability of committing a Type I error.
   b. No more than a 5 percent probability of committing a Type II error when $p = 0.04$.

Required:
1. Determine the minimum sample size that will meet the above constraints.
2. State the associated decision rule.

Problem 18.10

You have been asked to conduct a sampling study on the extent of participation in a public affairs program entitled "Summerthing" which is sponsored by the city and funded out of tax revenues. The city council has indicated that unless participation is at least 40 percent, they intend to discontinue the program. Upon further questioning, you feel that the council's willingness to commit Type I and Type II errors is aptly described by the following constraints:

a. No more than a 10 percent chance of committing a Type I error when $p = 0.40$.

b. No more than a 20 percent chance of committing a Type II error when $p = 0.30$.

Required:

1. Determine the minimum sample size that will meet the above constraints.
2. State the associated decision rule.

Problem 18.11

The accrediting body for graduate schools of business considers a program to be weak if more than 20 percent of the students admitted to the program fail to meet the following three requirements:

ATGSB score	> 450
Cumulative undergraduate quality point	> 2.5
Undergraduate quality point for junior and senior years	> 2.75

The accrediting body gives the institution the benefit of the doubt. It is presumed that an institution meets the specified standards unless evidence is observed to the contrary. The decision rule to be used must possess the following error characteristics:

a. No more than a 10 percent chance of committing a Type I error when $p = 0.20$.

b. No more than a 10 percent chance of committing a Type II error when $p = 0.25$.

Required:

1. Determine the minimum sample size that meets the above constraints.
2. State the associated decision rule.

Problem 18.12

An instructor designs an exam that he believes 90 percent of all students should be able to finish in the allotted time. Only 32 of a class of 40 complete the exam. Utilizing a critical probability of 10 percent, determine if the sample evidence is consistent or inconsistent with the original hypothesis.

Problem 18.13

Brotherbrooks Clothiers must reduce the number of sales people by one. The choice has been narrowed to two salesmen, Clark Softsell and John Hardsell. It has been decided to select a random sample from the sales slips of each of the salesmen. If the sample evidence demonstrates beyond a reasonable doubt that the average sales of one is greater than the other, the salesman with the higher average sales is to be retained. The sample evidence is summarized below:

	Clark Softsell	*John Hardsell*
Sample size	$n_1 = 50$	$n_2 = 50$
Sample mean	$\bar{X}_1 = \$43$	$\bar{X}_2 = \$37$
Sample standard deviation	$\hat{\sigma}_{x_1} = \sqrt{\dfrac{\sum (X_1 - \bar{X}_1)^2}{n_1 - 1}} = \$14$	$\hat{\sigma}_{x_2} = \sqrt{\dfrac{\sum (X_2 - \bar{X}_2)^2}{n_2 - 1}} = \$12$

Test the null hypothesis that there is no difference between the average sales of the two salesmen using a critical probability of 10 percent.

Problem 18.14

The personnel office claims that the rate of retention on new hires is greater for female employees who are married at the time of initial employment than it is for female employees who are single. Samples selected at random from company files revealed the following statistics:

	Female employees married at the time of initial employment	Female employees single at the time of initial employment
Sample size	$n_1 = 40$	$n_2 = 40$
Number still employed 2 years after initial hiring date	22	28

Utilizing the above sample evidence, test the null hypothesis that $p_1 - p_2 \geq 0$ using a critical probability of 10 percent.

Problem 18.15

A golfing supplies distributor has been trying to persuade the local pro to carry his brand of golfballs. He claims that recent refinements will enable an individual to lengthen his average drive by at least 10 yards. The pro indicates that he will hit 100 of his regular ball and 100 of the new "super ball." If the evidence is inconsistent with the null hypothesis, $\mu_{x_1} - \mu_{x_2} \leq 5$ yards at a 10 percent critical probability, the pro has agreed to carry the new line. The results of the test are as follows:

Super ball	Regular ball
$n_1 = 100$	$n_2 = 100$
$\bar{X}_1 = 237$ yards	$\bar{X}_2 = 230$ yards

$$\hat{\sigma}_{x_1} = \sqrt{\frac{\sum (X_1 - \bar{X}_1)^2}{n_1 - 1}} = 6 \text{ yards} \qquad \hat{\sigma}_{x_2} = \sqrt{\frac{\sum (X_2 - \bar{X}_2)^2}{n_2 - 1}} = 5 \text{ yards}$$

Using the above sample evidence, test the null hypothesis, $\mu_{x_1} - \mu_{x_2} \leq 5$ yards.

Problem 18.16

Ceramic Supplies, Inc., is considering one of two processes for producing greenware. A sample of 100 castings from process 1 disclosed 12 defectives. Examination of 100 castings by process 2 revealed 28 defectives. Using a critical probability of 0.01, test the null hypothesis that there is no difference between the proportion of defectives produced by the two processes.

Problem 18.17

The Green Thumb Tree Nursery guarantees all plantings that it performs. The two partners disagree as to the planting procedure that should be followed. A random sample of the plantings performed by each of the partners disclosed the following:

	Jack Green	Tom Thumb
No. of plantings	100	100
No. of successes	86	94

Using a critical probability of 0.05, test the null hypothesis that there is no difference between the proportion of successful plantings for the two techniques.

Problem 18.18

A large metropolitan transit authority has initiated a safety training program for bus drivers. The program has been administered to a representative group of bus drivers selected at random from company files. Before initiating the program on a company-wide basis, they wish to evaluate whether or not the program has resulted in improved performance. A sampling study generated the following statistics:

	Bus drivers participating in the safety program	Bus drivers not participating in the safety program
Number	100	100
Number involved in an accident or a customer complaint within the last 3 months	5	11

Using a critical probability of 0.10 test the null hypothesis that participation in the safety program does not result in a difference in the proportion of drivers who are involved in an accident or complaint.

Problem 18.19

All new automobiles are equipped with a buzzer system that emits a signal until the driver and right front seat passenger (if any) have fastened their seatbelts. Ways exist to bypass the buzzer system so that individuals can drive without their seatbelts fastened and not have to listen to the annoying buzz. Proponents of this system have claimed that more than 80 percent of the seatbelt systems will be operated correctly. A random sample of 100 new car owners is to be selected to test the hypothesis that $p \geq 80$ percent. The critical probability is established at 5 percent.

1. State the decision rule by specifying the appropriate critical value separating the rejection and acceptance regions.
2. State the decision rule in terms of values of z (i.e., standard error units).
3. Given that the sample was comprised of 74 who utilized the system correctly and 26 who disconnected the system, should the hypothesis be accepted or rejected?
4. What is the probability of committing a Type I error when $p = 0.82$?
5. What is the probability of committing a Type II error when $p = 0.75$?

Problem 18.20

A child psychologist has been studying the ability of children of different ages to understand abstract concepts. Forty-nine children of age 4 and 64 children of age 5 were asked to listen to an explanation of ten concepts and then asked to respond to questions designed to determine the number of concepts they understood. The results were as follows:

	4-year-olds	5-year-olds
$\overline{X}$	3.2	3.5
$\hat{\sigma}_x$	0.7	0.8

Using a critical probability of 1 percent, test the hypothesis that there is no difference between 4- and 5-year-olds in understanding concepts.

Problem 18.21

Referring to the sample results in Problem 18.20, construct a 99 percent confidence interval estimate of the difference between the population means.

THE χ^2 DISTRIBUTION

A game encountered at amusement parks involves tossing a 12-sided die, each side being imprinted with the name of a different month. The implication is that each month has an equal opportunity of turning up. If this is true, the appropriate theoretical probability distribution describing the pattern of expected outcomes is the uniform distribution. Suppose that the following pattern of outcomes is observed over 2016 plays of the game.

Month	Number of times turned up
January	173
February	162
March	180
April	172
May	170
June	168
July	157
August	167
September	175
October	183
November	176
December	133

The expected pattern associated with a uniform distribution is for each month to turn up 168 times. The question is whether the discrepancy between the actual outcomes and the expected outcomes is one that could reasonably be observed given that the universe is truly uniform. Questions of this type are answered within the framework of a theoretical distribution called the χ^2 distribution (pronounced chi-square) introduced in 1900 by Karl Pearson.

19.1 GENERAL PROPERTIES OF THE χ^2 DISTRIBUTION

The χ^2 distribution describes the expected pattern of behavior for the sum of a specified number of squares of independent normally distributed variables, each of which has been selected from a universe characterized by a mean equal to zero and standard deviation equal to one. Thus, the variable U is χ^2 distributed, where $U = X_1{}^2 + X_2{}^2 + X_3{}^2 + \cdots + X_n{}^2$ and $X_i{}^2$ represents the square of a randomly selected observation from a universe with zero mean and unit standard deviation.

The specific pattern that emerges for a χ^2 distribution depends on the number of independent $X_i{}^2$ terms comprising the sum. The number of independent $X_i{}^2$ terms is referred to as the number of degrees of freedom. There is a distinct χ^2 distribution associated with each level of degrees of freedom. All χ^2 distributions are characterized by a mean equal to ϕ and a variance equal to 2ϕ, where ϕ is defined as the number of degrees of freedom. In general, χ^2 distributions associated with a small number of degrees of freedom are skewed to the right; the fewer the degrees of freedom, the greater the degree of skewness. As the number of degrees of freedom increases, the associated χ^2 distributions become more symmetrical, approaching the pattern of the normal distribution as the limiting case. It is considered legitimate to use a normal approximation with mean equal to ϕ and standard deviation equal to $\sqrt{2\phi}$ for χ^2 distributions associated with 30 or more degrees of freedom.[1]

In the illustration, if one were to repeat the experiment involving 2016 tosses of the die an infinite number of times, the observed frequency for each month would cluster around the expected frequency for that month in the pattern of a normal curve. Let f_o represents the observed frequency for a month on a particular performance of the experiment and f_e the expected frequency for that month over many performances of the experiment. One can then say that the statistic $f_o - f_e$ is normally distributed with mean equal to zero. Karl Pearson has demonstrated that dividing by $\sqrt{f_e}$ is equivalent to standardizing the statistic $f_o - f_e$. It follows that the statistic $(f_o - f_e)/\sqrt{f_e}$ is normally distributed with mean equal to zero and standard deviation equal to one. Squaring the statistic $(f_o - f_e)/\sqrt{f_e}$ generates $(f_o - f_e)^2/f_e$, which is the square of a normally distributed variable with mean equal to zero and standard deviation equal to one. One of these terms is generated for each category of outcomes comprising the experiment in question. Thus, one of these terms will be generated for each of the 12 months. The behavior of the sum of a number of these terms can be described by the appropriate member of the χ^2 probability distribution family. The appropriate χ^2 distribution is

---

[1] In practice, it has been determined that the quantity $\sqrt{2\chi^2}$ provides a better approximation to normality than χ^2 itself for values of n of 30 or more. The distribution of $\sqrt{2\chi^2}$ has a mean equal to $\sqrt{2\phi - 1}$ and a standard deviation equal to one.

determined by the number of degrees of freedom associated with the experiment under investigation.

In the illustrative example, there are only 11 independent observations because of the restriction that the sum of the expected frequencies must equal the sum of the observed frequencies. Upon determination of 11 of the $f_o - f_e$ statistics, the twelfth is automatically determined. Thus, the appropriate distribution is the χ^2 distribution associated with 11 degrees of freedom, which is designated as $\chi^2(11)$.

Table I in the Appendix contains selected percentage points for χ^2 distributions associated with various degrees of freedom. Each row represents a separate χ^2 distribution. The column headings indicate the percentage (in decimal form) lying in the right-hand tail of the associated theoretical χ^2 distribution. The entries in the rows refer to the values of χ^2 beyond which the percentage specified at the top of the column lies. A segment of Table I is reproduced as Table 19.1.

Table 19.1 Percentage points of the χ^2 distribution

ϕ \ P	0.995	0.99	0.975	0.95	0.90	0.75	0.50	0.25	0.10	0.05	0.025	0.01	0.005
11	2.60	3.05	3.82	4.57	5.58	7.58	10.34	13.70	17.28	19.68	21.9	24.7	26.8

Referring to Figure 19.1, which depicts the χ^2 distribution associated with 11 degrees of freedom, note that one would expect to observe a value of χ^2 equal to or greater than 10.34, 50 percent of the time; one would expect to observe a value of χ^2 equal to or greater than 17.28, 10 percent of the time; and one would expect to observe a value of χ^2 equal to or greater than 19.68, 5 percent of the time.

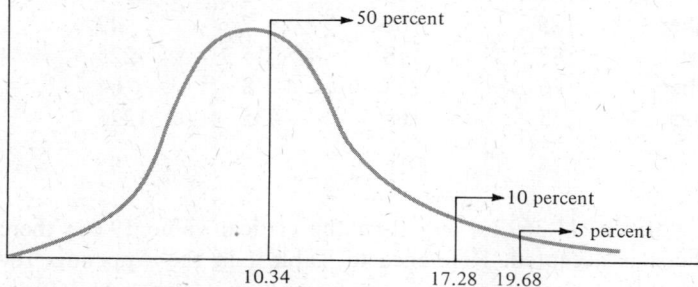

Figure 19.1 χ^2 Distribution associated with $\phi = 11$.

19.2 TESTS OF "GOODNESS OF FIT"

Tests of "goodness of fit" involve the computation of a χ^2 statistic from the observed frequencies and a comparison of the value so obtained with a previously determined critical value. The critical value represents a χ^2 value that the analyst would not reasonably expect to be exceeded due to chance selection in sampling. Tests of goodness of fit are one-tailed tests with the entire rejection region located in the right-hand tail. It is advisable to be suspicious, however, when extremely small values of χ^2 are obtained. Perfect or near-perfect fit may signal manipulation of the data selected for inclusion in the test.

Procedure

1. State the null hypothesis. In the illustrative example, the null hypothesis is that the sample observations have been drawn from a uniform distribution.

2. Specify the critical probability desired. In the illustrative example, the critical probability will be established at 5 percent. This reflects a willingness to employ a decision rule that will lead to rejection of a *true* null hypothesis 5 percent of the time.

3. Determine the appropriate χ^2 distribution and generate the decision rule. To determine the appropriate χ^2 distribution, it is necessary to determine the number of degrees of freedom associated with the experiment under investigation. In the illustrative example, ϕ is equal to 11.

Referring to Table I in the Appendix, the decision rule is determined to be:

Accept the null hypothesis if the observed $\chi^2 \leq 19.68$.
Reject the null hypothesis if the observed $\chi^2 > 19.68$.

4. Compute the χ^2 statistic and compare to the critical value.

Month	Observed frequency f_o	Expected frequency f_e	$f_o - f_e$	$(f_o - f_e)^2$	$\dfrac{(f_o - f_e)^2}{f_e}$
January	173	168	5	25	0.149
February	162	168	−6	36	0.214
March	180	168	12	144	0.857
April	172	168	4	16	0.095
May	170	168	2	4	0.024
June	168	168	0	0	0.000
July	157	168	−11	121	0.720
August	167	168	−1	1	0.006
September	175	168	7	49	0.292
October	183	168	15	225	1.339
November	176	168	8	64	0.381
December	133	168	−35	1225	7.292
	2016	2016			11.369

The χ^2 statistic, 11.369, is less than the critical value 19.68; therefore, the null hypothesis is accepted. Reference to Table I in the Appendix reveals that one would expect to observe a $\chi^2(11)$ equal to or greater than 11.369 significantly more frequently than 25 percent of the time. The aggregate discrepancy between the expected frequencies and those actually observed is one that could reasonably be attributed to chance selection in sampling. Acceptance of the null hypothesis does not necessarily mean that the universe is a uniform distribution. It simply indicates that the statistical evidence observed is not inconsistent with such a possibility.

Additional Considerations

1. A minimum of 50 observations is required if the theoretical χ^2 distribution is to provide a reasonable approximation of the expected sampling distribution.

2. The expected frequency for each category should be at least 5. On occasion it is necessary to combine categories in order to achieve this objective. When the expected frequency for a category is less than 5, the χ^2 approximation may be poor. If f_e is quite small, a slight discrepancy between f_o and f_e tends to be magnified out of proportion when the $(f_o - f_e)^2/f_e$ statistic is computed. To avoid this problem, the convention of a minimum f_e for each category has been adopted.

3. In general, the number of degrees of freedom may be interpreted as the number of unconstrained expected frequencies. In "goodness of fit" problems, this is equivalent to the number of classes or categories minus one degree of freedom for each restriction imposed on the expected frequency distribution.

> a. One degree of freedom is always lost due to the restriction that the sum of the expected frequencies must equal the sum of the observed frequencies, namely, $\sum f_e = \sum f_o$.
>
> b. In addition, a degree of freedom is lost for each parameter of the expected frequency distribution that is estimated from the sample observations.

The number of degrees of freedom is thus determined as follows:

$$\phi = n - r - 1$$

where

> ϕ represents the number of degrees of freedom
>
> n represents the number of categories or classes
>
> r represents the number of parameters that must be estimated from the sample observations
>
> 1 represents the degree of freedom lost due to the restriction that $\sum f_o = \sum f_e$.

EXAMPLE 19.1

In the illustrative example involving the uniform distribution, no universe parameters were estimated. Therefore,

$$\phi = n - r - 1$$
$$= 12 - 0 - 1$$
$$= 11$$

EXAMPLE 19.2

To test the goodness of fit of a Poisson distribution, one must have the parameter λ in order to compute the expected frequencies.

If λ is given, then

$$\phi = n - r - 1$$
$$= n - 0 - 1$$
$$= n - 1$$

If $\overline{X}$ is used to estimate λ, then

$$\phi = n - r - 1$$
$$n - 1 - 1$$
$$= n - 2$$

EXAMPLE 19.3

To test the goodness of fit for a binomial distribution, one must know n and p. The requirement $\sum f_o = \sum f_e$ is equivalent to specifying n.

If p is given, then

$$\phi = n - r - 1$$
$$= n - 0 - 1$$
$$= n - 1$$

If p is estimated from the sample observations, then

$$\phi = n - r - 1$$
$$= n - 1 - 1$$
$$= n - 2$$

EXAMPLE 19.4

To test the goodness of fit for a normal distribution, one must know μ and σ in order to compute the expected frequencies.

If μ and σ are specified, then

$$\phi = n - r - 1$$
$$= n - 0 - 1$$
$$= n - 1$$

If $\bar{X}$ and $\hat{\sigma}_x$ are used to estimate μ and σ, then

$$\phi = n - r - 1$$
$$= n - 2 - 1$$
$$= n - 3$$

Additional Illustrations

EXAMPLE 19.5[2]

The local law enforcement authorities have been advised that weekday robberies for suburban towns of similar income and population characteristics are Poisson distributed with a mean rate of occurrence of 0.8 robberies per day. The following pattern was observed when records covering the last 100 days were examined.

Number of robberies reported in one day	Number of days observed
0	61
1	28
2	9
3	0
4	2
5 or more	0
	100

Given the above pattern, conduct a χ^2 test to evaluate the legitimacy of the claim that the occurrence of weekday robberies is Poisson distributed

---

[2] If the reader is unfamiliar with the Poisson distribution, examples 19.5 and 19.6 should be omitted.

with a mean rate of occurrence equal to 0.8 robberies per day. Use a critical probability of 5 percent.

Procedure:

1. State the null hypothesis. The null hypothesis is that the pattern of robberies for the universe is Poisson distributed with a mean equal to 0.8 robberies per day.

2. Generate the expected frequencies. The expected frequencies are generated by multiplying the probability of observing zero successes, one success, two success, and so on (as provided by Table D in the Appendix for $\lambda = 0.8$) by the total number of observations (i.e., 100).

Number of robberies reported	Probability of occurrence	Expected frequency
0	0.4493	45
1	0.3595	36
2	0.1438	14
3	0.0383	4
4	0.0077	1
5 or more	0.0014	0
		100

3. Compute the χ^2 statistic.

Number of robberies reported	Observed frequency f_o	Expected frequency f_e	$f_o - f_e$	$(f_o - f_e)^2$	$\dfrac{(f_o - f_e)^2}{f_e}$
0	61	45	16	256	5.69
1	28	36	-8	64	1.78
2	9	14	-5	25	1.79
3 or more	2	5	-3	9	1.80
	100	100			11.06

Note that the category "3 or more" was obtained by combining a number of outcomes. Groupings such as this are frequently required in order to comply with the rule that the expected frequency for each cell must be equal to at least 5.

4. Establish the decision rule. λ, the mean rate of occurrence, is given and was not estimated from sample data. Therefore, the only restriction imposed is that $\sum f_o = \sum f_e$. There are four classes or categories. The number of degrees of freedom is equal to 3.

$$\chi^2_{0.05}(3) = 7.81$$

that is, 5 percent of the χ^2 values will have a value ≥ 7.81 in the χ^2 distribution associated with three degrees of freedom.

The decision rule is:

If $\chi^2 \leq 7.81$, accept the null hypothesis.
If $\chi^2 > 7.81$, reject the null hypothesis.

5. Application of the decision rule: $11.06 > 7.81$; therefore, the null hypothesis is rejected. The observed value of χ^2 is too large to be reasonably attributed to chance.

EXAMPLE 19.6[3]

Test that the distribution of robberies per day is Poisson distributed, but instead of using $\lambda = 0.8$, estimate λ from the sample data. Use a 5 percent critical probability.

Procedure:

1. Compute the arithmetic mean of the sample to provide an estimate of λ.

Number of robberies reported in one day	Number of days observed	
X	f_o	$f_o \cdot X$
0	61	0
1	28	28
2	9	18
3	0	0
4	2	8
	100	54

$$\bar{X} = \frac{\sum (f_o X)}{\sum (f_o)} = \frac{54}{100} = 0.54$$

$\hat{\lambda} \doteq \bar{X}$ or 0.54

2. State the null hypothesis. The null hypothesis is that the pattern of robberies for the universe is Poisson distributed with a mean equal to 0.5 robberies per day (0.5 is the closest value to 0.54 in Table D in the Appendix).

3. Generate the expected frequencies.

Number of robberies reported	Probability of occurrence	Expected frequency
0	0.6065	61
1	0.3033	30
2	0.0758	8
3	0.0126	1
4	0.0016	0
5 or more	0.0002	0
		100

4. Compute the χ^2 statistic.

Number of robberies reported	Observed frequency f_o	Expected frequency f_e	$f_o - f_e$	$(f_o - f_e)^2$	$\dfrac{(f_o - f_e)^2}{f_e}$
0	61	61	0	0	0
1	28	30	-2	4	0.13
2 or more	11	9	2	4	0.44
					0.57

5. Establish the decision rule. Two restrictions are imposed on the data. $\lambda = \bar{X}$ and $\sum f_o = \sum f_e$. There are three classes or categories. The number of degrees of freedom is equal to one. ($\phi = n - r - 1 = 3 - 1 - 1 = 1$).

$$\chi^2_{0.05}(1) = 3.84$$

---

[3] If the reader is unfamiliar with the Poisson distribution, example 19.6 should be omitted.

The decision rule is:

If $\chi^2 \leq 3.84$, accept the null hypothesis.
If $\chi^2 > 3.84$, reject the null hypothesis.

6. Application of the decision rule: $0.57 < 3.84$; therefore, the null hypothesis is accepted. The computed value of χ^2 is one that could reasonably be expected to be observed if the universe were Poisson distributed with $\lambda = 0.5$.

EXAMPLE 19.7

Measurement of the diameters of 200 gears selected from the output of a machine process generated the following pattern of observations:

Gear diameter (in.)	Observed frequency
Less than 2.497	1
2.497–2.498	7
2.498–2.499	33
2.499–2.500	64
2.500–2.501	59
2.501–2.502	29
2.502–2.503	5
2.503 or more	2
	200

Calculate χ^2 and test the null hypothesis that the universe from which the sample data has been selected is normally distributed with a mean equal to 2.500 in. and a standard deviation equal to 0.001 in. Use a critical probability of 5 percent.
Procedure:

1. Generate the expected frequencies.

Gear diameter	Probability[4] of occurrence	Expected frequency (probability × 200)
Less than 2.497	0.0014	0
2.497–2.498	0.0214	5
2.498–2.499	0.1359	27
2.499–2.500	0.3413	68
2.500–2.501	0.3413	68
2.501–2.502	0.1359	27
2.502–2.503	0.0214	5
2.503 or more	0.0014	0
		200

[4] Derived from the table of areas. To illustrate:

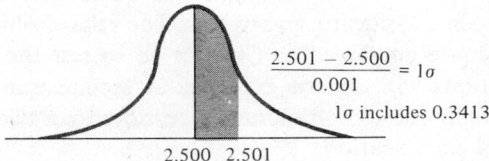

$$\frac{2.501 - 2.500}{0.001} = 1\sigma$$

1σ includes 0.3413

2.500 2.501

2. Compute the χ^2 statistic.

Gear diameter (in.)	Observed frequency f_o	Expected frequency f_e	$f_o - f_e$	$(f_o - f_e)^2$	$\dfrac{(f_o - f_e)^2}{f_e}$
Less than 2.498	8	5	3	9	1.80
2.498–2.499	33	27	6	36	1.33
2.499–2.500	64	68	−4	16	0.24
2.500–2.501	59	68	−9	81	1.19
2.501–2.502	29	27	2	4	0.15
2.502 or more	7	5	2	4	0.80
					5.51

3. Establish the decision rule. μ and σ are given and are not estimated from the sample data. The only restriction imposed is that $\sum f_o = \sum f_e$. There are six classes or categories. The number of degrees of freedom is equal to 5.

$$\chi^2_{0.05}(5) = 11.07$$

The decision rule is:

If $\chi^2 \leq 11.07$, accept the null hypothesis.
If $\chi^2 > 11.07$, reject the null hypothesis.

4. Application of the decision rule: $5.51 < 11.07$; therefore, the null hypothesis should be accepted. The pattern of observations observed in the sample is consistent with the results one would expect from a normally distributed universe with mean equal to 2.500 in. and standard deviation equal to 0.001 in.

19.3 χ^2 TEST OF INDEPENDENCE

In Chapter 18, one of the illustrative examples dealt with the hypothesis that no difference exists between the incidence of knee injuries associated with the use of a football-type shoe and the incidence of knee injuries when a soccer-type shoe is used. Alternatively, this hypothesis could have been expressed as follows: Occurrence of a knee injury and type of shoe worn are independent events; that is, the probability of incurring a knee injury is in no way affected by the type of shoe worn.

The technique introduced in Chapter 18 for testing the hypothesis that there is no difference between two universe proportions is one approach for evaluating such a hypothesis. An alternative equivalent test, called a χ^2 test of independence between classificatory variables, could also have been conducted. The χ^2 technique has an additional advantage in that it is capable of extension to situations involving more than two categories for each basis of classification.

In the following section, the knee-injury example will be used to illustrate a χ^2 test of independence between classificatory variables. The relationship between this technique and the procedure employed in Chapter 18 to test the difference between two sample proportions will also be explored. A second example illustrates the extension of the technique to situations where more than two variables are employed in each basis of classification.

EXAMPLE 19.8

The number of observations falling in each category are entered in a matrix referred to as an r by c contingency table, where r represents the number of rows and c represents the number of columns. In the illustration, whether or not a football player incurred a knee injury is the row basis of classification. The column basis of classification is whether the player wore a soccer-type or a conventional football-type shoe.

Contingency table

	Soccer-type shoe	Conventional football-type shoe	
Knee injury	36	92	128
No knee injury	364	308	672
	400	400	

The row and column totals reflect the total number of observations entered in each of the marginal categories. If the hypothesis of independence is true, then $p_s = p_f = p$, where p_s is the probability of a knee injury if a soccer-type shoe is worn, p_f is the probability of a knee injury if a conventional football-type shoe is worn, and p is the probability of a knee injury without specification as to the type of shoe worn.

The common proportion, p, can be estimated by combining the two sets of observations and determining the proportion of knee injuries for the entire sample. In the illustration, the common proportion equals 0.16.

$$\hat{p} = \frac{\begin{array}{c}\text{number of} \\ \text{knee injuries} \\ \text{with soccer-} \\ \text{type shoe}\end{array} + \begin{array}{c}\text{number of} \\ \text{knee injuries} \\ \text{with football-} \\ \text{type shoe}\end{array}}{\begin{array}{c}\text{total number} \\ \text{of observations}\end{array}} = \frac{36 + 92}{800} = \frac{128}{800} = 0.16$$

As in any χ^2 test, it is necessary to generate expected frequencies to compare with the observed frequencies. Two alternative approaches can be used.

Approach 1: To generate the expected frequencies for the first row, multiply the sum of column 1 and the sum of column 2 by $\hat{p}$.

$$(f_e)_{11} = (\hat{p})(\textstyle\sum \text{col. 1}) = (0.16)(400) = 64$$
$$(f_e)_{12} = (\hat{p})(\textstyle\sum \text{col. 2}) = (0.16)(400) = 64$$

To generate the expected frequencies for the second row, multiply the sum of column 1 and the sum of column 2 by $(1 - \hat{p})$.

$$(f_e)_{21} = (1 - \hat{p})(\textstyle\sum \text{col 1}) = (0.84)(400) = 336$$
$$(f_e)_{22} = (1 - \hat{p})(\textstyle\sum \text{col. 2}) = (0.84)(400) = 336$$

Approach 2: In general, the expected frequency for the ijth cell, $(f_e)_{ij}$, can be obtained by multiplying the marginal frequency for the ith row by the marginal frequency for the jth column and then dividing the product by the total number of observations comprising the contingency table.

$$(f_e)_{ij} = \frac{\sum (\text{row}_i) \times \sum (\text{col.}_j)}{\text{total number of observations}}$$

This is equivalent to multiplying the joint probability of two independent events by the total number of observations.

To illustrate:

$$(f_e)_{ij} = P\begin{pmatrix}\text{knee} \\ \text{injury}\end{pmatrix} \cap \begin{pmatrix}\text{soccer-} \\ \text{type shoe}\end{pmatrix}\begin{pmatrix}\text{total number} \\ \text{of observations}\end{pmatrix}$$

$$= P\begin{pmatrix}\text{knee} \\ \text{injury}\end{pmatrix} \cdot P\begin{pmatrix}\text{soccer-} \\ \text{type shoe}\end{pmatrix} \cdot \begin{pmatrix}\text{total number} \\ \text{of observations}\end{pmatrix}$$

$$= \left(\frac{\begin{array}{c}\text{number of} \\ \text{knee injuries}\end{array}}{\begin{array}{c}\text{total number} \\ \text{of observations}\end{array}}\right) \cdot \left(\frac{\begin{array}{c}\text{total number} \\ \text{wearing soccer-} \\ \text{type shoes}\end{array}}{\begin{array}{c}\text{total number} \\ \text{of observations}\end{array}}\right)\left(\begin{array}{c}\text{total number} \\ \text{of observations}\end{array}\right)$$

$$= \frac{\begin{pmatrix}\text{number of} \\ \text{knee injuries}\end{pmatrix}\begin{pmatrix}\text{total number} \\ \text{wearing soccer-} \\ \text{type shoes}\end{pmatrix}}{\begin{array}{c}\text{total number} \\ \text{of observations}\end{array}}$$

The cancelling process results in the simplified computational format:

$$(f_e)_{11} = \frac{\sum (\text{row}_i) \cdot \sum (\text{col.}_j)}{\text{total number of observations}}$$

To illustrate:

$$(f_e)_{11} = \frac{\sum (\text{row 1}) \cdot \sum (\text{col. 1})}{\text{total number of observations}} = \frac{128 \times 400}{800} = 64$$

$$(f_e)_{12} = \frac{\sum (\text{row 1}) \cdot \sum (\text{col. 2})}{\text{total number of observations}} = \frac{128 \times 400}{800} = 64$$

$$(f_e)_{21} = \frac{\sum (\text{row 2}) \cdot \sum (\text{col. 1})}{\text{total number of observations}} = \frac{672 \times 400}{800} = 336$$

$$(f_e)_{22} = \frac{\sum (\text{row 2}) \cdot \sum (\text{col. 2})}{\text{total number of observations}} = \frac{672 \times 400}{800} = 336$$

The expected frequency for each cell is entered in parentheses below the observed frequency for that cell.

Contingency table

	Soccer-type shoe	Conventional football-type shoe	
Knee injury	36 (64)	92 (64)	128
No knee injury	364 (336)	308 (336)	672
	400	400	

The conventional χ^2 test is then applied.

Procedure:

1. Compute the χ^2 statistic.

Cell (row, column)	Observed frequency f_o	Expected frequency f_e	$f_o - f_e$	$(f_o - f_e)^2$	$\dfrac{(f_o - f_e)^2}{f_e}$
1, 1	36	64	-28	784	12.25
1, 2	92	64	$+28$	784	12.25
2, 1	364	336	$+28$	784	2.33
2, 2	308	336	-28	784	2.33
					29.16

2. Establish the decision rule. The general rule for determining the number of degrees of freedom when dealing with a contingency table is $(r - 1)(c - 1)$, where r represents the number of rows and c represents the number of columns. Substituting:

$$(r - 1)(c - 1) = (2 - 1)(2 - 1) = 1$$

The one degree of freedom reflects the fact that although four expected frequencies are generated, as soon as one has been determined, the values for the remaining three are simultaneously determined as well. This is due to three independent restrictions imposed on the expected frequencies:

1. The total of the expected frequencies must equal the total of the observed frequencies.
2. The total *expected* frequency for the first row must equal the total *observed* frequency for the first row. (The total expected frequency for the second row must also equal the total observed frequency for the second row. This is not an independent restriction, however, since this constraint is automatically implied by constraints 1 and 2.)
3. The total *expected* frequency for the first column must equal the total *observed* frequency for the first column. (Similarly, the total expected frequency for the second column must equal the total observed frequency for the second column. Since this is already implied by restrictions 1 and 3, this does not constitute an independent restriction.)

Given the critical probability of 10 percent employed in Chapter 18, the critical value is

$$\chi^2_{0.10}(1) = 2.71$$

The decision rule is:

If $\chi^2 \leq 2.71$, accept the hypothesis of independence.
If $\chi^2 > 2.71$, reject the hypothesis of independence.

3. Application of the decision rule: $29.16 > 2.71$; therefore, the hypothesis of independence is rejected. The discrepancy between the expected frequencies and the observed frequencies is too great to be reasonably attributed to chance selection in sampling.

19.4 THE RELATIONSHIP BETWEEN THE χ^2 TEST OF INDEPENDENCE AND THE TEST OF THE NULL HYPOTHESIS THAT THERE IS NO DIFFERENCE BETWEEN TWO UNIVERSE PROPORTIONS

Test of the null hypothesis that there is no difference between the two universe proportions: Given a critical probability of 10 percent:

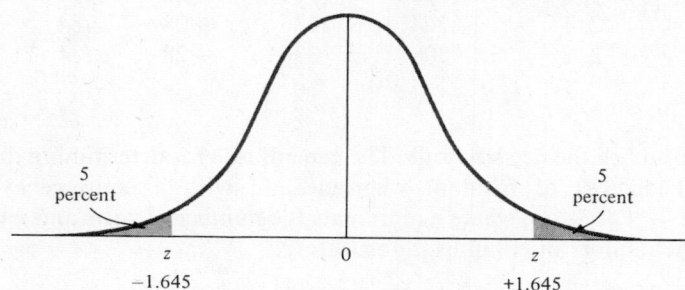

If the observed difference between the sample proportions differs from zero by more than 1.645 standard errors in either direction, the null hypothesis is rejected.

χ^2 test of independence: χ^2 deals with the square of normally distributed standardized values or z^2 values.

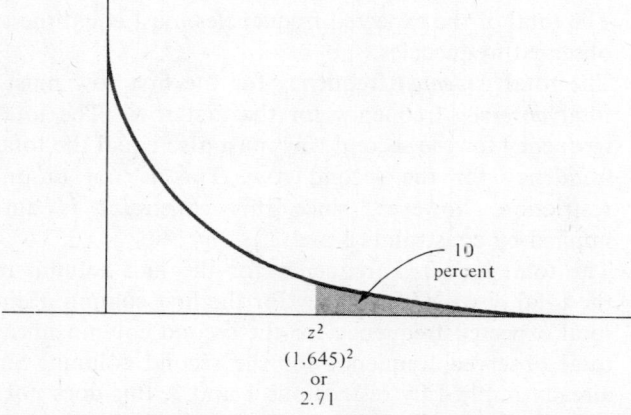

The probability of observing a $\chi^2 > 2.71$ is equivalent to observing a $z > |1.645|$.

EXAMPLE 19.9

The contingency table below summarizes the results obtained in a study
conducted by Consumer Studies, Inc., with respect to the performance of
four competing brands of toothpaste.

Contingency table

	Brand A	Brand B	Brand C	Brand D	
No cavities	9	13	17	11	50
One to five cavities	63	70	85	82	300
More than five cavities	28	37	48	37	150
	100	120	150	130	

Test the hypothesis that incidence of cavities is independent of the brand of
toothpaste used. Use a critical probability of 5 percent.
Procedure:

1. Generate the expected frequencies.

Cell	Computation of expected frequency	Cell	Computation of expected frequency
1, 1	$\dfrac{(50)(100)}{500} = 10$	2, 1	$\dfrac{(300)(100)}{500} = 60$
1, 2	$\dfrac{(50)(120)}{500} = 12$	2, 2	$\dfrac{(300)(120)}{500} = 72$
1, 3	$\dfrac{(50)(150)}{500} = 15$	2, 3	$\dfrac{(300)(150)}{500} = 90$
1, 4	$\dfrac{(50)(130)}{500} = 13$	2, 4	$\dfrac{(300)(130)}{500} = 78$

Cell	Computation of expected frequency
3, 1	$\dfrac{(150)(100)}{500} = 30$
3, 2	$\dfrac{(150)(120)}{500} = 36$
3, 3	$\dfrac{(150)(150)}{500} = 45$
3, 4	$\dfrac{(150)(130)}{500} = 39$

The expected frequencies are entered in the contingency table in parentheses below the observed frequency for each cell.

Contingency table

	Brand A	Brand B	Brand C	Brand D	
No cavities	9 (10)	13 (12)	17 (15)	11 (13)	50
One to five cavities	63 (60)	70 (72)	85 (90)	82 (78)	300
More than five cavities	28 (30)	37 (36)	48 (45)	37 (39)	150
	100	120	150	130	

2. Compute the χ^2 statistic.

Cell	f_o	f_e	$f_o - f_e$	$(f_o - f_e)^2$	$\dfrac{(f_o - f_e)^2}{f_e}$
1, 1	9	10	−1	1	0.1000
1, 2	13	12	+1	1	0.0833
1, 3	17	15	+2	4	0.2667
1, 4	11	13	−2	4	0.3077
2, 1	63	60	+3	9	0.1500
2, 2	70	72	−2	4	0.0556
2, 3	85	90	−5	25	0.2778
2, 4	82	78	+4	16	0.2051
3, 1	28	30	−2	4	0.1333
3, 2	37	36	+1	1	0.0278
3, 3	48	45	+3	9	0.2000
3, 4	37	39	−2	4	0.1026

$$1.9099$$

3. Establish the decision rule. There are six degrees of freedom.

$$(r - 1)(c - 1) = (3 - 1)(4 - 1) = (2)(3) = 6$$

The six restrictions are as follows: three of the four column totals are independently specified; two of the three row totals are independently specified; and the total number of expected frequencies must equal the total number of observations.

Twelve expected frequencies less six restrictions leaves six degrees of freedom.

$$\chi^2_{0.05}(6) = 12.59$$

The decision rule is:

If $\chi^2 \leq 12.59$, accept the hypothesis of independence.
If $\chi^2 > 12.59$, reject the hypothesis of independence.

4. Application of the decision rule: $1.9099 < 12.59$; therefore, the hypothesis of independence is accepted. Acceptance reflects the fact that the computed value of χ^2 is one that could reasonably be expected to occur if the probability of falling into each of the cavity categories is the same regardless of the brand of toothpaste used.

19.5 SAMPLING DISTRIBUTION OF THE VARIANCE[5]

The sampling distribution of the statistic ns^2/σ^2 will be χ^2 distributed with $n - 1$ degrees of freedom provided the samples of size n are drawn from a normally distributed universe.

n represents sample size
s^2 is the variance of the sample

$$s^2 = \frac{\sum (X - \bar{X})^2}{n}$$

σ^2 is the variance of the universe

Generation of a Confidence Interval Estimate of σ^2

In general,

$$L < \frac{ns^2}{\sigma^2} < U$$

where

L = lower confidence limit for the statistic ns^2/σ^2
U = upper confidence limit for the statistic ns^2/σ^2

If a 95 percent confidence interval is desired, L and U become

$L = \chi^2_{0.975}(n - 1)$
$U = \chi^2_{0.025}(n - 1)$

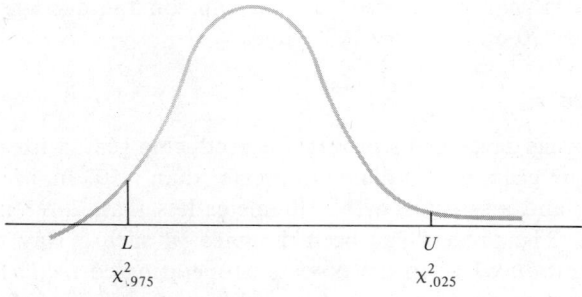

Substituting:

$$\chi^2_{0.975} < \frac{ns^2}{\sigma^2} < \chi^2_{0.025}$$

---

[5] This section may be omitted without loss of continuity.

Dividing by ns^2:

$$\frac{\chi^2_{0.975}}{ns^2} < \frac{1}{\sigma^2} < \frac{\chi^2_{0.025}}{ns^2}$$

Inverting:

$$\frac{ns^2}{\chi^2_{0.025}} < \sigma^2 < \frac{ns^2}{\chi^2_{0.975}}$$

The above expression is the general expression for a 95 percent confidence interval of the universe variance.

EXAMPLE 19.10

Suppose that a sample of size 25 generates a sample variance equal to 32 $[s^2 = \sum (X - \bar{X})^2/n]$. Generate a *90 percent* confidence interval estimate of the universe variance.

In general, a *90 percent* confidence interval estimate takes the following form:

$$\frac{ns^2}{\chi^2_{0.05}} < \sigma^2 < \frac{ns^2}{\chi^2_{0.95}}$$

$$\chi^2_{0.05}(24) = 36.4$$

$$\chi^2_{0.95}(24) = 13.85$$

Substituting:

$$\frac{(25)(32)}{36.4} < \sigma^2 < \frac{(25)(32)}{13.85}$$

$$\frac{800}{36.4} < \sigma^2 < \frac{800}{13.85}$$

$$21.70 < \sigma^2 < 57.76$$

"I am 90 percent confident that my statement is correct when I say that 'The universe variance has a value between 21.70 and 57.76'." The justification for such a statement is that the procedure used will, on the average, generate correct statements "90 out of every 100" times.

Test of a Hypothesis

Uniform Gear Works has developed a process for producing gears with an average diameter of 2.5 in. Any gears with a diameter greater than 2.503 in. are too loose to perform properly, and any gears with a diameter less than 2.497 in. are too tight to fit the shaft. The process has been designed in such a way that 99.73 percent of the gears produced when the process is operating correctly fall within the tolerances specified for usable gears (the value of a standard deviation is equal to 0.001 in. when the process is operating according to specifications). A sample of size 20 selected from the universe has generated a sample standard deviation equal to 0.0013 in. Test the hypothesis that the universe standard deviation is no greater than 0.001 in. Use a 5 percent critical probability.

Null hypothesis: $\sigma \leq 0.001$
Alternate hypothesis: $\sigma > 0.001$

Procedure: Compute the value of the statistic ns^2/σ^2. Compare the value ns^2/σ^2 to the critical value of χ^2 associated with $n - 1$ degrees of freedom.

1. Generate the decision rule.

$$\chi^2_{0.05}(19) = 30.1$$

The decision rule:

If $ns^2/\sigma^2 \leq 30.1$, accept the null hypothesis.
If $ns^2/\sigma^2 > 30.1$, reject the null hypothesis.

2. Compute ns^2/σ^2.

$$\frac{ns^2}{\sigma^2} = \frac{(20)(0.0013)^2}{(0.001)^2} = \frac{(20)(0.00000169)}{0.000001} = 20(1.69) = 33.80$$

3. Apply the decision rule: $33.80 > 30.1$; therefore, the null hypothesis is rejected. The value of 33.8 for ns^2/σ^2 is too large to be reasonably attributed to chance selection in sampling. The analyst would conclude that the process is not operating according to specifications.

TECHNICAL NOTE 11

The Yates Correction for Continuity

The computed value of χ^2 is a discrete variable because the observed frequencies from which the χ^2 statistic is generated are restricted to integer values. On the other hand, the entries in the χ^2 table are generated from a continuous theoretical distribution. When the total number of independent squares is large, the actual computed distribution of χ^2 values will closely approximate such a continuous distribution. However, when the number of independent squares entering into the computation of χ^2 is small, the approximation is not as accurate and the computed value of χ^2 tends to be overstated. In χ^2 tests involving *one degree of freedom*, the Yates correction for continuity is often employed to adjust for this problem.

The Yates correction for continuity takes the following form:

$$\chi^2 = \frac{\sum (|f_o - f_e| - \frac{1}{2})^2}{f_e}$$

$|f_o - f_e|$ is the absolute difference between the observed and the expected frequency. Note that the absolute difference between f_o and f_e is used to ensure consistency in the direction of correction for each term. Applying the Yates correction to illustrative example 19.8 yields the following results:

Cell	f_o	f_e	$|f_o - f_e|$	$|f_o - f_e| - \frac{1}{2}$	$(|f_o - f_e| - \frac{1}{2})^2$	$\dfrac{(|f_o - f_e| - \frac{1}{2})^2}{f_e}$
1, 1	36	64	28	27.5	756.25	11.8164
1, 2	92	64	28	27.5	756.25	11.8164
2, 1	364	336	28	27.5	756.25	2.2507
2, 2	308	336	28	27.5	756.25	2.2507
						28.1342

Although χ^2 is reduced slightly, 28.1342 is still significantly greater than the critical value 2.71. Introduction of the Yates correction for continuity is usually restricted to situations where the computed χ^2 value and the critical value are close and/or the expected frequencies for some of the cells are quite small.

PROBLEMS

Problem 19.1

In Chapter 2, 112 observations from the daily time cards of assembly workers in a Richmond, Virginia, plant were structured into the following frequency distribution:

Daily wage rate ($)	M	f
9–11	10	4
11–13	12	8
13–15	14	6
15–17	16	25
17–19	18	15
19–21	20	22
21–23	22	8
23–25	24	9
25–27	26	5
27–29	28	6
29–31	30	1
31–33	32	3
		112

The mean and standard deviation of the distribution were computed in Tables 3.3(b) and 4.9, respectively.

$$\bar{X} = \frac{\sum (fM)}{n} = \$19.16 \qquad \hat{\sigma}_x = \sqrt{\frac{\sum f(M - \bar{X})^2}{n - 1}} = \$5.00$$

Test the null hypothesis that the universe from which the above sample data has been selected is normally distributed. Use a critical probability of 5 percent. Estimate the universe parameters μ and σ from the sample data. (Recall that the expected frequency for each classification must be ≥ 5. Combine classes to achieve this where necessary.)

Procedure:

1. Generate the expected frequencies.
2. Establish the decision rule.
3. Compute the χ^2 statistic.
4. Apply the decision rule.

Problem 19.2

It has been hypothesized that the pattern of arrivals for 5-min intervals during the hours 9:00 a.m. to 11:00 a.m. at the Registry of Motor Vehicles office is Poisson distributed with a mean arrival rate of 1.7 arrivals per *5-min interval* (i.e., $\lambda = 1.7$). The following sample evidence has been collected.

Number of arrivals in a 5-min interval	Frequency of occurrence
0	41
1	36
2	17
3	5
4	1
5 or more	0
	100

1. Test the null hypothesis that the universe from which the above sample observations were obtained is Poisson distributed with a mean arrival rate of 1.7 arrivals per 5-min interval (i.e., $\lambda = 1.7$). Use a critical probability of 10 percent.

   Procedure:
   a. Generate the expected frequencies.
   b. Establish the decision rule.
   c. Compute the χ^2 statistic.
   d. Apply the decision rule.

2. Presume that the parameter λ is not known and must be estimated from the sample data. Conduct a χ^2 test of the goodness of fit of a Poisson distribution under these conditions.

Problem 19.3

Repeated observations on the outcomes of rolling a six-sided die revealed the following pattern of outcomes.

Face of die observed	Frequency
1	56
2	67
3	51
4	57
5	63
6	66

Using a 5 percent critical probability, test the hypothesis that the die is a fair die (i.e., evenly balanced).

Problem 19.4

An experiment consists of flipping a coin seven times and recording the number of heads observed. Four hundred performances of the experiment yielded the following pattern of outcomes:

No. of heads observed	No. of occasions the given number of heads was observed
0	33
1	99
2	126
3	92
4	39
5	9
6	1
7	1

1. Using a 5 percent critical probability, test the hypothesis that the coin is a fair coin.
2. Using a 5 percent critical probability, test the hypothesis that the coin is weighted such that heads will turn up 80 percent of the time.
3. Estimate p from the sample data. Using a 5 percent critical probability, test the hypothesis that the coin is weighted such that heads will turn up $\hat{p}$ percent of the time, where $\hat{p}$ is the sample estimate of p. Recall that $\mu_x = np$ and that $\overline{X} = \sum (fX)/\sum (f)$, therefore $\hat{p} = \overline{X}/n$, where n refers to the number of trials on each performance of the experiment.

Problem 19.5

In Problem 18.14, the null hypothesis $p_1 - p_2 \geq 0$ was tested using the sampling distribution of the difference between two sample proportions. A critical probability of 10 percent was used.
1. Design a χ^2 test of independence for the proportions p_1 and p_2 using a critical probability of 10 percent. State the decision rule and apply it.
2. How does the null hypothesis in part 1 differ from that employed in Problem 18.14?
3. Show the relationship between the critical value of χ^2 employed in the decision rule developed in part 1 and the value of z associated with the critical value in Problem 18.14.

Problem 19.6

In Problem 18.16, the null hypothesis that there is no difference between the proportion of defectives produced by two processes for casting ceramic molds was tested using the sampling distribution of the difference between two sample proportions. A critical probability of 0.01 was used.
1. Design a χ^2 test of independence for the proportion of defective castings with the two processes using a critical probability of 1 percent. State the decision rule and apply it.
2. Indicate the similarity between the null hypothesis tested in part 1 and the null hypothesis tested in Problem 18.16.
3. Demonstrate the relationship between the critical value of χ^2 associated with the decision rule employed in part 1 and the value of z associated with the critical value in Problem 18.16.

Problem 19.7

In Problem 18.17, the null hypothesis that there is no difference between the proportion of successful plantings for two alternative techniques employed by the Green Thumb Tree Nursery was tested using the sampling distribution of the difference between two sample proportions. A critical probability of 0.05 was used.
1. Design a χ^2 test of independence for the proportion of successful plantings with the two techniques using a critical probability of 5 percent. State the decision rule and apply it.
2. Indicate the similarity between the null hypothesis tested in part 1 and the null hypothesis tested in Problem 18.17.
3. Demonstrate the relationship between the critical value of χ^2 associated with the decision rule in part 1 and the value of z associated with the critical value in Problem 18.17.

Problem 19.8

In Problem 18.18, the null hypothesis that participation in a safety program does not result in a reduction in the proportion of drivers who are involved in an accident or complaint was tested using the sampling distribution of the difference between two sample proportions. A critical probability of 0.10 was used.

1. Design a χ^2 test of independence for the proportion of bus drivers involved in either an accident or a complaint for the two categories:
   a. drivers who participated in a safety program
   b. drivers who did not participate in a safety program
   Use a critical probability of 10 percent. State the decision rule and apply it.
2. How does the null hypothesis in part 1 differ from that employed in Problem 18.18?
3. Show the relationship between the critical value of χ^2 employed in the decision rule developed in part 1 and the value of z associated with the critical value in Problem 18.18.

Problem 19.9

The State Lottery Commission, in an effort to determine factors that distinguish participants from nonparticipants, conducted a study leading to the development of the following tables:

1.

	Engaged in some form of betting last year	Did not engage in some form of betting last year	
Never intends to play	46	24	70
Intends to be an infrequent player	33	117	150
Intends to be a regular player	1	79	80
	80	220	300

2.

	Does not consider the lottery a form of gambling	Does consider the lottery a form of gambling	
Never intends to play lottery	45	25	70
Infrequent player	80	70	150
Regular player	30	50	80
	155	145	300

3.

	Catholic	Jewish	Protestant	Other	None	
Never plays	20	6	34	4	6	70
Plays	120	20	68	12	10	230
	140	26	102	16	16	300

4.

	Did not play the numbers last year	Did play the numbers last year	
Never plays	1	69	70
Infrequent player	15	135	150
Regular player	30	50	80
	46	254	300

5.

	Male	Female	
Never plays	53	17	70
Plays	190	40	230
	243	57	300

6.

	Annual income						
	$3,000	$6,000	$10,000	$15,000	$20,000	$30,000	
Never plays	9	10	12	25	10	4	70
Plays	11	30	73	60	40	16	230
	20	40	85	85	50	20	300

Using a critical probability of 5 percent, test the null hypothesis that there is no relationship between the frequency with which an individual intends to purchase a lottery ticket and:
1. Whether or not he engaged in betting of some form last year.
2. Whether or not he considers the lottery a form of gambling.
3. His religious identification.
4. Whether or not he played the numbers last year.
5. Sex.
6. Income level.

Problem 19.10

A study to determine whether a relationship exists between academic major and achievement in the statistics course revealed the following pattern of observations.

	Accounting	Finance	Management	Marketing	
A	8	5	7	4	24
B	22	20	23	13	78
C	38	32	33	18	121
D	13	17	15	15	60
F	2	4	5	6	17
	83	78	83	56	300

Using a critical probability of 5 percent, test the null hypothesis that there is no relationship between academic major and achievement in the statistics course.

Problem 19.11

A recent development on college campuses has been the instructor evaluation form. To aid in the interpretation of the results of such evaluations, a study was made to determine whether any relationship existed between the stage of an individual's program and his attitude with respect to whether the academic work load was lighter than it should be, at the appropriate level, or heavier than it should be. A stratified random sample yielded the following results:

	Sophomore	Junior	Senior	
Believed work load is lighter than it should be	5	8	11	24
Believed work load is at the appropriate level	30	35	40	105
Believed work load is heavier than it should be	25	17	9	51
	60	60	60	180

1. Using a critical probability of 1 percent, test the null hypothesis that there is no relationship between the stage of an individual's program and his attitude with respect to the appropriateness of the academic work load.
2. What is the implication of a critical probability of 1 percent with respect to the likelihood of committing a Type I error?

Problem 19.12

The personnel office collected the following statistics in an effort to determine whether or not the degree of participation in student activities while in college had any relationship to level of achievement by members of the company's sales force. A sample selected from company records yielded the following.

	Low-level sales performance	Average sales performance	High-level sales performance	
No participation in student activities	18	18	4	40
Moderate participation in student activities	10	20	10	40
Active participation in student activities	2	12	6	20
	30	50	20	100

Using a critical probability of 10 percent, test the null hypothesis that there is no relationship between the degree of participation in student activities while in college and the level of achievement by members of the sales force.

Problem 19.13

1. Suppose that a sample of size 31 generates a sample variance equal to $32 [S^2 = \sum (X - \bar{X})^2 / n]$.
   a. Generate a 90 percent confidence interval estimate of the universe variance.
   b. What equivalent statement can be made with respect to the value of the universe standard deviation?
2. Suppose that a sample of size 61 generates a sample variance equal to $32 [S^2 = \sum (X - \bar{X})^2 / n]$.
   a. Generate a 90 percent confidence interval estimate of the universe variance.
   b. What equivalent statement can be made with respect to the value of the universe standard deviation?
3. Suppose that a sample of size 81 generates a sample variance equal to $32 [S^2 = \sum (X - \bar{X})^2 / n]$.
   a. Generate a 90 percent confidence interval estimate of the universe variance.
   b. What equivalent statement can be made with respect to the value of the universe standard deviation?

Problem 19.14

I. Snoop, an accountant for the firm, Merrill, Pierce and Kidder, is investigating the stocks and bonds sales records for 60 new recruits. He is particularly interested in determining whether or not a relationship exists between sales volume and educational background. The investigation revealed the following data:

	Less than $500,000 sales	$500,000 or more sales	Totals
College graduate	10	20	30
Not a college graduate	20	10	30
Totals	30	30	60

Using χ^2 analysis and a critical probability of 5 percent, determine whether sufficient evidence exists to support the claim that sales volume is dependent on educational level.

Problem 19.15

Bess Beyer, purchasing agent for Wino Vineyards in California, is interested in evaluating the effectiveness of three competitive brands of fertilizer. The vineyards were subdivided into 48 equivalent plots. Each brand of fertilizer was employed in 16 plots over the trial period. The following results were obtained:

Yield in 1000 wine gallons

Brand of fertilizer	Under 2	At least 2	Totals
A	10	6	16
B	8	8	16
C	6	10	16
Totals	24	24	

Employing a χ^2 analysis and a 1 percent critical probability, test whether yield is independent of the brand of fertilizer used. Briefly explain your answer in terms of the dependence or independence of these two bases of classification.

Problem 19.16

Indicate the nature of the null hypothesis employed in χ^2 tests of goodness of fit.

Problem 19.17

Under what condition(s) is it appropriate to use the Yates correction for continuity in a χ^2 test?

Problem 19.18

Suppose that you compute the χ^2 value in a test of independence for disposable income levels and consumption expenditure levels. This value is then compared to the table value of χ^2 associated with the specified degrees of freedom and critical probability. Explain precisely the conclusion that would be reached if the computed value exceeded the table value.

THE F DISTRIBUTION: ANALYSIS OF VARIANCE

20.1 TEST OF THE HYPOTHESIS OF EQUALITY OF POPULATION MEANS

U.N. Associates is a subcontractor specializing in the assembly of engine components. Three alternative techniques for training employees to perform a particular piecework operation are currently under consideration. Five workers were randomly selected for training in each technique. At the conclusion of a reasonable training period, an evaluation of each worker's performance was obtained. The results of the evaluation are summarized in Table 20.1.

The hypothesis to be tested is that no difference exists between the true mean output level that would be realized with adoption of one of the training techniques as compared to that which would be realized with adoption of any of the others. Stated more formally,

Table 20.1 Daily output

Training Technique 1		Training Technique 2		Training Technique 3	
Worker 1	90	Worker 6	97	Worker 11	88
Worker 2	85	Worker 7	85	Worker 12	96
Worker 3	105	Worker 8	102	Worker 13	86
Worker 4	88	Worker 9	95	Worker 14	91
Worker 5	92	Worker 10	101	Worker 15	79
Sum	460	Sum	480	Sum	440
$\bar{X}_1 = \frac{460}{5} = 92$		$\bar{X}_2 = \frac{480}{5} = 96$		$\bar{X}_3 = \frac{440}{5} = 88$	

the null hypothesis is $\mu_1 = \mu_2 = \mu_3$. The alternate hypothesis is that μ_1, μ_2, and μ_3 are not all equal. In Chapter 19, the χ^2 test of independence was introduced to test for the existence of a significant difference "between" two or more sample proportions. The hypothesis tested was that the population proportions were identical for the populations from which the samples had been drawn. Similarly, a technique is available for conducting a test to determine if a significant difference exists "between" more than two sample means. The technique is called *analysis of variance*, and the test is conducted within the framework of a sampling distribution that is a member of the F distribution family.[1] The nature of the F distribution will be discussed later in the chapter.

The test procedure involves the generation of two independent estimates of the unknown universe variance from the sample data. If the hypothesis is true, both estimates will be unbiased estimators of the unknown universe variance. A statistic is then formed from the ratio of the two estimates:

$$F = \frac{S_1{}^2}{S_2{}^2}$$

The sampling distribution of the $S_1{}^2/S_2{}^2$ statistic is described by a member of the F distribution family provided that three conditions are met:

1. The population from which each sample mean has been drawn is normally distributed.
2. There is equality of the population variances; that is,

   $$\sigma_1{}^2 = \sigma_2{}^2 = \sigma_3{}^2$$

3. The null hypothesis is true, namely,

   $$\mu_1 = \mu_2 = \mu_3$$

Slight departures with respect to the first two conditions can be tolerated without negating the usefulness of the technique. When such departures occur, the probability of rejecting a true hypothesis is usually greater than the stipulated level of significance. A technique occasionally utilized to offset such distortions is the substitution of a reduced critical probability.

A critical value is established. This is a value that the ratio would not be expected to exceed, say, more than 5 percent of the time, provided that the null hypothesis is true. If the null hypothesis is not true, the effect will be an increase

---

[1] The F is in honor of R. A. Fisher, who developed the technique during the 1920s.

in the numerator of the ratio and hence an increase in the magnitude of the ratio itself. If the computed value of the ratio exceeds the critical value, the null hypothesis is rejected and the alternate hypothesis, namely, that μ_1, μ_2, and μ_3 are not all equal, is accepted. The procedure for conducting such a test and the underlying rationale will be explored within the context of the illustrative example. It is interesting to note that the hypothesized equality of population means is tested by analyzing variances. At first glance, this appears to be a contradiction. The justification for the approach will become evident shortly.

Procedure:

1. State the hypothesis:

   Null hypothesis: $\mu_1 = \mu_2 = \mu_3$.
   Alternate hypothesis: μ_1, μ_2, and μ_3 are not all equal.

2. Generate the estimates of the unknown common universe variance.

σ_w^2, An Estimate Based on Within-Sample Variation

Given the equality of σ_1^2, σ_2^2, and σ_3^2 and that the populations they represent are normally distributed, an unbiased estimate of the common universe variance can be generated from any one of the three samples.

In general,

$$\bar{X}_j = \frac{\sum\limits_{j=1}^{n} (X_{ij})}{n}$$

$$S_j^2 = \frac{\sum\limits_{i=1}^{n} (X_{ij} - \bar{X}_j)^2}{n-1}$$

where

X_{ij} represents the output of the ith worker trained by the jth technique
$\bar{X}_j$ represents the mean output of the five workers trained by the jth technique
S_j^2 represents the sample variance for the jth technique computed as an unbiased estimator of the unknown universe variance

For training technique 1:

X_{i1}	$X_{i1} - \bar{X}_1$	$(X_{i1} - \bar{X}_1)^2$
90	-2	4
85	-7	49
105	$+13$	169
88	-4	16
92	0	0
460		238

$$\bar{X}_1 = \frac{\sum\limits_{i=1}^{n} (X_{i1})}{n} = \frac{460}{5} = 92$$

$$S_1^2 = \frac{\sum\limits_{i=1}^{n} (X_{i1} - \bar{X}_1)^2}{n-1} = \frac{238}{4} = 59.50$$

For training technique 2:

X_{i2}	$X_{i2} - \bar{X}_2$	$(X_{i2} - \bar{X}_2)^2$
97	$+1$	1
85	-11	121
102	$+6$	36
95	-1	1
101	$+5$	25
480		184

$$\bar{X}_2 = \frac{\sum_{i=1}^{n} (X_{i2})}{n} = \frac{480}{5} = 96$$

$$S_2^2 = \frac{\sum_{i=1}^{n} (X_{i2} - \bar{X}_2)^2}{n - 1} = \frac{184}{4} = 46$$

For training technique 3:

X_{i3}	$X_{i3} - \bar{X}_3$	$(X_{i3} - \bar{X}_3)^2$
88	0	0
96	$+8$	64
86	-2	4
91	$+3$	9
79	-9	81
440		158

$$\bar{X}_3 = \frac{\sum_{i=1}^{n} (X_{i3})}{n} = \frac{440}{5} = 88$$

$$S_3^2 = \frac{\sum_{i=1}^{n} (X_{i3} - \bar{X}_3)^2}{n - 1} = \frac{158}{4} = 39.5$$

Although any one of the three would constitute an unbiased estimator of the unknown common universe variance, an even better estimator can be generated by taking their mean value:

$$S_w^2 = \frac{S_1^2 + S_2^2 + S_3^2}{3} = \frac{59.50 + 46 + 39.5}{3} = \frac{145}{3} = 48.33$$

This estimate is called the within-sample estimate of the universe variance, because its computation is based on the variation observed *within* each sample. Note that the within-sample estimate provides an unbiased estimate of the true unknown common universe variance whether the null hypothesis is true or not. The common universe variance reflects the fact that the myriad of independent random influences operating *within* each of the categories has affected the variation within each category in a similar fashion. This is given by the constraint that $\sigma_1^2 = \sigma_2^2 = \sigma_3^2$. Thus, σ_w^2 can be treated as an estimate of the extent of variation that is attributable to chance selection in sampling from populations that are identical in shape (normal) and dispersion (common variance).

In general,

$$S_w{}^2 = \frac{\sum d_1{}^2 + \sum d_2{}^2 + \cdots + \sum d_k{}^2}{(n_1 - 1) + (n_2 - 1) + \cdots + (n_k - 1)}$$

Note that one degree of freedom is lost for each of the sample estimates. Thus, the total number of degrees of freedom is equal to $K(n - 1)$, where K represents the number of samples and n represents the common sample size.

An alternative approach for generating the value of $S_w{}^2$ involves the use of the following formula[2]:

$$S_w{}^2 = \frac{\sum\limits_{j=1}^{k} \sum\limits_{i=1}^{n} (X_{ij} - \overline{X}_j)^2}{K(n - 1)}$$

The double summation indicates that one first determines $\sum (X_{ij} - \overline{X}_j)^2$ for each j, allowing only the i subscript to vary, and then sums the resulting subtotals for $j = 1$ through k.

Procedure:

1. Determine the deviation of each observation from its sample mean.

$$X_{ij} - \overline{X}_j$$

2. Square the deviations.

$$(X_{ij} - \overline{X}_j)^2$$

3. Sum the squared deviations.

$$\sum\limits_{j=1}^{k} \sum\limits_{i=1}^{n} (X_{ij} - \overline{X}_j)^2$$

4. Divide by the number of degrees of freedom.

$$\frac{\sum\limits_{j=1}^{k} \sum\limits_{i=1}^{n} (X_{ij} - X_j)^2}{K(n - 1)}$$

Verify that this formula also generates the value $S_w{}^2 = 48.33$.

$\sigma_b{}^2$, An Estimate Based on Between-Sample Variation

If the null hypothesis is true, namely, $\mu_1 = \mu_2 = \mu_3$, then $\overline{X}_1$, $\overline{X}_2$, and $\overline{X}_3$ are unbiased estimators of the common universe mean. The dispersion of sample means around the true universe mean is measured by the standard error of the mean.

$$\sigma_{\bar{x}} = \frac{\sigma_x}{\sqrt{n}}$$

Multiplying both sides by $\sqrt{n}$:

$$\sigma_{\bar{x}} \cdot \sqrt{n} = \sigma_x$$

---

[2] The formulas discussed in this chapter all assume equal sample size. If samples of unequal size are selected, weights must be introduced to reflect this fact.

Squaring both sides:

$$\sigma_x^2 = n\sigma_{\bar{x}}^2$$

An estimate of $\sigma_{\bar{x}}^2$ can be generated from the observed sample means. The sampling distribution of the mean is treated as a universe and the standard error of the mean as its standard deviation (which, of course, it is).

In general,

$$\bar{\bar{X}} = \frac{\bar{X}_1 + \bar{X}_2 + \cdots + \bar{X}_k}{K}$$

and

$$\hat{\sigma}_{\bar{x}}^2 = \frac{(\bar{X}_1 - \bar{\bar{X}})^2 + (\bar{X}_2 - \bar{\bar{X}})^2 + \cdots + (\bar{X}_k - \bar{\bar{X}})^2}{K - 1}$$

where $\hat{\sigma}_{\bar{x}}$ is an unbiased estimator of the unknown standard error of the mean.

An estimate of the unknown common universe variance is obtained by multiplying $\hat{\sigma}_{\bar{x}}^2$ by the common sample size.[3]

$$S_b^2 = n\hat{\sigma}_{\bar{x}}^2$$

$$\bar{\bar{X}} = \frac{\bar{X}_1 + \bar{X}_2 + \bar{X}_3}{K} = \frac{92 + 96 + 88}{3} = \frac{276}{3} = 92$$

$$\hat{\sigma}_{\bar{x}}^2 = \frac{(\bar{X}_1 - \bar{\bar{X}})^2 + (\bar{X}_2 - \bar{\bar{X}})^2 + (\bar{X}_3 - \bar{\bar{X}})^2}{K - 1}$$

$$= \frac{(92 - 92)^2 + (96 - 92)^2 + (88 - 92)^2}{3 - 1}$$

$$= \frac{0^2 + 4^2 + (-4)^2}{2} = \frac{16 + 16}{2} = \frac{32}{2}$$

$$= 16$$

$$S_b^2 = n\hat{\sigma}_{\bar{x}}^2$$

$$= 5(16)$$

$$= 80$$

Provided that the null hypothesis is true, S_b^2 is subject to exactly the same battery of random independent influences as S_w^2. Thus, their ratio should conform closely to one. Some variation is expected, of course, because S_b^2 and S_w^2 are both sample estimates. The amount of variation that is reasonably attributable to chance selection in sampling is a function of the number of degrees of freedom entering into the computation of S_w^2 and S_b^2, respectively. If the null hypothesis is not true, the magnitude of the estimate of the common universe variance provided by S_b^2 will be biased upward. The upward bias is attributable to the influence of

---

[3] Recall:

$$\sigma_{\bar{x}} = \frac{\sigma_x}{\sqrt{n}}$$

$$\sqrt{n}\sigma_{\bar{x}} = \sigma_x$$

$$n\sigma_{\bar{x}}^2 = \sigma_x^2$$

the difference in the magnitude of the true population mean for each of the categories. Thus, S_b^2, on the average, will overstate the value of the common universe variance. The greater the difference between the population means, the greater the resulting overstatement in the estimate formed by S_b^2.

A more formal approach to generating the value of S_b^2 involves the use of the following relationship:

$$\hat{\sigma}_{\bar{x}}^2 = \frac{\sum\limits_{j=1}^{k} (\bar{X}_j - \bar{\bar{X}})^2}{K - 1}$$

Recall that

$$S_b^2 = n\hat{\sigma}_{\bar{x}}^2$$

Substituting:

$$S_b^2 = \frac{n \sum\limits_{j=1}^{k} (\bar{X}_j - \bar{\bar{X}})^2}{K - 1}$$

Procedure:

1. Compute the overall mean.

$$\bar{\bar{X}} = \frac{\bar{X}_1 + \bar{X}_2 + \bar{X}_3}{K}$$

2. Determine the deviation of each sample mean from the overall mean.

$$\bar{X}_j - \bar{\bar{X}}$$

3. Square the deviations.

$$(\bar{X}_j - \bar{\bar{X}})^2$$

4. Sum the squared deviations over the K samples.

$$\sum\limits_{j=1}^{k} (\bar{X}_j - \bar{\bar{X}})^2$$

5. Multiply by the common sample size.

$$n \sum\limits_{j=1}^{k} (\bar{X}_j - \bar{\bar{X}})^2$$

6. Divide by the number of degrees of freedom.

$$\frac{n \sum\limits_{j=1}^{k} (\bar{X}_j - \bar{\bar{X}})^2}{K - 1}$$

Verify that this formula generates the value $S_b^2 = 80$.

3. Construct the F ratio.

$$F = \frac{S_b^2}{S_w^2}$$

$S_b{}^2$ is always placed in the numerator. Thus, if the null hypothesis is false, it will be reflected by a ratio that exceeds one by a greater amount than could reasonably be attributed to chance selection in sampling.

4. Generate the decision rule. To determine a value for the F ratio that could not reasonably be exceeded due to chance selection in sampling, it is necessary to know the pattern of behavior that the F ratio will exhibit when the null hypothesis is true. The F distribution developed by R. A. Fisher describes the behavior of the ratio of two χ^2 distributed variables, each being divided by its appropriate degrees of freedom. Technical Note 12 at the end of the chapter demonstrates that the ratio formed by $S_b{}^2/S_w{}^2$ is such a statistic.

The appropriate member of the F distribution is determined by the number of degrees of freedom entering into the formation of the numerator and the number of degrees of freedom entering into the formation of the denominator. Letting ϕ_n represent the number of degrees of freedom in the numerator and ϕ_d the number of degrees of freedom in the denominator, a different F distribution is specified for each distinct ϕ_n, ϕ_d pair. In general, for ϕ_n and $\phi_d > 2$, the F distribution is unimodal and positively skewed. As the number of degrees of freedom increases, the F distribution approaches the normal curve as the limiting case. F can take on values ranging from zero to plus infinity. Tables have been constructed that contain the value of F that would be exceeded 5 percent of the time due to chance selection in sampling for each of the F distributions. The number of degrees of freedom in the numerator, ϕ_n, is designated by the column, and the number of degrees of freedom in the denominator, ϕ_d, is designated by the row. The critical value of F is entered in the body of the table at the intersection of the appropriate row and column. Table J in the Appendix reproduces a portion of such a table. Table J also contains the critical values associated with a 1 percent level of significance. The first listed or smaller value is the value on the F scale to the right of which lies 0.05 of the area under the curve. The second listed or larger value is the value on the F scale to the right of which lies 0.01 of the area under the curve.

In the illustrative example, $\phi_d = K(n - 1) = 3(5 - 1) = 12$ and $\phi_n = K - 1 = 3 - 1 = 2$. Using a critical probability of 5 percent, the critical value for F is determined to be 3.88 from Table J in the Appendix. If a 1 percent critical probability had been stipulated, the critical value for F would have been 6.93. The segment of Table J referred to is reproduced as Figure 20.1.

ϕ_d (degrees of freedom, denominator)	ϕ_n (degrees of freedom, numerator)		
	2		
12	3.88 6.93		

Figure 20.1

The decision rule associated with a 5 percent critical probability is

If $S_b^2/S_w^2 \leq 3.88$, accept the null hypothesis, namely, $\mu_1 = \mu_2 = \mu_3$.
If $S_b^2/S_w^2 > 3.88$, reject the null hypothesis and accept the alternate hypothesis, namely, that μ_1, μ_2, and μ_3 are not all equal.

5. Apply the decision rule.

$$\frac{S_b^2}{S_w^2} = \frac{80}{48.25} = 1.66$$

$1.66 < 3.88$; therefore, the null hypothesis is accepted. A value of 1.66 is one that could reasonably be expected to be observed when $\mu_1 = \mu_2 = \mu_3$. It is, of course, further assumed that $\sigma_1^2 = \sigma_2^2 = \sigma_3^2$ and that the underlying populations are all normally distributed.

A convenient format for generating the F statistic is the tabular arrangement shown in Table 20.2. SSB represents the sum of the squared deviations generated from between-sample variation. SSW represents the sum of the squared deviations generated from within-sample variation.

Table 20.2

Source of the estimate of the universe variance	Sum of the squared deviations	Degrees of freedom	Mean square (estimate of the universe variance)	Ratio
Between-sample variation	$SSB = n \sum_{j=1}^{k}(\bar{X}_j - \bar{\bar{X}})^2$	$K - 1$	$S_b^2 = \dfrac{SSB}{K-1}$	
				$F = \dfrac{S_b^2}{S_w^2}$
Within-sample variation	$SSW = \sum_{j=1}^{k}\sum_{i=1}^{n}(X_{ij} - \bar{X}_j)^2$	$K(n-1)$	$S_w^2 = \dfrac{SSW}{K(n-1)}$	

20.2 SHORT-CUT FORMULAS

Once the underlying relationships have been grasped, the following short-cut formulas are typically employed. The primary advantage associated with the use of the short-cut formulas is the elimination of the necessity for computing deviations.

$$SSB = \frac{\sum_{j=1}^{k}(T_j^2)}{n} - \frac{T^2}{K(n)}$$

where

SSB represents the sum of the squared deviations generated from between-sample variation

T_j represents the sum of the observations for the jth category

T represents the sum of the observations over all samples

n represents the common sample size

K represents the number of samples

$K(n)$ represents the total number of observations over all samples

$$SSW = \sum_{j=1}^{k}\sum_{i=1}^{n}[(X_{ij})^2] - \frac{\sum_{j=1}^{k}(T_j)^2}{n}$$

where

SSW represents the sum of the squared deviations generated from within-sample variation

$\sum_{j=1}^{k} \sum_{i=1}^{n} [(X_{ij})^2]$ represents the sum obtained when each of the individual observations over all samples are squared and then summed

$$F = \frac{K(n-1)(SSB)}{(K-1)(SSW)}$$

$$\sum_{j=1}^{k} (T_j^2) = T_1^2 + T_2^2 + T_3^2$$

$$= (460)^2 + (480)^2 + (440)^2$$

$$= 211600 + 230400 + 193600$$

$$= 635600$$

$$T^2 = (T_1 + T_2 + T_3)^2$$

$$= (460 + 480 + 440)^2$$

$$= (1380)^2$$

$$= 1904400$$

$$SSB = \frac{\sum T_j^2}{n} - \frac{T^2}{Kn}$$

$$= \frac{635600}{5} - \frac{1904400}{15}$$

$$= 127120 - 126960$$

$$= 160$$

$$SSW = \sum_{j=1}^{k} \sum_{i=1}^{n} [(X_{ij})^2] - \frac{\sum_{j=1}^{k} (T_j)^2}{n}$$

$$= [(90)^2 + (85)^2 + (105)^2 + (88)^2 + (92)^2 + (97)^2$$

$$+ (85)^2 + (102)^2 + (95)^2 + (101)^2 + (88)^2 + (96)^2$$

$$+ (86)^2 + (91)^2 + (79)^2] - \frac{635600}{5}$$

$$= 127700 - \frac{635600}{5}$$

$$= 127700 - 127120$$

$$= 580$$

$$F = \frac{K(n-1)(SSB)}{(K-1)(SSW)} = \frac{(3)(5-1)(160)}{(3-1)(580)}$$

$$= \frac{12(160)}{2(580)} = \frac{6(160)}{580} = \frac{960}{580}$$

$$= 1.66$$

20.3 SUMMARY COMMENTS

Suppose that the test conducted in the illustrative example had generated a value for the $S_b{}^2/S_w{}^2$ ratio that exceeded the critical value of 3.88. The null hypothesis would have been rejected and the alternate hypothesis accepted—namely, that μ_1, μ_2, and μ_3 are not all equal. Given that the analyst has concluded that there is a significant difference, can he proceed one step further and identify the source of the significant difference? The answer, of course, depends on the quality of the experimental design. The objective of the study was to determine whether a difference in training technique leads to a difference in performance. The sampling experiment should be set up in such a fashion that all possible sources of variation other than training technique influence each category in an equivalent fashion. All variation observed can then be attributed to either training technique or chance selection in sampling. Careful preparation and implementation of the sampling design is essential to a useful analytical study.

The technique discussed in the preceding pages is referred to as *one-way analysis of variance*. One-way analysis of variance deals with investigations where the data is classified on the basis of a single characteristic. In the illustrative example, the single basis of classification was the training technique used. Analysis of variance can be extended to include more than one basis of classification. To illustrate, each worker could have been classified not only according to training technique but according to level of I.Q. as well. More than one source of variability other than chance selection in sampling is recognized in the analysis. The reader interested in exploring the extension of variance analysis to more sophisticated applications is referred to the following selected readings.

References
1. Guenther, William C., *Analysis of Variance*, Englewood Cliffs, N.J.: Prentice-Hall, 1964.
2. Kempthorne, Oscar, *The Design and Analysis of Experiments*, New York: Wiley, 1952.
3. Yamane, Taro, *Statistics—An Introductory Analysis*, New York: Harper & Row, 1967.

20.4 TESTING FOR EQUALITY OF VARIANCE

Recall that one of the assumptions employed in the test for a significant difference between more than two sample means was homoscedasticity.[4] The F distribution can be used to construct a test of the validity of the equal variance assumption. Consider the following two random samples:

---

[4] Homoscedasticity is nothing more than a technical term used to describe equality of variance.

Sample 1

Daily output of the
five workers trained
by technique 1

X_{i1}	$X_{i1} - \bar{X}_1$	$(X_{i1} - \bar{X}_1)^2$
90	-2	4
85	-7	49
105	+13	169
88	-4	16
92	0	0
460		238

$$\bar{X}_1 = \frac{460}{5} = 92$$

$$\sum (X_{i1} - \bar{X}_1)^2 = 238$$

$$\hat{\sigma}_{x_1}^2 = \frac{\sum (X_{i1} - \bar{X}_1)^2}{n_1 - 1} = \frac{238}{4} = 59.5$$

Sample 2

Daily output of the
nine workers trained
by technique 2

X_{i2}	$X_{i2} - \bar{X}_2$	$(X_{i2} - \bar{X}_2)^2$
97	+1	1
85	-11	121
102	+6	36
95	-1	1
101	+5	25
88	-8	64
93	-3	9
99	+3	9
104	+8	64
864		330

$$\bar{X}_2 = \frac{864}{9} = 96$$

$$\sum (X_{i2} - \bar{X}_2)^2 = 330$$

$$\hat{\sigma}_{x_2}^2 = \frac{\sum (X_{i2} - \bar{X}_2)^2}{n_2 - 1} = \frac{330}{8} = 41.25$$

The following statistic is formed by the ratio of the two sample estimates:

$$\frac{\hat{\sigma}_{x_1}^2}{\hat{\sigma}_{x_2}^2} = \frac{[\sum (X_{i1} - \bar{X}_1)^2]/(n_1 - 1)}{[\sum (X_{i2} - \bar{X}_2)^2]/(n_2 - 1)}$$

It is clear that the numerator and denominator are χ^2 distributed variables divided by their respective degrees of freedom. Thus, the sampling distribution of the statistic $\hat{\sigma}_{x_1}^2/\hat{\sigma}_{x_2}^2$ is F distributed with $n_1 - 1$ degrees of freedom in the numerator and $n_2 - 1$ degrees of freedom in the denominator. It is implicitly assumed that both samples have been selected from normally distributed populations.

Test the hypothesis that $\sigma_{x_1} = \sigma_{x_2}$ using a critical probability of 10 percent and a two-tailed test.

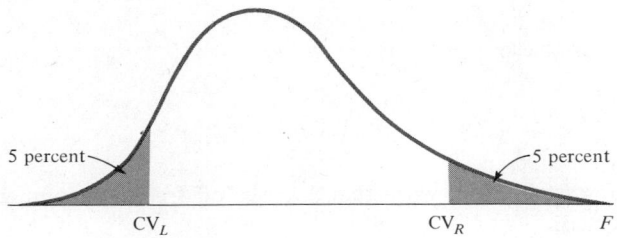

1. State the hypothesis.
   Null hypothesis: $\sigma_{x_1} = \sigma_{x_2}$
   Alternate hypothesis: $\sigma_{x_1} \neq \sigma_{x_2}$

2. Generate the decision rule. The F table in Appendix J is designed to provide the value beyond which 5 percent of the designated F distribution will lie (right-hand tail). The table can be used, however, to generate the left-hand critical value as well. The transformation required to accomplish this is illustrated below in part b.

Given an F distribution characterized by ϕ_1 degrees of freedom in the numerator and ϕ_2 degrees of freedom in the denominator:

a. The 5 percent right-tail critical value for F is determined as follows:

$$P(F > \mathrm{CV}_r \mid \phi_1, \phi_2) = 0.05$$

Substituting:

$$P(F > \mathrm{CV}_r \mid \phi_1 = 4, \phi_2 = 8) = 0.05$$

Referring to Table J in the Appendix:

$$\mathrm{CV}_r = 3.84$$

b. In general, to determine the critical value in the *left-hand tail* for an F distribution with ϕ_1 degrees of freedom in the numerator and ϕ_2 degrees of freedom in the denominator, locate the 5 percent critical value in the *right-hand tail* for the F distribution associated with ϕ_2 *degrees of freedom in the numerator and* ϕ_1 *degrees of freedom in the denominator*. The reciprocal of the value so obtained is the left-hand 5 percent critical value in the original F distribution.

The rationale underlying this relationship is as follows: Consider

$$F = \frac{U/\phi_1}{V/\phi_2}$$

where U is a χ^2 distributed variable with ϕ_1 degrees of freedom and V is a χ^2 distributed variable with ϕ_2 degrees of freedom.

$$P(F < CV_L) = 0.05$$

Inverting, we obtain

$$P\left(\frac{1}{F} > \frac{1}{CV_L}\right) = 0.05$$

Defining $1/F$ as F', we have

$$F' = \frac{1}{F} = \frac{V/\phi_2}{U/\phi_1}$$

Thus, F' is an F distribution with the degrees of freedom (numerator and denominator) reversed.

$$P\left(F' > \frac{1}{CV_L}\right) = 0.05$$

$1/CV_L = 5$ percent critical value right-hand tail in the F' distribution. Thus,

$$CV_L = \frac{1}{[5 \text{ percent critical value (right-hand tail) in the } F' \text{ distribution}]}$$

where CV_L is the 5 percent critical value (left-hand tail) in the original F distribution.

To illustrate:

$$P\left(F > \frac{1}{CV_L} \mid \underset{\text{num.}}{\phi_2 = 8}, \underset{\text{denom.}}{\phi_1 = 4}\right) = 0.05$$

$$\frac{1}{CV_L} = 6.04$$

$$CV_L = \frac{1}{6.04}$$

$$CV_L = 0.17$$

The decision rule is
If $0.17 \leq F \leq 3.84$, accept the null hypothesis, namely, $\sigma_{x_1}^2 = \sigma_{x_2}^2$.
If $F < 0.17$ or $F > 3.84$, reject the null hypothesis and accept the alternate hypothesis, namely, $\sigma_{x_1}^2 \neq \sigma_{x_2}^2$.
3. Compute the F statistic.

$$F = \frac{\hat{\sigma}_{x_1}^2}{\hat{\sigma}_{x_2}^2} = \frac{59.5}{41.25} = 1.44$$

4. Apply the decision rule. $0.17 < 1.44 < 3.84$; therefore, the null hypothesis is accepted. The F statistic of 1.44 is one that could reasonably be expected to occur due to chance selection in sampling when $\sigma_{x_1}^2 = \sigma_{x_2}^2$.

TECHNICAL NOTE 12

Demonstration that the ratio formed by $S_b{}^2/S_w{}^2$ is the ratio of two χ^2 distributed variables, each being divided by its appropriate degrees of freedom

$S_w{}^2$

$$\frac{X_{ij} - \bar{X}_j}{\sigma_x}$$

is a normally distributed variable with mean equal to zero and standard deviation equal to one.

$$\left(\frac{X_{ij} - \bar{X}_j}{\sigma_x}\right)^2 = \frac{(X_{ij} - \bar{X}_j)^2}{\sigma_x{}^2}$$

is the square of a normally distributed variable with mean equal to zero and standard deviation equal to one.

The expression

$$\frac{\sum_{j=1}^{k} \sum_{i=1}^{n} (X_{ij} - \bar{X}_j)^2}{\sigma_x{}^2}$$

represents a term that consists of the sum of squares of normally distributed variables each with zero mean and unit standard deviation. There are $K(n-1)$ independent squares comprising the numerator. Therefore, the expression

$$S_w{}^2 = \frac{\sum_{j=1}^{k} \sum_{i=1}^{n} (X_{ij} - \bar{X}_j)^2}{\sigma_x{}^2}$$

is χ^2 distributed with $K(n-1)$ degrees of freedom.

$S_b{}^2$

$$\frac{\bar{X}_j - \bar{\bar{X}}}{\sigma_{\bar{x}}}$$

is a normally distributed variable with mean equal to zero and standard deviation equal to one.

$$\left(\frac{\bar{X}_j - \bar{\bar{X}}}{\sigma_{\bar{x}}}\right)^2 = \frac{(\bar{X}_j - \bar{\bar{X}})^2}{\sigma_{\bar{x}}{}^2}$$

is the square of a normally distributed variable with mean equal to zero and standard deviation equal to one.

Substituting:

$$\frac{(\bar{X}_j - \bar{\bar{X}})^2}{\sigma_{\bar{x}}{}^2} = \frac{(\bar{X}_j - \bar{\bar{X}})^2}{(\sigma_x/\sqrt{n})^2} = \frac{(\bar{X}_j - \bar{\bar{X}})^2}{\sigma_x{}^2/n} = \frac{n(\bar{X}_j - \bar{\bar{X}})^2}{\sigma_x{}^2}$$

Summing over K samples:

$$\frac{\sum\limits_{j=1}^{k} n(\overline{X}_j - \overline{\overline{X}})^2}{\sigma_x^{\,2}}$$

n is a constant when sample size is the same for all samples, therefore the above expression can be written as

$$\frac{n \sum\limits_{j=1}^{k} (\overline{X}_j - \overline{\overline{X}})^2}{\sigma_x^{\,2}}$$

There are $K - 1$ *independent* squares comprising the expression, therefore the expression

$$S_b^{\,2} = \frac{n \sum\limits_{j=1}^{k} (\overline{X}_j - \overline{\overline{X}})^2}{\sigma_x^{\,2}}$$

is χ^2 distributed with $K - 1$ degrees of freedom.

$$\boldsymbol{F = \frac{S_b^{\,2}}{S_w^{\,2}}}$$

Dividing each of the above expressions by their respective degrees of freedom, we obtain

$$S_w^{\,2} = \frac{\sum\limits_{j=1}^{k} \sum\limits_{i=1}^{n} (X_{ij} - \overline{X}_j)^2}{\sigma_x^{\,2}} \div K(n - 1) = \frac{\sum\limits_{j=1}^{k} \sum\limits_{i=1}^{n} (X_{ij} - \overline{X}_j)^2}{(\sigma_x^{\,2})(K)(n - 1)}$$

and

$$S_b^{\,2} = \frac{n \sum\limits_{j=1}^{k} (\overline{X}_j - \overline{\overline{X}})^2}{\sigma_x^{\,2}} \div (K - 1) = \frac{n \sum\limits_{j=1}^{k} (\overline{X}_j - \overline{\overline{X}})^2}{\sigma_x^{\,2}(K - 1)}$$

Forming a ratio we obtain

$$\frac{\left[n \sum\limits_{j=1}^{k} (\overline{X}_j - \overline{\overline{X}})^2 \right] / [(\sigma_x)^2 (K - 1)]}{\left[\sum\limits_{j=1}^{k} \sum\limits_{i=1}^{n} (X_{ij} - \overline{X}_j)^2 \right] / [(\sigma_x)^2 (K)(n - 1)]}$$

Canceling out $(\sigma_x)^2$, the ratio becomes

$$\frac{\left[n \sum\limits_{j=1}^{k} (\overline{X}_j - \overline{\overline{X}})^2 \right] / (K - 1)}{\left[\sum\limits_{j=1}^{k} \sum\limits_{i=1}^{n} (X_{ij} - \overline{X}_j)^2 \right] / [K(n - 1)]}$$

The resulting expression is the F statistic, $S_b^{\,2}/S_w^{\,2}$. The above expression is also the ratio of two χ^2 distributed variables each being divided by its appropriate number of degrees of freedom. The sampling distribution of such a statistic is described by an F distribution. Therefore, $S_b^{\,2}/S_w^{\,2}$ is F distributed.

PROBLEMS

Problem 20.1

The director of an MBA program is interested in evaluating whether students with different academic backgrounds differ in quality of performance in the master's degree program. A random sample comprised of five students was selected for each category. Cumulative quality point averages at the end of the first year for the sample participants (business, engineering, liberal arts, and education) are recorded below:

Undergraduate degree

Business	Engineering	Liberal arts	Education
3.7	3.3	3.4	2.7
3.4	4.0	3.6	3.2
3.6	3.8	3.0	3.4
3.8	3.6	3.2	3.0
3.3	3.7	3.3	3.1

Utilizing the short-cut formulas, construct an analysis of variance table similar to Table 20.2 and test whether the differences among the four sample means are significant. Use a critical probability of 5 percent.

Problem 20.2

What risk is implied when a critical probability of 5 percent is used to test the null hypothesis, $\mu_1 = \mu_2 = \mu_3$?

Problem 20.3

An advertising firm conducted a study to determine whether different brands of instant coffee appealed to individuals with different intensities of coffee consumption. A sample of seven individuals was selected for each of the three major competing brands. Number of cups of coffee consumed per day were determined to be as follows:

Brand A (97 percent caffein free)	Brand B (moderate caffein content)	Brand C (heavy caffein content)
7	4	2
4	5	4
3	7	3
6	4	2
6	3	1
4	3	3
5	4	2

Utilizing the short-cut formulas, construct an analysis of variance table similar to Table 20.2 and test whether the differences among the three sample means are significant. Use a critical probability of 1 percent.

Problem 20.4

Marquis DeSade desired to test the effectiveness of three different techniques of teaching his sadistics course. Six students were randomly selected to be taught by each of the three techniques. Evaluations of performance are recorded.

Method A	Method B	Method C
96	87	79
91	93	86
75	82	96
84	78	83
87	95	92
82	76	85

Utilizing the short-cut formulas, construct an analysis of variance table similar to Table 20.2 and test whether the differences among the three sample means are significant. Use a critical probability of 5 percent.

Problem 20.5

What underlying assumptions are made in applying the *F* test in Problem 20.4?

Problem 20.6

You are interested in determining whether there is a difference in the average time lapse between placement of an order and its receipt (in work days) among the three vendors currently servicing your firm. A sample of size 5 was randomly selected from the freight billings for each of the three firms. The sampling results are presented below:

Tempus Fugit, Inc.	S. Lowe Molasses and Sons	O.K. Performance and Sons
7	11	9
8	7	7
8	9	7
5	8	6
6	10	7

Utilizing the short-cut formulas, construct an analysis of variance table similar to Table 20.2 and test whether the differences among the three sample means are significant. Use a critical probability of 1 percent.

Problem 20.7

Three alternative techniques for shipping eggs are under consideration. Five lots of eggs are shipped by each technique. The number of broken eggs observed in each of the 15 lots is recorded below:

Technique A	Technique B	Technique C
7	3	1
3	5	6
6	6	4
2	5	3
5	8	4

Utilizing the short-cut formulas, construct an analysis of variance table similar to Table 20.2 and test whether the differences among the three sample means are significant. Use a critical probability of 1 percent.

Problem 20.8

Tips earned by three waitresses working the same station on Friday evenings were observed for 12 consecutive Fridays, each waitress working the station every third Friday. The observations are summarized below:

Carol Davis	Jean Winn	Gloria Peckham
$27	$32	$28
35	37	32
42	39	37
29	35	34

Utilizing the short-cut formulas, construct an analysis of variance table similar to Table 20.2 and test whether the differences among the sample means are significant. Use a critical probability of 5 percent.

Problem 20.9

Given a ratio that is F distributed with ϕ_1 degrees of freedom in the numerator and ϕ_2 degrees of freedom in the denominator, where $\phi_1 = 8$ and $\phi_2 = 6$.
1. What value of the F ratio would you expect to be exceeded 5 percent of the time due to chance selection in sampling?
2. What value of the F ratio would you expect to be exceeded 1 percent of the time due to chance selection in sampling?

Problem 20.10

Test the hypothesis that $\sigma_{x_1} = \sigma_{x_2}$ using a critical probability of 10 percent and a two-tailed test, where $\hat{\sigma}_{x_1}^2 = 63.4$ and $\hat{\sigma}_{x_2}^2 = 40.38$. The sample sizes are $n_1 = 7$ and $n_2 = 6$, respectively.

Problem 20.11

Test the hypothesis that $\sigma_{x_1} = \sigma_{x_2}$ using a critical probability of 2 percent and a two-tailed test, where $\sigma_{x_1}^2 = 36.2$ and $\hat{\sigma}_{x_2}^2 = 27.5$. The sample sizes are $n_1 = 11$ and $n_2 = 7$, respectively.

Problem 20.12

Test the hypothesis that $\sigma_{x_1} = \sigma_{x_2}$ using a critical probability of 10 percent and a two-tailed test, where $\hat{\sigma}_{x_1}^2 = 120.6$ and $\hat{\sigma}_{x_2}^2 = 98$. The sample sizes are 11 and 17, respectively.

DECISION MAKING UNDER UNCERTAINTY: AN INTRODUCTION

21

Decisions must often be made in the face of an uncertain probabilistic environment. The decision maker must decide on a course of action to be pursued, knowing that the consequences associated with the selection of any given act will depend on future conditions and events, the outcomes of which are unknown at the time the decision must be made. The local hardware store, for example, is concerned with determining how many sump pumps to stock for the upcoming spring. If spring conditions are such that flooded cellars result, the optimal quantity to have stocked will be quite different than if conditions are such that minimal flooding occurs. If the decision maker were in a position to anticipate correctly the level of demand activity, the optimal stock action would be clear cut. Unfortunately, this is seldom the case. In fact, the decision maker is usually faced with a multiplicity of uncertainties with respect to the nature of the environment within which he must operate. The decision maker might be uncertain as to the cost of production of a new product under consideration, uncertain about the degree of market acceptability, uncertain as to the reactions of competitors, uncertain as to the impact of alternative sales promotion campaigns, uncertain as to the levels of demand activity associated with various price levels under consideration, and so on. Obviously, the existence of uncertainty complicates the decision-making process. It does not, however, negate the necessity of having to make a decision. A number of techniques have evolved that are designed to aid the decision maker in arriving at optimal decisions under conditions of uncertainty. Payoff table analysis, an elementary technique, is examined in this chapter. Chapters 22 and 23 introduce and explore some of the more sophisticated tools and techniques.

21.1 THE CONCEPT OF CONTRIBUTION

Consider the problem of Stayle Bakery Products, Inc.[1] Stayle Bakery Products, Inc., is attempting to determine the optimal number of loaves of its top-quality bread to produce on a daily basis. The top-quality bread sells for $0.38 a loaf, and company cost records have generated the information that variable costs totaling $0.25 are incurred with the production of each loaf. There is also a fixed cost associated with the production of bread—the firing and preheating of the ovens. If, after firing the ovens, it is decided not to produce anything, a non-recoverable cost of $2 is incurred. If Stayle Bakery arrives at the decision not to produce before the ovens are fired, the $2 cost will not be incurred. Thus, the $2 can be considered a fixed cost once the commitment to produce (and subsequent firing of the ovens) has taken place. On the basis of the above information, it can be said that Stayle Bakery Products, Inc., generates a contribution of $0.13 on each loaf sold toward the coverage of fixed costs and profit.

It is useful to examine the concept of contribution in greater detail before proceeding with the discussion of the decision facing Stayle Bakery Products, Inc. One of the underpinnings of the contribution concept is that even if nothing at all is produced, certain fixed costs continue to be incurred. For example, if a plant shutdown occurs because of a breakdown in labor negotiations, many of the costs of operating the organization continue (payments on fixed obligations, executive salaries, minimum upkeep expenses associated with maintenance of plant facilities, insurance, etc.).

To illustrate, suppose that a firm's fixed costs amount to $200 per production period. It produces a single product, variable costs of production per unit are $6, and the product sells for $10. The level of activity at which the firm would break even is that level of activity at which total revenue equals total cost.

$$\text{total revenue} = PQ$$
$$= \$10(Q)$$

where

P = price
Q = quantity produced and sold

$$\text{total cost} = TFC + (VC)Q$$
$$= \$200 + \$6Q$$

[1] It has been unfairly rumored that this firm specializes in serving campus cafeterias.

where

TFC = total fixed cost
VC = variable cost per unit
Q = quantity produced and sold

The break-even point occurs where total revenue equals total cost. Substituting and solving for Q:

$$\text{total revenue} = \text{total cost}$$
$$PQ = TFC + (VC)Q$$
$$\$10Q = \$200 + \$6Q$$
$$\$4Q = \$200$$
$$Q = 50 \text{ units}$$

Thus, the firm must produce and sell 50 units in order to break even.

Suppose the firm estimated that it would be able to sell only ten units during the next production period. Would it be preferable for the firm to produce the ten units or to shut down?

a. If the firm shut down (i.e., did not produce anything):

total revenue = $\$10Q = \$10(0)$ =   0
total cost
     fixed cost   = $200
     total variable cost = $(VC)Q = \$6(0)$ =   0 $200

     net loss   ($200)

b. If the firm produced ten units:

total revenue = $\$10Q = \$10(10)$ = $100
total cost
     fixed cost   = $200
     total variable cost = $(VC)(Q) = \$6(10)$ =   60 $260

     net loss   ($160)

Notice that although the firm still incurs a loss by operating and producing ten units, the magnitude of the loss is reduced from $200 to $160 because of the contribution of $4 generated by each of the ten units produced and sold. In general, if the price of the product exceeds the variable cost per unit, it is desirable to produce the product even though it involves operating at a loss. This is due to the fact that contribution will be generated toward the coverage of fixed costs, and as a result the loss will be less than if the firm shuts down and produces nothing.

The above policy is appropriate only for the period of time the economist refers to as the short run (i.e., until the firm is able to disassociate itself from its fixed commitments—the lease it up, etc.). At the break-even level, total contribution exactly equals total fixed cost. Verify this. At any level of operation above break-even, the contribution of $4 per unit sold is still generated, but with fixed costs covered, it accrues as profit. If the firm acts so as to maximize its total contribution over time, such an action is consistent with minimizing its losses if circumstances dictate that the firm has to operate below the break-even level. Such a policy is also consistent with maximization of profit in those situations where the break-even level of activity is exceeded. Thus, maximization of contribution is established as an all-purpose criterion for the firm.

21.2 GENERATION OF THE PROBABILITY DISTRIBUTION OF DEMAND

Meanwhile, back at the bakery, Stayle Bakery Products, Inc., very concerned with maintaining the integrity of its good name, has a policy that any bread left unsold at the end of the day is not to be sold even as day-old bread and is to be considered a total loss. The bakery is open seven days a week and has been unable to detect any pattern in daily sales that would lead them to expect the pattern of demand for any one day of the week to be different from that of any other. The bakery has been keeping historical records, tallying the number of loaves sold during each hour of the day as well as daily totals. On the days on which stockouts occurred, records were also kept as to the number of unfilled requests for premium bread. This additional information turned out to be very helpful, since historical records as to quantities sold will not depict the true pattern of demand if stockouts have occurred.

To obtain a correct entry for the level of demand on days when stockouts occurred, Stayle Bakery Products, Inc., had to combine the quantities sold and the number of unfilled requests. Fortunately, the records were kept in such a fashion that these adjustments could be readily made. Another approach sometimes used to adjust quantity-sold figures so as to reflect quantity demanded is to observe the time of day that stockout occurred and the percentage of daily demand typically sold by that hour of the day. For example, if the bakery had produced 60 loaves and stocked out at 3 p.m., and typically 75 percent of sales for any given day occurred by 3 p.m., then 60 loaves could be considered 75 percent of what total demand would have been for that day.

$$0.75 \text{ (total demand)} = 60 \text{ loaves}$$

$$\text{total demand} = \frac{60 \text{ loaves}}{0.75} \text{ or } 80 \text{ loaves}$$

Thus, the appropriate entry for the level of demand for that day is the adjusted estimate of total demand or 80 loaves. The important point to recognize is that unless appropriate upward adjustments are made on the days when stockouts occurred, a distortion in the projected pattern of demand will occur.

The relevant historical period on which to draw in generating the probability distribution of demand is another significant issue. The pattern of demand obtained is utilized as a forecast of expected future experience with respect to demand activity. Thus, any portion of the historical evidence generated under conditions considered nonrepresentative of the way conditions are expected to be in the future should be omitted.

After appropriate incorporation of the number of unfilled requests on days when stockouts occurred and the exclusion of any observations not relevant to a representative projection of the pattern of future demand activity, the historical pattern of demand depicted in Table 21.1 emerged. The artificial assumption has been made that the only levels of demand activity that can occur are 40, 50, 60, 70, and 80 loaves per day. Although totally unrealistic, such an assumption is convenient in that it allows the development of an understanding of the technique

Table 21.1 Stayle Bakery Products, Inc.: Historical pattern of demand for premium bread for the last 200 days

Number of loaves demanded	Number of days on which the given level of demand occurred	Probability of the given level of demand occurring on any given day (say, tomorrow)
40	20	0.10
50	40	0.20
60	80	0.40
70	40	0.20
80	20	0.10
Total	200	1.00

within a simple, uncomplicated framework. Once this has been accomplished, the assumption will be released and extensions will be examined that deal with the more realistic situation where demand can take on any discrete value over a wide range of possible demand levels.

In the case of Stayle Bakery Products, Inc., the firm is attempting to arrive at an optimal decision with respect to the number of loaves to produce in the face of uncertain information. The uncertain information takes the form of an inability to predict exact demand on a daily basis. It is not, however, as though the firm has no information at all. It is usually possible to generate the probability distribution of demand (i.e., the percentage of the time that various levels of demand activity are expected to occur).[2] The rub is the inability to predict which particular level of demand activity will actually be experienced on any one given particular day (say, tomorrow). Given that a probability distribution of demand can be generated, techniques have been developed that enable the decision maker to select an optimal action alternative given the state of his information. In the Stayle Bakery case, the state of the decision maker's information is a knowledge of the pattern of the probability distribution of demand.

21.3 THE CONDITIONAL PAYOFF TABLE

The initial phase of the analysis involves the development of a conditional payoff table (Table 21.2). A row is assigned to each possible state of nature that could conceivably occur (levels of demand activity in the Stayle Bakery Products, Inc., case). A column is assigned to each of the action alternatives under consideration (in the case of Stayle Bakery Products, Inc., the action alternatives are restricted to the production levels 40, 50, 60, 70, or 80 loaves of bread on a daily basis).

---

[2] The literature typically places the environment within which decisions are to be made into one of three categories.

1. Decision making under certainty: This is the situation where the decision maker knows exactly what is going to occur, and the appropriate action to take is readily ascertainable.
2. Decision making under risk: This is the situation where the decision maker is able to generate a probability distribution expressing his belief in the likelihood of various events. occurring although he cannot predict with certainty which event will actually occur.
3. Decision making under uncertainty: This is the situation where the decision maker is unable to generate a probability distribution describing the relative likelihood of occurrence of the events in question.

Table 21.2 Stayle Bakery Products, Inc.: Conditional profit table

	Action alternatives under consideration				
Levels of demand activity	Produce 40 loaves daily	Produce 50 loaves daily	Produce 60 loaves daily	Produce 70 loaves daily	Produce 80 loaves daily
40 loaves	3.20	0.70	−1.80	−4.30	−6.80
50 loaves	3.20	4.50	2.00	−0.50	−3.00
60 loaves	3.20	4.50	5.80	3.30	0.80
70 loaves	3.20	4.50	5.80	7.10	4.60
80 loaves	3.20	4.50	5.80	7.10	8.40

The entries in the cells of the body of the table are the conditional profits (positive or negative) associated with the joint occurrence of each of the conceivable act–event combinations. For example, the conditional profit that would result from producing 50 loaves on a day when the number of loaves demanded was 40 is entered in the cell at the intersection of the row "40 loaves demanded" and the column "50 loaves produced". The entry in each cell is obtained by comparing the total cost associated with the level of production and the total revenue generated by the quantity *sold*. (Remember that the quantity sold does not equal the quantity demanded given the occurrence of a stockout.)

In general, total cost is described by the following relationship:

$$TC = TFC + (VC)Q$$

where

TC = total cost
VC = average variable cost per unit
Q = quantity produced
TFC = total fixed cost

In the case of Stayle Bakery Products, Inc., total cost = $2 + ($0.25)Q$.

In general, total revenue is described by one of the following two relationships, depending on whether or not a stockout occurred.

TR = total revenue
P = price per unit
Q = quantity produced
D = quantity demanded

1. If there is no stockout; that is,

   if $Q \geq D$

   then $TR = PD$

2. If there is a stockout; that is,

   if $Q < D$

   then $TR = PQ$

In the case of Stayle Bakery Products, Inc. (Tables 21.3 and 21.4),

$TR = PD$ or   ($0.38)(D)$ if $Q \geq D$
$TR = PQ$ or   ($0.38)(Q)$ if $Q < D$

Table 21.3

40 loaves produced; 40 loaves demanded

Total revenue:	
$(\$0.38)(D) = (\$0.38)(40)$	$= \$15.20$
Less total cost:	
$\$2 + (\$0.25)(Q) = \$2 + \$0.25\,(40)$	$= \$12.00$
Conditional profit	$\$3.20$

Table 21.4

40 loaves produced; 50 loaves demanded

Total revenue:	
$(\$0.38)(Q) = (\$0.38)(40)$	$= \$15.20$
Less total cost:	
$\$2 + (\$0.25)Q = \$2 + (\$0.25)(40)$	$= \$12.00$
Conditional profit	$= \$3.20$

Note that when the quantity demanded exceeds the quantity supplied, the quantity sold is limited to the quantity produced. Upon examination of the conditional profit table, it is clear that once the quantity demanded exceeds the quantity produced, all remaining entries in the column are identical to that recorded for the cell where the quantity demanded equalled the quantity produced (Tables 21.5, 21.6, and 21.7).

Table 21.5

60 loaves produced; 50 loaves demanded

Total revenue:	
$(\$0.38)(D) = (\$0.38)(50)$	$= \$19.00$
Less total cost:	
$\$2 + (\$0.25)Q = \$2 + (\$0.25)(60)$	$= \$17.00$
Conditional profit	$= \$2.00$

Table 21.6

60 loaves produced; 60 loaves demanded

Total revenue:	
$(\$0.38)(Q) = (\$0.38)(60)$	$= \$22.80$
Less total cost:	
$\$2 + (\$0.25)(Q) = \$2 + (\$0.25)(60)$	$= \$17.00$
Conditional profit	$\$5.80$

Table 21.7

60 loaves produced; 70 loaves demanded

Total revenue:	
$(\$0.38)(Q) = (\$0.38)(60)$	$= \$22.80$
Less total cost:	
$\$2 + (\$0.25)(Q) = \$2 + (\$0.25)(60)$	$= \$17.00$
Conditional profit	$\$5.80$

The cell entries can also be generated by employing the following relationship:

$$\text{conditional profit} = \begin{bmatrix} \text{contribution} \\ \text{generated} \\ \text{on units} \\ \text{actually sold} \end{bmatrix} - \begin{array}{l} \text{fixed cost} \\ \text{incurred} \\ \text{to produce} \\ \text{the units} \\ \text{actually sold} \end{array} - \begin{bmatrix} \text{unrecovered} \\ \text{variable cost} \\ \text{on units} \\ \text{produced} \\ \text{but not sold} \end{bmatrix}$$

$$CP = [(P - VC)D - FC] - (VC)(Q - D) \qquad \text{if } Q > D$$
$$CP = [(P - VC)Q - FC] \qquad\qquad\qquad \text{if } Q \leq D$$

This approach is illustrated in table 21.8.

Table 21.8

60 loaves produced; 40 loaves demanded
$Q > D$

$$
\begin{aligned}
CP &= [(P - VC)D - FC] - (VC)(Q - D) \\
&= [(\$0.38 - \$0.25)(40) - \$2] - (\$0.25)(60 - 40) \\
&= [(\$0.13)(40) - \$2] - (\$0.25)(20) \\
&= (5.20 - 2) - \$5 \\
&= \$3.20 - 5 \\
&= -\$1.80
\end{aligned}
$$

The conditional profit table enables the analyst to observe the consequences associated with any given action alternative (level of production) and the various states of nature (levels of demand) with which it could be associated. The table does not, however, in and of itself indicate the optimal action alternative. Before the optimal action in the eyes of the decision maker can be determined, it is necessary for the decision maker to specify a criterion of choice.

21.4 ALTERNATIVE DECISION RULES

Three alternative decision rules will be examined: (1) maximin, a decision rule reflecting extreme pessimism on the part of the decision maker; (2) maximax, a decision rule reflecting extreme optimism on the part of the decision maker; and (3) selection of the action alternative with the maximum expected monetary payoff.

Maximin

The maximin decision rule operates under the assumption that the worst of all possible worlds is going to happen. It is designed to select the action alternative that will result in the most desirable consequence under these conditions (i.e., the action alternative that maximizes the minimum monetary payoff.)

Procedure:

1. Determine the least desirable consequence that can occur with respect to the action alternatives under consideration, that is, the minimum conceivable payoff (Table 21.9).

Table 21.9

Action alternatives	Least desirable consequence	State of nature associated with the least desirable consequence
Produce 40 loaves	3.20	40 loaves demanded
Produce 50 loaves	0.70	40 loaves demanded
Produce 60 loaves	− 1.80	40 loaves demanded
Produce 70 loaves	− 4.30	40 loaves demanded
Produce 80 loaves	− 6.80	40 loaves demanded

2. Select as the optimal action the action alternative that would cause the least desirable consequence or minimum payoff to be maximized.

If Stayle Bakery Products, Inc., were to employ the maximin criterion, the action alternative selected would be to produce 40 loaves. Production of 40 loaves maximizes the minimum payoff at $3.20.

Maximax

The maximax decision rule is based on extreme optimism—it operates under the assumption that the best of all possible worlds will happen and is designed to select the action alternative that will generate the greatest payoff under these conditions (i.e., the action alternative that maximizes the maximum monetary payoff).

Procedure:

1. Determine the most desirable consequence that can occur for each action alternative under consideration, that is, the maximum conceivable payoff (Table 21.10).

Table 21.10

Action alternative	Most desirable consequence	State of nature associated with the most desirable consequence
Produce 40 loaves	3.20	40 loaves demanded
Produce 50 loaves	4.50	50 loaves demanded
Produce 60 loaves	5.80	60 loaves demanded
Produce 70 loaves	7.10	70 loaves demanded
Produce 80 loaves	8.40	80 loaves demanded

2. Select as the optimal action the action alternative that causes the most desirable consequence to be maximized (i.e., maximizes the maximum payoff).

If Stayle Bakery Products, Inc., were to employ the maximax criterion, the action alternative selected would be to produce 80 loaves, which would generate a maximum payoff of $8.40.

Maximum Expected Monetary Profit

The decision rule employed under the criterion of maximum expected monetary payoff is to select the action alternative that would generate the maximum average profit per period if that action alternative were to be pursued consistently over time. This criterion is most appropriate in repetitive situations, where the action selected will be implemented many times and where the magnitude of the most undesirable consequence is small compared to the asset position of the firm, so that a string of bad luck would not cause severe financial difficulties. The willingness to employ this criterion implies that the decision maker plans to be around long enough for the probabilities to run true. In situations where the amount of "dough" involved is significant with respect to the asset position of the firm, the maximum expected monetary payoff criterion would no longer be appropriate. A technique designed to enable the decision maker to deal with situations where the maximum monetary payoff criterion is inappropriate is introduced in Chapter 23.

To compute the expected monetary profit associated with any given action alternative, multiply the conditional profit figure associated with each possible demand level by the probability of that particular level of demand occurring. The sum of these products is the expected monetary payoff. (When probabilities are used as weights, the sum of the weighted values is equal to the weighted average, because the sum of the probabilities or weights equals one.) Repeat the procedure for each action alternative. Select as the optimal act the action alternative that maximizes expected monetary payoff. The action that generates the maximum expected payoff (i.e., the largest average payoff per period) is also the act associated with the maximum expected total profit over any given period of time. Tables 21.11 through 21.14 illustrate the computation of expected monetary payoffs for the action alternatives under consideration by Stayle Bakery Products, Inc.

Table 21.11 Expected monetary payoff—Action alternative: Produce 70 loaves daily

Level of demand activity	Probability of a given level of demand activity occurring	Conditional monetary payoff	Computation of expected monetary payoff
40 loaves	0.10	−4.30	−0.43
50 loaves	0.20	−0.50	−0.10
60 loaves	0.40	+3.30	+1.32
70 loaves	0.20	+7.10	+1.42
80 loaves	0.10	+7.10	+0.71
Total	1.00		+3.45
			−0.53
			+2.92

Interpretation (Table 21.11): Over a period of 100 "days," if 70 loaves of bread were to be produced each day, it would be expected that:

On 10 of the 100 days, demand would be for 40 loaves. A loss of $4.30 would be incurred on each of those days.

On 20 of the 100 days, demand would be for 50 loaves with a loss of $0.50 incurred on each of the 20 days.

On 40 of the 100 days, demand would be for 60 loaves. A profit of $3.30 would be realized on each such occasion.

On 20 of the 100 days, demand would be for 70 loaves, with a $7.10 profit realized on each of the 20 days.

On the remaining 10 days, demand would be for 80 loaves, with a profit of $7.10 realized each time.

Averaging the firm's expected experience over the 100 days generates an expected profit of $2.92 per day.

In actuality, some departure from the expected demand pattern for the next 100 days would be expected, since that is not a sufficient period for the long-run probabilities to run true. As the number of days of implementation increased, the true probability pattern of demand would be expected to emerge. It is appropriate at this point to reflect on whether or not the projected probability distribution of demand is a reasonable estimate of the true probability distribution of demand. Great care must be exercised in generating an expected probability distribution of demand if the resultant analysis is to be meaningful in directing the decision maker toward the optimal action alternative. This, of course, is an issue to be raised and resolved within the context of each individual decision problem.

Clearly, there is no need to compute the expected monetary payoff for the action alternative, produce 40 loaves daily. The conditional monetary payoff is $3.20 regardless of the level of demand that actually materializes. Therefore,

$$EMP_{\text{produce } 40} = \$3.20$$

See Tables 21.12, 21.13, and 21.14 for computations of expected monetary payoffs for the action alternatives, produce 50, 60, and 80 loaves daily, respectively. Table 21.15 summarizes the expected monetary payoffs for the action alternatives under consideration.

According to the criterion, maximize expected monetary profit, the optimal action alternative is to produce 60 loaves daily. This would entail an average profit of $4.28 per day. Recall that the pessimistic maximin criterion would have led to the adoption of the action alternative, produce 40 loaves daily, and an

Table 21.12 Expected monetary payoff—Action alternative: Produce 50 loaves

Level of demand activity	Probability of a given level of demand activity occurring	Conditional monetary payoff	Computation of expected monetary payoff
40 loaves	0.10	0.70	0.07
50 loaves	0.20	4.50	0.90
60 loaves	0.40	4.50	1.80
70 loaves	0.20	4.50	0.90
80 loaves	0.10	4.50	0.45
Total	1.00		4.12

$$EMP_{\text{produce } 50} = \$4.12$$

Table 21.13 Expected monetary payoff—Action alternative: Produce 60 loaves

Level of demand activity	Probability of a given level of demand activity occurring	Conditional monetary payoff	Computation of expected monetary payoff
40 loaves	0.10	−1.80	−0.18
50 loaves	0.20	2.00	+0.40
60 loaves	0.40	5.80	+2.32
70 loaves	0.20	5.80	+1.16
80 loaves	0.10	5.80	+0.58
			4.46
			−0.18
Total	1.00		4.28

$\text{EMP}_{\text{produce } 60} = \$4.28$

Table 21.14 Expected monetary payoff—Action alternative: Produce 80 loaves

Level of demand activity	Probability of a given level of demand activity occurring	Conditional monetary payoff	Computation of expected monetary payoff
40 loaves	0.10	−6.80	−0.68
50 loaves	0.20	−3.00	−0.60
60 loaves	0.40	0.80	+0.32
70 loaves	0.20	4.60	+0.92
80 loaves	0.10	8.40	+0.84
Total	1.00		+2.08
			−1.28
			0.80

$\text{EMP}_{\text{produce } 80} = \$0.80$

Table 21.15 Summary: Expected monetary payoffs for the action alternatives under consideration by Stayle Bakery Products, Inc.

Action alternative	Expected monetary payoff
Produce 40 loaves	3.20
Produce 50 loaves	4.12
Produce 60 loaves	4.28
Produce 70 loaves	2.92
Produce 80 loaves	0.80

average profit per day of $3.20. Thus, ultraconservatism would have resulted in an average reduction in profit of $1.08 per day. The optimistic maximax criterion would have fared even worse: Producing 80 loaves daily would have resulted in an average profit of only $0.80 per day. This would have involved a reduction in average profits of $3.48 per day as compared to the optimal action alternative under the maximum expected monetary payoff criterion. Given that there are no negative payoffs sufficiently large to cause the firm financial hardship, and particularly if the decision is to be implemented repetitively, maximum expected monetary payoff is the logical criterion to use in selecting the optimal action alternative.

21.5 EXPECTED PROFIT UNDER CERTAINTY

If the firm possessed perfect information rather than operating under conditions of uncertainty, the level of demand activity occurring on any given day would be known to the decision maker prior to selection of the production level for that day. In effect, the firm would not only know the probability distribution of demand, but in addition would know the particular days on which the given levels of demand activity would occur. Given the possession of such information, the decision maker would vary the production level, always selecting the optimal action alternative associated with the known level of demand for the following day.

The conditional profit table describing the consequences of the various act–event combinations for the Stayle Bakery Product, Inc., case is reproduced as Table 21.16. To determine the optimal action alternative given a knowledge of the level of demand activity, simply survey the appropriate row in the conditional profit table. The action alternative associated with the largest monetary payoff in a row is the optimal act given the existence of that level of demand activity. In the case of Stayle Bakery Products, Inc., the optimal stock action for each level of demand activity is to produce the quantity demanded.[3] To determine the expected monetary payoff under conditions of certainty, one simply weights the optimal conditional profit figure associated with each demand level by the probability of that level of demand activity occurring (Table 21.17).

Table 21.16 Conditional profit table: Stayle Bakery Products, Inc.

	Action alternatives				
Levels of demand activity	Produce 40 loaves daily	Produce 50 loaves daily	Produce 60 loaves daily	Produce 70 loaves daily	Produce 80 loaves daily
40 loaves	3.20	0.70	−1.80	−4.30	−6.80
50 loaves	3.20	4.50	2.00	−0.50	−3.00
60 loaves	3.20	4.50	5.80	3.30	0.80
70 loaves	3.20	4.50	5.80	7.10	4.60
80 loaves	3.20	4.50	5.80	7.10	8.40

---

[3] It is not necessarily true that the optimal stock action is to produce the quantity demanded.

Table 21.17 Stayle Bakery Products, Inc.: Computation of expected monetary payoff under conditions of certainty

Level of demand activity	Probability of the level of demand activity occurring	Conditional monetary payoff under certainty (i.e., conditional monetary profit associated with the optimal action alternative given the level of demand activity)	Computation of expected profit under certainty
40 loaves	0.10	3.20	0.32
50 loaves	0.20	4.50	0.90
60 loaves	0.40	5.80	2.32
70 loaves	0.20	7.10	1.42
80 loaves	0.10	8.40	0.84
Total	1.00		5.80

$EMP_{under\ certainty} = \$5.80$

21.6 EXPECTED VALUE OF PERFECT INFORMATION

If the firm were able to determine the level of demand activity in advance and plan its production level accordingly, it could expect an average profit of $5.80 per day. Earlier analysis demonstrated that by consistently pursuing the optimal action alternative under conditions of uncertainty, an average profit of $4.28 per day would be realized. Clearly, uncertainty costs the firm money—an average of $1.52 per day. The $1.52 represents the extent to which average daily profits could be improved if Stayle Bakery Products, Inc., could shift from decision making under existing conditions of uncertainty to a situation characterized by perfect information (i.e., decision making under conditions of certainty). Thus, the $1.52 can be interpreted as the value of perfect information. In general, to compute the expected value of perfect information (or EVPI), deduct the expected monetary payoff associated with the optimal action alternative under conditions of uncertainty from the expected monetary payoff associated with decision making under conditions of certainty (Table 21.18).

Table 21.18 Computation of the expected value of perfect information

Expected monetary payoff under conditions of certainty	5.80
Less expected monetary payoff associated with the optimal action under conditions of uncertainty (i.e., produce 60 loaves)	4.28
EVPI	1.52

If perfect information were available, it would be worthwhile for Stayle Bakery Products, Inc., to purchase the information provided the price did not exceed $1.52 per day. Less than perfect information would command a lower price—the actual amount being determined by the difference between maximum expected monetary profit with the new information and the maximum expected monetary profit without the new information. Computation of the expected value of perfect information shows the maximum gain possible from efforts expended to reduce the uncertainty under which decisions must be made. The magnitude of the expected value of perfect information indicates the potential benefit from generating better information and is a useful input in deciding whether such an undertaking is merited.

21.7 MINIMIZATION OF EXPECTED OPPORTUNITY LOSS

Opportunity loss is defined as the profit forgone as the result of having selected other than the optimal action alternative given the occurrence of a particular state of nature. It is the difference between the profit actually obtained with the action alternative selected and the profit that could have been obtained had the optimal action alternative been selected.

Given demand for 50 loaves of bread, the optimal action alternative is to produce 50 loaves of bread. If the action alternative, produce 60 loaves of bread, had been selected, an opportunity loss of $2.50 would have been incurred. (The $2.50 is the difference between the payoff associated with the optimal action alternative, $4.50, and the payoff associated with the action alternative, produce 60 loaves, $2.00.) Opportunity losses can be considered as falling into two categories:

1. *Opportunity loss due to overstocking.* This category of opportunity loss refers to a reduction in profit due to incurring unnecessary nonrecoverable costs as a result of producing or stocking more units than one is able to sell. In the case of Stayle Bakery Products, Inc., the entire variable cost of production ($0.25 per unit) on any unsold units constitutes a nonrecoverable cost. If Stayle Bakery instituted a policy of selling day-old bread at a reduced price of, say, $0.15 per loaf, the opportunity loss due to overstocking would have been reduced by the "salvage value" and would be $0.10 per loaf. Any incremental nonrecoverable costs incurred as a result of overstocking constitute opportunity losses.

2. *Opportunity loss due to understocking.* This category of opportunity loss refers to a reduction in profit as a result of contribution forgone on sales that could have been made had the firm not stocked out. In the Stayle Bakery illustration, it is presumed that the only loss is the contribution forgone on unfilled requests. In actuality, one should consider not only the loss of contribution on the immediate sale but the impact on future lost sales as well. Contribution lost on other products that might have been purchased in conjunction with the unit with respect to which stockout occurred should also be considered.

In a fashion similar to the maximum expected monetary payoff approach, the initial phase of analysis involves construction of a conditional payoff table. The major difference in procedure is that instead of entering conditional monetary payoffs, the entries in the cells are conditional opportunity losses (i.e., the opportunity loss suffered given the joint occurrence of a state of nature and a given action alternative). Referring back to the conditional monetary payoff table, the opportunity losses can be generated by comparing the maximum payoff in each row (associated with the optimal action alternative given that state of nature) with the payoffs associated with each of the other action alternatives in that row.

520 CHAPTER 21 DECISION MAKING UNDER UNCERTAINTY: AN INTRODUCTION

Obviously, the opportunity loss would be zero for the optimal action alternative associated with each state of nature. In a conditional opportunity loss table (Table 21.19), entries to the right of a zero opportunity loss in any row represent opportunity losses due to overstocking, and entries to the left represent opportunity losses due to understocking.

Table 21.19 Stayle Bakery Products, Inc.: Conditional opportunity loss table

Levels of demand activity	Action alternatives				
	Produce 40 loaves daily	Produce 50 loaves daily	Produce 60 loaves daily	Produce 70 loaves daily	Produce 80 loaves daily
40 loaves	0	2.50	5.00	7.50	10.00
50 loaves	1.30	0	2.50	5.00	7.50
60 loaves	2.60	1.30	0	2.50	5.00
70 loaves	3.90	2.60	1.30	0	2.50
80 loaves	5.20	3.90	2.60	1.30	0

An alternative technique for generating the entries in the opportunity loss table can be illustrated by deriving the entries in the row associated with the level of demand activity, 60 loaves demanded. The optimal action alternative when demand is for 60 loaves is to have produced 60 loaves; thus, the opportunity loss entry in the produce 60 column is zero. The action alternatives, produce 70 and produce 80, involve opportunity losses due to overstocking and are generated as follows:

Produce 70 loaves: 10 loaves would have remained unsold at the end of the day, with a nonrecoverable cost of $0.25 each or a total opportunity loss of $2.50.

Produce 80 loaves: 20 loaves would have remained unsold at the end of the day, with a nonrecoverable cost of $0.25 each or a total opportunity loss of $5.00.

Selection of the action alternatives, produce 40 and produce 50 loaves, involve opportunity losses due to understocking and are computed as follows:

Produce 40 loaves: A contribution of $0.13 per loaf would have been lost on each of the 20 additional loaves that could have been sold had they been available, thus generating a total opportunity loss of $2.60.

Produce 50 loaves: A contribution of $0.13 per loaf would have been lost on each of the 10 additional loaves that could have been sold had they been available, resulting in a total opportunity loss of $1.30.

The remaining entries in the table are generated in a similar fashion by simply recognizing that there is a $0.13 opportunity loss associated with each unfilled request and a $0.25 opportunity loss incurred with each unit overstocked.

After the development of the conditional opportunity loss table, the next step is to determine the expected opportunity loss associated with each of the action alternatives. This is accomplished by multiplying each of the conditional opportunity losses in the column associated with a given action alternative by the probability of that conditional opportunity loss being incurred and then summing. The computations are illustrated in Tables 21.20 through 21.24 and summarized in Table 21.25.

Table 21.20 Stayle Bakery Products, Inc.: Expected opportunity loss—Action alternative, produce 40 loaves

Level of demand activity	Probability of a given level of demand activity occurring	Conditional opportunity loss	Computation of expected opportunity loss
40 loaves	0.10	0	0
50 loaves	0.20	1.30	0.26
60 loaves	0.40	2.60	1.04
70 loaves	0.20	3.90	0.78
80 loaves	0.10	5.20	0.52
	1.00		2.60

$EOL_{produce\ 40} = \$2.60$

Table 21.21 Stayle Bakery Products, Inc.: Expected opportunity loss—Action alternative, produce 50 loaves

Level of demand activity	Probability of a given level of demand activity occurring	Conditional opportunity loss	Computation of expected opportunity loss
40 loaves	0.10	2.50	0.25
50 loaves	0.20	0	0
60 loaves	0.40	1.30	0.52
70 loaves	0.20	2.60	0.52
80 loaves	0.10	3.90	0.39
	1.00		1.68

$EOL_{produce\ 50} = \$1.68$

Table 21.22 Stayle Bakery Products, Inc.: Expected opportunity loss—Action alternative, produce 60 loaves

Level of demand activity	Probability of a given level of demand activity occurring	Conditional opportunity loss	Computation of expected opportunity loss
40 loaves	0.10	5.00	0.50
50 loaves	0.20	2.50	0.50
60 loaves	0.40	0	0
70 loaves	0.20	1.30	0.26
80 loaves	0.10	2.60	0.26
	1.00		1.52

$EOL_{produce\ 60} = \$1.52$

Table 21.23 Stayle Bakery Products, Inc.: Expected opportunity loss—Action alternative, produce 70 loaves

Level of demand activity	Probability of a given level of demand activity occurring	Conditional opportunity loss	Computation of expected opportunity loss
40 loaves	0.10	7.50	0.75
50 loaves	0.20	5.00	1.00
60 loaves	0.40	2.50	1.00
70 loaves	0.20	0	0
80 loaves	0.10	1.30	0.13
	1.00		2.88

$EOL_{produce\ 70} = \$2.88$

Table 21.24 Stayle Bakery Products, Inc.: Expected opportunity loss—Action alternative, produce 80 loaves

Level of demand activity	Probability of a given level of demand activity occurring	Conditional opportunity loss	Computation of expected opportunity loss
40 loaves	0.10	10.00	1.00
50 loaves	0.20	7.50	1.50
60 loaves	0.40	5.00	2.00
70 loaves	0.20	2.50	0.50
80 loaves	0.10	0	0
	1.00		5.00

$EOL_{produce\ 80} = \$5.00$

Table 21.25 Stayle Bakery Products, Inc.: Summary of expected opportunity losses for the action alternatives under consideration

Action alternative	Expected opportunity loss
Produce 40 loaves daily	2.60
Produce 50 loaves daily	1.68
Produce 60 loaves daily	1.52
Produce 70 loaves daily	2.88
Produce 80 loaves daily	5.00

Expected opportunity loss is the average profit forgone as a result of pursuing a consistent action alternative instead of varying the action alternative so as always to generate the maximum payoff given a particular state of nature. (Of course, the latter action would require advance knowledge as to the state of nature and therefore would be possible only under conditions of certainty.) The most advantageous action alternative under conditions of uncertainty is the action that minimizes average profit forgone. This is the action alternative associated with the minimum expected opportunity loss. The minimum expected opportunity loss criterion and the maximum expected monetary payoff criterion identify the same optimal action alternative. *Maximizing expected monetary payoff is equivalent to minimizing expected opportunity loss.*

As Table 21.26 demonstrates, for any action alternative, the expected monetary payoff under certainty can be separated into the portion obtained by consistently pursuing that act under conditions of uncertainty and the portion that will be forgone. Identifying the action alternative with the maximum expected monetary payoff under uncertainty in the first column is identical to identifying the action alternative associated with the minimum expected opportunity loss under uncertainty in the second column. Maximizing the portion of the $5.80 obtained is identical to minimizing the portion of the $5.80 forgone.

Table 21.26 Relationship between expected monetary payoff under uncertainty, expected opportunity loss under uncertainty, and expected monetary payoff under certainty

Action alternative	Expected monetary payoff under uncertainty	+	Expected opportunity loss under uncertainty	=	Expected monetary payoff under certainty
Produce 40 loaves daily	3.20	+	2.60	=	5.80
Produce 50 loaves daily	4.12	+	1.68	=	5.80
Produce 60 loaves daily	4.28	+	1.52	=	5.80
Produce 70 loaves daily	2.92	+	2.88	=	5.80
Produce 80 loaves daily	0.80	+	5.00	=	5.80

The minimum expected opportunity loss also represents the expected value of perfect information (EVPI). Perfect information would ensure a zero opportunity loss and thus an improvement in expected profit per day equal to the minimum expected opportunity loss under conditions of uncertainty. This value (average profits forgone as a result of uncertainty) places an upper bound on the price the decision maker is willing to pay for improved information and serves to indicate whether the potential benefits warrant pursuing that possibility.

21.8 INCREMENTAL ANALYSIS

At this point it is appropriate to release the assumption that only certain discrete levels of demand can occur and replace it with a recognition that the level of demand activity can take on any value over the range of feasible demand levels. From the field of economics, the theory of the firm states that for the firm to maximize profits, it should produce that level of output that corresponds to the level of activity where marginal revenue equals marginal cost. This is known as the $MR = MC$ rule for profit maximization, and a graphical depiction of the relationships on which it is based is presented in Figure 21.1.

Marginal cost and marginal revenue intersect above the point on the horizontal axis labeled Q optimal. If the quantity produced were less than Q optimal, the amount added to revenue by producing another unit would exceed the amount added to cost. This would cause total profit to increase, and therefore the firm would have an incentive to expand production as long as $MR > MC$. Once the quantity produced exceeds Q optimal however, $MC > MR$, and more would be added to cost than to revenue by producing the last unit. Thus the firm would have an incentive to cut back production to the level Q optimal.

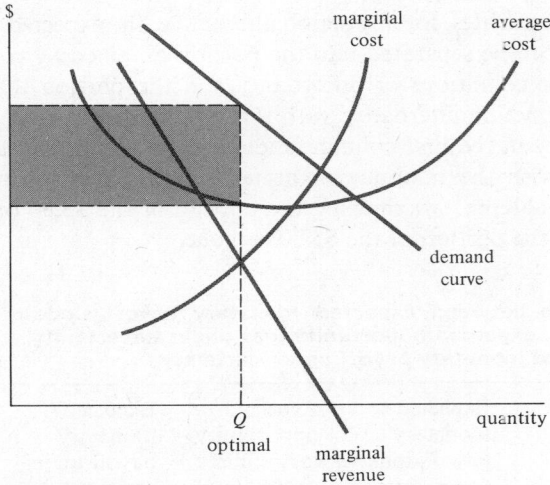

Figure 21.1

The $MR = MC$ rule is based on the assumption that the demand curve does not shift (i.e., that the only reason for a change in the quantity demanded is a change in the price charged for the product). This, of course, is an unrealistic assumption. Typically, the firm establishes a price—Stayle Bakery Products, Inc., for example, established a price of $0.38 a loaf. The quantity sold at that price then varies as the demand curve shifts (from day to day or period to period) because of many factors, some of which tend to inflate demand on any given day (company for dinner), some of which tend to deflate demand on any given day (it rains). Demand on any given day or the location of the demand curve is determined by the combined impact of all these influencing factors.

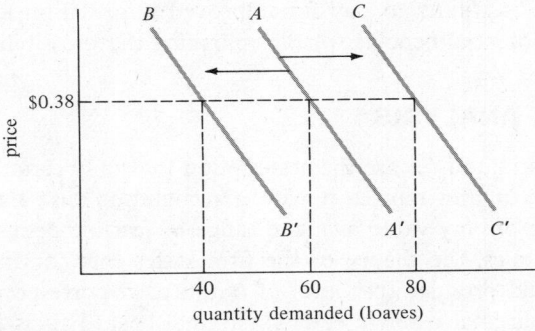

Figure 21.2

Figure 21.2 depicts Stayle Bakery Products', *average demand curve* as curve *AA'*. This curve can be considered as passing through the *average amount* demanded at each of the price levels referenced on the vertical axis. Thus, given a price of $0.38, for Stayle Bakery Products, one could say that sometimes demand is for as little as 40 loaves, sometimes for as much as 80 loaves, and over time averages about 60 loaves of bread demanded on a daily basis. The extent the demand curve shifts about the average demand curve can be reflected by a series of probability distributions recording the various levels of demand activity and the relative frequency with which they would be expected to occur for each price level.

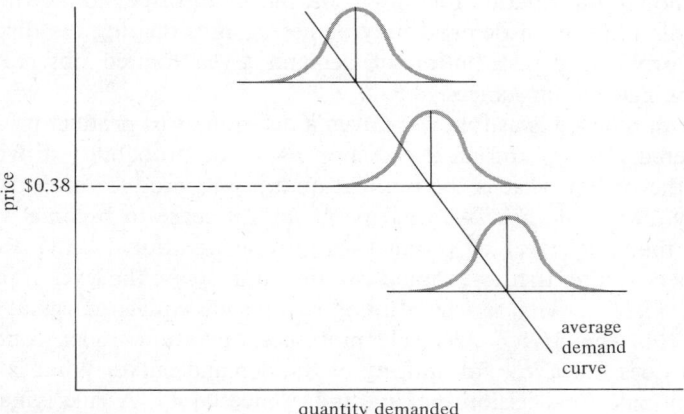

Figure 21.3

The influences underlying the variation in quantity demanded with the price held constant at, say, $0.38 in Figure 21.3 are simply reflections of the vagaries of consumer behavior and as such consist of the combined impact of a multitude of individual influences no one of which is significant enough to merit singling out. In the case of Stayle Bakery Products, Inc., the information utilized to develop the probability distribution of demand associated with a price of $0.38 should be based on those fluctuations in daily demand due to the many influences consistently operating on demand. Occasionally, there may be a change in a factor that exerts considerable influence on the overall level of demand and thereby brings about a *permanent* shift of the average demand curve although the pattern of demand remains the same. This is illustrated in Figure 21.4.

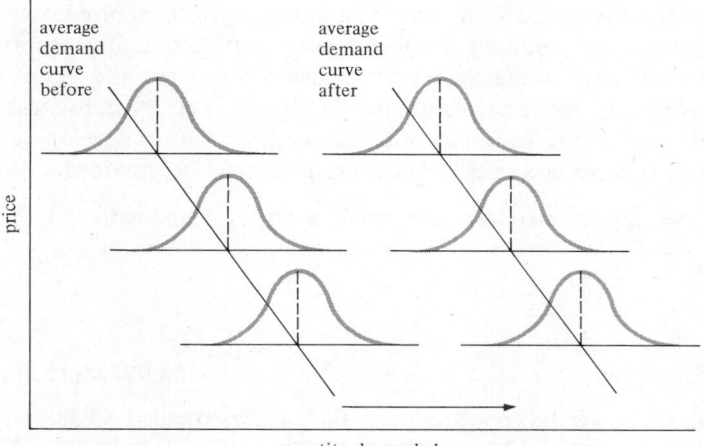

Figure 21.4

If a competitor significantly raised the price of his product relative to yours, it would be reflected by a permanent shift of the average demand curve for your product to the right, reflecting a greater demand on the average for your product at each and every price as a result of the changed circumstances. In such a case, the only information drawn upon to reflect the probability distribution of demand given the new situation should be those observations relating to the new circumstances—although information relating to the nature of dispersion in the previous probability distribution of demand may be useful in estimating the dispersion in the shifted probability distribution of demand given limited observation with respect to the new circumstances.

The question posed is as follows: Given a decision as to product price (known as administered pricing) and an expectation as to the probability distribution of demand at that price, what is the optimal quantity to produce on a daily basis? Equivalently, one could ask: At what point does it cease to become worthwhile to increase the daily level of production by one additional unit? Recall that economic theory states that one should continue to increase the level of production up to the level of activity at which marginal revenue equals marginal cost. The problem is that the $MR = MC$ rule presumes operation under conditions of certainty. It does not allow for shifting of the demand curve. What is needed is an equivalent rule for decision making under uncertainty. A rule is needed that will determine the optimal quantity to produce when faced with a shifting demand curve evidenced by a probability distribution of demand (Figure 21.5).

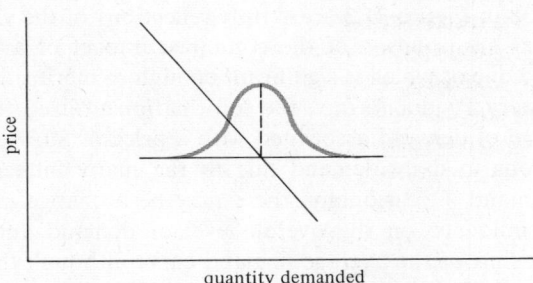

Figure 21.5

In general, as long as the amount of opportunity loss incurred over time as a result of not having produced the incremental unit (profit forgone) exceeds the opportunity loss that would have been incurred over time as a result of having produced the unit, the unit should be produced. This is consistent with the objective of minimizing expected opportunity loss. Before developing a rule to accomplish this, a set of symbols and definitions must be provided.

1. OL_o = opportunity loss incurred as a result of understocking by one unit
    $= P - VC$

where

   P = product price
   VC = increment in total cost as a result of producing one more unit

In the case of Stayle Bakery Products, Inc., price equalled $0.38, variable cost of production equalled $0.25, and therefore OL_u = $0.38 − $0.25 = $0.13.

2. OL_o = opportunity loss incurred as a result of overstocking by one unit
 = $VC - SV$

where

VC = increment in total cost as a result of producing one more unit
SV = salvage value, or the amount that can be recovered on any units left unsold at the end of the period

In the case of Stayle Bakery Products, Inc., there was no salvage value; therefore,

$$OL_o = VC - SV$$
$$= \$0.25 - 0$$
$$= \$0.25$$

If Stayle Bakery Products, Inc., introduced a policy of selling day-old bread at a reduced price of $0.15 a loaf, there would be a salvage value of $0.15 per loaf and OL_o would become $0.25 − $0.15 = $0.10.[4]

3. p represents the probability of selling the additional unit should the firm elect to produce or stock it. Another way of interpreting p is to think of it as the percentage of time daily demand would equal or exceed the new production level. Of course, p will take on different values depending on the level of production at which the firm is operating or planning to operate when the question is raised as to whether or not to produce an additional unit.

4. $1 - p$ represents the probability that the firm will be unable to sell the additional unit should it decide to produce and stock it (i.e., the percentage of the time that daily demand is expected to fall below the new production level).

Returning to Stayle Bakery Products, Inc., the assumption that demand for loaves of bread can take on only the values 40, 50, 60, 70, and 80 loaves will be replaced by the more realistic circumstance that daily demand for bread is normally distributed. Averaging the experience of Stayle Bakery over the last 200 days generated a mean demand of 60 loaves with a standard deviation of 7 loaves. Thus, given a price of $0.38 a loaf, the probability distribution of demand for Stayle Bakery Products, Inc., would appear as in Figure 21.6.

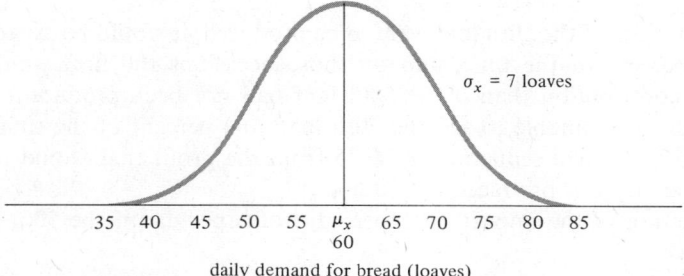

σ_x = 7 loaves

35   40   45   50   55 μ_x 65   70   75   80   85
                              60

daily demand for bread (loaves)

Figure 21.6

[4] Nonperishable items that are overstocked can be sold in subsequent periods (usually). In such instances, the variable cost of production would not represent an opportunity loss to the firm. The appropriate entry for OL_o in such an instance would be incremental cost incurred due to such things as increased insurance, increased storage requirements, increased working capital investment, etc.

Drawing on the tools of analysis referred to above, it can be determined whether it would be economically advisable for Stayle Bakery Products, Inc., to produce a 50th loaf of bread. (With questions of this sort, the analysis proceeds under the assumption that it has already been decided to produce at least 49 loaves.)

Recall that

$$OL_u = \$0.13$$
$$OL_o = \$0.25$$

Computation of the probability that the level of demand activity will be 50 loaves or more is as follows. Referring to Figure 21.7, the probability that $D \geq 50$ loaves is represented by the shaded area under the curve. From the maximum ordinate to 50 loaves is a distance of 10 loaves. One standard deviation unit is equal to 7 loaves, so a distance of 10 loaves is equivalent to a distance of 1.43 standard deviation units. Consultation of the table of areas under a normal curve reveals that 42.36 percent of the area under the curve lies between the maximum ordinate and a point 1.43 standard deviation units away. Combining this with the 50 percent that lies to the right of the mean, the following relationships are obtained:

$$p = \text{Prob}(D \geq 50 \text{ loaves}) = 0.4236 + 0.5000 = 0.9236$$
$$1 - p = \text{Prob}(D < 50 \text{ loaves}) = 1 - 0.9236 = 0.0764$$

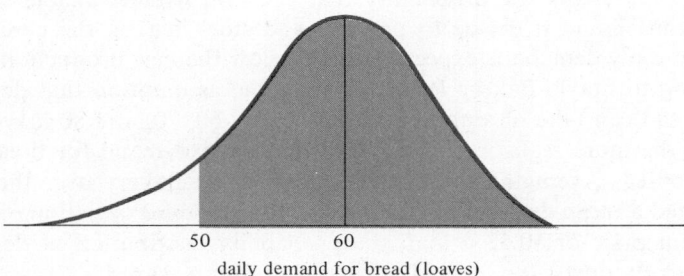

50   60
daily demand for bread (loaves)

Figure 21.7

Interpretation: If the 50th loaf were to be produced, it would be expected to be sold 92.36 percent of the time, and on those occasions the firm would obtain $0.13 more contribution than if the 50th loaf had not been produced. The firm would expect to be unable to sell the 50th loaf 7.64 percent of the time, and on those occasions incur a reduction of $0.25 from the profit that would have been made had the firm not produced the 50th loaf.

Computation of the impact on expected profits per day if the 50th loaf were to be produced:

$$
\begin{array}{lll}
0.9236(0.13) & = & 0.120068 \\
0.0764(-0.25) & = & -0.019100 \\
\hline
& & 0.100968
\end{array}
$$

Over time, as a result of producing the 50th loaf, average profits per day would be $0.101 higher than they would have been had the firm produced only 49 loaves. Clearly, the 50th loaf should have been produced.

The above procedure can be represented symbolically as follows:

$$(0.9236)(\$0.13) = p(OL_u) \qquad = EOL_u = \text{expected opportunity loss due to understocking}$$

$$(0.0764)(\$0.25) = (1 - p)(OL_o) = EOL_o = \text{expected opportunity loss due to overstocking}$$

Figure 21.8 describes the behavior of EOL_u and EOL_o. Note that at an extremely low level of operation, the probability of selling an additional unit is very high. As the level of operation rises, however, the probability of selling an additional unit declines. Thus, at the left of the graph, EOL_u or $p(OL_u)$ is virtually equal to OL_u. As the level of production increases, however, p declines and the value of EOL_u declines correspondingly. EOL_o behaves in just the opposite manner. At low levels of activity, $(1 - p)$ is virtually zero, but as the level of activity rises, $(1 - p)$ increases, causing the value of EOL_o to rise, gradually approaching the limiting value OL_o.

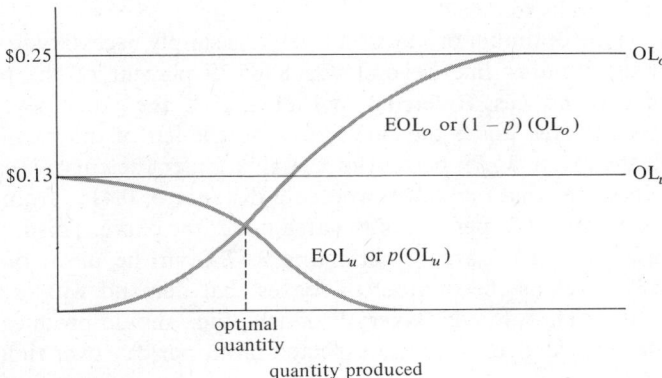

Figure 21.8

Interpretation: At any level of production below Q optimum, $EOL_u > EOL_o$. As a result, average profits over time would be greater as a result of stocking an additional unit than average profits over time would have been had the additional unit not been stocked. At any level of production above Q optimum, $EOL_u < EOL_o$. As a result, average profits over time would be less as a result of stocking an additional unit than average profits over time would have been had the additional unit not been stocked.

The only level of production at which no incentive exists to either expand or contract production is at Q optimum, which coincides with the production level where $EOL_u = EOL_o$. This condition is the equivalent rule under uncertainty to the $MR = MC$ rule. Thus, the rule is to produce up to the point where $EOL_u = EOL_o$. Substituting:

$$EOL_u = EOL_o$$
$$p(OL_u) = (1 - p)(OL_o)$$

Solving for p:

$$p(OL_u) = OL_o - p(OL_o)$$
$$p(OL_u) + p(OL_o) = OL_o$$
$$p(OL_u + OL_o) = OL_o$$

$$p = \frac{OL_o}{OL_u + OL_o}$$

The above formula generates the value of p that equalizes the expected opportunity loss due to overstocking and the expected opportunity loss due to understocking. Henceforth this value will be referred to as p optimum and will be designated by the symbol p_o. The firm should continue to expand the level of production as long as the probability of demand being equal to or exceeding that production level is greater than p_o or $OL_o/(OL_u + OL_o)$.

In the case of Stayle Bakery Products, Inc.:

$$p_o = \frac{OL_o}{OL_u + OL_o} = \frac{\$0.25}{\$0.13 + \$0.25} = \frac{\$0.25}{\$0.38} = 0.6579$$

Thus, Stayle Bakery Products, Inc., should continue expanding the level of production until the probability of daily demand being equal to or exceeding the level of production falls to 0.6579.

To determine the optimum production level, it is simply necessary to determine that point on the number line beyond which 65.79 percent of the probability distribution of demand lies. Referring to Figure 21.9, the point beyond which 65.79 percent of the area under the curve lies is to the left of the mean. Between that value and the mean, 15.79 percent of the area under the curve must lie. The table of areas indicates that one must proceed a distance of $0.41\sigma_x$ from the mean in order to encompass 15.79 percent of the area under the curve. This is equivalent to $(0.41)(7 \text{ loaves})$ or 2.87 loaves. Subtracting 2.87 from the mean of 60 loaves generates 57.13 loaves as the number of loaves that demand will exceed 65.79 percent of the time. Thus, Stayle Bakery Products, Inc., should produce 57 loaves of bread per day in order to maximize expected profit per day over time.

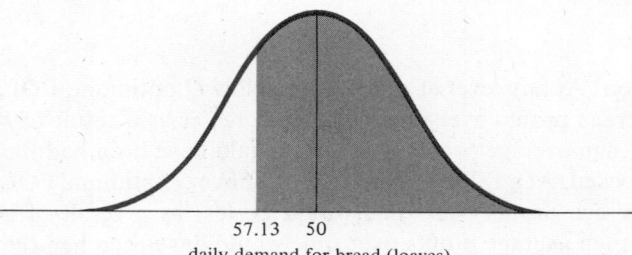

57.13　　50

daily demand for bread (loaves)

Figure 21.9

Suppose that Stayle Bakery Products, Inc., introduced the policy of selling as day-old bread any loaves left on hand at the end of a day. Presume that day-old bread and fresh bread serve different markets. Presume furthermore that the demand for day-old bread has always exceeded the supply—in fact, it is always sold by 10 a.m. of the next day. Moreover, day-old-bread customers never purchase fresh bread when they discover day-old bread is sold out, and fresh-bread customers never purchase day-old bread (unless they are planning on fondue, and even then they have to arrive before 10 a.m.; on those occasions they would be considered as part of the day-old-bread market). Under the revised circumstances, OL_o changes from \$0.25 to \$0.10.

$$(OL_o = VC - SV = \$0.25 - \$0.15)$$

$$p_o = \frac{OL_o}{OL_u + OL_o} = \frac{\$0.10}{\$0.13 + \$0.10} = \frac{\$0.10}{\$0.23} = 0.4348$$

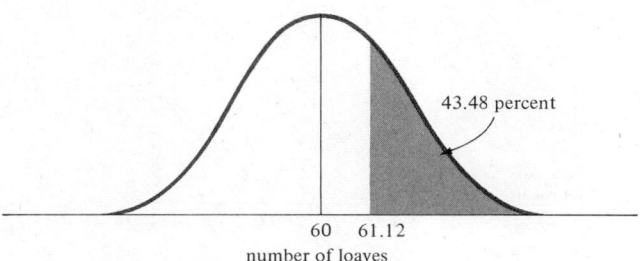

43.48 percent

60   61.12
number of loaves

Figure 21.10

Determination of the optimal stock level is shown in Figure 21.10. The table of areas indicates that one must proceed a distance of 0.16σ from the mean in order to encompass 6.52 percent of the area under the curve. This is equivalent to $(0.16)(7$ loaves$)$ or 1.12 loaves. The optimal stock level lies to the right of the mean when $p_o < 0.5000$, so 1.12 is added to the value of the universe mean. Given the adoption of the new policy, the optimal quantity to produce is 61 loaves. Notice that in both illustrations, the optimal quantity to produce does not correspond to the average quantity demanded. The only circumstance when the average quantity demanded is the optimal quantity to produce is when the penalty for overstocking by one unit is identical to the penalty for understocking by one unit. When $OL_o = OL_u$, $p_o = OL_o/(OL_o + OL_u) = 0.50$ and the optimal quantity to produce is the average quantity demanded.

PROBLEMS

Problem 21.1

Ty Coon is contemplating selling boxes of Christmas cards to pick up extra spending money. He has assembled a selling force of 100 urchins, well drilled in the art of door-to-door selling. The decumulative probability distribution depicted in Figure 21.11 describes Ty's assessment of the probability that any given urchin will sell x boxes or more. (Ty feels that if you've seen one urchin, you've seen them all.) To illustrate: A decumulative probability of 0.80 is associated with 122 boxes sold (i.e., Ty believes that 80 percent of the urchins will sell 122 boxes or more). The cards sell for \$5 a box. Each box costs Ty \$3.50, and the urchins are paid a \$0.50 commission on each box sold. Ty must order the boxes now, and intends to provide each urchin with the same number of boxes. Each urchin is on his own. An urchin who sells out completely cannot obtain more boxes even though other urchins may have boxes left over. Unsold boxes have no salvage value. The following approximation to the probability distribution was obtained by segmenting the decumulative probability distribution into deciles (see Figure 21.12).

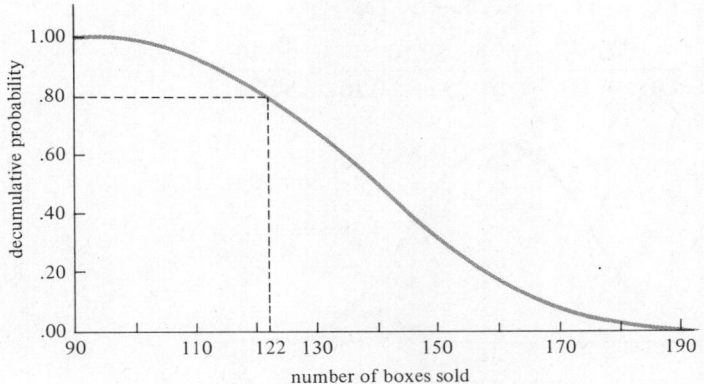

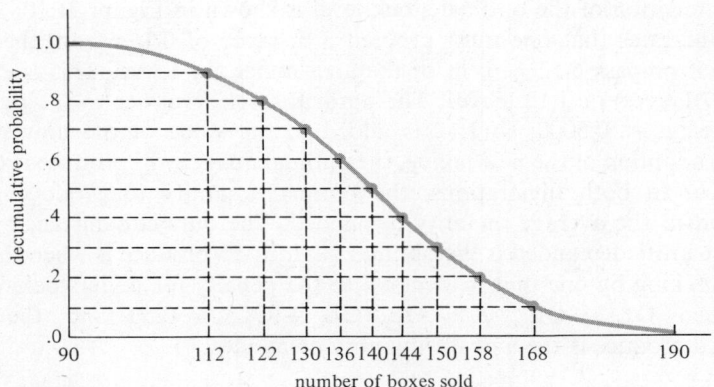

Figure 21.12

Number of boxes sold	Probability that the number sold will fall in the stipulated interval
90–112	0.10
112–122	0.10
122–130	0.10
130–136	0.10
136–140	0.10
140–144	0.10
144–150	0.10
150–158	0.10
158–168	0.10
168–190	0.10
	————
	1.00

For purposes of analysis, Ty restricted the demand possibilities to the bracket medians, assigning each a probability of 10 percent. Thus, the various states of nature considered are as follows:

Number of boxes sold	Probability of occurrence
101	0.10
117	0.10
126	0.10
133	0.10
138	0.10
142	0.10
147	0.10
154	0.10
163	0.10
179	0.10

Ty has decided to restrict his action alternatives to the following possibilities: Assign each urchin 110, 120, 130, 140, 150, or 160 boxes.
1. Construct a conditional profit table.
2. Determine the action alternative associated with maximizing expected profit per urchin.
3. Determine EVPI.
4. How is knowledge of the EVPI of value to the decision maker?

Problem 21.2

Referring to Problem 21.1:
1. Construct a conditional opportunity loss table.
2. Determine the action alternative associated with minimizing expected opportunity loss per urchin.
3. If a local testing agency guaranteed that they could predict with absolute certainty the sales performance for each urchin, what is the maximum amount you would be willing to pay for this service per urchin?

Problem 21.3

Referring to Problem 21.1 and using the marginal approach and the decumulative probability distribution of demand depicted in Figure 21.11, determine the optimal number of boxes to assign to the urchins.

Problem 21.4

Referring to Problem 21.1:
1. The decumulative probability distribution of demand depicted in Figure 21.11 refers to a probability density function that is normally distributed with a mean equal to 140 and a standard deviation equal to 20. Verify this.
2. Using the marginal approach and the normal curve description of demand, determine the optimal number of boxes to assign to the urchins.

Problem 21.5

Cumbersome Farms' local outlet desires to determine how many cartons of milk to stock on a daily basis. Historical data has generated the following pattern of demand.

Quantity demanded (no. of cartons)	Number of days on which given level of demand occurred
150	2
151	8
152	10
153	40
154	20
155	15
156	5
	——
	100

Assume that stock levels are restricted to the range 150–156 and that milk left unsold at the end of the day must be disposed of due to inadequate-refrigeration facilities. Milk costs $0.35 a carton to the retailer and sells for $0.50 a carton.

1. Construct a conditional profit table.
2. Determine the action alternative associated with the maximization of expected profit.
3. Determine EVPI.

Problem 21.6

Referring to Problem 21.5:

1. Construct a conditional opportunity loss table.
2. Determine the action alternative associated with the minimization of expected opportunity loss.

Problem 21.7

Referring to Problem 21.5:

1. Construct a decumulative probability distribution of demand.
2. Utilizing the marginal approach, determine the optimal number of cartons to stock on a daily basis.

Problem 21.8

Beck and Call, a news agency, is concerned with determining the optimal number of copies of the *Daily Clarion* to acquire for distribution on a daily basis. They have assessed a probability distribution of demand as follows:

No. of copies demanded	Percentage of time the given level of demand occurred
100	5
101	10
102	40
103	20
104	15
105	10
	——
	100

Papers remaining unsold at the end of the day may be returned for a credit of $0.02. Papers cost the distributor $0.07 and sell for $0.10.

1. Construct a conditional profit table.

2. Determine the action alternative associated with the maximization of expected profit.
3. Determine EVPI.

Problem 21.9

Referring to Problem 21.8:
1. Construct a conditional opportunity loss table.
2. Determine the action alternative associated with the minimization of expected opportunity loss.

Problem 21.10

Referring to Problem 21.8:
1. Construct a decumulative probability distribution of demand.
2. Utilizing the marginal approach, determine the optimal number of papers to order on a daily basis.

Problem 21.11

The probability distribution of demand is not necessarily obtained by observing the historical pattern of quantity sold. Discuss.

Problem 21.12[5]

B. Nomial, Inc., a chemical processing facility, cannot operate with less than 14 men. The permanent work force currently consists of 20 men and at no time have there been less than 14 men present. Accordingly, the plant has never had to shut down. Historically, the absentee rate has been approximately 5 percent, and the pattern of absenteeism has been as follows:

Number absent	Percentage of the time the given number of absences has occurred
0	36
1	38
2	19
3	6
4	1
5	0
6 or more	0
	100

The firm is considering a reduction in the size of the permanent work force. Company policy is to pay all workers (present and absent) even if the firm is unable to operate because more than six men are absent. All workers receive $20 per day. When the firm operates, a contribution (price-variable cost) of $600 per day is generated. For each reduction of one man in the size of the permanent work force, this figure increases by $20. If the firm is unable to operate, a fixed

[5] The idea for this problem originated with the case, Ashby Chemical Company, ICH 7C40 EA-C 472.

cost of $550 is incurred (including the $400 labor cost). This figure declines by $20 with each one-man reduction in the size of the labor force.

1. Construct a conditional contribution table.
2. Determine the action alternative associated with the maximization of expected contribution.
3. Determine EVPI.
4. Construct a conditional opportunity loss table.
5. Determine the action alternative associated with the minimization of expected opportunity loss.
6. Explain why the answers to parts 2 and 5 must be identical.
7. What is the expected contribution under conditions of *certainty*?
8. Would it be possible to obtain "certain" information in time for it to be of some use to the decision maker?
9. How much would it be worth per day for B. Nomial, Inc., to be able to operate under conditions of certainty? How might B. Nomial use this information?
10. Suppose a reduction in work size causes an increased awareness of the possibility of shutdown and therefore a reduction in the rate of absenteeism to 4 percent. How would this affect the optimal level for a permanent work force?
11. Repeat part 10, with a reduction in the absenteeism rate to 3 percent.
12. Repeat part 10, with a reduction in the absenteeism rate to 2 percent.
13. Suppose the reaction of the work force to reduced size is increased dissatisfaction and an increase in the absentee rate to 6 percent. How would this affect the optimal level for a permanent work force?

Problem 21.13

The techniques described in Chapter 21 may be applied to the cases Linmar Company (A), ICH 6C33 EA-C 467, and Linmar Company (B), ICH 6C34 EA-C 468. Copies of these cases may be obtained by writing to: Intercollegiate Case Clearing House, Harvard Business School, Soldiers Field Road, Boston, Massachusetts 02163.

Problem 21.14

Grossman's Bagel Emporium is attempting to determine the optimal daily production level for bagels with the objective of maximizing profit. Bagels sell for $0.20, and their production involves a variable expenditure of $0.08 each. Bagels not sold on any given day are simply disposed of. Analysis of past sales data has revealed that the average sale has been 372 bagels with a standard deviation of 25 bagels. Assume that daily sales are normally distributed. Using the marginal approach, what is the optimal daily production level?

$$p = \frac{OL_o}{OL_o + OL_u}$$

Problem 21.15

Given that a greeting card manufacturer believes that the likely sales volume for a particular card is normally distributed with a mean of 55,000 and a standard deviation of 6,000, determine the number of cards he should produce if he desires to maximize expected profit. Variable cost per card is $0.26, and the cards sell for $0.50 each. There is no salvage value, since he always scraps last year's line.

Problem 21.16

Daily sales of Allied Products, Inc., is normally distributed with a mean equal to 120 units and a standard deviation of 15 units. Units cost $12, sell for $15, and there is no salvage value. Units left unsold at the end of the day are discarded. What is the optimal daily stock?

Problem 21.17

Tenderhooks, Inc., has determined that its daily sales are normally distributed with a mean equal to 75 units and a standard deviation equal to 10 units. Units left unsold at the end of the day have a salvage value of $3. Units sell for $11 and cost the firm $7. What is the optimal daily stock?

Problem 21.18

Given that sales are normally distributed, what relationship must exist between OL_o and OL_u in order for the optimal daily stock level to be equal to the mean level of demand? Justify your answer.

DECISION TREE ANALYSIS: EXPECTED MONETARY VALUES

22

The decision maker needs a kit of tools that facilitate arriving at better decisions given the necessity of operating in an uncertain environment. Strategies comprise a series of interrelated sequential decisions representing the decision maker's response to a series of chance events. When the strategies under consideration are few and relatively uncomplicated, payoff table analysis is a useful technique. Payoff table analysis was introduced in Chapter 21. As the strategies under consideration increase in number and complexity, however, payoff table analysis becomes unwieldy. In such situations, a more sophisticated technique, decision tree analysis, should be used. The decision tree is a device that enables the decision maker to conceptualize chronologically "all the possible courses of action available, all the possible events which may affect the consequences of these actions, and all the

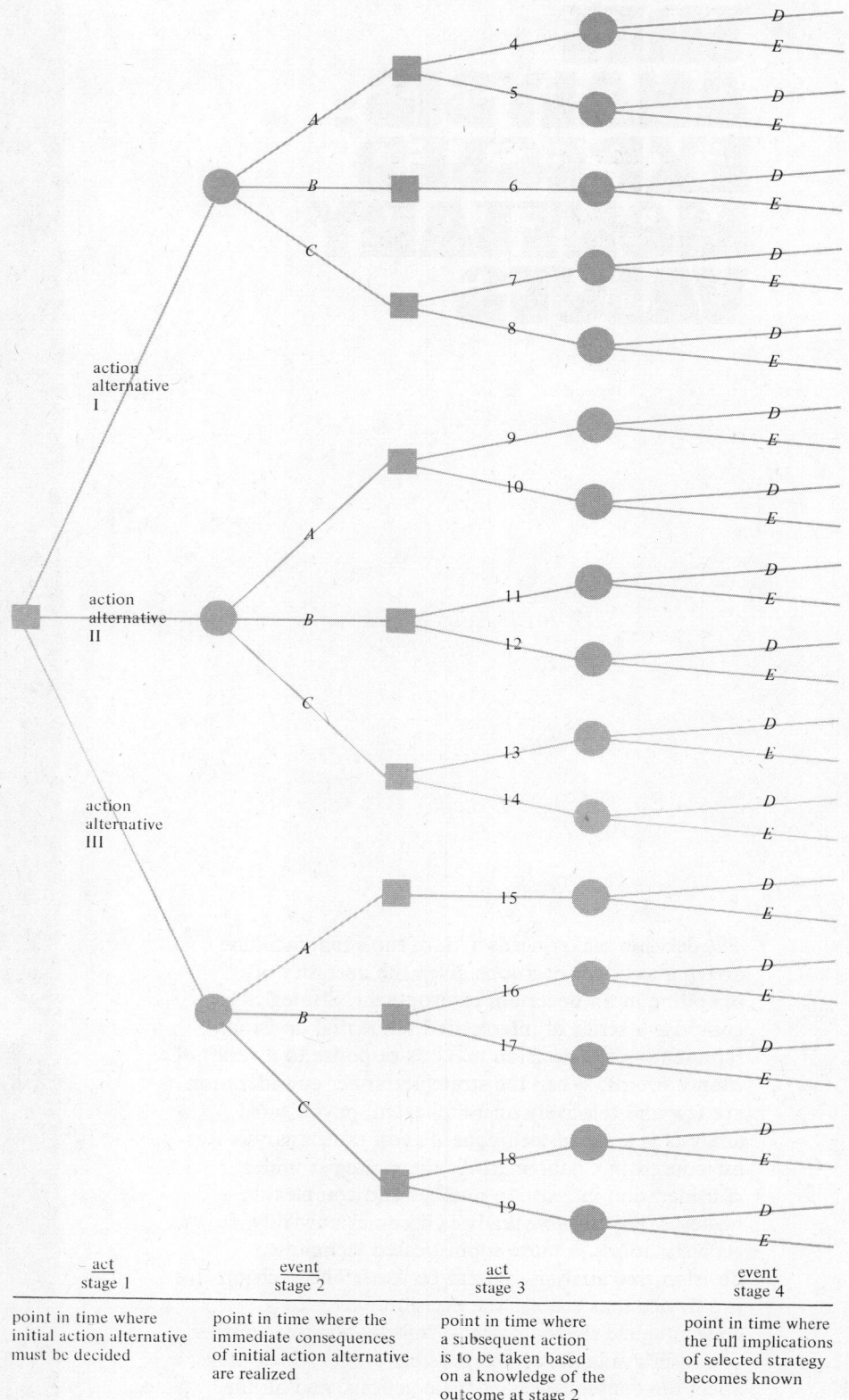

act stage 1	event stage 2	act stage 3	event stage 4
point in time where initial action alternative must be decided	point in time where the immediate consequences of initial action alternative are realized	point in time where a subsequent action is to be taken based on a knowledge of the outcome at stage 2	point in time where the full implications of selected strategy becomes known

Figure 22.1

relevant factual data required to evaluate each possible consequence (numerical evaluations of each outcome, probability of each event, etc.)."[1]

Figure 22.1 provides an illustrative example of a decision tree. At stage 1, the decision maker is faced with the necessity of deciding upon either action alternative I, action alternative II, or action alternative III. The immediate consequence associated with the selection of any of these alternatives, however, depends on whether A, B, or C eventually materializes. Unfortunately, as reflected by the ordering of the acts and events in the decision diagram, information as to which event has occurred will not be available at the point in time at which a course of action must be selected. The first two stages of the decision diagram, with the addition of the dollar consequences associated with arriving at each of the endpoints as of stage 2, are reproduced in Figure 22.2.

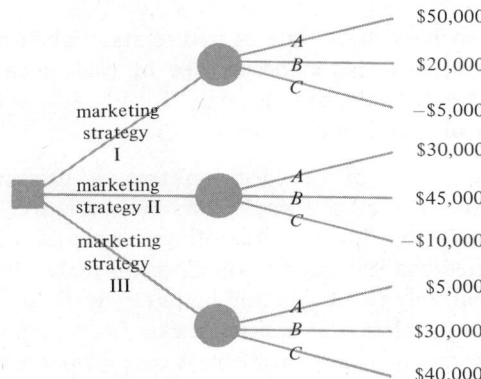

Figure 22.2

Consider the three action alternatives as alternative marketing strategies and the three events (A, B, and C) as referring to alternative descriptions of the nature of the market in which the strategies are to be implemented. The response to a particular marketing strategy depends on whether the true description of the market is state A, B, or C.

Given state of nature A, the most desirable immediate consequence is associated with marketing strategy I.
Given state of nature B, the most desirable immediate consequence is associated with marketing strategy II.
Given state of nature C, the most desirable immediate consequence is associated with marketing strategy III.

Unfortunately, the decision maker's knowledge is restricted to a subjective feeling about the likelihood of occurrence of these states of nature. While far from ideal, this limited information nevertheless can be useful to the decision maker in selecting among the available courses of action. Later in the chapter the manner in which this information can be utilized in arriving at an optimal decision under uncertainty will be demonstrated.

To perceive the magnitude of the dilemma facing the decision maker, it should be recognized that even if the decision maker did know in advance whether state of nature A, B, or C was going to materialize, it still would not be clear whether

[1] Robert Schlaifer, *Analysis of Decisions Under Uncertainty*, New York: McGraw-Hill, 1969, p. 4.

marketing strategy I, II, or III is the more desirable alternative. This is because the decision maker must still take subsequent action at stage 3, the consequences of which will not become known until stage 4. Events *D* and *E* at stage 4 might, for example, refer to whether or not a competitor elects to take some sort of retaliatory action.

The upshot is that the decision maker must consider not only the immediate consequences of any given action alternative but the ramifications of the various events and follow-up actions that might subsequently materialize as well. Summarizing, the decision maker must take into consideration the following factors in evaluating the desirability of any given action alternative:

1. States of nature (or events) and subsequent actions that could conceivably exert an influence on the consequences of selecting any of the action alternatives under consideration.
2. The likelihood of each such event or state of nature materializing.
3. The consequence associated with the occurrence of each such event in conjunction with the action alternative selected.
4. The relative desirability of each consequence.[2]

Without the analytical structure of the decision tree, a decision maker attempting to consider simultaneously all of these factors in association with each action alternative can be equated to a juggler attempting to keep 30 balls in the air simultaneously. Decision tree analysis enables the decision maker to consider each piece of the problem separately (to juggle one ball at a time). By gradually analyzing subsets of the overall problem, the decision maker is enabled ultimately to arrive at a solution to the overall aggregate problem. A case format, Component Supply, Inc., will be utilized as the vehicle for the development and illustration of the concepts and techniques of decision tree analysis.

Component Supply, Inc.

One of the products produced by Component Supply, Inc., is a small, transistorized component that serves as an integral part of the amplifier system of stereos. The part is produced in large volume and is sold to a number of manufacturers. It sells for $35 and involves a variable cost of production of $15. Until recently, it has been impossible to determine whether or not a particular transistorized component will work until it has actually been assembled into the completed amplifier system. In instances where the component turns out to be defective, Component Supply, Inc., has agreed to return the full purchase price of $35. In addition, Component Supply, Inc., has agreed to pay a penalty charge of $12 to reimburse the manufacturer for the inconvenience and loss of time involved in the replacement of the component. The transistorized component is quite delicate, and approximately 27 percent of all parts shipped in the past have subsequently proven to be defective. Recently, Component Supply, Inc., was approached by Prevention, Inc., a firm specializing in testing equipment. Prevention, Inc., proposed a testing device that would enable Component Supply to ascertain prior to shipment those parts that would later prove to be defective. A pilot model of the testing device has been

[2] In this chapter, it is assumed that the relative desirability of a given consequence is measured by the dollar magnitude of that outcome. In Chapter 23 this assumption will be released, and utility considerations will be introduced into the analysis.

developed, and Component Supply has subjected portions of its production over the past month to this test. Records have been kept, which indicate that 90 percent of the parts tested generated a positive reading. Upon shipment, it was determined that 20 percent of the parts recording a positive reading later proved to be defective. Those parts that were indicated as defective by the test were also shipped. Of the parts that had registered a negative reading, 90 percent subsequently were returned as defectives. Component Supply, Inc., is now in the process of deciding whether or not to rent the testing device and, if so, the maximum price they would be willing to pay (price is expressed as so much per part tested). Decision tree analysis will be utilized to generate a solution.

22.1 CHRONOLOGICAL DEPICTION OF THE POSSIBLE ACTS AND EVENTS

The first stage in the analysis is the development of a decision tree consisting of a chronological ordering of the various acts and events associated with the decision problem under investigation. Two types of branches or forks appear in the decision diagram or tree.

1. Act forks: Whenever the decision maker arrives at an act fork, the path to be followed is determined by the decision maker. An act fork is indicated by the placement of a square symbol at the base of the fork.
2. Event forks: Whenever the decision maker arrives at an event fork, the path the decision maker will follow is beyond his control. Direction of movement from an event fork is determined by chance. The decision maker is able to assign an expression of his attitude as to the likelihood of movement along each of these paths, but the actual path along which movement through the tree takes place is determined by chance. A chance or event fork is indicated by the placement of a circular symbol at the base of the fork.

The chronological depiction of the possible acts and events associated with the Component Supply, Inc., problem is presented in Figure 22.3. Normally it is not necessary to label the forks with numerical references. Labels have been included in Figure 22.3 to facilitate references to the decision tree.

The first act fork encountered, act fork $\boxed{1}$, technically offers the decision maker a choice as to whether or not to produce a part. It is presumed that this has already been decided in the affirmative. The second fork encountered, act fork $\boxed{2}$, offers the decision maker the option of continuing to operate as he has in the past or adopting the testing device. Should he elect to continue as before, the next fork encountered would be the trivial decision fork, ship or do not ship. Obviously, if the decision maker had no intention of shipping the parts, he never would have produced them in the first place. In such situations, trivial decision forks can be replaced by the obvious single action alternative, in this case, ship. Reflecting this policy, only the action alternative, ship, is included in the diagram. Upon shipment, the next information encountered deals with whether the part shipped is acceptable or defective. Historically, 73 percent of the parts have been acceptable and 27 percent defective. The status for any one part is determined by chance. This is reflected by the chance fork designated $\boxed{6}$.

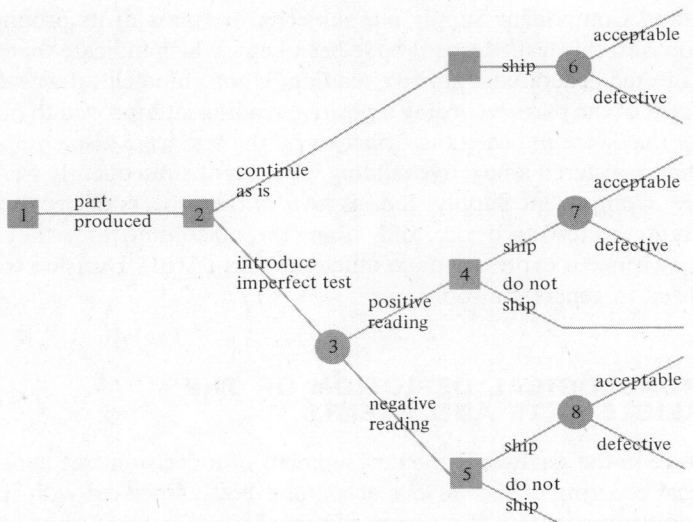

Figure 22.3 Skeletal framework of the decision tree.

Next return to act fork $\boxed{2}$ and examine the branch associated with adoption of the testing device. Given the adoption of the testing device, the next fork encountered is the event fork labeled ③. Event fork ③ reflects that the next piece of information made available to the decision maker will be whether the part tested records a positive or a negative reading. Upon receipt of this information, the decision maker must make another decision as reflected by act forks $\boxed{4}$ and $\boxed{5}$.

In actuality, act forks $\boxed{4}$ and $\boxed{5}$ are also somewhat trivial. If, after obtaining a positive reading, the decision maker decides not to ship (or after obtaining a negative reading, the decision maker decides to ship), he would be ignoring the test results. If he intends to ignore the test results, why bother to test? In spite of this, act forks $\boxed{4}$ and $\boxed{5}$ will be retained in the decision tree.

One reason they are retained is to provide additional opportunities for demonstrating the manner in which the decision tree is collapsed by a series of subset decisions into a final subset involving only the immediate action alternatives emanating from act fork $\boxed{2}$. The second reason that act forks $\boxed{4}$ and $\boxed{5}$ are retained is to illustrate that trivial forks are automatically taken care of in the process of analysis. Thus, if in doubt, leave the forks in question in the tree. The only disadvantage is a little more detail than would normally be required for a necessary and sufficient diagram.

22.2 ASSIGNING PARTIAL CASH FLOWS

Once the decision tree has been skeletonized by appropriate chronological sequencing[3] of the acts and events associated with the decision problem, the next

---

[3] It is important to recognize that chronological ordering is not on the basis of when an event occurs. Rather, it is based on where in the context of the decision process the *knowledge of the outcome* of the event becomes available to the decision maker.

stage is to put some flesh on the bones. In this context, flesh consists of partial cash flows and probabilities. They will be accorded separate treatment in the interest of reducing the complexity of the diagram.

There are dollar consequences associated with many of the acts and events encountered as one moves through the diagram. Each such consequence should be recorded on the diagram in association with the act or event with which it is realized. Such entries are referred to as *partial cash flows*. Figure 22.4 reproduces the skeletal framework of Figure 22.3 with the addition of partial cash flows. A $15 expenditure is associated with the production of each part. Whenever a part is shipped, a $35 receipt is recognized. Upon determination of a part being defective, the $35 receipt is negated and a penalty charge of $12 incurred. In addition a cost of T_I is incurred whenever a part is subjected to the test. The symbol T_I is used as a proxy for the unestablished cost of testing. If a specific dollar cost were to be stipulated, it would be entered in place of the generalized symbol, T_I.

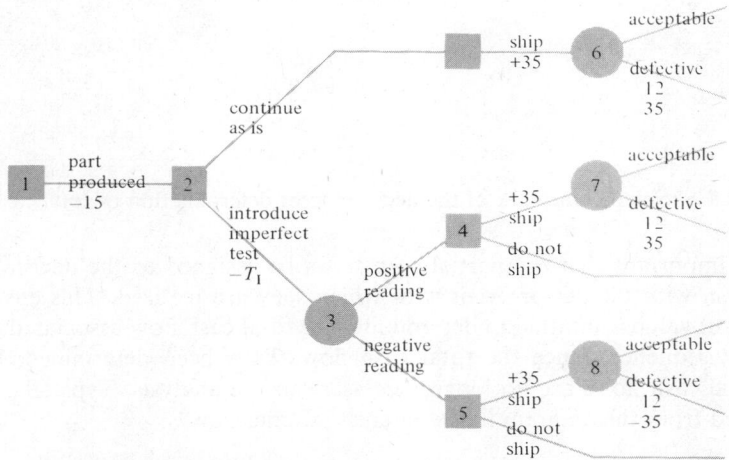

Figure 22.4 Skeletal framework of the decision tree: assignment of partial cash flows.

22.3 DETERMINATION OF THE TOTAL CASH FLOWS

Each path from the origin to an endpoint in the diagram describes an act–event sequence. Since the decision maker's concern is with ultimate consequences, the next logical step is to sum the partial cash flows encountered as one moves from the origin to the endpoint for each of the act–event sequences (Figure 22.5). Such totals are referred to as total cash flows or *terminal values*.[4]

---

[4] In this chapter, concern is restricted to situations where the relative desirability of each consequence is reflected by the dollar magnitude of that consequence. When this is the case, total cash flow and terminal value have the same meaning. When this assumption is released in Chapter 23, total cash flow and terminal value will not be synonymous. To illustrate: The value of concern at each endpoint may not be the total cash flow but rather the total cash position of the firm as of the point in time represented by the endpoint. Total cash position is equal to the sum of *total cash flow plus the beginning cash position*. Total cash position, rather than total cash flow, would then be assigned as the *terminal value*.

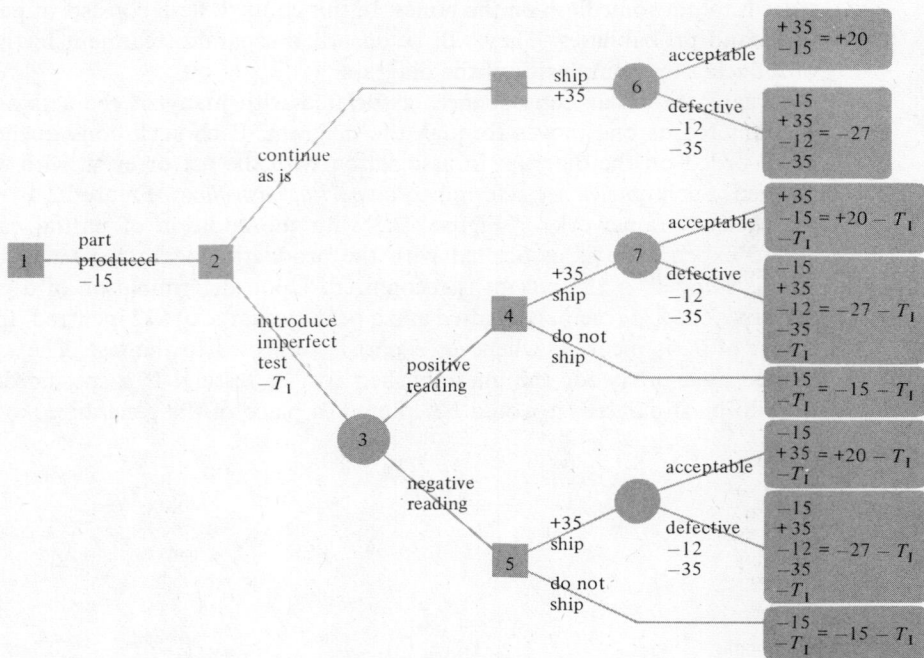

Figure 22.5 Skeletal framework of the decision tree: determination of total cash flows.

It is important that the partial cash flows be assigned to the decision tree in association with the act or event with which they are realized. This ensures that no relevant value is omitted in determining the total cash flow associated with any act–event sequence. Once the total cash flows have been determined, however, the partial cash flows are no longer necessary to the analysis. Typically, they are eliminated from the diagram in the interest of efficiency.

22.4 ASSIGNMENT OF PROBABILITIES TO THE EVENT FORKS

Probabilities must be assigned to each branch of each chance fork in the decision diagram. The probabilities assigned to the branches at any one chance fork must, of course, sum to one. A useful first step is to survey the diagram and list the probabilities needed:

event fork ③	P(positive reading)
	P(negative reading)
event fork ⑥	P(acceptable)
	P(defective)
event fork ⑦	P(acceptable | positive reading)
	P(defective | positive reading)
event fork ⑧	P(acceptable | negative reading)
	P(defective | negative reading)

Note that at event fork ⑥, the decision maker simply needs to know the proportion of all parts produced that will be acceptable and the proportion that will be defective.

At event fork ⑦, the population of interest is restricted to those parts that yield a positive reading. Of these, the decision maker needs to know the proportion that will ultimately prove acceptable and the proportion that will ultimately prove defective.

At event fork ⑧, the population of interest is restricted to those parts that yield a negative reading. Of these, the decision maker needs to know the proportion that will ultimately prove acceptable and the proportion that will ultimately prove defective.

The probabilities themselves are usually determined in one of three ways:

1. They are assessed directly from objective information at the decision maker's disposal (as they are in the illustrative example).
2. Some information on probabilities exists, but manipulation of the given information is required to obtain the probabilities required for analysis. Manipulation will often require the use of Bayes' rule.[5]
3. They are assigned subjectively by the decision maker, reflecting *his* strength of conviction with regard to the likelihood of occurrence of the various events.[6]

Referring to the problem description on pp. 542–543, the probabilities needed for the analysis are as follows:

event fork ③	P(positive reading)	= 0.90
	P(negative reading)	= 0.10
event fork ⑥	P(acceptable)	= 0.73
	P(defective)	= 0.27
event fork ⑦	P(acceptable | positive reading)	= 0.80
	P(defective | positive reading)	= 0.20
event fork ⑧	P(acceptable | negative reading)	= 0.10
	P(defective | negative reading)	= 0.90

Figure 22.6 depicts the decision tree skeleton with the flesh attached (total cash flows and probabilities). The decision tree is now ready for analysis.

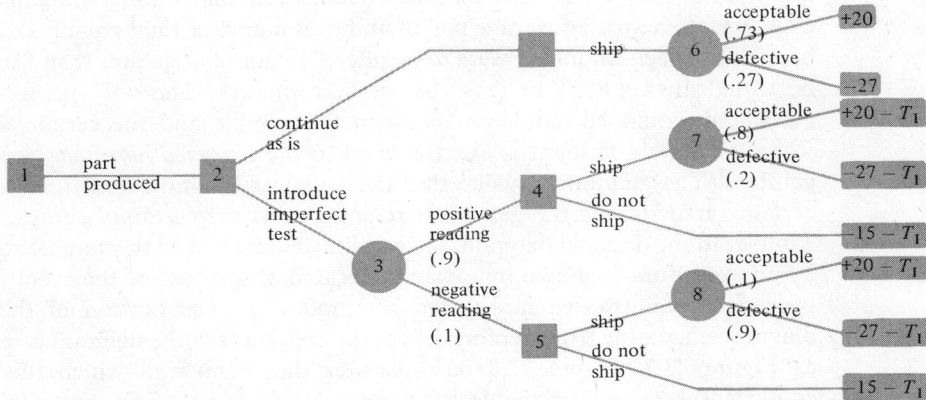

Figure 22.6 Skeletal framework of the decision tree: total cash flows and probabilities.

[5] This approach is required in the Technotronics Case, ICH 5C 30R EA-C 466R². The procedure for obtaining copies of this case is described at the end of the problem set for this chapter.

[6] The illustrative example in Chapter 23 provides an illustration of a situation where the probabilities are assigned subjectively by the decision maker.

22.5 THE TECHNIQUE OF BACKWARD INDUCTION

Once the decision tree has been established, the decision maker is in a position to obtain a solution to the aggregate overall problem by obtaining a series of solutions to subset problems by the technique of backward induction. The solution to the last subset problem will correspond to a solution to the overall aggregate problem represented by the decision tree.

The procedure: Analysis commences at the endpoints of the diagram and gradually folds the diagram back to the fork at the origin of the tree. Let us begin with event fork ⑦. Consider the decision maker as having worked his way through the decision tree up to the point designated as the base of event fork ⑦. Whenever the decision maker finds himself at the base of an event fork, he can consider himself faced with a gamble. In this case, the gamble consists of an 80 percent chance of ending up with $+\$20 - T_I$ and a 20 percent chance of ending up with $-\$27 - T_I$. It is the objective of this subset problem to determine the amount the decision maker would be willing to settle for in lieu of taking this gamble. Given that many such parts are produced, this is a gamble the decision maker would expect to encounter many times. Over time, his average payoff on these gambles would be equal to the mathematical expectation of the gamble. For event fork ⑦, the mathematical expectation is $\$10.60 - T_I$, computed as follows:

probability of occurrence	× conditional payoff	computation of expected monetary payoff
(0.8)	× $(+20 - T_I)$ =	$16 \quad - 0.8T_I$
(0.2)	× $(-27 - T_I)$ =	$- 5.40 - 0.2T_I$
		$\overline{10.60 - \quad T_I}$

Whenever faced with the gamble represented by event fork ⑦, the decision maker would always prefer the gamble if offered an alternative certain amount less than $\$10.60 - T_I$. This reflects a realization that taking the gamble leads, over time, to a greater average payoff and thus a greater total payoff. On the other hand, if the decision maker were to be offered an amount greater than $\$10.60 - T_I$, he would always elect to take that certain amount. The only point where the individual would be indifferent between the gamble and the certain amount is when the certain amount is exactly equal to the *expected monetary payoff* of the gamble.[7] The guaranteed value that the individual is indifferent to receiving for certain versus taking the gamble is referred to as the certainty equivalent to the gamble. In the decision diagram, the certainty equivalent to the gamble represented by an event fork is shown in a triangle located at the base of the event fork. This procedure is illustrated in Figures 22.7 and 22.8. The portion of the decision diagram emanating from act fork ④ to the endpoints of the diagram is reproduced as Figure 22.7. Figure 22.8 demonstrates the manner in which the certainty equivalent is entered on the decision tree.

---

[7] It is assumed that the individual has no aversion to nor unwarranted inclination toward gambling (i.e., that he is neither a risk averter nor a risk taker). Such individuals (risk averters and risk takers) are recognized and are taken into consideration in the more sophisticated model developed in Chapter 23.

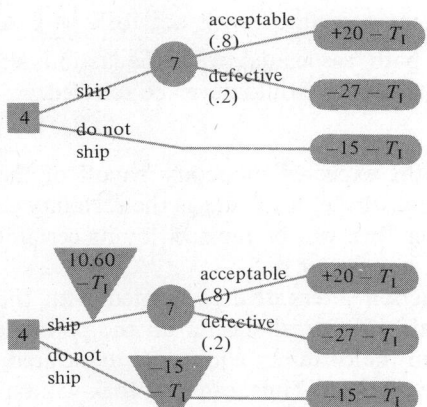

Figure 22.8

The certainty equivalent associated with the act "do not ship" is a certain $-\$15 - T_I$ and accordingly is entered in Figure 22.8. Given the decision maker's decisions on certainty equivalents, standing at the base of act fork $\boxed{4}$, he would consider either of the representations in Figure 22.9 as equivalent representations of the decision he must make. Clearly, "the number of balls the decision maker must juggle" in arriving at a determination of the optimal action alternative when standing at the base of act fork $\boxed{4}$ is considerably lessened when the choice is expressed in terms of certainty equivalents.

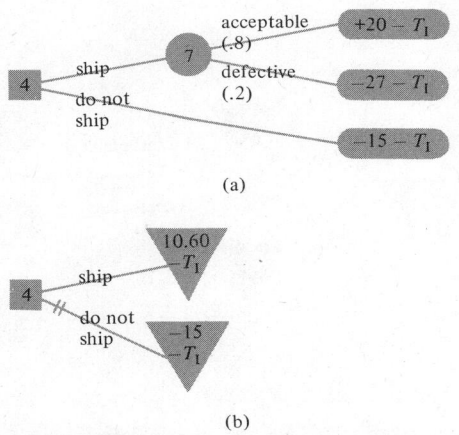

Figure 22.9(a) As originally depicted on the decision tree. **(b)** Reexpressed in terms of certainty equivalents.

The decision maker obviously would select the action alternative associated with the larger certainty equivalent. Therefore, given an arrival at the base of act fork $\boxed{4}$, the decision maker would always elect to ship the part. By so doing he would expect an average payoff of $\$10.60 - T_I$ over time. An arrival at act fork $\boxed{4}$ would always be followed by the action alternative, ship the part. Since the certainty equivalent associated with the action, ship the part, is $\$10.60 - T_I$, this becomes the certainty equivalent associated with act fork $\boxed{4}$. A pair of vertical slashes drawn across the path associated with the action alternative, do not ship, serves to indicate that that path would never be selected.

Summarizing the procedure:

At each event fork: Compute the expected monetary payoff of the gamble associated with the event fork. Substitute this value as the certainty equivalent to the gamble. Thus, each chance fork will be replaced by its certainty equivalent.

At each act fork: Select the action alternative associated with the highest certainty equivalent. Place a pair of vertical slashes on the paths associated with all acts not selected. Assign the certainty equivalent associated with the optimal action alternative to the act fork. Thus, each act fork will be replaced by the certainty equivalent associated with its optimal action alternative.

Backward induction involves a repetition of the above process, as the decision maker proceeds backward through the decision tree. Ultimately, this folding-back process will cause the decision maker to arrive at the action fork associated with the decision that must be made at the first stage of the decision tree analysis (i.e., act fork $\boxed{2}$). The decision tree reflecting a completed analysis is reproduced as Figure 22.10.

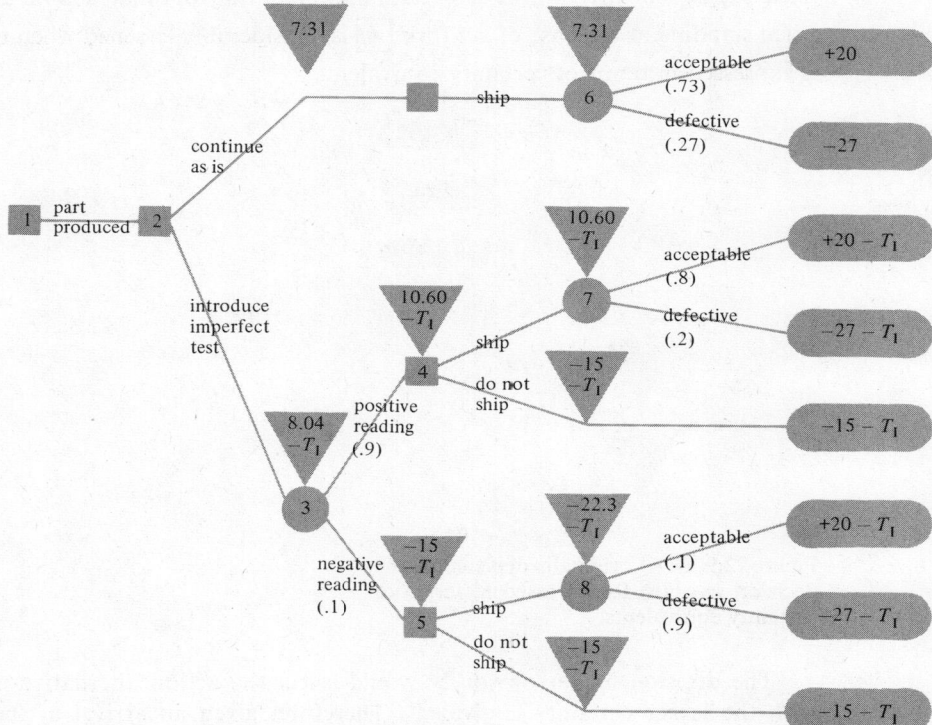

Figure 22.10 Completed decision tree analysis.

The certainty equivalent to the gamble represented by event fork $\boxed{7}$ and the certainty equivalent associated with act fork $\boxed{4}$ were computed above. (Recall that the certainty equivalent attached to an act fork is the certainty equivalent associated with the optimal action alternative at that act fork.) The derivation of the remaining certainty equivalents in the tree are as follows:

The expected monetary payoff for the gamble associated with event fork ⑧ is computed as follows:

(probability) ×	(conditional payoff)		computation of expected monetary payoff
(0.1)	×	$(+20 - T_I)$ =	$2 - 0.1T_I$
(0.9)	×	$(-27 - T_I)$ =	$-24.30 - 0.9T_I$

expected monetary payoff or certainty equivalent $\qquad -22.30 - \quad T_I$

Accordingly, the certainty equivalent is entered in the triangular symbol

$$\boxed{\begin{array}{c} -22.30 \\ -T_I \end{array}}$$

at the base of event fork ⑧.

For act fork $\boxed{4}$, the action alternative, ship the part, is associated with a certainty equivalent of $\$10.60 - T_I$. (If, upon arriving at act fork $\boxed{4}$, the decision maker always elects to ship the part, he can expect to average $\$10.60 - T_I$ over time as a payoff.) The action alternative, do not ship the part, is associated with a certainty equivalent of $-\$15 - T_I$. (If, upon arriving at act fork $\boxed{4}$, the decision maker elects *not* to ship the part, he will incur a guaranteed payoff of $-\$15 - T_I$.) The rational decision maker will always select the action alternative associated with the larger certainty equivalent. Accordingly, upon arriving at action fork $\boxed{4}$, the decision maker would elect to ship the part. The certainty equivalent associated with act fork $\boxed{4}$ is $\$10.60 - T_I$. A pair of vertical slashes is attached to the path associated with the action alternative, do not ship.

For act fork $\boxed{5}$, the action alternative, ship the part, is associated with a certainty equivalent of $-\$22.30 - T_I$. The action alternative, do not ship the part, is associated with a certainty equivalent of $-\$15 - T_I$. Selecting the action alternative associated with the larger certainty equivalent, the decision maker, upon arriving at act fork $\boxed{5}$, would elect not to ship the part (the lesser of two evils). Accordingly, act fork $\boxed{5}$ is assigned a certainty equivalent of $-\$15 - T_I$, and vertical slashes are attached to the path associated with the action alternative, ship the part.

For event fork ③, in the reduced diagram, the decision maker faces the following simplified gamble:

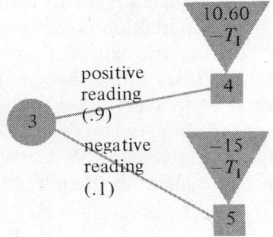

The above reduced diagram presents a substantial improvement over the more complicated choice presented prior to the employment of the technique of backward induction, that is,

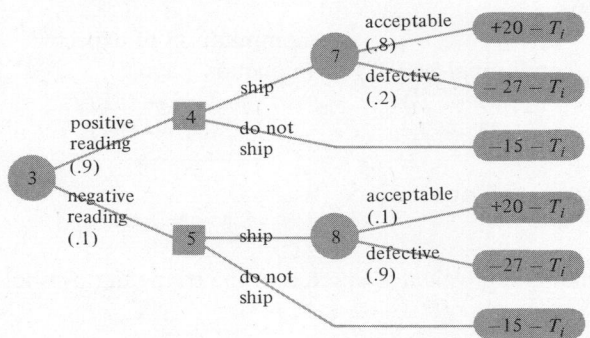

By compressing all the events, their probability of occurrence, and their consequences in association with each of the action alternatives under consideration into the two certainty equivalents, $10.60 - T_I$ at act fork $\boxed{4}$ and $-15 - T_I$ at act fork $\boxed{5}$, the gamble at event fork $\textcircled{3}$ is converted from a complicated set of consequences to a simplified equivalent gamble.

The expected monetary payoff for the gamble associated with event fork $\textcircled{3}$ is computed as follows:

			computation of expected
(probability)	× (conditional payoff)		monetary payoff
(0.9)	×	$(10.60 - T_I)$ =	$9.54 - 0.9T_I$
(0.1)	×	$(-15.00 - T_I)$ =	$-1.50 - 0.1T_I$

expected monetary payoff or certainty
equivalent $8.04 - T_I$

Accordingly, the certainty equivalent is entered in the triangular symbol

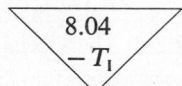

at the base of event fork $\textcircled{3}$.[8]

---

[8] It should be quite apparent from the diagram that the cost of the imperfect tester is reflected in every branch of every event fork and every act fork that follows the act: Institute the testing procedure. Accordingly, the computations could be made less cumbersome by simply inserting the symbol $(-T_I)$ in the certainty equivalent associated with event fork $\textcircled{3}$ and then suppressing this symbol for the analysis of certainty equivalents subsequent to the appearance of event fork $\textcircled{3}$ in the diagram. It should be noted that such suppression is legitimate only for an individual whose certainty equivalents to gambles are equal to the expected monetary value of those gambles. Such individuals are called averages players (they "play the averages"). In a similar fashion, the variable cost of production of $15 common to every act–event sequence on the tree could be totally ignored for an averages player without affecting the outcome of the analysis in any way. It bears repeating that this is legitimate only for an averages player. Why such an action is not legitimate for the nonaverages player will become evident in Chapter 23.

For event fork ⑥, the expected monetary payoff for the gamble associated with event fork ⑥ is computed as follows:

			computation of expected
(probability)	×	(conditional payoff)	monetary payoff
(0.73)	×	(+20)	=	14.60
(0.27)	×	(−27)	=	−7.29

expected monetary payoff or certainty
equivalent   7.31

Accordingly, the certainty equivalent is entered in the triangular symbol

$\overline{7.31}$

at the base of event fork ⑥.

For act fork ⟦2⟧, through the technique of backward induction, the decision maker has reduced the factors he must consider in deciding between the two action alternatives to a simple choice between two certainty equivalents, namely:

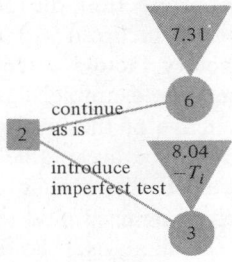

It is clear that the decision maker would prefer to introduce the testing device (imperfect as it may be), provided the rental fee is less than $0.73 per part. The $0.73 per part can be considered as the upper limit with which Component Supply, Inc., would enter into negotiations. Naturally, Component Supply, Inc., would hope to contract for a figure considerably less than this. If Prevention, Inc., (the firm offering the testing device) has undertaken a similar analysis, Component Supply, Inc., could be in for a hard bargaining session.

22.6 A THIRD ALTERNATIVE: A PERFECT TESTER

Suppose that Prevention, Inc., offered an additional proposal—the development and installation of a perfect tester. Installation of such a device would carry with it a guarantee that all acceptable parts would receive a positive reading and all defective parts a negative reading. Should a part be returned as a defective after having received a positive reading on the testing device, Component Supply, Inc., would be fully reimbursed for the nonrecoverable cost of production and the penalty fee. The previous decision tree analysis (as depicted in Figure 22.10) will be expanded to provide an evaluation of the merits of this new proposal. Figure 22.11 contains a skeletonized description of the perfect tester proposal, a recognition of the partial cash flows, and the determination of the total cash flow associated with each act–event sequence. The alternatives, continue as is and introduce the imperfect tester, having already been examined, are simply reproduced in their compressed form.

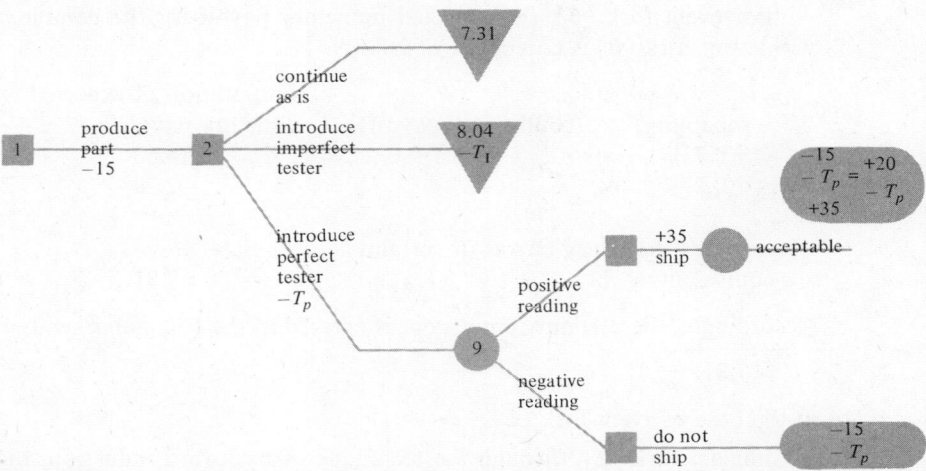

Figure 22.11 Skeletal framework of the decision tree with the inclusion of the option of a perfect tester: partial and total cash flows.

Partial cash flows consist of the $15 variable cost of production, the recognition of revenue of $35 with each part shipped, and a cost of T_p incurred with each part processed by the "perfect testing device." Nonmonetary factors often enter in and must be considered in an analysis of this type. For example, Component Supply, Inc., may feel that the improvement in the image of the firm associated with the ability to guarantee that all parts shipped will be acceptable is worth something. If so, this improvement should be reflected in the evaluation of the alternatives faced by the firm. Such factors have been recognized, and techniques exist for incorporating nonmonetary factors into the analysis.[9] To illustrate: One could ask how much Component Supply, Inc., would require as a dollar payoff to give up the improved image associated with the ability to guarantee that all parts shipped would be acceptable. This dollar amount could then be entered into the analysis as a proxy for the improved image. Such a substitution would enable the consequences of all the alternatives to be expressed in monetary values. The analysis would then be carried on as before. For our purposes, nonmonetary factors will be omitted from consideration.

A completed decision tree diagram is presented in Figure 22.12. Note that probability assessments have been assigned to the chance fork associated with a given part generating a positive or negative reading. Given the characteristics of a perfect tester, the probability of a positive reading is identical to the probability that a part produced will be acceptable, namely 0.73.

The certainty equivalent associated with the gamble described by event fork ⑨ is determined as follows:

(probability)	×	(conditional payoff)		computation of expected monetary payoff
(0.73)	×	$(+20 - T_p)$	=	$14.60 - 0.73T_p$
(0.27)	×	$(-15 - T_p)$	=	$-4.05 - 0.27T_p$

expected monetary payoff or certainty equivalent $\qquad\qquad 10.55 - \quad T_p$

[9] See Robert Schlaifer, *Analysis of Decisions Under Uncertainty*, New York: McGraw-Hill, 1969, Chap. 2.

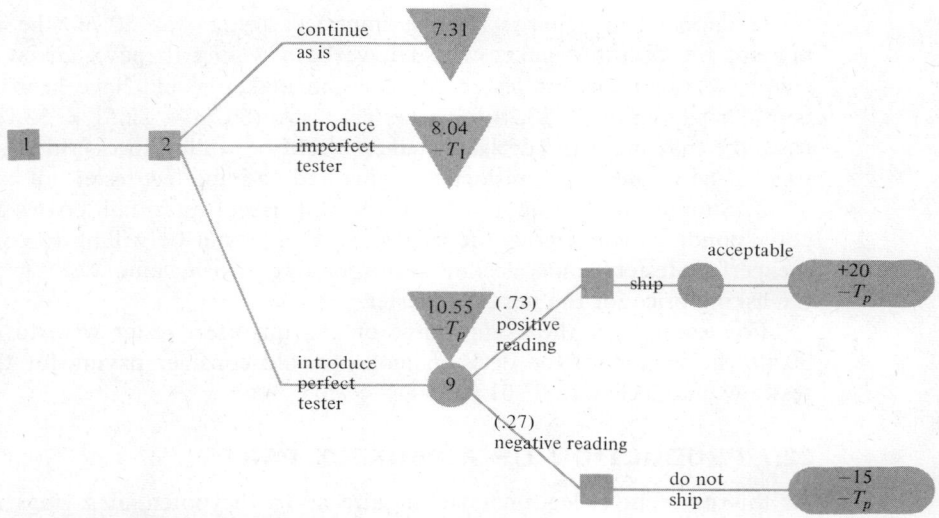

Figure 22.12 Completed decision tree analysis.

Determination of the optimal action alternative associated with act fork $\boxed{2}$: After folding back the decision tree diagram by the technique of backward induction, the decision alternatives associated with act fork $\boxed{2}$ appear as follows:

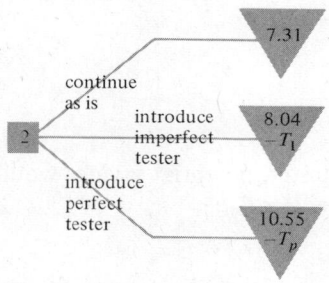

1. Given that the imperfect tester did not exist, the maximum amount the decision maker would consider paying for a perfect tester (per part) would be $3.24.

$10.55
−7.31
―――
$3.24

2. Given the existence of the imperfect tester, an additional constraint would be placed on the amount the decision maker would consider paying for a perfect tester.

$10.55
−8.04
―――
$2.51

The additional constraint would be that the decision maker would consider paying no more than $3.24 or $2.51 more than the charge for the imperfect tester, whichever was less. To illustrate—

a. Suppose the charge for the imperfect tester was $0.73 (the maximum amount the decision maker would have been willing to pay). Given the $0.73 charge for an imperfect tester, the decision maker would have been willing to consider paying up to $3.24 for a perfect tester ($0.73 + $2.51 = $3.24). In this case, the maximum the decision maker would be willing to consider is identical to what he would have considered paying had an imperfect tester not existed.

b. Should the asking price for the imperfect tester fall below $0.73, the corresponding ceiling price the decision maker would be willing to consider for the perfect tester would decline correspondingly to a value $2.51 greater than the asking price for the imperfect tester.

For example, if the asking price on the imperfect tester were to decline to $0.50, the maximum the decision maker would consider paying for the perfect tester would decline to $3.01 ($0.50 + $2.51).

22.7 PRODUCTION OF A PERFECT PART

At this point, it is legitimate to inquire as to the potential savings associated with the production of a perfect part, thereby eliminating the necessity for testing at all. If all parts produced were perfect, the firm would receive a payoff equal to $20 - C_I$, where C_I refers to the incremental variable cost of production necessary to bring about the production of a perfect part.

Determination of potential savings per part:

1. Assuming neither testing device existed, potential savings would amount to $12.69 per part less C_I.

$20.00
−7.31
———
$12.69

2. Given the existence of the imperfect tester, potential savings would amount to $11.96 plus the cost of the imperfect tester, less C_I.

$20.00
$-(+8.04 - T_I)$
—————
$11.96 + T_I$

$11.96 + T_I$ would amount to $12.69 only in the case where T_I is equal to the ceiling price of $0.73 per part.

3. Given the existence of the perfect tester, potential savings would amount to $9.45 plus the cost of the perfect tester, less C_I.

$20.00
$-(+10.55 - T_p)$
—————
$9.45 + T_p$

$9.45 + T_p$ would amount to $12.69 only in the case where T_p is equal to the ceiling price of $3.24.

22.8 SUMMARY

Decision tree analysis is an analytical device designed to provide for more effective utilization of the information and experience of the decision maker. The technique requires the decision maker to create a decision tree description of the acts and events that can exert a significant influence on the ultimate outcome of the immediate decision alternatives under consideration. This forces the decision maker to give more explicit consideration to these factors than might otherwise be the case. In addition, through the technique of backward induction, the decision maker is enabled to solve formally large, complex problems by arriving at a series of decisions on problems of a much smaller scope. The last stage in the folding-back process associated with backward induction creates a simple choice among the immediate action alternatives in which all the future ramifications of subsequent acts and events are compressed into a single evaluation referred to as a certainty equivalent. It is important to note that the decision tree does not supplant the need for judgment on the part of the decision maker, but rather provides a vehicle by which judgments can be reached in a more effective and efficient fashion.

In this chapter, it has been assumed throughout that the decision maker, when faced with a gamble, will assign to that gamble a certainty equivalent equal to the mathematical expectation of the consequences of that gamble. Such an assumption is legitimate when the decision under consideration is of a repetitive nature and where the overall consequences of selecting any given action alternative are relatively insignificant in comparison to the overall asset position of the decision maker. In such situations, the decision maker would expect to be around for many "plays of the game." He would thus be able to outlast a string of undesirable consequences and wait for the averages to run true. Under these circumstances, virtually all decision makers are "averages players." An averages player is one whose certainty equivalent to a gamble is equal to the mathematical expectation of the consequences associated with the gamble.

This assumption is no longer valid when considering decisions of a non-repetitive nature or where the dollar consequences associated with the outcomes of the various alternatives under consideration are quite large compared to the asset position of the decision maker. In decisions of this type, an encounter with a single undesirable consequence might eliminate the decision maker from the "game" permanently. To illustrate: Consider the following gamble.

A fair coin is to be flipped once. If a head is observed, I will pay you $5 million. If a tail is observed, you must pay me $1 million.

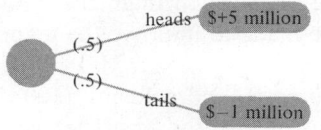

The mathematical expectation associated with the above gamble is $2 million.

			computation of expected
(probability)	×	(conditional payoff)	monetary payoff
(0.5)	×	(+$5 million)	=	+2.5 million
(0.5)	×	(−$1 million)	=	−0.5 million
			2 million

Most individuals, when given the option of taking a certain $1.5 million in lieu of the gamble, would not hesitate to opt for the certain $1.5 million (Howard Hughes excepted, of course). Such individuals are not averages players over the range of consequences encompassed by the gamble. The figure $1.5 million is $500,000 less than the expected monetary value of the gamble.

Suppose, moreover, that I indicate that should a tail be observed, payment of the $1 million will be required by sundown or an enforcer, "Black Pete," will be put on your trail (the latter being tantamount to swimming the East River with your feet encased in cement). Under these circumstances, you would undoubtedly be willing to pay to get out of the gamble. This is equivalent to assigning a negative certainty equivalent to the gamble. (Howard Hughes, of course, would still be an averages player—$1 million by sundown would hardly be anything for him to get excited about.)

The point of the above discussion is that often it is not legitimate to assume that the certainty equivalent to a gamble can be obtained by simply computing the mathematical expectation of the consequences associated with that gamble. For such an assumption to be valid, two prerequisites should be met (always the first and typically the second): (1) The dollar consequences associated with any of the alternatives under consideration should be insignificant with respect to the total asset position of the decision maker; and (2) the decision should be of a repetitive nature.

Although the above restrictions are quite narrow, they do not correspondingly limit decision tree analysis. A technique has been developed for incorporating the attitudes of nonaverage players into the analysis. The technique is called preference curve analysis. The introduction of the preference curve in Chapter 23 enables the technique of decision tree analysis to be extended to include the case of the nonaverages player.

CASE 1

Acme Foundry[10]

The Acme Foundry was having trouble with the quality of a particular type of castings, some of which were being produced with hidden flaws. The company had initiated a new test procedure in an attempt to overcome the problem, but management was beginning to question whether the test—which had proved far from perfect—was worth the expense. Some executives believed that the testing should be dropped, while others thought that a better procedure should be devised. Still others felt that this was looking at the problem through the wrong end of the telescope, and that more effort should be made to improve the manufacturing process.

Each casting cost $20 to make (variable cost), and sold for $100. Under an agreement between the foundry and its customers, the customer could return defective castings (which had no salvage value) for a full refund. In addition, Acme Foundry agreed to pay a penalty of $200 per defective casting to compensate the customer for the interruption of his machining operations. The customer had a continuing requirement for this particular casting. They took all that Acme could produce, but had several additional suppliers. Thus, the exact number of defectives in a single delivery had no effect beyond this $200 compensated loss per defective, since the missing units could be made up on the next order. The only true definition of a "defective" casting was whether the customer would reject it during his

[10] This case has been prepared and copyrighted, © 1964, revised © 1965, by Harbridge House, Inc., and is designed for teaching purposes only. The case does not necessarily indicate the policy or practice of any company or government agency and is not intended to illustrate either correct or incorrect, desirable or undesirable, management procedures. Where necessary, names and figures have been disguised. Reproduced with permission of Harbridge House, Inc.

machining operations—undiscovered internal flaws had no effect in this application.

The new test procedure had been in effect long enough to generate fairly reliable data on its characteristics. Approximately 80 percent of all castings passed the test; normally these were the only castings shipped to the customer. An average of about 10 percent of the castings had been returned by the customer as defective. As an experiment to learn more about the test, the foundry shipped out a few lots of castings that had failed the test. Approximately half of these were returned by the customer because of hidden flaws.

This test procedure, while not exactly reliable, was consistent. That is, castings that passed the test always passed a subsequent retest and those that failed always failed when retested.

Using decision tree analysis, generate answers to the questions posed below:

1. What is the maximum cost per casting manufactured that the foundry should be willing to pay for the imperfect test procedure?
2. What is the maximum cost per casting that Acme Foundry should be willing to pay for a perfect tester that would positively identify all (and only) those hidden flaws that would cause a defective casting, assuming that such a device could be provided?
3. Suppose the casting process could be changed so as to eliminate all troublesome hidden flaws. Suppose also that the imperfect tester cost $5 per casting, and that a more perfect tester was not available. How much increase over the present $20 manufacturing cost per casting could be justified by such an improved casting process?

CASE 2

The Valve Tester[11]

The reliability director at RRL (Rocket Research Laboratory) was talking to the program manager of the Laboratory's Skyprobe rocket system. He did not sound happy.

"I think it's ridiculous! Why don't we insist on an Engineering Change Notice to Stellar Equipment's contract and stop this whole argument about the LOX valve?" he asked impatiently. "We could just direct them to use our new plastic coating process on *all* those valves."

"I know how you feel," replied the program manager, "but that new coating we're using on each bad LOX valve costs us about $200 a unit. No doubt Stellar could do this for less during manufacture, but I still don't think it's worth it . . . certainly not on every single valve, good or bad. And furthermore, our experience with this new simulation tester is very encouraging. We've had only 4 percent bad valves among those that passed the test. It just doesn't make sense to spend $200 on every valve, just to catch those that slip by now."

"That's something else that bothers me," countered the reliability director. "I just can't see the point of this simulation tester. It's not only expensive—it's undependable. From what I've seen, it rejects about one-fourth of all the LOX valves, and still lets through some bad ones. For crying out loud, before we bothered with this test and simply installed *all* the valves, only 10 percent of them gave us any trouble. It seems to me that we're rejecting more good ones than bad ones, and that's a pretty foolish test."

[11] *Ibid.*

The Rocket Research Laboratory of the Military Space Agency had developed the Skyprobe rocket system for use in high-altitude measures of weather patterns and radioactivity intensity, and for various upper-atmosphere experiments. Normally, RRL scheduled several Skyprobe launches a week from its Southern Test Range, and this rate was expected to continue. The various subsystems and components supplied by the Skyprobe contractors were assembled and checked out by RRL personnel at the Southern Test Range. Among these was the LOX (liquid oxygen) valve, one of several components for which Stellar Equipment Corporation had contracts with RRL.

This valve produced according to specs prepared by RRL was delivered under a fixed-price contract at a cost of several hundred dollars per unit. It was designed to control the pressure of liquid oxygen in the Skyprobe rocket. The extremely cold and corrosive environment placed extraordinary demands on material perfection and close manufacturing tolerances. Slight surface cracks or out-of-round in the valve seats often caused leakage in the fueled rocket. When this happened, the rocket had to be defueled and the valve replaced—an operation estimated to cost about $1000, considering the evaporation loss of oxygen and the delay of launch and tracking crews, as well as the direct labor involved.

When careful incoming inspection at the Southern Test Range revealed that these bad valves had been well within the manufacturing specifications required of Stellar Equipment Corporation, it was generally agreed that this level of quality was the best obtainable using the current specifications. Serious difficulties had been encountered in repairing the bad valves until about three months ago, when RRL engineers applied a new type of plastic coating that had been developed for another project. This coating seemed to solve the problem perfectly—at least none of the coated valves failed when they were reinstalled in other Skyprobe rockets. The process and materials for the coating operation, however, were rather expensive—totaling about $200 per unit. Thus, the total cost incurred by the failure of a valve in the Skyprobe rocket was about $1200.

In an effort to reduce the incidence of bad LOX valves being installed, a simulation tester had been developed at the Southern Test Range. This tester consisted of a simple fixture into which the LOX valves were installed; liquid oxygen was fed first through the open valve and then against the closed valve. Although the pressures and temperatures in the tester were not as extreme as those in the Skyprobe rocket, the functioning of the valve could thus be approximately simulated.

Using decision tree analysis, generate answers to the questions posed below:

1. What portion of the LOX valves that are rejected by the simulation tester are actually good valves? What portion of the bad valves received at RRL now pass the simulation tester?
2. Assuming that the costs and quality experience given are valid, how much per valve is the simulation test worth relative to the alternative of installing all valves as received?
3. Again, assuming validity of the given data and assuming that the simulation tester costs $25 per valve, what is the maximum unit price that RRL could justify paying the manufacturer for the plastic coating on all valves?
4. What additional considerations and data should enter into the evaluation of the simulation tester and the improved manufacturing process?

CASE 3

The Spaceonics Corporation: quantitative analysis of a proposal strategy[12]

As president of Spaceonics Corporation, Harold Prentice was responsible for developing and carrying out company proposal strategy. Currently, Prentice was preparing for a review of the alternative courses of action Spaceonics might pursue in response to a NASA Request for Proposal (RFP) for research and development leading to design of a deep-space communications subsystem for Project Searcher. The Project Searcher prime contract had been awarded to Crest Astronautics a few months earlier. NASA planned to buy 12 to 18 unmanned satellites during the life of the program. The RFP indicated that the market for this communications subsystem might be expanded beyond Project Searcher requirements; this was the reason why the subsystem was being procured by NASA and supplied to Crest Astronautics as government-furnished equipment (GFE).

Prentice's immediate problem revolved about the fact that two of Spaceonics' most valuable men, Dr. Karl Hoffman, chief of a research section, and Harry Ryan, manager of the Radio Engineering Department, had each developed an approach to the new communications system. Prentice had asked both men to present their approaches to the top management committee for review before any decision was made on the NASA RFP. He also obtained figures from them on the anticipated net costs and gains that might be realized if either approach was adopted.

Background

The Spaceonics Corporation was founded in 1959 by a group of senior engineers from a large eastern electronics corporation. Most of its experience has been with communications systems for the space program, and nearly all of its $8 million annual volume consists of military or NASA cost-plus-fixed-fee or incentive-fee contracts.

In its brief corporate history, the company has won a well-deserved reputation for creative yet practical research, development, and manufacture. Among its contracts at the time of the Searcher RFP, for example, was one for the manufacture of a high-frequency radio (HFR) communications system for an earth-orbit Air Force observation satellite. Earlier, Spaceonics had been awarded one of the three R & D contracts for this system; but only Spaceonics was selected for production—substantially to the specifications it had developed. Preliminary tests of this HFR device had convinced Harry Ryan and other members of the Radio Engineering Department that the basic system could meet the requirements for the Project Searcher communications subsystem, with only one major modification.

In addition to the HFR satellite communications system contract, Spaceonics had nine other major projects in progress, all scheduled for completion within 8 to 15 months. The expected gross revenue for these contracts ranged from $200,000 to $2.5 million.

[12] *Ibid.*

The Management Review

Prentice opened the management review meeting with a brief synopsis of the three courses of action open to Spaceonics: (1) to propose a radically new application of laser beams, (2) to propose an improvement of the high-frequency radio communications system now being built for the observation satellite, or (3) not to propose at all. The committee agreed to review the alternatives in turn, and Prentice introduced Dr. Hoffman.

The research section chief explained that, theoretically, the potential application of laser beams to communications was well known, although no practical application had yet been made in this field. Dr. Hoffman, in a simplified version of a paper that he had recently presented before a professional society, argued convincingly for his theory. "If no unexpected problems arise," he concluded, "Spaceonics will be in an excellent position to get a very profitable manufacturing contract."

Ralph Drucker, marketing manager of Spaceonics, knew that other firms were showing interest in the application of lasers to communications problems. Accordingly, he asked Dr. Hoffman about Laser Laboratories, a leader in laser R & D work. Hoffman replied that, while Laser Labs was well known for the design and manufacture of basic laser devices, it did not have the capability to produce the entire communications system called for by the NASA RFP.

"I'm not so sure about that," Drucker answered. "You know, it's entirely possible that NASA might insist on a joint manufacturing contract between us and Laser Labs, if this concept ever gets to that stage."

"I think this laser thing is too risky," Harry Ryan added. "It involves a real state-of-the-art breakthrough. Besides, we already have a good HFR device that we can modify to fit this project requirement, and I think we'll have a better chance of getting both the development and production awards if we concentrate on it."

"Isn't it true, though," Dr. Hoffman asked, "that my laser concept would be more profitable to the firm?"

"Well, if it worked, it might make more money in the end," Ryan agreed. "The question is whether it will work. I think it's more sensible to propose on something we know more about, that has a better chance of being accepted, even if the returns are a little smaller."

"NASA will be more impressed by a bold new concept than by a modification to an existing device," Dr. Hoffman insisted. "After all, this is a research requirement, and the RFP only stated the functions the system should perform, not the techniques to achieve it."

"Let's look at the specific alternatives," Prentice broke in. "First, we can propose Karl Hoffman's laser approach, which is certainly creative and offers some substantial possible payoffs. On the other hand, we could stick with the proven HFR system that Harry Ryan advocates. If we get an R & D award, then logically we'd submit a production proposal based on the same concept. If we lose out on the R & D phase for the laser approach, I suppose we can still try to get a production award for whatever design approach NASA has selected. But I've yet to see a case where a loser at the R & D phase has been awarded a production contract. I really think that if we lose the R & D phase, we might as well forget it. And finally, of course, we can ignore this RFP and not propose at all. But if we do that we certainly won't be asked to bid on the manufacturing phase."

There was a general murmur of agreement; then the Spaceonics contract administrator spoke up. "Well, now, I'll admit this is a pretty exciting-sounding opportunity and something we could probably do quite well. But, after all, Crest Astronautics has several subcontractors working on subsystems of the satellite right now, and for all we know, one of those companies may have this program locked up. If that's the case, preparing a proposal could be a waste of time and money. What are the costs and the possible gains of these proposal alternatives, anyway?"

Prentice passed out a copy of the figures in question to each member of the committee (see Table 22.1). "These are the figures I worked out with Karl Hoffman and Harry Ryan in earlier conversations. As you can see, there are two columns. The first column shows the estimated cost of writing proposals, for both the present R & D phase and the follow-on manufacturing phase. I realize that frequently we can absorb a good share of such costs into our general overhead rate. But we're beginning to run into trouble with the government auditors in having all of these expenses allowed as part of overhead. What's more, we're locked in with our present overhead rate for several big contracts. What this boils down to is that for this RFP we should consider these proposal costs as coming out of company funds.

Table 22.1

	Costs of proposal preparation		Expected net fee (exclusive of proposal preparation expenses)
	Now (at the R & D phase)	Later (at the production phase)	
I. The laser approach			
Laser R & D effort	$22,000		$15,000
Laser production effort:			
a. All to Spaceonics		$60,000	$300,000
b. Half to Spaceonics, half to Laser Labs		$60,000	$150,000
II. The HFR approach			
HFR R & D effort	$10,000		$13,000
HFR production effort		$40,000 if win R & D $45,000 if lose R & D	$125,000

The other column is our estimate of the expected net fee that we'd realize from each of these possible contracts. We've had a lot of experience with this kind of R & D and small-scale production business, and practically always our net fee is close to 5 percent of the gross costs. I know the contracts sometimes state a higher or lower percentage, but by the time all adjustments, changes, and so on are considered we don't end up far from 5 percent. So these figures represent 5 percent of the gross costs for the various contracts."

After distributing his figures, Prentice went on, "As you can see, the R & D award by itself on the laser approach would involve a $7,000 net loss to us ($22,000 cost minus $15,000 gain equals $7,000 loss). Obviously, our only reason for getting into this situation is the hope of obtaining follow-on production work. If we landed this work, our figures indicate we'd make a really hefty gain, on the order of $300,000 or so—that is, 5 percent of a cost-plus-fixed-fee contract for about $6 million. But if the follow-on work had to be split up with Laser Labs, our gain would only be about half as much, or $150,000.

Our figures on the HFR approach show that we'd barely be ahead of the game if we got only the R & D phase ($13,000 net fee from R & D work minus $10,000 proposal cost equals a $3,000 overall gain). But really it's the shot at a follow-on production contract that we're after ($125,000 net fee for production, less $40,000 proposal cost, for an $85,000 overall gain from the production contract). We can see that lower proposal costs and potential gains are involved throughout this alternative, and hence less is at stake.

The dollar consequences of the do-nothing alternative would, naturally, be none."

Evaluation of the Uncertainties

Harold Prentice was reasonably sure that his figures considered all of the alternatives open to Spaceonics. "Well," he said, "I think we've got enough information to make a decision. Which approach do you think we should take?"

"I think it's quite likely that we'll win the contract if we use Ryan's HFR approach," one member said.

"But it's a lot more probable if we use the laser approach," a second man dissented.

"What do you mean by probable?" a third asked. "If you ask me, it's highly probable that a sub of Crest Astronautics is going to get the business, and we should save all those proposal dollars by not responding to this RFP."

For the next half-hour, the committee discussion centered on the uncertainties related to each of the alternatives. Toward noon Prentice remembered a recent article he had read on decision trees and decided to try a simple application of the idea.

"We could be here all day arguing about words," Prentice said. "The trouble is that these words describing the uncertainties involved have different meanings to each of us. I think we should try to quantify the chances of certain things happening, in terms of odds or probabilities. This might clear up some of the misunderstandings in this discussion."

Drucker countered with, "Really, Harold, this is carrying your engineering precision a little too far. We can't assign numerical odds to something as vague as this. After all, we aren't dealing with a horse race."

"Well, Ralph," Prentice replied, "I've been reading up on what seems like a useful way of looking at decisions like the one we're facing. It's called decision tree analysis, and I think we could use this new technique to help evaluate these alternatives. Now I've got some copies of an article that explains this approach. I suggest that we all have lunch, read the article, and come back about three o'clock this afternoon to continue our analysis."

After some half-joking remarks about reducing top management judgment to a magic formula, the group accepted Prentice's suggestion and the meeting adjourned.

When the committee reassembled later that day, Prentice asked for comments on the decision tree approach as a means of solving the proposal problem. Most members reacted positively. Accordingly, Prentice suggested that Albert Wallis, the Spaceonics comptroller, analyze the alternatives already discussed for this RFP and present his conclusions for review and assessment the next morning. "But first," he continued, "we've got to give Al some data to work with. Let's review our discussion of these alternatives and see if we can't pin down the uncertainties in these various situations. I realize this isn't a horse race, Ralph, or a dice game, where the bookies or the 'house' have worked out the percentages. But, after all, we've played this game of proposal-writing for some years now, and we should be able to set our odds pretty well. Lord knows, we've got to be gamblers in most any sort of business that's worthwhile. At least let's try to take calculated gambles if we can."

The committee nodded its agreement, and Prentice turned to the proposal alternatives.

The Laser Approach

After half an hour of discussion it was apparent that the committee thought the principal attraction of using the laser approach was also its chief drawback: No one was certain of NASA's current attitude toward a venture so inherently risky. Prentice pointed out that, in its last appropriation hearing, NASA had advanced the argument that its programs did much to promote technical break-throughs that later had wide application. Spaceonics' chief engineer, however, brought up the agency's recent budget problems, and quoted several high NASA officials as leaning toward proven hardware whenever possible. He emphasized that, in this particular case, a state-of-the-art advance would not be necessary to meet functional requirements if the HFR concept was used. After discussion of additional pros and cons, the committee finally agreed that Spaceonics had about one chance in five of winning an R & D contract based on the laser approach. (Dr. Hoffman protested that it was a misapplication of statistical theory to assign probability values to subjective feelings, but found little support for his view from other members of the committee.)

Most committee members believed that if Spaceonics was awarded an R & D contract for the laser concept, the company would be in a strong position to obtain the production contract. Ralph Drucker reminded the group, however, that NASA might insist on a joint production award with Laser Labs, with the consequent halving of Spaceonics' gain from the project.

"After all," he said, "NASA has traditionally preferred to go to well-known sources for such specialized technical components. Then, too, it seems to me, we must consider the possibility that even if we got the R & D award, we might not obtain any follow-on work."

This last possibility was discussed at some length with, at first, a wide range of opinions concerning the chances that this situation would occur. Several members of the committee said they thought Dr. Hoffman was too optimistic about the laser application's succeeding. It was also pointed out that NASA had often sponsored R & D work on interesting new techniques, only to revert to more conventional hardware for the actual production system. Eventually, however, the concensus was that the chances were about one in five that NASA would not award Spaceonics a production contract even if they did sponsor the firm's R & D laser approach. There was universal agreement, though, that if Spaceonics proposed for and lost the R & D phase on the laser concept, they would have no chance at getting a manufacturing contract, even for the HFR concept.

Prentice himself believed that the chances of the company's obtaining the entire production contract for laser systems (assuming that it won the R & D phase) were about 50–50. He explained that, allowing for the 20 percent possibility of getting no follow-on contract, this left about a 30 percent chance of a joint award with Laser Labs. These figures seemed reasonable to most of the group, although there remained some mild dissension in both directions from Prentice's figures.

The High-Frequency Radio Approach

Prentice then turned the discussion to Harry Ryan's modified HFR approach. Everyone agreed that this approach had a better chance of winning an R & D award than would the laser approach—but just how much better was hard to determine. NASA's attitude toward proven hardware was rehashed without providing much additional insight into the problem. After lengthy discussion, there was some agreement that Ryan's proposal would have about a one-third chance of being accepted by NASA.

The committee then estimated that the probability of obtaining a production contract if the company first won the development award would be about three chances out of five. Alternatively, the committee thought that Spaceonics had about one chance out of five of winning an HFR production contract without having won an R & D contract. These estimates reflected the awareness that several other firms had the technical ability to produce the HFR system, and they would certainly be considering this NASA RFP. (The cost of the production proposal would be $5000 greater without having won the R & D phase, as shown in Table 22.1.)

"Well," said Prentice, "I guess we have the figures Al Wallis should need. Now all he has to do is put them in the grinder and turn the crank! Seriously, though, Al's people should be able to run through the calculations in a couple of hours, so let's reconvene here tomorrow—say at ten o'clock—to review his efforts and make our decision."

Required:

1. Structure the above decision problem in the format of a decision tree. Display all action alternatives and events in appropriate chronological sequence.
2. Assign probabilities to all event forks, enter partial cash flows at the appropriate locations in the decision tree, and determine the total cash flows or terminal values.
3. Evaluate the decision tree under the assumption that Spaceonics is an averages player.
4. Would the decision change if the probability of obtaining a production contract given that the company first won a development contract on the HFR proposal is assessed at 0.8 rather than at 0.6? Support your answer.
5. Determine the value of the probability of obtaining a production contract given that the company first won a development contract on the HFR proposal that would cause the decision maker to be indifferent to the HFR and the laser approaches (holding everything else constant)? That is, what value for p would cause the certainty equivalent for both approaches to equal 8?

The approach outlined as follows can be used to determine how sensitive the optimal solution is to specified parameters in the decision problem. This technique is called sensitivity analysis.

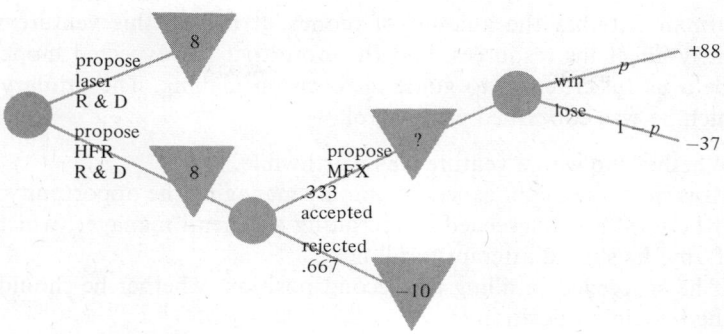

CASE 4[13]

David Kaufman, the owner and operator of a small executive recruiting agency, was somewhat puzzled by the complex proposal submitted to him by a German manufacturer and exporter of snowmobiles, Deutsche Schneemobil-fahrzeuge-fabrikenauslandsvertrieb Aktiengesellschaft (DSAG). DSAG was in the process of establishing a U.S. distributorship and was seeking help to fill the positions of general manager, sales manager, and service manager. It offered Kaufman a one-month exclusive opportunity to fill the position of general manager, and, if this was done successfully, the opportunity to proceed, still on an exclusive basis and for another one-month period, to fill either the position of sales manager or service manager—with the understanding that he would have to devote his full efforts to filling one position or the other. Then, only if he had succeeded in filling the second position, would DSAG be interested in having Kaufman continue for a third one-month exclusive period to fill the remaining position.

While this proposal seemed a bit unusual, Kaufman sensed that DSAG was serious and was not prepared to entertain counterproposals. However, because his agency was a one-man operation, Kaufman realized that he would have to determine if it would be reasonable for him to devote a full month to trying to fill each of these positions in light of his other commitments. As the first step in this determination, Kaufman estimated the fee that he would receive for filling each position and the total costs he could expect to incur (his own time, special advertising, telephone calls, and so forth) whether or not he was successful in filling the position. He then assessed his chances of filling each of the positions during the one-month period allowed. He felt that his chances of filling the third position (either sales manager or service manager) would be substantially better than that of filling the second position. This improved chance would follow from his increased knowledge about DSAG and the fact that he could point out to prospective hires that two of the three positions were already filled, indicating that DSAG was ready to start up the U.S. operation. The fees, costs, and probabilities of success estimated by Kaufman were as follows:

Position	Fee	Costs	Probabilities of success	
			If first	If second
General manager	$4000	$2000	0.50	—
Sales manager	$3000	$1400	0.50	0.80
Service manager	$2000	$1200	0.40	0.70

[13] *Ibid.*

Kaufman felt that the amount of money at risk in this venture would not particularly affect his resources, and therefore that the expected monetary value would be a useful criterion to guide his decision making. The primary questions with which he was concerned were as follows:

1. Whether the whole venture was worthwhile.
2. How much he might expect to gain by managing the opportunity optimally.
3. If he tried and succeeded in recruiting a general manager, which position, if any, he should attempt to fill next.
4. If he succeeded in filling the second position, whether he should try to fill the remaining position.

Utilizing decision tree analysis, generate answers to the questions posed above.

CASE 5

Colvario Incorporated has a present cash position of $1000. Colvario has three alternatives (projects A, B, and C) under consideration. His services are required for each and the projects are not postponable—therefore, he loses the opportunity to pursue the two projects he elects to turn down.

Project A: An outlay of $700 will be required. If completed on time, revenue of $1200 will be received. If not completed on time, a penalty of $100 per day delay will be deducted from the $1200 (with no more than $300 to be deducted under any circumstances). Colvario estimates the probability of delay as follows:

0 days delay	0.85
1 day delay	0.08
2 days delay	0.05
3 or more days delay	0.02
	1.00

Project B: An initial outlay of $800 will be required. Upon completion of the initial phase, Colvario will receive revenues of $1000. If completed in 2 days, Colvario has an option of pursuing a follow-up project that will involve a $200 expenditure on Colvario's part with an 80 percent chance of a $600 revenue receipt and a 20 percent chance of an $800 revenue receipt. If more than 2 days are required to complete the initial phase, he loses the option. Colvario feels he has a 50–50 chance of completing the initial phase of project B within the 2-day period.

Project C: Colvario's outlay will be his entire $1000. He assesses a two-thirds chance of success on project C, a one-third chance of no success. If the project works out successfully, he will receive a revenue receipt of $4000. If it works out unsuccessfully, he will lose everything.

Colvario has a fourth alternative, namely, leaving his money in the bank and earning $4 interest over this same period.

Required:

1. Construct a decision tree describing the above situation.
2. Evaluate it under the assumption that Colvario plays the averages.

ADDITIONAL CASES

Additional cases providing an opportunity to apply the techniques illustrated in this chapter are listed below:

1. Technotronics Corporation	ICH 5C 30R EA-C 466R[2]
2. Howe Properties	9-171-457 EA-C 1109
3. Warren Agency, Inc.	ICH 6C36 EA-C 476
4. J. D. Robinson Fertilizer and Explosives, Inc.	9-170-026 EA-C 808R
5. Everclear Plastics Company	ICH 9M 101 M-4

Copies of the above cases may be obtained by writing to: Intercollegiate Case Clearing House, Harvard Business School, Soldiers Field Road, Boston, Massachusetts, 02163.

DECISION TREE ANALYSIS: INTRODUCTION TO THE PREFERENCE CURVE

Suggested Approach to Chapter 23

Pages 571–583 present the underlying situation and
the supporting documentation for determination of
partial cash flows and probability assessments. The
main focus of the chapter is to describe a technique
(preference analysis) that enables the analyst to
introduce the decision maker's feelings about the
relative desirability of the outcomes as an integral part
of the analysis. Thus, one should not spend a great
deal of effort verifying the partial cash flows and
probability assessments. Simply read the first section to
provide an overview of the decision tree structure that
will serve as the base on which the discussion of
preference analysis will be developed.

23.1 A HYPOTHETICAL CASE ILLUSTRATION
INVOLVING THE DETERGENT INDUSTRY

A major ecological issue today is the phosphate detergent controversy. Due to the superb performance of phosphates as cleansing agents, phosphate detergents became an immediate success. Continuing research conducted by the detergent industry has yet to turn up an acceptable substitute. Many ecologists favor the removal of phosphates from detergents, claiming that phosphates have been identified as a major source of water pollution. It is their position that phosphates stimulate the growth of algae in bodies of water due to an overdose of nutrients. Extensive growth of algae results in a process called eutrophication. The process of eutrophication kills a body of water and its inhabitants by choking off the oxygen supply. Algae coats the surface, preventing the penetration of light and oxygen. Without light and oxygen, the lower algae layers die and decompose. Scientific research has revealed that 15 to 20 nutrients are required to support algae growth. These include carbon, nitrogen, phosphorous, potassium, silicon, trace metals, and vitamins. Absence of any of the critical nutrients will prevent algae growth. Those favoring the elimination of phosphates from detergents argue that phosphate reduction is the most feasible approach to combating eutrophication. Industry sources favor attacking one of the other nutrients (because of the outstanding properties of phosphates as a cleansing agent). Phosphate supporters also point out that only a small amount of phosphate is required to support algae growth and that phosphates are abundantly available from non-detergent sources. Many scientists believe that the answer lies in controlling carbon-rich organic wastes, primarily human wastes. Such controls could be obtained through the installation of modern sewage plants. Research is also underway to attempt to find a suitable phosphate substitute. One such ingredient, NTA, for which great hopes existed, has recently been ruled unsafe by the FDA.

Increasing public concern and governmental rumblings suggest that in the near future one of the following situations will materialize:

1. An acceptable phosphate substitute (an ecologically safe and effective cleansing agent) will be discovered. Such a discovery would sound the death knell for phosphate-based detergents and would lead to a significant increase in market share for the firm responsible for the discovery.

2. It will be conclusively demonstrated that the eutrophication process can be effectively and inexpensively controlled by regulating carbon wastes (or some other nonphosphate nutrient). This solution would place the burden on society rather than the detergent manufacturers. Considerable lobbying power would be essential if this approach were to be successful. The industry has agreed to provide such financial support (allocated in proportion to market share) should such a breakthrough occur. Such a development would not be expected to affect relative market shares. Note: If both an acceptable phosphate substitute and conclusive proof that the eutrophication process can be controlled economically by attacking an alternative nutrient exist, the industry is still committed to lobbying for the regulation of the nonphosphate nutrient. With the knowledge that a phosphate substitute exists and the bad press phosphates have received to date, however, the likelihood that such a lobbying effort would be successful is quite small.

3. If neither of the above breakthroughs occurs, it is expected that governmental prohibitions on the use of phosphates will be enacted. Given this, the industry would be forced to produce less effective detergent products. Little change in relative market shares would be anticipated.

Good Citizen, Inc., a major detergent manufacturer with 30 percent of the total market, is currently attempting to decide on the appropriate magnitude of expenditure to allocate to research and development efforts in this area. In the interest of simplification, it will be assumed that only three alternatives are under consideration, namely, $30 million, $35 million, and $40 million, In addition, a decision must be reached as to the proportion to be allocated to research and development oriented toward the discovery of a phosphate substitute and the proportion to be allocated toward uncovering an alternative nutrient to be controlled to prevent eutrophication. Again in the interest of simplicity, alternatives under consideration will be restricted to (1) 25 percent allocated to R & D oriented toward discovery of a phosphate substitute, and (2) 75 percent allocated to R & D oriented toward discovery of a phosphate substitute. It will also be assumed that if a breakthrough occurs, it will take place in exactly 2 years.

Given the above restrictions, the amounts that could be allocated to research and development in each area, together with Good Citizen, Inc.'s assessment of the probability of success, are displayed in Tables 23.1 and 23.2.

Table 23.1 Research and development oriented toward controlling eutrophication through an alternative nutrient

Dollar expenditure	Conditional probability of success
7.5 million (25 percent of 30 million)	0.25
8.75 million (25 percent of 35 million)	0.30
10 million (25 percent of 40 million)	0.35
22.5 million (75 percent of 30 million)	0.57
26.25 million (75 percent of 35 million)	0.60
30 million (75 percent of 40 million)	0.63

Table 23.2 Research and development oriented toward discovery of a phosphate substitute

Dollar expenditure	Conditional probability of success
7.5 million (25 percent of 30 million)	0.08
8.75 million (25 percent of 35 million)	0.10
10 million (25 percent of 40 million)	0.125
22.5 million (75 percent of 30 million)	0.28
26.25 million (75 percent of 35 million)	0.30
30 million (75 percent of 40 million)	0.32

Other expenditures:

1. If it can be conclusively demonstrated that the eutrophication process can be effectively and inexpensively controlled by regulating carbon wastes (or some other nonphosphate nutrient), Good Citizen, Inc.'s share of the lobbying effort will amount to $6 million.

2. If Good Citizen, Inc., discovers an acceptable phosphate substitute, a $15 million conversion cost is anticipated.

3. If no breakthroughs of any kind occur or if a competitor discovers an acceptable phosphate substitute, it is expected that phosphate detergents will be prohibited. Conversion to produce a nonphosphate detergent will cost $4.5 million.

Other considerations:

1. Total industry sales will be presumed to remain constant at $300 million. Good Citizen, Inc.'s market share of 30 percent amounts to sales of $90 million. With a 10 percent profit margin, Good Citizen, Inc.'s 30 percent market share generates an annual profit of $9 million. It would be a simple extension to expand the analysis to allow for the incorporation of alternative assumptions with respect to the growth pattern of industry sales. For the current analysis, it is assumed that total profit realized for the next 2 years will amount to $18 million. After this period, profit realized will depend on market share adjustments.

2. A decision must be reached with respect to the time horizon to be considered in evaluating the consequences associated with the action alternatives under consideration. In this instance, it is presumed that any consequences realized after the fifth year are not recognized in the analysis. If it is desired to recognize consequences realized beyond the fifth year, such data could readily be incorporated. It should also be noted that the time value of money has not been recognized in the analysis (i.e., the fact that a dollar received today is worth more than a dollar received 5 years from today). Such considerations will be left to more advanced treatments of decision tree analysis.

3. An important consideration is the criterion selected by the decision maker to reflect the consequences of arriving at the terminal point of any given act–event sequence. Depending on the situation and the interests of the decision maker, the criterion selected could be net cash flow, market price of common stock, net earnings, or net liquid assets, to name a few. Net liquid assets consist of the dollar value of those assets that would be expected to be converted into cash within 1 year less the dollar value of those liabilities that would be expected to be liquidated within a period of 1 year. The criterion of interest selected by Good Citizen, Inc., is the net liquid asset position 5 years hence. For any given endpoint in the decision diagram, the value of the criterion is computed in the fashion shown in Table 23.3.

It is presumed in Table 23.3 that the profit figures for years 1–5 reflect the aggregate change in net liquid assets over the 5-year time span exclusive of conversion costs, R & D expenditures, and lobbying expenditures. In actuality, a projection of cash flow, receivable and payable balances, and so on would be necessary to generate this figure.

4. For purposes of analysis, it is presumed that the likelihood that a competitor will discover a phosphate substitute as of the end of year 2 has been assessed at 30 percent. The likelihood that a competitor will develop an acceptable solution to the eutrophication problem involving regulation of a nonphosphate nutrient

Table 23.3 Computation of net liquid asset position 5 years hence

Current net liquid asset position	$50 million
Plus profits from detergent sales for the next 2 years (2 years at 9 million per year)	$18 million
Plus profits from detergent sales for years 3–5. (This figure will be determined by the market share that emerges as a result of R & D developments as of the end of year 2. This market share will be presumed constant for years 3–5.)	Dependent on market share
Less any conversion costs incurred as a result of changed circumstances	Either 15 million or 4.5 million
Less expenditure on R & D	Either $30 million, $35 million, or $40 million
Less expenditures incurred as a result of lobbying if conclusive proof can be provided that eutrophication can be effectively controlled by regulating an alternative nutrient	6 million if lobbying occurs
Net liquid asset position as of the end of the fifth year	

has been assessed at zero given the nature of competitors' research and development programs. Therefore, this alternative has been suppressed in Figure 23.1.

Figure 23.1 displays the various combinations of success (for Good Citizen, Inc., and its competitors) and the resulting market share Good Citizen, Inc., can anticipate under the given conditions for years 3–5. Each of the endpoints has been assigned a letter for reference purposes.

Probabilities have been assigned to the *terminal* chance forks in Figure 23.1 representing the assessment of Good Citizen, Inc., of the likelihood that a given market share will be realized under stipulated degrees of success in the research and development efforts of Good Citizen, Inc., and its competitors.

A word of explanation with respect to strategy selections by Good Citizen, Inc., as reflected by Figure 23.1:

1. If Good Citizen, Inc., is successful on both R & D projects, the best strategy is to hold back on the alternative eutrophication solution and introduce the phosphate substitute. (The existence of an alternative eutrophication solution would not be communicated to the industry.)
2. If Good Citizen, Inc., is successful in discovering an alternative eutrophication solution but unsuccessful in its search for a phosphate substitute, the existence of an alternative eutrophication solution should be communicated to the industry. If none of its competitors has discovered a phosphate substitute, it is for all intents and purposes certain that the lobbying effort will be successful. If a competitor has discovered a phosphate substitute, Good Citizen, Inc., feels that the chance of success on the lobbying effort is only 0.05. If the lobbying effort is unsuccessful, Good Citizen, Inc., will have to convert to an inferior nonphosphate-based detergent.

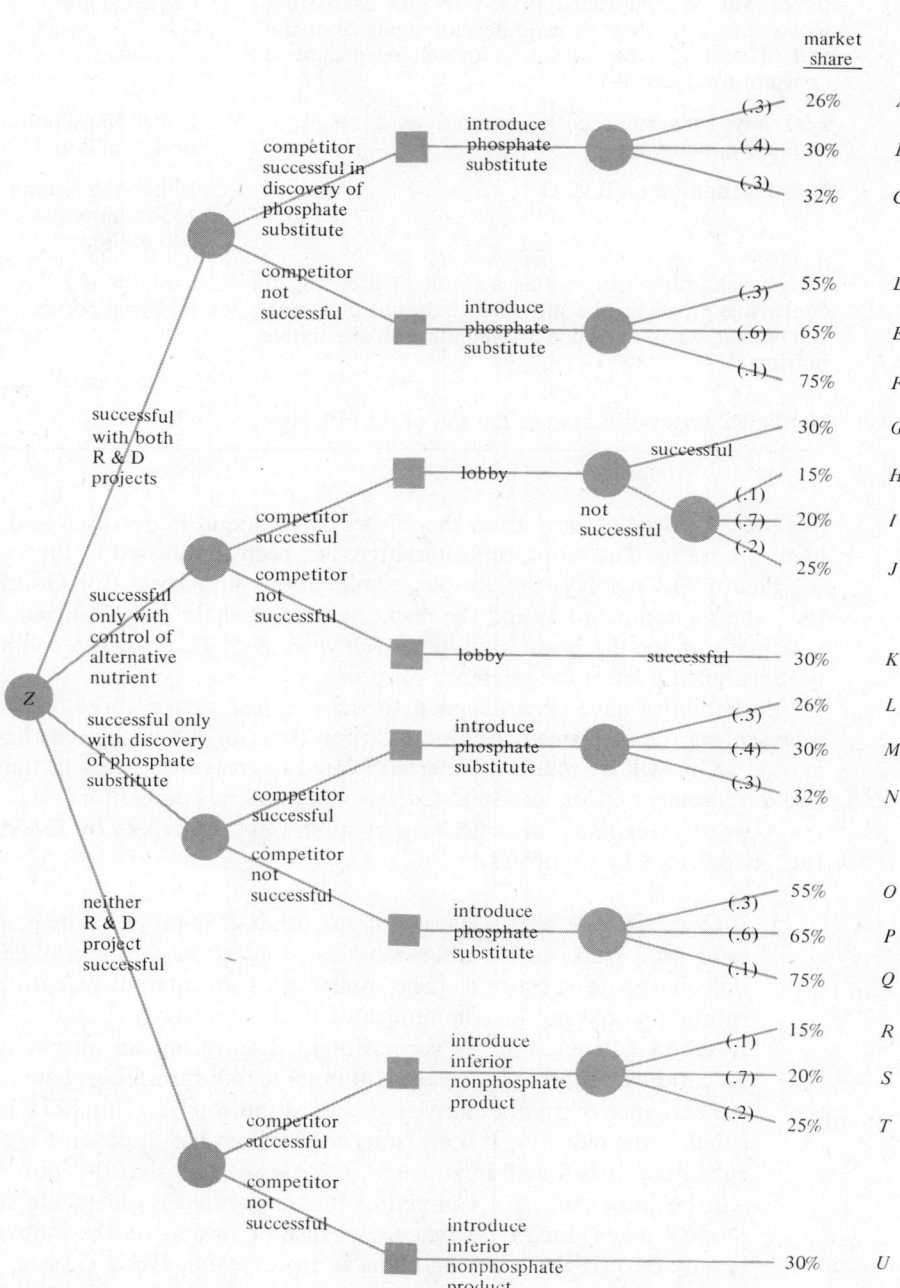

market
share

(.3)	26%	A
(.4)	30%	B
(.3)	32%	C
(.3)	55%	D
(.6)	65%	E
(.1)	75%	F
	30%	G
	15%	H
(.1)	20%	I
(.7)		
(.2)	25%	J
	30%	K
(.3)	26%	L
(.4)	30%	M
(.3)	32%	N
(.3)	55%	O
(.6)	65%	P
(.1)	75%	Q
(.1)	15%	R
(.7)	20%	S
(.2)	25%	T
	30%	U

Figure 23.1 Partial skeletonized decision diagram.

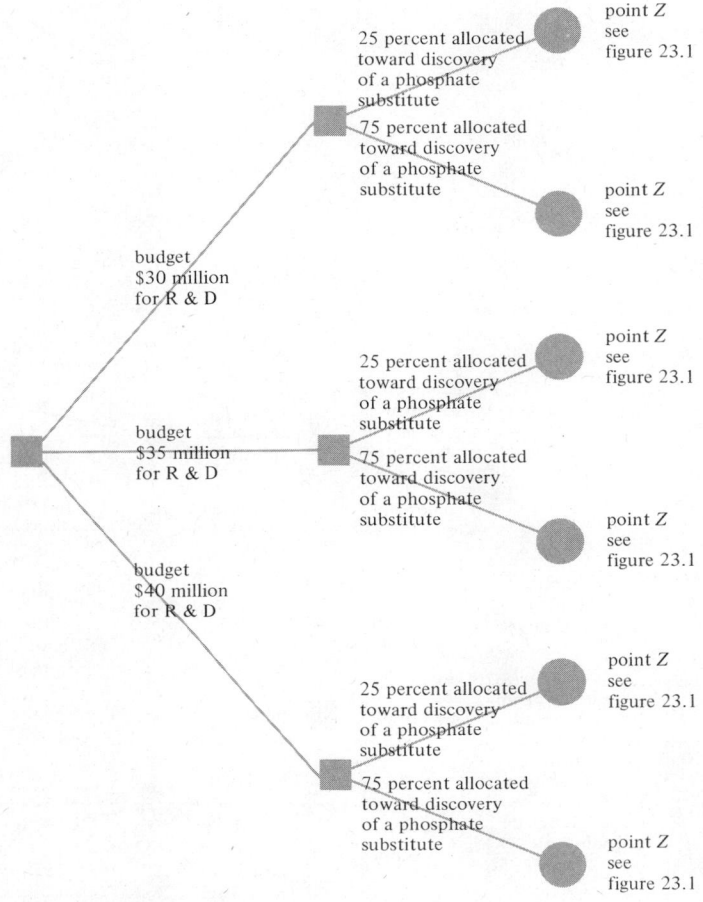

Figure 23.2

3. If Good Citizen, Inc., is unsuccessful on both R & D projects, it will have to convert to an inferior nonphosphate-based detergent regardless of the success or lack of success of its competitors in discovering a phosphate substitute.

The partial tree diagram emanating from point Z in Figure 23.1 is common to each of the action alternatives under consideration. The only difference is that the probabilities assigned to the chance forks emanating from point Z vary depending on the total dollar magnitude of the R & D budget and the manner in which it is allocated between the two programs. The total diagram reflecting the alternative actions available to Good Citizen, Inc., is presented in Figure 23.2.

In the following sections, attention is restricted to the branches emanating from the action alternative, budget $35 million for R & D. Ultimately, we shall return to Figure 23.2 and examine the decision tree in its entirety. The branches associated with the action alternative, budget $35 million for R & D, are reproduced in Figures 23.3(a) and 23.3(b) in skeletonized form.

The next step is the determination of the terminal value (total net liquid assets as of the end of the fifth year) for each of the endpoints. Following the format introduced in Table 23.3, the underlying computations are summarized in Table

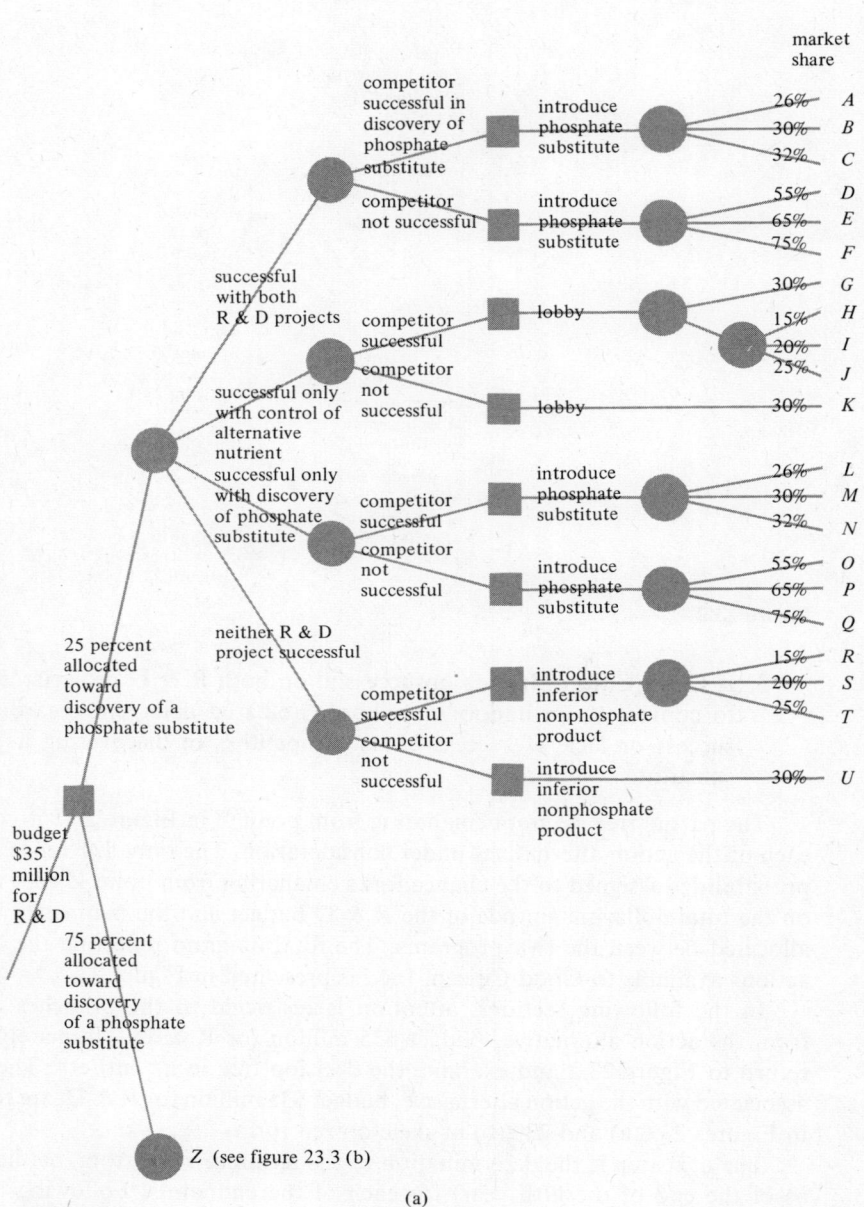

Figure 23.3(a)

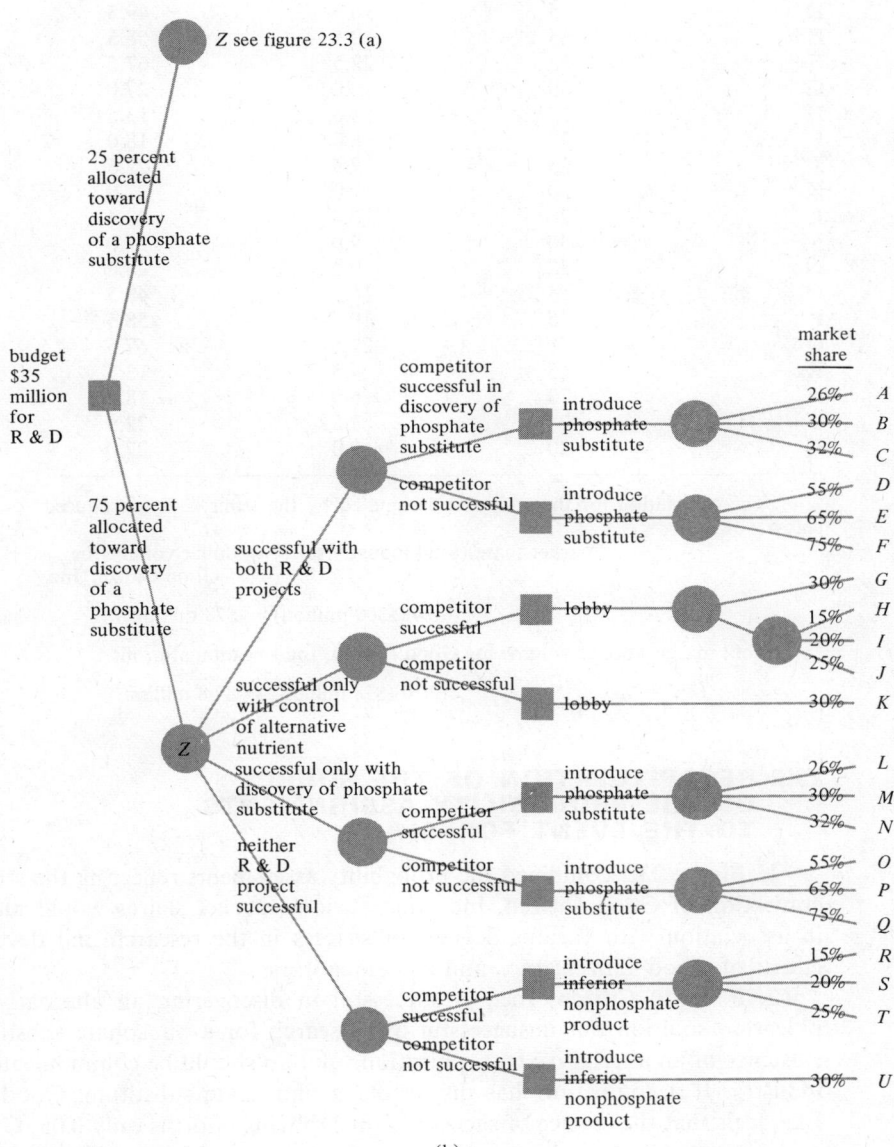

(b)

Figure 23.3(b)

23.5. The supporting computations underlying the determination of aggregate
profit realized for years 3–5 are presented in Table 23.4.

**Table 23.4 Determination of aggregate profit realized for years 3–5
(in millions)**

Designation of endpoint	Market share (percent)	Annual profit[a] in millions for years 3–5	Aggregate profit in millions for years 3–5
A	26	7.8	23.4
B	30	9.0	27.0
C	32	9.6	28.8
D	55	16.5	49.5
E	65	19.5	58.5
F	75	22.5	67.5
G	30	9.0	27.0
H	15	4.5	13.5
I	20	6.0	18.0
J	25	7.5	22.5
K	30	9.0	27.0
L	26	7.8	23.4
M	30	9.0	27.0
N	32	9.6	28.8
O	55	16.5	49.5
P	65	19.5	58.5
Q	75	22.5	67.5
R	15	4.5	13.5
S	20	6.0	18.0
T	25	7.5	22.5
U	30	9.0	27.0

[a] The calculation for the endpoint designated by the letter A is reproduced
below:

$$(\text{market share})(\text{total industry sales}) = \text{dollar volume for Good Citizen, Inc.}$$

$$(0.26)(\$300 \text{ million}) = \$78 \text{ million}$$

$$(\text{profit margin})(\text{dollar volume for Good Citizen, Inc.}) = \text{annual profit}$$

$$(0.10)(\$78 \text{ million}) = \$7.8 \text{ million}$$

23.2 RECAPITULATION OF THE SOURCE OF THE PROBABILITY ASSIGNMENTS TO THE EVENT FORKS

1. Figure 23.1 contained the probability assessments reflecting the strength of
conviction of Good Citizen, Inc., that various market shares would materialize
in association with various degrees of success in the research and development
efforts of Good Citizen, Inc., and its competitors.

2. If Good Citizen, Inc., is successful in discovering an alternative eutrophication solution but unsuccessful in its search for a phosphate substitute, the
existence of an alternative eutrophication solution should be communicated to the
industry. If a competitor has discovered a phosphate substitute, Good Citizen,
Inc., feels that the chance of success of the lobbying effort is only 0.05. Given that
a phosphate substitute has not been discovered, the lobbying effort is considered
certain to succeed.

Table 23.5 Computation of the underlying terminal values (total net liquid assets as of the end of the fifth year in millions)

Designation of endpoint	Current net liquid asset position	Plus profits from detergent sales years 1 and 2	Plus profits from detergent sales years 3–5 (see Table 23.4)	Less any conversion costs incurred	Less R & D expenditures	Less lobbying cost, if any	Terminal value of criterion net liquid assets as of end of fifth year
A	50	+18	+23.4	−15	−35	—	41.4
B	50	+18	+27.0	−15	−35	—	45.0
C	50	+18	+28.8	−15	−35	—	46.8
D	50	+18	+49.5	−15	−35	—	67.5
E	50	+18	+58.5	−15	−35	—	76.5
F	50	+18	+67.5	−15	−35	—	85.5
G	50	+18	+27.0	—	−35	−6	54.0
H	50	+18	+13.5	−4.5	−35	−6	36.0
I	50	+18	+18.0	−4.5	−35	−6	40.5
J	50	+18	+22.5	−4.5	−35	−6	45.0
K	50	+18	+27.0	—	−35	−6	54.0
L	50	+18	+23.4	−15	−35	—	41.4
M	50	+18	+27.0	−15	−35	—	45.0
N	50	+18	+28.8	−15	−35	—	46.8
O	50	+18	+49.5	−15	−35	—	67.5
P	50	+18	+58.5	−15	−35	—	76.5
Q	50	+18	+67.5	−15	−35	—	85.5
R	50	+18	+13.5	−4.5	−35	—	42.0
S	50	+18	+18.0	−4.5	−35	—	46.5
T	50	+18	+22.5	−4.5	−35	—	51.0
U	50	+18	+27.0	−4.5	−35	—	55.5

Table 23.6

$35 million allocated to R & D expenditure
25 percent or $8.75 million allocated to search for a phosphate substitute
75 percent or $26.25 million allocated to search for an alternative solution to eutrophication

Joint event	Probability of occurrence (See Tables 23.1 and 23.2)		
Good Citizen, Inc., successful with both R & D projects	$P\left(\begin{array}{l}\text{successful} \\ \text{with} \\ \text{phosphate} \\ \text{substitute}\end{array}\right) \cdot P\left(\begin{array}{l}\text{successful} \\ \text{in controlling} \\ \text{an alternative} \\ \text{nutrient}\end{array}\right)$	$= (0.1)(0.6)$	$= 0.06$
Good Citizen, Inc., successful only in controlling an alternative nutrient	$P\left(\begin{array}{l}\text{unsuccessful} \\ \text{with} \\ \text{phosphate} \\ \text{substitute}\end{array}\right) \cdot P\left(\begin{array}{l}\text{successful} \\ \text{in controlling} \\ \text{an alternative} \\ \text{nutrient}\end{array}\right)$	$= (0.9)(0.6)$	$= 0.54$
Good Citizen, Inc., successful only in discovering a phosphate substitute	$P\left(\begin{array}{l}\text{successful} \\ \text{with} \\ \text{phosphate} \\ \text{substitute}\end{array}\right) \cdot P\left(\begin{array}{l}\text{unsuccessful} \\ \text{in controlling} \\ \text{an alternative} \\ \text{nutrient}\end{array}\right)$	$= (0.1)(0.4)$	$= 0.04$
Good Citizen, Inc., unsuccessful in both R & D projects	$P\left(\begin{array}{l}\text{unsuccessful} \\ \text{with} \\ \text{substitute} \\ \text{phosphate}\end{array}\right) \cdot P\left(\begin{array}{l}\text{unsuccessful} \\ \text{in controlling} \\ \text{nutrient} \\ \text{an alternative}\end{array}\right)$	$= (0.9)(0.4)$	$= 0.36$
			$\overline{1.00}$

Table 23.7

$35 million allocated to R & D expenditure
75 percent or $26.25 million allocated to search for a phosphate substitute
25 percent or $8.75 million allocated to search for an alternative solution to eutrophication

Joint event	Probability of occurrence (see Tables 23.1 and 23.2)		
Good Citizen, Inc., successful with both R & D projects	$P\left(\begin{array}{l}\text{successful} \\ \text{with} \\ \text{phosphate} \\ \text{substitute}\end{array}\right) \cdot P\left(\begin{array}{l}\text{successful} \\ \text{in controlling} \\ \text{an alternative} \\ \text{nutrient}\end{array}\right)$	$= (0.3)(0.3)$	$= 0.09$
Good Citizen, Inc., successful only in controlling an alternative nutrient	$P\left(\begin{array}{l}\text{unsuccessful} \\ \text{with} \\ \text{phosphate} \\ \text{substitute}\end{array}\right) \cdot P\left(\begin{array}{l}\text{successful} \\ \text{in controlling} \\ \text{an alternative} \\ \text{nutrient}\end{array}\right)$	$= (0.7)(0.3)$	$= 0.21$
Good Citizen, Inc., successful only in discovering a phosphate substitute	$P\left(\begin{array}{l}\text{successful} \\ \text{with} \\ \text{phosphate} \\ \text{substitute}\end{array}\right) \cdot P\left(\begin{array}{l}\text{unsuccessful} \\ \text{in controlling} \\ \text{an alternative} \\ \text{nutrient}\end{array}\right)$	$= (0.3)(0.7)$	$= 0.21$
Good Citizen, Inc., unsuccessful in both R & D projects	$P\left(\begin{array}{l}\text{unsuccessful} \\ \text{with} \\ \text{phosphate} \\ \text{substitute}\end{array}\right) \cdot P\left(\begin{array}{l}\text{unsuccessful} \\ \text{in controlling} \\ \text{an alternative} \\ \text{nutrient}\end{array}\right)$	$= (0.7)(0.7)$	$= 0.49$
			$\overline{1.00}$

3. Good Citizen, Inc., has assigned a likelihood of 0.30 to a competitor
discovering a phosphate substitute as of the end of year 2.

4. The probability of occurrence of the joint events listed in Table 23.6 (and
Table 23.7) are a function of the dollar amounts allocated for research and
development (see Tables 23.1 and 23.2).

Insertion of the terminal values and the missing probabilities serves to flesh
out the skeletonized decision tree in Figures 23.3(a) and 23.3(b). The resultant
decision tree is depicted in Figures 23.4(a) and 23.4(b).

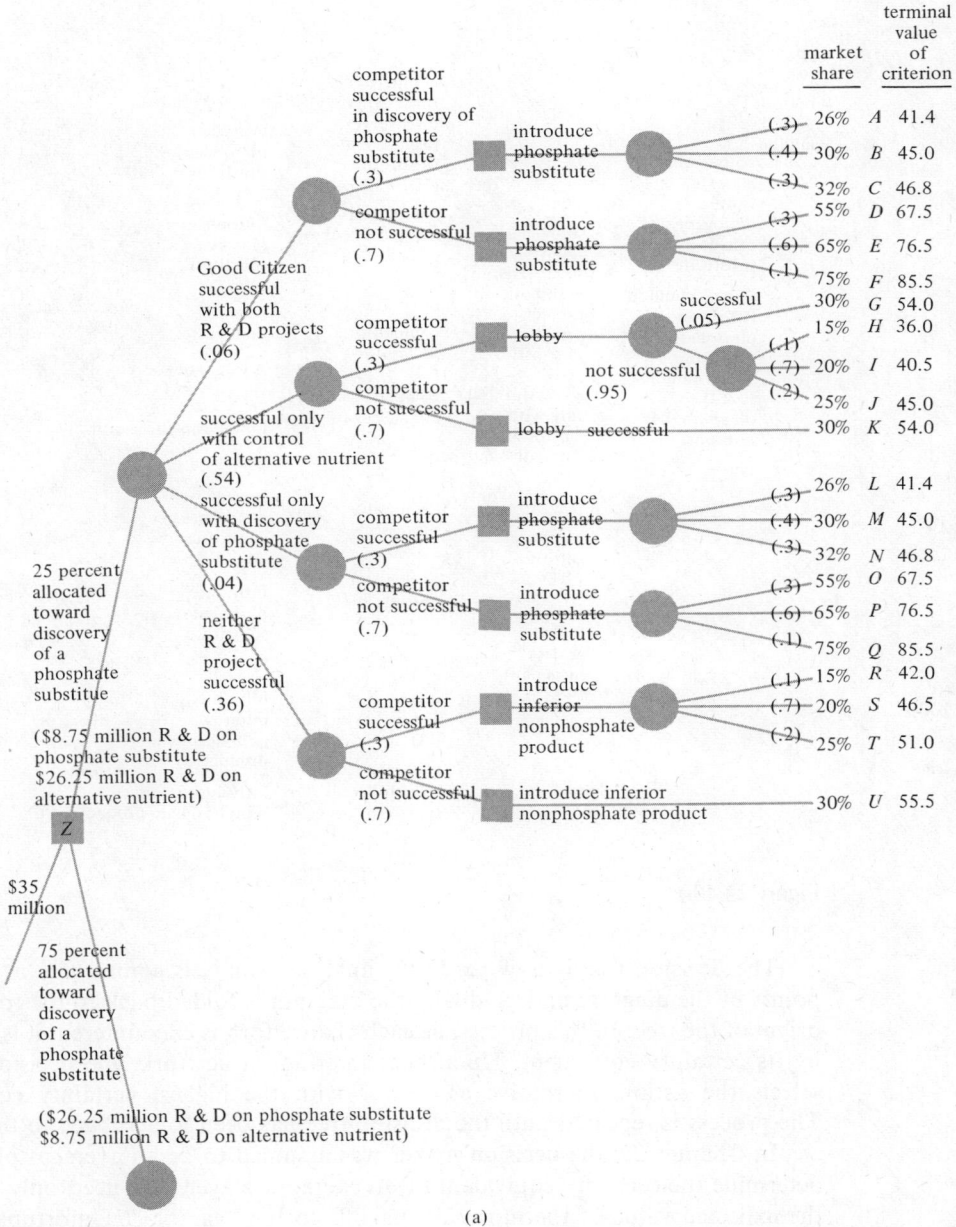

(a)

Figure 23.4(a)

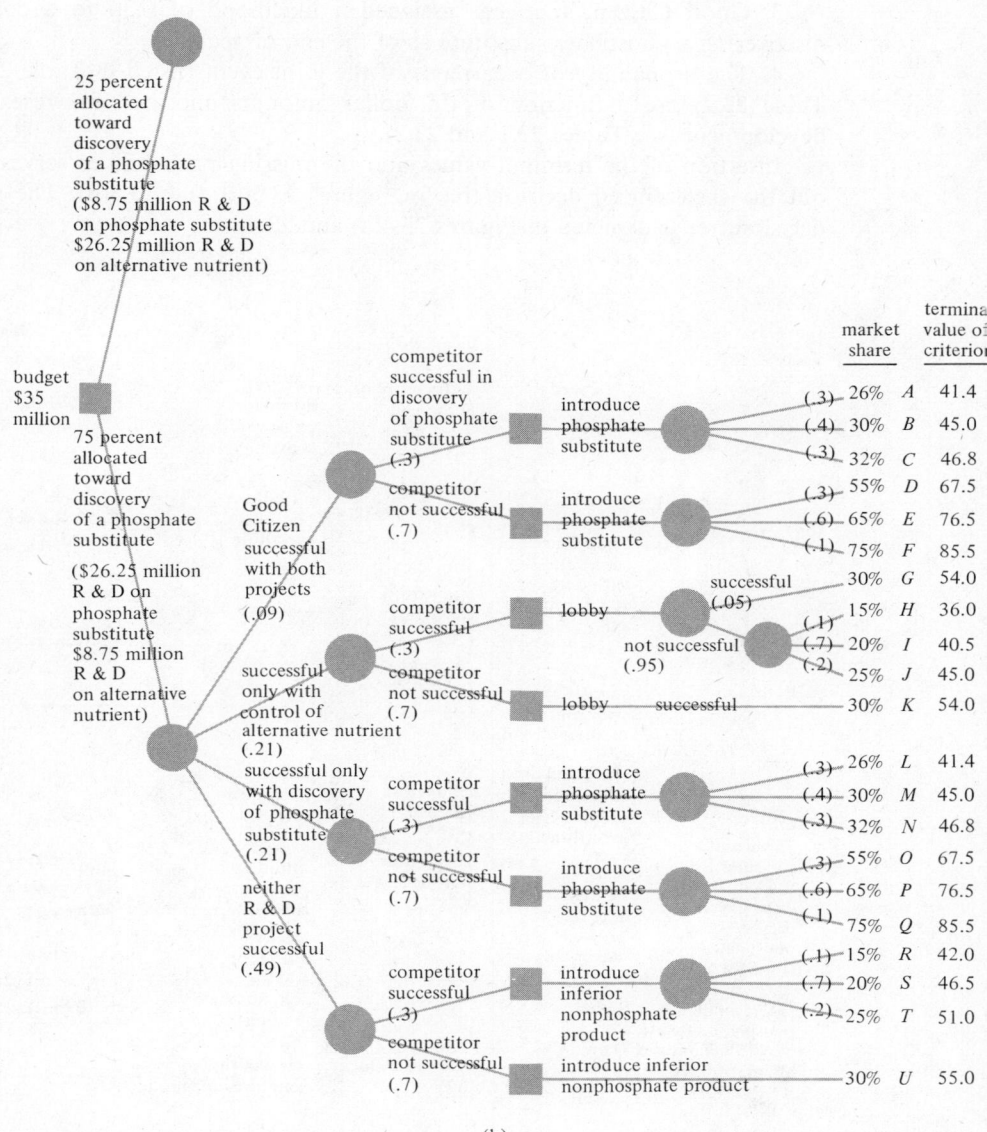

(b)

Figure 23.4(b)

The decision tree is now ready for analysis. Analysis commences at the endpoints of the diagram and gradually the diagram is folded back to the fork at the origin of the tree. In this process, as each chance fork is encountered, it is replaced by its certainty equivalent. Upon encountering an act fork, the decision maker selects the action alternative associated with the highest certainty equivalent. The process is repeated until the decision tree has been folded back to the origin.

In Chapter 22, the decision maker was assumed to be an averages player. To determine the certainty equivalent of an averages player, one need only compute the expected value of the monetary payoffs to the "gamble." Unfortunately, the assumption that the certainty equivalent to a gamble is equal to the mathematical expectation of the consequences of that gamble is no longer valid when considering gambles where the dollar consequences associated with the outcomes of the alternatives under consideration are quite large compared to the asset position of the decision maker. This is the situation in which we find Good Citizen, Inc.

An alternative method for determining the certainty equivalents must be found. One approach is to ask the decision maker to assess directly his certainty equivalents to the various gambles encountered.

One might inquire of the decision maker: What amount would you be indifferent to receiving for certain in lieu of taking a gamble that provides a 30 percent chance at $41.4 million, a 40 percent chance at $45.0 million, and a 30 percent chance at $46.8 million? The decision maker's response would constitute his certainty equivalent.

There are three major problems associated with this approach.

1. The decision maker will often find it difficult to assess his feelings about the desirability of a multibranched event fork. Such assessments require simultaneous consideration of the probability of occurrence of each event and the decision maker's attitude with respect to the relative desirability of the outcome associated with each event. Many find this a difficult task.
2. The evaluation process must be repeated for each event fork in the tree. Given numerous event forks, this can become a tedious undertaking.
3. Using this method, consistency in response is difficult to maintain from event fork to event fork. This is a serious problem. The overall responses throughout the tree should reflect a consistent attitude with respect to the relative desirability of the consequences associated with the alternatives under consideration.

Fortunately, an alternative approach is available. It is possible to develop an index of utility that reflects the relative desirability of each possible consequence over the range of consequences likely to be encountered by the decision maker. Certain advantages accrue if the utility index is constructed with the limiting values of 0.00 and 1.00. When a utility scale is constructed between the limiting values of 0.00 and 1.00, the resulting descriptive function is called a preference function or preference curve.

There are three major advantages associated with the preference curve approach.

1. Unlike the direct assessment approach, it enables the decision maker to concentrate solely on the relative desirability of the various outcomes. No consideration need be given to the likelihood of occurrence of those outcomes.
2. The preference curve approach enables the decision maker to express his attitude toward all possible outcomes under consideration simultaneously. This approach ensures that the individual assessments are consistent with the overall attitude of the decision maker with respect to the relative desirability of the various consequences.
3. The preference curve approach also simplifies the computational process involved in folding back a decision tree for an individual who is a non-averages player (i.e., a risk averter or a risk taker). This will be demonstrated subsequently.

23.3 PROCEDURE FOR DEVELOPMENT OF A PREFERENCE CURVE

The first decision involves the determination of the range of consequences over which the decision maker's attitude with respect to terminal values is to be assessed. Given an expenditure of $35 million for R & D, the lowest value of the terminal criterion encountered is $36 million and the highest is $85.5 million (see Table 23.5). Each outcome decreases by $5 million for the branch associated with an expenditure of $40 million for R & D. Thus, the overall low for the entire decision tree is $31 million. Similarly, each outcome increases by $5 million for the branch associated with an expenditure of $30 million. It follows that the overall high for the entire decision tree is $90.5 million. The range of consequences to be assessed by the preference curve must at least encompass these values. Arbitrarily, the range will be established with $30 million as the lower bound and $100 million as the upper bound. If the decision maker anticipates that other decisions will deal with consequences in the same "neighborhood," the range should be wide enough to encompass these values as well. The same preference curve continues to be applicable as long as the terminal values encountered fall within the range the preference curve is designed to serve and the decision maker's underlying attitude toward risk taking has not changed.

The two values selected as the limiting values of the range are referred to as *reference consequences*. The lower reference consequence is designated R_0 and the upper reference consequence as R_1. For Good Citizen, Inc., $30 million and $100 million are the designated values of R_0 and R_1, respectively. As will be demonstrated shortly, the decision maker's attitude with respect to the relative desirability of the consequences over the interval from R_0 to R_1 can be readily determined from the decision maker's responses to a series of questions. These questions take the form of easy-to-evaluate choices between a gamble and a certain event. The relative desirability of each consequence over the range of consequences is indicated by the assignment of a value called the *preference*. The preference is a relative ranking on a scale from zero to one.

Technically, the preference is considered to be the probability of occurrence the decision maker requires for the more desirable consequence in a reference gamble in order for the decision maker to be indifferent between that gamble and the associated terminal value recorded on the horizontal scale. A reference gamble is a gamble that provides some chance at R_1 and a complementary chance at R_0.

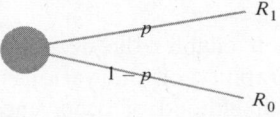

Two trivial illustrations of this relationship refer to the situations where the terminal values in question are R_0 and R_1.

For R_1 (or $100 million in the case of Good Citizen, Inc.): The question asked is what probability would the decision maker require for the outcome R_1 (i.e., $100 million) in a reference gamble in order for him to be indifferent between that gamble and a guarantee of R_1 for certain? Obviously, it must be 1.0.

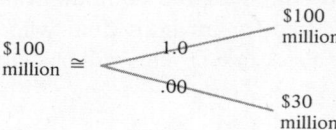

where $\cong$ is interpreted as "is considered equivalent to."
In general,

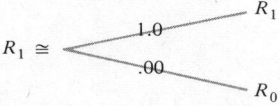

Clearly, as long as any chance exists of realizing the outcome $30 million, the decision maker would prefer the certain $100 million to the gamble. The only situation where the decision maker would be indifferent between the two would be where the gamble also promised a certain $100 million. This would be true only when there is a 1.0 chance of the outcome $100 million occurring. Thus, the preference for $100 million is 1.0. Note that the preference for R_1 must always equal 1.00 and therefore can be assigned directly.

For R_0 (or $30 million in the case of Good Citizen, Inc.): A similar line of reasoning discloses that the preference for R_0 will always be 0.00.

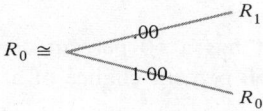

As long as there is any chance of obtaining the outcome R_1, no matter how small, the decision maker would prefer the gamble to the certain R_0. The only situation where the decision maker would be indifferent between the gamble and the certain event is when the probability that R_1 will occur is 0.0.

The above two illustrations are trivial in that the preferences for R_1 and R_0 will always be 1.0 and 0.0, respectively, and thus can be assigned automatically. Plotting the preferences for R_0 and R_1 provides two points on the preference curve. The plots are labeled CE_1 and CE_2 for identification (Figure 23.5).

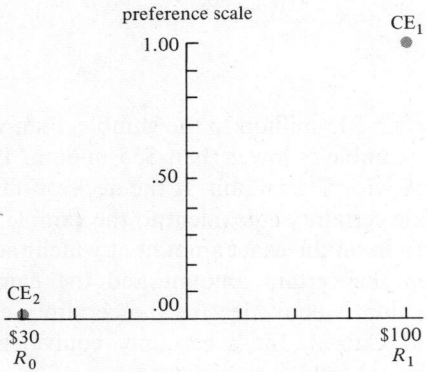

Figure 23.5

To obtain a third point on the curve, the decision maker generates his certainty equivalent to a reference gamble providing a 0.5 chance at R_1 and a 0.5 chance at R_0.
In general,

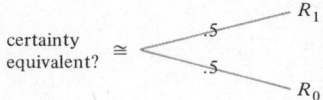

Specifically, for Good Citizen, Inc.,

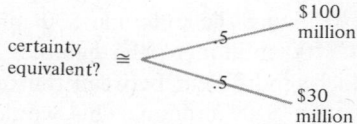

It is useful to couch the question in the format of a situation with which the decision maker is familiar. This aids the decision maker in assessing his certainty equivalent to a reference gamble such as the one posited above. To illustrate, the above question could alternatively have been posed as follows:

Consider yourself faced with having to choose from one of the following two alternatives.

Alternative 1: Introduce a new product that has a 50 percent chance of a payoff of $100 million and a 50 percent chance of a payoff of $30 million.

Alternative 2: Commit an equivalent amount of investment to a cost-reduction campaign guaranteed to generate savings of $55 million. (The $55 million is an arbitrarily selected figure.)

Schematically, the choice appears as follows:

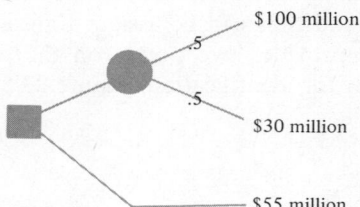

If the decision maker prefers the certain $55 million to the gamble, then you know that the certainty equivalent to the gamble is lower than $55 million. Lower the value accordingly and try again, say, with $35 million. If the decision maker now prefers the gamble, you know that the certainty equivalent to the gamble is greater than $35 million. Gradually, you zero in on the exact amount at which the decision maker becomes indifferent between the certain amount and the gamble. The certain amount obtained in this fashion is equivalent to the certainty equivalent to the gamble. Suppose that Good Citizen, Inc.'s certainty equivalent to this gamble is $45 million. Schematically,

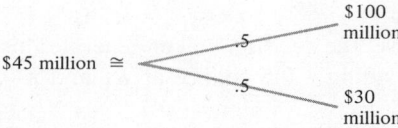

By definition, the preference for any given certainty equivalent is equal to the probability assigned to the more desirable consequence in a reference gamble. It follows that the decision maker's preference for $45 million is 0.50. This provides a third point on the preference curve. It is plotted in Figure 23.6 and is identified as CE_3.

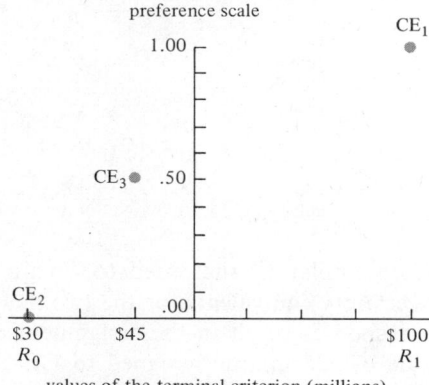

Figure 23.6

It is interesting to note that the certainty equivalent response that would have been supplied by an averages player would have been equal to the mathematical expectation of the reference consequences, namely:

$$0.5 \ (\$100 \ \text{million}) = \$50 \ \text{million}$$
$$0.5 \ (\$30 \ \text{million}) \ = \$15 \ \text{million}$$
$$\overline{\hspace{3cm}}$$
$$\$65 \ \text{million}$$

The difference between the certainty equivalent of an averages player and that of a nonaverages player with a lower certainty equivalent is referred to as the *risk premium*. The risk premium is a reflection of the extent of conservatism on the part of the decision maker. It can be considered as the reduction in expected monetary payoff the decision maker considers a fair tradeoff for avoiding risk. The greater the conservatism or unwillingness to bear risk, the greater the risk premium. Such individuals are referred to as *risk averters*. Occasionally, one will encounter a *risk taker* or gambler. Such individuals are characterized by a negative risk premium in that they would have to be offered a certainty equivalent in excess of the mathematical expectation of a gamble before they would be willing to forgo the gamble. In effect, such individuals enjoy risk and require compensation for relinquishing the opportunity to gamble. There are many variations within the ranks of risk takers and risk averters. This is a reflection of the varying extent to which they are smitten by the desire to gamble in the case of the risk taker or by the aversion to risk in the case of the risk averter. Between these extremes trods the "straight-and-narrow" averages player.[1]

___

[1] The averages player is literally "straight." Reference to Figure 23.8 discloses the preference curve for the averages player. It is always linear. The preference curve for a risk averter is always bowed to the left (concave to the origin)—the more conservative, the more bowed the curve. Conversely, risk takers or gamblers always have preference curves bowed to the right (convex). As you might have surmised, the greater the willingness to incur risk, the more bowed the curve.

Building on information already in hand, two additional points on the preference curve can be generated by determining the certainty equivalents associated with the following two nonreference gambles:

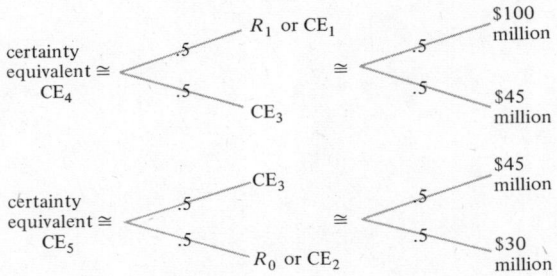

Utilizing a questioning procedure similar to that used to obtain CE_3, the decision maker can zero in on the certainty equivalents for the two new gambles. Presume that the decision maker's responses result in the assignment of a value of \$61.5 million to CE_4 and a value of \$35 million assigned to CE_5. Once the certainty equivalents are obtained, the next step is the generation of the preferences associated with these two terminal values.

For CE_4:

1. \$61.5 million is equivalent to

2. Recall that CE_3 is equivalent to

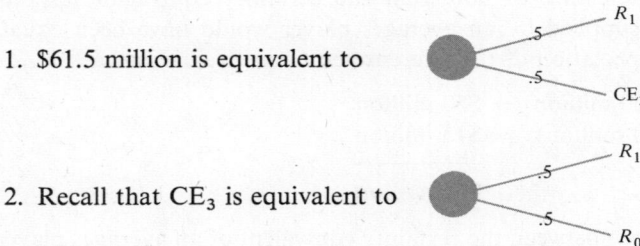

This equivalence reflects the fact that the decision maker is indifferent between the following two opportunities: (a) a 50 percent chance at CE_3, and (b) a 50 percent chance of having the opportunity to take a gamble providing a 50 percent chance at R_1 and a 50 percent chance at R_0. The decision maker visualizes these as equivalent opportunities. One can thus be substituted for the other without affecting the desirability of the original gamble.

3. Substituting, we obtain

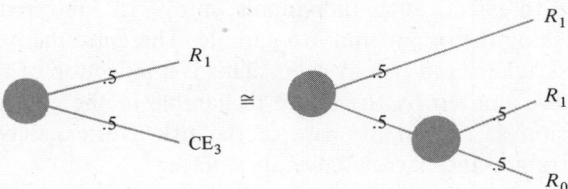

4. Note that in the two-stage gamble there are only two alternatives, namely, R_1 and R_0. This enables the gamble to be rewritten as an equivalent one-stage gamble reflecting the aggregate chance of arriving at outcome R_1 and outcome R_0. The probability of arriving at each endpoint in the two-stage gamble is obtained

by multiplying the probabilities one encounters as one moves from the origin to an endpoint.

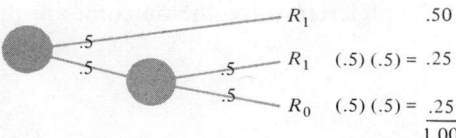

$$R_1 \qquad .50$$
$$R_1 \qquad (.5)(.5) = .25$$
$$R_0 \qquad (.5)(.5) = \underline{.25}$$
$$1.00$$

The combined probability of arriving at outcome R_1 equals $0.50 + 0.25 = 0.75$. The probability of arriving at R_0 is 0.25. Thus, an equivalent representation of the above gamble in the form of a one-stage gamble is

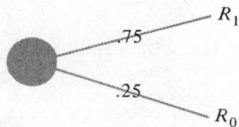

5. It follows that the following expressions are considered interchangeable by the decision maker:

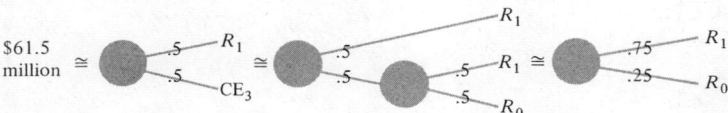

6. Eliminating the intermediate steps, one obtains the following equivalence:

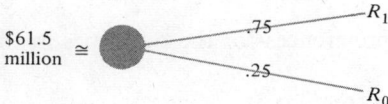

The nonreference gamble

$61.5 million $\cong$

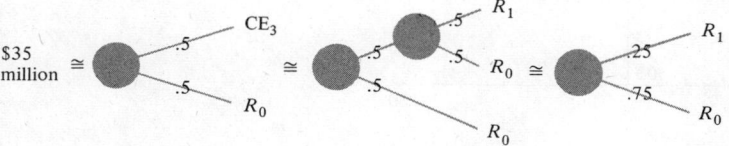

has now been transformed into an equivalent reference gamble. By definition, the preference for $61.5 million is equal to the probability required for the more desirable outcome in a *reference* gamble in order for the decision maker to be indifferent between the reference gamble and the certain amount.

It follows that the preference for CE_4 (i.e., $61.5 million) is 0.75. In a similar fashion, the preference for CE_5 (i.e., $35 million) is identified as 0.25.

In practice, a more efficient approach is used to determine the preference associated with any nonreference gamble. In effect, the above procedure for determining the probability of R_1 in the *reference* gamble is equivalent to finding the mathematical expectation of the preferences for the outcomes in the original *nonreference* gamble.

For CE_4:

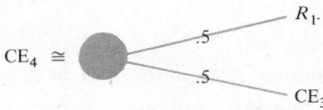

preference for R_1 = 1.0
preference for CE_3 = 0.5

Mathematical expectation of the preferences for the outcomes in the original nonreference gamble:

(0.5)(preference for R_1) = $(0.5)(1.0)$ = 0.50
(0.5)(preference for CE_3) = $(0.5)(0.5)$ = 0.25
—————
0.75

Thus, the preference for CE_4 is 0.75.

For CE_5:

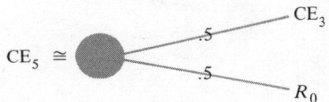

preference for CE_3 = 0.50
preference for R_0 = 0.00

Mathematical expectation of the preferences for the outcomes in the original nonreference gamble:

(0.5)(preference for CE_3) = $(0.5)(0.50)$ = 0.25
(0.5)(preference for R_0) = $(0.5)(0.00)$ = 0.00
—————
0.25

Thus, the preference for CE_5 is 0.25. Plotting the points for the preferences for CE_4 and CE_5 provides two additional points on the curve (Figure 23.7).

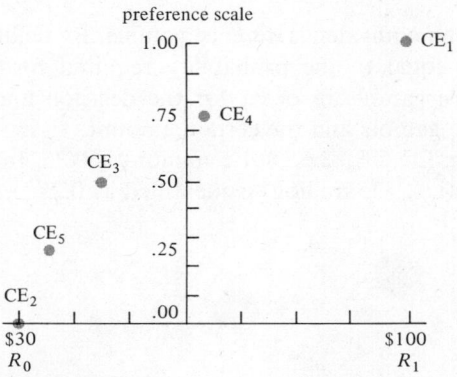

Figure 23.7

Additional points could be generated by building other nonreference gambles from the five known points on the curve. The gambles would take the following form:

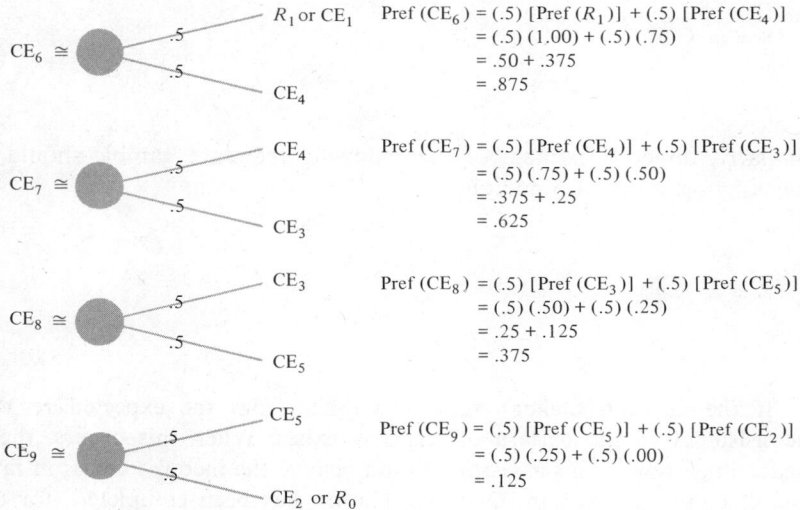

$$\text{Pref (CE}_6) = (.5) [\text{Pref} (R_1)] + (.5) [\text{Pref} (\text{CE}_4)]$$
$$= (.5) (1.00) + (.5) (.75)$$
$$= .50 + .375$$
$$= .875$$

$$\text{Pref (CE}_7) = (.5) [\text{Pref} (\text{CE}_4)] + (.5) [\text{Pref} (\text{CE}_3)]$$
$$= (.5) (.75) + (.5) (.50)$$
$$= .375 + .25$$
$$= .625$$

$$\text{Pref (CE}_8) = (.5) [\text{Pref} (\text{CE}_3)] + (.5) [\text{Pref} (\text{CE}_5)]$$
$$= (.5) (.50) + (.5) (.25)$$
$$= .25 + .125$$
$$= .375$$

$$\text{Pref (CE}_9) = (.5) [\text{Pref} (\text{CE}_5)] + (.5) [\text{Pref} (\text{CE}_2)]$$
$$= (.5) (.25) + (.5) (.00)$$
$$= .125$$

Each of the certainty equivalents would be obtained by a series of forced choices similar to those previously illustrated. The preferences for each gamble are obtained by finding the mathematical expectation of the preferences for the outcomes associated with the gamble (as illustrated above).

At this point the pattern should be clear. Through this process, the curve is filled out to whatever detail is desired. It is preferable, though not mandatory, to use 50–50 gambles because of the relative ease of assessment. Usually, the first five preferences computed are considered sufficient to enable identification of the preference curve (Table 23.8).

Table 23.8

Identification label	Dollar value of the terminal criterion	Preference
CE_2 or R_0	30 million	0.00
CE_5	35 million	0.25
CE_3	45 million	0.50
CE_4	61.5 million	0.75
CE_1 or R_1	100 million	1.00

Computer programs are available that fit a curve to the five points and print out the preference associated with values of the terminal criterion over the range of interest. The curve is typically subject to an additional constraint, namely, decreasing risk aversion as the value of the terminal value is increased. This reflects an increased willingness to undertake risk as the size of the asset base increases.

Verification of the validity of the curve as a correct representation of the attitude of the decision maker toward risk is essential. This is true for the first five points as well as for the curve derived from those points.

If the first five points are correct assessments, the decision maker should assign a certainty equivalent of $45 million to the following gamble:

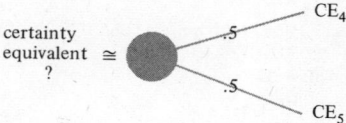

Similarly, direct assessment of the following reference gamble should result in the assignment of $61.5 million:

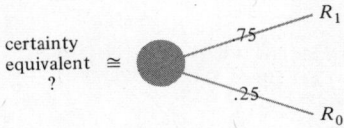

If the decision maker's responses differ from the expected responses, an inconsistency in the pattern of response exists. When this occurs, the decision maker must rework his assessments and resolve the inconsistencies in favor of an overall consistent pattern. Once verification has been completed (for the initial five points as well as the derived curve), the preference curve is ready for analysis. Figure 23.8 reproduces the preference curve generated for the decision maker in the Good Citizen, Inc., illustration based on his responses as summarized in Table 23.8. Figure 23.8 also displays a preference curve representing the responses

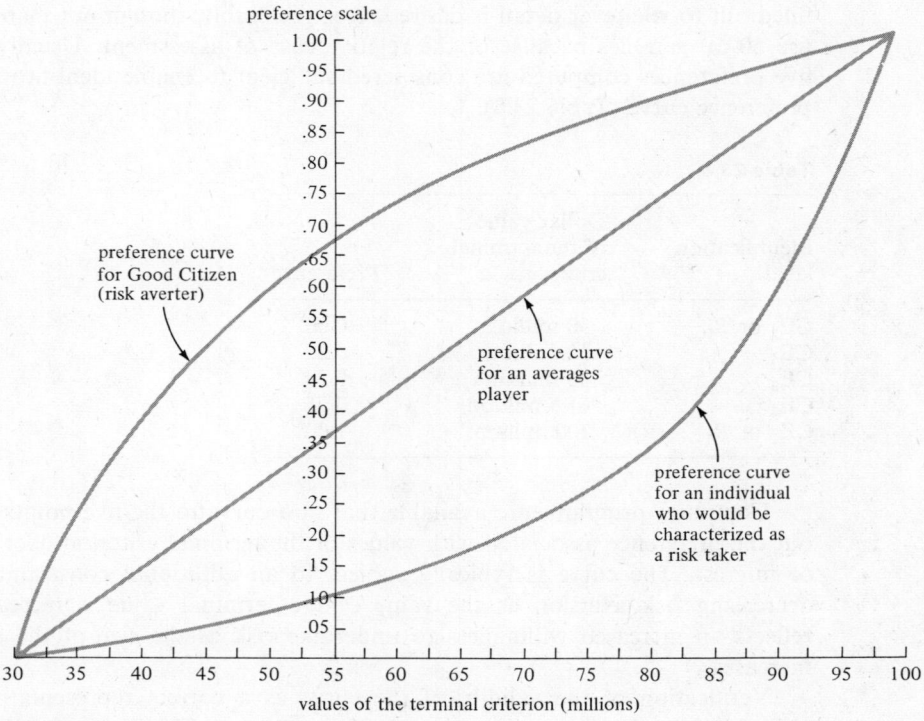

Figure 23.8

of an averages player and a curve describing the response pattern of an individual characterized as a risk taker.

23.4 APPLICATION OF THE PREFERENCE CURVE

The uppermost branch of Figure 23.4(a) reflects Good Citizen, Inc., achieving success with both R & D projects. This segment of Figure 23.4(a) is reproduced as Figure 23.9.

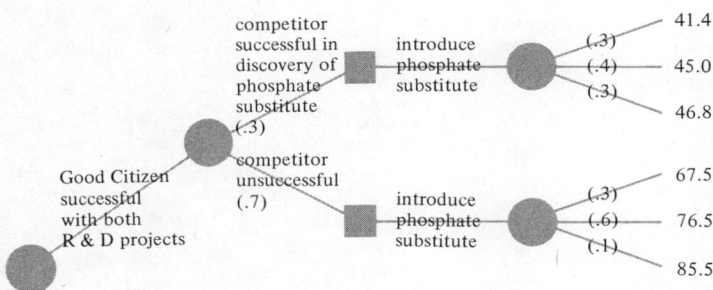

Figure 23.9

Terminal values are to be replaced by equivalent reference gambles. The preference associated with each terminal value is determined by raising a perpendicular at the terminal value in question and reading the corresponding preference from the vertical axis. Preferences obtained in this fashion from Figure 23.8 are displayed in Table 23.9.

Table 23.9

Terminal value (in millions of dollars)	Preference
41.4	42
45.0	50
46.8	55
67.5	80.5
76.5	87
85.5	93

The preference of 42 associated with the terminal value of 41.4 is interpreted as "the decision maker is indifferent between receiving \$41.4 million for certain and taking a gamble with a 42 percent chance at R_1 (i.e., \$100 million) and a 58 percent chance at R_0 (i.e., \$30 million)."

Figure 23.10 describes the same relationship as depicted in Figure 23.9. Terminal values have been replaced by their equivalent reference gambles. Each of the two-stage gambles in Figure 23.10 can be replaced by a one-stage gamble. This transformation is obtained by combining all the forks leading to the outcome R_1 into a single fork and all the forks leading to the outcome R_0 into a single fork. The resulting equivalent diagram is depicted as Figure 23.11.

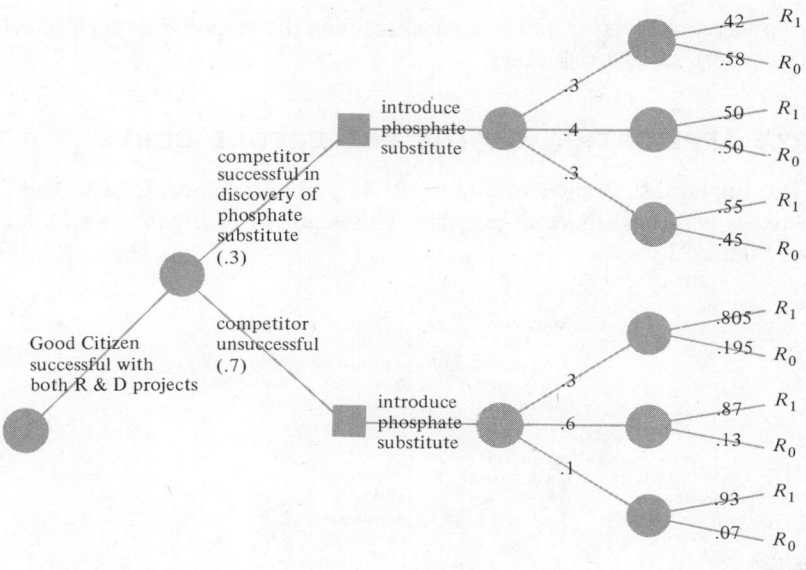

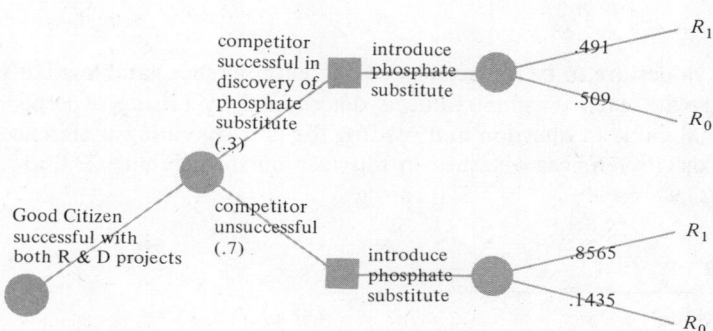

Figure 23.11

Upper fork:

probability (R_1) = (0.3)(0.42) + (0.4)(0.50) + (0.3)(0.55)
= 0.126 + 0.200 + 0.165 = 0.491

probability (R_0) = (0.3)(0.58) + (0.4)(0.50) + (0.3)(0.45)
= 0.174 + 0.200 + 0.135 = 0.509

Lower fork:

probability (R_1) = (0.3)(0.805) + (0.6)(0.87) + (0.1)(0.93)
= 0.2415 + 0.522 + 0.093 = 0.8565

probability (R_0) = (0.3)(0.195) + (0.6)(0.13) + (0.1)(0.07)
= 0.0585 + 0.078 + 0.007 = 0.1435

The reference gambles could be converted back into their corresponding equivalents in terms of the decision maker's criterion at each stage of the analysis.

To illustrate for the uppermost fork: Extend a perpendicular from the preference scale at 0.491 and read the corresponding value of the terminal value on the horizontal axis.

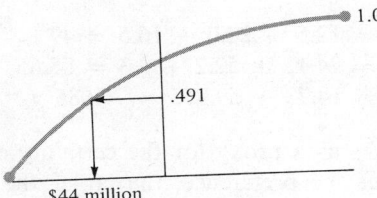

This is interpreted as "the decision maker considers a gamble with a 49.1 percent chance at R_1 and a 50.9 percent chance at R_0 equivalent to receiving $44 million for certain."

It is customary, however, to perform the entire analysis in terms of preferences. Conversion to original units of the terminal value does not occur until the final stage of the folding-back process.

The approach described above is useful as an aid in perceiving the underlying relationships involved and the manner in which they are manipulated. In practice, the folding-back process is performed much more efficiently by taking advantage of the following relationship: *The preference for any gamble can be obtained by simply computing the mathematical expectation of the preferences for the consequences associated with that gamble.* Figure 23.12 illustrates this relationship.

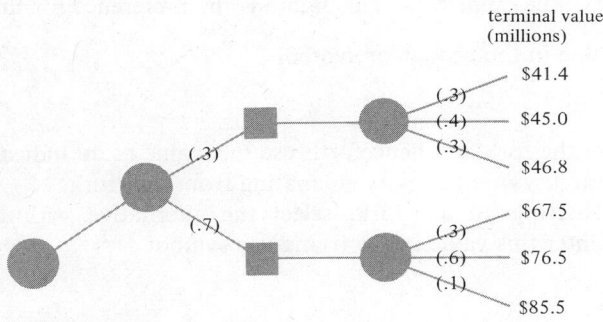

Figure 23.12

Substituting the decision maker's preferences for the terminal values, Figure 23.12 appears as depicted in Figure 23.13. Each event fork is replaced by the mathematical expectation of the preferences for the outcomes associated with that fork.

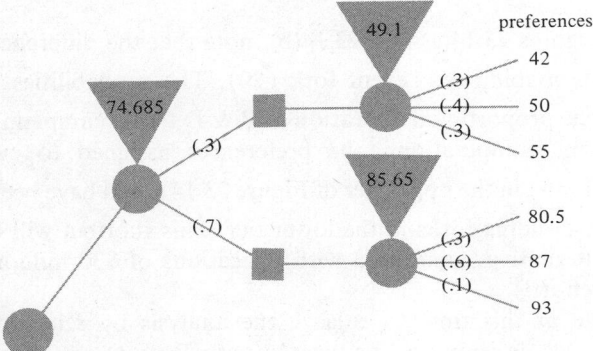

Figure 23.13

$$(0.3)(42) + (0.4)(50) + (0.3)(55) = 12.6 + 20.0 + 16.5 = 49.1$$
$$(0.3)(80.5) + (0.6)(87) + (0.1)(93) = 24.15 + 52.2 + 9.3 = 85.65$$
$$0.3(49.1) + (0.7)(85.65) = 14.73 + 59.955 = 74.685$$

The preference for the event fork serves as a proxy for the certainty equivalent associated with the gamble. The larger the preference, the larger the certainty equivalent.

Upon encountering an act fork, the decision maker should always select the alternative associated with the larger certainty equivalent. Obviously, when faced with two reference gambles, one would prefer the one that provides the greater chance at the more desirable consequence. This is equivalent to choosing the alternative with the larger preference. The process is repeated until the tree has been folded back to the origin.

Summarizing the Procedure

1. Replace each terminal value by its preference.
2. In the process of backward induction, replace each event fork encountered by the mathematical expectation of the preferences for the outcomes associated with the event fork. This figure is the preference for the gamble.

   Enter this value in the triangular symbol

   at the base of the fork and henceforth use this value as an indicator of the relative desirability of all activity emanating from that fork.
3. Upon encountering an act fork, select the alternative with the larger preference. Enter this value in the triangular symbol

   at the base of the fork.
4. Repeat steps 2 or 3 as they are appropriate until the tree has been folded back to the act fork at the origin.

A completed analysis for the portion of the overall decision tree displayed in Figures 23.4(a) and 23.4(b) is reproduced in Figures 23.14(a) and 23.14(b). Terminal values of the criterion have been replaced by the value of their preferences (see Table 23.10).

In comparing Figures 23.14(a) and 23.14(b), note that the difference is in the assignment of the probabilities at event fork ⟨20⟩. The probabilities of success vary depending on the proportional allocation of R & D to the competing projects. To reduce redundant computations, the preferences assigned to event forks ⟨6⟩, ⟨17⟩, ⟨18⟩, and ⟨19⟩ in the upper tier of Figure 23.14 could have been assigned directly to the corresponding forks in the lower tier. This shortcut will be utilized in evaluating the alternatives associated with allocations of $30 million and $40 million for R & D effort.

It is appropriate at this time to enlarge the analysis by reintroducing the portions of the overall decision tree associated with R & D allocations of $30 million and $40 million. Tables 23.11 and 23.12 develop the probabilities of success

Table 23.10

Endpoint	Terminal value (in millions of dollars)	Preference (for the risk averter described in Figure 23.8)
A	41.4	42
B	45.0	50
C	46.8	55
D	67.5	80.5
E	76.5	87
F	85.5	93
G	54.0	66
H	36.0	28
I	40.5	40
J	45.0	50
K	54.0	66
L	41.4	42
M	45.0	50
N	46.8	55
O	67.5	80.5
P	76.5	87
Q	85.5	93
R	42.0	44
S	46.5	54
T	51.0	61
U	55.5	68

Table 23.11

$30 million allocated to R & D expenditure
25 percent or $7.5 million allocated to search for a phosphate substitute
75 percent or $22.5 million allocated to search for an alternative solution to eutrophication

Joint event	Probability of occurrence (see Tables 23.1 and 23.2)
Good Citizen, Inc., successful with both R & D projects	$P\begin{pmatrix}\text{successful} \\ \text{with} \\ \text{phosphate} \\ \text{substitute}\end{pmatrix} \cdot P\begin{pmatrix}\text{successful} \\ \text{in controlling} \\ \text{an alternative} \\ \text{nutrient}\end{pmatrix} = (0.08)(0.57) = 0.0456$
Good Citizen, Inc., successful only in controlling an alternative nutrient	$P\begin{pmatrix}\text{unsuccessful} \\ \text{with} \\ \text{phosphate} \\ \text{substitute}\end{pmatrix} \cdot P\begin{pmatrix}\text{successful} \\ \text{in controlling} \\ \text{an alternative} \\ \text{nutrient}\end{pmatrix} = (0.92)(0.57) = 0.5244$
Good Citizen, Inc., successful only in discovering a phosphate substitute	$P\begin{pmatrix}\text{successful} \\ \text{with} \\ \text{phosphate} \\ \text{substitute}\end{pmatrix} \cdot P\begin{pmatrix}\text{unsuccessful} \\ \text{in controlling} \\ \text{an alternative} \\ \text{nutrient}\end{pmatrix} = (0.08)(0.43) = 0.0344$
Good Citizen, Inc., unsuccessful in both R & D projects	$P\begin{pmatrix}\text{unsuccessful} \\ \text{with} \\ \text{phosphate} \\ \text{substitute}\end{pmatrix} \cdot P\begin{pmatrix}\text{unsuccessful} \\ \text{in controlling} \\ \text{an alternative} \\ \text{nutrient}\end{pmatrix} = (0.92)(0.43) = 0.3956$

1.0000

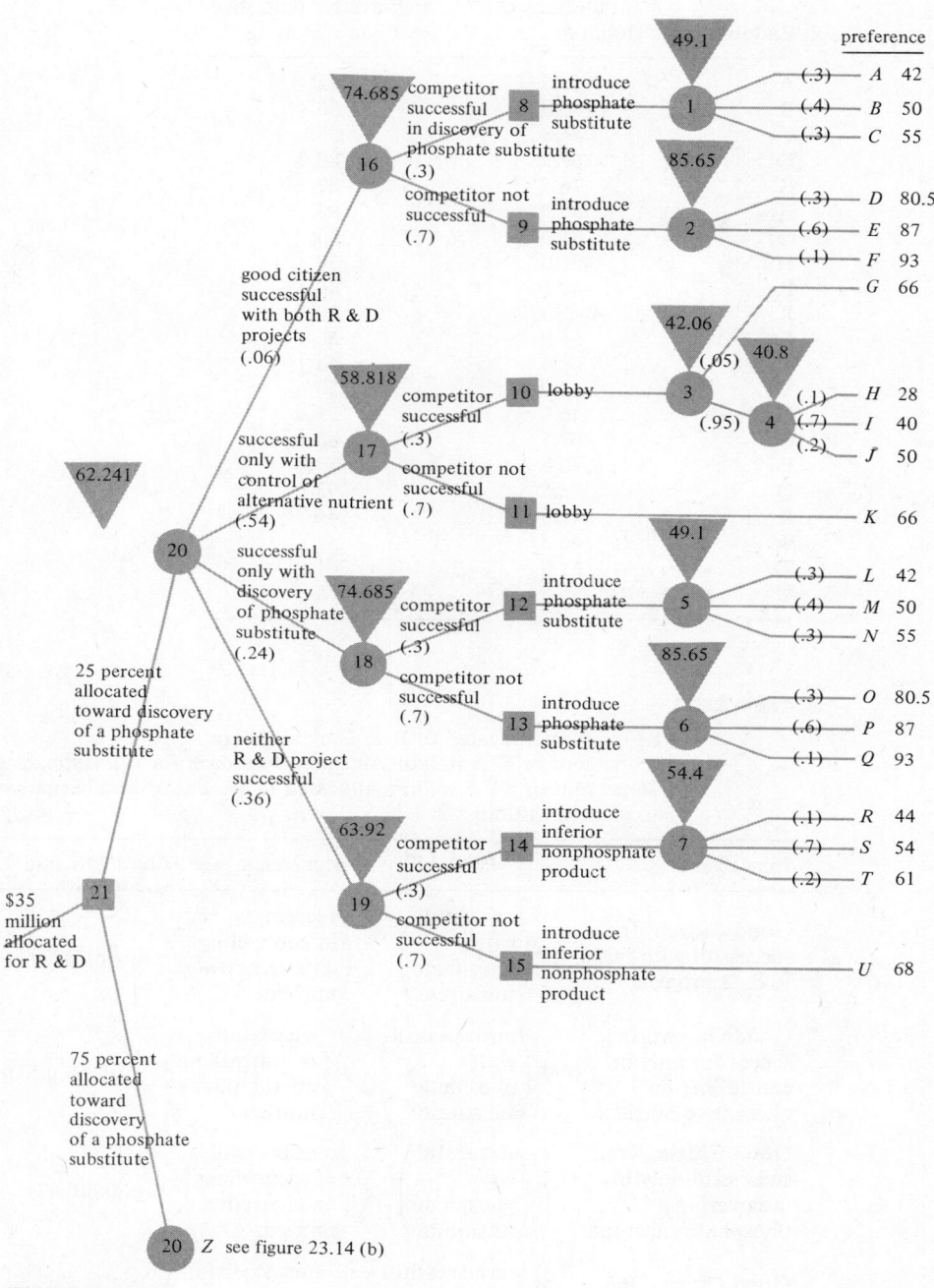

Figure 23.14(a)

(*a*)

associated with a \$30 million R & D expenditure. The terminal values and the
associated preferences are recorded in Table 23.13. Figure 23.15 displays a
completed analysis for the alternative—spend \$30 million for R & D.

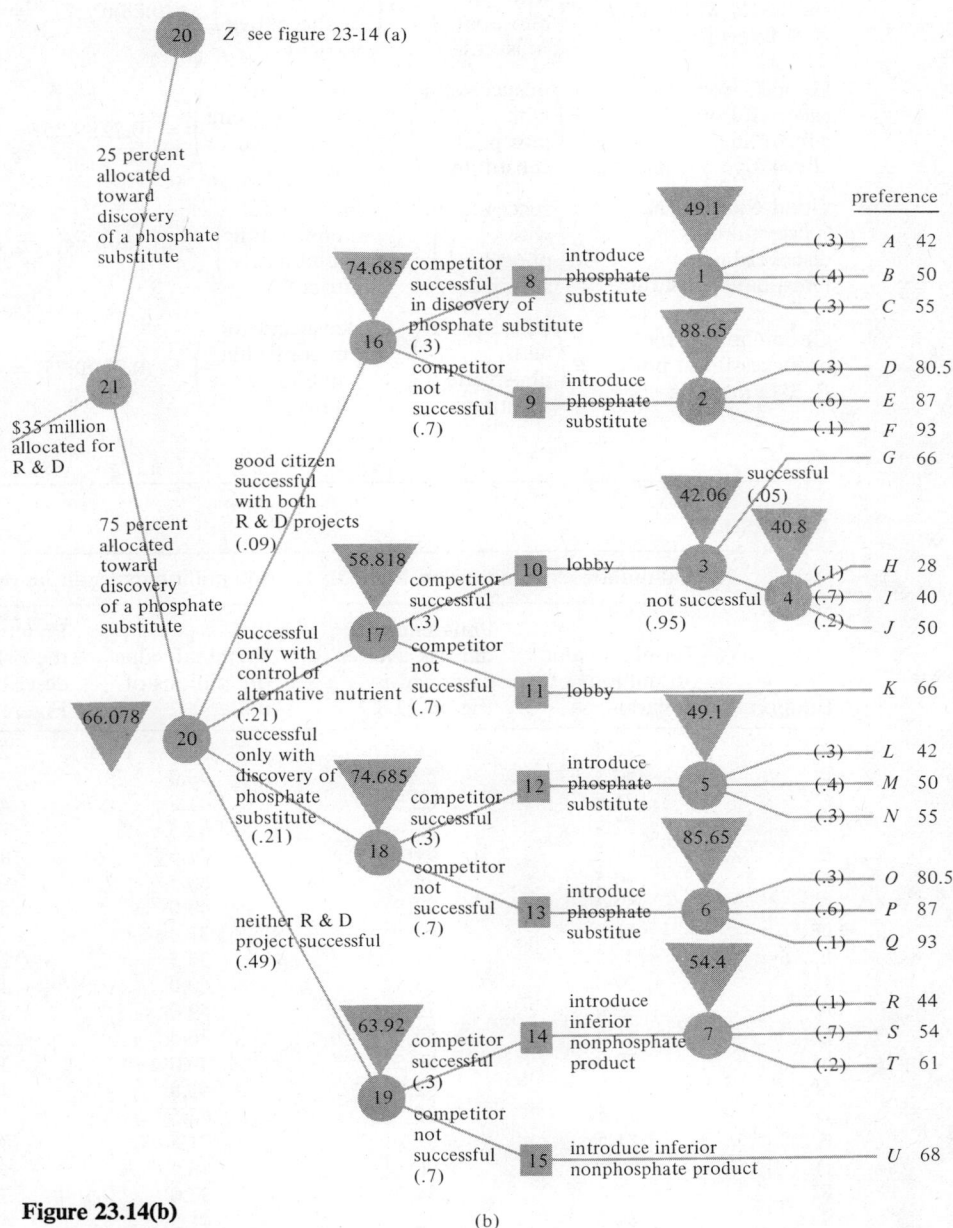

Figure 23.14(b)

(b)

Tables 23.14 and 23.15 develop the probabilities of success associated with a $40 million R & D expenditure. The terminal values and the associated preferences are recorded in Table 23.13. Figure 23.16 displays a completed analysis for the alternative, expend $40 million for R & D.

Table 23.12

$30 million allocated to R & D expenditure
75 percent or $22.5 million allocated to search for a phosphate substitute
25 percent or $7.5 million allocated to search for an alternative solution to eutrophication

Joint event	Probability of occurrence (see Tables 23.1 and 23.2)		
Good Citizen, Inc., successful with both R & D projects	$P\left(\begin{array}{c}\text{successful}\\\text{with}\\\text{phosphate}\\\text{substitute}\end{array}\right) \cdot P\left(\begin{array}{c}\text{successful}\\\text{in controlling}\\\text{an alternative}\\\text{nutrient}\end{array}\right)$	$= (0.28)(0.27)$	$= 0.07$
Good Citizen, Inc., successful only in controlling an alternative nutrient	$P\left(\begin{array}{c}\text{unsuccessful}\\\text{with}\\\text{phosphate}\\\text{substitute}\end{array}\right) \cdot P\left(\begin{array}{c}\text{successful}\\\text{in controlling}\\\text{an alternative}\\\text{nutrient}\end{array}\right)$	$= (0.72)(0.25)$	$= 0.18$
Good Citizen, Inc., successful only in discovering a phosphate substitute	$P\left(\begin{array}{c}\text{successful}\\\text{with}\\\text{phosphate}\\\text{substitute}\end{array}\right) \cdot P\left(\begin{array}{c}\text{unsuccessful}\\\text{in controlling}\\\text{an alternative}\\\text{nutrient}\end{array}\right)$	$= (0.28)(0.75)$	$= 0.21$
Good Citizen, Inc., unsuccessful in both R & D projects	$P\left(\begin{array}{c}\text{unsuccessful}\\\text{with}\\\text{phosphate}\\\text{substitute}\end{array}\right) \cdot P\left(\begin{array}{c}\text{unsuccessful}\\\text{in controlling}\\\text{an alternative}\\\text{nutrient}\end{array}\right)$	$= (0.72)(0.75)$	$= 0.54$

1.0000

Table 23.13

Endpoint	$30 million expenditure on R & D		$40 million expenditure on R & D	
	Terminal value[a] (in millions of dollars)	Preference for the risk averter described in Figure 23.8	Terminal value[a] (in millions of dollars)	Preference for the risk averter described in Figure 23.8
A	46.4	54	36.4	29
B	50.0	59.5	40.0	39
C	51.8	62.5	41.8	43
D	72.5	85	62.5	76
E	81.5	91	71.5	84
F	90.5	96	80.5	90
G	59.0	72.5	49.0	58.5
H	41.0	41	31.0	7
I	45.5	51	35.5	27
J	50.0	59.5	40.0	39
K	59.0	72.5	49.0	58.5
L	46.4	54	36.4	29
M	50.0	59.5	40.0	38
N	51.8	62.5	41.8	43
O	72.5	85	62.5	76
P	81.5	91	71.5	84
Q	90.5	96	80.5	90
R	47.0	54.5	37.0	31
S	51.5	62	41.5	42
T	56.0	68	46.0	53
U	60.5	73	50.5	60

[a] The terminal value is increased by $5 million for each outcome associated with an R & D expenditure of $30 million. For each outcome associated with an R & D expenditure of $40 million, the terminal value is decreased by $5 million.

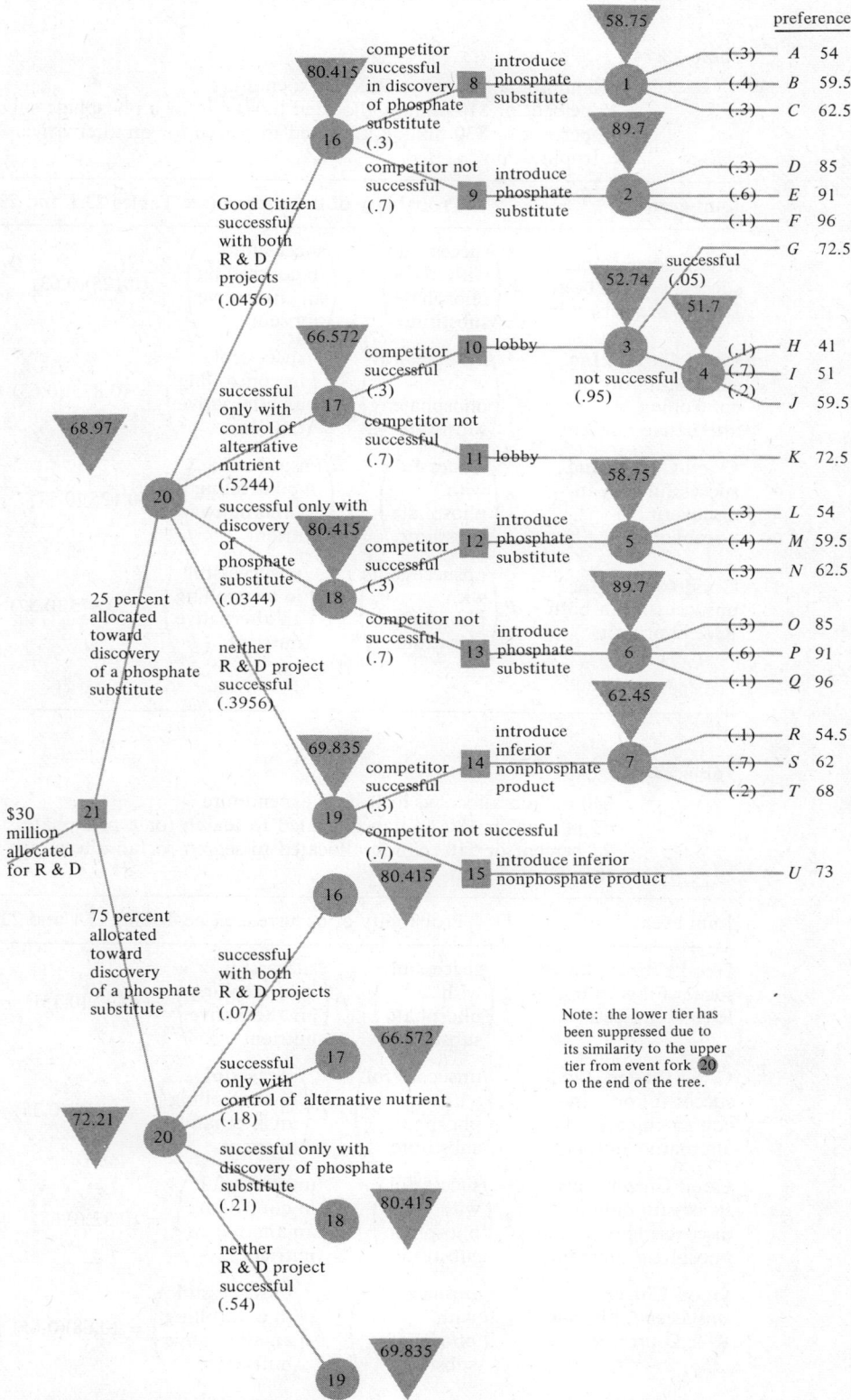

Figure 23.15 Partial decision tree: $30 million expenditure on R & D.

Table 23.14

$40 million allocated to R & D expenditure
25 percent or $10 million allocated to search for a phosphate substitute
75 percent or $30 million allocated to search for an alternative solution to eutrophication

Joint event		Probability of occurrence (see Tables 23.1 and 23.2)
Good Citizen, Inc., successful with both R & D projects	$P\begin{pmatrix}\text{successful}\\\text{with}\\\text{phosphate}\\\text{substitute}\end{pmatrix} \cdot P\begin{pmatrix}\text{successful}\\\text{in controlling}\\\text{an alternative}\\\text{nutrient}\end{pmatrix}$	$= (0.125)(0.63) = 0.07875$
Good Citizen, Inc., successful only in controlling an alternative nutrient	$P\begin{pmatrix}\text{unsuccessful}\\\text{with}\\\text{phosphate}\\\text{substitute}\end{pmatrix} \cdot P\begin{pmatrix}\text{successful}\\\text{in controlling}\\\text{an alternative}\\\text{nutrient}\end{pmatrix}$	$= (0.875)(0.63) = 0.55125$
Good Citizen, Inc., successful only in discovering a phosphate substitute	$P\begin{pmatrix}\text{successful}\\\text{with}\\\text{phosphate}\\\text{substitute}\end{pmatrix} \cdot P\begin{pmatrix}\text{unsuccessful}\\\text{in controlling}\\\text{an alternative}\\\text{nutrient}\end{pmatrix}$	$= (0.125)(0.37) = 0.04625$
Good Citizen, Inc., unsuccessful in both R & D projects	$P\begin{pmatrix}\text{unsuccessful}\\\text{with}\\\text{phosphate}\\\text{substitute}\end{pmatrix} \cdot P\begin{pmatrix}\text{unsuccessful}\\\text{in controlling}\\\text{an alternative}\\\text{nutrient}\end{pmatrix}$	$= (0.875)(0.37) = 0.32375$
		1.0000

Table 23.15

$40 million allocated to R & D expenditure
75 percent or $30 million allocated to search for a phosphate substitute
25 percent or $10 million allocated to search for an alternative solution to eutrophication

Joint event		Probability of occurrence (see Tables 23.1 and 23.2)
Good Citizen, Inc., successful with both R & D projects	$P\begin{pmatrix}\text{successful}\\\text{with}\\\text{phosphate}\\\text{substitute}\end{pmatrix} \cdot P\begin{pmatrix}\text{successful}\\\text{in controlling}\\\text{an alternative}\\\text{nutrient}\end{pmatrix}$	$= (0.32)(0.35) = 0.1120$
Good Citizen, Inc., successful only in controlling an alternative nutrient	$P\begin{pmatrix}\text{unsuccessful}\\\text{with}\\\text{phosphate}\\\text{substitute}\end{pmatrix} \cdot P\begin{pmatrix}\text{successful}\\\text{in controlling}\\\text{an alternative}\\\text{nutrient}\end{pmatrix}$	$= (0.68)(0.35) = 0.2380$
Good Citizen, Inc., successful only in discovering a phosphate substitute	$P\begin{pmatrix}\text{successful}\\\text{with}\\\text{phosphate}\\\text{substitute}\end{pmatrix} \cdot P\begin{pmatrix}\text{unsuccessful}\\\text{in controlling}\\\text{an alternative}\\\text{nutrient}\end{pmatrix}$	$= (0.32)(0.65) = 0.2080$
Good Citizen, Inc., unsuccessful in both R & D projects	$P\begin{pmatrix}\text{unsuccessful}\\\text{with}\\\text{phosphate}\\\text{substitute}\end{pmatrix} \cdot P\begin{pmatrix}\text{unsuccessful}\\\text{in controlling}\\\text{an alternative}\\\text{nutrient}\end{pmatrix}$	$= (0.68)(0.65) = 0.4420$
		1.0000

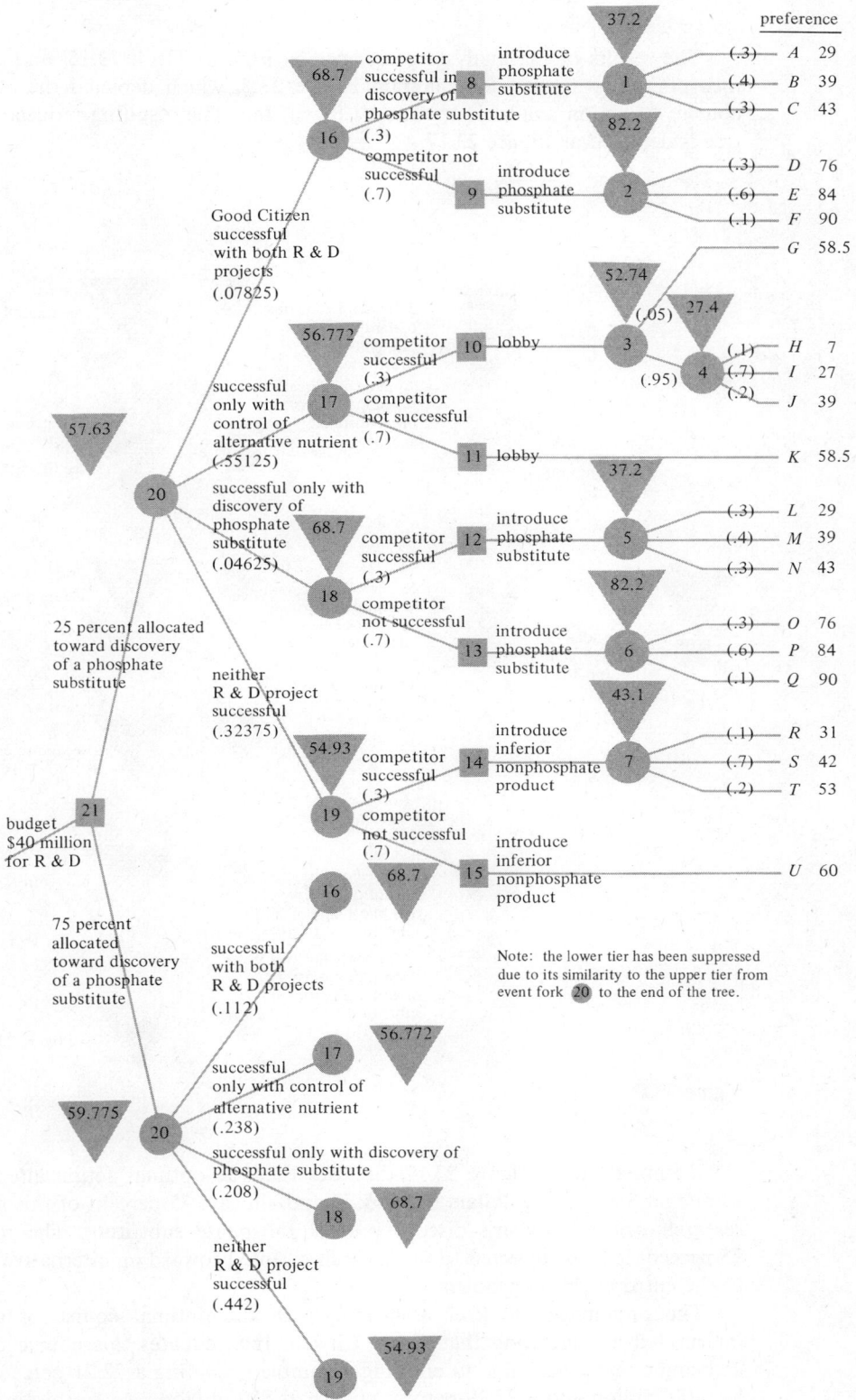

Figure 23.16 Partial decision tree: $40 million expenditure on R & D.

The results of the analyses performed in Figures 23.14, 23.15, and 23.16 are incorporated with decision diagram Figure 23.2, which depicted the alternative courses of action available to Good Citizen, Inc. The resulting reduced decision tree is depicted as Figure 23.17.

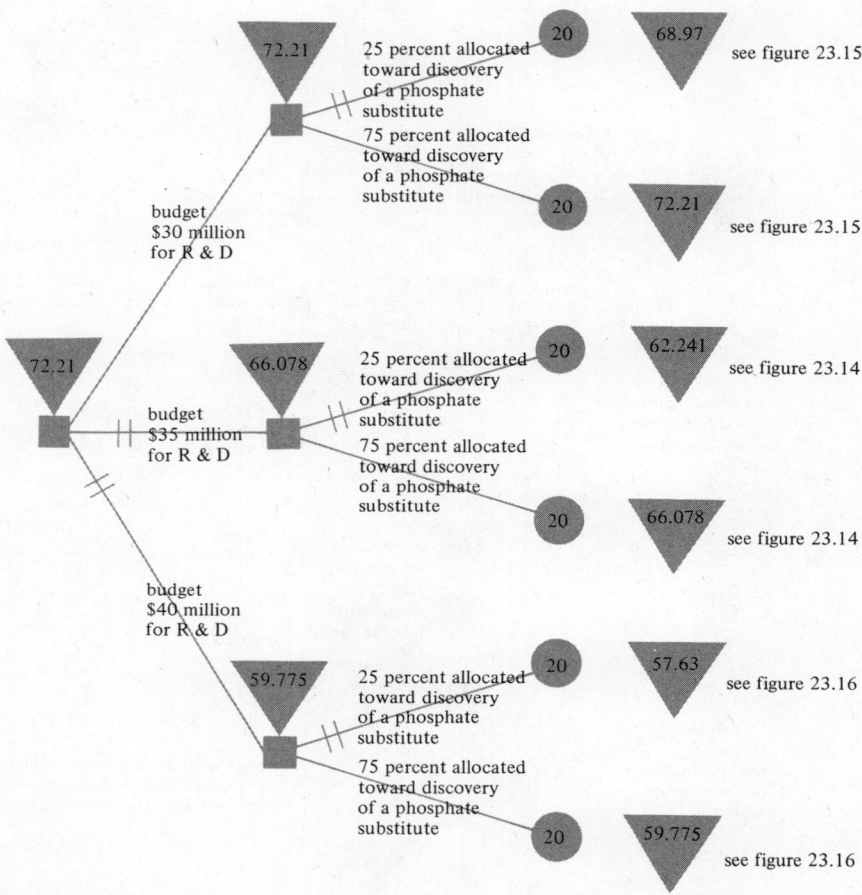

Figure 23.17

Examination of Figure 23.17 discloses that the optimal action alternative is to budget $30 million dollars for R & D and allocate 75 percent of this figure for research oriented toward discovery of a phosphate substitute. The remaining 25 percent is to be directed toward search oriented toward an alternative solution to the eutrophication problem.

The preference of 72.21 associated with the optimal course of action is interpreted as indicating that Good Citizen, Inc., equates this course of action as being equally desirable to entering a gamble providing a 72.21 percent chance at $100 million and a 27.79 percent chance at $30 million.

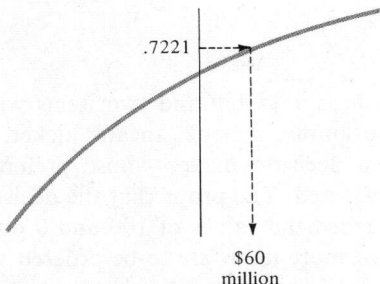

Referencing the preference curve discloses that the certainty equivalent associated with the optimal course of action is $60 million. The $60 million is interpreted as indicating that Good Citizen, Inc., is indifferent between pursuing the optimal course of action and some other alternative that would yield a guaranteed $60 million.

Decision tree analysis as an analytical technique enables the decision maker to separate out segments of a complicated, interrelated decision structure for individual attention. Creation of the decision tree structure itself requires determination of the act–event configurations likely to exert significant influence on the ultimate outcome of the immediate decision alternatives. When assigning probabilities to event forks, the decision maker need be concerned only with assessment of the likelihood of occurrence of a given event. It is essential that the implications associated with the occurrence of a given event be ignored when assessing probabilities. By the same token, when assigning preferences to the various consequences, the decision maker should be concerned only with the relative desirability of the consequences and should ignore the likelihood that any given event will occur. Upon completion of the above tasks, the technique of backward induction is utilized to fold back the decision tree through a series of simple decisions. Decision tree analysis is somewhat analogous to the approach utilized in completing a jigsaw puzzle—a complex task is accomplished one piece at a time.

Very complicated decision structures can be handled by this technique. Sophisticated extensions and insights into the successful application of this technique in complex decision structures is left to more advanced texts. The interested reader is referred to the following sources.

References

1. Schlaifer, Robert, *Analysis of Decisions Under Uncertainty*, New York: McGraw-Hill, 1969.
2. Raiffa, Howard, *Decision Analysis, Introductory Lectures on Choices Under Uncertainty*, Reading, Mass.: Addison-Wesley, 1968.
3. Hammond, John S., III, "Better Decisions with Preference Theory," *Harvard Business Review*, Nov.–Dec. 1967.
4. Brown, Rex. V., "Do Managers Find Decision Theory Useful?," *Harvard Business Review*, May–June 1970.

PROBLEMS

Problem 23.1

Locate three or four simple props, such as a $5 bill and two items whose retail value is of the same magnitude (for example, a book, theater ticket, penknife, or bottle of spirits). Then designate a decision maker whose preference scale (utility scale) over these items will be assessed. The props that the decision maker desires "most" and "least" will be assigned the values of 100 and 0 on the preference scale. The remaining one, two, or more items are to be ordered within this range from "best" to "worst." Finally, estimate the preference value that should be assigned to each of these items.

Problem 23.2

Bob Smith is faced with a gamble as shown below:

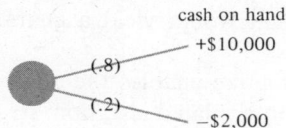

He is absolutely certain that the preference curve provided in Figure 23.18 is a correct depiction of his feelings about the relative desirability of the consequences measured on the horizontal axis. He claims that the preference curve indicates that his certainty equivalent to the gamble is $7600. Is this conclusion consistent with his preference curve? Not being particularly proficient at preference theory, he is not sure whether he has used the procedure correctly. What was his source of error, if any?

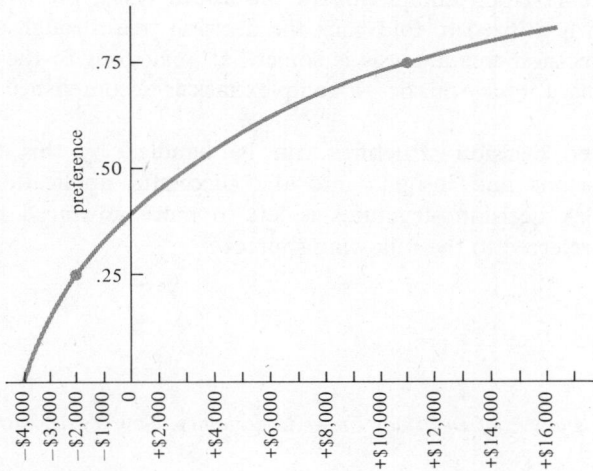

Figure 23.18

Problem 23.3

Demonstrate that the two event forks reproduced below are equivalent in the eyes of the decision maker whose preference curve is displayed in Figure 23.18 (within the error introduced by reading the preference curve).

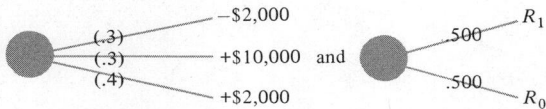

Problem 23.4

Intuitively justify the characteristic of decreasing risk aversion for a preference curve.

Problem 23.5

Under what circumstances is it appropriate to evaluate an individual's decision diagram employing values obtained from the individual's preference curve as opposed to utilizing the monetary values themselves? Is such a procedure in the *firm's* best interest? Discuss.

Problem 23.6

Describe the procedure you would employ to construct your personal preference curve between the two reference consequences $+\$10,000$ and $\$0$.

Problem 23.7

Given the preference curve provided in Figure 23.19 and the decision tree reproduced below, determine the answer to the following questions:
1. What is the decision maker's *preference* for act 3?
2. What is the decision maker's *certainty equivalent* for act 3?
3. What is the optimal strategy for the decision maker? (A strategy refers to a sequence of actions.)

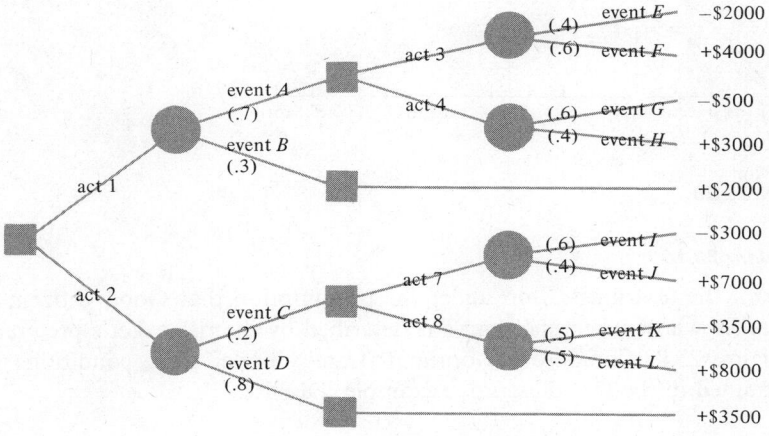

Problem 23.8

Under what circumstances is it legitimate to assume that an individual will be an averages player?

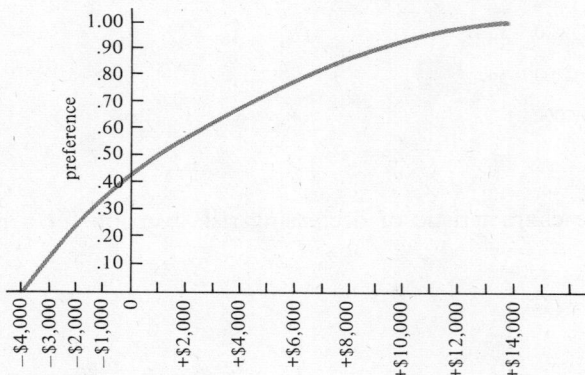

Figure 23.19

Problem 23.9

Presume that the preference curve provided in Figure 23.20 provides a correct description of Spaceonics Corporation's attitude toward the dollar consequences over the range −$100,000 to +$250,000.

1. Rework the Spaceonics case (case 3, Chapter 22) using the preference curve provided in Figure 23.20.
2. Referring to the analysis in part 1, discuss the implications of Spaceonics' attitude toward risk in terms of its influence on the selection of an optimal strategy.

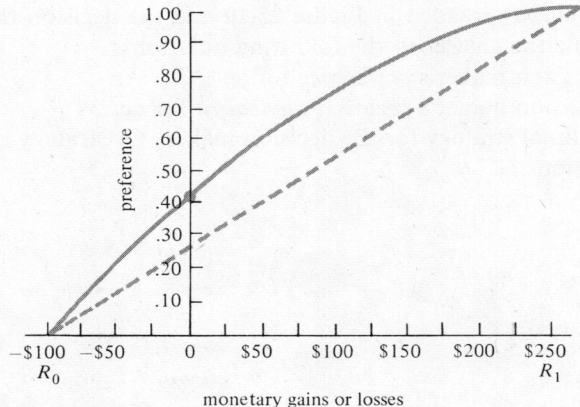

Figure 23.20

Problem 23.10

Rework the text illustration under the presumption that Good Citizen, Inc., is a risk taker whose preference curve is described by the risk taker's preference curve in Figure 23.8. Compare the optimal strategy under these conditions with that determined in the text illustrative example. Discuss.

Problem 23.11

Would it have been legitimate to have performed the analysis in the Good Citizen, Inc., case on the presumption that Good Citizen, Inc., was an averages player? Discuss.

Problem 23.12

The techniques described in Chapter 23 are utilized in the Edgartown Fisheries case, 9-170-001 EA-C 807R. Copies of this case may be obtained by writing to: Intercollegiate Case Clearing House, Harvard Business School, Soldiers Field Road, Boston, Massachusetts 02163.

Problem 23.13

The techniques described in Chapter 23 are illustrated in the case entitled "Mark Johnson," ICH 13C38 EA-C 736. This is an illustrative case describing the procedure an individual employed in developing his personal preference curve. Copies of this case may be obtained by writing the Intercollegiate Case Clearing House (see Problem 23.12).

TIME SERIES ANALYSIS: AN INTRODUCTION

24

A prime managerial responsibility is the design and
implementation of a strategy for the achievement of the
short-term and long-term objectives of the firm. Past
performance must be evaluated, and a forecast or
projection generated of future business activity.
Given a projection of the pattern and level of
future business activity, the desirability of alternative
strategies can then be investigated. Factors investigated
include the ability to meet projected sales activity levels
and the ability to maintain adequate but not excessive
inventory levels. Labor and material requirements
must be projected. Working capital needs must be
anticipated, and appropriate arrangements for
financing investigated. The suitability and timing of
capital expenditure projects must be evaluated. Finally,
once a strategy has been selected, control procedures
must be instituted to enable the firm to continually

reassess the validity of the original projection and the extent to which actual results correspond to expectations. If a departure from expectations occurs, appropriate corrective action must be undertaken.

Generation of a description of the pattern that has prevailed in the past is a useful first step in projecting the pattern of future activity. Information as to the pattern of past activity is typically recorded in the form of a time series. A time series consists of a set of chronological observations on the varying value of a statistical series recorded either at successive points in time or over successive periods of time.

Point data refers to the value of a stock variable recorded as of successive points in time. Typical of data of this type is the information recorded on a balance sheet:

Accounts receivable as of 12/31/71
Finished goods inventory as of 12/31/70
Retained earnings as of 6/30/71

Period data refers to the accumulated value of a flow variable recorded over successive periods of time. Typical of data of this type is the information recorded on an income statement:

Sales for the month of July 1970
Travel expense for the year 1971
Net income for the first quarter 1971

The behavior of a time series is a reflection of a myriad set of influences operating on the variable in question. These influences have traditionally been treated as subject to decomposition into the following four broad categories:

1. Secular trend
2. Seasonal variation
3. Cyclical fluctuations
4. Irregular influences

24.1 SECULAR TREND

Certain of the forces operating on a time series exert a consistently uplifting influence or a consistently depressing influence on the behavior of the series over time. Such forces tend to dominate the long-run underlying pattern of the series. Changes in technology, population growth, changes in the standard of living, and basic changes in consumer tastes are typical of the influences operating in this fashion. The nature of the influence exerted by such forces is captured by the trend component. The trend component describes the long-run pattern of growth and/or decline in the time series. Some typical secular trend patterns are depicted at the top of the following page.

24.2 SEASONAL VARIATION

Certain influences operating on the time series exert uplifting influences on certain occasions and depressing influences on other occasions in a repetitive, predictable pattern. This recurring pattern, which appears in a time series at given stated intervals and which tends to be constant with respect to both timing and amplitude, is referred to as seasonal variation. Constant with respect to timing means that

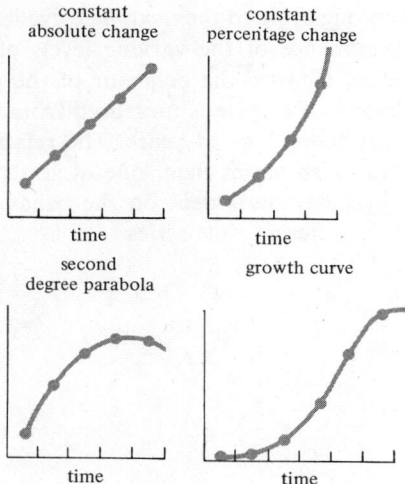

uplifting or depressing influences on the time series occur with approximately the same timing within each successive repetition of the seasonal pattern. Constant with respect to amplitude refers to the fact that the *percentage* uplift or *percentage* decline exerted on the time series at various stages of the seasonal pattern tends to remain the same for successive repetitions of the seasonal pattern. Most seasonal patterns are a reflection of underlying climatic conditions or customs:

> Beer consumption tends to be higher in the summer months.
> Suntan lotion sales are concentrated in the late spring and summer.
> Retail stores have heavy buying seasons around Easter, Christmas, and "back-to-school" time.
> Heating fuel sales peak in January and February and are negligible during the summer months.

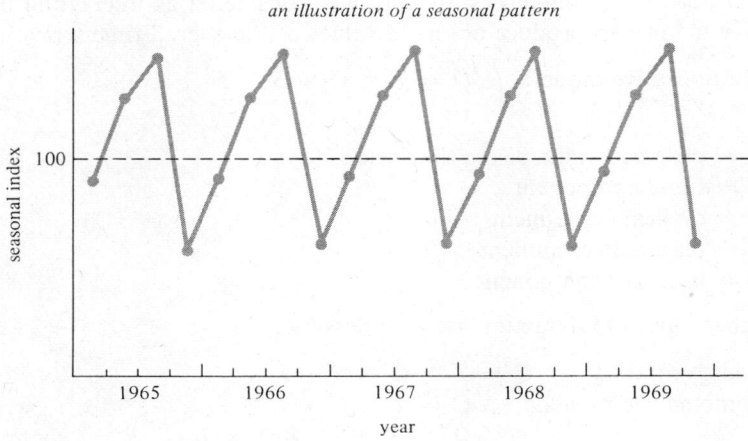

an illustration of a seasonal pattern

24.3 CYCLICAL FLUCTUATIONS

Cyclical fluctuations also refer to a recurring pattern that influences the behavior of the overall time series. The cyclical pattern differs from the seasonal, however, in that it is neither regular with respect to the duration of the pattern nor with respect to the timing or amplitude from one cycle to the next. The cyclical pattern for any given time series reflects the influence of the varying levels of business activity (expansion, prosperity, recession, etc.) on the behavior of the particular time series under investigation. The length of a cycle is measured from trough to trough. Cycle length has tended to vary from 1 to 14 years. The relative length of the stages (expansion, recession, etc.) also varies from one cycle to the next. The strength of the influence of the cyclical component on the behavior of the overall time series varies significantly for different time series.

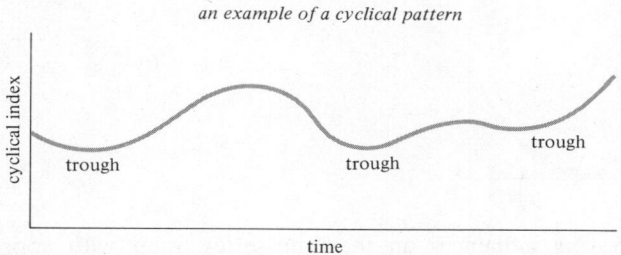

an example of a cyclical pattern

24.4 IRREGULAR INFLUENCES

Irregular influences fall into two categories:

1. Random influences: Irregular random influences refer to the myriad of minute environmental influences (some uplifting, some depressing) operating on a series at any one time—no one of which is significantly important in and of itself to warrant singling out for individual treatment.
2. The second category of irregular influence refers to nonrecurring influences that exert a significant one-time impact on the behavior of a time series and as such must be explicitly recognized. Included in this category are such events as floods, strikes, wars, and so on.

The above four influences are typically considered as interacting in a multiplicative manner to produce observed values of the overall time series.[1]

Multiplicative model: $\quad O = T \times C \times S \times I$

where

O = observed value of the time series
T = trend component
C = cyclical component
S = seasonal component
I = irregular component

Other types of interaction are also possible.

Additive model: $\quad O = T + C + S + I$
Combination models: $\quad O = T \times C \times S + I$
$\qquad\qquad\qquad\quad O = (T + C) \times S \times I$

and so on. Discussion in this text is restricted to the multiplicative model.

---

[1] There is no seasonal influence when dealing with annual data. Thus, for annual data, the multiplicative model becomes $O = T \times C \times I$.

Rather than examining the overall behavior of a time series, it is possible and frequently desirable to decompose the time series into its component elements for separate examination. This process is referred to as decomposition of the time series or time series analysis. Chapters 25 and 26 are devoted to an examination of these components. Secular trend is treated in Chapter 25. Seasonal variation, cyclical fluctuations, and irregular influences are examined in Chapter 26.

PROBLEMS

Problem 24.1

What is a time series?

Problem 24.2

Differentiate between point data and period data.

Problem 24.3

Time series are traditionally treated as subject to decomposition into four major components. What are these components? What are the distinctive characteristics of each of the components?

Problem 24.4

Seasonal variation refers to a pattern that is constant with respect to timing and amplitude. Explain what "constant with respect to timing and amplitude" means.

Problem 24.5

What are the two categories of irregular influence?

TIME SERIES ANALYSIS: SECULAR TREND

A description of secular trend is desirable for the following reasons:

1. To provide a description of the basic underlying pattern of behavior that has characterized the series in the past.

2. To provide a basis for extension of the basic underlying pattern of behavior of the past into the future. This procedure is called extrapolation. Extrapolation of secular trend reflects an implicit judgment that the underlying forces generating the past behavior of the series will continue to be operative in the same fashion in the future. If such an assumption is not legitimate, the analyst must evaluate the impact of the changes and attempt to predict the new pattern likely to emerge.

3. For purposes of elimination. By removing the

619

influence of trend from the overall time series, the behavior pattern of other components such as seasonal or cyclical are more readily discernable.

The dominant underlying trend influence in a time series is typically characterized by one of the following patterns:

1. Arithmetic straight line: An arithmetic straight-line trend line is characterized by constant *absolute* increase (or decrease) in the value of the trend component with the passage of each successive unit of time. Plotted on an arithmetic chart, it appears as a straight line.

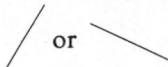

2. Logarithmic straight line: A logarithmic straight-line trend line is characterized by constant *percentage* increase (or decrease) in the value of the trend component with the passage of each successive unit of time. Plotted on an arithmetic chart, a logarithmic straight line is characterized by the following pattern:

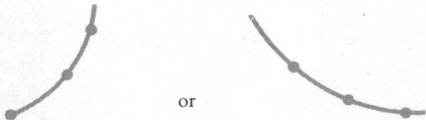

This is due to the fact that a constant percentage increase is associated with an increasing absolute increase with the passage of each successive unit of time. Plotted on a semilogarithmic chart, the logarithmic straight line appears as a straight line. Semilogarithmic charts are discussed in section 25.4.

3. Second-degree parabola: A second-degree parabola trend line can reflect a growth pattern that is increasing by increasing amounts, increasing by decreasing amounts, decreasing by increasing amounts, or decreasing by decreasing amounts. Moreover, the second-degree parabola is capable of a turning point. It can either reach a peak and turn down or reach a trough and turn up (one or the other). If two turning points appear appropriate for a given set of data, a cubic equation (a third-degree equation as opposed to a second-degree equation) should be fitted to the data. Typical patterns for second-degree parabolas plotted on an arithmetic chart are depicted below:

4. Growth curve: The growth curves reflect the pattern of growth of biological organisms. The long-run pattern of activity of many industries has tended to be characterized by such a pattern. Growth curves are characterized by slow initial growth, a period of rapid expansion, and a subsequent leveling off as market saturation takes place. A curve of this type, the Gompertz curve, is examined in section 25.6. Two general distinguishing characteristics of Gompertz curves are that the relative rate of growth declines at a constant rate, and the curve does not possess a turning point, asymptotically approaching an upper limit.

When plotted on an arithmetic chart, the Gompertz curve takes the shape of an elongated S:

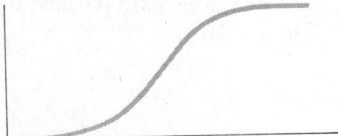

When plotted on a semilogarithmic chart, the Gompertz curve is concave downward:

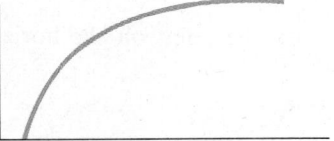

25.1 SELECTION OF THE LINE OF BEST FIT

The first step in describing secular trend is a determination of the basic pattern best reflecting the underlying long-run pattern in the series under investigation. This is accomplished by plotting the actual observations for the series on a chart. A visual judgment is then made on the basis of the observed pattern. It is suggested that a minimum of 20 years' observations be employed. To reduce the computational load and to focus attention on the underlying procedure, the illustrative example is restricted to 11 observations.

Table 25.1 Shipments of Naughty Pine (billions of board feet) 1952–1962

Year	Shipments
1952	10.15
1953	9.49
1954	9.41
1955	9.54
1956	8.74
1957	8.00
1958	8.44
1959	8.81
1960	8.29
1961	7.68
1962	7.98

The time series in Table 25.1 is plotted on an arithmetic chart (Figure 25.1). Shipments of Naughty pine is period data. Note the following conventions. The label for a period is placed below the middle of the period to which it refers. The value of the series for a period is plotted above the middle of the period to which it refers.[1]

---

[1] When plotting point data, the value of the variable is plotted directly above the *point* in time to which it refers.

25.2 ARITHMETIC LEAST-SQUARES LINE

Examination of Figure 25.1 discloses that an arithmetic straight line appears to provide an appropriate description of the long-run underlying pattern in the data. In general, the formula for a straight line takes the form

$$Y_c + a = bx$$

where

Y_c is the height of the line above the point on the horizontal axis associated with a given value of x.

a is the height of the line associated with the point on the horizontal axis assigned $x = 0$.

$$Y_c = a + bx$$
$$Y_c = a + b(0)$$
$$Y_c = a$$

b is the slope of the line or the change in the height of the line with a one-unit increase in the value of x. If b is positive, the height of the line rises as the value of x is increased. If b is negative, the height of the line declines with successive increases in the value of x.

x serves to locate points on the horizontal axis with reference to the distance and direction they deviate from the point on the horizontal axis associated with x equal to zero.

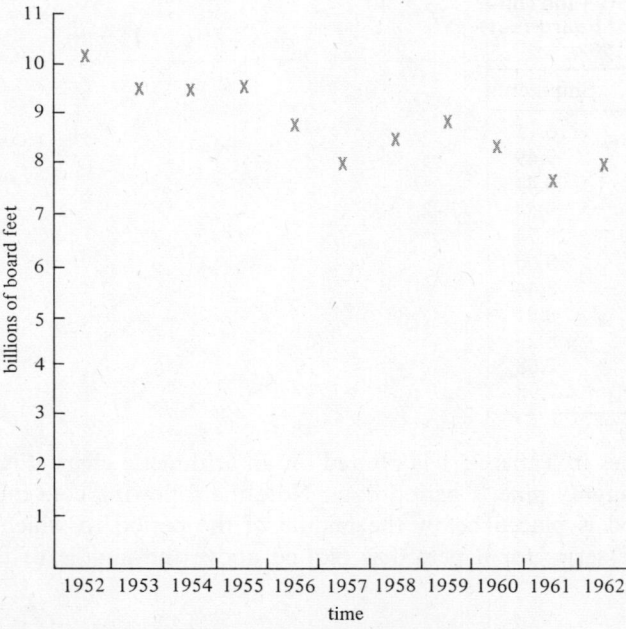

Figure 25.1 Shipments of Naughty Pine: 1952–1962.

$Y_c = a + bx$ is the general formula for any straight line. To identify a particular straight line, it is necessary to specify particular numerical values for a and b.

Previously, it was indicated that an arithmetic straight line would be fitted to the data. The problem now arises: How does one determine which of all the possible arithmetic straight lines is the one of best fit? To resolve this problem, it is necessary to establish a criterion that will indicate one of the lines as being preferable (more representative than the others in some sense). The criterion of best fit typically employed is the least-squares criterion.[2]

25.3 THE CONCEPT OF LEAST SQUARES

For each x value in the time series, there is (1) an observed value for the time series described by the symbol Y, and (2) a calculated value provided by the straight line fit to the data designated by the symbol Y_c. Some representative pairs of Y and Y_c values are plotted in Figure 25.2.

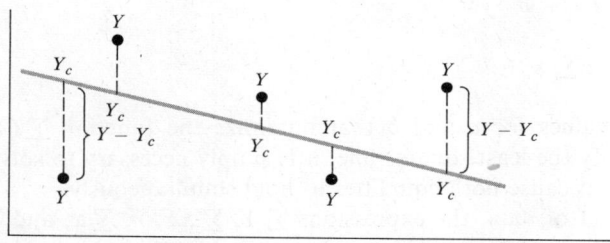

Figure 25.2

If the line is the arithmetic least-squares straight line, it possesses the following characteristics:

1. If positive signs are assigned to the deviations of the observed values of Y above the line and negative signs are assigned to the deviations of the observed values of Y lying below the line, the sum of the deviations will be equal to zero. That is,

$$\sum (Y - Y_c) = 0$$

2. If the deviations are squared and summed, the sum of the squared deviations from the least-squares line is less than the sum of the squared deviations would be if the procedure were to be repeated for any other line *of that type* fitted to the same set of data. That is,

$$\sum (Y - Y_c)^2 \text{ is at a minimum for lines of that type.}$$

---

[2] Other criteria of best fit are available. A simple graphic freehand fit is occasionally used. In such a case, the criterion of best fit is the subjective judgment of the individual fitting the freehand curve. As such, the curve fit for any given set of data would vary from one individual to the next. The least-squares criterion is an objective criterion in that anyone using this criterion to select an arithmetic straight line of best fit for a given set of data would identify the same line.

It is important to recognize that the least-squares criterion requires that the type of line desired be specified. Once this has been done, the technique generates the line of best fit of that type, that is, the least-squares line. Thus, for any given set of data, one could generate: (1) an arithmetic least-squares straight line, (2) a logarithmic least-squares straight line, and (3) a least-squares second-degree parabola. If one were to compute each of the above and determine which of the three has the lowest sum of the squared deviations, the least-squares criterion could also be used to indicate which *type* of line provides the best fit.

The least-squares criterion makes sense. The question is: How to implement it? One approach would be to construct all possible straight lines, compute the sum of the squared deviations from each, and select the line associated with the minimum sum as the line of best fit. Obviously, this approach is impractical. Fortunately, a system of simultaneous equations can be developed by differential calculus that allows the direct computation of the a and b coefficients identifying the least-squares arithmetic straight line for any given set of data. The equations, called the *normal equations*, are reproduced below[3]:

I $\qquad \sum Y = na + b \sum x$

II $\qquad \sum xY = a \sum x + b \sum x^2$

To determine the values for a and b that minimize the function $\sum (Y - Y_c)^2$ and therefore identify the least-squares line, it is simply necessary to solve for the values of a and b that cause both equalities to hold simultaneously.

For any given set of data, the expressions $\sum Y$, $\sum xY$, n, $\sum x$, and $\sum x^2$ are constants and can be determined. These values are generated for the data in the illustrative example in Table 25.2. $x = 0$ can be arbitrarily assigned to any of the

Table 25.2

	Y	x	x^2	xY
1952	10.15	0	0	0
1953	9.49	1	1	9.49
1954	9.41	2	4	18.82
1955	9.54	3	9	28.62
1956	8.74	4	16	34.96
1957	8.00	5	25	40.00
1958	8.44	6	36	50.64
1959	8.81	7	49	61.67
1960	8.29	8	64	66.32
1961	7.68	9	81	69.12
1962	7.98	10	100	79.80
	96.53	55	385	459.44

years. For all other years, the x value assigned serves to indicate the number of units (years in this case) that the middle of the associated year departs from the

[3] The derivation of the normal equations is provided in Technical Note 13 at the end of the chapter. The approach is intuitive and does not require a preliminary knowledge of differential calculus.

point in time assigned $x = 0$. In Table 25.2, 1952 has arbitrarily been selected as the year for which $x = 0$. The technique described is applicable, however, regardless of the year selected.

$$\text{I} \qquad \sum Y = na + b \sum x$$

$$\text{II} \qquad \sum xY = a \sum x + b \sum x^2$$

Substituting, we obtain two equations with two unknowns:

$$\text{I} \qquad 96.53 = 11a + 55b$$

$$\text{II} \qquad 459.44 = 55a + 385b$$

Multiplying both sides of equation I by 5 maintains the equality and creates a new equation designated IA:

$$\text{IA} \qquad 482.65 = 55a + 275b$$

Subtracting equation IA from equation II, a new equation designated II-IA is obtained. Again, the equality is maintained, since subtracting 482.65 from the right-hand member of equation II is equivalent to subtracting $55a + 275b$ from the left-hand member of equation II.

$$\text{II-IA} \qquad -23.21 = 110b$$

$$b = \frac{110}{-23.21}$$

$$b = -0.211$$

Substituting $b = -0.211$ in equation I and solving for a, we obtain

$$\begin{aligned}
\text{I} \qquad 96.53 &= 11a + 55b \\
96.53 &= 11a + 55(-0.211) \\
96.53 &= 11a - 11.605 \\
108.135 &= 11a \\
a &= 9.8304
\end{aligned}$$

The coefficients for the least-squares line are thus

$$a = 9.8304$$
$$b = -0.211$$

To obtain the value of the trend line for any given year, one need simply substitute the value of x for that year in the equation $Y_c = 9.8304 - 0.211x$. The equation generates the corresponding value of Y_c.

To illustrate: the x value for 1957 $= +5$.

$$\begin{aligned}
Y_{c_{1957}} &= 9.8304 - 0.211(5) \\
&= 9.8304 - 1.055 \\
&= 8.7754
\end{aligned}$$

Actually, the computation can be simplified considerably by employing the convention of assigning the x value $= 0$ to the *middle* of the period for which trend is to be computed.

Year	x
1952	-5
1953	-4
1954	-3
1955	-2
1956	-1
1957	0
1958	$+1$
1959	$+2$
1960	$+3$
1961	$+4$
1962	$+5$
	$\sum x = 0$

The reduction in computational effort is attributable to the fact that this convention causes $\sum x$ to equal zero. When $\sum x$ equals zero, the normal equations simplify as follows:

I
$$\sum Y = na + b \sum x$$
$$\sum Y = na + b(0)$$
$$\sum Y = na$$

$$a = \frac{\sum Y}{n}$$

II
$$\sum xY = a \sum x + b \sum x^2$$
$$\sum xY = a(0) + b \sum x^2$$
$$\sum xY = b \sum x^2$$

$$b = \frac{\sum xY}{\sum x^2}$$

This format permits the direct computation of the values of a and b.

Illustrating this approach with the previous example:

Year	Y	x	x^2	xY
1952	10.15	-5	25	-50.75
1953	9.49	-4	16	-37.96
1954	9.41	-3	9	-28.23
1955	9.54	-2	4	-19.08
1956	8.74	-1	1	-8.74
1957	8.00	0	0	0
1958	8.44	$+1$	1	$+8.44$
1959	8.81	$+2$	4	$+17.62$
1960	8.29	$+3$	9	$+24.87$
1961	7.68	$+4$	16	$+30.72$
1962	7.98	$+5$	25	$+39.90$
	96.53			-144.76
				$+121.55$
				-23.21

$$a = \frac{\sum Y}{n}$$

$$= \frac{96.53}{11}$$

$$= 8.7754$$

$$b = \frac{\sum xY}{\sum x^2}$$

$$= \frac{-23.21}{110}$$

$$= -0.211$$

$$Y_c = 8.7754 - 0.211(x)$$

This line is identical to the line calculated by the simultaneous equation approach. The a values differ (8.7754 versus 9.8304) because of the change in the year to which $x = 0$ was assigned.

When $x = 0$ in 1952, $Y_{c_{1952}} = a = 9.8304$. Letting $x = 0$ in 1957, $Y_{c_{1952}} \neq a$, but still equals 9.8304.

$$Y_{c_{1952}} = 8.7754 - 0.211(-5)$$

$$= 8.7754 + 1.055$$

$$= 9.8304$$

Obviously, this is the same line.[4]

If the trend value is desired for all years, one can either successively insert the appropriate x values and solve the equation, or utilize the following short-cut technique:

First compute directly the first and last values in the series by the formula.

$$Y_{c_{1952}} = a + bx \qquad\qquad Y_{c_{1962}} = a + bx$$
$$\quad= 8.7754 - 0.211(-5) \qquad\qquad = 8.7754 - 0.211(+5)$$
$$\quad= 8.7754 + 1.055 \qquad\qquad = 8.7754 - 1.055$$
$$\quad= 9.8304 \qquad\qquad\qquad = 7.7204$$

Year	Y	x	Y_c
1952	10.15	-5	9.8304
1953	9.49	-4	
1954	9.41	-3	
1955	9.54	-2	
1956	8.74	-1	
1957	8.00	0	
1958	8.44	$+1$	
1959	8.81	$+2$	
1960	8.29	$+3$	
1961	7.68	$+4$	
1962	7.98	$+5$	7.7204

[4] To avoid misinterpretation, the equation for secular trend should be accompanied by a legend. To illustrate:

$Y_c = 8.7754 - 0.211x$ $\qquad$ $Y_c = 9.8304 - 0.211x$
$x = 0$; June 30, 1957 $\qquad$ $x = 0$; June 30, 1952
x in terms of years $\qquad\qquad$ x in terms of years
Y in billions of board feet $\qquad$ Y in billions of board feet

Given that the trend is linear, the annual change in the height of the line is equal to a constant absolute amount. To generate the next trend value, simply add or subtract the constant absolute amount (depending on the sign) to or from the preceding value. The trend value in the last year (computed directly) can be used to verify the correctness of the calculations.

Trend value for 1952 =	9.8304	
	−0.211	
Trend value for 1953	9.6194	
	−0.211	
Trend value for 1954	9.4084	
	−0.211	
Trend value for 1955	9.1974	
	−0.211	
Trend value for 1956	8.9864	
	−0.211	
Trend value for 1957	8.7754	
	−0.211	
Trend value for 1958	8.5644	
	−0.211	
Trend value for 1959	8.3534	
	−0.211	
Trend value for 1960	8.1424	
	−0.211	
Trend value for 1961	7.9314	
	−0.211	
Trend value for 1962	7.7204	

It can be demonstrated that $\sum (Y - Y_c) = 0$ for the line $Y_c = 8.7754 - 0.211x$.

Year	Y	Y_c	$Y - Y_c$
1952	10.15	9.8304	+0.3196
1953	9.49	9.6194	−0.1294
1954	9.41	9.4084	+0.0016
1955	9.54	9.1974	+0.3426
1956	8.74	8.9864	−0.2464
1957	8.00	8.7754	−0.7754
1958	8.44	8.5644	−0.1244
1959	8.81	8.3534	+0.4566
1960	8.29	8.1424	+0.1476
1961	7.68	7.9314	−0.2514
1962	7.98	7.7204	+0.2596
	96.53	96.5294	+1.5276
			−1.5270
			+0.0004 $\simeq$ 0 Rounding error

Note also that $Y = Y_c$. The above relationships serve as checks.

25.4 LOGARITHMIC LEAST-SQUARES STRAIGHT LINE

Comprehension of the logarithmic least-squares straight line is facilitated by an analysis of the procedure utilized to construct a one-cycle semilogarithmic chart. The procedure for construction of a semilogarithmic chart is described below.

First, it is necessary to delineate the area within which the log cycle is to be generated. In the illustrative example, the cycle will be constructed within the 15 arithmetic subdivisions calibrated on the vertical axis of Figure 25.3. The number of subdivisions selected could have been 10, 20, or any other number. Fifteen arithmetic subdivisions represents an arbitrary choice.

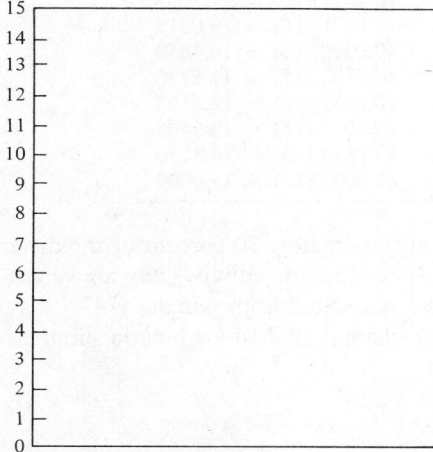

Figure 25.3

The number at the upper end of the vertical axis is assigned the log value 1.0000, which corresponds to the arithmetic number 10, and the number at the lower end of the vertical axis is assigned the log value 0.0000, which corresponds to the arithmetic number 1. The numbers 2, 3, 4, 5, 6, 7, 8, and 9 are then located on the vertical axis. They are not located in accordance with their arithmetic values but rather in accordance with the magnitude of their logarithms.

Logarithms of the numbers 1–10	
Number	Logarithm
1	0.0000
2	0.3010
3	0.4771
4	0.6021
5	0.6990
6	0.7782
7	0.8451
8	0.9031
9	0.9542
10	1.0000

As an aid to locating the numbers 1–10 on the vertical axis, consider the 15 arithmetic subdivisions as representing 100 percent of the scale and the logs of each of the numbers as representing the percentage of the distance above the base line that the line representing the associated arithmetic value should be drawn.

Arithmetic number	Logarithm of the number	Number of arithmetic subdivisions above the base line that the line representing the associated number should be located
1	0.0000	(0.0000)(15) = 0.0000
2	0.3010	(0.3010)(15) = 4.5150
3	0.4771	(0.4771)(15) = 7.1565
4	0.6021	(0.6021)(15) = 9.0315
5	0.6990	(0.6990)(15) = 10.4850
6	0.7782	(0.7782)(15) = 11.6730
7	0.8451	(0.8451)(15) = 12.6765
8	0.9031	(0.9031)(15) = 13.5465
9	0.9542	(0.9542)(15) = 14.3130
10	1.0000	(1.0000)(15) = 15.0000

The number 2 is thus located approximately 30 percent of the distance up the vertical axis or at approximately 4.5 arithmetic subdivisions above the base line. The line representing the number 3 is located approximately 47.7 percent of the distance up the vertical axis or approximately 7.16 arithmetic subdivisions above the baseline, and so on.

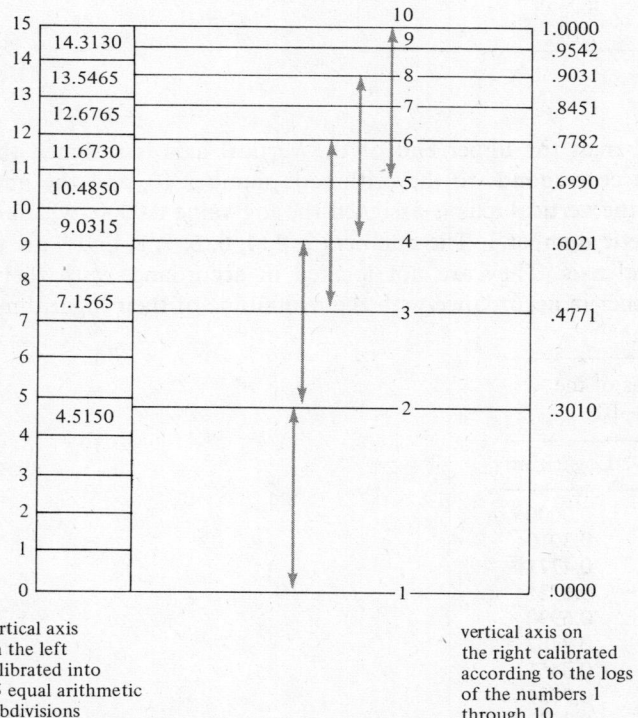

vertical axis on the left calibrated into 15 equal arithmetic subdivisions

vertical axis on the right calibrated according to the logs of the numbers 1 through 10

Figure 25.4 A one-cycle semilog scale superimposed on an arithmetic scale.

Figure 25.4 depicts a single-cycle semilog chart. A cycle is a segment of a vertical axis that has been calibrated according to the logarithmic relationship of the numbers 1–10.

Note that the vertical distance between the lines designated 2 and 1, the lines designated 4 and 2, the lines designated 6 and 3, the lines designated 8 and 4, and the lines designated 10 and 5 is identical. By exploring the underlying relationship that causes the equivalent vertical rise between each of the pairs of numbers listed above, the significance and interpretation of a semilog chart can be clearly demonstrated. The relationship between the pairs can be depicted as follows:

Second number in pair	Times	Constant multiple	=	First number in pair
1	×	2	=	2
2	×	2	=	4
3	×	2	=	6
4	×	2	=	8
5	×	2	=	10

Observe that in each case, the first number in the pair is twice the value of the second. Also note that the vertical distance between the numbers on the chart is identical for each pair. The above multiplications can also be expressed exponentially (to the base 10).

$$1 \times 2 = 10^{0.0000} \times 10^{0.3010} = 10^{0.3010} \quad \text{or} \quad 2$$
$$2 \times 2 = 10^{0.3010} \times 10^{0.3010} = 10^{0.6020} \quad \text{or} \quad 4$$
$$3 \times 2 = 10^{0.4771} \times 10^{0.3010} = 10^{0.7781} \quad \text{or} \quad 6$$
$$4 \times 2 = 10^{0.6021} \times 10^{0.3010} = 10^{0.9031} \quad \text{or} \quad 8$$
$$5 \times 2 = 10^{0.6990} \times 10^{0.3010} = 10^{1.0000} \quad \text{or} \quad 10$$

Note the following relationship: *Adding 0.3010 to the log of a number is equivalent to multiplying the original number by 2.* Stated in another way, adding 0.3010 to the log of a number always results in a 100 percent increase in the value of the number. *In general, adding a logarithmic amount to the log of a number is equivalent to multiplying the original number by the antilog of the logarithmic increment.*

The vertical distance between any two numbers on the graph will always be equal to the difference between their logs multiplied by the size of the scale within which the log cycle has been constructed. In general, a 100 percent increase is always depicted as a vertical rise equal to approximately 30 percent of the size of the scale within which the cycle has been constructed.

$$\log 27 = 1.4314$$
$$\text{add} \quad 0.3010$$
$$\overline{ 1.7324}$$

antilog 1.7324 is 54

In a similar fashion, adding 0.1761 to the log of a number is equivalent to a 50 percent increase. To demonstrate: Take any positive number greater than zero, proceed a distance up the vertical axis equal to 17.6 percent of the size of the scale assigned to the cycle, and the number corresponding to that location will be 50 percent greater than the original number. Similarly, it can be demonstrated that adding 0.1250 to the log of a number is equivalent to a 33 percent increase, and so on.

The above relationship also explains the increasing narrowness of the bands as one progresses up the cycle. The same arithmetic increment is equivalent to a smaller and smaller percentage increment as the size of the base increases. To illustrate:

From 1 to 2 is a 100 percent increase
From 2 to 3 is a 50 percent increase
From 3 to 4 is a 33 percent increase
From 4 to 5 is a 25 percent increase
From 5 to 6 is a 20 percent increase
    etc.

Although the vertical calibrations were determined on the basis of the differences between the logarithms of the numbers 1–10, the vertical axis can be used for any set of numbers as long as the numbers are related to each other in the same multiplicative fashion as the numbers 1–10. Thus, if the value at the bottom of a cycle is assigned the value 37, the label assigned to the line designated 2 must be 74 (2×37). The label assigned to the line designated 9 must be 333 (9×37), and the line designated as 10 must be labeled 370 (10×37). The difference between the logs of 370 and 333 is 0.0458.

log 370 = 2.5682
log 333 = 2.5224
———————
       0.0458

0.0458 indicates that approximately 4.6 percent of the vertical scale should lie between the value designated 370 and the value designated 333. The difference between the logs of 10 and 9 is also 0.0458.

log 10 = 1.0000
log 9 = 0.9542
———————
       0.0458

The explanation, of course, is that the percentage increase from 9 to 10 is identical to the percentage increase from 333 to 370. Thus, any set of labels can be attached to the vertical axis provided that their percentage relationship is the same as that for the numbers 1–10. The actual labels selected depend on the requirements established by the data. For example, if the data ranged from a low value of 47 to a high value of 382, a logical assignment might be a value of 40 to the bottom line in the cycle and a value of 400 to the top of the cycle. The value assigned to the top of the cycle must always be ten times the value assigned to the bottom of the cycle.

Suppose that the data ranged from a low of 72 to a high of 1024. No labeling system exists that would allow the data to be plotted within a single cycle, since the largest value is more than ten times the smallest. The solution is to use a two-cycle chart. A two-cycle chart consists of two cycles. The top line of the lower cycle serves as the bottom line of the upper cycle.

If the bottom line of the first cycle is labeled 1, the top line of the first cycle (which is also the bottom line of the second cycle) must be labeled 10, and the top line of the second cycle must be labeled 100. For data ranging from 72 to 1024, the values 20, 200, and 2000 might be assigned to the bottom of the first cycle, top of the first cycle, and top of the second cycle, respectively. Since the scale is based on percentage relationships, once the value has been established for the bottom line of the first cycle, all other labels are automatically determined. Additional cycles can be added if required. Semilog paper is available commercially. Thus, all the analyst typically need do is assign labels to the vertical axis consistent with the data to be plotted. Observations are plotted on a semilog chart in accordance with their arithmetic values as recorded on the vertical axis. The vertical scale is calibrated in such a fashion that the mere act of plotting converts the observations to their logarithmic relationship.

To illustrate: Consider the following four observations (period data):

1960	1500 lb
1961	3000 lb
1962	6000 lb
1963	12000 lb

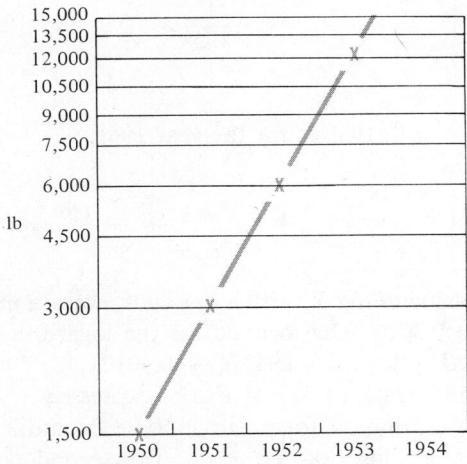

Figure 25.5

The four observations are plotted on a semilog chart in Figure 25.5. Note that when plotted on a semilog chart, the four values form a straight line. This is attributable to the fact that the vertical scale is calibrated in such a fashion that the plotted points reflect the logarithmic relationship of the numbers rather than their arithmetic relationship and that the values reflect a constant percentage increase with the passage of time. Any straight line can be described by the general formula $Y_c = a + bx$. In this case, it is the logarithms of the numbers that are linearly related. Accordingly, the line is described as follows:

$$\log Y_c = a + bx$$

where

> x once again serves to indicate the distance and direction the associated point in time departs from the point in time assigned the value $x = 0$.
>
> $\log Y_c$ is the logarithm of the trend-line value above the point in time indicated by the designated value of x. When $x = 0$, the formula becomes

$$\log Y_c = a + b(0)$$
$$\log Y_c = a$$

> a therefore is defined as the value of $\log Y_c$ (or the height of the line in log form) above the point in time represented by the value $x = 0$.
>
> b represents the change in the height of the line (change in the value of $\log Y_c$) with a one-unit increase in the value of x. In this case, $b = 0.3010$. Recall that adding a constant logarithmic increment is equivalent to continually multiplying by the antilog of that value. The antilog of 0.3010 is 2.

To describe the trend line depicted in Figure 25.5 by the formula $\log Y_c = a + bx$, a decision must be reached as to the point in time to be designated $x = 0$. Assume that $x = 0$ is assigned to the year 1961.

Year	x
1960	-1
1961	0
1962	$+1$
1963	$+2$

a is the value of $\log Y_c$, where $x = 0$; that is, for the year 1961,

$$Y_c = 3000$$
$$\log Y_c = \log 3000 = 3.4771$$
$$a = 3.4771$$

b is equal to the change in the value of $\log Y_c$ with a one-unit increase in the value of x. In this case, b is equal to 0.3010. The formula for the logarithmic straight line depicted in Figure 25.5 is thus $\log Y_c = 3.4771 + 0.3010x$.

The formula for a logarithmic straight line can also be expressed in arithmetic form. To rewrite the formula in arithmetic form, simply take the antilog of both sides. In the left-hand member, antilog $(\log Y_c) = Y_c$. The second term in the right-hand member in log form is bx where b is a logarithm and x is an arithmetic number. Multiplying a log by a number is equivalent to raising the antilog of the log to the power indicated by the number. Therefore,

$$bx = (\text{antilog } b)^x$$

Adding the logs of the two numbers is equivalent to multiplying the antilogs of the two logs. Therefore,

$$a + bx = (\text{antilog } a)(\text{antilog } b)^x$$

Thus, $\log Y_c = 3.4771 + 0.3010x$ can alternatively be expressed in arithmetic form as $Y_c = (3000)(2)^x$. Either specification generates the Y_c values. The logarithmic form is considered more convenient for computational purposes.

Once it has been decided that the basic long-run behavior pattern for a set of data is best described by constant percentage change (i.e., a logarithmic straight line), the next step is to determine the line of that type that best describes the behavior pattern of the data. Once again, the least-squares criterion is used to single out the line of best fit. To accomplish this, one need simply fit an arithmetic least-squares line to the logarithms of the original values. The logs of the Y values are substituted for the Y values.

Arithmetic least-squares straight line	Logarithmic least-squares straight line
$Y_c = a + bx$	$\log Y_c = a + bx$
where	where
$a = \dfrac{\sum Y}{n}$	$a = \dfrac{\sum (\log Y)}{n}$
$b = \dfrac{\sum xY}{\sum x^2}$	$b = \dfrac{\sum (x \log Y)}{\sum x^2}$

Recall from the discussion of the arithmetic least-squares line that the formulas indicated for a and b hold only for the special case where $x = 0$ has been assigned to the exact middle of the period to which trend is to be fit. When this is not the case, the values of a and b must be obtained by simultaneous solution of the system of normal equations. Substituting $\log Y$ for Y, the normal equations for the logarithmic least-squares straight line are

I $\quad \sum (\log Y) \quad = na + b \sum x$

II $\quad \sum (x \log Y) = a \sum x + b \sum x^2$

To illustrate the computational procedure, a logarithmic least-squares straight line will be fitted to the time series, shipments of Naughty pine in millions of board feet for the years 1951–1962. Note that in the illustrative example, the period for which trend is to be fitted is an even number of years (12) and that therefore the exact middle of the period for which trend is to be fit is December 31, 1956, the end of the sixth year.[5] The value of x assigned to each year serves to indicate the direction and extent (in time) that the *middle* of the year referred to departs from the point in time designated $x = 0$. A unit of x in this example represents the passage of 6 months (so designated in order to facilitate computation by working with whole numbers). The x value assigned to 1956 is thus -1 since the middle of 1956 is 6 months earlier than the end of 1956; $x = -3$ for the middle of 1955, which is a year and a half earlier, and so on (Table 25.3).

---

[5] An additional year's observations have been added in order to illustrate the procedure when dealing with an even number of years.

Table 25.3 Logarithmic least-squares straight line trend: shipments of Naughty Pine (millions of board feet) for the years 1951–1962

Year	Y	$\log Y$	x	x^2	$x \log Y$	$\log Y_c$	Y_c
1951	9.56	0.9805	-11	121	-10.7855	0.9966	9.92
1952	10.15	1.0065	-9	81	-9.0585	0.9872	9.71
1953	9.49	0.9773	-7	49	-6.8411	0.9778	9.50
1954	9.41	0.9736	-5	25	-4.8680	0.9684	9.30
1955	9.54	0.9795	-3	9	-2.9385	0.9590	9.10
1956	8.74	0.9415	-1	1	-0.9415	0.9496	8.90
1957	8.00	0.9031	$+1$	1	$+0.9031$	0.9402	8.71
1958	8.44	0.9263	$+3$	9	$+2.7789$	0.9308	8.53
1959	8.81	0.9450	$+5$	25	$+4.7250$	0.9214	8.35
1960	8.29	0.9186	$+7$	49	$+6.4302$	0.9120	8.17
1961	7.68	0.8854	$+9$	81	$+7.9686$	0.9026	7.99
1962	7.98	0.9020	$+11$	121	$+9.9220$	0.8932	7.82
		11.3393		572	-35.4331		
					32.7278		
					-2.7053		

$$a = \frac{\sum \log Y}{n} = \frac{11.3393}{12} = 0.9449$$

$$b = \frac{\sum (x \log Y)}{\sum x^2} = \frac{-2.7053}{572} = -0.0047$$

To determine the trend value for any year, insert the x value associated with that year in the formula $\log Y_c = 0.9449 - 0.0047x$, solve, and find the antilog. Illustrating the computation for the year 1951:

$$\log Y_{c_{1951}} = 0.9449 - 0.0047(-11)$$
$$= 0.9449 + 0.0517$$
$$= 0.9966$$
$$\text{antilog } 0.9966 = 9.92$$
$$Y_{c_{1951}} = 9.92$$

Similarly, for 1962:

$$\log Y_{c_{1962}} = 0.9449 - 0.0047(+11)$$
$$= 0.9449 - 0.0517$$
$$= 0.8932$$
$$\text{antilog } 0.8932 = 7.82$$
$$Y_{c_{1962}} = 7.82$$

Instead of inserting each x value and computing the associated value of $\log Y_c$, one can simply add the *annual* logarithmic increment to the $\log Y_c$ value of the previous year to obtain the consecutive values of $\log Y_c$ from 1951 through 1962. The annual logarithmic increment is obtained by multiplying the 6-months logarithmic increment by 2.

$$(-0.0047)(2) = -0.0094$$

1951	0.9966
	-0.0094
1952	0.9872
	-0.0094

1953	0.9778
	−0.0094
1954	0.9684
	−0.0094
1955	0.9590
	−0.0094
1956	0.9496
	−0.0094
1957	0.9402
	−0.0094
1958	0.9308
	−0.0094
1959	0.9214
	−0.0094
1960	0.9120
	−0.0094
1961	0.9026
	−0.0094
1962	0.8932

Using a calculator, insert the first value, 0.9966, and lock in the value 0.0094 as a constant subtraction. Each time the subtract button is pressed, the associated subtotal reading indicates the next annual log Y_c value. As a final step, the antilogs of the values are determined. The Y_c values can also be generated directly by working with the arithmetic representation of the logarithmic least-squares straight line. Note, log $Y_c = 0.9449 - 0.0047x$ is the logarithmic least-squares straight line in logarithmic form. To facilitate manipulation, the formula will be rewritten so that the minus sign is assigned to the x factor.

$$\log Y_c = 0.9449 - 0.0047x$$
$$= 0.9449 + (-1)(0.0047)(x)$$
$$= 0.9449 + 0.0047(-x)$$
$$\text{antilog } 0.9449 = 8.81$$
$$\text{antilog } 0.0047 = 1.011$$

Adding logs is equivalent to multiplying their antilogs. Multiplying a log by a numerical value is equivalent to raising its antilog to that power. Thus, log $Y_c = 0.9449 + (0.0047)(-x)$ can be expressed arithmetically as

$$Y_c = 8.81(1.011)^{-x}$$

An expression with a negative exponent can be equivalently written as the reciprocal of the same expression with a positive exponent. Therefore,

$$Y_c = 8.81(1.011)^{-x}$$

$$= 8.81\left(\frac{1}{1.011}\right)^x$$

$$= 8.81(0.9891)^x$$

The formula indicates that with each passage of a unit of x, the trend line value will be reduced to 98.91 percent of its previous value. This is the 6-months change. The annual change equals $(0.9891)^2$ or 0.97832. To generate each successive trend value, insert the initial trend value into the calculator, lock in the constant multiplier, 0.97832, press, and the new product is the next trend value, and so on. The values are already in final arithmetic form. Thus, the step of determining antilogs is eliminated.

25.5 LEAST-SQUARES SECOND-DEGREE PARABOLA

The general equation for a second-degree parabola is

$$Y_c = a + bx + cx^2$$

The expression "second-degree" refers to the fact that the equation contains the variable x raised to the second power. A second-degree parabola, when graphed, displays a geometric shape similar to those displayed in Figures 25.6(a) and 25.6(b). The family of parabolas displayed in Figure 25.6(a) all possess the common characteristic of a positive value for c (the coefficient of the x^2 term). They differ, however, in the extent to which the "arms" of the parabola are extended. The smaller the absolute value of c, the flatter the "arms." The larger the absolute value of c, the closer the arms approach the vertical. The parabolas depicted in Figure 25.6(b) have the common characteristic of a negative value for c.

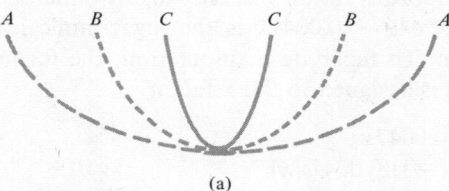

(a)

Figure 25.6(a)

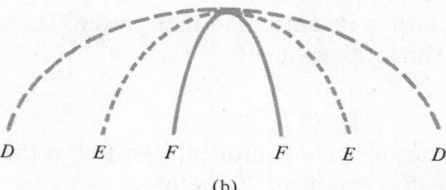

(b)

Figure 25.6(b)

The objective of the analyst is to select that segment of that parabola that will best represent the data. By applying the least-squares criterion, this objective can be achieved. As with the arithmetic least-squares straight line and the logarithmic least-squares straight line, a set of normal equations is generated.

Simultaneous solution of this set of equations provides the coefficients for a, b, and c associated with the second-degree parabola which when applied to the given set of data possesses the two characteristics: $\sum (Y - Y_c) = 0$ and $\sum (Y - Y_c)^2$ is at a minimum for lines of that type (second-degree parabolas) that could be fitted to the given set of data.

The three constants in the equation $Y_c = a + bx + cx^2$ may be interpreted in the following fashion:

a refers to the height of the parabola associated with that period (or point) in time for which x is designated as being equal to zero.

b refers to the slope of the curve at the point where $x = 0$ *only*.

c determines the amount (by its absolute value) and the direction (by its sign) of curvature away from the imaginary straight line formed by $Y_c = a + bx$ for the given values of a and b. Note that when $c = 0$, the equation $Y_c = a + bx + cx^2$ reduces to $Y_c = a + bx$, which is the formula for a straight line.

x refers, of course, to the number of units of time each given period lies away from the period of time assigned the designation $x = 0$. The sign indicates the direction of deviation.

Normal equations have been derived for the least-squares second-degree parabola in a similar fashion to their derivation for the arithmetic least-squares straight line. The function to be minimized is $F = \sum (Y - Y_c)^2$, where $Y_c = a + bx + cx^2$. This function is minimized when the partial derivatives with respect to a, b, and c are all equal to zero. That is,

$$\frac{\partial F}{\partial a} = 0$$

$$\frac{\partial F}{\partial b} = 0$$

$$\frac{\partial F}{\partial c} = 0$$

When $\partial F/\partial a = 0$, $\sum Y = na + b \sum x + c \sum x^2$.
When $\partial F/\partial b = 0$, $\sum xY = a \sum x + b \sum x^2 + c \sum x^3$.
When $\partial F/\partial c = 0$, $\sum x^2Y = a \sum x^2 + b \sum x^3 + c \sum x^4$.

The above three equations are referred to as the normal equations. When all three conditions hold simultaneously, we have the following system of three equations with three unknowns:

I $\qquad \sum Y = na + b \sum x + c \sum x^2$

II $\qquad \sum xY = a \sum x + b \sum x^2 + c \sum x^3$

III $\qquad \sum x^2Y = a \sum x^2 + b \sum x^3 + c \sum x^4$

The above three equations can be solved simultaneously for the values of a, b, and c. The least-squares second-degree parabola will be generated regardless of the point in time designated as $x = 0$. Once again, however, if x is set equal to zero at the exact middle of the period for which trend is to be fitted, a reduction in computational effort will result. In the case of the second-degree parabola, when $\sum x = 0$ and $\sum x^3 = 0$, the three normal equations simplify to

I $\sum Y = na + c \sum x^2$

II $\sum xY = b \sum x^2$

III $\sum x^2 Y = a \sum x^2 + c \sum x^4$

I Solving for a:

$$\sum Y = na + c \sum x^2$$

$$a = \frac{\sum Y - c \sum x^2}{n}$$

II Solving for b:

$$\sum xY = b \sum x^2$$

$$b = \frac{\sum x Y}{\sum x^2}$$

III Solving for c:

$$\sum x^2 Y = a \sum x^2 + c \sum x^4$$

Substituting:

$$a = \frac{\sum Y - c \sum x^2}{n}$$

$$\sum x^2 Y = \left(\frac{\sum Y - c \sum x^2}{n} \right) \sum x^2 + c \sum x^4$$

Multiplying both sides by n and removing the parentheses:

$$n \sum x^2 Y = \sum x^2 \sum Y - c(\sum x^2)^2 + nc \sum x^4$$

Collecting all terms not containing c in the left-hand member:

$$n \sum x^2 Y - \sum x^2 \sum Y = -c(\sum x^2)^2 + nc \sum x^4$$

Factoring c out in the right-hand member:

$$n \sum x^2 Y - \sum x^2 \sum Y = c[-(\sum x^2)^2 + n \sum x^4]$$

$$c = \frac{n \sum x^2 Y - \sum x^2 \sum Y}{n \sum x^4 - (\sum x^2)^2}$$

The three constants for the least-squares second-degree parabola are

$$a = \frac{\sum Y - c \sum x^2}{n}$$

$$b = \frac{\sum xY}{\sum x^2}$$

$$c = \frac{n \sum x^2 Y - \sum x^2 \sum Y}{n \sum x^4 - (\sum x^2)^2}$$

Table 25.4 displays the underlying computations for a least-squares second-degree parabola fitted to the data presented in Figure 25.7.

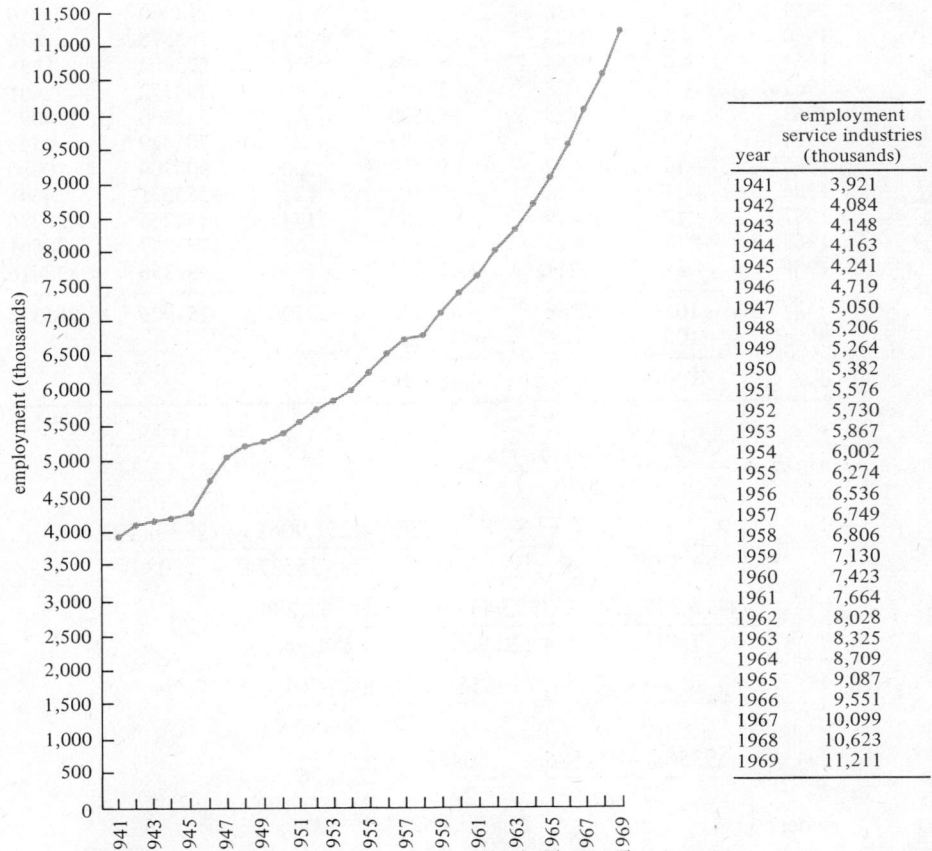

year	employment service industries (thousands)
1941	3,921
1942	4,084
1943	4,148
1944	4,163
1945	4,241
1946	4,719
1947	5,050
1948	5,206
1949	5,264
1950	5,382
1951	5,576
1952	5,730
1953	5,867
1954	6,002
1955	6,274
1956	6,536
1957	6,749
1958	6,806
1959	7,130
1960	7,423
1961	7,664
1962	8,028
1963	8,325
1964	8,709
1965	9,087
1966	9,551
1967	10,099
1968	10,623
1969	11,211

Figure 25.7 Employment—service industries: 1941–1969. (From U.S. Department of Labor.)

Table 25.4 Illustrative example: Computation of a least-squares second-degree parabola fitted to employment in the service industries, 1941–1969 (in thousands)

Year	x	Y	xY	x^2	x^2Y	x^4
1941	-14	3921	-54894	196	768516	38416
1942	-13	4084	-53092	169	690196	28561
1943	-12	4148	-49776	144	597312	20736
1944	-11	4163	-45793	121	503723	14641
1945	-10	4241	-42410	100	424100	10000
1946	-9	4719	-42471	81	382239	6561
1947	-8	5050	-40400	64	323200	4096
1948	-7	5206	-36442	49	255094	2401
1949	-6	5264	-31584	36	189504	1296
1950	-5	5382	-26910	25	134550	625
1951	-4	5576	-22304	16	89216	256
1952	-3	5730	-17190	9	51570	81
1953	-2	5867	-11734	4	23468	16
1954	-1	6002	-6002	1	6002	1
1955	0	6274	0	0	0	0
1956	$+1$	6536	$+6536$	1	6536	1
1957	$+2$	6749	$+13498$	4	26996	16
1958	$+3$	6806	$+20418$	9	61254	81
1959	$+4$	7130	$+28520$	16	114080	256
1960	$+5$	7423	$+37115$	25	185575	625
1961	$+6$	7664	$+45984$	36	275904	1296
1962	$+7$	8028	$+56196$	49	393372	2401
1963	$+8$	8325	$+66600$	64	532800	4096
1964	$+9$	8709	$+78381$	81	705429	6561
1965	$+10$	9087	$+90870$	100	908700	10000
1966	$+11$	9551	$+105061$	121	1155671	14641
1967	$+12$	10099	$+121188$	144	1454256	20736
1968	$+13$	10623	$+138099$	169	1795287	28561
1969	$+14$	11211	$+156954$	196	2197356	38416
	-105	193568	-481002	2030	14251906	255374
	$+105$		$+965420$			
	0		$+484418$			

$$b = \frac{\sum xY}{\sum x^2} = \frac{484,418}{2030} = 238.63$$

$$c = \frac{n \sum x^2Y - \sum x^2 \sum Y}{n \sum x^4 - (\sum x^2)^2} = \frac{(29)(14,251,906) - (2030)(193,568)}{(29)(255374) - (2030)^2}$$

$$= \frac{413,305,274 - 392,943,040}{7,405,846 - 4,120,900} = \frac{20,362,234}{3,284,946} = 6.20$$

$$a = \frac{\sum Y - c \sum x^2}{n} = \frac{193568 - 6.20(2030)}{29}$$

$$= \frac{193568 - 12586}{29} = \frac{180982}{29} = 6240.76$$

where

a = height of the parabola associated with the year designated as $x = 0$
b = slope at $x = 0$, that is, slope of the tangent to the parabola at $x = 0$
c = indicates the extent and direction of curvature

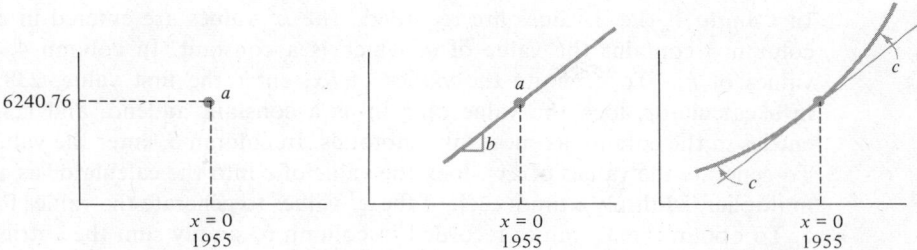

The following tabular format provides a convenient framework for calculating the trend values:

Year	Col. 1 x	Col. 2 x^2	Col. 3 a	Col. 4 bx	Col. 5 cx^2	Col. 6 Y_c
1941	−14	196	6240.76	238.63(x)	6.20(x^2)	
1942	−13	169	6240.76	238.63(x)	6.20(x^2)	
1943	−12	144	6240.76	238.63(x)	6.20(x^2)	
1944	−11	121	6240.76	238.63(x)	6.20(x^2)	

Table 25.5 Determination of the least-squares second-degree parabola trend values

Year	x	x^2	a	+	bx	+	cx^2	=	Y_c
1941	−14	196	6240.76		−3340.82		+1215.20	=	4115.14
1942	−13	169	6240.76		−3102.19				
1943	−12	144	6240.76		−2863.56				
1944	−11	121	6240.76		−2624.93				
1945	−10	100	6240.76		−2386.30				
1946	−9	81	6240.76		−2147.67				
1947	−8	64	6240.76		−1909.04		+396.80	=	4728.52
1948	−7	49	6240.76		−1670.41				
1949	−6	36	6240.76		−1431.78				
1950	−5	25	6240.76		−1193.15				
1951	−4	16	6240.76		−954.52		+99.20	=	5385.44
1952	−3	9	6240.76		−715.89				
1953	−2	4	6240.76		−477.26				
1954	−1	1	6240.76		−238.63				
1955	0	0	6240.76		0		0	=	6240.76
1956	+1	1	6240.76		+238.63				
1957	+2	4	6240.76						
1958	+3	9	6240.76						
1959	+4	16	6240.76		+954.52		+99.20		7294.48
1960	+5	25	6240.76						
1961	+6	36	6240.76						
1962	+7	49	6240.76						
1963	+8	64	6240.76		+1909.04		+396.80	=	8546.60
1964	+9	81	6240.76						
1965	+10	100	6240.76						
1966	+11	121	6240.76						
1967	+12	144	6240.76						
1968	+13	169	6240.76						
1969	+14	196	6240.76		+3340.82		+1215.20	=	10796.78

In column 1, the x values are recorded. The x^2 values are entered in column 2, column 3 contains the value of a, which is a constant. In column 4, enter the values of bx. To generate the values of bx, enter the first value, $238.63(-14)$, in a calculator, lock the value of b in as a constant addend, and generate the entries in the column as successive subtotals. In column 5, enter the values of cx^2. To generate the values of cx^2, lock the value of c into the calculator as a constant multiplier. Multiply c times each of the x^2 values to generate the values for column 5. To obtain the Y_c values recorded in column 6, simply sum the entries for that row in columns 3, 4, and 5 (Table 25.5).

25.6 THE GOMPERTZ CURVE—A GROWTH CURVE

Previous curves examined have been characterized by constant absolute change (the arithmetic least-squares line), by constant percentage change (the logarithmic least-squares line), and by the ability to reach a turning point and change direction (the second-degree parabola). The distinguishing feature of the Gompertz curve is that it continuously approaches but never reaches an upper limiting value. Moreover, the rate of growth declines at a constant rate. When plotted on an arithmetic chart, a Gompertz curve appears as an elongated S. When plotted on a semilogarithmic chart, the Gompertz curve is concave to the horizontal axis.

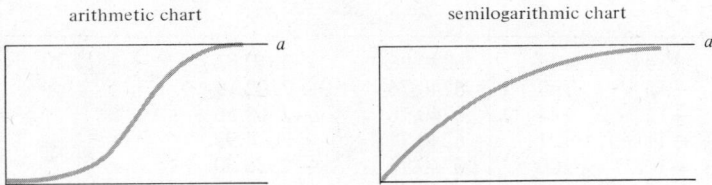

The formula for the Gompertz curve is

$$Y_c = ab^{c^x}$$

This relationship is usually expressed in logarithmic form to facilitate computation. In logarithmic form, the formula for the Gompertz curve appears as

$$\log Y_c = \log a + (\log b)c^x$$

Log a is the logarithm of the limiting value that is continuously approached but never reached (see the diagrams above). $(\log b)c^x$ is the difference between the log of the limiting value a and the log of the trend value for the time period in question. $(\log b)c^x$ is always negative. x represents the number of units of time each given period departs from the period of time designated $x = 0$. When fitting a Gompertz curve, $x = 0$ is always assigned to the first time period.

A Gompertz curve is not a least-squares curve. Instead, the curve is fitted to a time series by fitting the data at only a few selected points. The underlying computations are performed in such a fashion as to fit a representative point for the first third, middle third, and last third of the data.

The formula for the Gompertz curve contains three constants. The values for the constants are determined as follows:

$$c^n = \frac{S_3 - S_2}{S_2 - S_1}$$

$$\log b = \frac{(S_2 - S_1)(c - 1)}{(c^n - 1)^2}$$

$$\log a = \frac{1}{n}\left(S_1 - \frac{S_2 - S_1}{c^n - 1}\right)$$

The time series is divided into three subperiods (n refers to the number of years in a subperiod):

S_1 = sum of the logs of the observed Y values for the first third of the data
S_2 = sum of the logs of the observed Y values for the middle third of the data
S_3 = sum of the logs of the observed Y values for the last third of the data

The Gompertz curve presumes a declining rate of growth. Thus, unless $(S_3 - S_2)$ is less than $(S_2 - S_1)$, the Gompertz curve should not be fitted to the data.

Table 25.6 displays the underlying computations for a Gompertz curve fitted to the data presented in Figure 25.8.

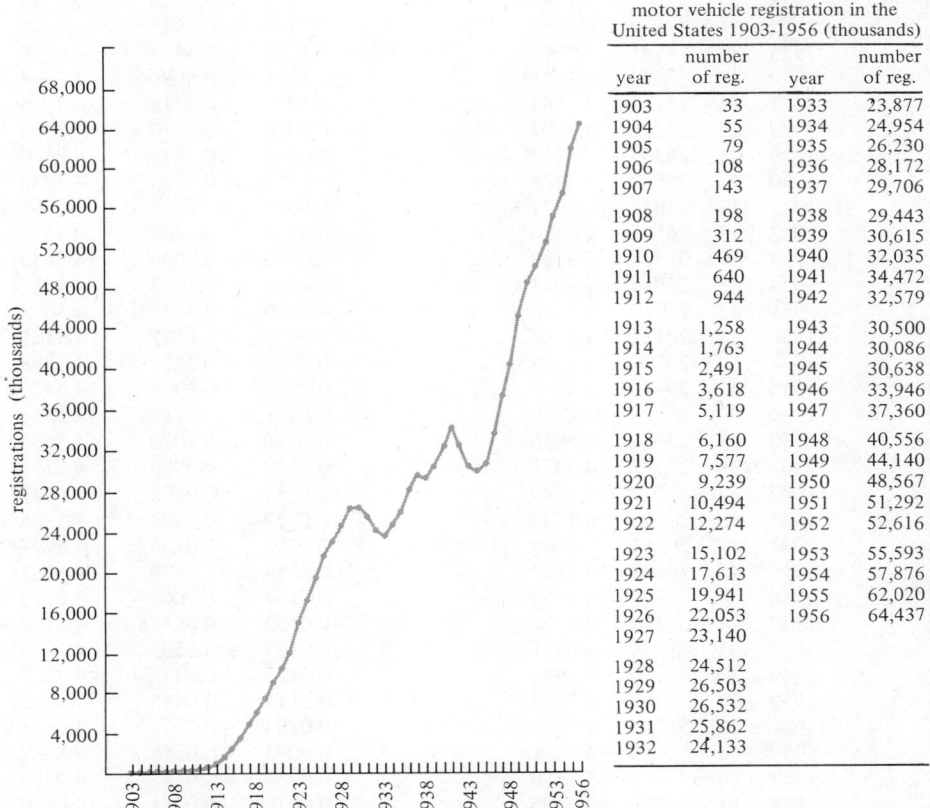

motor vehicle registration in the United States 1903-1956 (thousands)			
year	number of reg.	year	number of reg.
1903	33	1933	23,877
1904	55	1934	24,954
1905	79	1935	26,230
1906	108	1936	28,172
1907	143	1937	29,706
1908	198	1938	29,443
1909	312	1939	30,615
1910	469	1940	32,035
1911	640	1941	34,472
1912	944	1942	32,579
1913	1,258	1943	30,500
1914	1,763	1944	30,086
1915	2,491	1945	30,638
1916	3,618	1946	33,946
1917	5,119	1947	37,360
1918	6,160	1948	40,556
1919	7,577	1949	44,140
1920	9,239	1950	48,567
1921	10,494	1951	51,292
1922	12,274	1952	52,616
1923	15,102	1953	55,593
1924	17,613	1954	57,876
1925	19,941	1955	62,020
1926	22,053	1956	64,437
1927	23,140		
1928	24,512		
1929	26,503		
1930	26,532		
1931	25,862		
1932	24,133		

Figure 25.8 Motor vehicle registrations in the United States: 1903–1956. (From *Automobile Facts and Figures*, U.S. Bureau of Public Roads, 1961.)

Table 25.6 Fitting a Gompertz curve to motor vehicle registrations in the United States (thousands)

Year (1)	x (2)	Y (3)	log Y (Col. 4)	Subperiod totals (Col. 5)	c^x (6)	$(\log b)c^x$ (7)	Col. 7 + log a (8)	Y_c (9)
1903	0	33	1.5185		1.0000	−3.6415	1.0318	11
1904	1	55	1.7404		0.9093	−3.3112	1.3621	23
1905	2	79	1.8976		0.8268	−3.0108	1.6625	46
1906	3	108	2.0334		0.7518	−2.7377	1.9356	86
1907	4	143	2.1553		0.6836	−2.4893	2.1840	153
1908	5	198	2.2967		0.6216	−2.2636	2.4097	257
1909	6	312	2.4942		0.5652	−2.0582	2.6151	412
1910	7	469	2.6712		0.5139	−1.8714	2.8019	634
1911	8	640	2.8062		0.4673	−1.7017	2.9716	937
1912	9	944	2.9750		0.4249	−1.5473	3.1260	1337
1913	10	1258	3.0997		0.3864	−1.4071	3.2662	1846
1914	11	1763	3.2462		0.3514	−1.2796	3.3937	2476
1915	12	2491	3.3964		0.3195	−1.1635	3.5098	3234
1916	13	3618	3.5585		0.2905	−1.0579	3.6154	4125
1917	14	5119	3.7092		0.2642	−0.9621	3.7112	5143
1918	15	6160	3.7896		0.2403	−0.8751	3.7982	6283
1919	16	7577	3.8795		0.2185	−0.7957	3.8776	7544
1920	17	9239	3.9656	$S_1 = 51.2332$	0.1987	−0.7236	3.9497	8906
1921	18	10494	4.0209		0.1807	−0.6580	4.0153	10360
1922	19	12274	4.0890		0.1643	−0.5983	4.0750	11890
1923	20	15102	4.1790		0.1494	−0.5440	4.1293	13470
1924	21	17613	4.2458		0.1358	−0.4945	4.1788	15090
1925	22	19941	4.2998		0.1235	−0.4497	4.2236	16730
1926	23	22053	4.3435		0.1123	−0.4089	4.2644	18380
1927	24	23140	4.3644		0.1021	−0.3718	4.3015	20020
1928	25	24512	4.3894		0.0928	−0.3379	4.3354	21650
1929	26	26503	4.4233		0.0844	−0.3073	4.3660	23230
1930	27	26532	4.4238		0.0767	−0.2793	4.3940	24770
1931	28	25862	4.4127		0.0697	−0.2538	4.4195	26270
1932	29	24133	4.3826		0.0634	−0.2309	4.4424	27700
1933	30	23877	4.3880		0.0576	−0.2098	4.4635	29070
1934	31	24954	4.3971		0.0524	−0.1908	4.4825	30370
1935	32	26230	4.4188		0.0476	−0.1733	4.5000	31620
1936	33	28172	4.4498		0.0433	−0.1577	4.5156	32780
1937	34	29706	4.4728		0.0394	−0.1435	4.5298	33870
1938	35	29443	4.4690	$S_2 = 78.1697$	0.0358	−0.1304	4.5429	34910
1939	36	30615	4.4859		0.0326	−0.1187	4.5546	35860
1940	37	32035	4.5056		0.0296	−0.1078	4.5655	36770
1941	38	34472	4.5375		0.0269	−0.0980	4.5753	37610
1942	39	32579	4.5129		0.0245	−0.0892	4.5841	38380
1943	40	30500	4.4843		0.0222	−0.0808	4.5925	39130
1944	41	30086	4.4784		0.0202	−0.0736	4.5997	39780
1945	42	30638	4.4863		0.0184	−0.0670	4.6063	40390
1946	43	33946	4.5308		0.0167	−0.0608	4.6125	40970
1947	44	37360	4.5724		0.0152	−0.0554	4.6179	41490
1948	45	40556	4.6081		0.0138	−0.0503	4.6230	41980
1949	46	44140	4.6448		0.0125	−0.0455	4.6278	42440
1950	47	48567	4.6853		0.0114	−0.0415	4.6318	42840
1951	48	51292	4.7000		0.0104	−0.0379	4.6354	43190
1952	49	52616	4.7211		0.0095	−0.0346	4.6387	43520
1953	50	55593	4.7250		0.0086	−0.0313	4.6420	43850
1954	51	57876	4.7625		0.0078	−0.0284	4.6449	44150
1955	52	62020	4.7925		0.0071	−0.0259	4.6474	44400
1956	53	64437	4.8091	$S_3 = 83.0425$	0.0065	−0.0237	4.6496	44630

DETERMINATION OF THE CONSTRAINTS
(Supporting computations are presented in Table 25.6.)

$$c^n = \frac{S_3 - S_2}{S_2 - S_1}$$

$$= \frac{83.0425 - 78.1697}{78.1697 - 51.2332}$$

$$= \frac{4.8728}{26.9365} = 0.1809$$

$$\log c = \frac{\log 0.1809}{n} = \frac{9.2574 - 10}{18} = \frac{179.2574 - 180}{18}$$

$$= 9.9587 - 10$$

$$\text{antilog } 9.9587 - 10 = 0.9093$$
$$c = 0.9093$$

$$\log b = \frac{(S_2 - S_1)(c - 1)}{(c^n - 1)^2}$$

$$= \frac{(78.1697 - 51.2332)(0.9093 - 1.000)}{(0.1809 - 1.0000)^2}$$

$$= \frac{(26.9365)(-0.0907)}{(-0.8191)^2} = \frac{-2.44314055}{0.67092481}$$

$$= -3.6415$$

$$\log a = \frac{1}{n}\left(S_1 - \frac{S_2 - S_1}{c^n - 1}\right) = \frac{1}{18}\left(51.2332 - \frac{26.9365}{-0.8191}\right)$$

$$= \frac{1}{18}(51.2332 + 32.8855)$$

$$= \frac{84.1187}{18}$$

$$= 4.6733$$
$$\text{antilog } 4.6733 = 47130$$
$$a = 47130$$

A convenient approach for generating the trend values for the Gompertz curve is to utilize the tabular format provided in Table 25.6.

1. Calculate the value of c^x for each year and insert these values in column 6. When $x = 0$, $c^x = 1.000$. Each succeeding year's value can be generated by multiplying the value for the preceding year by the value of c.
2. Multiply the value of c^x for each year by the value of $\log b$. Enter these values in column 7. Column 7 thus contains the term $(\log b)c^x$.
3. Add the value of $\log a$ to each of the values in column 7. Record the result in column 8. Thus, column 8 contains $\log a + (\log b)c^x$. This is $\log Y_c$ or the value of the trend in log form.
4. Determine the antilog of the values in column 8. This result, recorded in column 9, provides the value of the trend.

25.7 PROJECTION OF THE TREND LINE

To project or extrapolate trend, simply insert the value of x associated with the period of interest in the trend-line equation and solve for Y_c. To illustrate, refer to the arithmetic least-squares straight-line trend line fitted to shipments of Naughty pine (billions of board feet) for the years 1952–1962.

The trend equation:

$Y_c = 8.7754 - 0.211x$
$x = 0$; June 30, 1957
x in terms of years
Y in billions of board feet

The Y_c value for 1972 is generated by substituting the value for x corresponding to June 30, 1972, in the trend equation. Substituting $x = +15$, Y_c is determined to be 5.6104 billion board feet.

$$
\begin{aligned}
Y_{c_{1972}} &= 8.7754 - 0.211x \\
&= 8.7754 - 0.211(+15) \\
&= 8.7754 - 3.165 \\
&= 5.6104
\end{aligned}
$$

Interpreting this value, note that the value of Y_c is only a prediction of the long-term secular component of the time series. The actual observation for 1972 will reflect the combined influence of trend, cyclical, and irregular elements. There is, of course, no seasonal component in annual data. Furthermore, it is important to recognize that 5.6104 billion board feet is an accurate prediction of the value of long-term secular trend in 1972 only so long as the long-term growth pattern in the series continues to be characterized by an average increment of 0.211 billion board feet per year.

25.8 ELIMINATION OF THE TREND COMPONENT

Figure 25.9(a) displays the original observed values of Y (shipments of Naughty pine in billions of board feet) for the years 1951–1962 (see Table 25.1). The pattern of long-run secular trend as described by the arithmetic least-squares straight-line equation $Y_c = 8.7754 - 0.211x$ is superimposed as a dotted line in Figure 25.9(a). Given that the components are related in a multiplicative manner and that the trend component is accurately measured by the equation $Y_c = 8.7754 - 0.211x$, the influence of cyclical and irregular influences on the time series can be isolated by dividing the observed value for each time period by the value of the trend component for the same period.[6] Figure 25.9(b) displays the original observed values of Y adjusted for the removal of the trend component.

To illustrate for 1962:

$$
O_{1962} = T_{1962} \times C_{1962} \times I_{1962}
$$

---

[6] If the data consisted of monthly observations, the residual would also reflect a seasonal influence, i.e., $C_{JULY, 1962} \times S_{JULY} \times I_{JULY, 1962}$.

Dividing:

$$\frac{O_{1962}}{T_{1962}} = \frac{\cancel{T_{1962}} \times C_{1962} \times I_{1962}}{\cancel{T_{1962}}} = C_{1962} \times I_{1962}$$

$$= \frac{7.98}{7.7204} = 1.03$$

The residual of 1.03 is interpreted as indicating that the influence of cyclical and irregular influences caused the observed value for 1962 to be 3 percent greater than it would have been had secular trend been the only influence on the data. Referring to Figure 25.9(b), note that the effect of dividing each observation by its trend component is equivalent to assigning a trend component of one to each of the observations. The objective in removing the trend component is to enable the analyst to better observe the nature of the impact of the other component elements, namely, cyclical, seasonal if not dealing with annual data, and irregular [Figure 25.9(b)].

Table 25.7 Determination of the residual ratios

Year	Observed value	Trend value	Residual ratio Y/Y_c
1952	10.15	9.8304	1.034
1953	9.49	9.6194	0.986
1954	9.41	9.4084	1.000
1955	9.54	9.1974	1.037
1956	8.74	8.9864	0.973
1957	8.00	8.7754	0.912
1958	8.44	8.5644	0.985
1959	8.81	8.3534	1.055
1960	8.29	8.1424	1.018
1961	7.68	7.9314	0.970
1962	7.98	7.7204	1.034

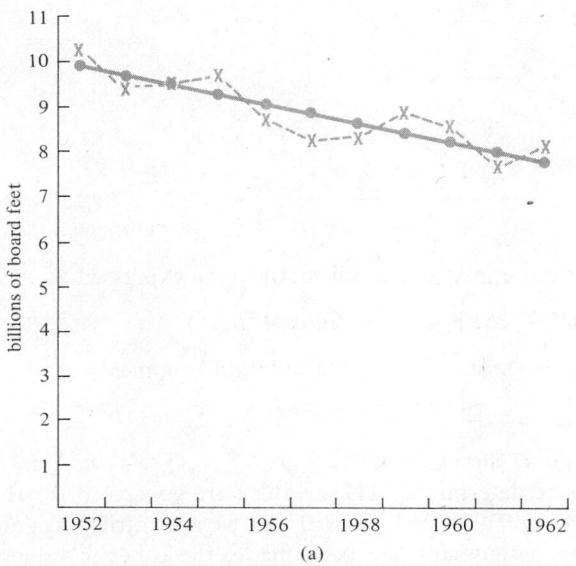

Figure 25.9(a) Shipments of Naughty Pine: 1952–1962.

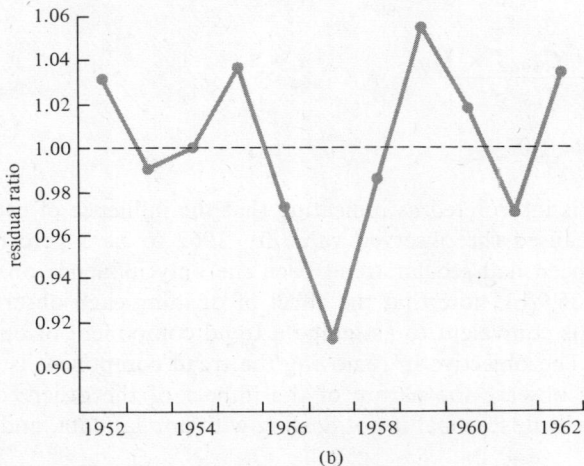

Figure 25.9(b) Shipments of Naughty Pine adjusted to remove trend component: 1952–1962.

TECHNICAL NOTE 13

Derivation of the normal equations

An understanding of calculus is not a prerequisite for this section.
    The objective is to minimize the function,

$$F = \sum (Y - Y_c)^2 \qquad \text{i.e., the sum of the squared deviations}$$
where $Y_c = a + bx$

Substituting, the objective can be restated as:

    minimize the function $\qquad F = \sum (Y - a - bx)^2$

Expanding the expression $(Y - a - bx)^2$, we obtain

$$
\begin{array}{l}
Y \;-\; a \;-\; bx \\
Y \;-\; a \;-\; bx \\
\hline
Y^2 \;-\; aY \;-\; bxY \\
 \;-\; aY \;+\; a^2 \;+\; abx \\
 \;-\; bxY \;+\; abx \;+\; b^2 x^2 \\
\hline
Y^2 \;-\; 2aY \;-\; 2bxY \;+\; a^2 \;+\; 2abx \;+\; b^2 x^2
\end{array}
$$

Thus, the function to be minimized can alternatively be expressed as

$$F = \sum (Y^2 - 2aY - 2bxY + a^2 + 2abx + b^2 x^2)$$

Expressing the summation term by term, the function becomes

$$F = \sum Y^2 - 2a \sum Y - 2b \sum xY + na^2 + 2ab \sum x + b^2 \sum x^2$$

For any given set of data, the expressions $\sum Y^2$, $\sum Y$, $\sum xY$, n, $\sum x$, and $\sum x^2$ are constants and can be determined. These values are generated for the data in the illustrative example in Table 25.8. $x = 0$ can be arbitrarily assigned to any of the years. Once this assignment has been made, the other x values serve to indicate the number of units (years in this case), the middle of the other years depart from the middle of the year assigned $x = 0$. In Table 25.8, 1952 has arbitrarily been selected as the year for which $x = 0$. The technique described is applicable, however, regardless of the year selected. Substituting, the function becomes

$$F = 853.2665 - 2a(96.53) - 2b(459.44) + 11a^2 + 2ab(55) + b^2(385)$$

To determine the sum of the squared deviations of any given straight line from the observed Y values, it is simply necessary to insert the values for a and b into the expression and solve for the value of F.

When $a = 1$ and $b = 1$, $F = 247.3265$.
When $a = 2$ and $b = 1$, $F = 197.2665$.
When $a = 3$ and $b = 1$, $F = 169.2065$.
When $a = 2$ and $b = 2$, $F = 653.3865$.

Table 25.8

Year	Y	x	x^2	xY	Y^2
1952	10.15	0	0	0	103.0225
1953	9.49	1	1	9.49	90.0601
1954	9.41	2	4	18.82	88.5481
1955	9.54	3	9	28.62	91.0116
1956	8.74	4	16	34.96	76.3876
1957	8.00	5	25	40.00	64.0000
1958	8.44	6	36	50.64	71.2336
1959	8.81	7	49	61.67	77.6161
1960	8.29	8	64	66.32	68.7241
1961	7.68	9	81	69.12	58.9824
1962	7.98	10	100	79.80	63.6804
	96.53	55	385	459.44	853.2665

The above relationships are depicted graphically in Figure 25.10. One can think of each possible point on the "floor" of the diagram as representing the parameters of a particular straight line. For example, the point on the "floor" designated (2, 1) refers to the particular straight line, $Y_c = 2 + 1x$. Directly

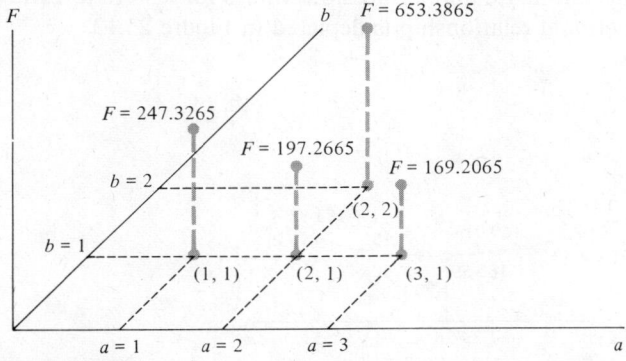

Figure 25.10

above that point on the "floor" is plotted the value of F, namely, the sum of the squared deviations of the 11 observed values of Y from that particular straight line. If this procedure were to be repeated for each and every possible (a, b) combination, the plotted values for F would form a smooth shape similar to that depicted in Figure 25.11.

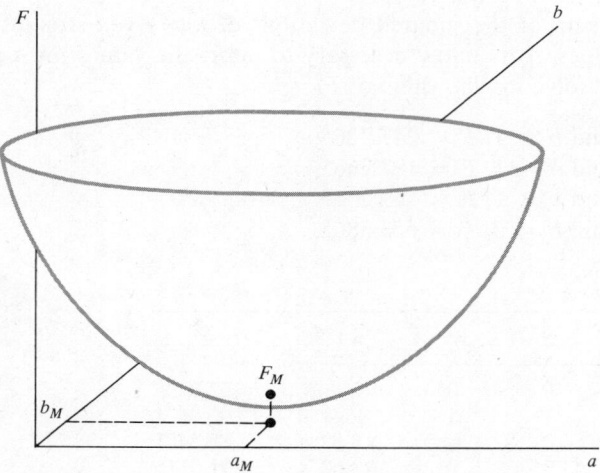

Figure 25.11

Given the geometric shape depicted in Figure 25.11, the minimum point of the function can be determined visually. This point, F_M, corresponds to the minimum sum of the squared deviations. By dropping a perpendicular to the "floor" and reading from the a axis and the b axis the associated values of a and b, the parameters of the least-squares line can be identified. The least-squares line would be

$$Y_c = a_M + b_M x$$

Although the geometric approach would work theoretically, the underlying computations required to obtain the values for F would be prohibitive. We can, however, capitalize on the nature of the functional relationship to determine the values for a_M and b_M directly. Consider the values for F that would be observed if b were held constant at 1.0 and all possible values for a were to be investigated. The resulting functional relationship is depicted in Figure 25.12.

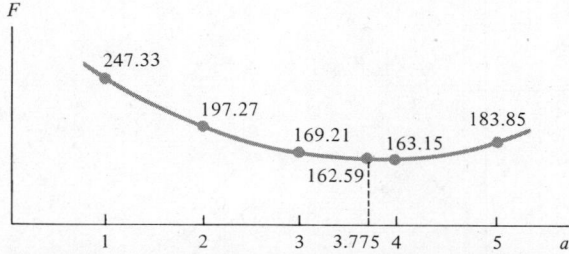

Figure 25.12 Sum of squared deviations for various values of a given $b = 1.0$.

Given any particular value of a, the effect of an infinitesimal increase in the value of a on the value of the function is described by the slope of the tangent to the function at that point. Referring to Figure 25.13: When a is equal to one, an infinitesimal increase in the value of a will cause the value of the function F to decrease. This is reflected by the negative slope of the tangent to the function at the value 247.33. When a is equal to 5, an infinitesimal increase in the value of a will cause the value of the function F to increase. This is reflected by the positive slope of the tangent to the function at the value 183.85. It is only when a is equal to 3.775 that an infinitesimal increase in the value of a causes the value of the function neither to increase nor to decline. This is due to the fact that at this point the function is changing from decreasing to increasing and is at its minimum point. This is reflected by the zero slope of the tangent to the function at the value 162.59.

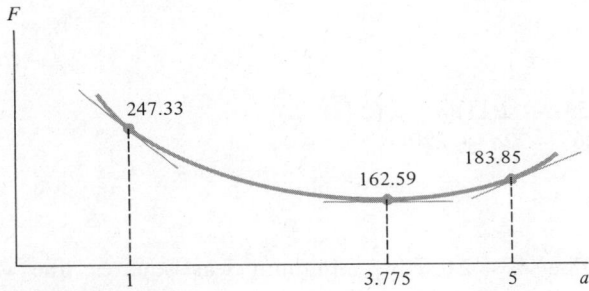

Figure 25.13

In general, whenever the tangent to the function has a negative slope, further improvement can be obtained by increasing the value assigned to the coefficient a. Whenever the tangent to the function has a positive slope, further improvement can be obtained by decreasing the value assigned to the coefficient a. It is only when the tangent to the function has a slope equal to zero that no further improvement can be obtained by changing the value of a, *given that the value of b is held constant*. Thus, the least-squares line associated with the specified value of b can be identified by determining the value of a associated with the tangent with a zero slope. The slope of the tangent to the function at any given point is the geometric equivalent to the derivative of the function. Using differential calculus, the partial derivative of the function F with respect to a is specified without proof to be[7]

$$\frac{\partial F}{\partial a} = -2 \sum Y + 2na + 2b \sum x$$

where

$$F = \sum Y^2 - 2a \sum Y - 2b \sum xY + na^2 + 2ab \sum x + b^2 \sum x^2$$

$\partial F/\partial a$ represents the ratio of the change in the value of F to an infinitesimal change in the value of a.

---

[7] A partial derivative describes the rate of change in the value of a function with respect to an infinitesimal change in the specified variable with all but the specified variable held constant. In this case, the specified variable is a and the variable b is held constant.

Given a specific value for b and the knowledge that the slope of the tangent to the function is zero when F is at its minimum (and therefore $\partial F/\partial a = 0$), the above equation can be solved for the value of a that corresponds to the least-squares line given a particular value for b.

When $b = 1.0$:

$$\frac{\partial F}{\partial a} = -2 \sum Y + 2na + 2b \sum x$$

$$
\begin{aligned}
0 &= -2(96.53) + 2(11)a + 2(1)(55) \\
0 &= -193.06 + 22a + 110 \\
-22a &= -83.06 \\
a &= 3.775
\end{aligned}
$$

Note that this corresponds to the minimum point on the function depicted in Figure 25.13.

When $b = 2.0$:

$$
\begin{aligned}
0 &= -2(96.53) + 2(11)(a) + (2)(2)(55) \\
0 &= -193.06 + 22a + 220 \\
-22a &= 26.94 \\
a &= -1.225
\end{aligned}
$$

Thus, given the value $b = 2.0$, the minimum least-squares line would be $Y_c = -1.225 + 2.0x$.

Our objective, however, is not to determine local least-squares lines (for given values of b) but rather to determine the global or overall least-squares line (considering all possible combinations of a and b). The global least-squares line corresponds to the point on the function where both the tangent to the function with respect to an infinitesimal change in the value of a *and* the tangent to the function with respect to an infinitesimal change in the value of b are characterized by a zero slope. As long as one or the other has a nonzero slope, it indicates that a lower sum of the squared deviations can be obtained by making an appropriate adjustment in the value of a or b. Geometrically, this relationship is depicted in Figure 25.14. In the language of calculus, this is equivalent to stating that the partial derivative of F with respect to a and the partial derivative of F with respect to b must both be equal to zero. It is stated without proof that

$$\frac{\partial F}{\partial a} = -2 \sum Y + 2na + 2b \sum x$$

$$\frac{\partial F}{\partial b} = -2 \sum xY + 2a \sum x + 2b \sum x^2$$

Both of the above expressions must be equal to zero in order to obtain the global least-squares line. One or the other equal to zero is not sufficient.

Thus, the following relationships hold for the least-squares line:

I $0 = -2 \sum Y + 2na + 2b \sum x$

II $0 = -2 \sum xY + 2a \sum x + 2b \sum x^2$

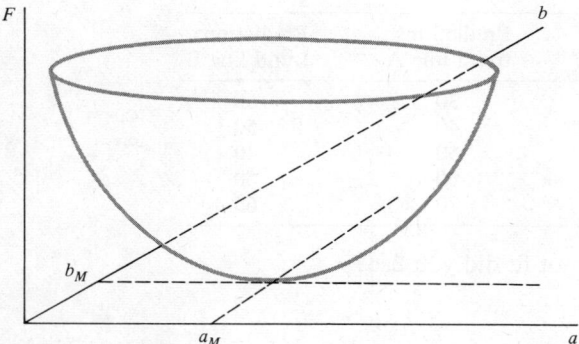

Figure 25.14

The previous equations can be simplified by multiplying both sides by $\frac{1}{2}$; they then become

I $0 = -\sum Y + na + b \sum x$ or $\sum Y = na + b \sum x$
II $0 = -\sum xY + a \sum x + b \sum x^2$ or $\sum xY = a \sum x + b \sum x^2$

In this final form, they are referred to as the normal equations:

I $\sum Y = na + b \sum x$
II $\sum xY = a \sum x + b \sum x^2$

The first equality must hold if $\partial F/\partial a$ is equal to zero, and the second must hold if $\partial F/\partial b$ is equal to zero. Thus, if one desires the values for a and b that will minimize the function $\sum (Y - Y_c)^2$, it is simply necessary to determine the values for a and b that will simultaneously solve both equations. The values $\sum Y$, n, $\sum x$, $\sum xY$, and $\sum x^2$ can be readily determined from the observed data (see Table 25.8). Thus, we have two equations with two unknowns, and the values a and b (the two unknowns) are readily ascertained.

PROBLEMS

Problem 25.1

1. Differentiate between point and period data.
2. Explain clearly the difference in plotting and interpreting *point* data and *period* data on a chart.

Problem 25.2

What are the two distinguishing characteristics of a least-squares line?

Problem 25.3

For the same set of data, one could compute an arithmetic least-squares straight line *and* a least-squares second-degree parabola. How can one have two different least-squares lines for the same set of data?

Problem 25.4

1. You have been given two trend lines and you have computed the predicted values they give for a 5-year period. Determine which one provides the best fit to the data.

Year	Actual values	Predictions trend line A	Predictions trend line B
1967	40	30	60
1968	50	40	50
1969	80	50	40
1970	70	60	50
1971	70	70	60

2. What criterion of fit did you use?

Problem 25.5

$Y_c = a + bx$ is the equation for the arithmetic least-squares straight line. Explain specifically what Y_c, a, b, and x represent in this formula.

Problem 25.6

$Y_c = a + bx + cx^2$ is the equation for the least-squares second-degree parabola trend line. Explain specifically what Y_c, a, b, c, and x represent in this formula.

Problem 25.7

Discuss thoroughly the similarities and the dissimilarities of the arithmetic and logarithmic least-squares lines.

Problem 25.8

1. What does the letter b represent in the formula for the arithmetic least-squares straight line?
2. How does this differ from the meaning of b for the logarithmic least-squares straight line?

Problem 25.9

In calculating the arithmetic least-squares straight-line trend, what difference in interpretation is given to the value of b for an even number of years as compared to an odd number of years?

Problem 25.10

Contrast the meaning of the symbol b in the formulas for
1. the arithmetic least-squares straight line
2. the logarithmic least-squares straight line
3. the least-squares second-degree parabola.

Problem 25.11

How does an arithmetic least-squares straight line differ from a logarithmic least-squares straight line with respect to interpretation?

Problem 25.12

In a two-cycle semilog chart, the value at the top of the second cycle is always
(1) _____ times the value of the item at the bottom of the first cycle and
(2) _____ times the value of the item at the bottom of the second cycle.

Problem 25.13

You can determine the absolute value of a point plotted on a semilogarithmic chart by reading on the vertical axis the corresponding value to the height of the point. Why, then, are semilogarithmic charts frowned upon when you want to make comparisons between the absolute values of items plotted on a chart?

Problem 25.14

A machine with an expected life of 7 years is to be depreciated at a uniform rate of diminishing value. The original cost of the machine was $10,000, and it is estimated that it will be worth $200 at the end of 7 years. Using semilog paper, graph the relationship described above. From the graph, estimate the credit to the reserve for depreciation for the second year.

Problem 25.15

Two members of the controller's staff computed an arithmetic least-squares straight-line trend for some corporate staff data for 7 years, 1960–1966:

Jones	$Y_c = 130 + 10X$
Wilson	$Y_c = 100 + 10X$

Both checked their trend lines and observed that they generated the same Y_c values.
  1. Is this possible? If so, how? If not, why not?
  2. Compute the trend value for 1964 for both lines. If you cannot, what additional information would you need in order to accomplish this task?

Problem 25.16

Given the following set of data:

Year	Y
1961	120
1962	85
1963	108
1964	107
1965	104
1966	112
1967	122

  1. Generate the arithmetic least-squares straight-line trend equation.
  2. Determine the value of the arithmetic least-squares straight-line trend for the year 1966.
  3. Generate the logarithmic least-squares straight-line trend equation.
  4. Determine the value of the logarithmic least-squares straight-line trend for the year 1962.

Problem 25.17

Given the following set of data:

Year	Y
1961	7
1962	11
1963	18
1964	32
1965	50
1966	61
1967	75
1968	82

1. Generate the arithmetic least-squares straight-line trend equation.
2. Determine the value of the arithmetic least-squares straight-line trend for the year 1962.
3. Generate the logarithmic least-squares straight-line trend equation.
4. Determine the value of the logarithmic least-squares straight-line trend for the year 1968.

Problem 25.18

1. From the following set of data, compute the numerical values for the *formula* for the logarithmic least-squares straight-line based on the observations from 1957 to 1965.

Annual production of electricity by electric utilities in the United States (in billions of kilowatt hours) for the years 1957–1965

Year	Annual production
1957	632
1958	645
1959	710
1960	753
1961	792
1962	852
1963	914
1964	984
1965	1055

Source: Federal Power Commission.

2. Use the formula constructed in part 1 to predict annual production of electricity by electric utilities in the United States for the year *1967 only*.
3. Would you expect the predicted value according to the trend line for 1967 to be equal to the actual value for 1967? *Why* or *why not*?

Problem 25.19

Given the following time series data:

Earned degrees: Bachelor's and first professional degrees, United States 1958–1959 to 1968–1969 (in thousands)

Academic year	Total	Men	Women
1958–1959	382.2	253.9	128.3
1959–1960	391.6	254.4	137.2
1960–1961	398.2	254.7	143.5
1961–1962	416.9	260.8	156.1
1962–1963	446.4	273.3	173.1
1963–1964	497.6	298.3	199.3
1964–1965	533.9	318.1	215.8
1965–1966	555.6	331.1	224.5
1966–1967	594.9	355.3	239.6
1967–1968	671.6	392.8	278.8
1968–1969	760.0	437.0	324.0

Source: *Yearbook of Higher Education*, 1971. Academic Media, Orange, New Jersey.

1. Plot the three time series on a chart.
2. Visually evaluate the three series and decide for each which of the following would provide the best fit:
   a. arithmetic least-squares straight line
   b. logarithmic least-squares straight line
   c. least-squares second-degree parabola
3. For the time series, total earned degrees:
   a. Determine the trend equation for the arithmetic least-squares straight line.
   b. Using the equation derived in part a, determine the Y_c values.
   c. Determine the trend equation for the logarithmic least-squares straight line.
   d. Using the equation derived in part b, determine the Y_c values.
   e. Determine the trend equation for the least-squares second-degree parabola.
   f. Using the equation derived in part e, determine the Y_c values.
   g. Compute the $\sum (Y - Y_c)^2$ for each of the three least-squares lines. Using the least-squares criterion, which of the three provides the line of best fit?
4. Repeat steps a–g as described in part 3 for the time series, earned degrees—men.
5. Repeat steps a–g as described in part 3 for the time series, earned degrees—women.

Problem 25.20

Given:

Estimated average charges (current dollars) per full-time undergraduate resident degree credit students in institutions of higher education, United States 1958–1959 to 1968–1969

Academic year	Tuition and required fees, public ($)	Tuition and required fees, private ($)
1958–1959	191	738
1959–1960	200	794
1960–1961	209	850
1961–1962	218	906
1962–1963	222	944
1963–1964	234	1012
1964–1965	243	1088
1965–1966	257	1153
1966–1967	275	1233
1967–1968	292	1326
1968–1969	314	1443

Source: *Yearbook of Higher Education*, 1971. Academic Media, Orange, New Jersey.

1. Plot the two time series on a chart.
2. Visually evaluate the two series and decide for each which of the following would provide the best fit:
   a. arithmetic least-squares straight line
   b. logarithmic least-squares straight line

    c. least-squares second-degree parabola.

3. For the time series, tuition and required fees—public:

    a. Determine the trend equation for the line selected as providing the best fit in part 1.

    b. Using the equation derived in part a, determine the Y_c values.

4. Repeat steps a and b as described in part 3 for the time series, tuition and required fees—private.

5. Referring to the chart in part 1 and the trend equations generated in parts 3a and 4a, what conclusions, if any, can be drawn?

Problem 25.21

Given:

Genuine Parts, Inc.: Sales Revenue (in millions of dollars) 1957–1969

Year	Sales revenue
1957	48.1
1958	56.5
1959	70.4
1960	75.0
1961	80.5
1962	90.3
1963	96.7
1964	120.3
1965	171.5
1966	190.6
1967	212.0
1968	275.3
1969	303.5

Source: *Moody's Industrial Manual*, 1970.

1. Plot the time series on a chart.

2. Visually evaluate the plots on the chart and decide which of the following would provide the best fit:

    a. arithmetic least-squares straight line

    b. logarithmic least-squares straight line

    c. least-squares second-degree parabola

3. a. Determine the trend equation for the arithmetic least-squares straight line.

    b. Using the equation derived in part a, determine the Y_c values.

    c. Determine the trend equation for the logarithmic least-squares straight line.

    d. Using the equation derived in part c, determine the Y_c values.

    e. Determine the trend equation for the least-squares second-degree parabola.

    f. Using the equation derived in part e, determine the Y_c values.

    g. Compute the $\sum (Y - Y_c)^2$ for each of the three least-squares lines. Using the least-squares criterion, which of the three provides the line of best fit?

Problem 25.22

Given:

Trans World Airlines: Operating Revenue (millions of dollars) 1947–1970

Year	Operating revenue
1947	78.5
1948	101.1
1949	105.6
1950	117.1
1951	144.9
1952	160.5
1953	187.1
1954	203.5
1955	217.3
1956	240.2
1957	263.5
1958	284.7
1959	348.3
1960	378.1
1961	362.9
1962	400.9
1963	476.5
1964	575.0
1965	671.8
1966	699.6
1967	873.2
1968	948.1
1969	1089.4
1970	1150.3

Source: *Moody's Transportation Manual*, 1970.

1. Plot the time series on a chart.
2. Determine the trend equation for a least-squares second-degree parabola fitted to the data.
3. Using the equation derived in part 2, determine the Y_c values.
4. Plot the Y_c values on the graph and connect them with a smooth curve.
5. Explain why a Gompertz curve cannot be fitted to the data.

TIME SERIES ANALYSIS: SEASONAL VARIATION AND CYCLICAL FLUCTUATIONS

26

26.1 SEASONAL VARIATION

The recurring pattern that appears in a time series at given stated intervals and that tends to be constant with respect to both timing and amplitude is referred to as seasonal variation. The interval within which seasonal variation typically is measured is the calendar year, and the seasonal pattern describes the monthly variation in the impact of the seasonal influence. It should be emphasized, however, that the concept of seasonal variation is equally applicable to periods of shorter duration. To illustrate: Over a series of weekly intervals, one can observe a repeatable pattern of daily variation in the number of lines of advertising copy in the daily newspaper. Advertising lineage is typically lowest on Monday and peaks with the Thursday edition. Similarly, over a series of 24-hour intervals, one can observe a repeatable pattern of hourly variation in the demand placed on telephone trunk lines. Moreover, the pattern differs depending on the day of the week (Mondays exhibit a different pattern than Saturdays, etc.).

A description of seasonal variation may be desirable for any of the following reasons:

1. Knowledge of the seasonal pattern is extremely useful for purposes of short-term planning. The seasonal pattern is quite stable from period to period, and as such the impact of seasonal influence can be predicted with a high degree of confidence.[1] Thus, the decision maker can project the *pattern* of man-hour requirements, raw material needs, receivable balances, cash inflows and outflows, the need for short-term financing, and so on with a certain degree of security. The *level* (as opposed to the pattern) of these receipts and expenditures is controlled by the relatively long-run underlying forces reflected in the trend and cyclical components. The behavior of the longer-run components is considerably more difficult to forecast. The farther into the future the period in question, the greater the degree of difficulty. Once the long-run forecast has been formulated, however, the short-term pattern associated with seasonal variation is fairly predictable.

2. Given a knowledge of the existing seasonal pattern, management can investigate ways to smooth out the observed variation in the level of activity. One technique is to add a product with a complementary selling season to the product line. To illustrate: A firm currently producing power lawnmowers could even out production and sales activity by adding a line of snowblowers. Another approach to combating an undesirable seasonal pattern involves institution of promotional campaigns to stimulate demand during slow periods of activity. Illustrations of this type of behavior include such activities as lowering hotel rates in the off season and offering different rate structures for long-distance telephone calls in nonpeak as opposed to peak hours of activity.

3. For purposes of evaluation: Removal of the influence of seasonal variation often enables the influence of longer-run forces such as trend and cyclical fluctuations to become more readily discernable.

26.2 INTERPRETATION OF A SEASONAL INDEX NUMBER

A seasonal index number indicates the extent to which a given month's observed value is expected to be inflated or deflated due to seasonal influences. Consider a firm that sells 3600 units a year. Assuming no seasonal pattern (and for the moment ignoring trend, cyclical, and irregular influences), the 3600 units would be evenly distributed over the 12-month period. Dividing 3600 by 12 generates average monthly sales of 300 units per month. Ignoring trend, cyclical, and irregular influences:

1. If no seasonal pattern exists, the seasonal index number for each of the 12 months is 100, indicating that the expected value for each month is equal to 100 percent of average monthly sales.

2. If the seasonal index number assigned to September is 110, it indicates that the level of activity for the month of September is expected to be

---

[1] Where the seasonal pattern is gradually changing over time, techniques are available that provide a changing seasonal pattern to accommodate the changing underlying circumstances.

10 percent greater than average monthly sales for the year (the figure that would be expected to be observed each and every month had there been no seasonal pattern).

Allowing for trend, cyclical, and irregular influences: If the seasonal index number assigned to July is 82, it indicates that the July level of activity typically will be 82 percent of the value that would be expected to be observed if only trend, cyclical, and irregular influences were operative. To illustrate: Assuming trend, cyclical, seasonal, and irregular influences operate in a multiplicative manner, the observed value for July 1970 can be described by the following relationship:

$$O_{\text{July 1970}} = T_{\text{July 1970}} \times C_{\text{July 1970}} \times S_{\text{July}} \times I_{\text{July 1970}}$$

Assume that the combined influence of trend, cyclical, and irregular influences result in projected sales for July 1970 of 321 units. After incorporating the influence of seasonal variation, the sales for July 1970 would be projected at 82 percent of this figure or approximately 263 units.

$$
\begin{aligned}
O_{\text{July 1970}} &= (321 \text{ units}) \times S_{\text{July}} \\
&= (321 \text{ units}) (0.82) \\
&= 263 \text{ units}
\end{aligned}
$$

A number of techniques have been developed for generating seasonal index numbers. Our attention will be restricted to two of these, namely, (1) the ratio to trend method, and (2) the ratio to the 12 months centered moving average method.

26.3 THE RATIO TO TREND METHOD

The procedure:

1. Compute the trend value for each period for which there is an observation.
2. Divide each observation (each month of each year) by the corresponding value of the trend component for the equivalent period of time. Multiply the resulting quotient by 100. The resulting values are referred to as ratios to trend (short for ratio of the original observation to the trend component). The series, hog receipts at Chicago Public Stockyards (in thousands) for the years 1960–1962, will be used as an illustrative example. The supporting computations are presented in Table 26.1. It is usually recommended that a minimum of 7 years' observations be used. In the interest of computational simplicity, the illustrative example is restricted to observations for 3 years. In the illustrative example, it is assumed that trend is described by an arithmetic least-squares straight-line fit to the data.
3. *Collect all the ratios to trend by month.* The result is depicted in Table 26.2. Note that all the Januarys are in the first column, all the Februarys in the second column, and so on.
4. *Obtain a representative ratio to trend for each of the 12 months.* Eliminate from consideration any ratios that appear to be unduly influenced by significant irregular influences. The mean of the ratios that remain is then used as the representative ratio to trend for the month.
5. *The typical representative ratio for each month is then divided by the average ratio to trend for the entire period.* The 12 resulting quotients (multiplied by 100) comprise the set of seasonal index numbers. By definition, the 12 seasonal index numbers must average to 100. If the computation does not produce this result,

Table 26.1 Hog receipts at Chicago Public Stockyards in thousands, 1960–1962

Year	Month	Y	$x = \frac{1}{2}$ month	xY	Y_c or trend	Ratio to trend
1960	Jan.	195	-35	-6825	152.692	127.71
	Feb.	158	-33	-5214	152.408	103.67
	Mar.	151	-31	-4681	152.124	99.26
	Apr.	149	-29	-4321	151.840	98.13
	May	163	-27	-4401	151.556	107.55
	June	150	-25	-3750	151.272	99.16
	July	119	-23	-2737	150.988	78.81
	Aug.	135	-21	-2835	150.704	89.58
	Sept.	130	-19	-2470	150.420	86.42
	Oct.	150	-17	-2550	150.136	99.91
	Nov.	169	-15	-2535	149.852	112.78
	Dec.	178	-13	-2314	149.568	119.01
1961	Jan.	153	-11	-1683	149.284	102.49
	Feb.	127	-9	-1143	149.000	85.23
	Mar.	143	-7	-1001	148.716	96.16
	Apr.	137	-5	-685	148.432	92.30
	May	179	-3	-537	148.148	120.82
	June	144	-1	-144	147.864	97.39
	July	120	$+1$	$+120$	147.580	81.31
	Aug.	121	$+3$	$+363$	147.296	82.15
	Sept.	116	$+5$	$+580$	147.012	78.91
	Oct.	162	$+7$	$+1134$	146.728	110.41
	Nov.	167	$+9$	$+1503$	146.444	114.04
	Dec.	159	$+11$	$+1749$	146.160	108.78
1962	Jan.	174	$+13$	$+2262$	145.876	119.28
	Feb.	131	$+15$	$+1965$	145.592	89.98
	Mar.	148	$+17$	$+2516$	145.308	101.85
	Apr.	126	$+19$	$+2394$	145.024	86.88
	May	151	$+21$	$+3171$	144.740	104.32
	June	147	$+23$	$+3381$	144.456	101.76
	July	121	$+25$	$+3025$	144.172	83.93
	Aug.	119	$+27$	$+3213$	143.888	82.70
	Sept.	109	$+29$	$+3161$	143.604	75.90
	Oct.	164	$+31$	$+5084$	143.320	114.43
	Nov.	177	$+33$	$+5841$	143.036	123.75
	Dec.	176	$+35$	$+6160$	142.752	123.29
				-49826		
				$+47622$		
				-2204		

$$a = \frac{\sum Y}{N}$$

$$= \frac{5318}{36}$$

$$= 147.722$$

$$b = \frac{\sum xY}{\sum x^2}$$

$$= \frac{-2204}{15540}$$

$$= -0.142$$

$$Y_c = a + bx$$

$$= 147.722 - 0.142x$$

When $x = -35$,
$$Y_c = 152.692$$

When $x = +35$,
$$Y_c = 142.752$$

All the other values are obtained by adding the monthly increment to the preceding month. In the illustration, the monthly increment is $2b$ or -0.284.

Source: Department of Agriculture.

the resulting index numbers must be leveled. This is accomplished by dividing each seasonal index number by the average value of the 12 seasonal index numbers and multiplying by 100.

Illustration of the computation for the month of January:

$$\text{average ratio for Jan.} = \frac{349.48}{3} = 117$$

$$\text{average ratio for the entire period} = \frac{3600.05}{36} = 100$$

$$\text{seasonal index number for January} = \frac{117}{100}(100) = 117$$

Table 26.2 Computation of the seasonal index by averaging the ratios to trend

Year	Month												
	Jan.	Feb.	Mar.	Apr.	May	June	July	Aug.	Sept.	Oct.	Nov.	Dec.	Total
1960	127.71	103.67	99.26	98.13	107.55	99.16	78.81	89.58	86.42	99.91	112.78	119.01	1221.99
1961	102.49	85.23	96.16	92.30	120.82	97.39	81.31	82.15	78.91	110.41	114.04	108.78	1169.99
1962	119.28	89.98	101.85	86.88	104.32	101.76	83.93	82.70	75.90	114.43	123.75	123.29	1208.07
Σ	349.48	278.88	297.27	277.31	332.69	298.31	244.05	254.43	241.23	324.75	350.57	351.08	3600.05
Average ratio for month	117	93	99	92	111	100	81	85	80	108	117	117	
Average ratio for entire period	117	93	99	92	111	100	81	85	80	108	117	117	

By coincidence, the average ratio to trend for the entire period is exactly equal to 100 for the illustrative example. Thus, the last step, the leveling of the seasonal index numbers to ensure that they average to 100, is unnecessary.

Underlying Rationale of the Ratio to Trend Method

The first step in the procedure involves dividing the observed value of the variable for each month of each year by the trend component as of that same period of time.

$$\frac{\text{observed value, Jan. 1960}}{\text{trend, Jan. 1960}} = \frac{\widehat{T_{\text{Jan. 1960}}} \times C_{\text{Jan. 1960}} \times I_{\text{Jan. 1960}} \times S_{\text{Jan.}}}{\widehat{T_{\text{Jan. 1960}}}}$$

$$= 1.0 \times C_{\text{Jan. 1960}} \times I_{\text{Jan. 1960}} \times S_{\text{Jan.}}$$

In effect, the computation has the impact of converting the trend value for each month of each year to the common value 1.0. Graphically, the adjustment has the following impact:

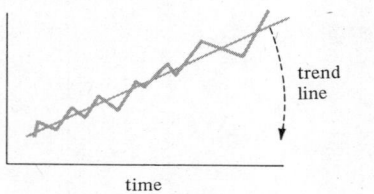 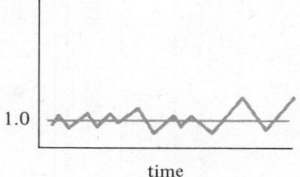

In the final step in the procedure, the representative ratio to trend for each of the 12 months is divided by the average ratio to trend for the entire period. This has the following effect:

$$\frac{\text{average ratio to trend for Jan.}}{\text{average ratio to trend for the entire period}} = \frac{1.0 \times \begin{array}{c}\text{average} \\ \text{cyclical for} \\ \text{Jan.}\end{array} \times \begin{array}{c}\text{average} \\ \text{seasonal for} \\ \text{Jan.}\end{array} \times \begin{array}{c}\text{average} \\ \text{irregular} \\ \text{Jan.}\end{array}}{1.0 \times \begin{array}{c}\text{average} \\ \text{cyclical for} \\ \text{entire period}\end{array} \times \begin{array}{c}\text{average} \\ \text{seasonal for} \\ \text{entire period}\end{array} \times \begin{array}{c}\text{average} \\ \text{irregular for} \\ \text{entire period}\end{array}}$$

average irregular value for Jan. = 1.0

Sometimes the irregular influence exerts an uplifting influence on the series and sometimes it exerts a depressing influence. Over time, these effects should average out. In addition, any irregular influence exerting significant influence on an observation should have been eliminated.

average irregular value for entire period = 1.0

for the same reason as above.

average seasonal value for the entire period = 1.0

By definition, the average of the seasonal index numbers for a year should always equal 1.0 (or 100 in whole-number form). Clearly, the average of the seasonal index numbers for a number of years should also average to 1.0.

The average seasonal value for January should equal the typical uplifting or depressing influence exerted on the series during the month of January due to seasonal influences. Therefore,

$$\frac{\text{average ratio to trend for Jan.}}{\text{average ratio to trend for entire period}} = \frac{1.0 \times \text{average cyclical for Jan.} \times \text{average seasonal for Jan.} \times 1.0}{1.0 \times \text{average cyclical for entire period} \times 1.0 \times 1.0}$$

Quite clearly, this method of computation will generate a true seasonal index only when

average cyclical Jan. = average cyclical for entire period

This will only be true when the cyclical pattern is characterized by negligible amplitude.

Thus, the ratio to trend method has the weakness of requiring negligible cyclical amplitude if a reasonable approximation to the true seasonal index is to be generated.

26.4 THE RATIO TO THE 12-MONTH CENTERED MOVING AVERAGE METHOD

The procedure:

1. Compute a 12-month moving total column (column 3 in Table 26.3). The moving total represents the sum of the observations for 12 consecutive months. The moving total is advanced by dropping the observation for the first month in the 12-month moving total and adding the observation of the next month encountered. To illustrate:

total for the 12 months			
7/1959–6/1960		=	2061
subtract	7/1959	=	−176
			1885
add	7/1960	=	+119
total for the 12 months			
8/1959–7/1960		=	2004
subtract	8/1959	=	−141
			1863
add	8/1960	=	+135
total for the 12 months			
9/1959–8/1960		=	1998

etc.

Table 26.3 Computation of ratio to 12-month centered moving average: hog receipts at Chicago Public Stockyards in thousands, 1960–1962

Col. 1	Col. 2	Col. 3	Col. 4	Col. 5	Col. 6
Year and month (middle of the month)	Receipts (middle of the month)	12-month moving total (end of sixth month)	12-month moving average (end of sixth month)	Centered 12-month moving average (middle of the month)	Ratio to the 12-month moving average (middle of the month)
1959					
July	176				
Aug.	141				
Sept.	165				
Oct.	177				
Nov.	214				
Dec.	222				
1960		2061	171.7		
Jan.	195			169.4	115 Jan.
		2004	167.0		
Feb.	158			166.8	95 Feb.
		1998	166.5		
Mar.	151			165.1	92 Mar.
		1963	163.6		
Apr.	149			162.5	92 Apr.
		1936	161.3		
May	163			159.5	102 May
		1891	157.6		
June	150			155.8	96 June
		1847	153.9		
July	119			152.2	78 July
		1805	150.4		
Aug.	135			149.1	91 Aug.
		1774	147.8		
Sept.	130			147.5	88 Sept.
		1766	147.2		
Oct.	150			146.7	102 Oct.
		1754	146.2		
Nov.	169			146.9	115 Nov.
		1770	147.5		
Dec.	178			147.3	121 Dec.
		1764	147.0		
1961					
Jan.	153			147.1	104 Jan.
		1765	147.1		
Feb.	127			146.5	87 Feb.
		1751	145.9		
Mar.	143			145.4	98 Mar.
		1737	144.8		
Apr.	137			145.3	94 Apr.
		1749	145.8		
May	179			145.7	123 May
		1747	145.6		
June	144			144.8	99 June
		1728	144.0		

Table 26.3 *(continued)*

Col. 1	Col. 2	Col. 3	Col. 4	Col. 5	Col. 6
Year and month (middle of the month)	Receipts (middle of the month)	12-month moving total (end of sixth month)	12-month moving average (end of sixth month)	Centered 12-month moving average (middle of the month)	Ratio to the 12-month moving average (middle of the month)
				144.9	83 July
		1749	145.8		
Aug.	121			146.0	83 Aug.
		1753	146.1		
Sept.	116			146.3	79 Sept.
		1758	146.5		
Oct.	162			146.1	111 Oct.
		1747	145.6		
Nov.	167			144.5	116 Nov.
		1719	143.3		
Dec.	159			143.4	111 Dec.
		1722	143.5		
1962					
Jan.	174			143.6	121 Jan.
		1723	143.6		
Feb.	131			143.5	91 Feb.
		1721	143.4		
Mar.	148			143.1	103 Mar.
		1714	142.8		
Apr.	126			142.9	88 Apr.
		1716	143.0		
May	151			143.4	105 May
		1726	143.8		
June	147			144.6	102 June
		1743	145.3		
July	121			144.8	84 July
		1732	144.3		
Aug.	119			144.5	82 Aug.
		1736	144.7		
Sept.	109			144.5	75 Sept.
		1732	144.3		
Oct.	164			145.4	113 Oct.
		1757	146.4		
Nov.	177			146.1	121 Nov.
		1749	145.8		
Dec.	176			144.9	121 Dec.
		1728	144.0		
1963					
Jan.	163				
Feb.	135				
Mar.	144				
Apr.	151				
May	143				
June	126				

Plotted on a graph, the 12-month total (being period data) is located directly over the end of the sixth month. The 12-month total is recorded in column 3 between the designations for the sixth and seventh months in the moving total to reflect this relationship.

2. Divide each of the 12-month moving totals by 12 to convert to a 12-month moving average (column 4). The 12-month moving average is the representative value for the 12-month period. Thus, if plotted on a graph, the 12-month moving average is also plotted directly over the end of the sixth month in the moving total. The 12-month moving average is recorded in column 4 between the designations for the sixth and seventh months in the moving total to reflect this relationship.

3. Center the moving averages by obtaining the mean of the 12-month moving average centered at the beginning of a month and the 12-month moving average centered at the end of the same month. If plotted, the mean of these two values would always be located above the middle of the month. This convention is reflected by the location of the entries in column 5 of Table 26.3.

4. Divide the actual observed value for each month by the 12-month centered moving average for that same month. Both values are considered as being located at the middle of the month. Thus, the resulting ratio reflects the relative magnitudes of the two series as of the same point in time. The resulting quotient (multiplied by 100) is referred to as the ratio to the 12-month centered moving average.

5. Collect all the January ratios and select a representative January ratio. The modified mean is often used to obtain the representative value (see Chapter 3, p. 59). Repeat the procedure for each month. The result is presented in Table 26.4.

Table 26.4 Computation of seasonal index by averaging ratios to moving average

	Jan.	Feb.	Mar.	Apr.	May	June	July	Aug.	Sept.	Oct.	Nov.	Dec
1960	115	95	92	92	102	96	78	91	88	102	115	121
1961	104	87	98	94	123	99	83	83	79	111	116	111
1962	121	91	103	88	105	102	84	82	75	113	121	121
Σ	340	273	293	274	330	297	245	256	242	326	352	353
Average	113	91	98	91	110	99	82	85	81	109	117	118

6. Level the 12 representative values such that their sum equals 1200 or their mean value equals 100. This has the effect of generating the seasonal index numbers expressed in conventional form (i.e., expressed to the base 100).

Before adjustment, the representative ratios sum to 1194 and average to 99.5. The index is then leveled by expressing each of the 12 seasonal index numbers as a percentage of their average value, 99.5. After leveling, the index numbers reflect the percentage variation around the average value 100 due to seasonal influences rather than around the average value 99.5.

	Jan.	Feb.	Mar.	Apr.	May	June
Index leveled	113.6	91.5	98.5	91.5	110.5	99.5

	July	Aug.	Sept.	Oct.	Nov.	Dec.
Index leveled	82.4	85.4	81.4	109.5	117.6	118.6

Underlying Rationale of the Ratio to the 12-Month Centered Moving Average Method

The actual observation for a given month (say, June 1960) reflects the combined influence of

$$T_{\substack{\text{Middle of} \\ \text{June 1960}}} \times C_{\substack{\text{Middle of} \\ \text{June 1960}}} \times S_{\text{June}} \times I_{\substack{\text{Middle of} \\ \text{June 1960}}}$$

When a 12-month moving average is constructed, 12 of these products are averaged together (one for each month).

$$
\begin{array}{cccc}
T_{\text{June 1960}} & \times\ C_{\text{June 1960}} & \times\ S_{\text{June}} & \times\ I_{\text{June 1960}} \\
T_{\text{July 1960}} & \times\ C_{\text{July 1960}} & \times\ S_{\text{July}} & \times\ I_{\text{July 1960}} \\
T_{\text{Aug. 1960}} & \times\ C_{\text{Aug. 1960}} & \times\ S_{\text{Aug.}} & \times\ I_{\text{Aug. 1960}} \\
\vdots & \vdots & \vdots & \vdots \\
T_{\text{May 1961}} & \times\ C_{\text{May 1961}} & \times\ S_{\text{May}} & \times\ I_{\text{May 1961}} \\
\hline
\end{array}
$$

$\sum/12$	$\times$	$\sum/12$	$\times\ \sum/12\ \times$	$\sum/12$	$=$ 12-month moving average
average trend value for the 12-month period		average cyclical value for the 12-month period			

seasonal pattern is always averaged out when 12 consecutive monthly values are averaged together $1200/12 = 100$
uplifting and depressing random irregular influences tend to be averaged out. Therefore $\text{Ave}(I) = 1.0$

Examining the trend component: If trend is linear, the average trend value and the individual trend value as of the end of the sixth month are approximately equal (graphs A and B).

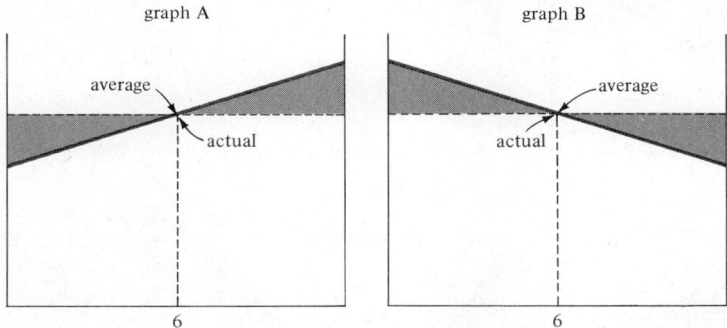

If trend is parabolic, the portion of the second-degree parabola assigned to any one 12-month period does not depart significantly from a straight line (graphs C and D). The one exception is when the parabola changes direction during the 12-month period (as illustrated in graphs E and F).

In graph E, trend as of the end of the sixth month would be understated if represented by the average trend figure. In graph F, it would be overstated. If

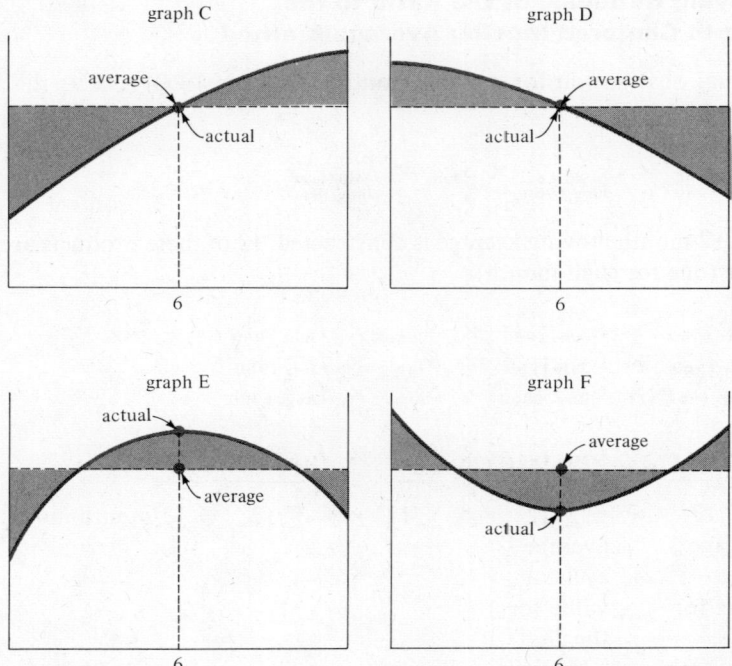

either of the latter two cases occur, however, the probability of elimination in
the computation of the modified mean in step 5 is extremely high.

Similar graphical arguments can be presented to demonstrate that cyclical
as of the end of the sixth month tends to be closely approximated by the cyclical
component in the 12-month moving average.

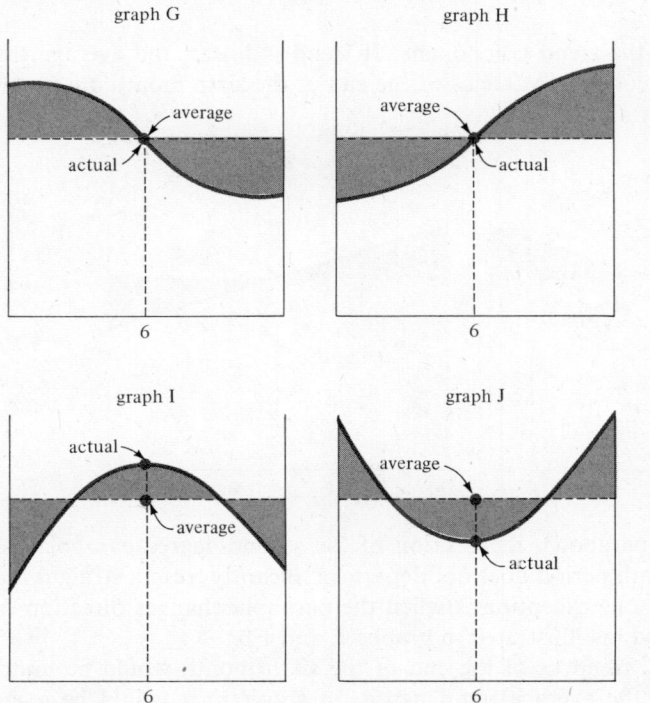

The only exception would be in a case such as graph I or graph J, where a cyclical turning point is experienced during the 12-month period. Once again, however, should either of those situations occur, the probability of elimination is extremely high when computing the modified mean in step 5.

The 12-month moving average can thus be treated as representing *only* the influences of trend and cyclical as of the end of the sixth month. The irregular and seasonal influences are averaged out in the computational process.

$$\text{12-month moving average} = T_{\text{end of the sixth month}} \times C_{\text{end of the sixth month}} \times 1.0 \times 1.0$$
$$= T_{\text{end of the sixth month}} \times C_{\text{end of the sixth month}}$$

In order to place the original observations and the moving average into the same time sequence, it is necessary to center the moving average. The result of averaging two consecutive moving averages is as follows:

$$\text{end of June} = T_{\text{end of June}} \times C_{\text{end of June}}$$
$$\text{end of July} = T_{\text{end of July}} \times C_{\text{end of July}}$$

12-month moving average:

$$\frac{\substack{\text{12-month moving} \\ \text{average centered} \\ \text{on June 30}} + \substack{\text{12-month moving} \\ \text{average centered} \\ \text{on July 30}}}{2} = T_{\text{middle of July}} \times C_{\text{middle of July}}$$

Thus, a 12-month centered moving average can be considered as representing trend and cyclical as of the middle of the month on which it is centered.

It follows that when one computes the ratio to the 12-month centered moving average, one in effect performs the following division:

$$\frac{\substack{\text{Observed value for} \\ \text{month of June 1960}}}{\substack{\text{12-month centered moving} \\ \text{average June 1960}}} = \frac{T_{\substack{\text{middle} \\ \text{of June}}} \times C_{\substack{\text{middle} \\ \text{of June}}} \times S_{\text{June}} \times I_{\substack{\text{middle} \\ \text{of June}}}}{T_{\substack{\text{middle} \\ \text{of June}}} \times C_{\substack{\text{middle} \\ \text{of June}}}}$$
$$= S_{\text{June}} \times I_{\text{middle of June}}$$

The next step is the selection of a representative June ratio, using the modified mean. Eliminate the three largest and three smallest ratios out of, say, 10 ratios. This removes from consideration any ratios containing significant trend and/or cyclical components that slipped past the earlier screening process. In addition, any ratios unduly influenced by unusual irregular influences also are eliminated. Obtain the arithmetic mean of the remaining ratios. This process averages out any random irregular influences and leaves a pure seasonal pattern as a residual.

26.5 CYCLICAL FLUCTUATIONS

The Residual Approach

Trend adjusted to reflect the influence of seasonal variation (i.e., $T \times S$) is referred to as *statistical normal*. Dividing the observed value for a given month and year by the value of statistical normal as of that same month and year isolates cyclical and irregular influences.

To illustrate:

$$\frac{O_{\text{June 1960}}}{\text{Statistical normal}_{\text{June 1960}}} = \frac{T_{\text{June 1960}} \times C_{\text{June 1960}} \times S_{\text{June}} \times I_{\text{June 1960}}}{T_{\text{June 1960}} \times S_{\text{June}}}$$
$$= C_{\text{June 1960}} \times I_{\text{June 1960}}$$

A 5-month moving average is fitted to the resulting ratios. The 5-month moving average centered on a given month is interpreted as the index of cyclical activity for that month. It is assumed that irregular influences are averaged out in the process of computing the 5-month moving average. This technique for generating a cyclical index is referred to as the residual approach. The computational process is illustrated in Table 26.5 using the series, hog receipts at Chicago Public Stockyards, 1960–1962. Trend values are obtained from Table 26.1, and the seasonal index numbers are obtained from Table 26.4 (after adjustment). The resulting indices of cyclical activity are depicted in Figure 26.1.

Table 26.5 Computation of indices of cyclical activity: hog receipts at Chicago Public Stockyards in thousands 1960–1962

Year	Month	Y	Y_c or trend	Seasonal	Statistical normal	Ratio to statistical normal	5-month moving total	5-month moving average
1960	Jan.	195	152.692	1.136	173.458	1.12		
	Feb.	158	152.408	0.915	139.453	1.13		
	Mar.	151	152.124	0.985	149.842	1.01	5.30	1.06
	Apr.	149	151.840	0.915	138.934	1.07	5.18	1.04
	May	163	151.556	1.105	167.469	0.97	5.01	1.00
	June	150	151.272	0.995	150.516	1.00	5.05	1.01
	July	119	150.988	0.824	124.414	0.96	5.04	1.01
	Aug.	135	150.704	0.854	128.701	1.05	4.98	1.00
	Sept.	130	150.420	0.814	122.442	1.06	4.94	0.99
	Oct.	150	150.136	1.095	164.399	0.91	4.98	1.00
	Nov.	169	149.852	1.176	176.226	0.96	4.83	0.97
	Dec.	178	149.568	1.186	177.388	1.00	4.70	0.94
1961	Jan.	153	149.284	1.136	169.587	0.90	4.77	0.95
	Feb.	127	149.000	0.915	136.335	0.93	4.82	0.96
	Mar.	143	148.716	0.985	146.485	0.98	4.91	0.98
	Apr.	137	148.432	0.915	135.815	1.01	4.99	1.00
	May	179	148.148	1.105	163.704	1.09	5.05	1.01
	June	144	147.864	0.995	147.125	0.98	5.03	1.01
	July	120	147.580	0.824	121.606	0.99	4.99	1.00
	Aug.	121	147.296	0.854	125.790	0.96	4.91	0.98
	Sept.	116	147.012	0.814	119.668	0.97	4.90	0.98
	Oct.	162	146.728	1.095	160.667	1.01	4.83	0.97
	Nov.	167	146.444	1.176	172.218	0.97	4.92	0.98
	Dec.	159	146.160	1.186	173.346	0.92	4.93	0.99
1962	Jan.	174	145.876	1.136	165.715	1.05	4.95	0.99
	Feb.	131	145.592	0.915	133.217	0.98	4.93	0.99
	Mar.	148	145.308	0.985	143.128	1.03	4.96	0.99
	Apr.	126	145.024	0.915	132.697	0.95	4.94	0.99
	May	151	144.740	1.105	159.938	0.95	4.98	1.00
	June	147	144.456	0.995	143.734	1.03	4.92	0.98
	July	121	144.172	0.824	118.798	1.02	4.90	0.98
	Aug.	119	143.888	0.854	122.880	0.97	5.00	1.00
	Sept.	109	143.604	0.814	116.894	0.93	5.02	1.00
	Oct.	164	143.320	1.095	156.935	1.05	5.04	1.01
	Nov.	177	143.036	1.176	168.210	1.05		
	Dec.	176	142.752	1.186	169.304	1.04		

Source: Department of Agriculture.

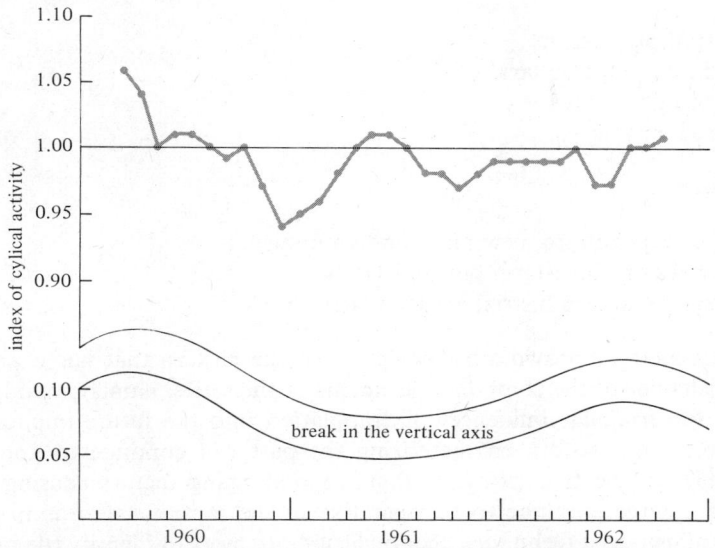

Figure 26.1 An index of cyclical activity: hog receipts at Chicago
public stockyards: 1960–1962. (From Table 26.5.)

Cyclical influence is a long-term influence. A considerably longer period than
3 years should be used to compute a cyclical index. Due to the irregularity from
one cycle to the next, the analyst cannot extrapolate future cyclical activity by
simply extending the historically observed pattern. Nevertheless, an investigation
of the historically observed pattern does provide information on amplitude (i.e.,
how sensitive the series is to cyclical influences) and the timing of turning points
in the cyclical pattern. The decision maker can attempt to predict turning points
in the series under investigation by determining other series that tend to lead the
series under investigation in the timing of turning points in the level of activity
due to cyclical influences. The National Bureau of Economic Research has
conducted many studies of this nature and has compiled lists of series that tend
to lead, coincide, or lag behind the general business cycle in the timing of cyclical
turning points.[2] The status of these indicators is published monthly by the Bureau
of the Census in the *Business Conditions Digest*. No one of these indicators is a
perfect predictor, but considered jointly they can provide valuable information
to the business planner. Examples of leading, coincident, and lagging series are
provided below:

Leading series:

  Change in consumer installment debt
  Corporation profits after taxes
  New building permits, private housing units
  New orders, durable goods
  Contracts and orders, plant and equipment

[2] G. Moore and J. Shiskin, *Indicators of Business Expansions and Contractions*, New York:
National Bureau of Economic Research, 1967.

Coincident series:

> Industrial production
> GNP in constant dollars
> Sales of retail stores
> Personal income

Lagging series:

> Business expenditure, new plant and equipment
> Bank rates on short-term business loans
> Commercial and industrial loans outstanding

Time series analysis provides a description of the pattern that has characterized the past behavior of the component elements of the series, namely, trend, cyclical, seasonal, and irregular influences. Extrapolation into the future implies a belief that the behavior pattern characterizing the past will continue to apply in the future. This will be true provided that the underlying factors causing the past pattern to emerge continue to be operative in the same fashion as in the past. A more sophisticated technique, econometrics, attempts to identify the underlying explanatory factors and the nature of the reaction of the variable in question to changes in the magnitude of the underlying explanatory variables. Given the projected status of the underlying explanatory factors and the relationship between these factors and the variable to be predicted, econometric techniques are used to provide an estimate of the value of the variable of interest. Econometric techniques are beyond the scope of this text. The interested reader is referred to the following sources.

References

1. Brennan, Michael, *Preface to Econometrics*, 2nd ed., Cincinnati: Southwestern Publishing Co., 1966.
2. Johnston, John, *Econometric Methods*, New York: McGraw-Hill, 1963.
3. Malinvaud, Edmond, *Statistical Methods of Econometrics*, New York: American Elsevier, 1970.

The reader interested in a more detailed treatment of time series analysis is referred to:

4. Brown, Robert G., *Smoothing, Forecasting and Prediction of Discrete Time Series*, Englewood Cliffs, N.J.: Prentice-Hall, 1963.

PROBLEMS

Problem 26.1

Exactly what kind of information is conveyed by a seasonal index number, for example, a seasonal index number for January of 120?

Problem 26.2

Explain how the ratio to trend method of calculating a seasonal index eliminates trend, cyclical, and irregular factors.

Problem 26.3

What underlying assumption must correspond to reality for a particular time series if the ratio to trend method of calculating a seasonal index is not to result in a distortion of the true seasonal pattern?

Problem 26.4

Discuss ways in which a business decision maker might utilize a seasonal index.

Problem 26.5

Thoroughly explain the theory behind the ratio to the 12-month centered moving average method of computing seasonal index numbers. Specifically explain how trend, cyclical, and irregular influences are removed.

Problem 26.6

What is "statistical normal"?

Problem 26.7

If the total of the 12 index numbers does not total to 1200, an adjustment must be made. Explain the nature of the adjustment and why it is necessary.

Problem 26.8

Forecasting the impact of the seasonal pattern is much more certain than forecasting the impact of the cyclical pattern. Why?

Problem 26.9

The long-run secular trend of sales for your firm is described by the equation

$$Y_c = \$80,000 + \$2,000X$$

where X represents the passage of a given number of months and $X = 0$ for June 1972. The seasonal pattern is described by the following set of indices:

Jan.	140	May	70	Sept.	95
Feb.	110	June	85	Oct.	100
Mar.	95	July	90	Nov.	110
Apr.	90	Aug.	95	Dec.	120

1. Using the above information, determine the value of "statistical normal" for March 1974.
2. Assuming that the seasonal indices and the trend equation are correct descriptions of reality, would you expect the observed value of sales for March 1974 to correspond to the value computed in part 1? Why or why not?

Problem 26.10

Assume:

  a. Secular trend of sales for the next calendar year is described by the equation

$$Y_c = \$100,000 + 3000X$$

  where $X = 0$ for January and X in the equation represents the passage of a given number of months.

b. The seasonal pattern of sales is described by the following set of indices:

Jan.	60	May	100	Sept.	140
Feb.	70	June	110	Oct.	120
Mar.	80	July	120	Nov.	100
Apr.	90	Aug.	130	Dec.	80

c. Cyclical influence for the next calendar year is *predicted* as follows:

Jan.	101	May	103	Sept.	107
Feb.	102	June	105	Oct.	108
Mar.	102	July	105	Nov.	108
Apr.	103	Aug.	106	Dec.	109

d. Trend, seasonal, and cyclical are related in a multiplicative manner.

e. Dollar sales in any month are converted to cash in the following pattern:

40 percent realized in the month of the sale
30 percent realized in the month following the sale
20 percent realized in the second month following the sale
10 percent realized in the third month following the sale

f. Sales for October, November, and December of the current year were as follows:

Oct.	$75,000
Nov.	$95,000
Dec.	$110,000

Required:
1. Project the cash inflow from sales for the 12 months comprising the next calendar year.
2. How might the above information prove useful?

Problem 26.11

Sales for September were $200,000 and sales for October are $180,000. A business acquaintance commented that October must have been a bad month for you. Given that the seasonal index number for September is 100 and that for October is 70, react to his comments.

Problem 26.12

1. Using the following values, calculate the ratio to the centered moving average for the second quarter of 1966.

Output

Year	First quarter (Jan.–Mar.)	Second quarter (Apr.–June)	Third quarter (July–Sept.)	Fourth quarter (Oct.–Dec.)
1964	30	100	220	100
1965	50	120	230	90
1966	45	115	246	94
1967	55	125	235	105
1968	60	130	225	115

2. Interpret the value obtained in the computation in part 1.

Problem 26.13

H. G. Wilson and Sons: quarterly ratios to trend

Year	First quarter (Jan.–Mar.)	Second quarter (Apr.–June)	Third quarter (July–Sept.)	Fourth quarter (Oct.–Dec.)
1965	120	400	80	200
1966	75	270	120	155
1967	120	660	150	270
1968	200	800	280	320

Assume that the entries in the table above were generated in order to determine seasonal indices by the ratio to trend method.

1. Describe the procedure by which the above values were obtained.
2. Complete the calculations leading to the generation of the seasonal indices.
3. Given that the forecast for next year is 400,000 units, project how sales will be broken down on a quarterly basis.

Problem 26.14

Jones, Inc., Average monthly output

Year	First quarter (Jan.–Mar.)	Second quarter (Apr.–June)	Third quarter (July–Sept.)	Fourth quarter (Oct.–Dec.)
1966	80	105	160	182
1967	84	107	145	180
1968	92	98	158	175
1969	78	110	154	180
1970	65	112	159	178

Trend has been calculated using the above data by the arithmetic least-squares method. The value of a in the least-squares formula (as of the middle of 1955) is 130.225, while the value of the *annual b* is -0.6. Determine the ratio to trend ordinate for the last quarter of 1968.

Problem 26.15

Justify the employment of a modified mean to determine a representative ratio to trend ordinate for each month.

Problem 26.16

Total sales of gas in billions of therms by months

Year	Jan.	Feb.	Mar.	Apr.	May	June	July	Aug.	Sept.	Oct.	Nov.	Dec.
1960	10.3	10.2	10.8	8.6	6.9	6.2	5.6	5.6	5.6	6.1	7.5	9.5
1961	11.4	10.8	9.4	9.0	7.5	6.3	5.7	5.7	5.8	6.7	8.0	10.0
1962	12.3	11.5	11.1	9.6	7.3	6.3	6.0	6.1	6.3	6.9	8.5	10.5
1963	12.5	12.7	11.8	9.1	8.0	7.1	6.7	6.6	6.5	7.0	8.1	11.2
1964	13.8	12.6	12.2	10.8	8.6	7.6	7.1	7.1	7.1	8.0	9.2	11.8
1965	13.0	13.3	13.1	11.7	8.7	7.7	7.5	7.3	7.3	8.2	9.8	12.0
1966	13.6	14.7	13.7	11.7	10.1	8.6	8.0	7.8	8.0	8.9	10.7	12.8
1967	15.0	14.3	14.3	11.8	10.6	9.1	8.3	8.2	8.3	9.2	11.5	14.1
1968	16.7	15.9	15.6	12.4	11.0	9.8	9.1	8.8	8.8	9.6	11.9	15.3
1969	17.8	16.4	16.4	13.8	11.2	10.1	9.9	9.6	9.5	10.3	12.9	16.1

Source: 1970 Gas Facts, American Gas Association.

Determine monthly seasonal indices for total sales of gas, utilizing the ratio to the 12-month centered moving average method.

Problem 26.17

Northwest Airlines, Inc. revenue passenger miles (in millions): quarterly data

Year	First quarter (Jan.–Mar.)	Second quarter (Apr.–June)	Third quarter (July–Sept.)	Fourth quarter (Oct.–Dec.)
1963	498.6	590.2	715.5	656.6
1964	663.0	793.6	898.4	793.9
1965	810.2	994.6	1217.4	1084.5
1966	1151.9	1425.3	1099.1[a]	1398.5
1967	1479.5	1730.1	1970.5[a]	1619.9
1968	1678.1	1903.8	1943.2	1601.4
1969	1624.8	1937.7	2214.3	1704.2
1970	1680.8	1979.1	1023.8[a]	726.6[a]

[a] Strike period.
Source: *Moody's Transportation Manual*, 1971.

1. Determine quarterly seasonal indices for revenue passenger miles, utilizing the ratio to the *four-quarter* centered moving average method.
2. What problems do the strike periods cause? How did you handle these problems?
3. Do you think the ratio to trend method would provide an undistorted representation of the true seasonal indices for airline revenue passenger miles? Why or why not?

Problem 26.18

U.S. production of beer (in millions of barrels)

Year	First quarter (Jan.–Mar.)	Second quarter (Apr.–June)	Third quarter (July–Sept.)	Fourth quarter (Oct.–Dec.)
1958	20.3	25.7	24.6	19.5
1959	19.9	26.9	26.5	19.8
1960	20.9	27.4	27.4	19.7
1961	21.4	27.0	26.4	20.3
1962	21.7	28.1	26.3	20.7
1963	22.3	28.6	27.6	22.1
1964	24.1	29.2	29.8	22.8
1965	24.9	30.5	29.2	23.7
1966	25.5	31.5	31.3	24.8
1967	27.2	33.2	30.3	25.8
1968	27.7	33.7	33.5	27.5
1969	28.8	32.9	36.7	29.0
1970	30.7	38.3	34.7	29.4

Source: Internal Revenue Service.

Determine quarterly seasonal indices for U.S. production of beer, utilizing the ratio to trend method. Fit trend to the quarterly data with an arithmetic least-squares straight line.

Problem 26.19

Newspaper advertising lineage (52 cities) in the United States in millions of lines

Year	First quarter (Jan.–Mar.)	Second quarter (Apr.–June)	Third quarter (July–Sept.)	Fourth quarter (Oct.–Dec.)
1962	637.1	730.5	676.3	754.4
1963	626.7	752.9	690.3	786.7
1964	669.0	788.0	712.7	803.7
1965	694.6	823.8	772.0	874.1
1966	753.3	880.4	816.7	903.8
1967	753.0	873.5	786.0	885.2
1968	749.8	863.2	820.6	947.6
1969	811.2	929.4	861.5	983.1
1970	778.0	901.5	838.1	926.1

Source: Survey of Current Business.

Determine quarterly seasonal indices for newspaper advertising lineage, utilizing the ratio to the *four-quarter* centered moving average method.

Problem 26.20

Production of creamery butter in factories in the United States in millions of pounds

Year	First quarter (Jan.–Mar.)	Second quarter (Apr.–June)	Third quarter (July–Sept.)	Fourth quarter (Oct.–Dec.)
1962	424.1	465.0	316.9	331.1
1963	390.7	439.0	294.7	295.2
1964	396.1	441.4	292.4	311.4
1965	397.6	418.2	261.7	247.3
1966	292.3	335.4	228.3	256.1
1967	325.4	374.9	263.6	258.9
1968	315.9	350.9	247.9	250.1
1969	306.8	337.9	238.0	238.4
1970	298.5	338.2	246.4	252.2

Source: Crop Reporting Board, U.S.D.A.

Determine quarterly seasonal indices for production of creamery butter utilizing the ratio to the *four-quarter* centered moving average method.

Problem 26.21

What underlying assumptions must correspond to reality for a particular time series if the residual method is to provide an appropriate description of the cyclical pattern?

Problem 26.22

What impact would an overstatement in the estimates of secular trend have on the cyclical indices obtained by the residual method?

Problem 26.23

To what do the labels "leading," "coincident," and "lagging" refer when discussing cyclical indices?

Problem 26.24

Referring to Problem 26.17:
1. Deseasonalize the data by dividing each of the observed quarterly values by the appropriate quarterly seasonal index.
2. Plot the deseasonalized time series on a chart.
3. Visually evaluate the series and decide which of the following would provide the best description of secular trend:
   a. arithmetic least-squares line
   b. logarithmic least-squares line
   c. least-squares second-degree parabola
4. Determine the equation for the trend line selected in part 3.
5. Compute the value of trend for each quarter of the years 1963–1970.
6. Determine the cyclical indices by the residual method.
7. Interpret the results obtained in part 6.

Problem 26.25

Referring to Problem 26.18:
1. Using the seasonal indices and secular trend values generated in Problem 26.18, determine the cyclical indices by the residual method.
2. Interpret the results obtained in part 1.

Problem 26.26

Referring to Problem 26.19:
1. Deseasonalize the data by dividing each of the observed quarterly values by the appropriate quarterly seasonal index.
2. Plot the deseasonalized time series on a chart.
3. Visually evaluate the series and decide which of the following would provide the best description of secular trend:
   a. arithmetic least-squares line
   b. logarithmic least-squares line
   c. least-squares second-degree parabola
4. Determine the equation for the trend line selected in part 3.
5. Compute the value of trend for each quarter of the years 1962–1970.
6. Determine the cyclical indices by the residual method.
7. Interpret the results obtained in part 6.

SIMPLE REGRESSION AND CORRELATION ANALYSIS

A farming cooperative has undertaken an investigation to determine the nature of the relationship between yield per acre of a particular crop and the intensity of application of a fertilizer compound. A technique referred to as regression analysis is to be used to generate an estimate of the relationship. In regression analysis, one hypothesizes that a relationship exists between the variable to be predicted and one or more explanatory variables. To illustrate: One might hypothesize that the variable Y is a function of the variables X, Z, and W:

$$Y = f(X, Z, W)$$

Multiple regression analysis, which allows for more than one explanatory variable, is treated in Chapter 28. In this chapter, discussion is restricted to simple regression analysis, which involves only one explanatory variable:

$$Y = f(X)$$

27

$y_c = a + bX$

y = dependent

X = independent

(Regression analysis is concerned with specification of the nature of the relationship and determination of the values of the coefficients in the regression function.) Discussion will be restricted to linear regression functions.[1] In general, linear relationships take the form $Y_c = a + bX$. Regression analysis is used to specify the value of the coefficients of the regression function, namely, a and b. The regression function or regression line is the line of average relationship between the two variables. The regression line indicates the *average* value of Y associated with a particular value of X, where Y is defined as the dependent variable and X is defined as the independent variable. The independent variable is the variable whose value is given, and the dependent variable is the variable whose value is to be predicted. In the illustrative example, the regression line provides an estimate of the average yield per acre associated with a given intensity of fertilizer application.

To facilitate comprehension of regression analysis and the role of the regression line, let us assume that the regression function is known and appears as represented in Figure 27.1.[2] In accordance with custom, values of X, the independent variable, are recorded on the horizontal axis, and values of Y, the dependent variable, are recorded on the vertical axis.

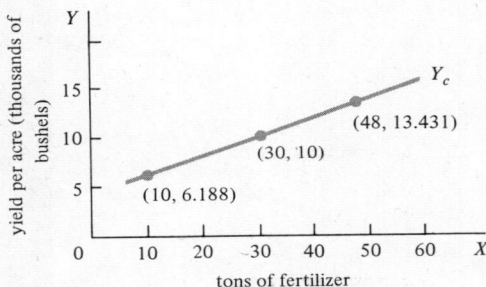

Figure 27.1

Interpretation: Given a fertilizer application of 10 tons, at $X = 10$, raise a perpendicular until it intersects the regression line. The associated value on the vertical axis indicates that, *on the average*, a yield of 6188 bushels is associated with a fertilizer application of 10 tons.[3] This is not to be interpreted as indicating that each acre tract administered a fertilizer application of 10 tons will yield 6188 bushels. Some will yield more than 6188 bushels and others less. If all performances were to be averaged together, however, the estimated *average* performance of acre tracts with a fertilizer application of 10 tons would be 6188 bushels. A similar interpretation applies to the average value of Y associated with each of the other X values. In effect, there is a probability distribution of possible values of Y

---

[1] Other relationships are possible, of course. If the analyst concludes that the relationship is curvilinear (say, parabolic), the general form of the regression function is $Y_c = a + bX + cX^2$, and regression analysis is used to specify the value of the coefficients a, b, and c.

[2] The computational procedure underlying the regression line depicted in Figure 27.1 is presented later in the chapter.

[3] Similarly, an average yield of 10,000 bushels per acre is associated with a fertilizer application of 30 tons, and an average yield of 13,431 bushels per acre is associated with a fertilizer application of 48 tons.

associated with each feasible magnitude of fertilizer application. These probability distributions are referred to as *conditional probability distributions* to reflect the fact that the particular probability distribution referred to is conditional upon the value of X with which it is associated. The means of these distributions are referred to as *conditional means*. Recall the notation μ_Y previously used when referring to the universe mean of the Y values. A new symbol, $\mu_{Y.X}$, is now introduced. The symbol $\mu_{Y.X}$ represents the mean of the Y values associated with the value of X indicated to the right of the period and thus symbolizes a conditional mean.

When $X = 10$, $\mu_{Y.X} = \mu_{Y.10} = 6.188$.
When $X = 30$, $\mu_{Y.X} = \mu_{Y.30} = 10$.
When $X = 48$, $\mu_{Y.X} = \mu_{Y.48} = 13.431$.

By definition, the regression line passes through each of the conditional means. Thus, the conditional means are the values on the regression line. The dispersion in each of the conditional probability distributions is measured by the distribution's standard deviation. Consistent with previous definitions, these are referred to as *conditional standard deviations*. They are designated $\sigma_{Y.X}$ when referring to the universe and $\hat{\sigma}_{Y.X}$ when estimated from a sample.

In the following section, construction of the regression line and measurement of the improvement in the estimation of the value of Y attributable to use of the regression function will be discussed. The analysis is built on the following assumptions: (1) The regression line is a straight line, (2) all the conditional probability distributions of Y are normally distributed, and (3) all the conditional standard deviations are equal. The regression line and the associated conditional probability distributions of Y appear as depicted in Figure 27.2.

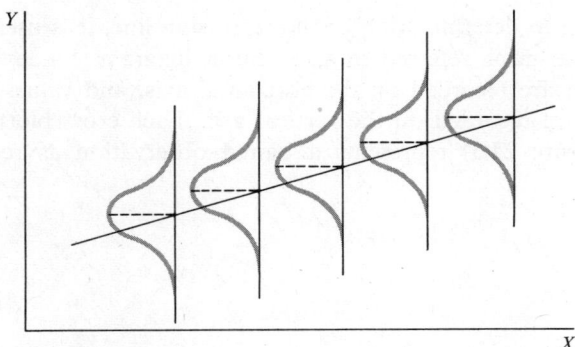

Figure 27.2

In actuality, the regression line is usually unknown and must be estimated from sample data.

$\mu_{Y.X} = \alpha + \beta X$ is the general form of the unknown linear regression line characterizing the relationship between X and Y for the universe. α and β are the unknown parameters of the universe regression line and specify the height and slope of a linear function.

$\overline{Y}_X = a + bX$ is the general form of the *estimate* of the regression line, where a and b are estimates of the universe parameters α and β derived from the sample data.

In the illustrative example, it is assumed that a sample is randomly selected from areas presently under cultivation and fertilizer is applied to these areas with varying intensity. To reduce the computational burden, it is assumed that the sample consists of eight tracts of land and yields the data recorded in Table 27.1.

Table 27.1 *indipendent   Dependent (random)*

Tract	Fertilizer in tons X	Yield per acre (thousands of bushels) Y
A	10	6
B	20	7
C	25	9
D	30	13
E	30	10
F	35	9
G	42	14
H	48	12

The sample data consists of observations on eight tracts of land, where each observation represents a random selection from the conditional probability distribution of Y associated with its specified value for X. The Y observations are not conditional means. They are random observations drawn from the conditional probability distributions associated with $X = 10$, $X = 20$, $X = 30$, $X = 35$, $X = 42$, and $X = 48$.

27.1 DETERMINATION OF THE REGRESSION LINE

As a preliminary step to determination of the regression line, it is useful to plot the observations on a graph referred to as a scatter diagram. Values of X, the independent variable, are recorded on the horizontal axis, and values of Y, the dependent variable, are recorded on the vertical axis. Each cross plotted on the scatter diagram (Figure 27.3) represents a paired observation as recorded in Table 27.1.

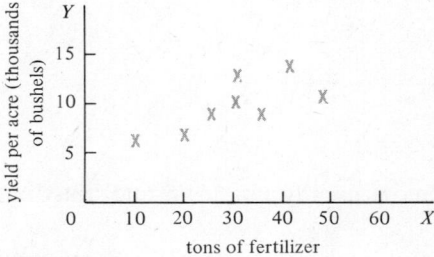

Figure 27.3

Plotting the observations on a scatter diagram enables the analyst to judge visually the *type* of line that best describes the pattern of the relationship. In this chapter, discussion is limited to linear regression lines. Examination of Figure 27.3 discloses that the assumption of linearity appears to be legitimate. The next step involves the determination of the particular straight line that best fits the plotted points on the scatter diagram. The criterion of best fit employed is the least-squares criterion.[4] The least-squares line possesses the following two characteristics: $\sum (Y - Y_c) = 0$, and $\sum (Y - Y_c)^2$ is less when computed from the arithmetic least-squares straight line than it would be if computed from any other arithmetic straight-line fit to the data.

The formula for the arithmetic least-squares straight line is $Y_c = a + bx$, where a and b are generated by the following formulas[5]:

$$a = \frac{\sum x^2 \cdot \sum Y - \sum x \cdot \sum xY}{n \cdot \sum x^2 - (\sum x)^2}$$

$$b = \frac{n \cdot \sum xY - \sum x \cdot \sum Y}{n \cdot \sum x^2 - (\sum x)^2}$$

Least-squares formula

a = height of regression line
b = slope

Referring to the formulas:

Y_c is the estimated mean value of Y associated with the value of small x substituted in the regression equation. Thus, the Y_c values are the conditional means, $\overline{Y}_x$.

a is the height of the regression line above the point on the large X axis associated with small x equal to zero.

x indicates the direction and number of units any given value of large X departs from the value of large X associated with small x equal to zero. In the special case where the value of large X associated with a is equal to zero, the deviations (small x's) are identical to the large X values. This has the advantage of eliminating a computational step—the large X values

---

[4] The conceptual basis for derivation of the least-squares line is developed in Technical Note 13 at the end of Chapter 25. In Chapter 25, time is treated as the independent variable.

[5] The above formulas are derived from the normal equations described in Technical Note 13, Chapter 25. The normal equations are reproduced below:

I $\sum Y = na + b\sum x$
II $\sum xY = a\sum x + b\sum x^2$ *NORMAL EQUATIONS*

Solving equation I for a, we obtain

$$a = \frac{\sum Y - b\sum x}{n}$$

Substituting for a in equation II and solving for b, we obtain

$$\sum xY = \left(\frac{\sum Y - b\sum x}{n}\right)\sum x + b\sum x^2$$

$$n \cdot \sum xY = \sum x \cdot \sum Y - b\sum x \cdot \sum x + b \cdot n\sum x^2$$
$$n \cdot \sum xY - \sum x \cdot \sum Y = b \cdot n\sum x^2 - b(\sum x)^2$$
$$n \cdot \sum xY - \sum x \cdot \sum Y = b[n\sum x^2 - (\sum x)^2]$$
$$b = \frac{n \cdot \sum xY - \sum x \cdot \sum Y}{n \cdot \sum x^2 - (\sum x)^2}$$

Substituting the above expression for b into equation I and solving for a generates the expression

$$a = \frac{\sum x^2 \cdot \sum Y - \sum x \cdot \sum xY}{n \cdot \sum x^2 - (\sum x)^2}$$

can be used directly as the small x values.[6] In all other cases, it is necessary to compute the small x values.

b is the change in the height of the regression line (or the change in the magnitude of the conditional mean) with a one-unit increase in the value of x.

n represents the number of paired observations.

The computation of the coefficients of the regression function for the illustrative example. letting small $x = 0$ where large $X = 0$, is illustrated in Table 27.2.

Table 27.2 Computation of the regression line where small $x = 0$ is associated with large $X = 0$

Tract	X Tons of fertilizer x	Y Yield per acre (thousands of bushels)	x^2	xY
A	10	6	100	60
B	20	7	400	140
C	25	9	625	225
D	30	13	900	390
E	30	10	900	300
F	35	9	1225	315
G	42	14	1764	588
H	48	12	2304	576
	240	80	8218	2594

Summations

$$\sum x = 240 \qquad \sum xY = 2594 \qquad \bar{X} = \frac{\sum X}{n} = \frac{240}{8} = 30$$

$$\sum Y = 80 \qquad \sum x^2 = 8218 \qquad \bar{Y} = \frac{\sum Y}{n} = \frac{80}{8} = 10$$

Solving for a[a]

$$a = \frac{\sum x^2 \cdot \sum Y - \sum x \cdot \sum xY}{n \cdot \sum x^2 - (\sum x)^2} = \frac{(8218)(80) - (240)(2594)}{8(8218) - (240)^2} = \frac{657440 - 622560}{65744 - 57600} = \frac{34880}{8144}$$

$$= 4.282$$

Solving for b [note that the denominator for both a and b is equal to $n \cdot \sum x^2 - (\sum x)^2$]:

$$b = \frac{n \cdot \sum xY - \sum x \sum Y}{n \cdot \sum (x^2) - (\sum x)^2} = \frac{8(2594) - (240)(80)}{8144} = \frac{20752 - 19200}{8144} = \frac{1552}{8144} = 0.1906$$

[a] Provided that small $x = 0$ is associated with large $X = 0$, a can alternatively be computed:
$$a = \bar{Y} - b\bar{X} = 10 - 0.1906(30) = 10 - 5.718 = 4.282$$

Once the values for a and b have been determined, the regression function is used to generate the conditional mean associated with any given value of x. Tract A: Fertilizer application equal to 10 tons:

$$Y_c = a + bx$$
$$= 4.282 + 0.1906x$$
$$= 4.282 + 0.1906(10)$$
$$= 4.282 + 1.906$$
$$= 6.188$$

6,188 bushels

[6] An alternative formula for the computation of a *when the value of large X associated with a is equal to zero* is $a = \bar{Y} - b\bar{X}$, where $\bar{Y}$ is the mean of the Y values and $\bar{X}$ is the mean of the X values.

Thus, $\overline{Y}_X = 6.188$ when $x = 10$.

Interpretation: It is estimated that a fertilizer application of 10 tons is associated with an average yield of 6188 bushels per acre.

Tract B: Fertilizer application equal to 20 tons:

$$Y_c = a + bx$$
$$= 4.282 + 0.1906(20)$$
$$= 4.282 + 3.812$$
$$= 8.094$$

Thus, $\overline{Y}_X = 8.094$ when $x = 20$.

Tract C: Fertilizer application equal to 25 tons:

$$Y_c = a + bx$$
$$= 4.282 + 0.1906(25)$$
$$= 4.282 + 4.765$$
$$= 9.047$$

Thus, $\overline{Y}_X = 9.047$ when $x = 25$.

Tract D: Fertilizer application equal to 30 tons:

$$Y_c = a + bx$$
$$= 4.282 + 0.1906(30)$$
$$= 4.282 + 5.718$$
$$= 10.00$$

Thus, $\overline{Y}_X = 10.00$ when $x = 30$.

The computation for tract E is identical to the computation for tract D.

Tract F: Fertilizer application equal to 35 tons:

$$Y_c = a + bx$$
$$= 4.282 + 0.1906(35)$$
$$= 4.282 + 6.671$$
$$= 10.953$$

Thus, $\overline{Y}_X = 10.953$ when $x = 35$.

Tract G: Fertilizer application equal to 42 tons:

$$Y_c = a + bx$$
$$= 4.282 + 0.1906(42)$$
$$= 4.282 + 8.0052$$
$$= 12.287$$

Thus, $\overline{Y}_X = 12.287$ when $x = 42$.

Tract H: Fertilizer application equal to 48 tons:

$$Y_c = a + bx$$
$$= 4.282 + 0.1906(48)$$
$$= 4.282 + 9.149$$
$$= 13.431$$

Thus, $\overline{Y}_X = 13.431$ when $x = 48$.

An interesting relationship is demonstrated by the computation for tract D. When the value of large X is equal to the value of the mean of the X values (in the illustrative example, $\overline{X} = 30$), the computed value of Y_c is equal to the arithmetic mean of the Y values (in the illustrative example, $\overline{Y} = 10$). Summarizing, *the height of the regression line above the point on the large X axis equal* *in value to the mean of the observed X values is always equal to the mean of the observed Y values.* In effect, perpendiculars raised at $\overline{X}$ on the X axis and at $\overline{Y}$ on the Y axis intersect on the regression line.

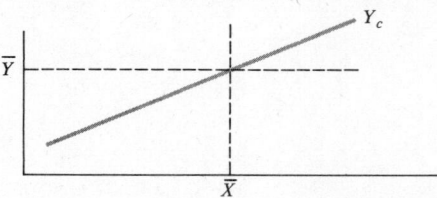

This relationship can be used to locate the height of the regression line. If *a*, the height of the regression line, is located above that point on the X axis equal in value to the arithmetic mean of the observed X values, then x is set equal to zero, where $X = \overline{X}$. Locating *a* in the manner described above has other ramifications as well. Small x will now indicate the distance and direction the observed large X values depart from the point on the large X axis equal in value to $\overline{X}$. The sum of the deivations of the individual observations from their arithmetic mean is always equal to zero. Given that $\sum x$ equals zero, the formulas for *a* and *b* simplify as follows:

$$a = \frac{\sum x^2 \cdot \sum Y - \sum x \cdot \sum xY}{n \cdot \sum x^2 - (\sum x)^2} \qquad b = \frac{n \cdot \sum xY - \sum x \cdot \sum Y}{n \cdot \sum x^2 - (\sum x)^2}$$

$$= \frac{\sum x^2 \cdot \sum Y - (0) \cdot \sum xY}{n \cdot \sum x^2 - (0)^2} \qquad = \frac{n \cdot \sum xY - (0) \cdot \sum Y}{n \cdot \sum x^2 - (0)^2}$$

$$= \frac{\sum x^2 \cdot \sum Y}{n \cdot \sum x^2} \qquad = \frac{n \cdot \sum xY}{n \cdot \sum x^2}$$

$$= \frac{\sum Y}{n} \qquad = \frac{\sum xY}{\sum x^2}$$

The regression function identified is identical to the one previously obtained. The only difference is that its height is located at a different point on the function. To demonstrate the equivalence, the regression line is computed in Table 27.3, reflecting the assignment of $x = 0$ to the point on the X axis equal to the arithmetic mean of the observed X values.

The computations of the Y_c values using the function $Y_c = 10 + 0.1906x$ (where $x = 0$ is associated with $X = \overline{X}$) appear below. A quick comparison with the computations performed on pp. 690–691 verify that this is indeed the same regression function previously identified as $Y_c = 4.282 + 0.1906x$ (where $x = 0$ was associated with $X = 0$). The *b* value is identical for both lines since the slope of the line is unchanged.

$X = 10; x = -20$	$X = 20; x = -10$
$Y_c = 10 + 0.1906(-20)$	$Y_c = 10 + 0.1906(-10)$
$= 10 - 3.812$	$= 10 - 1.906$
$= 6.188$	$= 8.094$
$X = 25; x = -5$	$X = 30; x = 0$
$Y_c = 10 + 0.1906(-5)$	$Y_c = 10 + 0.1906(0)$
$= 10 - 0.953$	$= 10 + 0$
$= 9.047$	$= 10$

$$X = 35; x = +5$$
$$Y_c = 10 + 0.1906(+5)$$
$$= 10 + 0.953$$
$$= 10.953$$

$$X = 42; x = +12$$
$$Y_c = 10 + 0.1906(+12)$$
$$= 10 + 2.287$$
$$= 12.287$$

$$X = 48; x = +18$$
$$Y_c = 10 + 0.1906(+18)$$
$$= 10 + 3.4308$$
$$= 13.431$$

Table 27.3 Computation of the regression line where small $x = 0$ is associated with large $X = \bar{X}$

Tract	Tons of fertilizer X	Yield per acre (thousands of bushels) Y	$X - \bar{X}$ x	xY	x^2
A	10	6	−20	−120	400
B	20	7	−10	−70	100
C	25	9	−5	−45	25
D	30	13	0	0	0
E	30	10	0	0	0
F	35	9	+5	+45	25
G	42	14	+12	+168	144
H	48	12	+18	+216	324
	240	80	+35	+429	1018
			−35	−235	
			0	+194	

$$a = \frac{\sum Y}{n} = \frac{80}{8} = 10 \qquad \bar{X} = 30$$

$$b = \frac{\sum xY}{\sum x^2} = \frac{194}{1018} = 0.1906$$

27.2 EVALUATION OF THE STRENGTH OF THE REGRESSION RELATIONSHIP[7]

Suppose that one were asked to provide an estimate of yield per acre with no information as to the extensiveness of fertilizer application. The logical choice as a representative value would be $\bar{Y}$ or the mean of the Y values for all eight tracts of land observed. The effectiveness of $\bar{Y}$ as a representative value is determined by the extent to which the individual observations are dispersed in the overall distribution of Y values. Dispersion in the distribution of the Y values is measured by the standard deviation of the Y values or σ_Y. Unfortunately, σ_Y is unknown. An unbiased estimate of σ_Y can be generated, however, from the sample data utilizing the formula $\hat{\sigma}_Y = \sqrt{\sum (Y - \bar{Y})^2/(n - 1)}$. The computation of $\hat{\sigma}_Y$ for the illustrative example is presented in Table 27.4.

[7] Strength as used in this context refers to the extent of improvement in the precision of the best estimate of performance when $\bar{Y}_x$, rather than Y, is used as the estimator.

Table 27.4

Tract	Yield per acre (in thousands of bushels) Y	$Y - \bar{Y}$	$(Y - \bar{Y})^2$
A	6	−4	16
B	7	−3	9
C	9	−1	1
D	13	+3	9
E	10	0	0
F	9	−1	1
G	14	+4	16
H	12	+2	4
	80	+9	56
		−9	
		0	

$$\bar{Y} = \frac{\sum Y}{n} = \frac{80}{8} = 10$$

$$\hat{\sigma}_Y = \sqrt{\frac{\sum (Y - \bar{Y})^2}{n - 1}} = \sqrt{\frac{56}{7}} = \sqrt{8} = 2.83$$

$\hat{\sigma}_y$ = std dev of y

Graphically, the precision of $\bar{Y}$ as an estimator of individual performance can be displayed on the scatter diagram by the interval $\bar{Y} \pm \hat{\sigma}_Y$, as displayed in Figure 27.4. Sixty-eight percent of the individual observations have a value within plus or minus one standard deviation of the mean. Thus the interval $\bar{Y} \pm \hat{\sigma}_Y$ can be interpreted as containing 68 percent of the individual Y values.

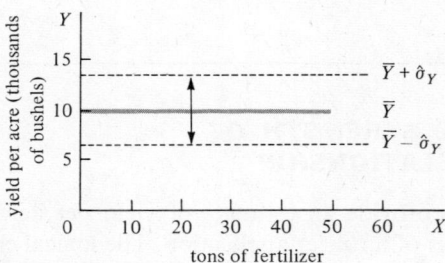

Figure 27.4

Given the specification of a particular value of X, the logical choice as a representative value becomes $\bar{Y}_X$ as furnished by the regression line. The effectiveness of a given $\bar{Y}_X$ as a representative value is measured by the extent to which the individual observations comprising the conditional probability distribution cluster around the conditional mean. Dispersion of the Y values comprising a conditional probability distribution is measured by $\sigma_{Y.X}$. Recall that the analysis presumes that the conditional standard deviations are all equal. Given this assumption, the dispersion of the individual Y values around their respective conditional means is equivalent to the dispersion of *all* the individual Y values around the regression line. The dispersion of the individual Y values around the

$\sigma_{YX} = std\ error\ of\ the\ estimate$

regression line is measured by the *standard error of the estimate*. The standard error of the estimate is analogous to the standard deviation. The major difference is that the standard error of the estimate is the square root of the average squared deviation around a *line*, whereas the standard deviation is the square root of the average squared deviation around *a point*, the mean. Given the assumption that the conditional standard deviations are all equal, the standard error of the estimate is equivalent to this common conditional standard deviation and is in fact designated by the same symbol, $\sigma_{Y.X}$.

Referring to Figure 27.5, if one were to add and subtract the value of a conditional standard deviation (i.e., the standard error of the estimate) to the value of the conditional mean for the conditional probability distribution associated with a fertilizer application of 10 tons, one would obtain the Y values designated A and B. The area contained under the curve in the interval between A and B comprises 68 percent of the total area under the curve. Recall that relative area under the curve is proportional to relative frequency. It follows that 68 percent of the yield per acre performances associated with a fertilizer application of 10 tons will have a value between A and B. Similar reasoning generates the values C and D for the probability distribution associated with a fertilizer application of 30 tons, and the values E and F for the conditional probability distribution associated with a fertilizer application of 48 tons. All the conditional standard deviations are equal. Thus, A, C, and E lie an equal vertical distance above the regression line. Similarly, B, D, and F lie an equal vertical distance below the regression line. If the points lying one conditional standard deviation above the regression line are connected and the points lying one conditional standard deviation below the regression line are connected, one obtains two lines parallel to the regression line and equidistant from it at all points.

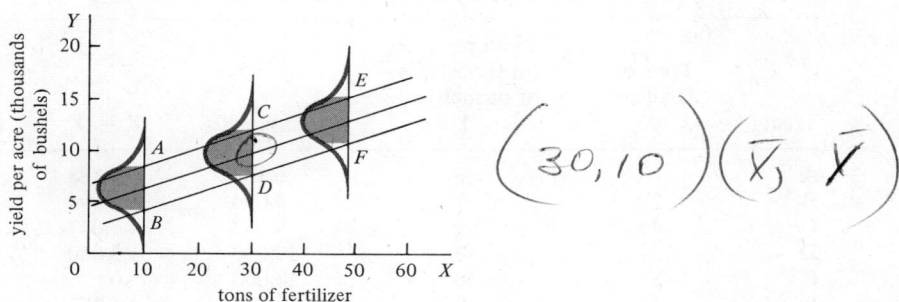

Figure 27.5

Between the two bands fall 68 percent of individual performances. The width of the band or interval $(\mu_{Y.X} \pm \sigma_{Y.X})$ provides an indication of the extent to which individual performances are dispersed around their respective conditional means (i.e., the regression line). All other things being equal, the wider the interval, the less precise the mean as a predictor of individual performance. The narrower the interval, the more precise the mean as a predictor of individual performance.

The extent of improvement in the precision of the estimated value of Y when $\overline{Y}_X$, rather than $\overline{Y}$, is used as the estimator can be observed by comparing the magnitude of $\sigma_{Y.X}$ with the magnitude of σ_Y. The smaller the value of $\sigma_{Y.X}$ relative to the value of σ_Y, the greater the improvement in the precision of the estimate.

The universe parameter, $\sigma_{Y.X}$, is unknown and also must be estimated from the sample data. The unbiased estimator of $\sigma_{Y.x}$ takes the form

$$\hat{\sigma}_{Y.x} = \sqrt{\frac{\sum (Y - Y_c)^2}{(n - m)}}$$ *unbiased estimator of $\sigma_{y.x}$*

where

Y represents the observed value of the dependent variable.

Y_c represents the expected value of the dependent variable as estimated by the regression line. The value of X must be specified in order to determine Y_c.

n refers to the total number of paired observations.

m refers to the number of degrees of freedom lost. In general, one degree of freedom is lost for each constant that must be estimated from the sample data. In the case of a linear relationship, the constants a and b must be generated from the sample data before any deviations from the regression line can be computed. Thus, the number of degrees of freedom lost is equal to two. Had the regression function been parabolic, three degrees of freedom would have been lost, since three constants, a, b, and c, must be determined from the sample data.

df = 2

The extent of improvement in the precision of the estimated value of Y when Y_X, rather than $\overline{Y}$, is used as the estimator is evaluated by comparing the magnitude of the following measures:

(1.78)

$$\hat{\sigma}_Y = \sqrt{\frac{\sum (Y - \overline{Y})^2}{n - 1}} \quad \text{versus} \quad \hat{\sigma}_{Y.x} = \sqrt{\frac{\sum (Y - Y_c)^2}{n - m}}$$

2.83 vs.

$\hat{\sigma}_Y$, previously computed in Table 27.4, is equal to 2.83. The underlying computation for the determination of $\hat{\sigma}_{Y.x}$ is presented in Table 27.5.

Table 27.5 Underlying computation for the determination of $\hat{\sigma}_{X.Y}$

Tract	Tons of fertilizer X	Yield per acre (in thousands of bushels) Y	Y_c (see p. 690)	$Y - Y_c$	$(Y - Y_c)^2$
A	10	6	6.188	−0.188	0.035344
B	20	7	8.094	−1.094	1.196836
C	25	9	9.047	−0.047	0.002209
D	30	13	10.000	+3.000	9.000000
E	30	10	10.000	0	0
F	35	9	10.953	−1.953	3.814209
G	42	14	12.287	+1.713	2.934369
H	48	12	13.431	−1.431	2.047761
				−4.713	19.030728
				+4.713	
				0	

$$\hat{\sigma}_{Y.x} = \sqrt{\frac{\sum (Y - Y_c)^2}{n - m}} = \sqrt{\frac{19.030728}{8 - 2}}$$

$$= \sqrt{\frac{19.030728}{6}} = \sqrt{3.171788}$$

$$= 1.78$$

Thus, the improvement in the precision of the estimate as provided by the regression line is indicated by the reduction in dispersion from 2.83 to 1.78.

A visual comparison of the extent of improvement in the precision of estimates provided by the regression line as compared to the estimate provided by the mean of *all* the Y values can be obtained by plotting the intervals $\bar{Y} \pm \hat{\sigma}_Y$ and $Y_c \pm \hat{\sigma}_{Y.X}$ on the scatter diagram. The diagrammatic representation of this relationship is depicted in Figure 27.6. The supporting computations are presented in Table 27.6.

Estimate

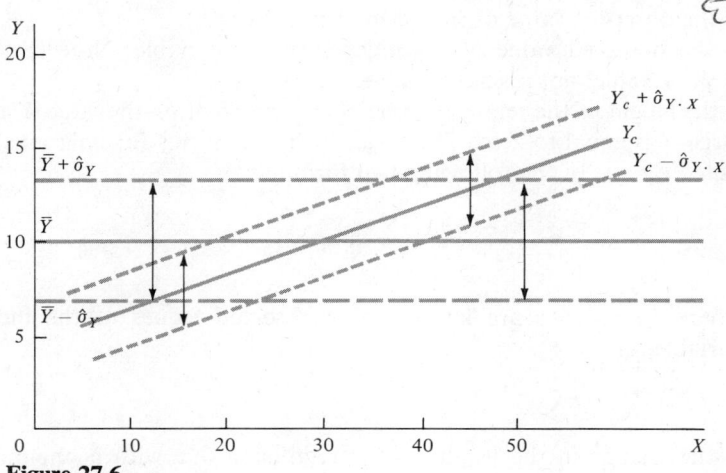

Figure 27.6

Table 27.6

Tract	X	Y_c	$Y_c - \hat{\sigma}_{Y.X}$	$Y_c + \hat{\sigma}_{Y.X}$
A	10	6.188	4.408	7.968
B	20	8.094	6.314	9.874
C	25	9.047	7.267	10.827
D, E	30	10.000	8.220	11.780
F	35	10.953	9.173	12.733
G	42	12.287	10.507	14.067
H	48	13.431	11.651	15.211

$\bar{Y}$	$\bar{Y} - \hat{\sigma}_Y$	$\bar{Y} + \hat{\sigma}_Y$
10	7.17	12.83

27.3 SHORT-CUT FORMULAS FOR THE DETERMINATION OF $\hat{\sigma}_Y$ AND $\hat{\sigma}_{Y.X}$

A major disadvantage associated with the computation of $\hat{\sigma}_{Y.X}$ by the formula $\sqrt{\sum (Y - Y_c)^2/(n - m)}$ is the necessity of computing the Y_c value associated with each observed X value. If one had 100 X observations, 100 Y_c values would have to be computed. An alternative formula for the computation of $\hat{\sigma}_{Y.X}$ exists which does not require the preliminary step of computing the Y_c values. The alternate and recommended formula is

$$\hat{\sigma}_{Y.X} = \sqrt{\frac{\sum Y^2 - a \sum Y - b \sum XY}{n - m}}$$

where

Y is the observed value of the dependent Y variable.

X is the observed value of the independent X variable. Note that this is a large X value, not a small x value.

a is the height of the regression line above the point on the large X axis where large X is equal to zero. The a value utilized in this formula must be computed by one of the following two formulas:

$$a = \frac{\sum X^2 \cdot \sum Y - \sum X \cdot \sum XY}{n \cdot \sum X^2 - (\sum X)^2}$$

where all the X's are large X's or observed values of the independent variable, or

$$a = \overline{Y} - b\overline{X}$$

b is the change in the height of the regression line with a one-unit change in the value of X.

All the summations and the values for a and b required for the formula

$$\hat{\sigma}_{Y.X} = \sqrt{\frac{\sum Y^2 - a \sum Y - b \sum XY}{n - m}}$$

with the exception of $\sum Y^2$ will have previously been determined provided the constants in the regression function have been computed by the formulas

$$a = \frac{\sum X^2 \cdot \sum Y - \sum X \cdot \sum XY}{n \cdot \sum X^2 - (\sum X)^2} \quad \text{or} \quad \overline{Y} - b\overline{X}$$

$$b = \frac{n \cdot \sum XY - \sum X \cdot \sum Y}{n \cdot \sum X^2 - (\sum X)^2}$$

where X is a large X and refers to the observed values of the independent variable.

These computations, previously described in Table 27.2, are reproduced together with the additional computation of $\sum Y^2$ in Table 27.7.

A short-cut formula for the computation of $\hat{\sigma}_Y$ also exists. The short-cut formula for $\hat{\sigma}_Y$ takes the form

$$\hat{\sigma}_Y = \sqrt{\frac{n \cdot \sum Y^2 - (\sum Y)^2}{n(n - 1)}}$$

Use of this formula eliminates the step of having to compute the deviations, $Y - \overline{Y}$. All the summations required for substitution into this formula have already been generated in Table 27.7.

Table 27.7 Supporting computations

Tract	Tons of fertilizer X	Yield per acre (thousands of bushels) Y	X^2	XY	Y^2
A	10	6	100	60	36
B	20	7	400	140	49
C	25	9	625	225	81
D	30	13	900	390	169
E	30	10	900	300	100
F	35	9	1225	315	81
G	42	14	1764	588	196
H	48	12	2304	576	144
	240	80	8218	2594	856

Summations:

$\sum X = 240$ $\sum X^2 = 8218$ $\sum Y^2 = 856$

$\sum Y = 80$ $\sum XY = 2594$

$$a = \frac{\sum X^2 \cdot \sum Y - \sum X \cdot \sum XY}{n \cdot \sum X^2 - (\sum X)^2} = 4.282$$

$$b = \frac{n \cdot \sum XY - \sum X \cdot \sum Y}{n \cdot \sum X^2 - (\sum X)^2} = 0.1906$$

The computation of $\hat{\sigma}_{Y.x}$ and $\hat{\sigma}_Y$ by the short-cut formulas is illustrated below:

$$\hat{\sigma}_{Y.x} = \sqrt{\frac{\sum Y^2 - a \sum Y - b \sum XY}{n - m}}$$

$$= \sqrt{\frac{856 - (4.282)(80) - (0.1906)(2594)}{8 - 2}}$$

$$= \sqrt{\frac{856 - 342.560 - 494.4164}{6}}$$

$$= \sqrt{\frac{19.0236}{6}}$$

$$= \sqrt{3.1706}$$

$$= 1.78$$

$$\hat{\sigma}_Y = \sqrt{\frac{n \cdot \sum Y^2 - (\sum Y)^2}{n(n-1)}}$$

$$= \sqrt{\frac{(8)(856) - (80)^2}{(8)(7)}}$$

$$= \sqrt{\frac{6848 - 6400}{56}}$$

$$= \sqrt{\frac{448}{56}}$$

$$= \sqrt{8}$$

$$= 2.83$$

27.4 INFERENCES WITH RESPECT TO THE UNIVERSE PARAMETER, β

The universe parameter, β, indicates the change in the expected value of the dependent variable with a one-unit change in the independent variable. In the context of the illustrative example, β refers to the anticipated change in expected yield per acre with a one-unit change (1 ton) in the extensiveness of fertilizer application. The analysis generated an estimate of β, namely, $b = 0.1906$. The analyst can conduct tests of hypotheses concerning the value of β or construct confidence interval estimates of the value of β by techniques analogous to those utilized for universe means in Chapters 13 and 15. In this case, the relevant sampling distribution is the sampling distribution of b (Figure 27.7). It is stated without proof that this distribution is normally distributed with a mean equal to β.

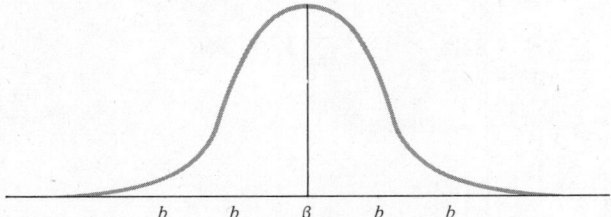

Figure 27.7 Estimated sampling distribution of β.

The standard error of b when estimated from sample data is determined as follows:

$$\hat{\sigma}_b = \frac{\hat{\sigma}_{Y.X}}{\sqrt{\sum (X - \overline{X})^2}}$$

Examination of the formula for $\hat{\sigma}_b$ discloses the following relationships:

1. All other things being equal, $\sum (X - \overline{X})^2$ varies directly with the portion of the range of the independent variable investigated. The greater the range of the independent variable investigated, the greater the term $\sum (X - \overline{X})^2$. The greater the term $\sum (X - \overline{X})^2$, the smaller the value of $\hat{\sigma}_b$.
2. $\hat{\sigma}_{Y.X}$ reflects the variability in the dependent variable Y within the conditional probability distributions. All other things being equal, the larger the estimated value of the standard error of the estimate, $\hat{\sigma}_{Y.X}$, the larger the value of $\hat{\sigma}_b$.
3. All other things being equal, the larger the sample size, the larger the expression $\sum (X - \overline{X})^2$. The larger $\sum (X - \overline{X})^2$, the smaller $\hat{\sigma}_b$. Thus, $\hat{\sigma}_b$ and sample size are inversely related.

Confidence interval estimates of β take the following form:
For small samples:

$$b \pm t\hat{\sigma}_b$$

where t refers to the number of standard errors associated with the desired level of confidence for the distribution associated with $n - m$ degrees of freedom.
For large samples:

$$b \pm z\hat{\sigma}_b$$

where z refers to the number of standard errors associated with the desired level of confidence in a normal distribution.

Tests of hypotheses with respect to β take the following form:
For small samples, compute the expression

$$t = \frac{|b - \beta|}{\hat{\sigma}_b}$$

Compare the calculated value of t with the value of t associated with $n - m$ degrees of freedom and the specified level of significance. If the computed value of t exceeds this value, reject the hypothesis.
For large samples, compute the expression

$$z = \frac{|b - \beta|}{\hat{\sigma}_b}$$

Compare the calculated value of z with the value of z associated with the specified level of significance. If the computed value of z exceeds this value, reject the hypothesis.
The calculation of $\hat{\sigma}_b$ for the illustrative example is performed in Table 27.8. This value is then used to construct a 95 percent confidence interval estimate of β and to test the null hypothesis that $\beta = 0$ at the 5 percent and 1 percent critical probability levels.

Table 27.8 Calculation of $\hat{\sigma}_b$

Tract	X	$X - \bar{X}$	$(X - \bar{X})^2$
A	10	-20	400
B	20	-10	100
C	25	-5	25
D	30	0	0
E	30	0	0
F	35	$+5$	25
G	42	$+12$	144
H	48	$+18$	324
	240		1018

$$\bar{X} = \frac{\sum X}{n} = \frac{240}{8} = 30$$

$$\sum (X - \bar{X})^2 = 1018$$

$$\hat{\sigma}_{Y.X} = 1.78$$

$$\hat{\sigma}_b = \frac{\hat{\sigma}_{Y.X}}{\sqrt{\sum (X - \bar{X})^2}} = \frac{1.78}{\sqrt{1018}}$$

$$= \frac{1.78}{31.91} = 0.0558$$

Determination of a 95 percent confidence interval estimate of β:

$n = 8$; degrees of freedom $= 6$; $t_{0.025} = 2.447$
$b \pm t_{0.025}\hat{\sigma}_b$
$0.1906 \pm 2.447(0.0558)$
0.1906 ± 0.1365
$0.0541 \leftrightarrow 0.3271$

One would be 95 percent confident that the statement is correct when it is stated that the value of β is between 0.0541 and 0.3271, where β represents the change in expected value of $\mu_{Y \cdot X}$ with a one-unit increase in the value of X.

Test of the null hypothesis that $\beta = 0$ at the 5 percent critical probability level:

$$t = \frac{|b - \beta|}{\hat{\sigma}_b} = \frac{|0.1906 - 0|}{0.0558} = 3.42$$

$t_{0.025} = 2.447$

One would expect to observe a value of b that differed from 0 by more than 2.447 standard errors only 5 percent of the time. The value 3.42 exceeds 2.447. Therefore, the sample evidence would be considered inconsistent with the null hypothesis at the 5 percent critical probability level and the null hypothesis would be rejected.

Test of the null hypothesis that $\beta = 0$ at the 1 percent critical probability level:

$$t = \frac{|b - \beta|}{\hat{\sigma}_b} = \frac{|0.1906 - 0|}{0.0558} = 3.42$$

$t_{0.005} = 3.707$

One would expect to observe a value of b that differed from 0 by more than 3.707 standard errors only 1 percent of the time. The calculated t of 3.42 is less than 3.707. Therefore, the sample evidence would be considered consistent with the null hypothesis at the 1 percent critical probability level and the null hypothesis would be accepted.

27.5 CORRELATION ANALYSIS

In regression analysis, the variable designated as the independent variable is treated as predetermined and the objective is the specification of the linear function or regression line describing the relationship between the independent and dependent variable. The regression line indicates the average value of the dependent variable associated with a specified value of the independent variable. The objective of correlation analysis, on the other hand, is to evaluate the extent to which covariation exists among the variables under investigation. Two variables are said to be correlated when a change in the value of one of the variables tends to be associated with a consistent corresponding change in the value of the other. Correlation analysis requires a more demanding set of assumptions than regression analysis.

Assumptions:

1. Unlike regression analysis, where the independent variable can be pre-determined, *both* the independent and the dependent variables are presumed to be random variables in correlation analysis.
2. In regression analysis, the conditional probability distributions of the dependent variable Y given specified values of the independent variable X are all assumed to be normally distributed. No assumption is required with respect to the distribution of the independent variable X. In correlation analysis, it is assumed that *both* the conditional probability distributions of Y given specified values of X and the conditional probability distributions of X given specified values of Y are normally distributed. When X and Y are jointly normally distributed, it is referred to as a bivariate normal distribution.

3. In regression analysis, $\sigma_{Y.X}$ is assumed to be equal for all the conditional probability distributions of Y given X. In correlation analysis, $\sigma_{Y.X}$ is assumed to be equal for all the conditional probability distributions of Y given X *and* $\sigma_{X.Y}$ is assumed to be equal for all the conditional probability distributions of X given Y.

4. As previously indicated, discussion in this text is restricted to the case where the relationship between the value of the independent variable and the average value of the dependent variable is described by a linear regression function. In the case of correlation analysis, it makes no difference which variable is designated as the independent variable. The measure of correlation (or the measure of covariation) is the same whether Y is regressed on X or X is regressed on Y.

Provided that the conditions described above are met, the regression line utilized in correlation analysis is computed in exactly the same fashion as the regression line previously computed in the discussion of regression analysis.

The extent to which correlation between the variables may or may not exist can be roughly evaluated by observing the nature of the scattered points representing the paired observations on a scatter diagram. Good correlation is said to exist when the relationship is so strong that one can almost predict the value of one of the variables provided the value of the other variable is known. Good correlation is associated with the situation where the observed values of the variables cluster closely around the regression line. Poor correlation is associated with the situation where the observed values of the variable are widely dispersed around the regression line. Two variables can be either positively or negatively correlated. An example of each is provided below:

1. Positive correlation: Two variables are said to be positively correlated when small values of Y are associated with small values of X and large values of Y are associated with large values of X. Positive correlation indicates a *direct* relationship between the variables. Whether or not it is good positive correlation depends on the extent of scatter of the plotted points around the regression line. Positive correlation is associated with a regression line with a positive slope. A scatter diagram depicting positive correlation is shown below:

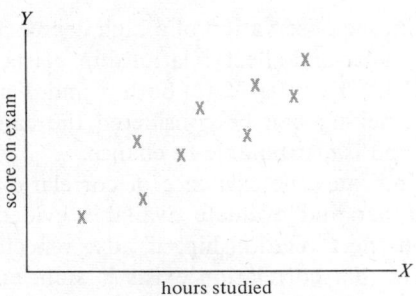

2. Negative correlation: Two variables are said to be negatively correlated when small values of X tend to be associated with large values of Y and large values of X tend to be associated with small values of Y. Negative correlation indicates an *inverse* relationship between the variables. Whether or not it is good negative correlation depends on the extent of scatter of the plotted points around the regression line. The terms negative and positive correlation are not indicative

of good and bad but rather are indicative of the direction of the relationship (i.e., direct or inverse). A scatter diagram depicting negative correlation is shown below:

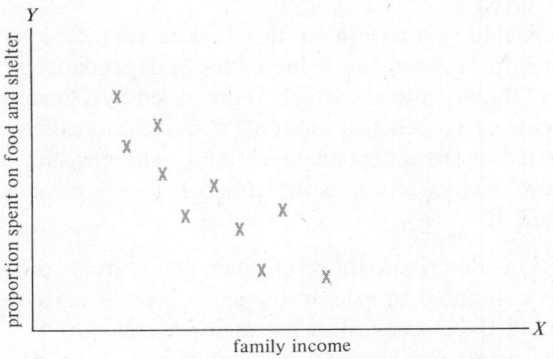

3. No correlation: When all values of X are associated with small, medium, and large values of Y, it is said that no correlation exists between the two variables. A scatter diagram depicting no correlation is shown below:

It does not necessarily follow from the observation of a high degree of correlation between two variables that a cause-and-effect relationship exists. Possible explanations include: (1) X causes Y, (2) Y causes X, (3) both X and Y are caused by the same underlying factors but neither can be considered the cause of the other, and (4) the relationship observed is attributable to chance.

The statistical evidence merely indicates the existence of correlation, not the reason for it. The analyst must gather and evaluate available evidence in an attempt to determine the cause-and-effect relationship, if any, reflected in the observed data. Occasionally, the fact that correlation exists is sufficient for the analyst's purpose. An example of such a situation is where the analyst merely desires to evaluate the extent to which variation in one of the variables is associated with variation in the other. In other instances, the causal factor is of utmost importance. For example, is it legitimate to claim that an increase in cigarette consumption will cause an increase in the likelihood of lung cancer? Medical research supports this contention.

27.6 THE COEFFICIENT OF DETERMINATION

A measure of the extent to which X and Y are correlated can be generated by determining the proportion of the total variation in Y that is associated with variation in X.

Each observed Y value can be expressed as follows:

$$Y = Y_c + (Y - Y_c)$$

Thus, variation in the value of Y can be treated as the net influence of variation in these two components.

$$\Delta Y = \Delta Y_c + \Delta(Y - Y_c)$$

$Y_c = a + bX$ and a and b are constants. Therefore, any change in Y_c is attributable to a change in the value of X.

The second term, $\Delta(Y - Y_c)$, represents the change in Y_c attributable to factors other than variation in the value of X. Thus,

ΔY represents the total variation in Y

ΔY_c represents the portion of the variation in Y attributable to variation in X

$\Delta(Y - Y_c)$ represents the portion of the variation in Y attributable to factors other than variation in X

A measure of the total variability in Y is provided by the variance of the original Y values.

$$\sigma_Y{}^2 = \frac{\sum (Y - \mu_Y)^2}{N}$$

A measure of the portion of the variation in Y attributable to factors other than variation in X is provided by the variance of the conditional probability distributions of Y given specified values of X. This is measured by the square of the standard error of the estimate or the average squared deviation around the regression line

$$\sigma_{Y.X}^2 = \frac{\sum (Y - \mu_{Y.X})^2}{N}$$

A measure of the portion of the variation in Y attributable to variation in X is provided by the average squared deviation of the conditional means around their average value. The mean of the conditional means is μ_Y.

$$\sigma_{\mu_{Y.X}}^2 = \frac{\sum (\mu_{Y.X} - \mu_Y)^2}{N}$$

It can be demonstrated that

$$\underset{\substack{\text{total} \\ \text{variation} \\ \text{in } Y}}{\sigma_Y{}^2} = \underset{\substack{\text{portion of the} \\ \text{variation in } Y \\ \text{attributable to} \\ \text{variation in } X}}{\sigma_{\mu_{Y.X}}^2} + \underset{\substack{\text{portion of the variation} \\ \text{in } Y \text{ attributable to} \\ \text{variation in factors} \\ \text{other than } X}}{\sigma_{Y.X}^2}$$

Table 27.9 Illustration of the relationship

$$\sigma_Y^2 = \sigma_{Y.x}^2 + \sigma_{\mu_{Y.x}}^2$$

Tract	X	Y	$\mu_{Y.x}$	$Y - \mu_Y$	$(Y - \mu_Y)^2$	$Y - \mu_{Y.x}$	$(Y - \mu_{Y.x})^2$	$\mu_{Y.x} - \mu_Y$	$(\mu_{Y.x} - \mu_Y)^2$
A	10	6	6.188	−4	16	−0.188	0.035344	−3.812	14.531344
B	20	7	8.094	−3	9	−1.094	1.196836	−1.906	3.632836
C	25	9	9.047	−1	1	−0.047	0.002209	−0.953	−0.908209
D	30	13	10.000	+3	9	+3.000	9.000000	0	0
E	30	10	10.000	0	0	0	0	0	0
F	35	9	10.953	−1	1	−1.953	3.814209	0.953	0.908209
G	42	14	12.287	+4	16	+1.713	2.934369	2.287	5.230369
H	48	12	13.431	+2	4	−1.431	2.047761	3.431	11.771761
		80	80	+9	56	−4.713	19.030728	−6.671	36.982728
				−9		+4.713		+6.671	
				0		0		0	

μ_Y always equals $\bar{\mu}_{Y.x}$ $\begin{cases} \mu_Y = \dfrac{\sum Y}{N} = \dfrac{80}{8} = 10 \\[2mm] \bar{\mu}_{Y.x} = \dfrac{\sum \mu_{Y.x}}{N} = \dfrac{80}{8} = 10 \end{cases}$

$$\sigma_{Y.x}^2 = \frac{\sum (Y - \mu_{Y.x})^2}{N} = \frac{19.030728}{8} = 2.38$$

$$\sigma_{\mu_{Y.x}}^2 = \frac{\sum (\mu_{Y.x} - \mu_Y)^2}{N} = \frac{36.982728}{8} = 4.62$$

$$\sigma_Y^2 = \frac{\sum (Y - \mu_Y)^2}{N} = \frac{56}{8} = 7$$

$$\sigma_Y^2 = \sigma_{Y.x}^2 + \sigma_{\mu_{Y.x}}^2$$

$$7 = 2.38 + 4.62$$

This relationship is illustrated in Table 27.9. For this example, the eight observations comprising the illustrative example are treated as the universe rather than as a sample. Dividing both sides of the equation by σ_Y^2, we obtain

$$1 \quad = \quad \frac{\sigma_{\mu_{Y.X}}^2}{\sigma_Y^2} \quad + \quad \frac{\sigma_{Y.X}^2}{\sigma_Y^2}$$

| 100 percent | = | percentage of the variation in Y associated with variation in X | + | percentage of the variation in Y associated with factors other than X |

The percentage of variation in Y associated with variation in X is referred to as the coefficient of determination and is designated by the symbol ρ^2 (rho square). Substituting, we obtain

$$1 = \rho^2 + \frac{\sigma_{Y.X}^2}{\sigma_Y^2}$$

Solving for ρ^2:

$$\rho^2 = 1 - \frac{\sigma_{Y.X}^2}{\sigma_Y^2}$$

In effect, one determines the percentage of the variation in Y *not* associated with variation in X and then subtracts this value from 1 to obtain ρ^2, the percentage of variation in Y that is associated with variation in X. ρ^2 can also be computed directly using the formula $\rho^2 = \sigma_{\mu_{Y.X}}^2/\sigma_Y^2$. The disadvantage of the direct approach is that the $\mu_{Y.X}$ associated with each of the paired observations must be determined in order to compute $\sigma_{\mu_{Y.X}}^2$. Thus, the coefficient of determination is generally computed using the formula $\rho^2 = 1 - \sigma_{Y.X}^2/\sigma_Y^2$.

Of course, the universe parameters $\sigma_{Y.X}^2$ and σ_Y^2 are usually not known and must be estimated from the sample data. The unbiased estimators of each of these values, having been previously generated in the discussion of regression analysis, are reproduced below:

$$\hat{\sigma}_{Y.X}^2 = \frac{\sum (Y - \overline{Y}_X)^2}{n - m}$$

$$\hat{\sigma}_Y^2 = \frac{\sum (Y - \overline{Y})^2}{n - 1}$$

Provided that the estimates of $\hat{\sigma}_{Y.X}^2$ and $\hat{\sigma}_Y^2$ are used, the resulting coefficient of determination obtained is itself an estimate of ρ^2. The estimated coefficient of determination obtained from a sample is represented by the symbol r^2. The computation of r^2 for the illustrative example is presented below:

$$r^2 = 1 - \frac{\hat{\sigma}_{Y.X}^2}{\hat{\sigma}_Y^2}$$

$$= 1 - \frac{3.17195}{8}$$

$$= 1 - 0.396$$

$$= 0.604$$

Interpretation: On the basis of the sample, it is estimated that approximately 60 percent of the variation in Y is associated with or "explained by" variation in X. The remaining 40 percent is attributable to other factors. This, of course, is just a point estimate of the universe parameter. Inferential statements with respect to the value of the universe parameter are discussed later in the chapter.

The extent to which correlation exists between the two variables can be roughly visualized by plotting the two bands $\overline{Y} \pm \hat{\sigma}_Y$ and $Y_c \pm \hat{\sigma}_{Y.X}$ on the scatter diagram. The diagram depicting this relationship for the illustrative example is reproduced as Figure 27.8.

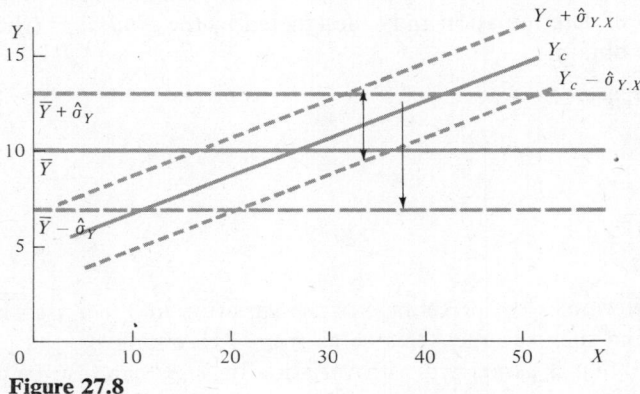

Figure 27.8

In general, the narrower the band $Y_c \pm \hat{\sigma}_{Y.X}$ relative to the band $\overline{Y} \pm \hat{\sigma}_Y$, the greater the percentage of total variation "explained" by variation in X and accordingly the greater the value of the coefficient of determination. The visual representation provided by Figure 27.8 is only a rough approximation of the extent of correlation. The bands reflect the relative magnitude of $\hat{\sigma}_{Y.X}$ versus $\hat{\sigma}_Y$, whereas the percentage of variation in Y *not associated* with variation in X is indicated by the ratio of the squares of these values or $\hat{\sigma}_{Y.X}^2/\hat{\sigma}_Y^2$. To obtain a precise estimate of the extent of correlation, one must compute the coefficient of determination.

No Correlation

Provided that the value of X is unknown, the best estimate of an unknown individual performance is $\overline{Y}$. The precision of this estimate is indicated by $\hat{\sigma}_Y$. Given the specification of a particular value of X, the best estimate of an unknown individual performance is provided by $\overline{Y}_X$ or Y_c. The precision of this estimate is measured by $\hat{\sigma}_{Y.X}$. In the case where no correlation exists between the variables, the regression line has a zero slope and $\overline{Y}_X = \overline{Y}$ for all values of X. Under these conditions, the measure of dispersion $\hat{\sigma}_Y$, which measures the variation of the individual observations around $\overline{Y}$, is equal in value to $\hat{\sigma}_{Y.X}$, which measures the variation of the individual observations around the regression line. Knowledge of the value of X in no way improves one's ability to estimate the value of Y. The best estimate is still $\overline{Y}(= \overline{Y}_X)$, and the precision of the estimate is not improved. The band $\overline{Y}_X \pm \hat{\sigma}_{Y.X}$ is as wide as the band $\overline{Y} \pm \hat{\sigma}_Y$.

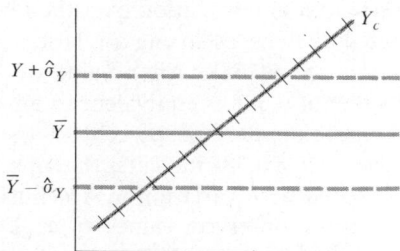

The coefficient of determination when no correlation exists is equal to zero:

$$r^2 = 1 - \frac{\hat{\sigma}_{Y.X}^2}{\hat{\sigma}_Y{}^2}$$

When there is no correlation, $\hat{\sigma}_Y{}^2 = \hat{\sigma}_{Y.X}^2$ and all of the variation in Y is still unexplained.[8] The r^2 of zero indicates that none of the variation in Y is associated with or "explained" by variation in X.

$$r^2 = 1 - 1$$
$$= 0$$

Perfect Correlation

When perfect correlation exists between two variables, all the observations fall directly on the regression line. For any given conditional probability distribution of Y, all the observed values of Y coincide in value with $\overline{Y}_X$ (and for any given conditional probability distribution of X, all the observed values of X coincide in value with $\overline{X}_Y$). The dispersion of the Y values around their conditional means as measured by $\hat{\sigma}_{Y.X}$ is therefore zero. In the case of perfect correlation, knowledge of the value of X enables the analyst to predict the unknown value of an individual Y observation with absolute certainty.

[8] When the sample size is small, $\hat{\sigma}_Y{}^2$ will not be equal to $\hat{\sigma}_{Y.X}^2$ because $\hat{\sigma}_Y{}^2$ is based on $n - 1$ degrees of freedom and $\hat{\sigma}_{Y.X}^2$ is based on $n - 2$ degrees of freedom. When the sample size is large, they are approximately equal.

The coefficient of determination when perfect correlation exists is equal to one:

$$r^2 = 1 - \frac{\hat{\sigma}_{Y.X}^2}{\hat{\sigma}_Y^2}$$

$$= 1 - \frac{0}{\hat{\sigma}_Y^2}$$

$$= 1 - 0$$

$$= 1$$

An r^2 of 1 indicates that all of the variation in Y is associated with or "explained by" variation in X. In the diagram above, it is perfect *positive* correlation, as indicated by the positive slope of the regression line.

27.7 THE COEFFICIENT OF CORRELATION

An alternative measure frequently used as an index of the degree of correlation between two variables is the square root of the coefficient of determination. This value is called the coefficient of correlation.

For the universe: the coefficient of correlation $= \sqrt{\rho^2} = \rho$
For the sample: the coefficient of correlation $= \sqrt{r^2} = r$

The coefficient of correlation for the illustrative example is

$$r = \sqrt{r^2}$$

$$= \sqrt{1 - \frac{\hat{\sigma}_{Y.X}^2}{\hat{\sigma}_Y^2}}$$

$$= \sqrt{0.604}$$

$$= +0.777$$

The sign is determined from the sign assigned to the coefficient b in the regression equation. The coefficient of correlation ranges in value from 0 to ± 1. The sign indicates whether correlation is positive or negative (i.e., whether the nature of covariation is direct or inverse). The extent of correlation is indicated by the magnitude of the coefficient. The coefficient of correlation is an abstract number and as such must be interpreted within the scale of possible values that could occur. In the illustrative example, the coefficient of correlation was equal to $+0.777$. This value is an indication of relative strength of correlation over the range from 0 to $+1$. However, it cannot be interpreted as twice as strong a relationship as that indicated by a coefficient of correlation of $+0.388$. An r value of $+0.777$ is equivalent to an r^2 of 0.60, and an r value of 0.388 is equivalent to an r^2 of 0.15. Thus, an r of $+0.777$ indicates that approximately 60 percent of the variation in Y is associated with variation in X, and an r of 0.388 indicates that approximately 15 percent of the variation in Y is associated with variation in X. Thus, one must be quite careful in attempting to interpret observed values of r. Despite the increased awkwardness of interpretation associated with the use of r, it is often preferred to r^2, since the sampling distribution of r is more conducive to the construction of inferential statements than the sampling distribution of the statistic, r^2. The procedure for construction of inferential statements with respect to r is discussed in Section 20.9.

27.8 THE PRODUCT MOMENT METHOD

The coefficient of correlation for the universe can be computed directly by the following relationship:

$$\rho = \frac{N(\sum XY) - (\sum X)(\sum Y)}{\sqrt{N(\sum X^2) - (\sum X)^2}\sqrt{N(\sum Y^2) - (\sum Y)^2}}$$

When estimated from a sample, the formula becomes

$$r = \frac{n(\sum XY) - (\sum X)(\sum Y)}{\sqrt{n(\sum X^2) - (\sum X)^2}\sqrt{n(\sum Y^2) - (\sum Y)^2}}$$

This is equivalent to the r that would have been obtained by the following computation:

$$r = \sqrt{1 - \frac{\sum(Y - \bar{Y}_x)^2/n}{\sum(Y - \bar{Y})^2/n}}$$

A correction factor, $(n - 1)/(n - m)$, must be introduced when the sample size is small,[9] in order to generate an unbiased estimator of ρ. The $n - m$ is associated with the use of $\sum(Y - \bar{Y}_x)^2/(n - m)$ as an unbiased estimator of $\sigma_{Y.x}^2$, and the $n - 1$ is associated with the use of $\sum(Y - \bar{Y})^2/(n - 1)$ as an unbiased estimator of σ_Y^2. As the sample size increases, the ratio $(n - 1)/(n - m)$ approaches one, and the correction factor can be ignored. The correction factor, when r is computed by the product moment formula, takes the following form:

$$\text{adjusted } r = \sqrt{1 - (1 - r^2)\frac{n - 1}{n - 2}}$$

All the summations required by the product moment formula have previously been computed for the illustrative example and are recorded in Table 27.7. Substituting, we obtain

$$r = \frac{n(\sum XY) - (\sum X)(\sum Y)}{\sqrt{n(\sum X^2) - (\sum X)^2}\sqrt{n(\sum Y^2) - (\sum Y)^2}}$$

$$= \frac{8(2594) - (240)(80)}{\sqrt{8(8218) - (240)^2}\sqrt{8(856) - (80)^2}}$$

$$= \frac{20752 - 19200}{\sqrt{65744 - 57600}\sqrt{6848 - 6400}}$$

$$= \frac{1552}{\sqrt{8144}\sqrt{448}}$$

$$= \frac{1552}{\sqrt{3648512}}$$

$$= \frac{1552}{1910}$$

$$= 0.81$$

[9] n and m are defined on p. 696.

$$\text{adjusted } r = \sqrt{1 - (1 - (0.81)^2)\left(\frac{8 - 1}{8 - 2}\right)}$$

$$= \sqrt{1 - (1 - 0.6561)(\tfrac{7}{6})}$$

$$= \sqrt{1 - (0.3439)(\tfrac{7}{6})}$$

$$= \sqrt{1 - 0.4012}$$

$$= \sqrt{0.5988}$$

$$= +0.77$$

Note that the adjusted r is identical in value to the r obtained by the formula $r = \sqrt{1 - \hat{\sigma}_{Y.X}^2/\hat{\sigma}_Y^2}$. The product moment formula is particularly advantageous when all that is desired is a measure of correlation. The value of r can be computed directly by the product moment method and does not require the determination of the regression line, the standard deviation of Y, or the standard error of the estimate.

27.9 INFERENTIAL STATEMENTS[10] WITH RESPECT TO ρ

The sampling distribution of r has a mean equal to ρ and a standard error of r equal to $\sqrt{(1 - r^2)/(n - 2)}$. It is approximately t distributed with $n - 2$ degrees of freedom *provided the value of ρ is approximately equal to zero*. As the value of the universe parameter, ρ, departs from zero, the sampling distribution of r becomes increasingly skewed. Fisher discovered that when ρ departs from zero, the statistic formed by the transformation $z^* = \tfrac{1}{2}\log_e[(1 + r)/(1 - r)]$ is approximately normally distributed with a mean equal to $\tfrac{1}{2}\log_e[(1 + \rho)/(1 - \rho)]$ and a standard deviation equal to $1/\sqrt{n - 3}$. Thus, one can construct inferential statements and tests of hypotheses with respect to ρ utilizing normal curve analysis by conducting such analyses with the transformed variable z^*. A table providing values of r associated with specified values of z^* is reproduced in the Appendix as Table K.

When $\rho \cong 0$, confidence interval statements take the form

$$r \pm t\sigma_r$$

where t refers to the number of standard errors associated with the specified level of confidence for a t distribution with $n - 2$ degrees of freedom and

$$\sigma_r = \sqrt{\frac{1 - r^2}{n - 2}}$$

Tests of hypotheses take the form: Compare the value of t associated with the specified critical probability for a t distribution with $n - 2$ degrees of freedom with the statistic

$$t = \frac{|r - \text{hypothesized } \rho|}{\sqrt{(1 - r^2)/(n - 2)}}$$

---

[10] This section may be omitted without loss of continuity.

If the calculated value of t is less than or equal to the critical value of t, accept the null hypothesis that $\rho =$ hypothesized ρ. If the calculated value of t exceeds the specified value of t, reject the null hypothesis.

When $\rho \neq 0$, confidence interval statements take the form

$$z^* \pm \varkappa\sigma_{z^*}$$

where

$z^* = \frac{1}{2}\log_e[(1 + r)/(1 - r)]$
$\varkappa =$ number of standard errors associated with the specified level of confidence for a normal distribution

$\sigma_{z^*} = 1/\sqrt{n - 3}$

Tests of hypotheses take the form: Compare the value of $\varkappa$ associated with the specified critical probability for a normal distribution with the statistic

$$\varkappa = \frac{|z^* - \text{hypothesized } z^*|}{\sigma_{z^*}}$$

If the calculated value of $\varkappa$ is less than or equal to the critical value of $\varkappa$, accept the null hypothesis that $z^* =$ hypothesized z^*, where the hypothesized z^* is

$$z^* = \frac{1}{2}\log_e[(1 + \text{hypothesized } \rho)/(1 - \text{hypothesized } \rho)]$$

Otherwise reject the null hypothesis.

Illustrative example: Construction of a 95 percent confidence interval estimate for ρ utilizing the data in the illustrative example.

$n = 8$
$r = +0.77$
$\varkappa = 1.96$

Referring to Table K in the Appendix, when $r = 0.77$, the value of $z^* = 1.02$.
95 percent confidence interval:

$$z^* \pm \varkappa\sigma_{z^*}$$

$$1.02 \pm 1.96\frac{1}{\sqrt{5}}$$

$$1.02 \pm 1.96\frac{1}{2.236}$$

$$1.02 \pm 1.96(0.447)$$

$$1.02 \pm 0.88$$

$$0.14 \leftrightarrow 1.90$$

Thus, one could be 95 percent confident that one's statement was correct when stating that the value of z^* corresponding to ρ is somewhere between 0.14 and 1.90.
Converting z^* values to r values:

$z^* = 0.14$ is associated with $r = 0.1391$.
$z^* = 1.90$ is associated with $r = 0.9562$.

Given the above transformation, one could be 95 percent confident that one's statement was correct when stating that the value of ρ is somewhere in the interval 0.1391 to 0.9562.

Converting r values to r^2 values, confidence interval statements can be generated with respect to ρ^2.

$r = 0.1391$ is associated with $r^2 = 0.02$
$r = 0.9562$ is associated with $r^2 = 0.91$.

The statement: One could be 95 percent confident that one's statement was correct when stating that the percentage of the variation in Y associated with variation in X is somewhere between 2 and 91 percent.

Obviously, a much larger sample size is required if meaningful confidence interval estimates are desired.

The discussion in this chapter has dealt with analyses that can be performed when the underlying assumptions correspond to reality. A number of tests have been developed to evaluate whether or not the assumptions are met by a given set of data. Interested readers are referred to advanced treatments of regression and correlation analysis for a discussion of these tests and the impact on the analysis of violation of the underlying assumptions.

TECHNICAL NOTE 14

Confidence interval statements with respect to the forecasted value of Y given the stipulation of a specific value of X

Given that the value of $\mu_{Y.x}$ is known:

Where $\hat{\sigma}_{Y.x}$ has been estimated from a small sample: One could state that "95 percent of the individual observations comprising a conditional probability distribution have a value within $t_{0.025}\hat{\sigma}_{Y.x}$ of $\mu_{Y.x}$, where $\hat{\sigma}_{Y.x}$ is subject to variation from sample to sample."

Where $\hat{\sigma}_{Y.x}$ has been estimated from a large sample: One could state that "95 percent of the individual observations comprising a conditional probability distribution have a value within $1.96\hat{\sigma}_{Y.x}$ of $\mu_{Y.x}$, where the value of $\hat{\sigma}_{Y.x}$ can be considered constant from sample to sample."

$\mu_{Y.x}$ is typically unknown, however, and it is necessary to use a point estimate, $\overline{Y}_x$, as generated by the regression equation. $\overline{Y}_x$ is subject to sampling error, and this must be incorporated when constructing confidence statements with respect to the forecasted value of Y given a specific value for X. The sampling error in $\overline{Y}$ is attributable to the sampling error associated with the estimates of α and β generated by the regression equation.

The standard error of a forecasted Y value, when α and β have been estimated from sample data, takes the following form:

$$\hat{\sigma}_{Y|x} = \hat{\sigma}_{Y.x}\sqrt{1 + \frac{1}{n} + \frac{(X - \overline{X})^2}{\sum(X - \overline{X})^2}}$$

The height of the regression function is located at the point where $X = \overline{X}$. At this point only the estimating error attributable to a is operative. Thus, where

$X = \bar{X}$, $\hat{\sigma}_{Y|X}$ reduces to the following form:

$$\hat{\sigma}_{Y|X} = \hat{\sigma}_{Y.X}\sqrt{1 + \frac{1}{n}}$$

The correction factor $\sqrt{1 + 1/n}$ reflects the impact of the sampling error attributable to a on the standard error of a forecasted Y value.

As one departs from the point where the height of the regression function has been located, the sampling error attributable to b becomes operative. The farther the departure from the point where $X = \bar{X}$, the greater the impact of the sampling error associated with the estimate of β on the value of $\hat{\sigma}_{Y|X}$. This impact is reflected by the third term in the correction factor, $(X - \bar{X})^2/\sum(X - \bar{X})^2$. The net impact is that the greater the departure of the specified value of X from the point where $X = \bar{X}$, the wider the confidence interval with respect to the forecasted value of Y. As the sample size increases, both $1/n$ and $(X - \bar{X})^2/\sum(X - \bar{X})^2$ approach zero. Thus, when n is quite large, $\hat{\sigma}_{Y|X} = \hat{\sigma}_{Y.X}$.

The computation of the 95 percent confidence intervals for forecasted values of Y associated with the values of X in the illustrative example is illustrated in Table 27.10. The confidence intervals are depicted superimposed on the scatter diagram in Figure 27.9.

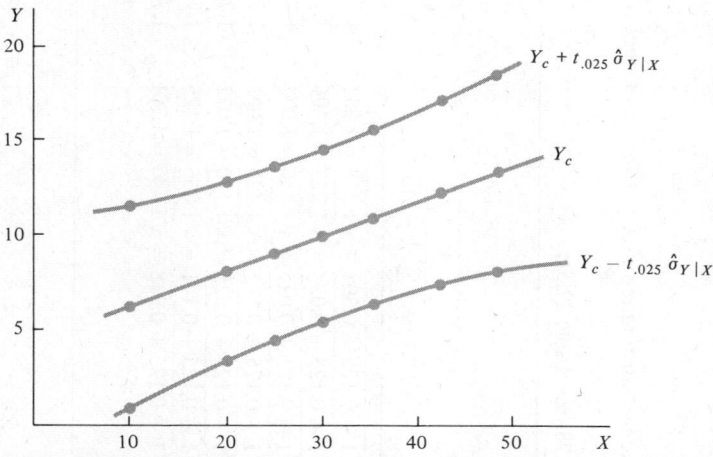

Figure 27.9 Ninety-five percent confidence band for forecasted Y given a stipulated value of X.

TECHNICAL NOTE 15

Confidence interval statements with respect to $\mu_{Y.X}$ given the stipulation of a specific value of X

Given that the value of $\mu_{Y.X}$ is known:

Where $\hat{\sigma}_{Y.X}$ has been estimated from a small sample: One could state that "95 percent of the sample means comprising the conditional sampling distribution of the mean associated with the specified value X and the sample

Table 27.10 Computation of the 95 percent confidence interval for the forecasted values of Y for the values of X specified in the illustrative example

$n = 8$; $\phi = 6$; $\hat{\sigma}_{Y.x} = 1.78$ (see Table 27.5); $\bar{X} = 30$

Tract	X	Y_c	$X - \bar{X}$	$(X - \bar{X})^2$	$\sqrt{1 + \dfrac{1}{n} + \dfrac{(X - \bar{X})^2}{\sum(X - \bar{X})^2}}$	$\hat{\sigma}_{Y\vert x}$ or $\hat{\sigma}_{Y.x}\sqrt{1 + \dfrac{1}{n} + \dfrac{(X - \bar{X})^2}{\sum(X - \bar{X})^2}}$	$t_{0.025}\sigma_{Y\vert x}$ = $2.447\hat{\sigma}_{Y\vert x}$	$Y_c - t_{0.025}\hat{\sigma}_{Y\vert x}$ = $Y_c - 2.447\hat{\sigma}_{Y\vert x}$	$Y_c + t_{0.025}\hat{\sigma}_{Y\vert x}$ = $Y_c + 2.447\hat{\sigma}_{Y\vert x}$
A	10	6.188	−20	400	$\sqrt{1 + 0.125 + 0.393} = \sqrt{1.518} = 1.232$	$1.78(1.232) = 2.193$	5.366	0.822	11.554
B	20	8.094	−10	100	$\sqrt{1 + 0.125 + 0.098} = \sqrt{1.223} = 1.106$	$1.78(1.106) = 1.969$	4.818	3.276	12.912
C	25	9.047	−5	25	$\sqrt{1 + 0.125 + 0.025} = \sqrt{1.150} = 1.072$	$1.78(1.072) = 1.908$	4.669	4.378	13.716
D, E	30	10.000	0	0	$\sqrt{1 + 0.125 + 0} = \sqrt{1.125} = 1.061$	$1.78(1.061) = 1.889$	4.622	5.378	14.622
F	35	10.953	+5	25	$\sqrt{1 + 0.125 + 0.025} = \sqrt{1.150} = 1.072$	$1.78(1.072) = 1.908$	4.669	6.284	15.622
G	42	12.287	+12	144	$\sqrt{1 + 0.125 + 0.141} = \sqrt{1.266} = 1.125$	$1.78(1.125) = 2.002$	4.899	7.388	17.186
H	48	13.431	+18	324	$\sqrt{1 + 0.125 + 0.318} = \sqrt{1.443} = 1.201$	$1.78(1.201) = 2.138$	5.232	8.199	18.663
				1018					

size n have a value within $t_{0.025}\hat{\sigma}_{Y.x}/\sqrt{n}$ of $\mu_{Y.x}$. The value assigned to $\hat{\sigma}_{Y.x}$ would vary from sample to sample."

Where $\hat{\sigma}_{Y.x}$ has been estimated from a large sample: One could state that "95 percent of the sample means comprising the conditional sampling distribution of the mean associated with the specified value X and the sample size n have a value within $1.96\hat{\sigma}_{Y.x}/\sqrt{n}$ of $\mu_{Y.x}$, where the value $\hat{\sigma}_{Y.x}$ can be considered constant from sample to sample."

$\mu_{Y.x}$ is typically unknown, however, and it is necessary to estimate its value by the regression equation. The regression estimate is subject to sampling error in the estimation of α and β, and this must be incorporated when generating confidence interval estimates of the value of $\mu_{Y.x}$ for a specified value of X. The standard error of the conditional mean when α and β have been estimated from sample data takes the following form:

$$\hat{\sigma}_{\bar{Y}|x} = \hat{\sigma}_{Y.x}\sqrt{\frac{1}{n} + \frac{(X - \bar{X})^2}{\sum(X - \bar{X})^2}}$$

As in the case of the forecasted value of Y given the stipulation of a specific value of X, the confidence interval will be narrowest where $X = \bar{X}$ and widen as the X value of interest departs farther and farther from this point. This is attributable to the fact that any error in the estimation of the slope of the line will have an increased impact the greater the departure of the X value of interest from the point at which the height of the regression line is estimated, namely, where $X = \bar{X}$.

The computation of the 95 percent confidence intervals for $\mu_{Y.x}$ associated with the values of X in the illustrative example is illustrated in Table 27.11. The confidence intervals are depicted in Figure 27.10.

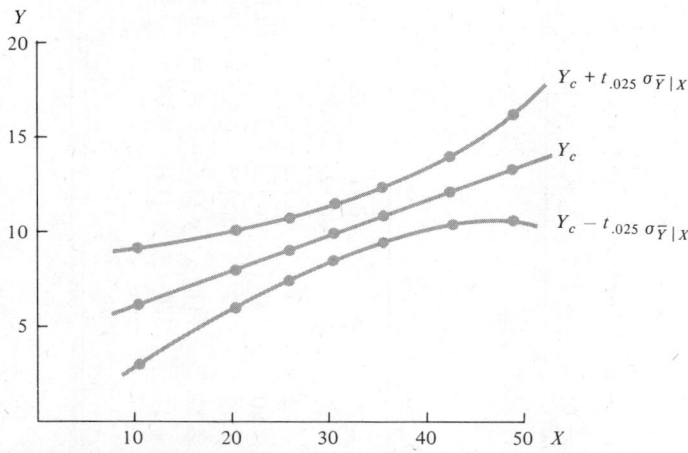

Figure 27.10 Ninety-five percent confidence band for $\mu_{Y.x}$.

PROBLEMS

Problem 27.1

1. What is a regression line? What information does it convey?
2. Does the regression line in and of itself provide sufficient information to

Table 27.11 Computation of the 95 percent confidence interval of $\mu_{Y\cdot x}$ for the values of X specified in the illustrative example

$n = 8$; $\phi = 6$; $\hat{\sigma}_{Y\cdot x} = 1.78$; $\bar{X} = 30$

Tract	X	Y_c	$X - \bar{X}$	$(X - \bar{X})^2$	$\sqrt{\dfrac{1}{n}+\dfrac{(X-\bar{X})^2}{\sum(X-\bar{X})^2}}$	$\hat{\sigma}_{\bar{Y}\|x}$ or $\hat{\sigma}_{Y\cdot x}\sqrt{\dfrac{1}{n}+\dfrac{(X-\bar{X})^2}{\sum(X-\bar{X})^2}}$	$t_{0.25}\hat{\sigma}_{\bar{Y}\|x}$ $2.447\hat{\sigma}_{\bar{Y}\|x}$	$Y_c - t_{0.025}\hat{\sigma}_{\bar{Y}\|x}$	$Y_c + t_{0.025}\hat{\sigma}_{\bar{Y}\|x}$
A	10	6.188	−20	400	$\sqrt{0.125 + 0.393} = \sqrt{0.518} = 0.72$	$1.78(0.72) = 1.282$	3.137	3.051	9.325
B	20	8.094	−10	100	$\sqrt{0.125 + 0.098} = \sqrt{0.223} = 0.47$	$1.78(0.47) = 0.837$	2.048	6.046	10.142
C	25	9.047	−5	25	$\sqrt{0.125 + 0.025} = \sqrt{0.150} = 0.39$	$1.78(0.39) = 0.694$	1.698	7.349	10.745
D,E	30	10.000	0	0	$\sqrt{0.125 + 0} = \sqrt{0.125} = 0.35$	$1.78(0.35) = 0.623$	1.524	8.476	11.524
F	35	10.953	+5	25	$\sqrt{0.125 + 0.025} = \sqrt{0.150} = 0.39$	$1.78(0.39) = 0.694$	1.698	9.255	12.651
G	42	12.287	+12	144	$\sqrt{0.125 + 0.141} = \sqrt{0.266} = 0.52$	$1.78(0.52) = 0.926$	2.266	10.021	14.533
H	48	13.431	+18	324	$\sqrt{0.125 + 0.318} = \sqrt{0.443} = 0.67$	$1.78(0.67) = 1.193$	2.919	10.512	16.350
				1018					

enable the analyst to evaluate improved predictive ability? If not, what additional information would he like to have?

Problem 27.2

What is a conditional standard deviation?

Problem 27.3

What information is conveyed by b in the regression equation $\bar{Y}_x = a + bx$?

Problem 27.4

Once the regression line has been obtained, it is desirable to evaluate the extent to which the conditional means are representative predictors of individual performance as compared to relying on the arithmetic mean of all the Y values. Thoroughly explain the role that the standard error of the estimate and the standard deviation of the Y values play in such an evaluation.

Problem 27.5

Draw scatter diagrams depicting the following situations:
  a. good positive correlation
  b. good negative correlation
  c. poor negative correlation
  d. poor positive correlation
  e. no correlation

On each scatter diagram, roughly indicate the location of the following measures:
  1. the regression line
  2. $\bar{Y} \pm \hat{\sigma}_Y$
  3. $\bar{Y}_x \pm \hat{\sigma}_{Y.x}$

Problem 27.6

A chicken farmer desires to determine the impact on gain in weight associated with variation in the grain ration. To study the impact, the chicken farmer randomly selected 200 chickens, which were divided into ten groups of 20 chickens each. The first group received a ration of X oz, the second group a ration of $X + 1$ oz, the third group a ration of $X + 2$ oz, and so on in successive 1-oz increments. The gain in weight for each chicken was then determined upon completion of the typical growing period. The farmer has heard that statistical methods are available that will enable him to estimate the weight gain attributable to incremental grain rations. Moreover, he also has heard that it is possible to evaluate the accuracy of those estimates. Drawing on your understanding of regression analysis, explain, in terms the chicken farmer would understand, the statistical techniques to which he refers. Be sure to include the regression line, $\bar{Y}_x$, $\hat{\sigma}_{Y.x}$, $\hat{\sigma}_Y$, and r^2.

Problem 27.7

If the standard error of the estimate, $\sigma_{Y.x}$, is equal in value to the standard deviation of the Y values, σ_Y, what would be the value of the coefficient of determination, and why?

Problem 27.8

Is it possible for $\sigma_{Y.X}$ to be greater in value than σ_Y? Why or why not?

Problem 27.9

What does it indicate in terms of the strength of the *association* of one variable with another if all of the estimated values of the conditional means generated by the regression function ($\overline{Y}_X$) are identical in value to the corresponding original observations (Y)?

Problem 27.10

The standard error of the estimate in one correlation study was $1250. The standard error of the estimate in a second, unrelated study was observed to be $750. In which of the two studies was correlation higher? Justify your answer.

Problem 27.11

In calculating the standard error of the estimate, the formula used involves dividing the sum of the squared deviations from the regression line by $n - 2$ instead of by n. Thoroughly explain *why* one divides by $n - 2$ instead of by n when computing the value of the standard error of the estimate.

Problem 27.12

You wish to determine the relationship between the annual production of pig iron and the annual number of pigs slaughtered. A sample of five observations revealed the following information:

Year	Production of pig iron (millions of tons)	No. of pigs slaughtered (ten thousands)
A	8	16
B	7	12
C	6	10
D	5	8
E	9	20

1. Letting pig iron production be the dependent variable, construct a regression line relating the production of pig iron and the number of pigs slaughtered.
2. Estimate the proportion of the variation in pig iron production that is associated with the variation in factors *other than* the number of pigs slaughtered.
3. Determine the value of the coefficient of determination. Does this value make sense considering the nature of the two variables (slaughter of pigs and production of pig iron)?
4. Determine the value of the coefficient of correlation. Interpret the result.

Problem 27.13

A company has developed an aptitude test to be used for evaluating the potential performance of applicants for sales positions. Hopefully, a high test score will indicate possession of the traits conducive to high sales performance, and a low test score will indicate lack of the traits. The question is: Does the test accomplish what it purports to do? To investigate this question, the company selected a sample of their present employees and gave them the test. The company then compared the test scores with the associated sales performance of these individuals to see if high scores tended to be associated with high sales, and so on. The sample results (based on eight salesmen) are reproduced below:

Salesman	Test score	Sales performance (thousands of dollars)
Jones	35	60
Wilson	80	70
Smith	75	90
Lewis	90	90
Pierce	65	70
Blue	50	80
Allen	70	60
Murray	95	120

1. Construct a scatter diagram.
2. Compute the least-squares linear regression line.
3. Determine the values of $\hat{\sigma}_Y$ and $\hat{\sigma}_{Y.X}$.
4. Discuss the role the values computed in step 3 play in evaluating the improvement in the ability to predict when $\overline{Y}_X$, rather than $\overline{Y}$, is employed as the best estimate.
5. Calculate the coefficient of correlation using the values obtained in step 3.
6. Calculate the coefficient of correlation by the product moment method. Make the appropriate adjustment for the degrees of freedom.
7. Calculate the coefficient of determination.
8. Employing a 5 percent critical probability, test the hypothesis that the true value of the universe parameter β is zero.
9. Construct a 90 percent confidence interval estimate of the value of the universe parameter, β.

Problem 27.14

To facilitate price determination for a newly developed product, a number of pilot studies were conducted in market areas that were virtually identical with respect to income level, degree of education, density of population, and so on. The product was test marketed in each of ten areas, with the only dimension allowed to vary being the price of the product. Sales volumes observed in association with each price level are recorded below:

Area	Price	Sales volume (in units)
A	0.10	12,000
B	0.15	10,000
C	0.20	7,000
D	0.25	6,000
E	0.30	5,000
F	0.35	4,000
G	0.40	3,500
H	0.45	2,000
I	0.50	2,500
J	0.55	2,000

1. Construct a scatter diagram.
2. Determine the least-squares linear regression line.
3. Estimate the volume associated with a price of $0.12.
4. Estimate the volume associated with a price of $0.50. Discuss the validity of this estimate.
5. Determine the values of $\hat{\sigma}_Y$ and $\hat{\sigma}_{Y.X}$.
6. Discuss the role the values computed in step 5 play in evaluating the improvement in the ability to predict when $\overline{Y}_X$, rather than $\overline{Y}$, is employed as the best estimate.
7. Employing a critical probability of 1 percent, test the hypothesis that the true value of the universe parameter β is zero.
8. Construct a 90 percent confidence interval estimate of the value of the universe parameter, β.

Problem 27.15

A jobshop printing outfit desires to estimate the relationship between the number of copies produced by an offset printing technique and the associated direct labor cost. Eight orders were observed and the following information obtained:

	Based on a single page	
Order	Number of copies	Total direct labor cost
1	200	1.80
2	50	1.20
3	300	2.70
4	100	1.40
5	20	1.10
6	80	1.50
7	150	1.70
8	60	1.40

1. Construct a scatter diagram.
2. Compute the least-squares linear regression line.
3. Estimate the total direct labor cost associated with 250 copies.
4. Calculate the value of the coefficient of correlation by the formula

$$ r = \sqrt{1 - \frac{\hat{\sigma}_{Y.X}}{\hat{\sigma}_Y}} $$

5. Calculate the coefficient of correlation by the product moment method. Incorporate the appropriate adjustment for degrees of freedom.
6. Calculate the coefficient of determination.
7. Employing a z^* transformation, construct a 90 percent confidence interval estimate of the universe parameter, ρ.
8. Employing a critical probability of 1 percent, test the hypothesis that the true value of the universe parameter β is zero.
9. Construct a 95 percent confidence interval estimate of the value of the universe parameter, β.

Problem 27.16

A firm is currently involved in negotiating a penalty charge for defective components. It is believed that the time lost in disassembling and reassembling the product in which the components are used is a linear function of the number of defective components encountered. A random sample of ten such subassemblies revealed the following information:

Job number	Number of defective components	Time lost due to disassembly and reassembly (in hours)
201	1	2.3
372	3	6.4
126	4	8.0
240	2	3.9
302	2	4.0
184	4	4.6
76	3	4.2
420	5	8.0
311	2	3.4
276	1	2.6
		42.0

1. Construct a scatter diagram.
2. Compute the least-squares linear regression line.
3. Company policy is to charge a fixed base amount if any defectives are encountered and then a variable penalty charge for each defective encountered.
   a. What should the fixed base amount be?
   b. What should the variable penalty charge per defective component encountered be?
4. Determine the values of $\hat{\sigma}_Y$ and $\hat{\sigma}_{Y.X}$.
5. Discuss the role the values computed in step 4 play in evaluating the improvement in the ability to predict when $\overline{Y}_x$, rather than $\overline{Y}$, is employed as the best estimate.
6. Calculate the coefficient of correlation using the values obtained in step 4.
7. Calculate the coefficient of correlation by the product moment method. Incorporate the appropriate adjustment for the degrees of freedom.
8. Calculate the coefficient of determination.
9. Employing a z^* transformation, construct a 95 percent confidence interval estimate of the universe parameter, ρ.
10. Employing a 5 percent critical probability, test the hypothesis that the true value of the universe parameter β is zero.
11. Construct a 90 percent confidence interval estimate of the value of the universe parameter β.

Problem 27.17

Most computer installations have a regression-correlation routine as part of the software package. One need simply supply the X and Y values and the computer program will perform the necessary underlying computations. The output from a typical program applied to the illustrative example in Chapter 27 is reproduced in Printout 27.1.

```
                          TABLE OF RESIDUALS
             CASE NO.    Y VALUE      Y ESTIMATE      RESIDUAL
               1         6.00000       6.18861        -.18861
               2         7.00000       8.09430       -1.09430
               3         9.00000       9.04715         -.04715
               4        13.00000      10.00000        3.00000
               5        10.00000      10.00000          .00000
               6         9.00000      10.95285       -1.95285
               7        14.00000      12.28684        1.71316
               8        12.00000      13.43026       -1.43026


VARIABLE    MEAN     STANDARD      CORRELATION    REGRESSION    STD. ERROR     COMPUTED
NO.                  DEVIATION       X VS Y       COEFFICIENT   OF REG.COEF.   T VALUE
 2        30.00000   12.05939        .81252         .19057        .05582       3.41421
DEPENDENT
 1        10.00000    2.82843


                    INTERCEPT                  4.28291

                    MULTIPLE CORRELATION        .81252

                    (ADJUSTED FOR DEGREES OF FREEDOM)
                                                .77689

                    STD. ERROR OF ESTIMATE     1.78089

                    R-SQUARED                   .66019

                    (ADJUSTED FOR DEGREES OF FREEDOM)
                                                .60355
```

Printout 27.1

1. The regression line is specified by the numerical values for *a* and *b*. In the computer printout, the *a* value is designated as the intercept and the *b* value as the regression coefficient. Locate these values on the printout and compare them to the values computed in Table 27.2.
2. Justify the use of the label "intercept" for the *a* coefficient.
3. The precision of the regression line estimates as compared to $\overline{Y}$ is evaluated by comparing $\hat{\sigma}_{Y.X}$ (standard error of the estimate) and $\hat{\sigma}_Y$ (standard deviation of the dependent variable). Locate these values on the printout and compare them to the values computed on p. 699.
4. The coefficient of correlation is indicated in the printout under the column heading, CORRELATION X VS Y. In the case of a single independent variable, the coefficient of multiple correlation is identical to the coefficient of simple correlation (i.e., CORRELATION X VS Y). Verify that these values are equal on the printout.
5. Compare the value indicated under the column heading CORRELATION X VS Y with the value of *r* computed by the product moment method on p. 711. Note that the value on the printout has not allowed for degrees of freedom. Using the formula on p. 711, make the appropriate adjustment.
6. The coefficient of determination is indicated on the printout as R-SQUARED. The value of R-SQUARED (ADJUSTED FOR DEGREES OF FREEDOM) appears directly below. Compare the latter value to the value of r^2 computed in the text on p. 707.

7. Using the formula on p. 711, derive the value of R-SQUARED (AD-JUSTED FOR DEGREES OF FREEDOM) from the value of R-SQUARED.

8. The standard error of b is indicated under the column heading STD. ERROR OF REG. COEF. Compare this value to the value computed in Table 27.8.

9. Discuss the significance of the value 3.41421 indicated under the column heading COMPUTED T VALUE. Refer to p. 701 in the text.

10. Compare the Y estimate values in the table of residuals with the values computed for $\bar{Y}_x$ on pp. 690–691.

11. The $Y - \bar{Y}_x$ values are entered in the residuals column. Total these values. Is the sum obtained in accordance with expectations?

Problem 27.18

What is the reason for instituting a z^* transformation?

Problem 27.19

Given $r^2 = 0.8448$.

1. Assuming a sample size of 28 and using a z^* transformation, what statement can be made with respect to ρ^2 with 95 percent confidence? (That is, construct a 95 percent confidence interval estimate of the universe parameter ρ^2.)

2. Repeat part 1 assuming a sample size of 53.

3. Repeat part 1 assuming a sample size of 103.

4. How is the confidence interval estimate affected by the increase in sample size?

Problem 27.20

A coefficient of correlation of $r = 0.9121$ has been obtained from a sample consisting of 103 pairs of observations. Using the z^* transformation, what statement can you make with respect to r^2 with 95 percent confidence? (That is, using the z^* transformation, construct a 95 percent confidence interval for the universe parameter, ρ^2.)

Problem 27.21

In attempting to evaluate the strength of the relationship between score on an aptitude test and subsequent sales performance, you have calculated a coefficient of correlation of $+0.82$. The computation was based on a set of 67 individuals. (You have a test score and a sales performance observation for each individual.) What *specific numerical* statement can be made with 90 percent confidence that the statement is correct with respect to the percentage of the observed variation in Y that is associated with variation in X? All computations must be shown.

MULTIPLE REGRESSION AND CORRELATION ANALYSIS

28

In Chapter 27, the nature of the relationship between the dependent variable, yield per acre, and the single independent variable, intensity of fertilizer application, was investigated. A simple linear regression function was fitted to the sample evidence. Comparison of the standard error of the estimate and the standard deviation of the Y values revealed that specification of the intensity of fertilizer application enabled the analyst to generate more precise estimates of yield per acre than otherwise would be the case. It is a natural and logical step for the analyst to attempt to improve further his ability to forecast by expanding the regression function to include additional independent variables. The technique for dealing with more than one independent variable is referred to as multiple regression analysis.

28.1 MULTIPLE REGRESSION ANALYSIS

Consider the following extension of the illustrative example in Chapter 27. Presume that in addition to yield per acre and the extensiveness of fertilizer application, the sample also provides an index of the original suitability of the soil prior to application of the fertilizer. To reduce the computational burden, once again only eight tracts of land will be considered. The sample observations are summarized in Table 28.1.

Table 28.1

Tract	Yield per acre (thousands of bushels)	Fertilizer in tons	Index indicating the original suitability of the soil
A	6	10	4
B	7	20	3
C	9	25	5
D	13	30	9
E	10	30	5
F	9	35	3
G	14	42	8
H	12	48	3

Regression analysis is concerned with specification of the nature of the relationship and determination of the values of the coefficients in the regression function. As before, discussion is restricted to linear regression functions. With two independent variables and one dependent variable and the imposed restriction of a linear relationship, the regression function is linear in two dimensions or a plane.

In Chapter 27, when dealing with a two-variable linear relationship, Y was used to represent the dependent variable and X the independent variable. When dealing with a relationship in more than two variables, it is convenient to change the notation in the following fashion: All variables, including the dependent variable, are designated X. Each variable is identified by its subscript. The subscript 1 is typically assigned to the dependent variable. In accordance with this custom, the notation utilized in this chapter will be as follows:

X_1 represents yield per acre
X_2 represents fertilizer application in tons
X_3 represents the index indicating the original suitability of the soil

The general form of a linear regression function associated with one dependent variable and two independent variables (i.e., a plane) for the universe is

$$\mu_{1.23} = \alpha_{1.23} + \beta_{12.3}X_2 + \beta_{13.2}X_3$$

where

$\mu_{1.23}$ is the conditional mean of X_1 associated with a specified value of X_2 and X_3. To illustrate: Suppose that $\mu_{1.23} = 10$ when $X_2 = 30$ and $X_3 = 5$. This indicates that when the original quality of the soil is assigned an index rating of 5 and 30 tons of fertilizer are applied, the *expected* yield per acre is 10,000 bushels.

$\alpha_{1.23}$ is the conditional mean of X_1 associated with $X_2 = 0$ and $X_3 = 0$. $\alpha_{1.23}$ is the X_1 intercept.

$\beta_{12.3}$ and $\beta_{13.2}$ are referred to as *partial regression coefficients*.

$\beta_{12.3}$ is the change in the value of $\mu_{1.23}$ associated with a one-unit change in the value of X_2. The value of all variables indicated to the right of the period in $\beta_{12.3}$ are presumed to remain constant. $\beta_{12.3}$ is the slope of the plane with respect to the X_2 axis. Thus, if the original quality of the soil is held constant and fertilizer application is increased by one unit, $\beta_{12.3}$ is the associated change in the value of $\mu_{1.23}$.

$\beta_{13.2}$ is the change in the value of $\mu_{1.23}$ associated with a one-unit change in the value of X_3. The value of all variables indicated to the right of the period in $\beta_{13.2}$ are presumed to remain constant. $\beta_{13.2}$ is the slope of the plane with respect to the X_3 axis. Thus, if the fertilizer application is held constant and the index rating of the original quality of the soil is incremented by one unit, $\beta_{13.2}$ is the associated change in the value of $\mu_{1.23}$.[1]

The analysis is built on the following assumptions:

1. The conditional means are linearly related. In other words, the regression function is a plane.
2. All the conditional probability distributions of X_1 are normally distributed. There is a separate conditional probability distribution associated with each conceivable pair of X_2, X_3 values.
3. All the conditional standard deviations are equal. $\sigma_{1.23}$ is a measure of the variability in X_1 when the values of X_2 and X_3 are held constant. This measure of dispersion in the conditional probability distributions is presumed to be the same for all combinations of X_2 and X_3.

The universe regression function is typically unknown and must be estimated from sample data. The sample estimate takes the form

$$\overline{X}_{1.23} = a_{1.23} + b_{12.3}X_2 + b_{13.2}X_3$$

The sample data consists of observations on eight tracts of land. Each observation consists of a random selection from a conditional probability distribution of X_1, associated with a specified value of X_2 *and* a specified value of X_3. To facilitate visualization of the regression fitting process, it is useful to plot the observations on a scatter diagram. With one dependent and two independent variables, the scatter diagram is three dimensional. Each cross placed in three dimensional space represents a trivariate observation (Figure 28.1).

---

[1] The partial regression coefficients cannot be interpreted as stated above in cases where the independent variables are highly correlated. In such cases, the influence attributable to each of the independent variables cannot be distinguished. This problem is called multicollinearity. Where multicollinearity exists, one of the independent variables should be eliminated. The techniques for detecting the existence of multicollinearity and the procedures for dealing with this problem will be left to more advanced treatments of multiple regression and correlation analysis.

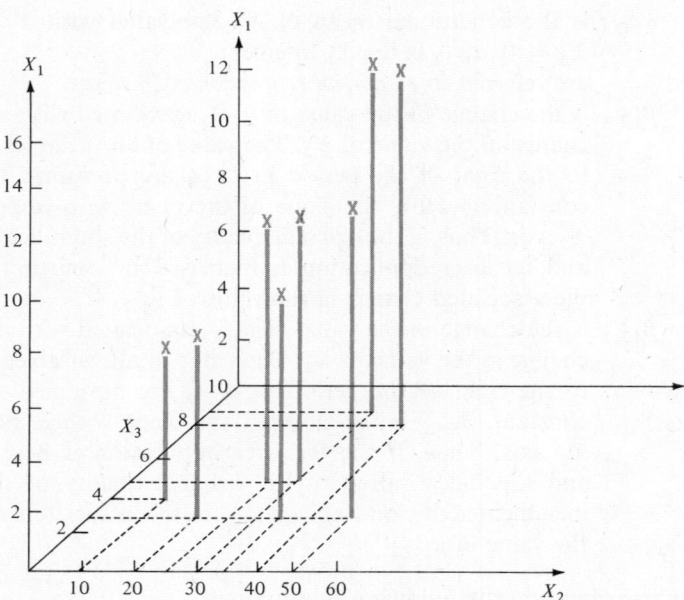

Figure 28.1

The least-squares criterion is utilized to determine the coefficients of the regression function. In effect, one determines the values for the coefficients $a_{1.23}$, $b_{12.3}$, and $b_{13.2}$ that minimize the function

$$F = \sum (X_1 - a_{1.23} - b_{12.3}X_2 - b_{13.2}X_3)^2$$

The coefficients that minimize the sum of the squared deviations are identical to the coefficients that solve the following system of *normal equations*[2]:

I $\qquad \sum X_1 = na_{1.23} + b_{12.3} \sum X_2 + b_{13.2} \sum X_3$

II $\qquad \sum X_1 X_2 = a_{1.23} \sum X_2 + b_{12.3} \sum X_2^{\,2} + b_{13.2} \sum X_2 X_3$

III $\qquad \sum X_1 X_3 = a_{1.23} \sum X_3 + b_{12.3} \sum X_2 X_3 + b_{13.2} \sum X_3^{\,2}$

The summations required by the normal equations for the illustrative example are generated in Table 28.2. Substituting, the normal equations become

I $\qquad 80 = 8a_{1.23} + 240b_{12.3} + 40b_{13.2}$

II $\qquad 2594 = 240a_{1.23} + 8218b_{12.3} + 1230b_{13.2}$

III $\qquad 432 = 40a_{1.23} + 1230b_{12.3} + 238b_{13.2}$

The system now consists of three equations with three unknowns. Multiplying equation I by 5 and subtracting from equation III, we obtain equation IIIA:

III $\qquad 432 = 40a_{1.23} + 1230b_{12.3} + 238b_{13.2}$

$-(I \times 5) \qquad 400 = 40a_{1.23} + 1200b_{12.3} + 200b_{13.2}$

IIIA $= (III - 5I) \qquad 32 = \qquad\qquad 30b_{12.3} + 38b_{13.2}$

---

[2] It can be demonstrated that equation I holds when $\partial F/\partial a_{1.23} = 0$; that equation II holds when $\partial F/\partial b_{12.3} = 0$; and that equation III holds when $\partial F/\partial b_{13.2} = 0$. Thus, when all three equations hold, all the partials are equal to zero and the function is minimized. There will be one normal equation for each constant in the regression equation. Thus, a system of four normal equations would be required if the regression equation took the form $\bar{X}_{1.234} = a_{1.234} + b_{12.34}X_2 + b_{13.24}X_3 + b_{14.23}X_4$.

Table 28.2 Computation of the summations required by the normal equations

Tract	X_1	X_2	X_3	$X_2{}^2$	$X_3{}^2$	X_1X_2	X_1X_3	X_2X_3	$X_1{}^2$
A	6	10	4	100	16	60	24	40	36
B	7	20	3	400	9	140	21	60	49
C	9	25	5	625	25	225	45	125	81
D	13	30	9	900	81	390	117	270	169
E	10	30	5	900	25	300	50	150	100
F	9	35	3	1225	9	315	27	105	81
G	14	42	8	1764	64	588	112	336	196
H	12	48	3	2304	9	576	36	144	144
	80	240	40	8218	238	2594	432	1230	856

$\sum X_1 = 80$ $\qquad \sum X_3 = 40$ $\qquad \sum X_3{}^2 = 238$ $\qquad \sum X_1X_3 = 432$

$\sum X_2 = 240$ $\qquad \sum X_2{}^2 = 8218$ $\qquad \sum X_1X_2 = 2594$ $\qquad \sum X_2X_3 = 1230$

Multiplying equation I by 30 and subtracting from equation II, we obtain equation IIA:

$$\begin{array}{ll} \text{II} & 2594 = 240a_{1.23} + 8218b_{12.3} + 1230b_{13.2} \\ -(\text{I} \times 30) & 2400 = 240a_{1.23} + 7200b_{12.3} + 1200b_{13.2} \\ \hline \text{IIA} = (\text{II} - 30\text{I}) & 194 = \phantom{240a_{1.23} +} 1018b_{12.3} + 30b_{13.2} \end{array}$$

Equations IIIA and IIA form a system of two equations with two unknowns:

$$\begin{array}{lll} \text{IIIA} & 32 = & 30b_{12.3} + 38b_{13.2} \\ \text{IIA} & 194 = & 1018b_{12.3} + 30b_{13.2} \end{array}$$

Multiplying equation IIIA by 3 and equation IIA by -3.8 and adding, we obtain

$$\begin{array}{ll} 3(\text{IIIA}) & 96 = 90b_{12.3} + 114b_{13.2} \\ -3.8(\text{IIA}) & -737.2 = -3868.4b_{12.3} - 114b_{13.2} \\ \hline & -641.2 = -3778.4b_{12.3} \end{array}$$

$$b_{12.3} = \frac{-641.2}{-3778.4}$$

$$b_{12.3} = 0.1697$$

Substituting $b_{12.3} = 0.1697$ in equation IIIA, we obtain

$$\begin{array}{ll} \text{IIIA} & 32 = 30b_{12.3} + 38b_{13.2} \\ & 32 = 30(0.1697) + 38b_{13.2} \\ & 32 = 5.0910 + 38b_{13.2} \\ & 26.909 = 38b_{13.2} \\ & b_{13.2} = 0.7081 \end{array}$$

Substituting $b_{12.3} = 0.1697$ and $b_{13.2} = 0.7081$ in equation I, we obtain

$$\begin{array}{ll} \text{I} & 80 = 8a_{1.23} + 240b_{12.3} + 40b_{13.2} \\ & 80 = 8a_{1.23} + 240(0.1697) + 40(0.7081) \\ & 80 = 8a_{1.23} + 40.7280 + 28.324 \\ & 80 = 8a_{1.23} + 69.052 \\ & 8a_{1.23} = 10.948 \\ & a_{1.23} = 1.3685 \end{array}$$

Thus, the coefficients of the regression function are

$$a_{1.23} = 1.3685$$
$$b_{12.3} = 0.1697$$
$$b_{13.2} = 0.7081$$

Once the values for the coefficients have been determined, the regression function can be used to generate the conditional mean associated with any specified combination of X_2 and X_3.

Tract A: Fertilizer application equal to 10 tons and a soil index rating of 4

$$\begin{aligned}\overline{X}_{1.23} &= a_{1.23} + b_{12.3}X_2 + b_{13.2}X_3\\ &= 1.3685 + 0.1697(10) + 0.7081(4)\\ &= 1.3685 + 1.697 + 2.8324\\ &= 5.8979\end{aligned}$$

Interpretation: It is estimated that a soil index rating of 4 and a fertilizer application of 10 tons is associated with an average yield of 5898 bushels per acre.

The supporting computations for the determination of $\overline{X}_{1.23}$ for the remaining seven tracts are reproduced in Table 28.3. Referring to tract E, note that when the values of X_2 and X_3 substituted into the regression function are equal to their respective means, the value of $\overline{X}_{1.23}$ is equal to the mean of the X_1 values. Short-cut formulas for the computation of the regression line that capitalize on this relationship are available. The objective of this chapter is to provide a conceptual understanding of multiple regression and correlation analysis. Short-cut computational approaches will be left to other texts.[3]

Table 28.3 Determination of $\overline{X}_{1.23}$ for tracts B–H

Tract	X_2	X_3	$\overline{X}_{1.23} = 1.3685 + 0.1697X_2 + 0.7081X_3$
B	20	3	1.3685 + 0.1697(20) + 0.7081(3) 1.3685 + 3.394 + 2.1243 6.8868
C	25	5	1.3685 + 0.1697(25) + 0.7081(5) 1.3685 + 4.2425 + 3.5405 9.1515
D	30	9	1.3685 + 0.1697(30) + 0.7081(9) 1.3685 + 5.091 + 6.3729 12.8324
E	30	5	1.3685 + 0.1697(30) + 0.7081(5) 1.3685 + 5.091 + 3.5405 10.000
F	35	3	1.3685 + 0.1697(35) + 0.7081(3) 1.3685 + 5.9395 + 2.1243 9.4323
G	42	8	1.3685 + 0.1697(42) + 0.7081(8) 1.3685 + 7.1274 + 5.6648 14.1607
H	48	3	1.3685 + 0.1697(48) + 0.7081(3) 1.3685 + 8.1456 + 2.1243 11.6384

[3] Charles T. Clark and Lawrence L. Schkade, *Statistical Methods for Business Decisions*, Cincinnati, Ohio: Southwestern Publishing Company, 1969, pp. 625–630.

28.2 EVALUATION OF THE STRENGTH OF THE REGRESSION RELATIONSHIP

The extent of improvement in the precision of the estimated value of X_1 when $\overline{X}_{1.23}$, rather than $\overline{X}_1$, is used as the estimator can be observed by comparing the magnitude of $\sigma_{1.23}$ with the magnitude of σ_1. The smaller the value of $\sigma_{1.23}$ relative to the value of σ_1, the greater the improvement in the precision of the estimate. In general, when there are K *independent* variables, the extent of improvement in the precision of the estimated value of X_1 when $\overline{X}_{1.2345\ldots(k+1)}$, rather than $\overline{X}_1$, is used as the estimator can be observed by comparing the magnitude of $\sigma_{1.234\ldots(k+1)}$ with the magnitude of σ_1. Dispersion of the X_1 values comprising a given conditional probability distribution is measured by $\sigma_{1.234\ldots(k+1)}$. Given that the conditional standard deviations are all equal, the dispersion of the individual X_1 values around their respective conditional means is equivalent to the dispersion of *all* the individual X_1 values around the regression equation. The dispersion of the individual X_1 values around the regression equation is measured by the *standard error of the estimate*.

$$\sigma_{1.234\ldots(k+1)} = \sqrt{\frac{\sum (X_1 - \mu_{1.234\ldots(k+1)})^2}{N}}$$

The universe parameter, $\sigma_{1.234\ldots(k+1)}$, is unknown and must be estimated from the sample data. The unbiased estimator of $\sigma_{1.234\ldots(k+1)}$ takes the form

$$\hat{\sigma}_{1.234\ldots(k+1)} = \sqrt{\frac{\sum (X_1 - \overline{X}_{1.234\ldots(k+1)})^2}{n - m}}$$

where m refers to the number of degrees of freedom lost. In general, one degree of freedom is lost for each constant that must be estimated from the sample data. In the illustrative example, three parameters must be estimated from the sample data, $a_{1.23}$, $b_{12.3}$, and $b_{13.2}$. Thus, there are three degrees of freedom lost. In general, for each additional independent variable added to the regression equation, one additional degree of freedom will be lost.

An unbiased estimator of σ_1, the standard deviation of the X_1 values, was presented in Chapter 27. Recall:

$$\hat{\sigma}_1 = \sqrt{\frac{\sum (X_1 - \overline{X}_1)^2}{n - 1}}$$

Comparison of $\hat{\sigma}_1$ and $\sigma_{1.23}\ldots(k+1)$ for the illustrative example

The computation of $\hat{\sigma}_1$, previously performed in Table 27.4, is reproduced in Table 28.4. The computation of $\hat{\sigma}_{1.234\ldots(k+1)}$ is displayed in Table 28.5. In the illustrative example, the number of independent variables is equal to two. Therefore, $(k + 1) = (2 + 1) = 3$ and $\hat{\sigma}_{1.234\ldots(k+1)} = \hat{\sigma}_{1.23}$.

The improvement in the precision of the estimate as provided by the regression equation is indicated by the reduction in dispersion from 2.83 to 0.289. It is useful to compare the stepwise improvement in precision since additional independent variables are included in the regression function.

Estimator	Precision of the estimate
$\bar{X}_1$	$\hat{\sigma}_1$ = 2.83
$\bar{X}_{1.2}$	$\hat{\sigma}_{1.2}$ = 1.78
where X_2 indicates	
fertilizer application	
$\bar{X}_{1.23}$	$\hat{\sigma}_{1.23}$ = 0.289
where X_3 measures	
original suitability of soil	

Typically, the order in which independent variables are added to the regression equation is determined by evaluating the extent to which precision of the estimate of the dependent variable is improved with the addition of each of the independent variables under consideration. The independent variable associated with the greatest improvement in precision is the next variable added to the regression function. One cannot add independent variables without limit, however. An independent variable will not be considered for inclusion unless its introduction results in a reduction of the sum of the squared error term, $\sum (X_1 - \bar{X}_{1.234\ldots(k+1)})^2$. Even this is not sufficient. The introduction of an additional independent variable results in the loss of an additional degree of freedom, causing the divisor of the term $\sum (X_1 - \bar{X}_{1.234\ldots(k+1)})^2/(n-m)$ to become smaller. Thus, the reduction in the sum of the squared error term must be more than sufficient to offset the impact of a smaller divisor before the associated additional independent variable can be considered for inclusion in the regression function.

A visual comparison of the extent of improvement in the precision of estimates provided by the regression function as compared to the estimate provided by $\bar{X}_1$ can be obtained by plotting the intervals $\bar{X}_1 \pm \hat{\sigma}_1$ and $\bar{X}_{1.23} \pm \hat{\sigma}_{1.23}$ on the scatter diagram. This relationship is depicted in Figure 28.2. Note that the intervals

Table 28.4 Computation of $\hat{\sigma}_1$

Tract	X_1 yield per acre (thousands of bushels)	$X_1 - \bar{X}_1$	$(X_1 - \bar{X}_1)^2$
A	6	-4	16
B	7	-3	9
C	9	-1	1
D	13	$+3$	9
E	10	0	0
F	9	-1	1
G	14	$+4$	16
H	12	$+2$	4
	80	$+9$ / -9 / 0	56

$$\bar{X}_1 = \frac{\sum X_1}{n} = \frac{80}{8} = 10 \qquad \hat{\sigma}_{x_1} = \sqrt{\frac{\sum (X_1 - \bar{X}_1)^2}{n-1}} = \sqrt{\frac{56}{7}} = \sqrt{8} = 2.83$$

Table 28.5 Computation of $\hat{\sigma}_{1.23}$

Tract	X_1 yield per acre (thousands of bushels)	$\bar{X}_{1.23}$ (see Table 21.3)	$X_1 - \bar{X}_{1.23}$	$(X_1 - \bar{X}_{1.23})^2$
A	6	5.8979	0.1021	0.01042441
B	7	6.8868	0.1132	0.01281424
C	9	9.1515	−0.1515	0.02295225
D	13	12.8324	0.1676	0.02808976
E	10	10.0000	—	—
F	9	9.4323	−0.4323	0.18688329
G	14	14.1607	−0.1607	0.02582449
H	12	11.6384	+0.3616	0.13075456
	80	80.0000	+0.7445 −0.7445	0.41774300
			0	

$$\hat{\sigma}_{1.23} = \sqrt{\frac{\sum (X_1 - \bar{X}_{1.23})^2}{n - m}}$$

$$= \sqrt{\frac{0.41774300}{8 - 3}}$$

$$= \sqrt{\frac{0.41774300}{5}}$$

$$= \sqrt{0.0835486}$$

$$= 0.289$$

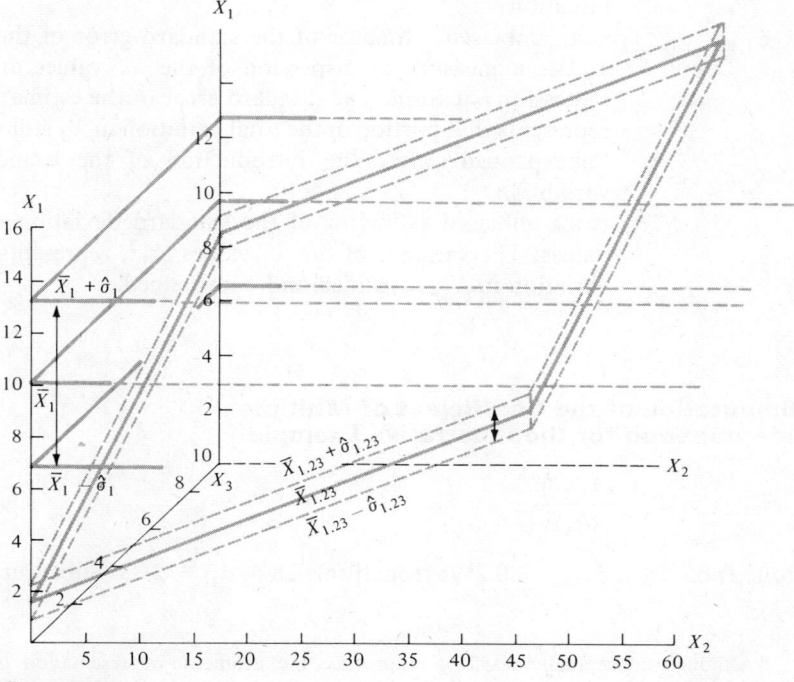

Figure 28.2

when depicted on the scatter diagram take the form of planes parallel to the plane formed by $\overline{X}_1$ and the plane formed by $\overline{X}_{1.23}$. When more than two independent variables are introduced, it is no longer possible to represent the improvement in precision geometrically.[4] We are not constrained, of course, with respect to calculation of the improvement in precision. If four independent variables were included in the regression equation, the improvement in precision would be evaluated by comparing

$$\hat{\sigma}_1 = \sqrt{\frac{\sum (X_1 - \overline{X}_1)^2}{n - 1}} \quad \text{and} \quad \hat{\sigma}_{1.2345} = \sqrt{\frac{\sum (X_1 - \overline{X}_{1.2345})^2}{n - 5}}$$

28.3 MULTIPLE CORRELATION ANALYSIS

The objective in multiple correlation analysis is to evaluate the extent to which variation in the dependent variable is associated with variation in the independent variables specified in the regression equation. The extent to which correlation exists is evaluated by the coefficient of multiple determination.

$$R^2_{1.23\ldots(k+1)} = 1 - \frac{(\hat{\sigma}_{1.23\ldots(k+1)})^2}{(\hat{\sigma}_1)^2}$$

where

$R^2_{1.23\ldots(k+1)}$ is interpreted as the percentage of variation in X_1 that is associated with or "explained by" the variation in the k independent variables indicated to the right of the period. A capital R^2 is used to designate the coefficient of *multiple* determination.

$\hat{\sigma}_{1.23\ldots(k+1)}$ is an unbiased estimator of the standard error of the estimate and is a measure of dispersion of the X_1 values around the regression equation. The standard error of the estimate squared represents that portion of the total variation in X_1 still remaining "unexplained" after the introduction of the k independent variables.

$\hat{\sigma}_1$ is an unbiased estimator of the standard deviation of the X_1 values. The variance of the X_1 values, $\hat{\sigma}_1{}^2$, represents the total variation in X_1, explained and unexplained.

Computation of the Coefficient of Multiple Determination for the Illustrative Example

$$R^2_{1.23} = 1 - \frac{(\hat{\sigma}_{1.23})^2}{(\hat{\sigma}_1)^2}$$

From Table 28.5, $\hat{\sigma}_{1.23} = 0.289$; from Table 28.4, $\hat{\sigma}_1 = 2.83$. Substituting,

---

[4] Although conceptually feasible, in practice the geometric representation of the two-independent-variable case is seldom employed.

$$R^2_{1.23} = 1 - \frac{(0.289)^2}{(2.83)^2}$$

$$= 1 - \frac{0.083521}{8}$$

$$= 1 - 0.0104$$

$$= 0.9896$$

Interpretation: 98.96 percent of the variation in X_1 is associated with variation in X_2 and X_3. The remaining 1.04 percent is associated with other factors not formally recognized in the regression equation.

28.4 COEFFICIENT OF MULTIPLE CORRELATION

The square root of the coefficient of multiple determination produces a measure analogous to the coefficient of correlation discussed in simple correlation analysis. This measure is called the coefficient of multiple correlation and is subject to the same interpretation as the simple coefficient of correlation (a somewhat awkward interpretation, as you may recall). There is one difference: There is no sign attached to the coefficient of multiple correlation. This is because the dependent variable may vary directly with some independent variables and inversely with others. Because both types may be present in the regression equation, no sign is attached to the coefficient of multiple correlation.

Computation of the coefficient of multiple correlation for the illustrative example:

$$R_{1.23} = \sqrt{R^2_{1.23}} = \sqrt{1 - \frac{(\hat{\sigma}_{1.23})^2}{(\hat{\sigma}_1)^2}}$$

$$= \sqrt{1 - \frac{(0.289)^2}{(2.83)^2}}$$

$$= \sqrt{0.9896}$$

$$= 0.99$$

28.5 COEFFICIENTS OF PARTIAL DETERMINATION

A useful measure that indicates the percentage of the still "unexplained" variation in X_1 that would be "explained" with the introduction of a given additional independent variable is referred to as the coefficient of partial determination. Simple coefficients of determination, as discussed in Chapter 27, are identical to zero-order partial coefficients of determination. The order of the partial is related to the number of independent variables in the regression equation prior to consideration of the independent variable being evaluated by the partial coefficient of determination.

Computation of the Zero-Order Partial Coefficients of Determination for the Illustrative Example

r_{12}^2

$$r_{12}^2 = 1 - \frac{(\hat{\sigma}_{1.2})^2}{(\hat{\sigma}_1)^2}$$

where

r_{12}^2 represents the proportion of the total variation in X_1 previously "unexplained" that will be "explained" by introduction of the independent variable, X_2. In the case of a zero-order partial coefficient, *all* of the total variation in X_1 is previously "unexplained."

$(\hat{\sigma}_1)^2$ represents the total variation in X_1.

$(\hat{\sigma}_{1.2})^2$ represents the total variation in X_1 still "unexplained" after introduction of the independent variable X_2 into the regression equation.

$$r_{12}^2 = 1 - \frac{(\hat{\sigma}_{1.2})^2}{(\hat{\sigma}_1)^2}$$

Substituting: From Table 27.5, $(\hat{\sigma}_{1.2})^2 = 3.17179$; from Table 27.4, $(\hat{\sigma}_1)^2 = 8$.

$$r_{12}^2 = 1 - \frac{3.17179}{8}$$

$$= 1 - 0.396$$

$$= 0.604$$

Interpretation: 60 percent of the variation in X_1 is associated with variation in X_2. The remaining 40 percent is associated with factors other than X_2 that are not presently recognized in the regression function.

r_{13}^2

$$r_{13}^2 = 1 - \frac{(\hat{\sigma}_{1.3})^2}{(\hat{\sigma}_1)^2}$$

The underlying computations leading to the determination of $\hat{\sigma}_{1.3}^2$ are presented in Table 28.6.

$$r_{13}^2 = 1 - \frac{4.857}{8}$$

$$= 1 - 0.607$$

$$= 0.393$$

Interpretation: 39 percent of the variation in X_1 is associated with variation in X_3. The remaining 61 percent is associated with factors other than X_3 that are not presently recognized in the regression function.

Computation of the First-Order Coefficients of Partial Determination for the Illustrative Example

$r_{13.2}^2$

$$r_{13.2}^2 = 1 - \frac{(\hat{\sigma}_{1.23})^2}{(\hat{\sigma}_{1.2})^2}$$

where

$r_{13.2}^2$ represents the proportion of the total variation in X_1 that remained unexplained after inclusion of X_2 in the regression equation, that will be explained with the introduction of the additional independent variable, X_3.

Table 28.6 Underlying computations leading to the determination of $\hat{\sigma}_{1.3}^2$

Tract	X_1	X_3	X_3^2	X_1X_3	X_1^2
A	6	4	16	24	36
B	7	3	9	21	49
C	9	5	25	45	81
D	13	9	81	117	169
E	10	5	25	50	100
F	9	3	9	27	81
G	14	8	64	112	196
H	12	3	9	36	144
	80	40	238	432	856

$$a_{13} = \frac{\sum X_3^2 \cdot \sum X_1 - \sum X_3 \cdot \sum X_1X_3}{n \cdot \sum X_3^2 - (\sum X_3)^2}$$

$$= \frac{(238)(80) - (40)(432)}{8(238) - (40)^2}$$

$$= \frac{19040 - 17280}{1904 - 1600}$$

$$= \frac{1760}{304}$$

$$= 5.789$$

$$b_{13} = \frac{n \cdot \sum X_1X_3 - \sum X_3 \cdot \sum X_1}{n \cdot \sum (X_3^2) - (\sum X_3)^2}$$

$$= \frac{8(432) - (40)(80)}{8(238) - (40)^2}$$

$$= \frac{3456 - 3200}{304}$$

$$= \frac{256}{304}$$

$$= 0.842$$

$$\bar{X}_{1.3} = a_{13} + b_{13}X_3$$

$$= 5.789 + 0.842X_3$$

$$\hat{\sigma}_{1.3}^2 = \frac{\sum X_1^2 - a_{13} \cdot \sum X_1 - b_{13} \sum X_1X_3}{n - 2}$$

$$= \frac{856 - 5.789(80) - 0.842(432)}{6}$$

$$= \frac{856 - 463.120 - 363.744}{6} = \frac{856 - 826.864}{6} = \frac{29.136}{6}$$

$$= 4.857$$

$(\hat{\sigma}_{1.2})^2$ represents the portion of the total variation in X_1 that remained unexplained after inclusion of X_2 in the regression equation.

$(\hat{\sigma}_{1.23})^2$ represents the portion of the total variation in X_1 that remained unexplained after inclusion of X_2 *and* X_3 in the regression equation.

Substituting: From Table 28.5, $(\hat{\sigma}_{1.23})^2 = 0.0835486$; from Table 27.5, $(\hat{\sigma}_{1.2})^2 = 3.17179$.

$$r^2_{13.2} = 1 - \frac{0.0835486}{3.17179}$$

$$= 1 - 0.026$$
$$= 0.974$$

Interpretation: 97.4 percent of the portion of the total variation in X_1 remaining "unexplained" after inclusion of X_2 in the regression equation will be explained by the additional inclusion of the independent variable X_3.

$r^2_{12.3}$

$$r^2_{12.3} = 1 - \frac{(\hat{\sigma}_{1.23})^2}{(\hat{\sigma}_{1.3})^2}$$

Substituting: From Table 28.5, $(\hat{\sigma}_{1.23})^2 = 0.0835486$; from Table 28.6, $(\hat{\sigma}_{1.3})^2 = 4.857$.

$$r^2_{12.3} = 1 - \frac{0.0835486}{4.857}$$

$$= 1 - 0.0172$$
$$= 0.9828$$

Interpretation: 98.28 percent of the portion of the total variation in X_1 remaining "unexplained" after inclusion of X_3 in the regression equation will be explained by the additional inclusion of the independent variable X_2.

28.6 RELATIONSHIP BETWEEN THE COEFFICIENT OF MULTIPLE DETERMINATION AND THE COEFFICIENTS OF PARTIAL DETERMINATION

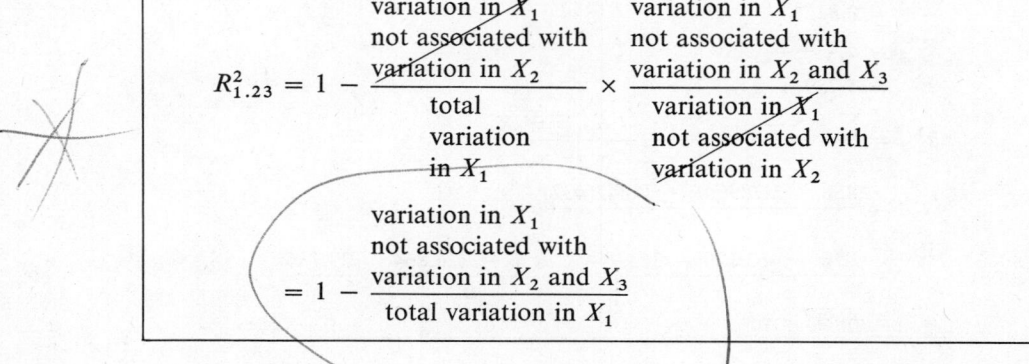

$$R^2_{1.23} = 1 - \frac{(\hat{\sigma}_{1.2})^2}{(\hat{\sigma}_1)^2} \times \frac{(\hat{\sigma}_{1.23})^2}{(\hat{\sigma}_{1.2})^2}$$

$$= 1 - \frac{(\hat{\sigma}_{1.23})^2}{(\hat{\sigma}_1)^2}$$

$$R^2_{1.23} = 1 - (1 - r^2_{12})(1 - r^2_{13.2})$$

where

$(1 - r^2_{12})$ represents the percentage of the variation in X_1 still "unexplained" posterior to the introduction of X_2 in the regression equation.

$(1 - r^2_{13.2})$ represents the percentage of the variation in X_1 still "unexplained" posterior to the introduction of X_2 that will still be "unexplained" posterior to the introduction of X_3 in the regression equation.

$(1 - r^2_{12})(1 - r^2_{13.2})$ represents the percentage of total variation in X_1 still "unexplained" after the introduction of X_2 and X_3 in the regression equation. Therefore,

$$1 - R^2_{1.23} = (1 - r^2_{12})(1 - r^2_{13.2})$$

and

$$R^2_{1.23} = 1 - (1 - r^2_{12})(1 - r^2_{13.2})$$

Referring to the illustrative example:

$$R^2_{1.23} = 1 - (1 - 0.604)(1 - 0.974)$$
$$= 1 - (0.396)(0.026)$$
$$= 1 - 0.010296$$
$$= 0.9897$$

This relationship can be extended to include any given number of independent variables. To illustrate: The coefficient of determination for three independent variables in general is expressed as

$$R^2_{1.234} = 1 - (1 - r^2_{12})(1 - r^2_{13.2})(1 - r^2_{14.23})$$

This is equivalent to

$$1 - \frac{\hat{\sigma}^2_{1.234}}{\hat{\sigma}_1^2}$$

28.7 THE COEFFICIENT OF PARTIAL CORRELATION

The square root of a coefficient of partial determination produces a measure called the coefficient of partial correlation. This measure is analogous to the coefficient of correlation and is subject to similar interpretation.

28.8 TESTS OF SIGNIFICANCE WITH RESPECT TO THE PARTIAL REGRESSION COEFFICIENTS

Most computer programs designed to perform multiple regression analysis provide for tests of significance with respect to the partial regression coefficients generated by the least-squares regression line. The sampling distribution of a given partial regression coefficient is normal with a mean equal to the value of the partial regression coefficient for the universe. The estimated sampling error of the partial regression coefficient is usually generated by matrix algebra and as such is beyond the scope of this text.[5] Suffice it to say that each partial regression coefficient is tested against the hypothesis that the true partial regression coefficient for the universe is equal to zero. The test takes the following form:

With respect to $b_{12.3}$:

$$t = \frac{b_{12.3} - 0}{\sigma_{b_{12.3}}}$$

This statistic is t distributed with $n - k - 1$ degrees of freedom, where k refers to the number of partial regression coefficients in the regression equation.[6] The computed value of t is then compared with the value of t associated with the specified critical probability to determine if the observed value of $b_{12.3}$ is significantly different from zero to warrant rejection of the null hypothesis that $\beta_{12.3} = 0$.

28.9 TESTS OF SIGNIFICANCE WITH RESPECT TO THE COEFFICIENT OF DETERMINATION

A similar test of significance can be applied to the coefficient of multiple determination. In this case, the test involves an application of analysis of variance.[7] It can be demonstrated that the following ratio is F distributed with $k - 1$ degrees of freedom in the numerator and $n - k$ degrees of freedom in the denominator, where k represents the total number of variables (independent and dependent).

$$F = \frac{(R^2_{1.23\ldots k})/(k - 1)}{(1 - R^2_{1.23\ldots k})/(n - k)}$$

The computed value of F is then compared with the value of F associated with the specified critical probability to determine if the observed value of $R^2_{1.23\ldots k}$ is significantly different from zero to justify rejection of the null hypothesis that the universe coefficient of multiple determination is zero.

---

[5] The interested reader is referred to Taro Yamane, *Statistics—An Introductory Analysis*, 2nd ed., New York: Harper & Row, 1967, pp. 774–792.

[6] In the illustrative example, there were two partial regression coefficients, $b_{12.3}$ and $b_{13.2}$. Therefore, the statistic is t distributed with $8 - 2 - 1$ or 5 degrees of freedom.

[7] Refer to Chapter 20.

28.10 TEST OF THE HYPOTHESIS THAT THE UNIVERSE COEFFICIENT OF MULTIPLE DETERMINATION IS EQUAL TO ZERO

A critical probability of 1 percent will be employed. In the illustrative example, there are eight observations and three variables. Therefore, the ratio formed by the statistic

$$F = \frac{(R^2_{1.23})/(k - 1)}{(1 - R^2_{1.23})/(n - k)}$$

will be F distributed with two degrees of freedom in the numberator ($\phi_1 = 2$) and five degrees of freedom in the denominator ($\phi_2 = 5$). Referring to the table, percentage points of the F distribution, the critical value of $F_{0.01}$ is observed to be 13.3.

ϕ_2 \ ϕ_1	2
5	13.3

Computation of the F statistic:

$$F = \frac{(R^2_{1.23})/(k - 1)}{(1 - R^2_{1.23})/(n - k)} = \frac{0.99/(3 - 1)}{0.01/(8 - 3)} = \frac{0.99/2}{0.01/5} = \frac{0.495}{0.002} = 247.5$$

247.5 > 13.3

The coefficient of multiple determination is clearly significantly different from zero, and the null hypothesis is accordingly rejected.

The interested reader is referred to the following texts for advanced treatments on the subject of regression and correlation analysis.

References

1. Clark, Charles T., and Schkade, Lawrence L., *Statistical Methods for Business Decisions*, Cincinnati, Ohio: Southwestern Publishing Company, 1969.
2. Croxton, Frederick E., and Cowden, Dudley J., *Applied General Statistics*, 2nd ed., Englewood Cliffs, N.J.: Prentice-Hall, 1955.
3. Ezekial, Mordecai, and Fox, Karl, *Methods of Correlation and Regression Analysis*, 3rd ed., New York: Wiley, 1969.
4. Yamane, Taro, *Statistics—An Introductory Analysis*, 2nd ed., New York: Harper & Row, 1967.

PROBLEMS

Problem 28.1

An instructor is concerned with determining the relationship between hours studied and performance in the statistics course (performance is measured by numerical grade score on the final exam). Each of the students also has an I.Q. test score, which the instructor felt should somehow be taken into consideration in the study. The instructor plans to use multiple regression and correlation analysis. Discuss the following statistical concepts and the information conveyed by each within the context of the situation described above (let X_1 = numerical grade score on final exam, X_2 = hours studied, and X_3 = I.Q. test score):

1. The simple linear regression function of performance regressed on hours studied, $X_1 = f(X_2)$.
2. The multiple linear regression function of performance regressed on hours studied and I.Q. test score, $X_1 = f(X_2, X_3)$.
3. $\hat{\sigma}_1$, $\hat{\sigma}_{1.2}$, and $\hat{\sigma}_{1.23}$.
4. The simple coefficient of determination, $r^2_{1.2}$.
5. The partial coefficient of determination, $r^2_{13.2}$.
6. The multiple coefficient of determination, $R^2_{1.23}$.

Problem 28.2

Explain why the sequence of entry of the independent variables into the stepwise regression function does not necessarily correspond to the ranking obtained by looking at the simple coefficients of correlation.

Problem 28.3

Clearly differentiate between the two expressions $R^2_{1.456}$ and $r^2_{14.56}$.

Problem 28.4

What information is conveyed by a coefficient of partial determination? For example,

$$r^2_{12.34} = 1 - \frac{\hat{\sigma}^2_{1.234}}{\hat{\sigma}^2_{1.34}}$$

Problem 28.5

Given the general form of the linear regression function,

$$\overline{X}_{1.234} = a_{1.234} + b_{12.34}X_2 + b_{13.24}X_3 + b_{14.23}X_4$$

interpret and define the following symbols:

1. $\overline{X}_{1.234}$
2. $a_{1.234}$
3. $b_{13.24}$

Problem 28.6

Why would one expect the value of the regression coefficient b_{12} in the regression function $\overline{X}_{1.2} = a_{12} + b_{12}X_2$ to be different from the value of the regression coefficient $b_{12.3}$ in the regression function $\overline{X}_{1.23} = a_{1.23} + b_{12.3}X_2 + b_{13.2}X_3$?

Problem 28.7

The coefficient of simple correlation and the coefficients of partial correlation have either a positive or negative sign, indicating whether the nature of the relationship is direct or inverse. Explain why there is no sign attached to the coefficient of multiple correlation.

Problem 28.8

Most computer solutions involving the use of a stepwise multiple correlation algorithm provide, as part of the output, a correlation matrix. A correlation matrix (Printout 28.1) indicates the simple coefficient of correlation for each of the variables with each of the others (including itself). The variables associated with a given coefficient of correlation can be determined by referencing the number of the row and the number of the column with which it is associated. For example, $r_{32} = 0.2552$ is located in the cell at the intersection of the third row and the second column. Note that the cell associated with the second row and the third column is blank. This is because $r_{23} = r_{32}$, and specification of one or the other is sufficient.

Printout 28.1

```
CORRELATION MATRIX
  1.0000
   .5481   1.0000
   .2137    .2552   1.0000
 -0.4949    .1006    .0280   1.0000
   .7818    .4073    .1539  -0.4235   1.0000
   .6979    .4540    .2229  -0.1513    .5765   1.0000
   .1669  -0.1751  -0.1878  -0.5972    .3695  -0.1323   1.0000
   .0104  -0.0486  -0.5796  -0.0902  -0.3034    .1332  -0.3215   1.0000
```

1. Why are all the entries on the diagonal of the correlation matrix equal to one?
2. Why are all the cells to the right of the diagonal blank?
3. Which of the other seven variables has the greatest degree of correlation if X_1 is the dependent variable?
4. Which of the other seven variables has the greatest degree of correlation if X_7 is the dependent variable?
5. What is the significance of the negative signs in the correlation matrix?

Problem 28.9

Most computer installations have a multiple regression-correlation routine as part of the software package. One need simply supply the X_1, X_2, X_3, ... values. The computer program causes the necessary underlying computations to be performed. The output from a typical stepwise multiple linear regression program applied to the illustrative example in Chapter 28 is reproduced in Printout 28.2.

Printout 28.2

```
                        CORRELATION MATRIX

                  VARIABLE     1          2          3
                  NUMBER

                     1       1.000       .813       .694
                     2                  1.000       .153
                     3                             1.000

     STEP NUMBER    1
     VARIABLE ENTERED    2

     MULTIPLE R            .8125
     STD. ERROR OF EST.   1.7809


             VARIABLES IN EQUATION              .    VARIABLES NOT IN EQUATION
                           STD. ERROR   COMPUTED  .
     VARIABLE    COEFFICIENT OF REG.COEF. T VALUE  .    VARIABLE   PARTIAL CORR.
                                                   .
       (CONSTANT    4.28291 )                      .
        2             .19057    .05582    3.41421   .       3          .98896
```

```
     STEP NUMBER    2
     VARIABLE ENTERED    3

     MULTIPLE R            .9963
     STD. ERROR OF EST.   .2890

             VARIABLES IN EQUATION                 .
                           STD. ERROR   COMPUTED   .
     VARIABLE    COEFFICIENT OF REG.COEF. T VALUE  .

       (CONSTANT    1.36830 )                      .
        2             .16970    .00917   18.51306   .
        3             .70813    .04744   14.92532   .

             ANALYSIS OF VARIANCE FOR THE REGRESSION

                            DEGREES
     SOURCE OF VARIATION   OF FREEDOM SUM OF SQUARES   MEAN SQUARE    F RATIO
     ATTRIBUTABLE TO REGRESSION   2       55.582          27.791      332.634
     DEVIATION FROM REGRESSION    5         .418            .084
        TOTAL                     7



     SUMMARY TABLE

          VARIABLE        MEAN      STANDARD DEVIATION
             1         10.00000        2.82843
             2         30.00000       12.05938
             3          5.00000        2.32993

       STEP      VARIABLE       MULTIPLE        INCREASE
      NUMBER  ENTERED REMOVED   R        RSQ     IN RSQ


         1       2               .8125    .6602    .6602
         2       3               .9963    .9925    .3324
```

1. What information is conveyed by the correlation matrix?
2. In a stepwise regression program, independent variables are introduced one by one on the basis of their incremental explanatory power. Which of the independent variables is the first to enter the stepwise regression function and why? Refer to the correlation matrix in answering this question.

3. Referring to step 1 of the printout, determine the numerical coefficients of the regression function, $\bar{X}_{1.2} = a_{12} + b_{12}X_2$.

   $a_{12} =$
   $b_{12} =$

   Verify the values of a_{12} and b_{12} by referring to the values computed in Chapter 27, Table 27.2.

4. The precision of the regression line estimates (i.e., the various $\bar{X}_{1.2}$) as compared to $\bar{X}_1$ is evaluated by comparing $\hat{\sigma}_{1.2}$ (the standard error of the estimate) and $\hat{\sigma}_1$ (the standard deviation of the independent variable). Locate the value of $\hat{\sigma}_1$ in the summary statistics and the value of $\hat{\sigma}_{1.2}$ in step 1 of the printout. Verify these values by comparing them to the values recorded on p. 734.

5. The coefficient of multiple correlation is labeled multiple R on the printout. In step 1, R_{12} equals 0.8125. The program computes the multiple R on the basis of universe relationships and accordingly must be adjusted by the degrees of freedom if the analysis is based on sample data.

$$\text{adjusted } R_{12} = \sqrt{1 - (1 - R_{12}^2)\left(\frac{n-1}{n-2}\right)}$$

   Adjust the printout value and compare the result to the value of r_{12} as computed by the formula $r_{12} = \sqrt{1 - \hat{\sigma}_{1.2}^2/\hat{\sigma}_1{}^2}$ on p. 710.

6. Note that the printout separates the variables that are in the regression equation from those that are not. The program then evaluates the variables not yet included in the regression equation to determine the one that explains the greatest percentage of the remaining unexplained variation in the dependent variable. The variable that meets this description is the one that possesses the highest coefficient of partial correlation. In the illustrative example, the only variable not included in the regression equation is X_3. The partial correlation coefficient ($r_{13.2}$) is 0.98896. Locate this value on the printout. What information does this value convey?

7. Determine the values of $R_{1.23}$, R_{12}, and $r_{13.2}$, and demonstrate that the coefficient of multiple determination in step 2 of the printout is related to the coefficient of multiple determination in step 1 of the printout and the coefficient of partial determination ($r_{13.2}$)^2 in the manner described below:

$$\begin{pmatrix}\text{percentage of the} \\ \text{variation in } X_1 \\ \text{associated with} \\ \text{variation in } X_2 \\ \text{and } X_3\end{pmatrix} = \begin{pmatrix}\text{percentage} \\ \text{associated} \\ \text{with} \\ \text{variation} \\ \text{in } X_2\end{pmatrix} + \begin{pmatrix}\text{percentage} \\ \text{not} \\ \text{associated} \\ \text{with} \\ \text{variation} \\ \text{in } X_2\end{pmatrix}\begin{pmatrix}\text{percentage of the} \\ \text{portion of the} \\ \text{variation in } X_1 \text{ not} \\ \text{associated with} \\ \text{variation in } X_2 \\ \text{that is associated} \\ \text{with variation} \\ \text{in } X_3\end{pmatrix}$$

$$R_{1.23}^2 = R_{12}^2 + (1 - R_{12}^2)(r_{13.2})^2$$

8. a. Referring to step 2 of the printout, write out the regression equation,

$$\bar{X}_{1.23} = a_{1.23} + b_{12.3}X_2 + b_{13.2}X_3$$

  b. Interpret each of the coefficients in the equation.
  c. Are the values of $b_{12.3}$ and $b_{13.2}$ significant?
9. In step 2, $R_{1.23}$ equals 0.9963. The program computes the multiple R on the basis of universe relationships and, accordingly, it must be adjusted by the degrees of freedom if the analysis is based on sample data.

$$\text{adjusted } R_{1.23} = \sqrt{1 - (1 - R_{1.23}^2)\left(\frac{n-1}{n-3}\right)}$$

Adjust the printout value and compare the result to the value of $R_{1.23}$ computed by the formula $R_{1.23} = \sqrt{1 - \hat{\sigma}_{1.23}^2/\hat{\sigma}_1^2}$ on p. 737. Slight differences are attributable to rounding.

10. The computation entitled ANALYSIS OF VARIANCE FOR THE REGRESSION on the printout, that is,

$$F = \frac{\text{mean squares (attributable to regression)}}{\text{mean squares (deviation from regression)}} = \frac{27.791}{0.084} = 332.6$$

is equivalent to the computation represented on p. 743,

$$F = \frac{R_{1.23}^2/(k-1)}{(1 - R_{1.23}^2)/(n-k)}$$

They are equivalent procedures for conducting tests of significance with respect to the coefficient of determination. Obtain the value of $R_{1.23}^2$ from the printout and demonstrate that the F ratio obtained under either procedure is the same. Slight differences are attributable to rounding.

Problem 28.10

Referring to Problem 27.13, the company has decided to introduce an additional explanatory variable in an effort to explain the variation in sales performance that still remains after the score on the aptitude test has been taken into account. They decided that some index of effort expended would be appropriate. Accordingly, for each salesman, an additional piece of data was gathered—the ratio of the mileage on his company car to the total mileage projected for adequate coverage of his sales territory. The total data set is summarized below:

Salesman	Sales performance (thousands of dollars) X_1	Test score X_2	Effort index X_3
Jones	60	35	0.9
Wilson	70	80	0.3
Smith	90	75	1.2
Lewis	90	90	0.5
Pierce	70	65	0.4
Blue	80	50	1.6
Allen	60	70	0.1
Murray	120	95	2.0

1. Determine the values of $a_{1.23}$, $b_{12.3}$, and $b_{13.2}$ for the least-squares linear regression line.

$$\overline{X}_{1.23} = a_{1.23} + b_{12.3}X_2 + b_{13.2}X_3$$

2. Determine the values of $\hat{\sigma}_1$ and $\hat{\sigma}_{1.23}$.
3. Discuss the role the values computed in part 2 play in evaluating the improvement in the ability to predict when $\overline{X}_{1.23}$, rather than $\overline{X}_1$, is employed as the best estimate.
4. Determine the value of the coefficient of multiple determination (i.e., $R^2_{1.23}$). What information does this measure convey?
5. Determine the value of the coefficient of multiple correlation (i.e., $R_{1.23}$). What information does this measure convey?
6. Determine the coefficient of partial determination, $r^2_{12.3}$. What information does this measure convey?

Problem 28.11

An enterprising college professor is entertaining the idea of employing a regression function to predict the final average of the students in his graduate statistics course in lieu of giving a final exam. He collected the following data on 15 students from the previous class.

Student	Final average X_1	Numerical score on midterm X_2	Admission test for graduate school of business — ATGSB overall score X_3	Admission test for graduate school of business — ATGSB math score X_4	Index indicating professor's evaluation of student's interest and dedication X_5	number of classes attended X_6
1	72	79	458	32	6	8
2	85	88	510	40	8	11
3	87	79	525	43	8	11
4	96	92	535	51	10	10
5	88	91	480	41	8	11
6	81	85	466	43	7	9
7	99	96	610	49	10	11
8	67	74	495	35	4	10
9	75	81	510	37	6	10
10	83	87	535	42	9	9
11	92	95	580	46	8	11
12	96	97	602	45	10	11
13	87	89	535	38	9	11
14	94	92	540	43	9	10
15	81	85	585	36	9	10

He then submitted the data to a stepwise multiple regression-correlation program. (If you are unfamiliar with the concept of a stepwise multiple linear regression program, see Problem 28.9.) A copy of the resulting printout appears as Printout 28.3.

Printout 28.3

```
CORRELATION MATRIX

VARIABLE      1          2          3          4          5          6
 NUMBER

   1        1.000       .902       .659       .864       .892       .567
   2                   1.000       .644       .732       .822       .477
   3                              1.000       .514       .688       .534
   4                                         1.000       .691       .418
   5                                                    1.000       .391
   6                                                               1.000
```

```
STEP NUMBER    1
VARIABLE ENTERED.    2

MULTIPLE R                  .9022
STD. ERROR OF EST.       4.1600

            VARIABLES IN EQUATION          .  VARIABLES NOT IN EQUATION
                        STD. ERROR   COMPUTED .
VARIABLE    COEFFICIENT OF REG.COEF.  T VALUE .  VARIABLE    PARTIAL CORR.
                                               .
   (CONSTANT  -21.52330 )                      .
   2           1.22584    .16250     7.54342   .     3           .23478
                                               .     4           .69405
                                               .     5           .61127
                                               .     6           .35951
```

```
STEP NUMBER    2
VARIABLE ENTERED    4

MULTIPLE R                  .9506
STD. ERROR OF EST.       3.1172

            VARIABLES IN EQUATION          .  VARIABLES NOT IN EQUATION
                        STD. ERROR   COMPUTED .
VARIABLE    COEFFICIENT OF REG.COEF.  T VALUE .  VARIABLE    PARTIAL CORR.
                                               .
   (CONSTANT  -15.66585 )                      .
   2            .78896    .17872     4.41446   .     3           .24784
   4            .78011    .23360     3.33956   .     5           .64449
                                               .     6           .39178
```

```
STEP NUMBER    3
VARIABLE ENTERED    5

MULTIPLE R                  .9714
STD. ERROR OF EST.       2.4894

            VARIABLES IN EQUATION          .  VARIABLES NOT IN EQUATION
                        STD. ERROR   COMPUTED .
VARIABLE    COEFFICIENT OF REG.COEF.  T VALUE .  VARIABLE    PARTIAL CORR.
                                               .
   (CONSTANT   2.81084 )                       .
   2            .45447    .18625     2.44014   .     3           .02689
   4            .65677    .19170     3.42604   .     6           .53837
   5           1.96390    .70250     2.79559   .
```

Printout 28.3 (*Continued*)

```
STEP NUMBER     4
VARIABLE ENTERED    6

MULTIPLE R              .9798
STD. ERROR OF EST.     2.2003
```

VARIABLE	COEFFICIENT	STD. ERROR OF REG.COEF.	COMPUTED T VALUE		VARIABLE	PARTIAL CORR.
		VARIABLES IN EQUATION			VARIABLES NOT IN EQUATION	
(CONSTANT	-3.58130)					
2	.37450	.16930	2.21201		3	-.21728
4	.61603	.17063	3.61039			
5	2.00220	.62119	3.22320			
6	1.44639	.71595	2.02023			

```
STEP NUMBER     5
VARIABLE ENTERED    3

MULTIPLE R              .9808
STD. ERROR OF EST.     2.2639
```

VARIABLE	COEFFICIENT	STD. ERROR OF REG.COEF.	COMPUTED T VALUE		VARIABLE	PARTIAL CORR.
		VARIABLES IN EQUATION			VARIABLES NOT IN EQUATION	
(CONSTANT	-.86565)					
2	.38555	.17498	2.20336			
3	-.01293	.01936	-.66781			
4	.61038	.17576	3.47269			
5	2.18129	.69312	3.14705			
6	1.64012	.79171	2.07161			

ANALYSIS OF VARIANCE FOR THE REGRESSION

SOURCE OF VARIATION	DEGREES OF FREEDOM	SUM OF SQUARES	MEAN SQUARE	F RATIO
ATTRIBUTABLE TO REGRESSION	5	1163.607	232.721	45.408
DEVIATION FROM REGRESSION	9	46.126	5.125	

TABLE OF RESIDUALS

CASE NO.	Y VALUE	Y ESTIMATE	RESIDUAL
1	72.00000	69.41140	2.58860
2	85.00000	86.37490	-1.37490
3	87.00000	84.54215	2.45785
4	96.00000	97.03045	-1.03045
5	88.00000	88.52982	-.52982
6	81.00000	82.15679	-1.15679
7	99.00000	98.02225	.97775
8	67.00000	67.75403	-.75403
9	75.00000	75.84225	-.84225
10	83.00000	85.78792	-2.78792
11	92.00000	91.83090	.16910
12	95.00000	95.06973	-.06973
13	87.00000	87.33774	-.33774
14	94.00000	89.90150	4.09850
15	81.00000	82.34818	-1.34818

SUMMARY TABLE

VARIABLE	MEAN	STANDARD DEVIATION
1	85.53333	9.23567
2	87.33333	6.84175
3	531.06667	47.12371
4	41.40000	5.23450
5	8.06667	1.70392
6	10.20000	.34112

STEP NUMBER	VARIABLE ENTERED	REMOVED	MULTIPLE R	RSQ	INCREASE IN RSQ
1	2		.9022	.8140	.8140
2	4		.9506	.9036	.0896
3	5		.9714	.9436	.0400
4	6		.9798	.9600	.0163
5	3		.9808	.9619	.0019

1. a. What information is provided by the correlation matrix?
   b. If only one independent variable were to be used, which would it be, and why?
   c. Indicate the independent variable that would be the second choice, third choice, fourth choice, and last choice.
   d. Refer to the stepwise multiple regression printout. Note the order in which the independent variables were introduced into the regression equation. Explain why this order differs from that indicated in the answers to parts b and c.
2. Referring to the printout:
   a. If no information is available with respect to the values of the *independent variables*, what value would be employed as the best estimate of the value of the dependent variable for an individual observation? Indicate the precision associated with such an estimate. Justify the fashion with which you indicated the precision of the estimate.
   b. Given a value for the independent variable designated X_2, what value would be employed as the best estimate of the value of the dependent variable? Indicate the precision associated with such an estimate. Justify the fashion with which you indicated the precision of the estimate.
   c. Given the knowledge of the value of the independent variable designated X_2, is the precision of the estimate improved? Justify your answer.
3. Referring to the stepwise multiple regression printout, step 2:
   a. Explain the meaning of the multiple R value of 0.9506.
   b. Explain the meaning of the values entered in the partial correlation column.
   c. Indicate the relationship between the multiple R value of 0.9714 in step 3 of the printout and the values discussed in parts a and b.
   d. Derive the value 0.9714 from the information referred to in parts a and b.
4. Referring to step 5 in the stepwise multiple regression printout:
   a. Write out the regression equation and interpret the meaning of each of the coefficients in the equation.
   b. Which of the regression coefficients are "significant"? Support your answer.
   c. Construct a 95 percent confidence interval estimate of $\beta_{16.2345}$.
5. Referring to step 5 in the stepwise multiple regression printout:
   a. Substitute the appropriate numerical values into the following formula and solve for $R_{1.23456}^2$.

$$1 - \frac{\hat{\sigma}_{1.23456}^2}{\hat{\sigma}_1^{\,2}}$$

   b. The value for $R_{1.23456}^2$ computed in part a differs from the value on the printout because the computer program computes the value of $R_{1.23456}^2$ on the basis of universe relationships. This figure should be adjusted by the degrees of freedom if the analysis is based on sample data. The adjustment is as follows:

$$\text{adjusted } R_{1.23456}^2 = 1 - (1 - R_{1.23456}^2)\left(\frac{n-1}{n-6}\right)$$

Adjust the printout value and compare the result to the value obtained in part a. Slight differences are attributable to rounding.

6. Clearly differentiate between the two expressions $R^2_{1.23456}$ and $r^2_{13.2456}$ (both numerically and in terms of their interpretation).

7. Is the coefficient of multiple correlation in step 5 of the printout (i.e., $R_{1.23456}$) "significant"? Support your answer.

8. Assume the instructor grades on the following basis:

      90–100 receives an A

      80–90   receives a B (interpreted as 80 up to but not including 90)

      70–80   receives a C

   Less than 70 receives an F

   a. If the model were to be used in lieu of giving a final exam, how many of the students would have received the grade they deserved?

   b. If the students knew the parameters of the model, how would it affect their behavior pattern?

Problem 28.12

A real estate firm collected the following data as input into a stepwise multiple regression program with the objective of trying to generate a predictor of the market value of properties to serve as a general guide to their clientele.

Obser-vation	Market value X_1	Number of bedrooms X_2	Acreage X_3	Tax rate (100 percent evaluation) X_4	Distance from shopping center (miles) X_5	Square footage of home X_6	Commuting time to an urban center (min) X_7	Dollars spent per pupil in school system X_8
1	38,000	4	0.5	35	0.5	1500	35	850
2	20,000	3	0.3333	53	0.5	1000	45	750
3	35,000	3	0.5	41	2.0	1900	20	920
4	32,500	3	0.5	58	3.0	2400	25	712
5	40,000	4	0.5	37	10.0	2400	45	762
6	27,000	4	0.5	101	0.5	2500	10	835
7	32,000	3	1.25	39	2.0	1600	35	850
8	33,000	3	1.0	59	3.0	2640	35	890
9	75,000	4	1.0	35	12.0	2500	60	605
10	30,000	3	0.5	56	1.0	1250	15	900
11	49,000	4	0.3333	37	1.0	2700	25	1050
12	23,500	4	0.75	101	0.5	1200	5	805
13	20,000	3	0.125	45	0.5	1000	75	680
14	50,000	4	18.0	55	5.0	2500	20	500
15	75,000	4	1.0	34	8.0	3000	35	1000

A copy of the resulting printout is reproduced as Printout 28.4. (If you are unfamiliar with the concept of a stepwise multiple linear regression program, see Problem 28.9.)

Printout 28.4

```
CORRELATION MATRIX

VARIABLE    1        2        3        4        5        6        7        8
NUMBER

   1      1.000     .548     .214    -.495     .782     .698     .167     .010
   2               1.000     .255     .101     .407     .454    -.175    -.049
   3                        1.000     .028     .154     .223    -.188    -.580
   4                                 1.000    -.424    -.151    -.597    -.090
   5                                          1.000     .576     .369    -.304
   6                                                   1.000    -.132     .133
   7                                                            1.000    -.322
   8                                                                     1.000
```

```
STEP NUMBER    1
VARIABLE ENTERED    5

MULTIPLE R              .7818
STD. ERROR OF EST.   11132.6307

                                                .
          VARIABLES IN EQUATION                 .   VARIABLES NOT IN EQUATION
                                                .
                          STD. ERROR   COMPUTED .
VARIABLE    COEFFICIENT   OF REG.COEF.  T VALUE .   VARIABLE   PARTIAL CORR.
                                                .
   (CONSTANT 26891.02564 )                      .
   5         3568.37607    789.37388   4.52051  .      2          .40318
                                                .      3          .15152
                                                .      4         -.29000
                                                .      6          .48520
                                                .      7         -.21041
                                                .      8          .41694



STEP NUMBER    2
VARIABLE ENTERED    6

MULTIPLE R              .8383
STD. ERROR OF EST.   10131.9749
                                                .
          VARIABLES IN EQUATION                 .   VARIABLES NOT IN EQUATION
                          STD. ERROR   COMPUTED .
VARIABLE    COEFFICIENT   OF REG.COEF.  T VALUE .   VARIABLE   PARTIAL CORR.
                                                .
   (CONSTANT 11483.41529 )                      .
   5         2594.15349    879.20492   2.95057  .      2          .31189
   6            9.28342      4.82962   1.92218  .      3          .08219
                                                .      4         -.40446
                                                .      7          .01313
                                                .      8          .28008
STEP NUMBER    3
VARIABLE ENTERED    4

MULTIPLE R              .8668
STD. ERROR OF EST.    9678.2867
                                                .
          VARIABLES IN EQUATION                 .   VARIABLES NOT IN EQUATION
                          STD. ERROR   COMPUTED .
VARIABLE    COEFFICIENT   OF REG.COEF.  T VALUE .   VARIABLE   PARTIAL CORR.
                                                .
   (CONSTANT 21815.27378 )                      .
   4         -194.40435    132.53905  -1.46677  .      2          .50246
   5         2030.02197    923.71411   2.19767  .      3          .12797
   6           10.13914      4.65011   2.18041  .      7         -.25799
                                                .      8          .16873
```

Printout 28.4 (*Continued*)

```
STEP NUMBER     4
VARIABLE ENTERED     2

MULTIPLE R          .9023
STD. ERROR OF EST.     8776.2362
```

		VARIABLES IN EQUATION			VARIABLES NOT IN EQUATION	
VARIABLE	COEFFICIENT	STD. ERROR OF REG.COEF.	COMPUTED T VALUE		VARIABLE	PARTIAL CORR.
(CONSTANT	-3919.13039)					
2	10054.54082	5471.16877	1.83773		3	.05785
4	-266.24214	126.38403	-2.10661		7	-.20720
5	1521.08424	882.22031	1.72415		8	.15119
6	7.97172	4.37856	1.82063			

```
STEP NUMBER     5
VARIABLE ENTERED     7

MULTIPLE R          .9067
STD. ERROR OF EST.     9050.2658
```

		VARIABLES IN EQUATION			VARIABLES NOT IN EQUATION	
VARIABLE	COEFFICIENT	STD. ERROR OF REG.COEF.	COMPUTED T VALUE		VARIABLE	PARTIAL CORR.
(CONSTANT	6105.92096)					
2	9467.52255	5717.10315	1.65600		3	.02067
4	-311.64814	148.63533	-2.09673		8	.09577
5	1799.28014	1009.63463	1.78211			
6	6.64677	4.97350	1.33644			
7	-118.51079	186.51582	-.63539			

ANALYSIS OF VARIANCE FOR THE REGRESSION

SOURCE OF VARIATION	DEGREES OF FREEDOM	SUM OF SQUARES	MEAN SQUARE	F RATIO
ATTRIBUTABLE TO REGRESSION	5	3406667527.154	681333505.431	8.318
DEVIATION FROM REGRESSION	9	737165806.179	81907311.798	

```
SUMMARY TABLE
```

VARIABLE	MEAN	STANDARD DEVIATION
1	38666.66667	17204.30456
2	3.53333	.51640
3	1.78611	4.49611
4	52.40000	21.71504
5	3.30000	3.76924
6	2006.00000	686.16741
7	32.33333	18.69556
8	807.26667	145.05441

STEP NUMBER	VARIABLE ENTERED	REMOVED	MULTIPLE R	RSQ	INCREASE IN RSQ
1	5		.7818	.6112	.6112
2	6		.8383	.7027	.0915
3	4		.8668	.7514	.0486
4	2		.9023	.8141	.0628
5	7		.9067	.8221	.0080

1. a. What information is provided by the correlation matrix?
   b. If only one independent variable were to be used, which would it be, and why?
   c. Indicate the independent variable that would be the second choice, third choice, fourth choice, fifth choice, sixth choice, and last choice.
   d. Refer to the stepwise multiple regression printout. Note the order in which the independent variables were introduced into the regression equation. Explain why this order differs from that indicated in the answers to parts b and c.
2. Referring to the printout:
   a. If no information is available with respect to the values of the *independent variables*, what value would be employed as the best estimate of the value of the dependent variable for an individual observation? Indicate the precision associated with such an estimate. Justify the fashion with which you indicated the precision of the estimate.
   b. Given a value for the independent variable designated X_5, what value would be employed as the best estimate of the value of the dependent variable? Indicate the precision associated with such an estimate. Justify the fashion with which you indicated the precision of the estimate.
   c. Given the knowledge of the value of the independent variable designated X_5, is the precision of the estimate improved? Justify your answer.
3. Referring to the stepwise multiple regression printout, step 3:
   a. Explain the meaning of the multiple R value of 0.8668.
   b. Explain the meaning of the values entered in the partial correlation column.
   c. Indicate the relationship between the multiple R value of 0.9023 in step 4 of the printout and the values discussed in parts a and b.
   d. Derive the value 0.9023 from the information referred to in parts a and b.
4. Referring to step 3 in the stepwise multiple regression printout:
   a. Write out the regression equation and interpret the meaning of each of the coefficients in the equation.
   b. Which of the regression coefficients are "significant"? Support your answer.
   c. Construct a 95 percent confidence interval estimate of $\beta_{16.45}$.
5. Referring to step 3 in the stepwise multiple regression printout:
   a. Substitute the appropriate numerical values into the following formula and solve for $R^2_{1.456}$.

$$1 - \frac{\hat{\sigma}^2_{1.456}}{\hat{\sigma}_1{}^2}$$

   b. The value for $R^2_{1.456}$ computed in part a differs from the value on the printout because the computer program computes the value of $R^2_{1.456}$ on the basis of universe relationships. This figure should be adjusted by the degrees of freedom if the analysis is based on sample data. The adjustment is as follows:

$$\text{adjusted } R^2_{1.456} = 1 - (1 - R^2_{1.456})\left(\frac{n-1}{n-4}\right)$$

Adjust the printout value and compare the result to the value obtained in part a. Slight differences are attributable to rounding.

6. Clearly differentiate between the two expressions $R^2_{1.456}$ and $r^2_{14.56}$ (both numerically and in terms of their interpretation).

7. Is the coefficent of multiple correlation in step 5 of the printout ($R_{1.24567}$) "significant"? Support your answer.

Problem 28.13

The firm of Huckster, Dun, and Finagle, Attorneys-at-Law, specializes in divorce cases. They have decided to generate a regression function designed to estimate monthly alimony payments given the client's monthly income, the wife's monthly income, and an index of the client's guilt. They collected data on ten of their former clients.

Client	Monthly alimony payment (excluding child support) X_1	Monthly income of client X_2	Wife's monthly income X_3	Guilt index (the number of times per week the client screws the cap on the toothpaste) X_4
1	600	2000	250	3
2	250	1200	500	1
3	800	1600	50	9
4	300	1000	100	4
5	450	900	20	7
6	700	1750	100	9
7	900	3000	200	1
8	550	1200	10	14
9	450	900	30	11
10	550	1100	20	10

A copy of the printout of a stepwise multiple regression-correlation program applied to this data is reproduced as Printout 28.5. If you are unfamiliar with the concept of a stepwise multiple linear regression program, see Problem 28.9.

1. a. What information is conveyed by the correlation matrix?
   b. If only one independent variable were to be used, which would it be, and why?
   c. Indicate the independent variable that would be the second choice, and the one that would be chosen last.
   d. Refer to the stepwise multiple regression printout. Note the order in which the independent variables were introduced into the regression equation. Explain why this order differs from that indicated in the answers to parts b and c.

2. Referring to the printout:
   a. If no information is available with respect to the values of the independent variables, what value would be employed as the best estimate of the value of the dependent variable for an individual observation? Indicate the precision associated with such an estimate. Justify the fashion with which you indicated the precision of the estimate.
   b. Given a value for the independent variable designated X_2, what value would be employed as the best estimate of the value of the dependent

Printout 28.5

```
CORRELATION MATRIX

VARIABLE      1          2          3          4
 NUMBER

    1       1.000       .780      -.268       .079
    2                  1.000       .285      -.484
    3                             1.000      -.781
    4                                         1.000
```

```
STEP NUMBER    1
VARIABLE ENTERED    2

   MULTIPLE R            .7804
   STD. ERROR OF EST.   136.6698

            VARIABLES IN EQUATION            .  VARIABLES NOT IN EQUATION
                         STD. ERROR   COMPUTED .
   VARIABLE    COEFFICIENT  OF REG.COEF.  T VALUE .    VARIABLE    PARTIAL CORR.

   (CONSTANT   195.68180 )
    2             .24527      .06947    3.53052 .       3          -.81805
                                                .       4           .83512

STEP NUMBER    2
VARIABLE ENTERED    4

MULTIPLE R              .9390
STD. ERROR OF EST.    80.3687

            VARIABLES IN EQUATION            .  VARIABLES NOT IN EQUATION
                         STD. ERROR   COMPUTED .
   VARIABLE    COEFFICIENT  OF REG.COEF.  T VALUE .    VARIABLE    PARTIAL CORR.

   (CONSTANT  -127.58429 )
    2             .33602      .04668    7.19781 .       3          -.50308
    4           27.58111     6.86647    4.01678

STEP NUMBER    3
VARIABLE ENTERED    3

MULTIPLE R              .9548
STD. ERROR OF EST.    75.0229

            VARIABLES IN EQUATION            .  VARIABLES NOT IN EQUATION
                         STD. ERROR   COMPUTED .
   VARIABLE    COEFFICIENT  OF REG.COEF.  T VALUE .    VARIABLE    PARTIAL CORR.

   (CONSTANT    11.74158 )
    2             .32523      .04422    7.35527
    3            -.37684      .26428   -1.42588
    4           16.65955     9.98764    1.56802


            ANALYSIS OF VARIANCE FOR THE REGRESSION

                           DEGREES
   SOURCE OF VARIATION   OF FREEDOM  SUM OF SQUARES   MEAN SQUARE   F RATIO
   ATTRIBUTABLE TO REGRESSION   3     348479.430     116159.810    20.638
   DEVIATION FROM REGRESSION    6      33770.570       5628.428
```

```
SUMMARY TABLE

      VARIABLE         MEAN         STANDARD DEVIATION
         1          555.00000         206.08790
         2         1465.00000         655.76503
         3          128.00000         153.82530
         4            6.90000           4.45845

   STEP       VARIABLE        MULTIPLE            INCREASE
  NUMBER   ENTERED  REMOVED     R        RSQ      IN RSQ


    1         2                .7804    .6091     .6091
    2         4                .9390    .8817     .2726
    3         3                .9548    .9117     .0299
```

variable? Indicate the precision associated with such an estimate. Justify the fashion with which you indicated the precision of the estimate.

c. Given the knowledge of the value of the independent variable designated X_2, is the precision of the estimate improved? Justify your answer.

3. Referring to step 1 in the stepwise multiple regression printout:
   a. Explain the meaning of the multiple R value of 0.7804.
   b. Explain the meaning of the values entered in the partial correlation column.
   c. Indicate the relationship between the multiple R value of 0.9390 in step 2 of the printout and the values discussed in parts a and b.
   d. Derive the value 0.9390 from the information referred to in parts a and b.

4. Referring to step 3 in the stepwise multiple regression printout:
   a. Write out the regression equation and interpret the meaning of each of the coefficients in the equation.
   b. Which of the regression coefficients are "significant"? Support your answer.
   c. Construct a 95 percent confidence interval estimate of $\beta_{12.34}$.

5. Referring to step 3 in the stepwise multiple regression printout:
   a. Substitute the appropriate numerical values into the following formula and solve for $R^2_{1.234}$.

$$1 - \frac{\hat{\sigma}^2_{1.234}}{\hat{\sigma}_1{}^2}$$

   b. The value for $R^2_{1.234}$ computed in part a differs from the value on the printout because the computer program computes the value of $R^2_{1.234}$ on the basis of universe relationships. This figure should be adjusted by the degrees of freedom if the analysis is based on sample data. The adjustment is as follows:

$$\text{adjusted } R^2_{1.234} = 1 - (1 - R^2_{1.234})\left(\frac{n-1}{n-4}\right)$$

   Adjust the printout value and compare the result to the value obtained in part a. Slight differences are attributable to rounding.

6. Clearly differentiate between the two expressions $R^2_{1.234}$ and $r^2_{13.24}$ (both numerically and in terms of their interpretation).

7. Is the coefficient of multiple correlation in step 3 of the printout (i.e., $R_{1.234}$) "significant"? Support your answer.

INDEX NUMBERS

Index numbers are measures that serve to indicate the relative change in magnitude of the observations in a series from time to time or from place to place. The value of the series as of a particular period in time (or as in a particular geographical location) is selected as the basis of comparison. This period of time is referred to as the base period. The index number for the base period is always 100. The value of the index numbers assigned to other periods of time reflect the relative change in the magnitude of the series *as compared to* the value of the series in the base period.

To illustrate:

A chronological comparison: dollar sales—ABC Company (base period = 1965)

Year	Index number
1965	100
1967	113
1970	120

Interpretation: Dollar sales in 1967 are 13 percent higher than or 113 percent of dollar sales in 1965. Similarly, one can observe that dollar sales in 1970 have increased 20 percent over dollar sales in the base period.

Note: It is incorrect to compare the index numbers for 1967 and 1970 and conclude, because there is a difference of 7 percentage points in their index numbers, that dollar sales in 1970 are 7 percent higher than dollar sales in 1967. The true increase is

$$\frac{120 - 113}{113} (100) \quad \text{or} \quad \frac{7}{113} \ (100)$$

which equals 6.2 percent. Direct interpretation of the index numbers is valid only when comparison is made with the base period (i.e., an index number of 100).

A spatial comparison: ABC Company—dollar sales by branch, 1967 (base location = Atlanta)

Branch location	Index number
Chicago	90
Atlanta	100
New York	110

Interpretation: Dollar sales in New York are 110 percent of dollar sales in Atlanta, and dollar sales in Chicago are 90 percent of dollar sales in Atlanta.

By utilizing index numbers as opposed to raw data, comparison is simplified considerably. It is easier to compare relative change to a base of 100 than to a value such as $152.62. Many series of interest to the general public are provided in the form of index numbers. Three of the more widely recognized are the Consumer Price Index, the Wholesale Price Index, and the Index of Industrial Production. The Consumer Price Index (the inverse of which measures changes in the purchasing power of the dollar) is employed in many collective bargaining contracts to adjust wage levels automatically to reflect changes in the cost of living. In addition, index numbers are available that describe the behavior of such series as imports, exports, construction costs, stock market prices, and farm production, to name a few. These series are extremely useful to analysts in their efforts to assess past economic activity, evaluate current economic performance, and generate predictions of the level of future activity. Another advantage of index numbers is that management is often willing to disseminate information in index number form that it would otherwise be unwilling to make available. Index numbers allow management to suppress the absolute magnitude of the data. Only the relative change that has taken place is revealed.

When an index number is used to describe the relative change in a single series, it is referred to as a *simple index number*. The computation of a simple index number is described below and illustrated in Table 29.1.

Procedure:

1. Select the period in time that is to serve as the base period.
2. Divide each of the observations comprising the series by the observation for the base period. Multiply the quotient by 100.

Table 29.1 Computation of a simple index number series depicting the relative change in annual dollar sales for the Gotham Company (in thousands) (base year = 1967)

Year	Sales ($)	Computation of index number	Index number
1966	190	(190/200)(100)	95
1967	200	(200/200)(100)	100
1968	216	(215/200)(100)	108
1969	240	(240/200)(100)	120
1970	270	(270/200)(100)	135

Index numbers are also used to summarize in a single figure the net impact of the changes in a number of related series. When an index number is used to summarize the overall net relative change in a group of related series, the resulting index number is referred to as a *composite index number*. There are four major alternative constructions for composite index numbers. Each will be examined and its relative merits discussed, but first it is useful to examine some of the basic problems encountered in index number construction.

29.1 BASIC PROBLEMS ENCOUNTERED IN INDEX NUMBER CONSTRUCTION

1. Definition of the purpose for which the index number is to be constructed: The importance of an appropriate statement of purpose cannot be overemphasized. The choice of items to be included, the base year, and the weighting system to be employed are all predicated on a clear definition of the purpose for which the index is to be used.

2. Determination of the items to be included: The items to be included are determined by the basic purpose the index number is to serve. The judgment of the analyst plays an important role in the determination of the items selected. The analyst typically attempts to include all the important items (in his judgment) and to select a representative sample from the others. It is difficult to make appropriate adjustments for changes in quality from one period to another. Accordingly, preference is shown for those commodities whose physical characteristics are not subject to change from period to period.

3. Selection of the base period: The base period is the period with which all other periods are compared. In general, the period that is to be compared to the base period is referred to as the *given-year* period. The base year selected should be considered normal in that the observations for that period should not be unusually high or low It is also desirable that the base period be relatively recent since it is difficult to relate to conditions as they existed in a period in the distant past. Moreover, with the passage of time, the composition and relative importance of the items comprising the index may change to such an extent that the selection of a base period in the distant past no longer would be meaningful. Precisely for this reason, agencies of the Federal government have periodically updated the base period of indices they publish. Base periods used have been the average of the four years 1935–1939, the three years 1947–1949, the three years 1957–1959, and, with the most recent revision, the single year 1967. The justification for utilizing the average of a number of years is that the averaging process tends to cancel out year-to-year variation.

4. Determination of the weighting system to be employed: The relative influence each of the commodities comprising the composite index is to play in the generation of an overall representative index number must be determined. In general, the relative importance of the prices of a group of commodities is measured by the quantities sold, produced, imported, and so on of the products in question. Thus, quantity weights are employed when constructing a price index. In a similar fashion, when the index to be constructed is a quantity index, the prices of the products comprising the index indicate the relative importance of the quantities sold, produced, and so on, and thus prices are employed as weights. Only the variable of interest should be allowed to change so that any change that occurs can be attributed solely to price change in a price index (or quantity change in a quantity index). The issues involved in the selection of an appropriate set of weights will be discussed in conjunction with the examination of the alternative constructions for composite index numbers.

29.2 METHODS OF CONSTRUCTION FOR COMPOSITE INDEX NUMBERS

There are four basic alternative constructions. Each will be illustrated in the context of a price index. The underlying procedure can readily be extended to a quantity index since the computational process is virtually identical for price and quantity indices—the roles of prices and quantities are simply interchanged. The four methods to be examined are (1) the relative of aggregates method, (2) the average of relatives method, (3) the relative of weighted aggregates method, and (4) the average of weighted relatives method.

The following symbols will be employed in the discussion of the construction of composite index numbers:

P_b = prices in the base period
P_g = prices in the given period (the period being compared to the base period)
Q_b = quantities in the base period
Q_g = quantities in the given period
N = number of different series comprising the index
I_g = index number or relative for the given period

29.3 THE RELATIVE OF AGGREGATES METHOD

The name of the method provides an apt description of the underlying computational procedure. The computational procedure is summarized as follows:

1. Compute an aggregate or sum of the prices (or quantities if it is a quantity index) of the commodities comprising the index for each year.
2. Divide each of the aggregates obtained in step 1 by the aggregate obtained for the base year. Multiply the resulting quotient by 100.

Formulas:

Relative of aggregates price index:

$$I_g = \frac{\sum P_g}{\sum P_b}(100)$$

Relative of aggregates quantity index:

$$I_g = \frac{\sum Q_g}{\sum Q_b}(100)$$

The computation of a relative of aggregates prices index is illustrated in Table 29.2.

Table 29.2 Relative of aggregates price index (base period Jan. 1967)

Commodity	Unit of measurement	Price Jan. 1967	Price Jan. 1970	Price Jan. 1973
Bread	Loaf	0.30	0.32	0.34
Milk	Quart	0.20	0.25	0.27
Butter	Pound	0.60	0.64	0.71
Aggregate		1.10	1.21	1.32

$$I_{1967} = \frac{1.10}{1.10}(100) = 100$$

$$I_{1970} = \frac{1.21}{1.10}(100) = 110$$

$$I_{1973} = \frac{1.32}{1.10}(100) = 120$$

The relative of aggregates method is easy to employ and easy to understand. It has the disadvantage of implicitly assuming that the relative importance of the commodities in the overall index is reflected by the magnitude of their prices (or quantities in the case of a quantity index). Thus, butter exerts the most influence on the relative of aggregate index number computed above. This would be incorrect, since only 1 lb of butter is consumed per week, whereas milk consumption consists of 7 quarts. Moreover, the aggregates obtained are a function of the units in which the commodities are quoted. Thus, if milk had been quoted at half-gallon prices, an entirely different set of indices would have resulted. Another disadvantage of the relative of aggregates method is that only a measure of relative overall change is provided. It does not allow comparison of the relative change for the separate components comprising the index. One further qualification: The units of measurement must be comparable in order to use the relative of aggregates method to compute a quantity index. A relative of aggregates quantity index cannot be computed for the above example, since the units of measurement (loaves, quarts, and pounds) are not comparable and therefore cannot be aggregated.

29.4 THE AVERAGE OF RELATIVES METHOD

The name of the average of relatives method provides an apt description of the underlying computational procedure. Summarizing the procedure:

1. Obtain a relative for each commodity for each year by dividing the price (or quantity) for the given year by the price (or quantity) for the base year.

2. For each year, the relatives are summed and their average obtained by dividing by the number of relatives. The method of averaging customarily employed is the arithmetic mean. Occasionally the geometric mean will be used.

Formulas:
Average of relatives price index:

$$I_g = \frac{\sum [(P_g/P_b)(100)]}{N}$$

Average of relatives quantity index:

$$I_g = \frac{\sum [(Q_g/Q_b)(100)]}{N}$$

The computation of an average of relatives price index is illustrated in Table 29.3.

Table 29.3 Average of relatives price index (base period Jan. 1967)

Commodity	Unit of measure-ment	Price Jan. 1967	Price Jan. 1970	Price Jan. 1973	Price relatives Jan. 1967	Price relatives Jan. 1970	Price relatives Jan. 1973
Bread	Loaf	0.30	0.32	0.34	100	107	113
Milk	Quart	0.20	0.25	0.27	100	125	135
Butter	Pound	0.60	0.64	0.71	100	107	118
Total					300	339	366
Index number					100	113	122

The average of relatives method has the advantage of permitting comparison of the relative change for each of the component commodities as well as generating an overall aggregate index. Furthermore, the average of relatives method can be used to compute a quantity index when the units of measurement of the commodities comprising the index are not comparable. It has the disadvantage of implicitly assuming that the relative importance of the commodities in the overall index is reflected by the relative magnitude of their relatives. Neither the average of relatives nor the relative of aggregates method incorporates a weighting system designed to give each of the component commodities its proper relative importance. Because of the lack of such a system, the first two methods are seldom employed in practice. The next two methods examined are similar to the first two methods with the addition of an appropriately superimposed system of weighting the component commodities in accordance with their relative importance in the overall index.

29.5 THE RELATIVE OF WEIGHTED AGGREGATES METHOD

The relative importance of the prices of a group of commodities are best reflected by the quantities sold, produced, imported, and so on of the products in question. (In the case of a quantity index, the prices of the products comprising the index indicate the relative importance of the quantities sold, produced, and so on, and thus prices are employed as weights.) The weights selected must be held constant, so that the index will reflect only changes in price (or quantity in the case of a quantity index.) Possible weighting systems include the following:

1. A relative of weighted aggregates index employing base-year weights is called a *Laspeyres* index.
2. A relative of weighted aggregates index employing given-year weights is referred to as a *Paasche* index. A major disadvantage of a Paasche index is that for a continuing series, the weights would have to be redetermined each year and the entire series recomputed utilizing the new weighting system.
3. A relative of weighted aggregates index employing weights from a period other than the base year or the given year is referred to as a *fixed weight* relative of weighted aggregates index. It is not uncommon for the weights to be generated from a period other than the one selected as the base period.

The relative importance of the commodities comprising an index will typically change with the passage of years. To reflect this, indices covering an extensive period of years periodically have the weighting system revised (often at 10-year intervals).

Procedure for the computation of a relative of weighted aggregates index number:

1. Determine the appropriate weighting factor for each commodity comprising the index.
2. Multiply the price (or quantity) of each commodity for each year by the appropriate weighting factor.
3. Sum the products obtained in step 2 for each year to obtain the weighted aggregates.
4. Divide the weighted aggregate for each year by the weighted aggregate of the year selected as the base year.

Formulas:
Relative of weighted aggregates *price* index using *base year* weights:

$$I_g = \frac{\sum (P_g Q_b)}{\sum (P_b Q_b)}(100)$$

Note that since identical quantities are being employed as weights in both numerator and denominator, the index reflects only the relative change in the prices of the items comprising the index.

Relative of weighted aggregates *price* index using *given-year* weights:

$$I_g = \frac{\sum (P_g Q_g)}{\sum (P_b Q_g)}(100)$$

Relative of weighted aggregates *price* index using *fixed weights*:

$$I_g = \frac{\sum (P_g Q_f)}{\sum (P_b Q_f)}(100)$$

Relative of weighted aggregates *quantity* index using *base-year* weights:

$$I_g = \frac{\sum (P_b Q_g)}{\sum (P_b Q_b)} (100)$$

The computation of a relative of weighted aggregates price index is illustrated in Table 29.4.

Table 29.4 Relative of weighted aggregates price index (base period Jan. 1967—base-year weights)

Commodity	Unit of measure-ment	Quantity consumed per week (Jan. 1967) Q_b	Price Jan. 1967	Price Jan. 1970	Price Jan. 1973	$P_{67}Q_{67}$	$P_{70}Q_{67}$	$P_{73}Q_{67}$
Bread	Loaf	2	0.30	0.32	0.34	0.60	0.64	0.68
Milk	Quart	7	0.20	0.25	0.27	1.40	1.75	1.89
Butter	Pound	1	0.60	0.64	0.71	0.60	0.64	0.71
						2.60	3.03	3.28

$$I_{1967} = \frac{2.60}{2.60} (100) = 100$$

$$I_{1970} = \frac{3.03}{2.60} (100) = 117$$

$$I_{1973} = \frac{3.28}{2.60} (100) = 126$$

The relative of weighted aggregates method has the advantage of introducing weights, thereby providing each commodity with its proper relative importance relative to the other commodities comprising the index. The major disadvantage is that it does not provide for comparison of relative change for the separate components comprising the index.

29.6 THE AVERAGE OF WEIGHTED RELATIVES METHOD

The average of weighted relatives method is similar to the average of relatives method except that an explicit weighting system is introduced.
Procedure:

1. Obtain a relative for each commodity for each year by dividing the price (or quantity) for the given year by the price (or quantity) for the base year.
2. Determine the appropriate weighting factor for the relatives for each commodity comprising the index. The weighting factor customarily employed when relatives are used is the total value expended, consumed, produced, and so on for each commodity in some representative period. Total value is obtained by multiplying price times quantity for the period selected. If base-year values $(P_b Q_b)$ are employed as weights, the average of weighted relatives price index will yield identical index numbers to the Laspeyres relative of weighted aggregates price index. Similarly, if given-year

values $(P_g Q_g)$ are employed as weights, the average of weighted relatives price index will yield identical index numbers to the Paasche relative of weighted aggregates price index. Algebraically, the methods are equivalent. The same operations are performed in each method. Only the order in which they are performed differs.

3. Weight the relative for each commodity by multiplying the relative for the given commodity by the value weight for that commodity.
4. Sum the products obtained in step 3 and obtain the index number for the given year by dividing the sum of the products obtained for the given year by the sum of the products generated for the base period. When multiplied by 100, this figure provides the index number for the given year.

Formulas:
Average of weighted relatives price index employing base-year weights:

$$I_g = \frac{\sum \{(P_b Q_b)[(P_g/P_b)100]\}}{\sum \{(P_b Q_b)[(P_b/P_b)100]\}}(100)$$

Average of weighted relatives price index employing given-year weights:

$$I_g = \frac{\sum \{(P_g Q_g)[(P_g/P_b)100]\}}{\sum \{(P_g Q_g)[(P_b/P_b)100]\}}(100)$$

Average of weighted relatives quantity index employing base-year weights:

$$I_g = \frac{\sum (P_b Q_b)[(Q_g/Q_b)100]}{\sum \{(P_b Q_b)[(Q_b/Q_b)100]\}}(100)$$

The computation of an average of weighted relatives price index is illustrated in Table 29.5.

Table 29.5 Average of weighted relatives price index (base period Jan. 1967—base-year weights $P_b Q_b$)

Commodity	Unit of measure- ment	Quantity consumed per week (Jan. 1967) Q_b	Price Jan. 1967	Price Jan. 1973	Value weights $P_b Q_b$	Price relatives Jan. 1967	Price relatives Jan. 1973	Weighted relatives Jan. 1967	Weighted relatives Jan. 1973
Bread	Loaf	2	0.30	0.34	0.60	100	113	60	67.8
Milk	Quart	7	0.20	0.27	1.40	100	135	140	189.0
Butter	Pound	1	0.60	0.71	0.60	100	118	60	70.8
								260	327.6

$$I_{1967} = \frac{260}{260}(100) = 100$$

$$I_{1973} = \frac{327.6}{260}(100) = 126$$

The average of weighted relatives method has the advantage of not only employing weights and thereby providing each item with its proper relative importance, but it also permits a comparison of the relative change for each of the component commodities comprising the index. It does involve a more complex computation than the relative of weighted aggregates method and therefore should be used only when a comparison of the component commodities comprising the index is desired.

29.7 SHIFTING THE BASE OF AN INDEX

If the analyst desires to compare two series of index numbers that do not share a common base period, it is necessary to shift them to a common base. Another instance where shifting the base of an index number series is desirable is when the existing base period needs to be moved to a more recent time period in order to allow comparison with a period with which it is easy to identify. *Provided that the weighting system is not changed*, the process of shifting the base is performed by simply dividing the existing index numbers by the index number for the period selected as the new base period and multiplying by 100 (Table 29.6). The relationship, of course, is unchanged. Only the base with which comparison is made is changed. If the weighting system is also changed, the entire series of index numbers must be recomputed.

Table 29.6 Shifting the base of an index number (original base period Jan. 1967; revised base period Jan. 1970)

Period	Original index number series (base period = Jan. 1967)	Revised index number series (base period = Jan. 1970)
Jan. 1967	100	$(100/117)(100) = 85$
Jan. 1970	117	$(117/117)(100) = 100$
Jan. 1973	126	$(126/117)(100) = 108$

29.8 DEFLATION OF A DOLLAR VALUE SERIES

A price index is occasionally employed to deflate a dollar value series. Deflation is performed by dividing the dollar value figure by the corresponding value of the appropriate price index as of the same period in time. The impact is to wash out the change in dollar value that is attributable to change in price. In effect, one is left with a series that reflects the value of each year's output when priced at the prices observed in the period selected as the base period for the price index used to deflate the dollar value series. When expressed in original dollar figures, the series is referred to as being stated in current dollars. After deflation, the resulting series is referred to as being stated in constant dollars.

29.9 COMMON INDEX NUMBERS

Consumer Price Index

The Consumer Price Index is constructed by the Bureau of Labor Statistics, U.S. Department of Labor. This index measures the average change in prices of goods and services purchased by families of urban wage earners and clerical workers and by single persons living alone. Approximately 400 goods and services are covered by the index. They are divided into such subcategories as food, housing, apparel and upkeep, transportation, and health and recreation. The single year

1967 is currently the standard reference base period for all general-purpose index numbers prepared by Federal agencies. This base period was adopted in March 1970, and replaced the 1957–1959 base period that had been in use since 1960. Expenditure weights employed relate to the period 1960–1961. The reciprocal of this price index provides an index displaying the purchasing power of money.

Referring to Table 29.7, the Consumer Price Index rose from 100 in 1967 to 116.3 in 1970. Alternatively stated, one could say that the purchasing power of the dollar fell from 1.00 in 1967 to 0.89 in 1970. In other words, in 1970, a dollar would be required to purchase the equivalent of what would have cost $0.89 in 1967.

Table 29.7

Consumer price index		Purchasing power of dollar	
Year	Index	Year	Index
1966	97.2	1966	1.029
1967	100	1967	1.000
1968	104.2	1968	0.960
1969	109.8	1969	0.911
1970	116.3	1970	0.890

Source: U.S. Department of Labor, Bureau of Labor Statistics.

Wholesale Price Index

The Wholesale Price Index is also published by the Bureau of Labor Statistics, U.S. Department of Labor. This series is designed to measure relative change in the prices for all commodities sold in primary markets in the United States (i.e., the first significant large-volume commercial transaction for each commodity). The index is currently comprised of approximately 2450 commodity price series. As of January 1967, the weights employed are the values of the net shipments of the commodities derived from the industrial census of 1963 and other data. The single year 1967 is the base period employed as of March 1970.

Federal Reserve Board Index of Industrial Production

The Federal Reserve Board Index of Industrial Production is the most widely employed and widely recognized quantity index in the United States. It measures changes in the physical volume of output of manufacturing, mining, and utilities. Component indices are prepared for 25 major industrial groups and 175 subgroups. Further information on this and other index numbers can be obtained by writing the appropriate government agency responsible for preparation.

PROBLEMS

Problem 29.1

What advantages does the average of relatives method possess as compared to the relative of aggregates method for computing an index number?

Problem 29.2

Indicate three advantages possessed by the average of weighted relatives method as contrasted with the relative of aggregates method for computing an index number.

Problem 29.3

1. Compute a price index for the following set of data by the relative of weighted aggregates method using 1967 as the base year. Use base-year weights.

Commodity	Unit	Price 1967	Price 1970	Quantity 1967	Quantity 1970
A	pound	0.10	0.15	80	200
B	ton	10.00	9.00	10	5
C	gallon	0.80	1.00	50	75
D	dozen	0.20	0.25	100	100

2. What additional information would have been provided (in words, not numbers) if the average of weighted relatives method had been employed?
3. Why can one not compute the relative of aggregates quantity index for the set of data in part 1?

Problem 29.4

What common disadvantage characterizes the average of relatives and the relative of aggregates methods of index number construction?

Problem 29.5

What basic factors should be taken into consideration in deciding whether to use the average of weighted relatives or the relative of weighted aggregates method of index number construction?

Problem 29.6

For the following set of data, compute a quantity index number employing the relative of weighted aggregates method of index number construction. Let 1968 be the base year and use base-year weights.

Commodity	Unit	Price 1968	Quantity 1968	Price 1970	Quantity 1970
A	pound	1.00	200	2.50	190
B	pound	0.80	100	0.60	120
C	pound	0.75	6	1.25	5

Problem 29.7

Repeat Problem 29.6 using given-year weights.

Problem 29.8

What is the primary disadvantage associated with the employment of given-year weights?

Problem 29.9

Differentiate between the Laspeyres relative of weighted aggregates method of index number construction and the Paasche relative of weighted aggregates method of index number construction. Indicate the difference in terms of what each of these methods of index number construction measures.

Problem 29.10

For the following set of data, compute the index number that not only *correctly* indicates the overall percentage price change but also indicates the percentage price change for each of the component series as well. Let 1967 be the base year. Use base-year weights.

Commodity	Unit	Price 1967	Quantity 1967	Price 1970	Quantity 1970
A	pound	1.00	200	2.50	190
B	pound	0.80	100	0.60	120
C	pound	0.75	6	1.25	5

Problem 29.11

Is there such a thing as an "unweighted" composite index number?

Problem 29.12

What "implicit" weighting system characterizes the relative of aggregates method of index number construction?

Problem 29.13

What "implicit" weighting system characterizes the average of relatives method of index number construction?

Problem 29.14

Under what circumstances would it be considered desirable to shift the base of a series of index numbers?

Problem 29.15

It is preferable to employ commodities whose physical characteristics are not subject to change from time period to time period or from place to place when constructing an index number series. Why is this true?

Problem 29.16

Given the following set of data:

Commodity	Unit	Price 1968	Quantity 1968	Price 1970	Quantity 1970	Price 1972	Quantity 1972
A	pound	0.10	10	0.12	10	0.15	10
B	pound	0.10	20	0.08	22	0.08	23
C	pound	0.30	5	0.33	3	0.39	2

1. Compute a relative of aggregates price index number series using 1968 as the base year.
2. Shift the index number series obtained in part 1 to a base year of 1970.
3. Compute a relative of aggregates quantity index number series using 1968 as the base year.
4. Compute an average of relatives price index number series using 1968 as the base year.
5. Compute an average of relatives quantity index number series using 1968 as the base year.

Problem 29.17

Referring to the data in Problem 29.16:

1. Compute a relative of weighted aggregates price index number for the two years 1968 and 1970 using base-year weights. Use 1968 as the base year.
2. Shift the index number series obtained in part 1 to a base year of 1970.
3. Compute a relative of weighted aggregates price index number for the two years 1968 and 1970 using given-year (i.e., 1970) weights.
4. Compute a relative of weighted aggregates price index number for the three years 1968, 1970, and 1972 using given-year (i.e., 1972) weights. Why can one not just add on to the series generated in part 3?
5. Compute a relative of weighted aggregates price index number for the year 1972, employing base-year weights. Use 1968 as the base year.
6. Referring to part 5, explain why one can just add on to the series generated in solving part 1 to obtain the index number for the additional year requested in part 5.

Problem 29.18

Referring to the data in Problem 29.16:

1. Compute a relative of weighted aggregates quantity index number using base-year weights. Use 1968 as the base year.
2. Compute a relative of weighted aggregates quantity index number using given-year (i.e., 1972) weights. Use 1968 as the base year.

Problem 29.19

Referring to the data in Problem 29.16:

1. Compute an average of weighted relatives price index number using base-year weights. Use 1968 as the base year. Explain why the index number obtained is identical to that obtained in Problem 29.17, part 1.
2. Compute an average of weighted relatives price index number using given-year (i.e., 1972) weights. Use 1968 as the base year. Explain why the index number obtained is not identical to that obtained in Problem 29.17, part 4.
3. Compute an average of weighted relatives quantity index number using base-year weights. Use 1968 as the base year.
4. Compute an average of weighted relatives quantity index number using given-year (i.e., 1972) weights. Use 1968 as the base year.

APPENDIX

Table A *The concept of a logarithm: table of mantissas*

A.1 RULES OF OPERATION FOR EXPONENTIAL EXPRESSIONS

Rule 1: Multiplication of Exponential Expressions

$$a^3 \cdot a^8 = a^{11}$$
$$(a \cdot a \cdot a)(a \cdot a \cdot a \cdot a \cdot a \cdot a \cdot a \cdot a) = a^{11}$$

The basic rule for multiplication of exponential expression is as follows: *When multiplying exponential expressions with the same base, add the exponents and attach to the common base.*

Rule 2: Division with Exponential Expressions

$$\frac{a^8}{a^3} = a^5 \quad \text{or} \quad \frac{\cancel{a} \cdot \cancel{a} \cdot \cancel{a} \cdot a \cdot a \cdot a \cdot a \cdot a}{\cancel{a} \cdot \cancel{a} \cdot \cancel{a}} = a^5$$

The basic rule for division with exponential expressions is as follows: *Subtract the exponent of the divisor from the exponent of the dividend and attach to the common base.*

Note the following special cases:

Case A.

$$\frac{a^7}{a^7} = a^{7-7} = a^0 \qquad \text{also} \quad \frac{\cancel{a^7}}{\cancel{a^7}} = 1$$

The above is a simple proof that any expression to the zero power is equivalent to that expression divided by itself and therefore equal to 1.

Case B.

$$\frac{a^3}{a^5} = \frac{\cancel{a} \cdot \cancel{a} \cdot \cancel{a}}{\cancel{a} \cdot \cancel{a} \cdot \cancel{a} \cdot a \cdot a} = \frac{1}{a^2}$$

Applying the rule:

$$\frac{a^3}{a^5} = a^{3-5} = a^{-2}$$

Thus,

$$a^{-2} = \frac{1}{a^2}$$

An expression with a negative exponent can thus be written equivalently as the reciprocal of the same expression with a positive exponent.

Rule 3: Raising an Exponential Expression to a Power

$$(a^3)^4 = a^{12}$$
$$(a^3)^4 = a^3 \cdot a^3 \cdot a^3 \cdot a^3 = a^{12}$$

The basic rule for raising an exponential expression to a power is as follows: *Multiply the external exponent* (i.e., the exponent or power to which the exponential expression is to be raised) *by the internal exponent* (the exponent attached to the base in the exponential expression being raised to the power) *and attach the product to the common base.*

Rule 4: Extracting the Root of an Exponential Expression

$$\sqrt[4]{a^8}$$

Extracting the fourth root is equivalent to raising the expression to the one-fourth power.

$$(a^8)^{1/4} = a^{8/4} = a^2$$

Quite clearly, a^2 is the value that when multiplied times itself four times generates the value a^8. The basic rule for extracting the root from exponential expressions is as follows: *Divide the exponent on the base of the expression that is to have the root extracted by the value of the root to be extracted and attach to the common base.*

$$\sqrt[7]{a^2} = a^{2/7}$$
$$\sqrt[7]{a^{14}b^7} = a^{14/7}b^{7/7} = a^2b$$

The preceding four rules constitute the basic rules for performing operations with exponential expressions.

Operations with *common* logarithms refer to operations performed with exponential expressions all to the base 10.

To illustrate: Arithmetically, $10 \cdot 30 = 300$. The same result can be obtained by converting 10 and 30 into exponential expressions and performing the multiplication operation using exponential operations. Recall that the common logarithm of a number is the power to which 10 must be raised in order to generate that number. Thus,

$$\log_{10} 10 = 1.0000$$

The above expression states that the power to which 10 must be raised in order to generate the number 10 is 1. Similarly,

$$\log_{10} 30 = 1.4771 \quad \text{or} \quad 10^{1.4771} = 30$$

Thus,

$$10 \cdot 30 = 10^{1.0000} \cdot 10^{1.4771}$$

To perform the multiplication exponentially (or logarithmically given the base of 10), simply add the exponents (logarithms) and attach to the common base.

$$10^{1.0000} \cdot 10^{1.4771} = 10^{1.0000+1.4771} = 10^{2.4771}$$

The answer in exponential (or logarithmic form) is $10^{2.4771}$. To convert the answer to original number form, the number 10 is raised to the indicated power. This procedure is referred to as obtaining the antilog.

Procedure for generating an antilog: Every log has two parts—a whole-number part and a fractional part. The whole-number part is called the characteristic and the fractional part is called the mantissa. In the illustrative example, the log is 2.4771, 2 is the characteristic, and .4771 is the mantissa. The function of the mantissa is to indicate the sequence of numerical digits that identify the particular number. The function of the characteristic is to indicate the location of the decimal point in the sequence of numerical digits.

A table of mantissas would indicate that .4771 corresponds to the sequence of numerical digits, 30000. The characteristic locates the decimal point and indicates whether the number is 3; 30; 3000; 0.300; 0.0300; and so on.

If the characteristic is positive, it is always one unit less than the number of numerical digits to the left of the decimal point. In this case it is 2, so there must be three places to the left of the decimal point; thus the number is 300.

If the characteristic is negative, it is always one unit greater than the number of zeros between the decimal point and the first significant digit. Convention is to express negative exponents in the following form:

Negative characteristic	Customary mode of expression
-3	7. mantissa $-$ 10
-8	2. mantissa $-$ 10

Thus, the expression $6.4771 - 10$ indicates that the sequence of numerical digits is 30000 and that there are three zeros between the decimal point and the first significant digit. Thus, $10^{6.4771-10} = 0.0003$.

A quick survey of the rules for performing operations logarithmically discloses that they are identical to the rules for performing operations exponentially, with the word logarithm substituted for the word exponent. The word logarithm serves to indicate that one is referring to the subset of exponential expressions that are to the base 10.

Rules for performing operations logarithmically:

1. Multiplication: Obtain the logs of the numbers and *add* the logs. Determine the antilog of the sum to return the answer to original arithmetic form.

TABLE A Table of Mantissas

	0	1	2	3	4	5	6	7	8	9
10	.0000	.0043	.0086	.0128	.0170	.0212	.0253	.0294	.0334	.0374
11	.0414	.0453	.0492	.0531	.0569	.0607	.0645	.0682	.0719	.0755
12	.0792	.0828	.0864	.0839	.0934	.0969	.1004	.1033	.1072	.1106
13	.1139	.1173	.1206	.1239	.1271	.1303	.1335	.1367	.1399	.1430
14	.1461	.1492	.1523	.1553	.1584	.1614	.1644	.1673	.1703	.1732
15	.1761	.1790	.1818	.1847	.1875	.1903	.1931	.1959	.1987	.2014
16	.2041	.2068	.2095	.2122	.2148	.2175	.2201	.2227	.2253	.2279
17	.2304	.2330	.2355	.2380	.2405	.2430	.2455	.2480	.2504	.2529
18	.2553	.2577	.2601	.2625	.2648	.2672	.2695	.2718	.2742	.2765
19	.2788	.2810	.2833	.2856	.2878	.2900	.2923	.2945	.2967	.2989
20	.3010	.3032	.3054	.3075	.3096	.3118	.3139	.3160	.3181	.3201
21	.3222	.3243	.3263	.3234	.3304	.3324	.3345	.3365	.3385	.3404
22	.3424	.3444	.3464	.3483	.3502	.3522	.3541	.3560	.3579	.3598
23	.3617	.3636	.3655	.3674	.3692	.3711	.3729	.3747	.3766	.3784
24	.3802	.3820	.3838	.3856	.3874	.3892	.3909	.3927	.3945	.3962
25	.3979	.3997	.4014	.4031	.4048	.4065	.4082	.4099	.4116	.4133
26	.4150	.4166	.4183	.4200	.4216	.4232	.4249	.4265	.4281	.4298
27	.4314	.4330	.4346	.4362	.4378	.4393	.4409	.4425	.4440	.4456
28	.4472	.4487	.4502	.4518	.4533	.4548	.4564	.4579	.4594	.4609
29	.4624	.4639	.4654	.4669	.4683	.4698	.4713	.4728	.4742	.4757
30	.4771	.4786	.4800	.4814	.4829	.4843	.4857	.4871	.4886	.4900
31	.4914	.4928	.4942	.4955	.4969	.4983	.4997	.5011	.5024	.5038
32	.5051	.5065	.5079	.5092	.5105	.5119	.5132	.5145	.5159	.5172
33	.5185	.5198	.5211	.5224	.5237	.5250	.5263	.5276	.5289	.5302
34	.5315	.5328	.5340	.5353	.5366	.5378	.5391	.5403	.5416	.5428
35	.5441	.5453	.5465	.5478	.5490	.5502	.5515	.5527	.5539	.5551
36	.5563	.5575	.5587	.5599	.5611	.5623	.5635	.5647	.5658	.5670
37	.5682	.5694	.5705	.5717	.5729	.5740	.5752	.5763	.5775	.5786
38	.5798	.5809	.5821	.5832	.5843	.5855	.5866	.5877	.5888	.5899
39	.5911	.5922	.5933	.5944	.5955	.5966	.5977	.5988	.5999	.6010
40	.6021	.6031	.6042	.6053	.6064	.6075	.6085	.6096	.6107	.6117
41	.6128	.6138	.6149	.6160	.6170	.6180	.6191	.6201	.6212	.6222
42	.6232	.6243	.6253	.6263	.6274	.6284	.6294	.6304	.6314	.6325
43	.6335	.6345	.6355	.6365	.6375	.6385	.6395	.6405	.6415	.6425
44	.6435	.6444	.6454	.6464	.6474	.6484	.6493	.6503	.6513	.6522
45	.6532	.6542	.6551	.6561	.6571	.6580	.6590	.6599	.6609	.6618
46	.6628	.6637	.6546	.6656	.6665	.6675	.6684	.6693	.6702	.6712
47	.6721	.6730	.6739	.6749	.6758	.6767	.6776	.6785	.6794	.6803
48	.6812	.6821	.6830	.6839	.6848	.6857	.6866	.6875	.6884	.6893
49	.6902	.6911	.6920	.6928	.6937	.6946	.6955	.6964	.6972	.6981
50	.6990	.6998	.7007	.7016	.7024	.7033	.7042	.7050	.7059	.7067
51	.7076	.7084	.7093	.7101	.7110	.7118	.7126	.7135	.7143	.7152
52	.7160	.7168	.7177	.7185	.7193	.7202	.7210	.7218	.7226	.7235
53	.7243	.7251	.7259	.7267	.7275	.7284	.7292	.7300	.7308	.7316
54	.7324	.7332	.7340	.7348	.7356	.7364	.7372	.7380	.7388	.7396
55	.7404	.7412	.7419	.7427	.7435	.7443	.7451	.7459	.7466	.7474

2. Division: *Subtract* the log of the divisor from the log of the dividend to obtain the log of the quotient. Determine the antilog of the quotient to return the answer to original arithmetic form.

3. To raise to a power: *Multiply* the logarithm of the number to be raised to the power by the numerical value of the power to which it is to be raised. Determine the antilog of the product to return the answer to original arithmetic form.

4. To extract a root: *Divide* the log of the number whose root is to be extracted by the numerical value of the root which is to be extracted. To return the answer to original arithmetic form, determine the antilog of the quotient.

TABLE A Table of Mantissas (*Continued*)

	0	1	2	3	4	5	6	7	8	9
56	.7482	.7490	.7497	.7505	.7513	.7520	.7528	.7536	.7543	.7551
57	.7559	.7566	.7574	.7582	.7589	.7597	.7604	.7612	.7619	.7627
58	.7634	.7642	.7649	.7657	.7664	.7672	.7679	.7686	.7694	.7701
59	.7709	.7716	.7723	.7731	.7738	.7745	.7752	.7760	.7767	.7774
60	.7782	.7789	.7796	.7803	.7810	.7818	.7825	.7832	.7839	.7846
61	.7853	.7860	.7868	.7875	.7882	.7889	.7896	.7903	.7910	.7917
62	.7924	.7931	.7938	.7945	.7952	.7959	.7966	.7973	.7980	.7987
63	.7993	.8000	.8007	.8014	.8021	.8028	.8035	.8041	.8048	.8055
64	.8062	.8069	.8075	.8082	.8089	.8096	.8102	.8109	.8116	.8122
65	.8129	.8136	.8142	.8149	.8156	.8162	.8169	.8176	.8182	.8189
66	.8195	.8202	.8209	.8215	.8222	.8228	.8235	.8241	.8248	.8254
67	.8261	.8267	.8274	.8280	.8287	.8293	.8299	.8306	.8312	.8319
68	.8325	.8331	.8338	.8344	.8351	.8357	.8363	.8370	.8376	.8382
69	.8388	.8395	.8401	.8407	.8414	.8420	.8426	.8432	.8439	.8445
70	.8451	.8457	.8463	.8470	.8476	.8482	.8488	.8494	.8500	.8506
71	.8513	.8519	.8525	.8531	.8537	.8543	.8549	.8555	.8561	.8567
72	.8573	.8579	.8585	.8591	.8597	.8603	.8609	.8615	.8621	.8627
73	.8633	.8639	.8645	.8651	.8657	.8663	.8669	.8675	.8681	.8686
74	.8692	.8698	.8704	.8710	.8716	.8722	.8727	.8733	.8739	.8745
75	.8751	.8756	.8762	.8768	.8774	.8779	.8785	.8791	.8797	.8802
76	.8808	.8814	.8820	.8825	.8831	.8837	.8842	.8848	.8854	.8859
77	.8865	.8871	.8876	.8882	.8887	.8893	.8899	.8904	.8910	.8915
78	.8921	.8927	.8932	.8938	.8943	.8949	.8954	.8960	.8965	.8971
79	.8976	.8982	.8987	.8993	.8998	.9004	.9009	.9015	.9020	.9025
80	.9031	.9036	.9042	.9047	.9053	.9058	.9063	.9069	.9074	.9079
81	.9085	.9090	.9096	.9101	.9106	.9112	.9117	.9122	.9128	.9133
82	.9138	.9143	.9149	.9154	.9159	.9165	.9170	.9175	.9180	.9186
83	.9191	.9196	.9201	.9206	.9212	.9217	.9222	.9227	.9232	.9238
84	.9243	.9248	.9253	.9258	.9263	.9269	.9274	.9279	.9284	.9289
85	.9294	.9299	.9304	.9309	.9315	.9320	.9325	.9330	.9335	.9340
86	.9345	.9350	.9355	.9360	.9365	.9370	.9375	.9380	.9385	.9390
87	.9395	.9400	.9405	.9410	.9415	.9420	.9425	.9430	.9435	.9440
88	.9445	.9450	.9455	.9460	.9465	.9469	.9474	.9479	.9484	.9489
89	.9494	.9499	.9504	.9509	.9513	.9518	.9523	.9528	.9533	.9538
90	.9542	.9547	.9552	.9557	.9562	.9566	.9571	.9576	.9581	.9586
91	.9590	.9595	.9600	.9605	.9609	.9614	.9619	.9624	.9628	.9633
92	.9638	.9643	.9647	.9652	.9657	.9661	.9666	.9671	.9675	.9680
93	.9685	.9689	.9694	.9699	.9703	.9708	.9713	.9717	.9722	.9727
94	.9731	.9736	.9741	.9745	.9750	.9754	.9759	.9763	.9768	.9773
95	.9777	.9782	.9786	.9791	.9795	.9800	.9805	.9809	.9814	.9818
96	.9823	.9827	.9832	.9835	.9841	.9845	.9850	.9854	.9859	.9863
97	.9868	.9872	.9877	.9881	.9886	.9890	.9894	.9899	.9903	.9908
98	.9912	.9917	.9921	.9926	.9930	.9934	.9939	.9943	.9948	.9952
99	.9956	.9961	.9965	.9969	.9974	.9978	.9983	.9987	.9991	.9996

TABLE B Squares and Square Roots

n	n^2	$\sqrt{n}$	$\sqrt{10n}$
1.00	1.0000	1.000000	3.162278
1.01	1.0201	1.004988	3.178050
1.02	1.0404	1.009950	3.193744
1.03	1.0609	1.014889	3.209361
1.04	1.0816	1.019804	3.224903
1.05	1.1025	1.024695	3.240370
1.06	1.1236	1.029563	3.255764
1.07	1.1449	1.034408	3.271085
1.08	1.1664	1.039230	3.286335
1.09	1.1881	1.044031	3.301515
1.10	1.2100	1.048809	3.316625
1.11	1.2321	1.053565	3.331666
1.12	1.2544	1.058301	3.346640
1.13	1.2769	1.063015	3.361547
1.14	1.2996	1.067708	3.376389
1.15	1.3225	1.072381	3.391165
1.16	1.3456	1.077033	3.405877
1.17	1.3689	1.081665	3.420526
1.18	1.3924	1.086278	3.435113
1.19	1.4161	1.090871	3.449638
1.20	1.4400	1.095445	3.464102
1.21	1.4641	1.100000	3.478505
1.22	1.4884	1.104536	3.492850
1.23	1.5129	1.109054	3.507136
1.24	1.5376	1.113553	3.521363
1.25	1.5625	1.118034	3.535534
1.26	1.5876	1.122497	3.549648
1.27	1.6129	1.126943	3.563706
1.28	1.6384	1.131371	3.577709
1.29	1.6641	1.135782	3.591657
1.30	1.6900	1.140175	3.605551
1.31	1.7161	1.144552	3.619392
1.32	1.7424	1.148913	3.633180
1.33	1.7689	1.153256	3.646917
1.34	1.7956	1.157584	3.660601
1.35	1.8225	1.161895	3.674235
1.36	1.8496	1.166190	3.687818
1.37	1.8769	1.170470	3.701351
1.38	1.9044	1.174734	3.714835
1.39	1.9321	1.178983	3.728270
1.40	1.9600	1.183216	3.741657
1.41	1.9881	1.187434	3.754997
1.42	2.0164	1.191638	3.768289
1.43	2.0449	1.195826	3.781534
1.44	2.0736	1.200000	3.794733
1.45	2.1025	1.204159	3.807887
1.46	2.1316	1.208305	3.820995
1.47	2.1609	1.212436	3.834058
1.48	2.1904	1.216553	3.847077
1.49	2.2201	1.220656	3.860052
1.50	2.2500	1.224745	3.872983
1.51	2.2801	1.228821	3.885872
1.52	2.3104	1.232883	3.898718
1.53	2.3409	1.236932	3.911521
1.54	2.3716	1.240967	3.924283
1.55	2.4025	1.244990	3.937004
1.56	2.4336	1.249000	3.949684
1.57	2.4649	1.252996	3.962323
1.58	2.4964	1.256981	3.974921
1.59	2.5281	1.260952	3.987480,
1.60	2.5600	1.264911	4.000000
1.61	2.5921	1.268858	4.012481
1.62	2.6244	1.272792	4.024922
1.63	2.6569	1.276715	4.037326
1.64	2.6896	1.280625	4.049691
1.65	2.7225	1.284523	4.062019
1.66	2.7556	1.288410	4.074310
1.67	2.7889	1.292285	4.086563
1.68	2.8224	1.296148	4.098780
1.69	2.8561	1.300000	4.110961
1.70	2.8900	1.303840	4.123106
1.71	2.9241	1.307670	4.135215
1.72	2.9584	1.311488	4.147288
1.73	2.9929	1.315295	4.159327
1.74	3.0276	1.319091	4.171331
1.75	3.0625	1.322876	4.183300
1.76	3.0976	1.326650	4.195235
1.77	3.1329	1.330413	4.207137
1.78	3.1684	1.334166	4.219005
1.79	3.2041	1.337909	4.230839
1.80	3.2400	1.341641	4.242641
1.81	3.2761	1.345362	4.254409
1.82	3.3124	1.349074	4.266146
1.83	3.3489	1.352775	4.277850
1.84	3.3856	1.356466	4.289522
1.85	3.4225	1.360147	4.301163
1.86	3.4596	1.363818	4.312772
1.87	3.4969	1.367479	4.324350

TABLE B Squares and Square Roots (*Continued*)

n	n²	√n	√10n
1.88	3.5344	1.371131	4.335897
1.89	3.5721	1.374773	4.347413
1.90	3.6100	1.378405	4.358899
1.91	3.6481	1.382027	4.370355
1.92	3.6864	1.385641	4.381780
1.93	3.7249	1.389244	4.393177
1.94	3.7636	1.392839	4.404543
1.95	3.8025	1.396424	4.415880
1.96	3.8416	1.400000	4.427189
1.97	3.8809	1.403567	4.438468
1.98	3.9204	1.407125	4.449719
1.99	3.9601	1.410674	4.460942
2.00	4.0000	1.414214	4.472136
2.01	4.0401	1.417745	4.483302
2.02	4.0804	1.421267	4.494441
2.03	4.1209	1.424781	4.505552
2.04	4.1615	1.428286	4.516636
2.05	4.2025	1.431782	4.527693
2.06	4.2436	1.435270	4.538722
2.07	4.2849	1.438749	4.549725
2.08	4.3264	1.442221	4.560702
2.09	4.3681	1.445683	4.571652
2.10	4.4100	1.449138	4.582576
2.11	4.4521	1.452584	4.593474
2.12	4.4944	1.456022	4.604346
2.13	4.5369	1.459452	4.615192
2.14	4.5796	1.462874	4.626013
2.15	4.6225	1.466288	4.636809
2.16	4.6656	1.469694	4.647580
2.17	4.7089	1.473092	4.658326
2.18	4.7524	1.476482	4.669047
2.19	4.7961	1.479865	4.679744
2.20	4.8400	1.483240	4.690416
2.21	4.8841	1.486607	4.701064
2.22	4.9284	1.489966	4.711688
2.23	4.9729	1.493318	4.722288
2.24	5.0176	1.496663	4.732864
2.25	5.0625	1.500000	4.743416
2.26	5.1076	1.503330	4.753946
2.27	5.1529	1.506652	4.764452
2.28	5.1984	1.509967	4.774935
2.29	5.2441	1.513275	4.785394
2.30	5.2900	1.516575	4.795832
2.31	5.3361	1.519868	4.806246
2.32	5.3824	1.523155	4.816638
2.33	5.4289	1.526434	4.827007
2.34	5.4756	1.529706	4.837355
2.35	5.5225	1.532971	4.847680
2.36	5.5696	1.536229	4.857983
2.37	5.6169	1.539480	4.868265
2.38	5.6644	1.542725	4.878524
2.39	5.7121	1.545962	4.888763
2.40	5.7600	1.549193	4.898979
2.41	5.8081	1.552417	4.909175
2.42	5.8564	1.555635	4.919350
2.43	5.9049	1.558846	4.929503
2.44	5.9536	1.562050	4.939636
2.45	6.0025	1.565248	4.949747
2.46	6.0516	1.568439	4.959839
2.47	6.1009	1.571623	4.969909
2.48	6.1504	1.574802	4.979960
2.49	6.2001	1.577973	4.989990
2.50	6.2500	1.581139	5.000000
2.51	6.3001	1.584298	5.009990
2.52	6.3504	1.587451	5.019960
2.53	6.4009	1.590597	5.029911
2.54	6.4516	1.593738	5.039841
2.55	6.5025	1.596872	5.049752
2.56	6.5536	1.600000	5.059644
2.57	6.6049	1.603122	5.069517
2.58	6.6564	1.606238	5.079370
2.59	6.7081	1.609348	5.089204
2.60	6.7600	1.612452	5.099020
2.61	6.8121	1.615549	5.108816
2.62	6.8644	1.618641	5.118594
2.63	6.9169	1.621727	5.128353
2.64	6.9696	1.624808	5.138093
2.65	7.0225	1.627882	5.147815
2.66	7.0756	1.630951	5.157519
2.67	7.1289	1.634013	5.167204
2.68	7.1824	1.637071	5.176872
2.69	7.2361	1.640122	5.186521
2.70	7.2900	1.643168	5.196152
2.71	7.3441	1.646208	5.205766
2.72	7.3984	1.649242	5.215362
2.73	7.4529	1.652271	5.224940
2.74	7.5076	1.655295	5.234501
2.75	7.5625	1.658312	5.244044

TABLE B Squares and Square Roots (*Continued*)

n	n^2	$\sqrt{n}$	$\sqrt{10n}$
2.76	7.6176	1.661325	5.253570
2.77	7.6729	1.664332	5.263079
2.78	7.7284	1.667333	5.272571
2.79	7.7841	1.670329	5.282045
2.80	7.8400	1.673320	5.291503
2.81	7.8961	1.676305	5.300943
2.82	7.9524	1.679286	5.310367
2.83	8.0089	1.682260	5.319774
2.84	8.0656	1.685230	5.329165
2.85	8.1225	1.688194	5.338539
2.86	8.1796	1.691153	5.347897
2.87	8.2369	1.694107	5.357238
2.88	8.2944	1.697056	5.366563
2.89	8.3521	1.700000	5.375872
2.90	8.4100	1.702939	5.385165
2.91	8.4681	1.705872	5.394442
2.92	8.5264	1.708801	5.403702
2.93	8.5849	1.711724	5.412947
2.94	8.6436	1.714643	5.422177
2.95	8.7025	1.717556	5.431390
2.96	8.7616	1.720465	5.440588
2.97	8.8209	1.723369	5.449771
2.98	8.8804	1.726268	5.458938
2.99	8.9401	1.729162	5.468089
3.00	9.0000	1.732051	5.477226
3.01	9.0601	1.734935	5.486347
3.02	9.1204	1.737815	5.495453
3.03	9.1809	1.740690	5.504544
3.04	9.2416	1.743560	5.513620
3.05	9.3025	1.746425	5.522681
3.06	9.3636	1.749286	5.531727
3.07	9.4249	1.752142	5.540758
3.08	9.4864	1.754993	5.549775
3.09	9.5481	1.757840	5.558777
3.10	9.6100	1.760682	5.567764
3.11	9.6721	1.763519	5.576737
3.12	9.7344	1.766352	5.585696
3.13	9.7969	1.769181	5.594640
3.14	9.8596	1.772005	5.603570
3.15	9.9225	1.774824	5.612486
3.16	9.9856	1.777639	5.621388
3.17	10.0489	1.780449	5.630275
3.18	10.1124	1.783255	5.639149
3.19	10.1761	1.786057	5.648008
3.20	10.2400	1.788854	5.656854
3.21	10.3041	1.791647	5.665686
3.22	10.3684	1.794436	5.674504
3.23	10.4329	1.797220	5.683309
3.24	10.4976	1.800000	5.692100
3.25	10.5625	1.802776	5.700877
3.26	10.6276	1.805547	5.709641
3.27	10.6929	1.808314	5.718391
3.28	10.7584	1.811077	5.727128
3.29	10.8241	1.813836	5.735852
3.30	10.8900	1.816590	5.744563
3.31	10.9561	1.819341	5.753260
3.32	11.0224	1.822087	5.761944
3.33	11.0889	1.824829	5.770615
3.34	11.1556	1.827567	5.779273
3.35	11.2225	1.830301	5.787918
3.36	11.2896	1.833030	5.796551
3.37	11.3569	1.835756	5.805170
3.38	11.4244	1.838478	5.813777
3.39	11.4921	1.841195	5.822371
3.40	11.5600	1.843909	5.830952
3.41	11.6281	1.846619	5.839521
3.42	11.6964	1.849324	5.848077
3.43	11.7649	1.852026	5.856620
3.44	11.8336	1.854724	5.865151
3.45	11.9025	1.857418	5.873670
3.46	11.9716	1.860108	5.882176
3.47	12.0409	1.862794	5.890671
3.48	12.1104	1.865476	5.899152
3.49	12.1801	1.868154	5.907622
3.50	12.2500	1.870829	5.916080
3.51	12.3201	1.873499	5.924525
3.52	12.3904	1.876166	5.932959
3.53	12.4609	1.878829	5.941380
3.54	12.5316	1.881489	5.949790
3.55	12.6025	1.884144	5.958188
3.56	12.6736	1.886796	5.966574
3.57	12.7449	1.889444	5.974948
3.58	12.8164	1.892089	5.983310
3.59	12.8881	1.894730	5.991661
3.60	12.9600	1.897367	6.000000
3.61	13.0321	1.900000	6.008328
3.62	13.1044	1.902630	6.016644
3.63	13.1769	1.905256	6.024948
3.64	13.2496	1.907878	6.033241

TABLE B Squares and Square Roots (*Continued*)

n	n^2	$\sqrt{n}$	$\sqrt{10n}$
3.65	13.3225	1.910497	6.041523
3.66	13.3956	1.913113	6.049793
3.67	13.4689	1.915724	6.058052
3.68	13.5424	1.918333	6.066300
3.69	13.6161	1.920937	6.074537
3.70	13.6900	1.923538	6.082763
3.71	13.7641	1.926136	6.090977
3.72	13.8384	1.928730	6.099180
3.73	13.9129	1.931321	6.107373
3.74	13.9876	1.933908	6.115554
3.75	14.0625	1.936492	6.123724
3.76	14.1376	1.939072	6.131884
3.77	14.2129	1.941649	6.140033
3.78	14.2884	1.944222	6.148170
3.79	14.3641	1.946792	6.156298
3.80	14.4400	1.949359	6.164414
3.81	14.5161	1.951922	6.172520
3.82	14.5924	1.954482	6.180615
3.83	14.6689	1.957039	6.188699
3.84	14.7456	1.959592	6.196773
3.85	14.8225	1.962142	6.204837
3.86	14.8996	1.964688	6.212890
3.87	14.9769	1.967232	6.220932
3.88	15.0544	1.969772	6.228965
3.89	15.1321	1.972308	6.236986
3.90	15.2100	1.974842	6.244998
3.91	15.2881	1.977372	6.252999
3.92	15.3664	1.979899	6.260990
3.93	15.4449	1.982423	6.268971
3.94	15.5236	1.984943	6.276942
3.95	15.6025	1.987461	6.284903
3.96	15.6816	1.989975	6.292853
3.97	15.7609	1.992486	6.300794
3.98	15.8404	1.994994	6.308724
3.99	15.9201	1.997498	6.316645
4.00	16.0000	2.000000	6.324555
4.01	16.0801	2.002498	6.332456
4.02	16.1604	2.004994	6.340347
4.03	16.2409	2.007486	6.348228
4.04	16.3216	2.009975	6.356099
4.05	16.4025	2.012461	6.363961
4.06	16.4836	2.014944	6.371813
4.07	16.5649	2.017424	6.379655
4.08	16.6464	2.019901	6.387488
4.09	16.7281	2.022375	6.395311
4.10	16.8100	2.024846	6.403124
4.11	16.8921	2.027313	6.410928
4.12	16.9744	2.029778	6.418723
4.13	17.0569	2.032240	6.426508
4.14	17.1396	2.034699	6.434283
4.15	17.2225	2.037155	6.442049
4.16	17.3056	2.039608	6.449806
4.17	17.3889	2.042058	6.457554
4.18	17.4724	2.044505	6.465292
4.19	17.5561	2.046949	6.473021
4.20	17.6400	2.049390	6.480741
4.21	17.7241	2.051828	6.488451
4.22	17.8084	2.054264	6.496153
4.23	17.8929	2.056696	6.503845
4.24	17.9776	2.059126	6.511528
4.25	18.0625	2.061553	6.519202
4.26	18.1476	2.063977	6.526868
4.27	18.2329	2.066398	6.534524
4.28	18.3184	2.068816	6.542171
4.29	18.4041	2.071232	6.549809
4.30	18.4900	2.073644	6.557439
4.31	18.5761	2.076054	6.565059
4.32	18.6624	2.078461	6.572671
4.33	18.7489	2.080865	6.580274
4.34	18.8356	2.083267	6.587868
4.35	18.9225	2.085665	6.595453
4.36	19.0096	2.088061	6.603030
4.37	19.0969	2.090454	6.610598
4.38	19.1844	2.092845	6.618157
4.39	19.2721	2.095233	6.625708
4.40	19.3600	2.097618	6.633250
4.41	19.4481	2.100000	6.640783
4.42	19.5364	2.102380	6.648308
4.43	19.6249	2.104757	6.655825
4.44	19.7136	2.107131	6.663332
4.45	19.8025	2.109502	6.670832
4.46	19.8916	2.111871	6.678323
4.47	19.9809	2.114237	6.685806
4.48	20.0704	2.116601	6.693280
4.49	20.1601	2.118962	6.700746
4.50	20.2500	2.121320	6.708204
4.51	20.3401	2.123676	6.715653
4.52	20.4304	2.126029	6.723095
4.53	20.5209	2.128380	6.730527

TABLE B Squares and Square Roots (*Continued*)

n	n²	√n	√10n
4.54	20.6116	2.130728	6.737952
4.55	20.7025	2.133073	6.745369
4.56	20.7936	2.135416	6.752777
4.57	20.8849	2.137756	6.760178
4.58	20.9764	2.140093	6.767570
4.59	21.0681	2.142429	6.774954
4.60	21.1600	2.144761	6.782330
4.61	21.2521	2.147091	6.789698
4.62	21.3444	2.149419	6.797058
4.63	21.4369	2.151743	6.804410
4.64	21.5296	2.154066	6.811755
4.65	21.6225	2.156386	6.819091
4.66	21.7156	2.158703	6.826419
4.67	21.8089	2.161018	6.833740
4.68	21.9024	2.163331	6.841053
4.69	21.9961	2.165641	6.848357
4.70	22.0900	2.167948	6.855655
4.71	22.1841	2.170253	6.862944
4.72	22.2784	2.172556	6.870226
4.73	22.3729	2.174856	6.877500
4.74	22.4676	2.177154	6.884766
4.75	22.5625	2.179449	6.892024
4.76	22.6576	2.181742	6.899275
4.77	22.7529	2.184033	6.906519
4.78	22.8484	2.186321	6.913754
4.79	22.9441	2.188607	6.920983
4.80	23.0400	2.190890	6.928203
4.81	23.1361	2.193171	6.935416
4.82	23.2324	2.195450	6.942622
4.83	23.3289	2.197726	6.949820
4.84	23.4256	2.200000	6.957011
4.85	23.5225	2.202272	6.964194
4.86	23.6196	2.204541	6.971370
4.87	23.7169	2.206808	6.978539
4.88	23.8144	2.209072	6.985700
4.89	23.9121	2.211334	6.992853
4.90	24.0100	2.213594	7.000000
4.91	24.1081	2.215852	7.007139
4.92	24.2064	2.218107	7.014271
4.93	24.3049	2.220360	7.021396
4.94	24.4036	2.222611	7.028513
4.95	24.5025	2.224860	7.035624
4.96	24.6016	2.227106	7.042727
4.97	24.7009	2.229350	7.049823
4.98	24.8004	2.231591	7.056912
4.99	24.9001	2.233831	7.063993
5.00	25.0000	2.236068	7.071068
5.01	25.1001	2.238303	7.078135
5.02	25.2004	2.240536	7.085196
5.03	25.3009	2.242766	7.092249
5.04	25.4016	2.244994	7.099296
5.05	25.5025	2.247221	7.106335
5.06	25.6036	2.249444	7.113368
5.07	25.7049	2.251666	7.120393
5.08	25.8064	2.253886	7.127412
5.09	25.9081	2.256103	7.134424
5.10	26.0100	2.258318	7.141428
5.11	26.1121	2.260531	7.148426
5.12	26.2144	2.262742	7.155418
5.13	26.3169	2.264950	7.162402
5.14	26.4196	2.267157	7.169379
5.15	26.5225	2.269361	7.176350
5.16	26.6256	2.271563	7.183314
5.17	26.7289	2.273763	7.190271
5.18	26.8324	2.275961	7.197222
5.19	26.9361	2.278157	7.204165
5.20	27.0400	2.280351	7.211103
5.21	27.1441	2.282542	7.218033
5.22	27.2484	2.284732	7.224957
5.23	27.3529	2.286919	7.231874
5.24	27.4576	2.289105	7.238784
5.25	27.5625	2.291288	7.245688
5.26	27.6676	2.293469	7.252586
5.27	27.7729	2.295648	7.259477
5.28	27.8784	2.297825	7.266361
5.29	27.9841	2.300000	7.273239
5.30	28.0900	2.302173	7.280110
5.31	28.1961	2.304344	7.286975
5.32	28.3024	2.306513	7.293833
5.33	28.4089	2.308679	7.300685
5.34	28.5156	2.310844	7.307530
5.35	28.6225	2.313007	7.314369
5.36	28.7296	2.315167	7.321202
5.37	28.8369	2.317326	7.328028
5.38	28.9444	2.319483	7.334848
5.39	29.0521	2.321637	7.341662
5.40	29.1600	2.323790	7.348469
5.41	29.2681	2.325941	7.355270
5.42	29.3764	2.328089	7.362065

TABLE B Squares and Square Roots (*Continued*)

n	n^2	$\sqrt{n}$	$\sqrt{10n}$
5.43	29.4849	2.330236	7.368853
5.44	29.5936	2.332381	7.375636
5.45	29.7025	2.334524	7.382412
5.46	29.8116	2.336664	7.389181
5.47	29.9209	2.338803	7.395945
5.48	30.0304	2.340940	7.402702
5.49	30.1401	2.343075	7.409453
5.50	30.2500	2.345208	7.416198
5.51	30.3601	2.347339	7.422937
5.52	30.4704	2.349468	7.429670
5.53	30.5809	2.351595	7.436397
5.54	30.6916	2.353720	7.443118
5.55	30.8025	2.355844	7.449832
5.56	30.9136	2.357965	7.456541
5.57	31.0249	2.360085	7.463243
5.58	31.1364	2.362202	7.469940
5.59	31.2481	2.364318	7.476630
5.60	31.3600	2.366432	7.483315
5.61	31.4721	2.368544	7.489993
5.62	31.5844	2.370654	7.496666
5.63	31.6969	2.372762	7.503333
5.64	31.8096	2.374868	7.509993
5.65	31.9225	2.376973	7.516648
5.66	32.0356	2.379075	7.523297
5.67	32.1489	2.381176	7.529940
5.68	32.2624	2.383275	7.536577
5.69	32.3761	2.385372	7.543209
5.70	32.4900	2.387467	7.549834
5.71	32.6041	2.389561	7.556454
5.72	32.7184	2.391652	7.563068
5.73	32.8329	2.393742	7.569676
5.74	32.9476	2.395830	7.576279
5.75	33.0625	2.397916	7.582875
5.76	33.1776	2.400000	7.589466
5.77	33.2929	2.402082	7.596052
5.78	33.4084	2.404163	7.602631
5.79	33.5241	2.406242	7.609205
5.80	33.6400	2.408319	7.615773
5.81	33.7561	2.410394	7.622336
5.82	33.8724	2.412468	7.628892
5.83	33.9889	2.414539	7.635444
5.84	34.1056	2.416609	7.641989
5.85	34.2225	2.418677	7.648529
5.86	34.3396	2.420744	7.655064
5.87	34.4569	2.422808	7.661593
5.88	34.5744	2.424871	7.668116
5.89	34.6921	2.426932	7.674634
5.90	34.8100	2.428992	7.681146
5.91	34.9281	2.431049	7.687652
5.92	35.0464	2.433105	7.694154
5.93	35.1649	2.435159	7.700649
5.94	35.2836	2.437212	7.707140
5.95	35.4025	2.439262	7.713624
5.96	35.5216	2.441311	7.720104
5.97	35.6409	2.443358	7.726578
5.98	35.7604	2.445404	7.733046
5.99	35.8801	2.447448	7.739509
6.00	36.0000	2.449490	7.745967
6.01	36.1201	2.451530	7.752419
6.02	36.2404	2.453569	7.758866
6.03	36.3609	2.455606	7.765307
6.04	36.4816	2.457641	7.771744
6.05	36.6025	2.459675	7.778175
6.06	36.7236	2.461707	7.784600
6.07	36.8449	2.463737	7.791020
6.08	36.9664	2.465766	7.797435
6.09	37.0881	2.467793	7.803845
6.10	37.2100	2.469818	7.810250
6.11	37.3321	2.471841	7.816649
6.12	37.4544	2.473863	7.823043
6.13	37.5769	2.475884	7.829432
6.14	37.6996	2.477902	7.835815
6.15	37.8225	2.479919	7.842194
6.16	37.9456	2.481935	7.848567
6.17	38.0689	2.483948	7.854935
6.18	38.1924	2.485961	7.861298
6.19	38.3161	2.487971	7.867655
6.20	38.4400	2.489980	7.874008
6.21	38.5641	2.491987	7.880355
6.22	38.6884	2.493993	7.886698
6.23	38.8129	2.495997	7.893035
6.24	38.9376	2.497999	7.899367
6.25	39.0625	2.500000	7.905694
6.26	39.1876	2.501999	7.912016
6.27	39.3129	2.503997	7.918333
6.28	39.4384	2.505993	7.924645
6.29	39.5641	2.507987	7.930952
6.30	39.6900	2.509980	7.937254
6.31	39.8161	2.511971	7.943551

TABLE B Squares and Square Roots (*Continued*)

n	n²	√n	√10n
6.32	39.9424	2.513961	7.949843
6.33	40.0689	2.515949	7.956130
6.34	40.1956	2.517936	7.962412
6.35	40.3225	2.519921	7.968689
6.36	40.4496	2.521904	7.974961
6.37	40.5769	2.523886	7.981228
6.38	40.7044	2.525866	7.987490
6.39	40.8321	2.527845	7.993748
6.40	40.9600	2.529822	8.000000
6.41	41.0881	2.531798	8.006248
6.42	41.2164	2.533772	8.012490
6.43	41.3449	2.535744	8.018728
6.44	41.4736	2.537716	8.024961
6.45	41.6025	2.539685	8.031189
6.46	41.7316	2.541653	8.037413
6.47	41.8609	2.543619	8.043631
6.48	41.9904	2.545584	8.049845
6.49	42.1201	2.547548	8.056054
6.50	42.2500	2.549510	8.062258
6.51	42.3801	2.551470	8.068457
6.52	42.5104	2.553429	8.074652
6.53	42.6409	2.555386	8.080842
6.54	42.7716	2.557342	8.087027
6.55	42.9025	2.559297	8.093207
6.56	43.0336	2.561250	8.099383
6.57	43.1649	2.563201	8.105554
6.58	43.2964	2.565151	8.111720
6.59	43.4281	2.567100	8.117881
6.60	43.5600	2.569047	8.124038
6.61	43.6921	2.570992	8.130191
6.62	43.8244	2.572936	8.136338
6.63	43.9569	2.574879	8.142481
6.64	44.0896	2.576820	8.148620
6.65	44.2225	2.578759	8.154753
6.66	44.3556	2.580698	8.160882
6.67	44.4889	2.582634	8.167007
6.68	44.6224	2.584570	8.173127
6.69	44.7561	2.586503	8.179242
6.70	44.8900	2.588436	8.185353
6.71	45.0241	2.590367	8.191459
6.72	45.1584	2.592296	8.197561
6.73	45.2929	2.594224	8.203658
6.74	45.4276	2.596151	8.209750
6.75	45.5625	2.598076	8.215838
6.76	45.6976	2.600000	8.221922
6.77	45.8329	2.601922	8.228001
6.78	45.9684	2.603843	8.234076
6.79	46.1041	2.605763	8.240146
6.80	46.2400	2.607681	8.246211
6.81	46.3761	2.609598	8.252272
6.82	46.5124	2.611513	8.258329
6.83	46.6489	2.613427	8.264381
6.84	46.7856	2.615339	8.270429
6.85	46.9225	2.617250	8.276473
6.86	47.0596	2.619160	8.282512
6.87	47.1969	2.621068	8.288546
6.88	47.3344	2.622975	8.294577
6.89	47.4721	2.624881	8.300602
6.90	47.6100	2.626785	8.306624
6.91	47.7481	2.628688	8.312641
6.92	47.8864	2.630589	8.318654
6.93	48.0249	2.632489	8.324662
6.94	48.1636	2.634388	8.330666
6.95	48.3025	2.636285	8.336666
6.96	48.4416	2.638181	8.342661
6.97	48.5809	2.640076	8.348653
6.98	48.7204	2.641969	8.354639
6.99	48.8601	2.643861	8.360622
7.00	49.0000	2.645751	8.366600
7.01	49.1401	2.647640	8.372574
7.02	49.2804	2.649528	8.378544
7.03	49.4209	2.651415	8.384510
7.04	49.5616	2.653300	8.390471
7.05	49.7025	2.655184	8.396428
7.06	49.8436	2.657066	8.402381
7.07	49.9849	2.658947	8.408329
7.08	50.1264	2.660827	8.414274
7.09	50.2681	2.662705	8.420214
7.10	50.4100	2.664583	8.426150
7.11	50.5521	2.666458	8.432082
7.12	50.6944	2.668333	8.438009
7.13	50.8369	2.670206	8.443933
7.14	50.9796	2.672078	8.449852
7.15	51.1225	2.673948	8.455767
7.16	51.2656	2.675818	8.461678
7.17	51.4089	2.677686	8.467585
7.18	51.5524	2.679552	8.473488
7.19	51.6961	2.681418	8.479387
7.20	51.8400	2.683282	8.485281

TABLE B Squares and Square Roots (*Continued*)

n	n^2	$\sqrt{n}$	$\sqrt{10n}$
7.21	51.9841	2.685144	8.491172
7.22	52.1284	2.687006	8.497058
7.23	52.2729	2.688866	8.502941
7.24	52.4176	2.690725	8.508819
7.25	52.5625	2.692582	8.514693
7.26	52.7076	2.694439	8.520563
7.27	52.8529	2.696294	8.526429
7.28	52.9984	2.698148	8.532292
7.29	53.1441	2.700000	8.538150
7.30	53.2900	2.701851	8.544004
7.31	53.4361	2.703701	8.549854
7.32	53.5824	2.705550	8.555700
7.33	53.7289	2.707397	8.561542
7.34	53.8756	2.709243	8.567380
7.35	54.0225	2.711088	8.573214
7.36	54.1696	2.712932	8.579044
7.37	54.3169	2.714774	8.584870
7.38	54.4644	2.716616	8.590693
7.39	54.6121	2.718455	8.596511
7.40	54.7600	2.720294	8.602325
7.41	54.9081	2.722132	8.608136
7.42	55.0564	2.723968	8.613942
7.43	55.2049	2.725803	8.619745
7.44	55.3536	2.727636	8.625543
7.45	55.5025	2.729469	8.631338
7.46	55.6516	2.731300	8.637129
7.47	55.8009	2.733130	8.642916
7.48	55.9504	2.734959	8.648699
7.49	56.1001	2.736786	8.654479
7.50	56.2500	2.738613	8.660254
7.51	56.4001	2.740438	8.666026
7.52	56.5504	2.742262	8.671793
7.53	56.7009	2.744085	8.677557
7.54	56.8516	2.745906	8.683317
7.55	57.0025	2.747726	8.689074
7.56	57.1536	2.749545	8.694826
7.57	57.3049	2.751363	8.700575
7.58	57.4564	2.753180	8.706320
7.59	57.6081	2.754995	8.712061
7.60	57.7600	2.756810	8.717798
7.61	57.9121	2.758623	8.723531
7.62	58.0644	2.760435	8.729261
7.63	58.2169	2.762245	8.734987
7.64	58.3696	2.764055	8.740709
7.65	58.5225	2.765863	8.746428
7.66	58.6756	2.767671	8.752143
7.67	58.8289	2.769476	8.757854
7.68	58.9824	2.771281	8.763561
7.69	59.1361	2.773085	8.769265
7.70	59.2900	2.774887	8.774964
7.71	59.4441	2.776689	8.780661
7.72	59.5984	2.778489	8.786353
7.73	59.7529	2.780288	8.792042
7.74	59.9076	2.782086	8.797727
7.75	60.0625	2.783882	8.803408
7.76	60.2176	2.785678	8.809086
7.77	60.3729	2.787472	8.814760
7.78	60.5284	2.789265	8.820431
7.79	60.6841	2.791057	8.826098
7.80	60.8400	2.792848	8.831761
7.81	60.9961	2.794638	8.837420
7.82	61.1524	2.796426	8.843076
7.83	61.3089	2.798214	8.848729
7.84	61.4656	2.800000	8.854377
7.85	61.6225	2.801785	8.860023
7.86	61.7796	2.803569	8.865664
7.87	61.9369	2.805352	8.871302
7.88	62.0944	2.807134	8.876936
7.89	62.2521	2.808914	8.882567
7.90	62.4100	2.810694	8.888194
7.91	62.5681	2.812472	8.893818
7.92	62.7264	2.814249	8.899438
7.93	62.8849	2.816026	8.905055
7.94	63.0436	2.817801	8.910668
7.95	63.2025	2.819574	8.916277
7.96	63.3616	2.821347	8.921883
7.97	63.5209	2.823119	8.927486
7.98	63.6804	2.824889	8.933085
7.99	63.8401	2.826659	8.938680
8.00	64.0000	2.828427	8.944272
8.01	64.1601	2.830194	8.949860
8.02	64.3204	2.831960	8.955445
8.03	64.4809	2.833725	8.961027
8.04	64.6416	2.835489	8.966605
8.05	64.8025	2.837252	8.972179
8.06	64.9636	2.839014	8.977750
8.07	65.1249	2.840775	8.983318
8.08	65.2864	2.842534	8.988882
8.09	65.4481	2.844293	8.994443

TABLE B Squares and Square Roots (*Continued*)

n	n²	√n	√10n
8.10	65.6100	2.846050	9.000000
8.11	65.7721	2.847806	9.005554
8.12	65.9344	2.849561	9.011104
8.13	66.0969	2.851315	9.016651
8.14	66.2596	2.853069	9.022195
8.15	66.4225	2.854820	9.027735
8.16	66.5856	2.856571	9.033272
8.17	66.7489	2.858321	9.038805
8.18	66.9124	2.860070	9.044335
8.19	67.0761	2.861818	9.049862
8.20	67.2400	2.863564	9.055385
8.21	67.4041	2.865310	9.060905
8.22	67.5684	2.867054	9.066422
8.23	67.7329	2.868798	9.071935
8.24	67.8976	2.870540	9.077445
8.25	68.0625	2.872281	9.082951
8.26	68.2276	2.874022	9.088454
8.27	68.3929	2.875761	9.093954
8.28	68.5584	2.877499	9.099451
8.29	68.7241	2.879236	9.104944
8.30	68.8900	2.880972	9.110434
8.31	69.0561	2.882707	9.115920
8.32	69.2224	2.884441	9.121403
8.33	69.3889	2.886174	9.126883
8.34	69.5556	2.887906	9.132360
8.35	69.7225	2.889637	9.137833
8.36	69.8896	2.891366	9.143304
8.37	70.0569	2.893095	9.148770
8.38	70.2244	2.894823	9.154234
8.39	70.3921	2.896550	9.159694
8.40	70.5600	2.898275	9.165151
8.41	70.7281	2.900000	9.170605
8.42	70.8964	2.901724	9.176056
8.43	71.0649	2.903446	9.181503
8.44	71.2336	2.905168	9.186947
8.45	71.4025	2.906888	9.192388
8.46	71.5716	2.908608	9.197826
8.47	71.7409	2.910326	9.203260
8.48	71.9104	2.912044	9.208692
8.49	72.0801	2.913760	9.214120
8.50	72.2500	2.915476	9.219544
8.51	72.4201	2.917190	9.224966
8.52	72.5904	2.918904	9.230385
8.53	72.7609	2.920616	9.235800
8.54	72.9316	2.922328	9.241212
8.55	73.1025	2.924038	9.246621
8.56	73.2736	2.925748	9.252027
8.57	73.4449	2.927456	9.257429
8.58	73.6164	2.929164	9.262829
8.59	73.7881	2.930870	9.268225
8.60	73.9600	2.932576	9.273618
8.61	74.1321	2.934280	9.279009
8.62	74.3044	2.935984	9.284396
8.63	74.4769	2.937686	9.289779
8.64	74.6496	2.939388	9.295160
8.65	74.8225	2.941088	9.300538
8.66	74.9956	2.942788	9.305912
8.67	75.1689	2.944486	9.311283
8.68	75.3424	2.946184	9.316652
8.69	75.5161	2.947881	9.322017
8.70	75.6900	2.949576	9.327379
8.71	75.8641	2.951271	9.332738
8.72	76.0384	2.952965	9.338094
8.73	76.2129	2.954657	9.343447
8.74	76.3876	2.956349	9.348797
8.75	76.5625	2.958040	9.354143
8.76	76.7376	2.959730	9.359487
8.77	76.9129	2.961419	9.364828
8.78	77.0884	2.963106	9.370165
8.79	77.2641	2.964793	9.375500
8.80	77.4400	2.966479	9.380832
8.81	77.6161	2.968164	9.386160
8.82	77.7924	2.969848	9.391486
8.83	77.9689	2.971532	9.396808
8.84	78.1456	2.973214	9.402127
8.85	78.3225	2.974895	9.407444
8.86	78.4996	2.976575	9.412757
8.87	78.6769	2.978255	9.418068
8.88	78.8544	2.979933	9.423375
8.89	79.0321	2.981610	9.428680
8.90	79.2100	2.983287	9.433981
8.91	79.3881	2.984962	9.439280
8.92	79.5664	2.986637	9.444575
8.93	79.7449	2.988311	9.449868
8.94	79.9236	2.989983	9.455157
8.95	80.1025	2.991655	9.460444
8.96	80.2816	2.993326	9.465728
8.97	80.4609	2.994996	9.471008
8.98	80.6404	2.996665	9.476286

TABLE B Squares and Square Roots (*Continued*)

n	n²	√n	√10n
8.99	80.8201	2.998333	9.481561
9.00	81.0000	3.000000	9.486833
9.01	81.1801	3.001666	9.492102
9.02	81.3604	3.003331	9.497368
9.03	81.5409	3.004996	9.502631
9.04	81.7216	3.006659	9.507891
9.05	81.9025	3.008322	9.513149
9.06	82.0836	3.009983	9.518403
9.07	82.2649	3.011644	9.523655
9.08	82.4464	3.013304	9.528903
9.09	82.6281	3.014963	9.534149
9.10	82.8100	3.016621	9.539392
9.11	82.9921	3.018278	9.544632
9.12	83.1744	3.019934	9.549869
9.13	83.3569	3.021589	9.555103
9.14	83.5396	3.023243	9.560335
9.15	83.7225	3.024897	9.565563
9.16	83.9056	3.026549	9.570789
9.17	84.0889	3.028201	9.576012
9.18	84.2724	3.029851	9.581232
9.19	84.4561	3.031501	9.586449
9.20	84.6400	3.033150	9.591663
9.21	84.8241	3.034798	9.596874
9.22	85.0084	3.036445	9.602083
9.23	85.1929	3.038092	9.607289
9.24	85.3776	3.039737	9.612492
9.25	85.5625	3.041381	9.617692
9.26	85.7476	3.043025	9.622889
9.27	85.9329	3.044667	9.628084
9.28	86.1184	3.046309	9.633276
9.29	86.3041	3.047950	9.638465
9.30	86.4900	3.049590	9.643651
9.31	86.6761	3.051229	9.648834
9.32	86.8624	3.052868	9.654015
9.33	87.0489	3.054505	9.659193
9.34	87.2356	3.056141	9.664368
9.35	87.4225	3.057777	9.669540
9.36	87.6096	3.059412	9.674709
9.37	87.7969	3.061046	9.679876
9.38	87.9844	3.062679	9.685040
9.39	88.1721	3.064311	9.690201
9.40	88.3600	3.065942	9.695360
9.41	88.5481	3.067572	9.700515
9.42	88.7364	3.069202	9.705668
9.43	88.9249	3.070831	9.710819
9.44	89.1136	3.072458	9.715966
9.45	89.3025	3.074085	9.721111
9.46	89.4916	3.075711	9.726253
9.47	89.6809	3.077337	9.731393
9.48	89.8704	3.078961	9.736529
9.49	90.0601	3.080584	9.741663
9.50	90.2500	3.082207	9.746794
9.51	90.4401	3.083829	9.751923
9.52	90.6304	3.085450	9.757049
9.53	90.8209	3.087070	9.762172
9.54	91.0116	3.088689	9.767292
9.55	91.2025	3.090307	9.772410
9.56	91.3936	3.091925	9.777525
9.57	91.5849	3.093542	9.782638
9.58	91.7764	3.095158	9.787747
9.59	91.9681	3.096773	9.792855
9.60	92.1600	3.098387	9.797959
9.61	92.3521	3.100000	9.803061
9.62	92.5444	3.101612	9.808160
9.63	92.7369	3.103224	9.813256
9.64	92.9296	3.104835	9.818350
9.65	93.1225	3.106445	9.823441
9.66	93.3156	3.108054	9.828530
9.67	93.5089	3.109662	9.833616
9.68	93.7024	3.111270	9.838699
9.69	93.8961	3.112876	9.843780
9.70	94.0900	3.114482	9.848858
9.71	94.2841	3.116087	9.853933
9.72	94.4784	3.117691	9.859006
9.73	94.6729	3.119295	9.864076
9.74	94.8676	3.120897	9.869144
9.75	95.0625	3.122499	9.874209
9.76	95.2576	3.124100	9.879271
9.77	95.4529	3.125700	9.884331
9.78	95.6484	3.127299	9.889388
9.79	95.8441	3.128898	9.894443
9.80	96.0400	3.130495	9.899495
9.81	96.2361	3.132092	9.904544
9.82	96.4324	3.133688	9.909591
9.83	96.6289	3.135283	9.914636
9.84	96.8256	3.136877	9.919677
9.85	97.0225	3.138471	9.924717
9.86	97.2196	3.140064	9.929753
9.87	97.4169	3.141656	9.934787

TABLE B Squares and Square Roots (*Continued*)

n	n^2	$\sqrt{n}$	$\sqrt{10n}$
9.88	97.6144	3.143247	9.939819
9.89	97.8121	3.144837	9.944848
9.90	98.0100	3.146427	9.949874
9.91	98.2081	3.148015	9.954898
9.92	98.4064	3.149603	9.959920
9.93	98.6049	3.151190	9.964939
9.94	98.8036	3.152777	9.969955
9.95	99.0025	3.154362	9.974969
9.96	99.2016	3.155947	9.979980
9.97	99.4009	3.157531	9.984989
9.98	99.6004	3.159114	9.989995
9.99	99.8001	3.160696	9.994999

TABLE C Binomial Distribution: Individual Terms $p(r) = c\binom{n}{r} p^r q^{n-r}$

N = 1

R	P	.01	.02	.03	.04	.05	.06	.07	.08	.09	.10		
0		.9900	.9800	.9700	.9600	.9500	.9400	.9300	.9200	.9100	.9000		1
1		.0100	.0200	.0300	.0400	.0500	.0600	.0700	.0800	.0900	.1000		0
		.99	.98	.97	.96	.95	.94	.93	.92	.91	.90	P	R

R	P	.11	.12	.13	.14	.15	.16	.17	.18	.19	.20		
0		.8900	.8800	.8700	.8600	.8500	.8400	.8300	.8200	.8100	.8000		1
1		.1100	.1200	.1300	.1400	.1500	.1600	.1700	.1800	.1900	.2000		0
		.89	.88	.87	.86	.85	.84	.83	.82	.81	.80	P	R

R	P	.21	.22	.23	.24	.25	.26	.27	.28	.29	.30		
0		.7900	.7800	.7700	.7600	.7500	.7400	.7300	.7200	.7100	.7000		1
1		.2100	.2200	.2300	.2400	.2500	.2600	.2700	.2800	.2900	.3000		0
		.79	.78	.77	.76	.75	.74	.73	.72	.71	.70	P	R

R	P	.31	.32	.33	.34	.35	.36	.37	.38	.39	.40		
0		.6900	.6800	.6700	.6600	.6500	.6400	.6300	.6200	.6100	.6000		1
1		.3100	.3200	.3300	.3400	.3500	.3600	.3700	.3800	.3900	.4000		0
		.69	.68	.67	.66	.65	.64	.63	.62	.61	.60	P	R

R	P	.41	.42	.43	.44	.45	.46	.47	.48	.49	.50		
0		.5900	.5800	.5700	.5600	.5500	.5400	.5300	.5200	.5100	.5000		1
1		.4100	.4200	.4300	.4400	.4500	.4600	.4700	.4800	.4900	.5000		0
		.59	.58	.57	.56	.55	.54	.53	.52	.51	.50	P	R

N = 2

R	P	.01	.02	.03	.04	.05	.06	.07	.08	.09	.10		
0		.9801	.9604	.9409	.9216	.9025	.8836	.8649	.8464	.8281	.8100		2
1		.0198	.0392	.0582	.0768	.0950	.1128	.1302	.1472	.1638	.1800		1
2		.0001	.0004	.0009	.0016	.0025	.0036	.0049	.0064	.0081	.0100		0
		.99	.98	.97	.96	.95	.94	.93	.92	.91	.90	P	R

R	P	.11	.12	.13	.14	.15	.16	.17	.18	.19	.20		
0		.7921	.7744	.7569	.7396	.7225	.7056	.6889	.6724	.6561	.6400		2
1		.1958	.2112	.2262	.2408	.2550	.2688	.2822	.2952	.3078	.3200		1
2		.0121	.0144	.0169	.0196	.0225	.0256	.0289	.0324	.0361	.0400		0
		.89	.88	.87	.86	.85	.84	.83	.82	.81	.80	P	R

R	P	.21	.22	.23	.24	.25	.26	.27	.28	.29	.30		
0		.6241	.6084	.5929	.5776	.5625	.5476	.5329	.5184	.5041	.4900		2
1		.3318	.3432	.3542	.3648	.3750	.3848	.3942	.4032	.4118	.4200		1
2		.0441	.0484	.0529	.0576	.0625	.0676	.0729	.0784	.0841	.0900		0
		.79	.78	.77	.76	.75	.74	.73	.72	.71	.70	P	R

R	P	.31	.32	.33	.34	.35	.36	.37	.38	.39	.40		
0		.4761	.4624	.4489	.4356	.4225	.4096	.3969	.3844	.3721	.3600		2
1		.4278	.4352	.4422	.4488	.4550	.4608	.4662	.4712	.4758	.4800		1
2		.0961	.1024	.1089	.1156	.1225	.1296	.1369	.1444	.1521	.1600		0
		.69	.68	.67	.66	.65	.64	.63	.62	.61	.60	P	R

R	P	.41	.42	.43	.44	.45	.46	.47	.48	.49	.50		
0		.3481	.3364	.3249	.3136	.3025	.2916	.2809	.2704	.2601	.2500		2
1		.4838	.4872	.4902	.4928	.4950	.4968	.4982	.4992	.4998	.5000		1
2		.1681	.1764	.1849	.1936	.2025	.2116	.2209	.2304	.2401	.2500		0
		.59	.58	.57	.56	.55	.54	.53	.52	.51	.50	P	R

N = 3

R	P	.01	.02	.03	.04	.05	.06	.07	.08	.09	.10		
0		.9703	.9412	.9127	.8847	.8574	.8306	.8044	.7787	.7536	.7290		3
1		.0294	.0576	.0847	.1106	.1354	.1590	.1816	.2031	.2236	.2430		2
2		.0003	.0012	.0026	.0046	.0071	.0102	.0137	.0177	.0221	.0270		1
3		.0000	.0000	.0000	.0001	.0001	.0002	.0003	.0005	.0007	.0010		0
		.99	.98	.97	.96	.95	.94	.93	.92	.91	.90	P	R

R	P	.11	.12	.13	.14	.15	.16	.17	.18	.19	.20		
0		.7050	.6815	.6585	.6361	.6141	.5927	.5718	.5514	.5314	.5120		3
1		.2614	.2788	.2952	.3106	.3251	.3387	.3513	.3631	.3740	.3840		2
2		.0323	.0380	.0441	.0506	.0574	.0645	.0720	.0797	.0877	.0960		1
3		.0013	.0017	.0022	.0027	.0034	.0041	.0049	.0058	.0069	.0080		0

TABLE C Binomial Distribution: Individual Terms $p(r) = c \binom{n}{r} p^r q^{n-r}$ (*Continued*)

		.89	.98	.97	.86	.85	.84	.83	.82	.81	.80	P	R
R	P	.21	.22	.23	.24	.25	.26	.27	.28	.29	.30		
0		.4930	.4746	.4565	.4390	.4219	.4052	.3890	.3732	.3579	.3430		3
1		.3932	.4015	.4091	.4159	.4219	.4271	.4316	.4355	.4386	.4410		2
2		.1045	.1133	.1222	.1313	.1406	.1501	.1597	.1693	.1791	.1890		1
3		.0093	.0106	.0122	.0138	.0156	.0176	.0197	.0220	.0244	.0270		0
		.79	.78	.77	.76	.75	.74	.73	.72	.71	.70	P	R

R	P	.31	.32	.33	.34	.35	.36	.37	.38	.39	.40		
0		.3285	.3144	.3008	.2875	.2746	.2621	.2500	.2383	.2270	.2160		3
1		.4428	.4439	.4444	.4443	.4436	.4424	.4406	.4382	.4354	.4320		2
2		.1989	.2089	.2189	.2289	.2389	.2488	.2587	.2686	.2783	.2880		1
3		.0298	.0328	.0359	.0393	.0429	.0467	.0507	.0549	.0593	.0640		0
		.69	.68	.67	.66	.65	.64	.63	.62	.61	.60	P	R

P	P	.41	.42	.43	.44	.45	.46	.47	.48	.49	.50		
0		.2054	.1951	.1852	.1756	.1664	.1575	.1489	.1406	.1327	.1250		3
1		.4282	.4239	.4191	.4140	.4084	.4024	.3961	.3894	.3823	.3750		2
2		.2975	.3069	.3162	.3252	.3341	.3428	.3512	.3594	.3674	.3750		1
3		.0689	.0741	.0795	.0852	.0911	.0973	.1038	.1106	.1176	.1250		0
		.59	.58	.57	.56	.55	.54	.53	.52	.51	.50	P	R

N = 4

R	P	.01	.02	.03	.04	.05	.06	.07	.08	.09	.10		
0		.9606	.9224	.8853	.8493	.8145	.7807	.7481	.7164	.6857	.6561		4
1		.0388	.0753	.1095	.1416	.1715	.1993	.2252	.2492	.2713	.2916		3
2		.0006	.0023	.0051	.0088	.0135	.0191	.0254	.0325	.0402	.0486		2
3		.0000	.0000	.0001	.0002	.0005	.0008	.0013	.0019	.0027	.0036		1
4		.0000	.0000	.0000	.0000	.0000	.0000	.0000	.0000	.0001	.0001		0
		.99	.98	.97	.96	.95	.94	.93	.92	.91	.90	P	R

P	P	.11	.12	.13	.14	.15	.16	.17	.18	.19	.20		
0		.6274	.5997	.5729	.5470	.5220	.4979	.4746	.4521	.4305	.4096		4
1		.3102	.3271	.3424	.3562	.3685	.3793	.3888	.3970	.4039	.4096		3
2		.0575	.0669	.0767	.0870	.0975	.1084	.1195	.1307	.1421	.1536		2
3		.0047	.0051	.0076	.0094	.0115	.0138	.0163	.0191	.0222	.0256		1
4		.0001	.0002	.0003	.0004	.0005	.0007	.0008	.0010	.0013	.0016		0
		.89	.98	.87	.86	.85	.84	.83	.82	.81	.80	P	R

P	P	.21	.22	.23	.24	.25	.26	.27	.28	.29	.30		
0		.3895	.3702	.3515	.3336	.3164	.2999	.2840	.2687	.2541	.2401		4
1		.4142	.4176	.4200	.4214	.4219	.4214	.4201	.4180	.4152	.4116		3
2		.1651	.1767	.1882	.1996	.2109	.2221	.2331	.2439	.2544	.2646		2
3		.0293	.0332	.0375	.0420	.0469	.0520	.0575	.0632	.0693	.0756		1
4		.0019	.0023	.0028	.0033	.0039	.0046	.0053	.0061	.0071	.0081		0
		.79	.78	.77	.76	.75	.74	.73	.72	.71	.70	P	R

R	P	.31	.32	.33	.34	.35	.36	.37	.38	.39	.40		
0		.2267	.2138	.2015	.1897	.1785	.1678	.1575	.1478	.1385	.1296		4
1		.4074	.4025	.3970	.3910	.3845	.3775	.3701	.3623	.3541	.3456		3
2		.2745	.2841	.2933	.3021	.3105	.3185	.3260	.3330	.3396	.3456		2
3		.0822	.0991	.0963	.1038	.1115	.1194	.1276	.1361	.1447	.1536		1
4		.0092	.0105	.0119	.0134	.0150	.0168	.0187	.0209	.0231	.0256		0
		.69	.68	.67	.66	.65	.64	.63	.62	.61	.60	P	R

P	P	.41	.42	.43	.44	.45	.46	.47	.48	.49	.50		
0		.1212	.1132	.1056	.0983	.0915	.0850	.0789	.0731	.0677	.0625		4
1		.3368	.3278	.3185	.3091	.2995	.2897	.2799	.2700	.2600	.2500		3
2		.3511	.3560	.3604	.3643	.3675	.3702	.3723	.3738	.3747	.3750		2
3		.1627	.1719	.1813	.1908	.2005	.2102	.2201	.2300	.2400	.2500		1
4		.0283	.0311	.0342	.0375	.0410	.0448	.0488	.0531	.0576	.0625		0
		.59	.58	.57	.56	.55	.54	.53	.52	.51	.50	P	R

N = 5

P	P	.01	.02	.03	.04	.05	.06	.07	.08	.09	.10		
0		.9510	.9039	.8587	.8154	.7738	.7339	.6957	.6591	.6240	.5905		5
1		.0480	.0922	.1328	.1699	.2036	.2342	.2618	.2866	.3086	.3280		4
2		.0010	.0038	.0082	.0142	.0214	.0299	.0394	.0498	.0610	.0729		3
3		.0000	.0001	.0003	.0006	.0011	.0019	.0030	.0043	.0060	.0081		2
4		.0000	.0000	.0000	.0000	.0000	.0001	.0001	.0002	.0003	.0004		1
		.99	.98	.97	.96	.95	.94	.93	.92	.91	.90	P	R

TABLE C Binomial Distribution: Individual Terms $p(r) = c\binom{n}{r} p^r q^{n-r}$ (*Continued*)

R	P	.11	.12	.13	.14	.15	.16	.17	.18	.19	.20		
0		.5584	.5277	.4984	.4704	.4437	.4182	.3939	.3707	.3487	.3277		5
1		.3451	.3598	.3724	.3829	.3915	.3983	.4034	.4069	.4089	.4096		4
2		.0853	.0981	.1113	.1247	.1382	.1517	.1652	.1786	.1919	.2048		3
3		.0105	.0134	.0166	.0203	.0244	.0289	.0338	.0392	.0450	.0512		2
4		.0007	.0009	.0012	.0017	.0022	.0028	.0035	.0043	.0053	.0064		1
5		.0000	.0000	.0000	.0001	.0001	.0001	.0001	.0002	.0002	.0003		0
		.89	.88	.87	.86	.85	.84	.83	.82	.81	.80	P	R

R	P	.21	.22	.23	.24	.25	.26	.27	.28	.29	.30		
0		.3077	.2887	.2707	.2536	.2373	.2219	.2073	.1935	.1804	.1681		5
1		.4090	.4072	.4043	.4003	.3955	.3898	.3834	.3762	.3685	.3601		4
2		.2174	.2297	.2415	.2529	.2637	.2739	.2836	.2926	.3010	.3087		3
3		.0578	.0648	.0721	.0798	.0879	.0962	.1049	.1138	.1229	.1323		2
4		.0077	.0091	.0108	.0126	.0146	.0169	.0194	.0221	.0251	.0283		1
5		.0004	.0005	.0006	.0008	.0010	.0012	.0014	.0017	.0021	.0024		0
		.79	.78	.77	.76	.75	.74	.73	.72	.71	.70	P	R

R	P	.31	.32	.33	.34	.35	.36	.37	.38	.39	.40		
0		.1564	.1454	.1350	.1252	.1160	.1074	.0992	.0916	.0845	.0778		5
1		.3513	.3421	.3325	.3226	.3124	.3020	.2914	.2808	.2700	.2592		4
2		.3157	.3220	.3275	.3323	.3364	.3397	.3423	.3441	.3452	.3456		3
3		.1418	.1515	.1613	.1712	.1811	.1911	.2010	.2109	.2207	.2304		2
4		.0319	.0357	.0397	.0441	.0488	.0537	.0590	.0646	.0706	.0768		1
5		.0029	.0034	.0039	.0045	.0053	.0060	.0069	.0079	.0090	.0102		0
		.69	.68	.67	.66	.65	.64	.63	.62	.61	.60	P	R

R	P	.41	.42	.43	.44	.45	.46	.47	.48	.49	.50		
0		.0715	.0656	.0602	.0551	.0503	.0459	.0418	.0380	.0345	.0313		5
1		.2484	.2376	.2270	.2164	.2059	.1956	.1854	.1755	.1657	.1563		4
2		.3452	.3442	.3424	.3400	.3369	.3332	.3289	.3240	.3185	.3125		3
3		.2399	.2492	.2583	.2671	.2757	.2838	.2916	.2990	.3060	.3125		2
4		.0834	.0902	.0974	.1049	.1128	.1209	.1293	.1380	.1470	.1563		1
5		.0116	.0131	.0147	.0165	.0185	.0206	.0229	.0255	.0282	.0313		0
		.59	.58	.57	.56	.55	.54	.53	.52	.51	.50	P	R

N = 6

R	P	.01	.02	.03	.04	.05	.06	.07	.08	.09	.10		
0		.9415	.8858	.8330	.7828	.7351	.6899	.6470	.6064	.5679	.5314		6
1		.0571	.1085	.1546	.1957	.2321	.2642	.2922	.3164	.3370	.3543		5
2		.0014	.0055	.0120	.0204	.0305	.0422	.0550	.0688	.0833	.0984		4
3		.0000	.0002	.0005	.0011	.0021	.0036	.0055	.0080	.0110	.0146		3
4		.0000	.0000	.0000	.0000	.0001	.0002	.0003	.0005	.0008	.0012		2
5		.0000	.0000	.0000	.0000	.0000	.0000	.0000	.0000	.0000	.0001		1
		.99	.98	.97	.96	.95	.94	.93	.92	.91	.90	P	R

R	P	.11	.12	.13	.14	.15	.16	.17	.18	.19	.20		
0		.4970	.4644	.4336	.4046	.3771	.3513	.3269	.3040	.2824	.2621		6
1		.3685	.3800	.3888	.3952	.3993	.4015	.4018	.4004	.3975	.3932		5
2		.1139	.1295	.1452	.1608	.1762	.1912	.2057	.2197	.2331	.2458		4
3		.0188	.0236	.0289	.0349	.0415	.0486	.0562	.0643	.0729	.0819		3
4		.0017	.0024	.0032	.0043	.0055	.0069	.0086	.0106	.0128	.0154		2
5		.0001	.0001	.0002	.0003	.0004	.0005	.0007	.0009	.0012	.0015		1
6		.0000	.0000	.0000	.0000	.0000	.0000	.0000	.0000	.0000	.0001		0
		.89	.88	.87	.86	.85	.84	.83	.82	.81	.80	P	R

R	P	.21	.22	.23	.24	.25	.26	.27	.28	.29	.30		
0		.2431	.2252	.2084	.1927	.1780	.1642	.1513	.1393	.1281	.1176		6
1		.3877	.3811	.3735	.3651	.3560	.3462	.3358	.3251	.3139	.3025		5
2		.2577	.2687	.2789	.2882	.2966	.3041	.3105	.3160	.3206	.3241		4
3		.0913	.1011	.1111	.1214	.1318	.1424	.1531	.1639	.1746	.1852		3
4		.0182	.0214	.0249	.0287	.0330	.0375	.0425	.0478	.0535	.0595		2
5		.0019	.0024	.0030	.0036	.0044	.0053	.0063	.0074	.0087	.0102		1
6		.0001	.0001	.0001	.0002	.0002	.0003	.0004	.0005	.0006	.0007		0
		.79	.78	.77	.76	.75	.74	.73	.72	.71	.70	P	R

R	P	.31	.32	.33	.34	.35	.36	.37	.38	.39	.40		
0		.1079	.0989	.0905	.0827	.0754	.0687	.0625	.0568	.0515	.0467		6
1		.2909	.2792	.2673	.2555	.2437	.2319	.2203	.2089	.1976	.1866		5
2		.3267	.3284	.3292	.3290	.3280	.3261	.3235	.3201	.3159	.3110		4
3		.1957	.2061	.2162	.2260	.2355	.2446	.2533	.2616	.2693	.2765		3
4		.0660	.0727	.0799	.0873	.0951	.1032	.1116	.1202	.1291	.1382		2
5		.0119	.0137	.0157	.0180	.0205	.0232	.0262	.0295	.0330	.0369		1
6		.0009	.0011	.0013	.0015	.0018	.0022	.0026	.0030	.0035	.0041		0
		.69	.68	.67	.66	.65	.64	.63	.62	.61	.60	P	R

R	P	.41	.42	.43	.44	.45	.46	.47	.48	.49	.50		
0		.0422	.0381	.0343	.0308	.0277	.0248	.0222	.0198	.0176	.0156		6
1		.1759	.1654	.1552	.1454	.1359	.1267	.1179	.1095	.1014	.0938		5

TABLE C Binomial Distribution: Individual Terms $p(r) = c\binom{n}{r} p^r q^{n-r}$ (Continued)

2	.3055	.2994	.2928	.2856	.2780	.2699	.2615	.2527	.2436	.2344		4
3	.2831	.2891	.2945	.2992	.3032	.3065	.3091	.3110	.3121	.3125		3
4	.1475	.1570	.1666	.1763	.1861	.1958	.2056	.2153	.2249	.2344		2
5	.0410	.0455	.0503	.0554	.0609	.0667	.0729	.0795	.0864	.0938		1
6	.0048	.0055	.0063	.0073	.0083	.0095	.0108	.0122	.0138	.0156		0
	.59	.58	.57	.56	.55	.54	.53	.52	.51	.50	P	R

N = 7

R	P	.01	.02	.03	.04	.05	.06	.07	.08	.09	.10		
0		.9321	.8681	.8080	.7514	.6983	.6485	.6017	.5578	.5168	.4783		7
1		.0659	.1240	.1749	.2192	.2573	.2897	.3170	.3396	.3578	.3720		6
2		.0020	.0076	.0162	.0274	.0406	.0555	.0716	.0886	.1061	.1240		5
3		.0000	.0003	.0008	.0019	.0036	.0059	.0090	.0128	.0175	.0230		4
4		.0000	.0000	.0000	.0001	.0002	.0004	.0007	.0011	.0017	.0026		3
5		.0000	.0000	.0000	.0000	.0000	.0000	.0000	.0001	.0001	.0002		2
		.99	.98	.97	.96	.95	.94	.93	.92	.91	.90	P	R

R	P	.11	.12	.13	.14	.15	.16	.17	.18	.19	.20		
0		.4423	.4047	.3773	.3479	.3206	.2951	.2714	.2493	.2288	.2097		7
1		.3827	.3901	.3946	.3965	.3960	.3935	.3891	.3830	.3756	.3670		6
2		.1419	.1596	.1769	.1936	.2097	.2248	.2391	.2523	.2643	.2753		5
3		.0292	.0363	.0441	.0525	.0617	.0714	.0816	.0923	.1033	.1147		4
4		.0036	.0049	.0066	.0086	.0109	.0136	.0167	.0203	.0242	.0287		3
5		.0003	.0004	.0006	.0008	.0012	.0016	.0021	.0027	.0034	.0043		2
6		.0000	.0000	.0000	.0000	.0001	.0001	.0001	.0002	.0003	.0004		1
		.89	.88	.87	.86	.85	.84	.83	.82	.81	.80	P	R

R	P	.21	.22	.23	.24	.25	.26	.27	.28	.29	.30		
0		.1920	.1757	.1605	.1465	.1335	.1215	.1105	.1003	.0910	.0824		7
1		.3573	.3468	.3356	.3237	.3115	.2989	.2860	.2731	.2600	.2471		6
2		.2850	.2935	.3007	.3067	.3115	.3150	.3174	.3186	.3186	.3177		5
3		.1263	.1379	.1497	.1614	.1730	.1845	.1956	.2065	.2169	.2269		4
4		.0336	.0389	.0447	.0510	.0577	.0648	.0724	.0803	.0886	.0972		3
5		.0054	.0066	.0080	.0097	.0115	.0137	.0161	.0187	.0217	.0250		2
6		.0005	.0006	.0008	.0010	.0013	.0016	.0020	.0024	.0030	.0036		1
7		.0000	.0000	.0000	.0000	.0001	.0001	.0001	.0001	.0002	.0002		0
		.79	.78	.77	.76	.75	.74	.73	.72	.71	.70	P	R

R	P	.31	.32	.33	.34	.35	.36	.37	.38	.39	.40		
0		.0745	.0672	.0606	.0546	.0490	.0440	.0394	.0352	.0314	.0280		7
1		.2342	.2215	.2090	.1967	.1848	.1732	.1619	.1511	.1407	.1306		6
2		.3156	.3127	.3088	.3040	.2985	.2922	.2853	.2778	.2698	.2613		5
3		.2363	.2452	.2535	.2610	.2679	.2740	.2793	.2838	.2875	.2903		4
4		.1062	.1154	.1248	.1345	.1442	.1541	.1640	.1739	.1838	.1935		3
5		.0286	.0326	.0369	.0416	.0466	.0520	.0578	.0640	.0705	.0774		2
6		.0043	.0051	.0061	.0071	.0084	.0098	.0113	.0131	.0150	.0172		1
7		.0003	.0003	.0004	.0005	.0006	.0008	.0009	.0011	.0014	.0016		0
		.69	.68	.67	.66	.65	.64	.63	.62	.61	.60	P	R

R	P	.41	.42	.43	.44	.45	.46	.47	.48	.49	.50		
0		.0249	.0221	.0195	.0173	.0152	.0134	.0117	.0103	.0090	.0078		7
1		.1211	.1119	.1032	.0950	.0872	.0798	.0729	.0664	.0604	.0547		6
2		.2524	.2431	.2336	.2239	.2140	.2040	.1940	.1840	.1740	.1641		5
3		.2923	.2934	.2937	.2932	.2918	.2897	.2867	.2830	.2786	.2734		4
4		.2031	.2125	.2216	.2304	.2388	.2468	.2543	.2612	.2676	.2734		3
5		.0847	.0923	.1003	.1086	.1172	.1261	.1353	.1447	.1543	.1641		2
6		.0196	.0223	.0252	.0284	.0320	.0358	.0400	.0445	.0494	.0547		1
7		.0019	.0023	.0027	.0032	.0037	.0044	.0051	.0059	.0068	.0078		0
		.59	.58	.57	.56	.55	.54	.53	.52	.51	.50	P	R

N = 8

R	P	.01	.02	.03	.04	.05	.06	.07	.08	.09	.10		
0		.9227	.8508	.7837	.7214	.6634	.6096	.5596	.5132	.4703	.4305		8
1		.0746	.1389	.1939	.2405	.2793	.3113	.3370	.3570	.3721	.3826		7
2		.0026	.0099	.0210	.0351	.0515	.0695	.0888	.1087	.1288	.1488		6
3		.0001	.0004	.0013	.0029	.0054	.0089	.0134	.0189	.0255	.0331		5
4		.0000	.0000	.0001	.0002	.0004	.0007	.0013	.0021	.0031	.0046		4
5		.0000	.0000	.0000	.0000	.0000	.0000	.0001	.0001	.0002	.0004		3
		.99	.98	.97	.96	.95	.94	.93	.92	.91	.90	P	R

R	P	.11	.12	.13	.14	.15	.16	.17	.18	.19	.20		
0		.3937	.3596	.3282	.2992	.2725	.2479	.2252	.2044	.1853	.1678		8
1		.3892	.3923	.3923	.3897	.3847	.3777	.3691	.3590	.3477	.3355		7
2		.1684	.1872	.2052	.2220	.2376	.2518	.2646	.2758	.2855	.2936		6
3		.0416	.0511	.0613	.0723	.0839	.0959	.1084	.1211	.1339	.1468		5
4		.0064	.0087	.0115	.0147	.0185	.0228	.0277	.0332	.0393	.0459		4
5		.0006	.0009	.0014	.0019	.0026	.0035	.0045	.0058	.0074	.0092		3

TABLE C Binomial Distribution: Individual Terms $p(r) = c\binom{n}{r} p^r q^{n-r}$ (*Continued*)

	.89	.88	.87	.86	.85	.84	.83	.82	.81	.80		
6	.0000	.0001	.0001	.0002	.0002	.0003	.0005	.0006	.0009	.0011		2
7	.0000	.0000	.0000	.0000	.0000	.0000	.0000	.0000	.0001	.0001		1
	.89	.88	.87	.86	.85	.84	.83	.82	.81	.80	P	R

R	P	.21	.22	.23	.24	.25	.26	.27	.28	.29	.30	
0		.1517	.1370	.1236	.1113	.1001	.0899	.0806	.0722	.0646	.0576	8
1		.3226	.3092	.2953	.2812	.2670	.2527	.2386	.2247	.2110	.1977	7
2		.3002	.3052	.3087	.3108	.3115	.3108	.3089	.3058	.3017	.2965	6
3		.1596	.1722	.1844	.1963	.2076	.2184	.2285	.2379	.2464	.2541	5
4		.0530	.0607	.0689	.0775	.0865	.0959	.1056	.1156	.1258	.1361	4
5		.0113	.0137	.0165	.0196	.0231	.0270	.0313	.0360	.0411	.0467	3
6		.0015	.0019	.0025	.0031	.0038	.0047	.0058	.0070	.0084	.0100	2
7		.0001	.0002	.0002	.0003	.0004	.0005	.0006	.0008	.0010	.0012	1
8		.0000	.0000	.0000	.0000	.0000	.0000	.0000	.0000	.0001	.0001	0
		.79	.78	.77	.76	.75	.74	.73	.72	.71	.70	P
												R

R	P	.31	.32	.33	.34	.35	.36	.37	.38	.39	.40	
0		.0514	.0457	.0406	.0360	.0319	.0281	.0249	.0218	.0192	.0168	8
1		.1847	.1721	.1600	.1484	.1373	.1267	.1166	.1071	.0981	.0896	7
2		.2904	.2835	.2758	.2675	.2587	.2494	.2397	.2297	.2194	.2090	6
3		.2609	.2668	.2717	.2756	.2786	.2805	.2815	.2815	.2806	.2787	5
4		.1465	.1569	.1673	.1775	.1875	.1973	.2067	.2157	.2242	.2322	4
5		.0527	.0591	.0659	.0732	.0808	.0888	.0971	.1058	.1147	.1239	3
6		.0118	.0139	.0162	.0188	.0217	.0250	.0285	.0324	.0367	.0413	2
7		.0015	.0019	.0023	.0028	.0033	.0040	.0048	.0057	.0067	.0079	1
8		.0001	.0001	.0001	.0002	.0002	.0003	.0004	.0004	.0005	.0007	0
		.69	.68	.67	.66	.65	.64	.63	.62	.61	.60	P
												R

R	P	.41	.42	.43	.44	.45	.46	.47	.48	.49	.50	
0		.0147	.0128	.0111	.0097	.0084	.0072	.0062	.0053	.0046	.0039	8
1		.0816	.0742	.0672	.0608	.0548	.0493	.0442	.0395	.0352	.0313	7
2		.1985	.1880	.1776	.1672	.1569	.1469	.1371	.1275	.1183	.1094	6
3		.2759	.2723	.2679	.2627	.2568	.2503	.2431	.2355	.2273	.2188	5
4		.2397	.2465	.2526	.2580	.2627	.2665	.2695	.2717	.2730	.2734	4
5		.1332	.1428	.1525	.1622	.1719	.1816	.1912	.2006	.2098	.2188	3
6		.0463	.0517	.0575	.0637	.0703	.0774	.0848	.0926	.1008	.1094	2
7		.0092	.0107	.0124	.0143	.0164	.0188	.0215	.0244	.0277	.0313	1
8		.0008	.0010	.0012	.0014	.0017	.0020	.0024	.0028	.0033	.0039	0
		.59	.58	.57	.56	.55	.54	.53	.52	.51	.50	P
												R

N = 9

R	P	.01	.02	.03	.04	.05	.06	.07	.08	.09	.10	
0		.9135	.8337	.7602	.6925	.6302	.5730	.5204	.4722	.4279	.3874	9
1		.0830	.1531	.2116	.2597	.2985	.3292	.3525	.3695	.3809	.3874	8
2		.0034	.0125	.0262	.0433	.0629	.0840	.1061	.1285	.1507	.1722	7
3		.0001	.0006	.0019	.0042	.0077	.0125	.0186	.0261	.0348	.0446	6
4		.0000	.0000	.0001	.0003	.0006	.0012	.0021	.0034	.0052	.0074	5
5		.0000	.0000	.0000	.0000	.0000	.0001	.0002	.0003	.0005	.0008	4
6		.0000	.0000	.0000	.0000	.0000	.0000	.0000	.0000	.0000	.0001	3
		.99	.98	.97	.96	.95	.94	.93	.92	.91	.90	P
												R

R	P	.11	.12	.13	.14	.15	.16	.17	.18	.19	.20	
0		.3504	.3165	.2855	.2573	.2316	.2082	.1869	.1676	.1501	.1342	9
1		.3897	.3884	.3840	.3770	.3679	.3569	.3446	.3312	.3169	.3020	8
2		.1927	.2119	.2295	.2455	.2597	.2720	.2823	.2908	.2973	.3020	7
3		.0556	.0674	.0800	.0933	.1069	.1209	.1349	.1489	.1627	.1762	6
4		.0103	.0138	.0179	.0228	.0283	.0345	.0415	.0490	.0573	.0661	5
5		.0013	.0019	.0027	.0037	.0050	.0066	.0085	.0108	.0134	.0165	4
6		.0001	.0002	.0003	.0004	.0006	.0008	.0012	.0016	.0021	.0028	3
7		.0000	.0000	.0000	.0000	.0000	.0001	.0001	.0001	.0002	.0003	2
		.89	.88	.87	.86	.85	.84	.83	.82	.81	.80	P
												R

R	P	.21	.22	.23	.24	.25	.26	.27	.28	.29	.30	
0		.1199	.1069	.0952	.0846	.0751	.0665	.0589	.0520	.0458	.0404	9
1		.2867	.2713	.2558	.2404	.2253	.2104	.1960	.1820	.1685	.1556	8
2		.3049	.3061	.3056	.3037	.3003	.2957	.2899	.2831	.2754	.2668	7
3		.1891	.2014	.2130	.2238	.2336	.2424	.2502	.2569	.2624	.2668	6
4		.0754	.0852	.0954	.1060	.1168	.1278	.1388	.1499	.1608	.1715	5
5		.0200	.0240	.0285	.0335	.0389	.0449	.0513	.0583	.0657	.0735	4
6		.0036	.0045	.0057	.0070	.0087	.0105	.0127	.0151	.0179	.0210	3
7		.0004	.0005	.0007	.0010	.0012	.0016	.0020	.0025	.0031	.0039	2
8		.0000	.0000	.0001	.0001	.0001	.0001	.0002	.0002	.0003	.0004	1
		.79	.78	.77	.76	.75	.74	.73	.72	.71	.70	P
												R

R	P	.31	.32	.33	.34	.35	.36	.37	.38	.39	.40	
0		.0355	.0311	.0272	.0238	.0207	.0180	.0156	.0135	.0117	.0101	9
1		.1433	.1317	.1206	.1102	.1004	.0912	.0826	.0747	.0673	.0605	8
2		.2576	.2478	.2376	.2270	.2162	.2052	.1941	.1831	.1721	.1612	7
3		.2701	.2721	.2731	.2729	.2716	.2693	.2660	.2618	.2567	.2508	6
4		.1820	.1921	.2017	.2109	.2194	.2272	.2344	.2407	.2462	.2508	5

TABLE C Binomial Distribution: Individual Terms $p(r) = c(^n_r)\, p^r q^{n-r}$ *(Continued)*

	.60	.61	.62	.63	.64	.65	.66	.57	.68	.69	
5	.0818	.0904	.0994	.1086	.1181	.1278	.1376	.1475	.1574	.1672	4
6	.0245	.0284	.0326	.0373	.0424	.0479	.0539	.0603	.0671	.0743	3
7	.0047	.0057	.0069	.0082	.0098	.0116	.0136	.0158	.0184	.0212	2
8	.0005	.0007	.0008	.0011	.0013	.0016	.0020	.0024	.0029	.0035	1
9	.0000	.0000	.0000	.0001	.0001	.0001	.0001	.0002	.0002	.0003	0
	.69	.68	.57	.66	.65	.64	.63	.62	.61	.60	P R

P	P	.41	.42	.43	.44	.45	.46	.47	.48	.49	.50	
0		.0087	.0074	.0064	.0054	.0046	.0039	.0033	.0028	.0023	.0020	9
1		.0542	.0494	.0431	.0383	.0339	.0299	.0263	.0231	.0202	.0176	8
2		.1506	.1402	.1301	.1204	.1110	.1020	.0934	.0853	.0776	.0703	7
3		.2442	.2369	.2291	.2207	.2119	.2027	.1933	.1837	.1739	.1641	6
4		.2545	.2573	.2592	.2601	.2600	.2590	.2571	.2543	.2506	.2461	5
5		.1769	.1863	.1955	.2044	.2128	.2207	.2280	.2347	.2408	.2461	4
6		.0819	.0990	.0983	.1070	.1160	.1253	.1348	.1445	.1542	.1641	3
7		.0244	.0279	.0318	.0360	.0407	.0458	.0512	.0571	.0635	.0703	2
8		.0042	.0051	.0060	.0071	.0083	.0097	.0114	.0132	.0153	.0176	1
9		.0003	.0004	.0005	.0006	.0008	.0009	.0011	.0014	.0016	.0020	0
		.59	.58	.57	.56	.55	.54	.53	.52	.51	.50	P R

N = 10

R	P	.01	.02	.03	.04	.05	.06	.07	.08	.09	.10	
0		.9044	.8171	.7374	.6648	.5987	.5386	.4840	.4344	.3894	.3487	10
1		.0914	.1667	.2281	.2770	.3151	.3438	.3643	.3777	.3851	.3874	9
2		.0042	.0153	.0317	.0519	.0746	.0988	.1234	.1478	.1714	.1937	8
3		.0001	.0008	.0026	.0058	.0105	.0168	.0248	.0343	.0452	.0574	7
4		.0000	.0000	.0001	.0004	.0010	.0019	.0033	.0052	.0078	.0112	6
5		.0000	.0000	.0000	.0000	.0001	.0001	.0003	.0005	.0009	.0015	5
6		.0000	.0000	.0000	.0000	.0000	.0000	.0000	.0000	.0001	.0001	4
		.99	.98	.97	.96	.95	.94	.93	.92	.91	.90	P R

P	P	.11	.12	.13	.14	.15	.16	.17	.18	.19	.20	
0		.3118	.2785	.2484	.2213	.1969	.1749	.1552	.1374	.1216	.1074	10
1		.3854	.3798	.3712	.3603	.3474	.3331	.3178	.3017	.2852	.2684	9
2		.2143	.2330	.2496	.2639	.2759	.2856	.2929	.2980	.3010	.3020	8
3		.0706	.0847	.0995	.1146	.1298	.1450	.1600	.1745	.1883	.2013	7
4		.0153	.0202	.0260	.0326	.0401	.0483	.0573	.0670	.0773	.0881	6
5		.0023	.0033	.0047	.0064	.0085	.0111	.0141	.0177	.0218	.0264	5
6		.0002	.0004	.0006	.0009	.0012	.0018	.0024	.0032	.0043	.0055	4
7		.0000	.0000	.0000	.0001	.0001	.0002	.0003	.0004	.0006	.0008	3
8		.0000	.0000	.0000	.0000	.0000	.0000	.0000	.0000	.0001	.0001	2
		.89	.88	.87	.86	.85	.84	.83	.82	.81	.80	P R

P	P	.21	.22	.23	.24	.25	.26	.27	.28	.29	.30	
0		.0947	.0834	.0733	.0643	.0563	.0492	.0430	.0374	.0326	.0282	10
1		.2517	.2351	.2188	.2030	.1877	.1730	.1590	.1456	.1330	.1211	9
2		.3011	.2984	.2942	.2885	.2816	.2735	.2646	.2548	.2444	.2335	8
3		.2134	.2244	.2343	.2429	.2503	.2563	.2609	.2642	.2662	.2668	7
4		.0993	.1108	.1225	.1343	.1460	.1576	.1689	.1798	.1903	.2001	6
5		.0317	.0375	.0439	.0509	.0584	.0664	.0750	.0839	.0933	.1029	5
6		.0070	.0088	.0109	.0134	.0162	.0195	.0231	.0272	.0317	.0368	4
7		.0011	.0014	.0019	.0024	.0031	.0039	.0049	.0060	.0074	.0090	3
8		.0001	.0002	.0002	.0003	.0004	.0005	.0007	.0009	.0011	.0014	2
9		.0000	.0000	.0000	.0000	.0000	.0000	.0001	.0001	.0001	.0001	1
		.79	.78	.77	.76	.75	.74	.73	.72	.71	.70	P R

R	P	.31	.32	.33	.34	.35	.36	.37	.38	.39	.40	
0		.0245	.0211	.0182	.0157	.0135	.0115	.0098	.0084	.0071	.0060	10
1		.1099	.0995	.0898	.0808	.0725	.0649	.0578	.0514	.0456	.0403	9
2		.2222	.2137	.1990	.1990	.1757	.1642	.1529	.1419	.1312	.1209	8
3		.2662	.2644	.2614	.2573	.2522	.2462	.2394	.2319	.2237	.2150	7
4		.2093	.2177	.2253	.2320	.2377	.2424	.2461	.2487	.2503	.2508	6
5		.1128	.1229	.1332	.1434	.1536	.1636	.1734	.1829	.1920	.2007	5
6		.0422	.0482	.0547	.0616	.0689	.0767	.0849	.0934	.1023	.1115	4
7		.0108	.0130	.0154	.0181	.0212	.0247	.0285	.0327	.0374	.0425	3
8		.0018	.0023	.0028	.0035	.0043	.0052	.0063	.0075	.0090	.0106	2
9		.0002	.0002	.0003	.0004	.0005	.0006	.0008	.0010	.0013	.0016	1
0		.0000	.0000	.0000	.0000	.0000	.0000	.0000	.0001	.0001	.0001	0
		.69	.68	.67	.66	.65	.64	.63	.62	.61	.60	P R

P	P	.41	.42	.43	.44	.45	.46	.47	.48	.49	.50	
0		.0051	.0043	.0036	.0030	.0025	.0021	.0017	.0014	.0012	.0010	10
1		.0355	.0312	.0273	.0238	.0207	.0180	.0155	.0133	.0114	.0098	9
2		.1111	.1017	.0927	.0843	.0763	.0688	.0619	.0554	.0494	.0439	8
3		.2058	.1963	.1865	.1765	.1665	.1564	.1464	.1364	.1267	.1172	7
4		.2503	.2488	.2462	.2427	.2384	.2331	.2271	.2204	.2130	.2051	6
5		.2087	.2162	.2229	.2289	.2340	.2383	.2417	.2441	.2456	.2461	5
6		.1209	.1304	.1401	.1499	.1536	.1692	.1786	.1878	.1966	.2051	4
7		.0480	.0540	.0604	.0673	.0746	.0824	.0905	.0991	.1080	.1172	3
8		.0125	.0147	.0171	.0198	.0229	.0263	.0301	.0343	.0389	.0439	2
9		.0019	.0024	.0029	.0035	.0042	.0050	.0059	.0070	.0083	.0098	1

TABLE C Binomial Distribution: Individual Terms $p(r) = c\binom{n}{r} p^r q^{n-r}$ (*Continued*)

1C	.0001	.0002	.0002	.0003	.0003	.0004	.0005	.0006	.0008	.0010		0
	.59	.58	.57	.56	.55	.54	.53	.52	.51	.50	P	R

N = 11

P	P	.01	.02	.03	.04	.05	.06	.07	.08	.09	.10		
0		.8953	.8007	.7153	.6382	.5688	.5063	.4501	.3996	.3544	.3138		11
1		.0995	.1798	.2433	.2925	.3293	.3555	.3727	.3823	.3855	.3835		10
2		.0050	.0183	.0376	.0609	.0867	.1135	.1403	.1662	.1906	.2131		9
3		.0002	.0011	.0035	.0076	.0137	.0217	.0317	.0434	.0566	.0710		8
4		.0000	.0000	.0002	.0006	.0014	.0028	.0048	.0075	.0112	.0158		7
5		.0000	.0000	.0000	.0000	.0001	.0002	.0005	.0009	.0015	.0025		6
6		.0000	.0000	.0000	.0000	.0000	.0000	.0000	.0001	.0002	.0003		5
		.99	.98	.97	.96	.95	.94	.93	.92	.91	.90	P	R

P	P	.11	.12	.13	.14	.15	.16	.17	.18	.19	.20		
0		.2775	.2451	.2161	.1903	.1673	.1469	.1288	.1127	.0985	.0859		11
1		.3773	.3676	.3552	.3408	.3248	.3078	.2901	.2721	.2541	.2362		10
2		.2332	.2507	.2654	.2774	.2866	.2932	.2971	.2987	.2980	.2953		9
3		.0865	.1025	.1190	.1355	.1517	.1675	.1826	.1967	.2097	.2215		8
4		.0214	.0280	.0356	.0441	.0536	.0638	.0748	.0864	.0984	.1107		7
5		.0037	.0053	.0074	.0101	.0132	.0170	.0214	.0265	.0323	.0388		6
6		.0005	.0007	.0011	.0016	.0023	.0032	.0044	.0058	.0076	.0097		5
7		.0000	.0001	.0001	.0002	.0003	.0004	.0006	.0009	.0013	.0017		4
8		.0000	.0000	.0000	.0000	.0000	.0000	.0001	.0001	.0001	.0002		3
		.89	.88	.87	.86	.85	.84	.83	.82	.81	.80	P	R

P	P	.21	.22	.23	.24	.25	.26	.27	.28	.29	.30		
0		.0748	.0650	.0564	.0489	.0422	.0364	.0314	.0270	.0231	.0198		11
1		.2187	.2017	.1854	.1697	.1549	.1408	.1276	.1153	.1038	.0932		10
2		.2907	.2845	.2768	.2680	.2581	.2474	.2360	.2242	.2121	.1998		9
3		.2318	.2407	.2481	.2539	.2581	.2608	.2619	.2616	.2599	.2568		8
4		.1232	.1358	.1482	.1603	.1721	.1832	.1937	.2035	.2123	.2201		7
5		.0459	.0536	.0620	.0709	.0803	.0901	.1003	.1108	.1214	.1321		6
6		.0122	.0151	.0185	.0224	.0268	.0317	.0371	.0431	.0496	.0566		5
7		.0023	.0030	.0039	.0050	.0064	.0079	.0098	.0120	.0145	.0173		4
8		.0003	.0004	.0006	.0008	.0011	.0014	.0018	.0023	.0030	.0037		3
9		.0000	.0000	.0001	.0001	.0001	.0002	.0002	.0003	.0004	.0005		2
		.79	.78	.77	.76	.75	.74	.73	.72	.71	.70	P	R

P	P	.31	.32	.33	.34	.35	.36	.37	.38	.39	.40		
0		.0169	.0144	.0122	.0104	.0088	.0074	.0062	.0052	.0044	.0036		11
1		.0834	.0744	.0662	.0587	.0518	.0457	.0401	.0351	.0306	.0266		10
2		.1874	.1751	.1630	.1511	.1395	.1284	.1177	.1075	.0978	.0887		9
3		.2526	.2472	.2408	.2335	.2254	.2167	.2074	.1977	.1876	.1774		8
4		.2269	.2326	.2372	.2406	.2428	.2438	.2436	.2423	.2399	.2365		7
5		.1427	.1533	.1636	.1735	.1830	.1920	.2003	.2079	.2148	.2207		6
6		.0641	.0721	.0806	.0894	.0985	.1080	.1176	.1274	.1373	.1471		5
7		.0206	.0242	.0283	.0329	.0379	.0434	.0494	.0558	.0627	.0701		4
8		.0046	.0057	.0070	.0085	.0102	.0122	.0145	.0171	.0200	.0234		3
9		.0007	.0009	.0011	.0015	.0018	.0023	.0028	.0035	.0043	.0052		2
10		.0001	.0001	.0001	.0001	.0002	.0003	.0003	.0004	.0005	.0007		1
		.69	.68	.67	.66	.65	.64	.63	.62	.61	.60	P	R

R	P	.41	.42	.43	.44	.45	.46	.47	.48	.49	.50		
0		.0030	.0025	.0021	.0017	.0014	.0011	.0009	.0008	.0006	.0005		11
1		.0231	.0199	.0171	.0147	.0125	.0107	.0090	.0076	.0064	.0054		10
2		.0801	.0721	.0646	.0577	.0513	.0454	.0401	.0352	.0308	.0269		9
3		.1670	.1566	.1462	.1359	.1259	.1161	.1067	.0976	.0888	.0806		8
4		.2321	.2267	.2206	.2136	.2060	.1978	.1892	.1801	.1707	.1611		7
5		.2258	.2299	.2329	.2350	.2360	.2359	.2348	.2327	.2296	.2256		6
6		.1569	.1664	.1757	.1846	.1931	.2010	.2083	.2148	.2206	.2256		5
7		.0779	.0861	.0947	.1036	.1128	.1223	.1319	.1416	.1514	.1611		4
8		.0271	.0312	.0357	.0407	.0462	.0521	.0585	.0654	.0727	.0806		3
9		.0063	.0075	.0090	.0107	.0126	.0148	.0173	.0201	.0233	.0269		2
10		.0009	.0011	.0014	.0017	.0021	.0025	.0031	.0037	.0045	.0054		1
11		.0001	.0001	.0001	.0001	.0002	.0002	.0002	.0003	.0004	.0005		0
		.59	.58	.57	.56	.55	.54	.53	.52	.51	.50	P	R

N = 12

P	P	.01	.02	.03	.04	.05	.06	.07	.08	.09	.10		
0		.8864	.7847	.6938	.6127	.5404	.4759	.4186	.3677	.3225	.2824		12
1		.1074	.1922	.2575	.3064	.3413	.3645	.3781	.3837	.3827	.3766		11
2		.0060	.0216	.0438	.0702	.0988	.1280	.1565	.1835	.2082	.2301		10
3		.0002	.0015	.0045	.0098	.0173	.0272	.0393	.0532	.0686	.0852		9
4		.0000	.0001	.0003	.0009	.0021	.0039	.0067	.0104	.0153	.0213		8
5		.0000	.0000	.0000	.0001	.0002	.0004	.0008	.0014	.0024	.0038		7
6		.0000	.0000	.0000	.0000	.0000	.0000	.0001	.0001	.0003	.0005		6
		.99	.98	.97	.96	.95	.94	.93	.92	.91	.90	P	R

TABLE C Binomial Distribution: Individual Terms $p(r) = c_r^n\, p^r q^{n-r}$ (*Continued*)

P	P	.11	.12	.13	.14	.15	.16	.17	.18	.19	.20		
0		.2470	.2157	.1880	.1637	.1422	.1234	.1069	.0924	.0798	.0687		12
1		.3663	.3529	.3372	.3197	.3012	.2821	.2627	.2434	.2245	.2062		11
2		.2490	.2647	.2771	.2863	.2924	.2955	.2960	.2939	.2897	.2835		10
3		.1026	.1203	.1380	.1553	.1720	.1876	.2021	.2151	.2265	.2362		9
4		.0285	.0369	.0464	.0569	.0683	.0804	.0931	.1062	.1195	.1329		8
5		.0056	.0081	.0111	.0148	.0193	.0245	.0305	.0373	.0449	.0532		7
6		.0008	.0013	.0019	.0028	.0040	.0054	.0073	.0096	.0123	.0155		6
7		.0001	.0001	.0002	.0004	.0006	.0009	.0013	.0018	.0025	.0033		5
8		.0000	.0000	.0000	.0000	.0001	.0001	.0002	.0002	.0004	.0005		4
9		.0000	.0000	.0000	.0000	.0000	.0000	.0000	.0000	.0000	.0001		3
		.89	.88	.87	.86	.85	.84	.83	.82	.81	.80	P	R

P	P	.21	.22	.23	.24	.25	.26	.27	.28	.29	.30		
0		.0591	.0507	.0434	.0371	.0317	.0270	.0229	.0194	.0164	.0138		12
1		.1885	.1717	.1557	.1407	.1267	.1137	.1016	.0906	.0804	.0712		11
2		.2756	.2663	.2558	.2444	.2323	.2197	.2068	.1937	.1807	.1678		10
3		.2442	.2503	.2547	.2573	.2581	.2573	.2549	.2511	.2460	.2397		9
4		.1460	.1589	.1712	.1828	.1936	.2034	.2122	.2197	.2261	.2311		8
5		.0621	.0717	.0818	.0924	.1032	.1143	.1255	.1367	.1477	.1585		7
6		.0193	.0236	.0285	.0340	.0401	.0469	.0542	.0620	.0704	.0792		6
7		.0044	.0057	.0073	.0092	.0115	.0141	.0172	.0207	.0246	.0291		5
8		.0007	.0010	.0014	.0018	.0024	.0031	.0040	.0050	.0063	.0078		4
9		.0001	.0001	.0002	.0003	.0004	.0005	.0007	.0009	.0011	.0015		3
10		.0000	.0000	.0000	.0000	.0000	.0001	.0001	.0001	.0001	.0002		2
		.79	.78	.77	.76	.75	.74	.73	.72	.71	.70	P	R

P	P	.31	.32	.33	.34	.35	.36	.37	.38	.39	.40		
0		.0116	.0098	.0082	.0068	.0057	.0047	.0039	.0032	.0027	.0022		12
1		.0628	.0552	.0484	.0422	.0368	.0319	.0276	.0237	.0204	.0174		11
2		.1552	.1429	.1310	.1197	.1088	.0986	.0890	.0800	.0716	.0639		10
3		.2324	.2241	.2151	.2055	.1954	.1849	.1742	.1634	.1526	.1419		9
4		.2349	.2373	.2384	.2382	.2367	.2340	.2302	.2254	.2195	.2128		8
5		.1688	.1737	.1879	.1963	.2039	.2106	.2163	.2210	.2246	.2270		7
6		.0885	.0981	.1079	.1180	.1281	.1382	.1482	.1580	.1675	.1766		6
7		.0341	.0396	.0456	.0521	.0591	.0666	.0746	.0830	.0918	.1009		5
8		.0096	.0116	.0140	.0168	.0199	.0234	.0274	.0318	.0367	.0420		4
9		.0019	.0024	.0031	.0038	.0048	.0059	.0071	.0087	.0104	.0125		3
10		.0003	.0003	.0005	.0006	.0008	.0010	.0013	.0016	.0020	.0025		2
11		.0000	.0000	.0000	.0001	.0001	.0001	.0001	.0002	.0002	.0003		1
		.69	.68	.67	.66	.65	.64	.63	.62	.61	.60	P	R

P	P	.41	.42	.43	.44	.45	.46	.47	.48	.49	.50		
0		.0018	.0014	.0012	.0010	.0008	.0006	.0005	.0004	.0003	.0002		12
1		.0148	.0126	.0106	.0090	.0075	.0063	.0052	.0043	.0036	.0029		11
2		.0567	.0502	.0442	.0388	.0339	.0294	.0255	.0220	.0189	.0161		10
3		.1314	.1211	.1111	.1015	.0923	.0836	.0754	.0676	.0604	.0537		9
4		.2054	.1973	.1886	.1794	.1700	.1602	.1504	.1405	.1306	.1208		8
5		.2294	.2285	.2276	.2256	.2225	.2184	.2134	.2075	.2008	.1934		7
6		.1851	.1931	.2003	.2068	.2124	.2171	.2208	.2234	.2250	.2256		6
7		.1103	.1198	.1295	.1393	.1489	.1585	.1678	.1768	.1853	.1934		5
8		.0479	.0542	.0611	.0684	.0762	.0844	.0930	.1020	.1113	.1208		4
9		.0148	.0175	.0205	.0239	.0277	.0319	.0367	.0418	.0475	.0537		3
10		.0031	.0038	.0046	.0056	.0068	.0082	.0098	.0116	.0137	.0161		2
11		.0004	.0005	.0006	.0008	.0010	.0013	.0016	.0019	.0024	.0029		1
12		.0000	.0000	.0000	.0001	.0001	.0001	.0001	.0001	.0002	.0002		0
		.59	.58	.57	.56	.55	.54	.53	.52	.51	.50	P	R

N = 13

P	P	.01	.02	.03	.04	.05	.06	.07	.08	.09	.10		
0		.8775	.7690	.6730	.5882	.5133	.4474	.3893	.3383	.2935	.2542		13
1		.1152	.2040	.2706	.3186	.3512	.3712	.3809	.3824	.3773	.3672		12
2		.0070	.0250	.0502	.0797	.1109	.1422	.1720	.1995	.2239	.2448		11
3		.0003	.0019	.0057	.0122	.0214	.0333	.0475	.0636	.0812	.0997		10
4		.0000	.0001	.0004	.0013	.0028	.0053	.0089	.0138	.0201	.0277		9
5		.0000	.0000	.0000	.0001	.0003	.0006	.0012	.0022	.0036	.0055		8
6		.0000	.0000	.0000	.0000	.0000	.0001	.0001	.0003	.0005	.0008		7
7		.0000	.0000	.0000	.0000	.0000	.0000	.0000	.0000	.0000	.0001		6
		.99	.98	.97	.96	.95	.94	.93	.92	.91	.90	P	R

P	P	.11	.12	.13	.14	.15	.16	.17	.18	.19	.20		
0		.2198	.1898	.1636	.1408	.1209	.1037	.0887	.0758	.0646	.0550		13
1		.3532	.3364	.3178	.2979	.2774	.2567	.2362	.2163	.1970	.1787		12
2		.2619	.2753	.2849	.2910	.2937	.2934	.2903	.2848	.2773	.2680		11
3		.1187	.1376	.1561	.1737	.1900	.2049	.2180	.2293	.2385	.2457		10
4		.0367	.0469	.0583	.0707	.0838	.0976	.1116	.1258	.1399	.1535		9
5		.0082	.0115	.0157	.0207	.0266	.0335	.0412	.0497	.0591	.0691		8
6		.0013	.0021	.0031	.0045	.0063	.0085	.0112	.0145	.0185	.0230		7
7		.0002	.0003	.0005	.0007	.0011	.0016	.0023	.0032	.0043	.0058		6
8		.0000	.0000	.0001	.0001	.0001	.0002	.0004	.0005	.0008	.0011		5
9		.0000	.0000	.0000	.0000	.0000	.0000	.0000	.0001	.0001	.0001		4
		.89	.88	.87	.86	.85	.84	.83	.82	.81	.80	P	R

TABLE C Binomial Distribution: Individual Terms $p(r) = c\binom{n}{r} p^r q^{n-r}$ *(Continued)*

P	P	.21	.22	.23	.24	.25	.26	.27	.28	.29	.30		
0		.0467	.0396	.0334	.0282	.0238	.0200	.0167	.0140	.0117	.0097		13
1		.1613	.1450	.1299	.1159	.1029	.0911	.0804	.0706	.0619	.0540		12
2		.2573	.2455	.2328	.2195	.2059	.1921	.1784	.1648	.1516	.1388		11
3		.2508	.2539	.2550	.2542	.2517	.2475	.2419	.2351	.2271	.2181		10
4		.1667	.1790	.1904	.2007	.2097	.2174	.2237	.2285	.2319	.2337		9
5		.0797	.0909	.1024	.1141	.1258	.1375	.1489	.1600	.1705	.1803		8
6		.0283	.0342	.0408	.0480	.0559	.0644	.0734	.0829	.0928	.1030		7
7		.0075	.0096	.0122	.0152	.0186	.0226	.0272	.0323	.0379	.0442		6
8		.0015	.0020	.0027	.0036	.0047	.0060	.0075	.0094	.0116	.0142		5
9		.0002	.0003	.0005	.0006	.0009	.0012	.0015	.0020	.0026	.0034		4
10		.0000	.0000	.0001	.0001	.0001	.0002	.0002	.0003	.0004	.0006		3
11		.0000	.0000	.0000	.0000	.0000	.0000	.0000	.0000	.0000	.0001		2
		.79	.78	.77	.76	.75	.74	.73	.72	.71	.70	P	Q

R	P	.31	.32	.33	.34	.35	.36	.37	.38	.39	.40		
0		.0080	.0066	.0055	.0045	.0037	.0030	.0025	.0020	.0016	.0013		13
1		.0469	.0407	.0351	.0302	.0259	.0221	.0188	.0159	.0135	.0113		12
2		.1265	.1148	.1037	.0933	.0836	.0746	.0663	.0586	.0516	.0453		11
3		.2084	.1981	.1874	.1763	.1651	.1538	.1427	.1317	.1210	.1107		10
4		.2341	.2331	.2307	.2270	.2222	.2163	.2095	.2018	.1934	.1845		9
5		.1893	.1974	.2045	.2105	.2154	.2190	.2215	.2227	.2226	.2214		8
6		.1134	.1239	.1343	.1446	.1546	.1643	.1734	.1820	.1898	.1968		7
7		.0509	.0583	.0662	.0745	.0833	.0924	.1019	.1115	.1213	.1312		6
8		.0172	.0206	.0244	.0288	.0336	.0390	.0449	.0513	.0582	.0656		5
9		.0043	.0054	.0067	.0082	.0101	.0122	.0146	.0175	.0207	.0243		4
10		.0008	.0010	.0013	.0017	.0022	.0027	.0034	.0043	.0053	.0065		3
11		.0001	.0001	.0002	.0002	.0003	.0004	.0006	.0007	.0009	.0012		2
12		.0000	.0000	.0000	.0000	.0000	.0000	.0001	.0001	.0001	.0001		1
		.69	.68	.67	.66	.65	.64	.63	.62	.61	.60	P	R

P	P	.41	.42	.43	.44	.45	.46	.47	.48	.49	.50		
0		.0010	.0008	.0007	.0005	.0004	.0003	.0003	.0002	.0002	.0001		13
1		.0095	.0079	.0066	.0054	.0045	.0037	.0030	.0024	.0020	.0016		12
2		.0395	.0344	.0298	.0256	.0220	.0188	.0160	.0135	.0114	.0095		11
3		.1007	.0913	.0823	.0739	.0660	.0587	.0519	.0457	.0401	.0349		10
4		.1750	.1653	.1553	.1451	.1350	.1250	.1151	.1055	.0962	.0873		9
5		.2189	.2154	.2108	.2053	.1989	.1917	.1838	.1753	.1664	.1571		8
6		.2029	.2080	.2121	.2151	.2169	.2177	.2173	.2158	.2131	.2095		7
7		.1410	.1506	.1600	.1690	.1775	.1854	.1927	.1992	.2048	.2095		6
8		.0735	.0818	.0905	.0996	.1089	.1185	.1282	.1379	.1476	.1571		5
9		.0284	.0329	.0379	.0435	.0495	.0561	.0631	.0707	.0788	.0873		4
10		.0079	.0095	.0114	.0137	.0162	.0191	.0224	.0261	.0303	.0349		3
11		.0015	.0019	.0024	.0029	.0036	.0044	.0054	.0066	.0079	.0095		2
12		.0002	.0002	.0003	.0004	.0005	.0006	.0008	.0010	.0013	.0016		1
13		.0000	.0000	.0000	.0000	.0000	.0000	.0001	.0001	.0001	.0001		0
		.59	.58	.57	.56	.55	.54	.53	.52	.51	.50	P	R

N = 14

P	P	.01	.02	.03	.04	.05	.06	.07	.08	.09	.10		
0		.8687	.7536	.6528	.5647	.4877	.4205	.3620	.3112	.2670	.2288		14
1		.1229	.2153	.2827	.3294	.3593	.3758	.3815	.3788	.3698	.3559		13
2		.0081	.0286	.0568	.0892	.1229	.1559	.1867	.2141	.2377	.2570		12
3		.0003	.0023	.0070	.0149	.0259	.0398	.0562	.0745	.0940	.1142		11
4		.0000	.0001	.0006	.0017	.0037	.0070	.0116	.0178	.0256	.0349		10
5		.0000	.0000	.0000	.0001	.0004	.0009	.0018	.0031	.0051	.0078		9
6		.0000	.0000	.0000	.0000	.0000	.0001	.0002	.0004	.0008	.0013		8
7		.0000	.0000	.0000	.0000	.0000	.0000	.0000	.0000	.0001	.0002		7
		.99	.98	.97	.96	.95	.94	.93	.92	.91	.90	P	R

R	P	.11	.12	.13	.14	.15	.16	.17	.18	.19	.20		
0		.1956	.1670	.1423	.1211	.1028	.0871	.0736	.0621	.0523	.0440		14
1		.3385	.3188	.2977	.2759	.2539	.2322	.2112	.1910	.1719	.1539		13
2		.2720	.2826	.2892	.2919	.2912	.2875	.2811	.2725	.2620	.2501		12
3		.1345	.1542	.1728	.1901	.2056	.2190	.2303	.2393	.2459	.2501		11
4		.0457	.0578	.0710	.0851	.0998	.1147	.1297	.1444	.1586	.1720		10
5		.0113	.0158	.0212	.0277	.0352	.0437	.0531	.0634	.0744	.0860		9
6		.0021	.0032	.0048	.0068	.0093	.0125	.0163	.0209	.0262	.0322		8
7		.0003	.0005	.0009	.0013	.0019	.0027	.0038	.0052	.0070	.0092		7
8		.0000	.0001	.0001	.0002	.0003	.0005	.0007	.0010	.0014	.0020		6
9		.0000	.0000	.0000	.0000	.0000	.0001	.0001	.0001	.0002	.0003		5
		.89	.88	.87	.86	.85	.84	.83	.82	.81	.80	P	R

R	P	.21	.22	.23	.24	.25	.26	.27	.28	.29	.30		
0		.0369	.0309	.0258	.0214	.0178	.0148	.0122	.0101	.0083	.0068		14
1		.1372	.1218	.1077	.0948	.0832	.0726	.0632	.0548	.0473	.0407		13
2		.2371	.2234	.2091	.1946	.1802	.1659	.1519	.1385	.1256	.1134		12
3		.2521	.2520	.2499	.2459	.2402	.2331	.2248	.2154	.2052	.1943		11
4		.1843	.1955	.2052	.2135	.2202	.2252	.2286	.2304	.2305	.2290		10
5		.0980	.1103	.1226	.1348	.1468	.1583	.1691	.1792	.1883	.1963		9
6		.0391	.0466	.0549	.0639	.0734	.0834	.0938	.1045	.1153	.1262		8
7		.0119	.0150	.0188	.0231	.0280	.0335	.0397	.0464	.0538	.0618		7

TABLE C Binomial Distribution: Individual Terms $p(r) = c\binom{n}{r} p^r q^{n-r}$ (*Continued*)

	.79	.78	.77	.76	.75	.74	.73	.72	.71	.70		
8	.0028	.0037	.0049	.0064	.0082	.0103	.0128	.0158	.0192	.0232		6
9	.0005	.0007	.0010	.0013	.0018	.0024	.0032	.0041	.0052	.0066		5
10	.0001	.0001	.0001	.0002	.0003	.0004	.0006	.0008	.0011	.0014		4
11	.0000	.0000	.0000	.0000	.0000	.0001	.0001	.0001	.0002	.0002		3
	.79	.78	.77	.76	.75	.74	.73	.72	.71	.70	P	R

R	P	.31	.32	.33	.34	.35	.36	.37	.38	.39	.40		
0		.0055	.0045	.0037	.0030	.0024	.0019	.0016	.0012	.0010	.0008		14
1		.0349	.0298	.0253	.0215	.0181	.0152	.0128	.0106	.0088	.0073		13
2		.1018	.0911	.0811	.0719	.0634	.0557	.0487	.0424	.0367	.0317		12
3		.1830	.1715	.1599	.1481	.1366	.1253	.1144	.1039	.0940	.0845		11
4		.2261	.2219	.2164	.2098	.2022	.1938	.1848	.1752	.1652	.1549		10
5		.2032	.2088	.2132	.2161	.2178	.2181	.2170	.2147	.2112	.2066		9
6		.1369	.1474	.1575	.1670	.1759	.1840	.1912	.1974	.2026	.2066		8
7		.0703	.0793	.0886	.0983	.1082	.1183	.1283	.1383	.1480	.1574		7
8		.0276	.0326	.0382	.0443	.0510	.0582	.0659	.0742	.0828	.0918		6
9		.0083	.0102	.0125	.0152	.0183	.0218	.0258	.0303	.0353	.0408		5
10		.0019	.0024	.0031	.0039	.0049	.0061	.0076	.0093	.0113	.0136		4
11		.0003	.0004	.0006	.0007	.0010	.0013	.0016	.0021	.0026	.0033		3
12		.0000	.0000	.0001	.0001	.0001	.0002	.0002	.0003	.0004	.0005		2
13		.0000	.0000	.0000	.0000	.0000	.0000	.0000	.0000	.0000	.0001		1
		.69	.68	.67	.66	.65	.64	.63	.62	.61	.60	P	R

R	P	.41	.42	.43	.44	.45	.46	.47	.48	.49	.50		
0		.0006	.0005	.0004	.0003	.0002	.0002	.0001	.0001	.0001	.0001		14
1		.0060	.0049	.0040	.0033	.0027	.0021	.0017	.0014	.0011	.0009		13
2		.0272	.0233	.0198	.0168	.0141	.0118	.0099	.0082	.0068	.0056		12
3		.0757	.0674	.0597	.0527	.0462	.0403	.0350	.0303	.0260	.0222		11
4		.1446	.1342	.1239	.1138	.1040	.0945	.0854	.0768	.0687	.0611		10
5		.2009	.1943	.1869	.1788	.1701	.1610	.1515	.1418	.1320	.1222		9
6		.2094	.2111	.2115	.2108	.2088	.2057	.2015	.1963	.1902	.1833		8
7		.1663	.1747	.1824	.1892	.1952	.2003	.2043	.2071	.2089	.2095		7
8		.1011	.1107	.1204	.1301	.1398	.1493	.1585	.1673	.1756	.1833		6
9		.0469	.0534	.0605	.0682	.0762	.0848	.0937	.1030	.1125	.1222		5
10		.0163	.0193	.0228	.0268	.0312	.0361	.0415	.0475	.0540	.0611		4
11		.0041	.0051	.0063	.0076	.0093	.0112	.0134	.0160	.0189	.0222		3
12		.0007	.0009	.0012	.0015	.0019	.0024	.0030	.0037	.0045	.0056		2
13		.0001	.0001	.0001	.0002	.0002	.0003	.0004	.0005	.0007	.0009		1
14		.0000	.0000	.0000	.0000	.0000	.0000	.0000	.0000	.0000	.0001		0
		.59	.58	.57	.56	.55	.54	.53	.52	.51	.50	P	R

N = 15

R	P	.01	.02	.03	.04	.05	.06	.07	.08	.09	.10		
0		.8601	.7386	.6333	.5421	.4633	.3953	.3367	.2863	.2430	.2059		15
1		.1303	.2261	.2938	.3388	.3658	.3785	.3801	.3734	.3605	.3432		14
2		.0092	.0323	.0636	.0988	.1348	.1691	.2003	.2273	.2496	.2669		13
3		.0004	.0029	.0085	.0178	.0307	.0468	.0653	.0857	.1070	.1285		12
4		.0000	.0002	.0008	.0022	.0049	.0090	.0148	.0223	.0317	.0428		11
5		.0000	.0000	.0001	.0002	.0006	.0013	.0024	.0043	.0069	.0105		10
6		.0000	.0000	.0000	.0000	.0000	.0001	.0003	.0006	.0011	.0019		9
7		.0000	.0000	.0000	.0000	.0000	.0000	.0000	.0001	.0001	.0003		8
		.99	.98	.97	.96	.95	.94	.93	.92	.91	.90	P	P

R	P	.11	.12	.13	.14	.15	.16	.17	.18	.19	.20		
0		.1741	.1470	.1238	.1041	.0874	.0731	.0611	.0510	.0424	.0352		15
1		.3228	.3006	.2775	.2542	.2312	.2090	.1878	.1678	.1492	.1319		14
2		.2793	.2870	.2903	.2897	.2856	.2787	.2692	.2578	.2449	.2309		13
3		.1496	.1696	.1880	.2044	.2184	.2300	.2389	.2452	.2489	.2501		12
4		.0555	.0694	.0843	.0998	.1156	.1314	.1468	.1615	.1752	.1876		11
5		.0151	.0208	.0277	.0357	.0449	.0551	.0662	.0780	.0904	.1032		10
6		.0031	.0047	.0069	.0097	.0132	.0175	.0226	.0285	.0353	.0430		9
7		.0005	.0008	.0013	.0020	.0030	.0043	.0059	.0081	.0107	.0138		8
8		.0001	.0001	.0002	.0003	.0005	.0008	.0012	.0018	.0025	.0035		7
9		.0000	.0000	.0000	.0000	.0001	.0001	.0002	.0003	.0005	.0007		6
10		.0000	.0000	.0000	.0000	.0000	.0000	.0000	.0000	.0001	.0001		5
		.89	.88	.87	.86	.85	.84	.83	.82	.81	.80	P	R

R	P	.21	.22	.23	.24	.25	.26	.27	.28	.29	.30		
0		.0291	.0241	.0198	.0163	.0134	.0109	.0089	.0072	.0059	.0047		15
1		.1162	.1018	.0889	.0772	.0668	.0576	.0494	.0423	.0360	.0305		14
2		.2162	.2010	.1858	.1707	.1559	.1416	.1280	.1150	.1029	.0916		13
3		.2490	.2457	.2405	.2336	.2252	.2156	.2051	.1939	.1821	.1700		12
4		.1986	.2079	.2155	.2213	.2252	.2273	.2276	.2262	.2231	.2186		11
5		.1161	.1290	.1416	.1537	.1651	.1757	.1852	.1935	.2005	.2061		10
6		.0514	.0606	.0705	.0809	.0917	.1029	.1142	.1254	.1365	.1472		9
7		.0176	.0220	.0271	.0329	.0393	.0465	.0543	.0627	.0717	.0811		8
8		.0047	.0062	.0081	.0104	.0131	.0163	.0201	.0244	.0293	.0348		7
9		.0010	.0014	.0019	.0025	.0034	.0045	.0058	.0074	.0093	.0116		6
10		.0002	.0002	.0003	.0005	.0007	.0009	.0013	.0017	.0023	.0030		5
11		.0000	.0000	.0000	.0001	.0001	.0002	.0002	.0003	.0004	.0006		4
12		.0000	.0000	.0000	.0000	.0000	.0000	.0000	.0000	.0001	.0001		3
		.79	.78	.77	.76	.75	.74	.73	.72	.71	.70	P	R

TABLE C Binomial Distribution: Individual Terms $p(r) = c(_r^n) p^r q^{n-r}$ (Continued)

P	P	.31	.32	.33	.34	.35	.36	.37	.38	.39	.40	
0		.0038	.0031	.0025	.0020	.0016	.0012	.0010	.0008	.0006	.0005	15
1		.0258	.0217	.0182	.0152	.0126	.0104	.0086	.0071	.0058	.0047	14
2		.0811	.0715	.0627	.0547	.0476	.0411	.0354	.0303	.0259	.0219	13
3		.1579	.1457	.1338	.1222	.1110	.1002	.0901	.0805	.0716	.0634	12
4		.2128	.2057	.1977	.1888	.1792	.1692	.1587	.1481	.1374	.1268	11
5		.2103	.2130	.2142	.2140	.2123	.2093	.2051	.1997	.1933	.1859	10
6		.1575	.1671	.1759	.1837	.1906	.1963	.2008	.2040	.2059	.2066	9
7		.0910	.1011	.1114	.1217	.1319	.1419	.1516	.1608	.1693	.1771	8
8		.0409	.0476	.0549	.0627	.0710	.0798	.0890	.0985	.1082	.1181	7
9		.0143	.0174	.0210	.0251	.0298	.0349	.0407	.0470	.0538	.0612	6
10		.0038	.0049	.0062	.0078	.0096	.0118	.0143	.0173	.0206	.0245	5
11		.0008	.0011	.0014	.0018	.0024	.0030	.0038	.0048	.0060	.0074	4
12		.0001	.0002	.0002	.0003	.0004	.0006	.0007	.0010	.0013	.0016	3
13		.0000	.0000	.0000	.0000	.0001	.0001	.0001	.0001	.0002	.0003	2
		.69	.68	.67	.66	.65	.64	.63	.62	.61	.60	P R

P	P	.41	.42	.43	.44	.45	.46	.47	.48	.49	.50	
0		.0004	.0003	.0002	.0002	.0001	.0001	.0001	.0001	.0000	.0000	15
1		.0038	.0031	.0025	.0020	.0016	.0012	.0010	.0008	.0006	.0005	14
2		.0185	.0156	.0130	.0108	.0090	.0074	.0060	.0049	.0040	.0032	13
3		.0558	.0489	.0426	.0369	.0318	.0272	.0232	.0197	.0166	.0139	12
4		.1163	.1061	.0963	.0869	.0780	.0696	.0617	.0545	.0478	.0417	11
5		.1778	.1691	.1598	.1502	.1404	.1304	.1204	.1106	.1010	.0916	10
6		.2060	.2041	.2010	.1967	.1914	.1851	.1780	.1702	.1617	.1527	9
7		.1840	.1900	.1949	.1987	.2013	.2028	.2030	.2020	.1997	.1964	8
8		.1279	.1376	.1470	.1561	.1647	.1727	.1800	.1864	.1919	.1964	7
9		.0691	.0775	.0863	.0954	.1048	.1144	.1241	.1338	.1434	.1527	6
10		.0288	.0337	.0390	.0450	.0515	.0585	.0661	.0741	.0827	.0916	5
11		.0091	.0111	.0134	.0161	.0191	.0226	.0266	.0311	.0361	.0417	4
12		.0021	.0027	.0034	.0042	.0052	.0064	.0079	.0096	.0116	.0139	3
13		.0003	.0004	.0006	.0008	.0010	.0013	.0016	.0020	.0026	.0032	2
14		.0000	.0000	.0001	.0001	.0001	.0002	.0002	.0003	.0004	.0005	1
		.59	.58	.57	.56	.55	.54	.53	.52	.51	.50	P R

N = 16

P	P	.01	.02	.03	.04	.05	.06	.07	.08	.09	.10	
0		.8515	.7238	.6143	.5204	.4401	.3716	.3131	.2634	.2211	.1853	16
1		.1376	.2363	.3040	.3469	.3706	.3795	.3771	.3665	.3499	.3294	15
2		.0104	.0362	.0705	.1084	.1463	.1817	.2129	.2390	.2596	.2745	14
3		.0005	.0034	.0102	.0211	.0359	.0541	.0748	.0990	.1198	.1423	13
4		.0000	.0002	.0010	.0029	.0061	.0112	.0183	.0274	.0385	.0514	12
5		.0000	.0000	.0001	.0003	.0008	.0017	.0033	.0057	.0091	.0137	11
6		.0000	.0000	.0000	.0000	.0001	.0002	.0005	.0009	.0017	.0028	10
7		.0000	.0000	.0000	.0000	.0000	.0000	.0000	.0001	.0002	.0004	9
8		.0000	.0000	.0000	.0000	.0000	.0000	.0000	.0000	.0000	.0001	8
		.99	.98	.97	.96	.95	.94	.93	.92	.91	.90	P R

P	P	.11	.12	.13	.14	.15	.16	.17	.18	.19	.20	
0		.1550	.1293	.1077	.0895	.0743	.0614	.0507	.0418	.0343	.0281	16
1		.3065	.2822	.2575	.2332	.2097	.1873	.1662	.1468	.1289	.1126	15
2		.2841	.2886	.2886	.2847	.2775	.2675	.2554	.2416	.2267	.2111	14
3		.1638	.1837	.2013	.2163	.2285	.2378	.2441	.2475	.2482	.2463	13
4		.0658	.0814	.0977	.1144	.1311	.1472	.1625	.1766	.1892	.2001	12
5		.0195	.0266	.0351	.0447	.0555	.0673	.0799	.0930	.1065	.1201	11
6		.0044	.0067	.0096	.0133	.0180	.0235	.0300	.0374	.0458	.0550	10
7		.0008	.0013	.0020	.0031	.0045	.0064	.0088	.0117	.0153	.0197	9
8		.0001	.0002	.0003	.0006	.0009	.0014	.0020	.0029	.0041	.0055	8
9		.0000	.0000	.0000	.0001	.0001	.0002	.0004	.0006	.0008	.0012	7
10		.0000	.0000	.0000	.0000	.0000	.0000	.0001	.0001	.0001	.0002	6
		.89	.88	.87	.86	.85	.84	.83	.82	.81	.80	P R

P	P	.21	.22	.23	.24	.25	.26	.27	.28	.29	.30	
0		.0230	.0188	.0153	.0124	.0100	.0081	.0065	.0052	.0042	.0033	16
1		.0979	.0847	.0730	.0626	.0535	.0455	.0385	.0325	.0273	.0228	15
2		.1952	.1792	.1635	.1482	.1336	.1198	.1068	.0947	.0835	.0732	14
3		.2421	.2359	.2279	.2185	.2079	.1964	.1843	.1718	.1591	.1465	13
4		.2092	.2162	.2212	.2242	.2252	.2243	.2215	.2171	.2112	.2040	12
5		.1334	.1464	.1586	.1699	.1802	.1891	.1966	.2026	.2071	.2099	11
6		.0650	.0757	.0869	.0984	.1101	.1218	.1333	.1445	.1551	.1649	10
7		.0247	.0305	.0371	.0444	.0524	.0611	.0704	.0803	.0905	.1010	9
8		.0074	.0097	.0125	.0158	.0197	.0242	.0293	.0351	.0416	.0487	8
9		.0017	.0024	.0033	.0044	.0058	.0075	.0096	.0121	.0151	.0185	7
10		.0003	.0005	.0007	.0010	.0014	.0019	.0025	.0033	.0043	.0056	6
11		.0000	.0001	.0001	.0002	.0002	.0004	.0005	.0007	.0010	.0013	5
12		.0000	.0000	.0000	.0000	.0000	.0001	.0001	.0001	.0002	.0002	4
		.79	.78	.77	.76	.75	.74	.73	.72	.71	.70	P R

P	P	.31	.32	.33	.34	.35	.36	.37	.38	.39	.40	
0		.0026	.0021	.0016	.0013	.0010	.0008	.0006	.0005	.0004	.0003	16
1		.0190	.0157	.0130	.0107	.0087	.0071	.0058	.0047	.0038	.0030	15
2		.0639	.0555	.0480	.0413	.0353	.0301	.0255	.0215	.0180	.0150	14

TABLE C Binomial Distribution: Individual Terms $p(r) = c\binom{n}{r} p^r q^{n-r}$ (*Continued*)

R	.69	.68	.67	.66	.65	.64	.63	.62	.61	.60	R
3	.1341	.1220	.1103	.0992	.0888	.0790	.0699	.0615	.0538	.0468	13
4	.1958	.1855	.1766	.1662	.1553	.1444	.1333	.1224	.1118	.1014	12
5	.2111	.2107	.2088	.2054	.2008	.1949	.1879	.1801	.1715	.1623	11
6	.1739	.1818	.1885	.1940	.1982	.2010	.2024	.2024	.2010	.1983	10
7	.1116	.1222	.1326	.1428	.1524	.1615	.1698	.1772	.1835	.1889	9
8	.0564	.0647	.0735	.0827	.0923	.1022	.1122	.1222	.1320	.1417	8
9	.0225	.0271	.0322	.0379	.0442	.0511	.0586	.0666	.0753	.0840	7
10	.0071	.0089	.0111	.0137	.0167	.0201	.0241	.0286	.0335	.0392	6
11	.0017	.0023	.0030	.0038	.0049	.0062	.0077	.0095	.0117	.0142	5
12	.0003	.0004	.0006	.0008	.0011	.0014	.0019	.0024	.0031	.0040	4
13	.0000	.0001	.0001	.0001	.0002	.0003	.0003	.0005	.0005	.0008	3
14	.0000	.0000	.0000	.0000	.0000	.0000	.0000	.0001	.0001	.0001	2
	.69	.68	.67	.66	.65	.64	.63	.62	.61	.60	P
											R

R P	.41	.42	.43	.44	.45	.46	.47	.48	.49	.50	
0	.0002	.0002	.0001	.0001	.0001	.0001	.0000	.0000	.0000	.0000	16
1	.0024	.0019	.0015	.0012	.0009	.0007	.0005	.0004	.0003	.0002	15
2	.0125	.0103	.0085	.0069	.0056	.0046	.0037	.0029	.0023	.0018	14
3	.0405	.0349	.0299	.0254	.0215	.0181	.0151	.0126	.0104	.0085	13
4	.0915	.0821	.0732	.0649	.0572	.0501	.0436	.0378	.0325	.0278	12
5	.1526	.1426	.1325	.1224	.1123	.1024	.0929	.0837	.0749	.0667	11
6	.1944	.1894	.1833	.1762	.1684	.1600	.1510	.1416	.1319	.1222	10
7	.1930	.1959	.1975	.1978	.1969	.1947	.1912	.1867	.1811	.1746	9
8	.1509	.1596	.1676	.1749	.1812	.1865	.1908	.1939	.1958	.1964	8
9	.0932	.1027	.1124	.1221	.1318	.1413	.1504	.1591	.1672	.1746	7
10	.0453	.0521	.0594	.0672	.0755	.0842	.0934	.1028	.1124	.1222	6
11	.0172	.0206	.0244	.0288	.0337	.0391	.0452	.0518	.0589	.0667	5
12	.0050	.0062	.0077	.0094	.0115	.0139	.0167	.0199	.0236	.0278	4
13	.0011	.0014	.0018	.0023	.0029	.0036	.0046	.0057	.0070	.0085	3
14	.0002	.0002	.0003	.0004	.0005	.0007	.0009	.0011	.0014	.0018	2
15	.0000	.0000	.0000	.0000	.0001	.0001	.0001	.0001	.0002	.0002	1
	.59	.58	.57	.56	.55	.54	.53	.52	.51	.50	P
											R

N = 17

P P	.01	.02	.03	.04	.05	.06	.07	.08	.09	.10	
0	.8429	.7093	.5958	.4996	.4181	.3493	.2912	.2423	.2012	.1668	17
1	.1447	.2461	.3133	.3539	.3741	.3790	.3726	.3582	.3383	.3150	16
2	.0117	.0402	.0775	.1180	.1575	.1935	.2244	.2492	.2677	.2800	15
3	.0006	.0041	.0120	.0246	.0415	.0618	.0844	.1083	.1324	.1556	14
4	.0000	.0003	.0013	.0036	.0076	.0138	.0222	.0330	.0458	.0605	13
5	.0000	.0000	.0001	.0004	.0010	.0023	.0044	.0075	.0118	.0175	12
6	.0000	.0000	.0000	.0000	.0001	.0003	.0007	.0013	.0023	.0039	11
7	.0000	.0000	.0000	.0000	.0000	.0000	.0001	.0002	.0004	.0007	10
8	.0000	.0000	.0000	.0000	.0000	.0000	.0000	.0000	.0000	.0001	9
	.99	.98	.97	.96	.95	.94	.93	.92	.91	.90	P
											R

P P	.11	.12	.13	.14	.15	.16	.17	.18	.19	.20	
0	.1379	.1138	.0937	.0770	.0631	.0516	.0421	.0343	.0278	.0225	17
1	.2898	.2638	.2381	.2131	.1893	.1671	.1466	.1279	.1109	.0957	16
2	.2865	.2878	.2846	.2775	.2673	.2547	.2402	.2245	.2081	.1914	15
3	.1771	.1963	.2126	.2259	.2359	.2425	.2460	.2464	.2441	.2393	14
4	.0766	.0937	.1112	.1287	.1457	.1617	.1764	.1893	.2004	.2093	13
5	.0246	.0332	.0432	.0545	.0668	.0801	.0939	.1081	.1222	.1361	12
6	.0061	.0091	.0129	.0177	.0236	.0305	.0385	.0474	.0573	.0680	11
7	.0012	.0019	.0039	.0045	.0065	.0091	.0124	.0164	.0211	.0267	10
8	.0002	.0003	.0006	.0009	.0014	.0022	.0032	.0045	.0062	.0084	9
9	.0000	.0000	.0001	.0002	.0003	.0004	.0006	.0010	.0015	.0021	8
10	.0000	.0000	.0000	.0000	.0000	.0001	.0001	.0002	.0003	.0004	7
11	.0000	.0000	.0000	.0000	.0000	.0000	.0000	.0000	.0000	.0001	6
	.89	.88	.87	.86	.85	.84	.83	.82	.81	.80	P
											R

P P	.21	.22	.23	.24	.25	.26	.27	.28	.29	.30	
0	.0182	.0146	.0118	.0094	.0075	.0060	.0047	.0038	.0030	.0023	17
1	.0822	.0702	.0597	.0505	.0426	.0357	.0299	.0248	.0206	.0169	16
2	.1747	.1584	.1427	.1277	.1136	.1005	.0883	.0772	.0672	.0581	15
3	.2322	.2234	.2131	.2016	.1893	.1765	.1634	.1502	.1372	.1245	14
4	.2161	.2205	.2228	.2228	.2209	.2170	.2115	.2044	.1961	.1868	13
5	.1493	.1617	.1730	.1830	.1914	.1982	.2033	.2067	.2083	.2081	12
6	.0794	.0912	.1034	.1156	.1276	.1393	.1504	.1608	.1701	.1784	11
7	.0332	.0404	.0485	.0573	.0668	.0769	.0874	.0982	.1092	.1201	10
8	.0110	.0143	.0181	.0226	.0279	.0338	.0404	.0478	.0558	.0644	9
9	.0029	.0040	.0054	.0071	.0093	.0119	.0150	.0186	.0228	.0276	8
10	.0006	.0009	.0013	.0018	.0025	.0033	.0044	.0058	.0074	.0095	7
11	.0001	.0002	.0002	.0004	.0005	.0007	.0010	.0014	.0019	.0026	6
12	.0000	.0000	.0000	.0001	.0001	.0001	.0002	.0003	.0004	.0006	5
13	.0000	.0000	.0000	.0000	.0000	.0000	.0000	.0000	.0001	.0001	4
	.79	.78	.77	.76	.75	.74	.73	.72	.71	.70	P
											R

P P	.31	.32	.33	.34	.35	.36	.37	.38	.39	.40	
0	.0018	.0014	.0011	.0009	.0007	.0005	.0004	.0003	.0002	.0002	17
1	.0139	.0114	.0093	.0075	.0060	.0048	.0039	.0031	.0024	.0019	16
2	.0500	.0428	.0364	.0309	.0260	.0218	.0182	.0151	.0125	.0102	15

TABLE C Binomial Distribution: Individual Terms $p(r) = c\binom{n}{r} p^r q^{n-r}$ *(Continued)*

	.69	.68	.67	.66	.65	.64	.63	.62	.61	.60		
3	.1123	.1007	.0898	.0795	.0701	.0614	.0534	.0463	.0398	.0341		14
4	.1766	.1659	.1547	.1434	.1320	.1208	.1099	.0993	.0892	.0796		13
5	.2063	.2030	.1982	.1921	.1849	.1767	.1677	.1582	.1482	.1379		12
6	.1854	.1910	.1952	.1979	.1991	.1988	.1970	.1939	.1895	.1839		11
7	.1309	.1413	.1511	.1602	.1685	.1757	.1818	.1868	.1904	.1927		10
8	.0735	.0831	.0930	.1032	.1134	.1235	.1335	.1431	.1521	.1606		9
9	.0330	.0391	.0458	.0531	.0611	.0695	.0784	.0877	.0973	.1070		8
10	.0119	.0147	.0181	.0219	.0263	.0313	.0368	.0430	.0498	.0571		7
11	.0034	.0044	.0057	.0072	.0090	.0112	.0138	.0168	.0202	.0242		6
12	.0008	.0010	.0014	.0018	.0024	.0031	.0040	.0051	.0065	.0081		5
13	.0001	.0002	.0003	.0004	.0005	.0007	.0009	.0012	.0016	.0021		4
14	.0000	.0000	.0000	.0001	.0001	.0001	.0002	.0002	.0003	.0004		3
15	.0000	.0000	.0000	.0000	.0000	.0000	.0000	.0000	.0000	.0001		2
	.69	.68	.67	.66	.65	.64	.63	.62	.61	.60	P	R

R	P	.41	.42	.43	.44	.45	.46	.47	.48	.49	.50		
0		.0001	.0001	.0001	.0001	.0000	.0000	.0000	.0000	.0000	.0000		17
1		.0015	.0012	.0009	.0007	.0005	.0004	.0003	.0002	.0002	.0001		16
2		.0094	.0068	.0055	.0044	.0035	.0028	.0022	.0017	.0013	.0010		15
3		.0290	.0246	.0207	.0173	.0144	.0119	.0097	.0079	.0064	.0052		14
4		.0706	.0622	.0546	.0475	.0411	.0354	.0302	.0257	.0217	.0182		13
5		.1276	.1172	.1070	.0971	.0875	.0784	.0697	.0616	.0541	.0472		12
6		.1773	.1697	.1614	.1525	.1432	.1335	.1237	.1138	.1040	.0944		11
7		.1936	.1932	.1914	.1883	.1841	.1787	.1723	.1650	.1570	.1484		10
8		.1692	.1748	.1805	.1850	.1883	.1903	.1910	.1904	.1886	.1855		9
9		.1169	.1266	.1361	.1453	.1540	.1621	.1694	.1758	.1812	.1855		8
10		.0650	.0733	.0822	.0914	.1008	.1105	.1202	.1298	.1393	.1484		7
11		.0287	.0338	.0394	.0457	.0525	.0599	.0678	.0763	.0851	.0944		6
12		.0100	.0122	.0149	.0179	.0215	.0255	.0301	.0352	.0409	.0472		5
13		.0027	.0034	.0043	.0054	.0068	.0084	.0103	.0125	.0151	.0182		4
14		.0005	.0007	.0009	.0012	.0016	.0020	.0026	.0033	.0041	.0052		3
15		.0001	.0001	.0001	.0002	.0003	.0003	.0005	.0006	.0008	.0010		2
16		.0000	.0000	.0000	.0000	.0000	.0000	.0001	.0001	.0001	.0001		1
		.59	.58	.57	.56	.55	.54	.53	.52	.51	.50	P	R

N = 18

R	P	.01	.02	.03	.04	.05	.06	.07	.08	.09	.10		
0		.8345	.6951	.5780	.4796	.3972	.3283	.2708	.2229	.1831	.1501		18
1		.1517	.2554	.3217	.3597	.3763	.3772	.3669	.3489	.3260	.3002		17
2		.0130	.0443	.0846	.1274	.1683	.2047	.2348	.2579	.2741	.2835		16
3		.0007	.0048	.0140	.0283	.0473	.0697	.0942	.1196	.1446	.1680		15
4		.0000	.0004	.0016	.0044	.0093	.0167	.0266	.0390	.0536	.0700		14
5		.0000	.0000	.0001	.0005	.0014	.0030	.0056	.0095	.0148	.0218		13
6		.0000	.0000	.0000	.0000	.0002	.0004	.0009	.0018	.0032	.0052		12
7		.0000	.0000	.0000	.0000	.0000	.0000	.0001	.0003	.0005	.0010		11
8		.0000	.0000	.0000	.0000	.0000	.0000	.0000	.0000	.0001	.0002		10
		.99	.98	.97	.96	.95	.94	.93	.92	.91	.90	P	R

R	P	.11	.12	.13	.14	.15	.16	.17	.18	.19	.20		
0		.1227	.1002	.0815	.0662	.0536	.0434	.0349	.0281	.0225	.0180		18
1		.2731	.2458	.2193	.1940	.1704	.1486	.1288	.1110	.0951	.0811		17
2		.2869	.2850	.2785	.2685	.2556	.2407	.2243	.2071	.1897	.1723		16
3		.1891	.2072	.2220	.2331	.2406	.2445	.2450	.2425	.2373	.2297		15
4		.0877	.1060	.1244	.1423	.1592	.1746	.1882	.1996	.2087	.2153		14
5		.0303	.0405	.0520	.0649	.0787	.0931	.1079	.1227	.1371	.1507		13
6		.0081	.0120	.0168	.0229	.0301	.0384	.0479	.0584	.0697	.0816		12
7		.0017	.0028	.0043	.0064	.0091	.0126	.0168	.0220	.0280	.0350		11
8		.0003	.0005	.0009	.0014	.0022	.0033	.0047	.0066	.0090	.0120		10
9		.0000	.0001	.0001	.0003	.0004	.0007	.0011	.0016	.0024	.0033		9
10		.0000	.0000	.0000	.0000	.0001	.0001	.0002	.0003	.0005	.0008		8
11		.0000	.0000	.0000	.0000	.0000	.0000	.0000	.0001	.0001	.0001		7
		.89	.88	.87	.86	.85	.84	.83	.82	.81	.80	P	R

R	P	.21	.22	.23	.24	.25	.26	.27	.28	.29	.30		
0		.0144	.0114	.0091	.0072	.0056	.0044	.0035	.0027	.0021	.0016		18
1		.0687	.0580	.0487	.0407	.0338	.0280	.0231	.0189	.0155	.0126		17
2		.1553	.1390	.1236	.1092	.0958	.0836	.0725	.0626	.0537	.0458		16
3		.2202	.2091	.1969	.1839	.1704	.1567	.1431	.1298	.1169	.1046		15
4		.2195	.2212	.2205	.2177	.2130	.2065	.1985	.1892	.1790	.1681		14
5		.1634	.1747	.1845	.1925	.1988	.2031	.2055	.2061	.2048	.2017		13
6		.0941	.1067	.1194	.1317	.1436	.1546	.1647	.1736	.1812	.1873		12
7		.0429	.0516	.0611	.0713	.0820	.0931	.1044	.1157	.1269	.1376		11
8		.0157	.0200	.0251	.0310	.0376	.0450	.0531	.0619	.0713	.0811		10
9		.0046	.0063	.0083	.0109	.0139	.0176	.0218	.0267	.0323	.0386		9
10		.0011	.0016	.0022	.0031	.0042	.0056	.0073	.0094	.0119	.0149		8
11		.0002	.0003	.0005	.0007	.0010	.0014	.0020	.0026	.0035	.0046		7
12		.0000	.0001	.0001	.0001	.0002	.0003	.0004	.0006	.0008	.0012		6
13		.0000	.0000	.0000	.0000	.0000	.0000	.0001	.0001	.0002	.0002		5
		.79	.78	.77	.76	.75	.74	.73	.72	.71	.70	P	R

R	P	.31	.32	.33	.34	.35	.36	.37	.38	.39	.40		
0		.0013	.0010	.0007	.0006	.0004	.0003	.0002	.0002	.0001	.0001		18

TABLE C Binomial Distribution: Individual Terms $p(r) = c\binom{n}{r} p^r q^{n-r}$ *(Continued)*

R	.69	.68	.67	.66	.65	.64	.63	.62	.61	.60	R
1	.0102	.0082	.0066	.0052	.0042	.0033	.0026	.0020	.0016	.0012	17
2	.0388	.0327	.0275	.0229	.0190	.0157	.0129	.0105	.0086	.0069	16
3	.0930	.0822	.0722	.0630	.0547	.0471	.0404	.0344	.0292	.0246	15
4	.1567	.1450	.1333	.1217	.1104	.0994	.0890	.0791	.0699	.0614	14
5	.1971	.1911	.1838	.1755	.1664	.1566	.1463	.1358	.1252	.1146	13
6	.1919	.1948	.1962	.1959	.1941	.1908	.1862	.1803	.1734	.1655	12
7	.1478	.1572	.1656	.1730	.1792	.1840	.1875	.1895	.1900	.1892	11
8	.0913	.1017	.1122	.1226	.1327	.1423	.1514	.1597	.1671	.1734	10
9	.0456	.0532	.0614	.0701	.0794	.0890	.0988	.1087	.1187	.1284	9
10	.0184	.0225	.0272	.0325	.0385	.0450	.0522	.0600	.0683	.0771	8
11	.0060	.0077	.0097	.0122	.0151	.0184	.0223	.0267	.0318	.0374	7
12	.0016	.0021	.0028	.0037	.0047	.0060	.0076	.0096	.0118	.0145	6
13	.0003	.0005	.0006	.0009	.0012	.0016	.0021	.0027	.0035	.0045	5
14	.0001	.0001	.0001	.0002	.0002	.0003	.0004	.0006	.0008	.0011	4
15	.0000	.0000	.0000	.0000	.0000	.0000	.0001	.0001	.0001	.0002	3
	.69	.68	.67	.66	.65	.64	.63	.62	.61	.60	P / R

P / R	.41	.42	.43	.44	.45	.46	.47	.48	.49	.50	R
0	.0001	.0001	.0000	.0000	.0000	.0000	.0000	.0000	.0000	.0000	18
1	.0009	.0007	.0005	.0004	.0003	.0002	.0002	.0001	.0001	.0001	17
2	.0055	.0044	.0035	.0028	.0022	.0017	.0013	.0010	.0008	.0006	16
3	.0206	.0171	.0141	.0116	.0095	.0077	.0062	.0050	.0039	.0031	15
4	.0536	.0464	.0400	.0342	.0291	.0246	.0206	.0172	.0142	.0117	14
5	.1042	.0941	.0844	.0753	.0666	.0586	.0512	.0444	.0382	.0327	13
6	.1569	.1477	.1380	.1281	.1181	.1081	.0983	.0887	.0795	.0708	12
7	.1869	.1833	.1785	.1726	.1657	.1579	.1494	.1404	.1310	.1214	11
8	.1786	.1825	.1852	.1864	.1864	.1850	.1822	.1782	.1731	.1669	10
9	.1379	.1469	.1552	.1628	.1694	.1751	.1795	.1828	.1843	.1855	9
10	.0862	.0957	.1054	.1151	.1248	.1342	.1433	.1519	.1598	.1669	8
11	.0436	.0504	.0578	.0658	.0742	.0831	.0924	.1020	.1117	.1214	7
12	.0177	.0213	.0254	.0301	.0354	.0413	.0478	.0549	.0626	.0708	6
13	.0057	.0071	.0089	.0109	.0134	.0162	.0196	.0234	.0278	.0327	5
14	.0014	.0018	.0024	.0031	.0039	.0049	.0062	.0077	.0095	.0117	4
15	.0003	.0004	.0005	.0006	.0009	.0011	.0015	.0019	.0024	.0031	3
16	.0000	.0000	.0001	.0001	.0001	.0002	.0002	.0003	.0004	.0006	2
17	.0000	.0000	.0000	.0000	.0000	.0000	.0000	.0000	.0000	.0001	1
	.59	.58	.57	.56	.55	.54	.53	.52	.51	.50	P / R

N = 19

R / P	.01	.02	.03	.04	.05	.06	.07	.08	.09	.10	R
0	.8262	.6812	.5606	.4604	.3774	.3086	.2519	.2051	.1666	.1351	19
1	.1586	.2642	.3294	.3645	.3774	.3743	.3602	.3389	.3131	.2852	18
2	.0144	.0485	.0917	.1367	.1787	.2150	.2440	.2652	.2787	.2852	17
3	.0008	.0056	.0161	.0323	.0533	.0778	.1041	.1307	.1562	.1796	16
4	.0000	.0005	.0020	.0054	.0112	.0199	.0313	.0455	.0618	.0798	15
5	.0000	.0000	.0002	.0007	.0018	.0038	.0071	.0119	.0183	.0266	14
6	.0000	.0000	.0000	.0001	.0002	.0006	.0012	.0024	.0042	.0069	13
7	.0000	.0000	.0000	.0000	.0000	.0001	.0002	.0004	.0008	.0014	12
8	.0000	.0000	.0000	.0000	.0000	.0000	.0000	.0001	.0001	.0002	11
	.99	.98	.97	.96	.95	.94	.93	.92	.91	.90	P / R

R / P	.11	.12	.13	.14	.15	.16	.17	.18	.19	.20	R
0	.1092	.0881	.0709	.0569	.0456	.0364	.0290	.0230	.0182	.0144	19
1	.2565	.2294	.2014	.1761	.1529	.1318	.1129	.0961	.0813	.0685	18
2	.2854	.2803	.2708	.2581	.2428	.2259	.2081	.1898	.1717	.1540	17
3	.1999	.2166	.2293	.2381	.2428	.2439	.2415	.2361	.2282	.2182	16
4	.0988	.1181	.1371	.1550	.1714	.1858	.1979	.2073	.2141	.2182	15
5	.0366	.0483	.0614	.0757	.0907	.1062	.1216	.1365	.1507	.1636	14
6	.0106	.0154	.0214	.0288	.0374	.0472	.0581	.0699	.0825	.0955	13
7	.0024	.0039	.0059	.0087	.0122	.0167	.0221	.0285	.0359	.0443	12
8	.0004	.0008	.0013	.0021	.0032	.0048	.0068	.0094	.0126	.0166	11
9	.0001	.0001	.0002	.0004	.0007	.0011	.0017	.0025	.0036	.0051	10
10	.0000	.0000	.0000	.0001	.0001	.0002	.0003	.0006	.0009	.0013	9
11	.0000	.0000	.0000	.0000	.0000	.0000	.0001	.0001	.0002	.0003	8
	.89	.88	.87	.86	.85	.84	.83	.82	.81	.80	P / R

R / P	.21	.22	.23	.24	.25	.26	.27	.28	.29	.30	R
0	.0113	.0089	.0070	.0054	.0042	.0033	.0025	.0019	.0015	.0011	19
1	.0573	.0477	.0396	.0326	.0268	.0219	.0178	.0144	.0116	.0093	18
2	.1371	.1212	.1064	.0927	.0803	.0692	.0592	.0503	.0426	.0358	17
3	.2065	.1937	.1800	.1659	.1517	.1377	.1240	.1109	.0985	.0869	16
4	.2196	.2185	.2151	.2096	.2023	.1935	.1835	.1726	.1610	.1491	15
5	.1751	.1849	.1928	.1986	.2023	.2040	.2036	.2013	.1973	.1916	14
6	.1086	.1217	.1343	.1463	.1574	.1672	.1757	.1827	.1880	.1916	13
7	.0536	.0637	.0745	.0858	.0974	.1091	.1207	.1320	.1426	.1525	12
8	.0214	.0270	.0334	.0406	.0487	.0575	.0670	.0770	.0874	.0981	11
9	.0069	.0093	.0122	.0157	.0198	.0247	.0303	.0366	.0436	.0514	10
10	.0018	.0026	.0036	.0050	.0066	.0087	.0112	.0142	.0178	.0220	9
11	.0004	.0006	.0009	.0013	.0018	.0025	.0034	.0045	.0060	.0077	8
12	.0001	.0001	.0002	.0003	.0004	.0006	.0008	.0012	.0016	.0022	7
13	.0000	.0000	.0000	.0000	.0001	.0001	.0002	.0002	.0004	.0005	6
14	.0000	.0000	.0000	.0000	.0000	.0000	.0000	.0000	.0001	.0001	5
	.79	.78	.77	.76	.75	.74	.73	.72	.71	.70	P / R

TABLE C Binomial Distribution: Individual Terms $p(r) = c\binom{n}{r} p^r q^{n-r}$ *(Continued)*

P	P	.31	.32	.33	.34	.35	.36	.37	.38	.39	.40	
0		.0009	.0007	.0005	.0004	.0003	.0002	.0002	.0001	.0001	.0001	19
1		.0074	.0059	.0046	.0036	.0029	.0022	.0017	.0013	.0010	.0008	18
2		.0299	.0249	.0206	.0169	.0138	.0112	.0091	.0073	.0058	.0046	17
3		.0762	.0664	.0574	.0494	.0422	.0358	.0302	.0253	.0211	.0175	16
4		.1370	.1249	.1131	.1017	.0909	.0806	.0710	.0621	.0540	.0467	15
5		.1846	.1764	.1672	.1572	.1468	.1360	.1251	.1143	.1036	.0933	14
6		.1935	.1936	.1921	.1890	.1844	.1785	.1714	.1634	.1546	.1451	13
7		.1615	.1692	.1757	.1808	.1844	.1865	.1870	.1860	.1835	.1797	12
8		.1088	.1195	.1298	.1397	.1489	.1573	.1647	.1710	.1760	.1797	11
9		.0597	.0687	.0782	.0880	.0980	.1082	.1182	.1281	.1375	.1464	10
10		.0268	.0323	.0385	.0453	.0528	.0608	.0694	.0785	.0879	.0976	9
11		.0099	.0124	.0155	.0191	.0233	.0280	.0334	.0394	.0460	.0532	8
12		.0030	.0039	.0051	.0066	.0083	.0105	.0131	.0161	.0196	.0237	7
13		.0007	.0010	.0014	.0018	.0024	.0032	.0041	.0053	.0067	.0085	6
14		.0001	.0002	.0003	.0004	.0006	.0008	.0010	.0014	.0018	.0024	5
15		.0000	.0000	.0000	.0001	.0001	.0001	.0002	.0003	.0004	.0005	4
16		.0000	.0000	.0000	.0000	.0000	.0000	.0000	.0000	.0001	.0001	3
		.69	.68	.67	.66	.65	.64	.63	.62	.61	.60	P R

P	P	.41	.42	.43	.44	.45	.46	.47	.48	.49	.50	
1		.0006	.0004	.0003	.0002	.0002	.0001	.0001	.0001	.0001	.0000	18
2		.0037	.0029	.0022	.0017	.0013	.0010	.0008	.0006	.0004	.0003	17
3		.0144	.0118	.0096	.0077	.0062	.0049	.0039	.0031	.0024	.0018	16
4		.0400	.0341	.0289	.0243	.0203	.0168	.0138	.0113	.0092	.0074	15
5		.0834	.0741	.0653	.0572	.0497	.0429	.0368	.0313	.0265	.0222	14
6		.1353	.1252	.1150	.1049	.0949	.0853	.0761	.0674	.0593	.0518	13
7		.1746	.1683	.1611	.1530	.1443	.1350	.1254	.1156	.1058	.0961	12
8		.1820	.1829	.1823	.1803	.1771	.1725	.1668	.1601	.1525	.1442	11
9		.1546	.1618	.1681	.1732	.1771	.1796	.1808	.1806	.1791	.1762	10
10		.1074	.1172	.1268	.1361	.1449	.1530	.1603	.1667	.1721	.1762	9
11		.0611	.0694	.0783	.0875	.0970	.1066	.1163	.1259	.1352	.1442	8
12		.0283	.0335	.0394	.0458	.0529	.0606	.0688	.0775	.0866	.0961	7
13		.0106	.0131	.0160	.0194	.0233	.0278	.0328	.0385	.0448	.0518	6
14		.0032	.0041	.0052	.0065	.0082	.0101	.0125	.0152	.0185	.0222	5
15		.0007	.0010	.0013	.0017	.0022	.0029	.0037	.0047	.0059	.0074	4
16		.0001	.0002	.0002	.0003	.0005	.0006	.0008	.0011	.0014	.0018	3
17		.0000	.0000	.0000	.0000	.0001	.0001	.0001	.0002	.0002	.0003	2
		.59	.58	.57	.56	.55	.54	.53	.52	.51	.50	P R

N = 20

P	P	.01	.02	.03	.04	.05	.06	.07	.08	.09	.10	
0		.8179	.6676	.5438	.4420	.3585	.2901	.2342	.1887	.1516	.1216	20
1		.1652	.2725	.3364	.3683	.3774	.3703	.3526	.3282	.3000	.2702	19
2		.0159	.0528	.0988	.1458	.1887	.2246	.2521	.2711	.2818	.2852	18
3		.0010	.0065	.0183	.0364	.0596	.0860	.1139	.1414	.1672	.1901	17
4		.0000	.0006	.0024	.0065	.0133	.0233	.0364	.0523	.0703	.0898	16
5		.0000	.0000	.0002	.0009	.0022	.0048	.0088	.0145	.0222	.0319	15
6		.0000	.0000	.0000	.0001	.0003	.0008	.0017	.0032	.0055	.0089	14
7		.0000	.0000	.0000	.0000	.0000	.0001	.0002	.0005	.0011	.0020	13
8		.0000	.0000	.0000	.0000	.0000	.0000	.0000	.0001	.0002	.0004	12
9		.0000	.0000	.0000	.0000	.0000	.0000	.0000	.0000	.0000	.0001	11
		.99	.98	.97	.96	.95	.94	.93	.92	.91	.90	P R

P	P	.11	.12	.13	.14	.15	.16	.17	.18	.19	.20	
0		.0972	.0776	.0617	.0490	.0388	.0306	.0241	.0189	.0148	.0115	20
1		.2403	.2115	.1844	.1595	.1368	.1165	.0986	.0829	.0693	.0576	19
2		.2822	.2740	.2618	.2466	.2293	.2109	.1919	.1730	.1545	.1369	18
3		.2093	.2242	.2347	.2409	.2428	.2410	.2358	.2278	.2175	.2054	17
4		.1099	.1299	.1491	.1666	.1821	.1951	.2053	.2125	.2168	.2182	16
5		.0435	.0567	.0713	.0868	.1028	.1189	.1345	.1493	.1627	.1746	15
6		.0134	.0193	.0266	.0353	.0454	.0566	.0689	.0819	.0954	.1091	14
7		.0033	.0053	.0080	.0115	.0160	.0216	.0282	.0360	.0448	.0545	13
8		.0007	.0012	.0019	.0030	.0046	.0067	.0094	.0128	.0171	.0222	12
9		.0001	.0002	.0004	.0007	.0011	.0017	.0026	.0038	.0053	.0074	11
10		.0000	.0000	.0001	.0001	.0002	.0004	.0006	.0009	.0014	.0020	10
11		.0000	.0000	.0000	.0000	.0000	.0001	.0001	.0002	.0003	.0005	9
12		.0000	.0000	.0000	.0000	.0000	.0000	.0000	.0000	.0001	.0001	8
		.89	.88	.87	.86	.85	.84	.83	.82	.81	.80	P R

P	P	.21	.22	.23	.24	.25	.26	.27	.28	.29	.30	
0		.0090	.0069	.0054	.0041	.0032	.0024	.0018	.0014	.0011	.0008	20
1		.0477	.0392	.0321	.0261	.0211	.0170	.0137	.0109	.0087	.0068	19
2		.1204	.1050	.0910	.0783	.0669	.0569	.0480	.0403	.0336	.0278	18
3		.1920	.1777	.1631	.1484	.1339	.1199	.1065	.0940	.0823	.0716	17
4		.2169	.2131	.2070	.1991	.1897	.1790	.1675	.1553	.1429	.1304	16
5		.1845	.1923	.1979	.2012	.2023	.2013	.1982	.1933	.1868	.1789	15
6		.1226	.1356	.1478	.1589	.1686	.1768	.1833	.1879	.1907	.1916	14
7		.0652	.0765	.0883	.1003	.1124	.1242	.1356	.1462	.1558	.1643	13
8		.0282	.0351	.0429	.0515	.0609	.0709	.0815	.0924	.1034	.1144	12
9		.0100	.0132	.0171	.0217	.0271	.0332	.0402	.0479	.0563	.0654	11
10		.0029	.0041	.0056	.0075	.0099	.0128	.0163	.0205	.0253	.0308	10
11		.0007	.0010	.0015	.0022	.0030	.0041	.0055	.0072	.0094	.0120	9

N = 20

TABLE C Binomial Distribution: Individual Terms $p(r) = c\binom{n}{r} p^r q^{n-r}$ (Continued)

r		.79	.78	.77	.76	.75	.74	.73	.72	.71	.70		R
12		.0001	.0002	.0003	.0005	.0008	.0011	.0015	.0021	.0029	.0039		8
13		.0000	.0000	.0001	.0001	.0002	.0002	.0003	.0005	.0007	.0010		7
14		.0000	.0000	.0000	.0000	.0000	.0000	.0001	.0001	.0001	.0002		6
		.79	.78	.77	.76	.75	.74	.73	.72	.71	.70	P	R

R	P	.31	.32	.33	.34	.35	.36	.37	.38	.39	.40		
0		.0006	.0004	.0003	.0002	.0002	.0001	.0001	.0001	.0001	.0000		20
1		.0054	.0042	.0033	.0025	.0020	.0015	.0011	.0009	.0007	.0005		19
2		.0229	.0188	.0153	.0124	.0100	.0080	.0064	.0050	.0040	.0031		18
3		.0619	.0531	.0453	.0383	.0323	.0270	.0224	.0185	.0152	.0123		17
4		.1181	.1062	.0947	.0839	.0738	.0645	.0559	.0482	.0412	.0350		16
5		.1698	.1599	.1493	.1384	.1272	.1161	.1051	.0945	.0843	.0746		15
6		.1907	.1881	.1839	.1782	.1712	.1632	.1543	.1447	.1347	.1244		14
7		.1714	.1770	.1811	.1836	.1844	.1836	.1812	.1774	.1722	.1659		13
8		.1251	.1354	.1450	.1537	.1614	.1678	.1730	.1767	.1790	.1797		12
9		.0750	.0849	.0952	.1056	.1158	.1259	.1354	.1444	.1526	.1597		11
10		.0370	.0440	.0516	.0598	.0686	.0779	.0875	.0974	.1073	.1171		10
11		.0151	.0188	.0231	.0280	.0336	.0398	.0467	.0542	.0624	.0710		9
12		.0051	.0066	.0085	.0108	.0136	.0168	.0206	.0249	.0299	.0355		8
13		.0014	.0019	.0026	.0034	.0045	.0058	.0074	.0094	.0118	.0146		7
14		.0003	.0005	.0006	.0009	.0012	.0016	.0022	.0029	.0038	.0049		6
15		.0001	.0001	.0001	.0002	.0003	.0004	.0005	.0007	.0010	.0013		5
16		.0000	.0000	.0000	.0000	.0000	.0001	.0001	.0001	.0002	.0003		4
		.69	.68	.67	.66	.65	.64	.63	.62	.61	.60	P	R

R	P	.41	.42	.43	.44	.45	.46	.47	.48	.49	.50		
1		.0004	.0003	.0002	.0001	.0001	.0001	.0001	.0000	.0000	.0000		19
2		.0024	.0018	.0014	.0011	.0008	.0006	.0005	.0003	.0002	.0002		18
3		.0100	.0080	.0064	.0051	.0040	.0031	.0024	.0019	.0014	.0011		17
4		.0295	.0247	.0206	.0170	.0139	.0113	.0092	.0074	.0059	.0046		16
5		.0656	.0573	.0496	.0427	.0365	.0309	.0260	.0217	.0180	.0148		15
6		.1140	.1037	.0936	.0839	.0746	.0658	.0577	.0501	.0432	.0370		14
7		.1585	.1502	.1413	.1318	.1221	.1122	.1023	.0925	.0830	.0739		13
8		.1790	.1768	.1732	.1683	.1623	.1553	.1474	.1388	.1296	.1201		12
9		.1658	.1707	.1742	.1763	.1771	.1763	.1742	.1708	.1661	.1602		11
10		.1268	.1359	.1446	.1524	.1593	.1652	.1700	.1734	.1755	.1762		10
11		.0801	.0895	.0991	.1089	.1185	.1280	.1370	.1455	.1533	.1602		9
12		.0417	.0486	.0561	.0642	.0727	.0818	.0911	.1007	.1105	.1201		8
13		.0178	.0217	.0260	.0310	.0366	.0429	.0497	.0572	.0653	.0739		7
14		.0062	.0078	.0098	.0122	.0150	.0183	.0221	.0264	.0314	.0370		6
15		.0017	.0023	.0030	.0038	.0049	.0062	.0078	.0098	.0121	.0148		5
16		.0004	.0005	.0007	.0009	.0013	.0017	.0022	.0028	.0036	.0046		4
17		.0001	.0001	.0001	.0002	.0002	.0003	.0005	.0006	.0008	.0011		3
18		.0000	.0000	.0000	.0000	.0000	.0000	.0001	.0001	.0001	.0002		2
		.59	.58	.57	.56	.55	.54	.53	.52	.51	.50	P	R

N = 25

R	P	.01	.02	.03	.04	.05	.06	.07	.08	.09	.10		
0		.7778	.6035	.4670	.3604	.2774	.2129	.1630	.1244	.0946	.0718		25
1		.1964	.3079	.3611	.3754	.3650	.3398	.3066	.2704	.2340	.1994		24
2		.0238	.0754	.1340	.1877	.2305	.2602	.2770	.2821	.2777	.2659		23
3		.0018	.0118	.0318	.0600	.0930	.1273	.1598	.1881	.2106	.2265		22
4		.0001	.0013	.0054	.0137	.0269	.0447	.0662	.0899	.1145	.1384		21
5		.0000	.0001	.0007	.0024	.0060	.0120	.0209	.0329	.0476	.0646		20
6		.0000	.0000	.0001	.0003	.0010	.0026	.0052	.0095	.0157	.0239		19
7		.0000	.0000	.0000	.0000	.0001	.0004	.0011	.0022	.0042	.0072		18
8		.0000	.0000	.0000	.0000	.0000	.0001	.0002	.0004	.0009	.0018		17
9		.0000	.0000	.0000	.0000	.0000	.0000	.0000	.0001	.0002	.0004		16
10		.0000	.0000	.0000	.0000	.0000	.0000	.0000	.0000	.0000	.0001		15
		.99	.98	.97	.96	.95	.94	.93	.92	.91	.90	P	R

R	P	.11	.12	.13	.14	.15	.16	.17	.18	.19	.20		
0		.0543	.0409	.0308	.0230	.0172	.0128	.0095	.0070	.0052	.0038		25
1		.1678	.1395	.1149	.0938	.0759	.0609	.0486	.0384	.0302	.0236		24
2		.2488	.2283	.2060	.1832	.1607	.1392	.1193	.1012	.0851	.0708		23
3		.2358	.2387	.2360	.2286	.2174	.2033	.1874	.1704	.1530	.1358		22
4		.1603	.1790	.1940	.2047	.2110	.2130	.2111	.2057	.1974	.1867		21
5		.0832	.1025	.1217	.1399	.1564	.1704	.1816	.1897	.1945	.1960		20
6		.0343	.0466	.0606	.0759	.0920	.1082	.1240	.1388	.1520	.1633		19
7		.0115	.0173	.0246	.0336	.0441	.0559	.0689	.0827	.0968	.1108		18
8		.0032	.0053	.0083	.0123	.0175	.0240	.0318	.0408	.0511	.0623		17
9		.0007	.0014	.0023	.0038	.0058	.0086	.0123	.0169	.0226	.0294		16
10		.0001	.0003	.0006	.0010	.0016	.0026	.0040	.0059	.0085	.0118		15
11		.0000	.0001	.0001	.0002	.0004	.0007	.0011	.0018	.0027	.0040		14
12		.0000	.0000	.0000	.0000	.0001	.0002	.0003	.0005	.0007	.0012		13
13		.0000	.0000	.0000	.0000	.0000	.0000	.0001	.0001	.0002	.0003		12
14		.0000	.0000	.0000	.0000	.0000	.0000	.0000	.0000	.0000	.0001		11
		.89	.88	.87	.86	.85	.84	.83	.82	.81	.80	P	R

R	P	.21	.22	.23	.24	.25	.26	.27	.28	.29	.30		
0		.0028	.0020	.0015	.0010	.0008	.0005	.0004	.0003	.0002	.0001		25
1		.0183	.0141	.0109	.0083	.0063	.0047	.0035	.0026	.0020	.0014		24

TABLE C Binomial Distribution: Individual Terms $p(r) = {}_c\binom{n}{r} p^r q^{n-r}$ (*Continued*)

R	.79	.78	.77	.76	.75	.74	.73	.72	.71	.70	R
2	.0585	.0479	.0389	.0314	.0251	.0199	.0157	.0123	.0096	.0074	23
3	.1192	.1035	.0891	.0759	.0641	.0537	.0446	.0367	.0300	.0243	22
4	.1742	.1606	.1463	.1318	.1175	.1037	.0906	.0785	.0673	.0572	21
5	.1945	.1903	.1836	.1749	.1645	.1531	.1408	.1282	.1155	.1030	20
6	.1724	.1789	.1828	.1841	.1828	.1793	.1736	.1661	.1572	.1472	19
7	.1244	.1369	.1482	.1578	.1654	.1709	.1743	.1754	.1743	.1712	18
8	.0744	.0869	.0996	.1121	.1241	.1351	.1450	.1535	.1602	.1651	17
9	.0373	.0463	.0562	.0669	.0781	.0897	.1013	.1127	.1236	.1336	16
10	.0159	.0209	.0269	.0338	.0417	.0504	.0600	.0701	.0808	.0916	15
11	.0058	.0080	.0109	.0145	.0189	.0242	.0302	.0372	.0450	.0536	14
12	.0018	.0026	.0038	.0054	.0074	.0099	.0130	.0169	.0214	.0268	13
13	.0005	.0007	.0011	.0017	.0025	.0035	.0048	.0066	.0088	.0115	12
14	.0001	.0002	.0003	.0005	.0007	.0010	.0015	.0022	.0031	.0042	11
15	.0000	.0000	.0001	.0001	.0002	.0003	.0004	.0006	.0009	.0013	10
16	.0000	.0000	.0000	.0000	.0000	.0001	.0001	.0002	.0002	.0004	9
17	.0000	.0000	.0000	.0000	.0000	.0000	.0000	.0000	.0001	.0001	8
	.79	.78	.77	.76	.75	.74	.73	.72	.71	.70	P R

P R	P	.31	.32	.33	.34	.35	.36	.37	.38	.39	.40	R
0		.0001	.0001	.0000	.0000	.0000	.0000	.0000	.0000	.0000	.0000	25
1		.0011	.0008	.0006	.0004	.0003	.0002	.0001	.0001	.0001	.0000	24
2		.0057	.0043	.0033	.0025	.0018	.0014	.0010	.0007	.0005	.0004	23
3		.0195	.0156	.0123	.0097	.0076	.0058	.0045	.0034	.0026	.0019	22
4		.0482	.0403	.0334	.0274	.0224	.0181	.0145	.0115	.0091	.0071	21
5		.0910	.0797	.0691	.0594	.0506	.0427	.0357	.0297	.0244	.0199	20
6		.1363	.1250	.1134	.1020	.0908	.0801	.0700	.0606	.0520	.0442	19
7		.1662	.1596	.1516	.1426	.1327	.1222	.1115	.1008	.0902	.0800	18
8		.1680	.1690	.1681	.1652	.1607	.1547	.1474	.1390	.1298	.1200	17
9		.1426	.1502	.1563	.1608	.1635	.1644	.1635	.1609	.1567	.1511	16
10		.1025	.1131	.1232	.1325	.1409	.1479	.1536	.1578	.1603	.1612	15
11		.0628	.0726	.0828	.0931	.1034	.1135	.1230	.1319	.1398	.1465	14
12		.0329	.0399	.0476	.0560	.0650	.0745	.0843	.0943	.1043	.1140	13
13		.0148	.0188	.0234	.0288	.0350	.0419	.0495	.0578	.0667	.0760	12
14		.0057	.0076	.0099	.0127	.0161	.0202	.0249	.0304	.0365	.0434	11
15		.0019	.0026	.0036	.0048	.0064	.0083	.0107	.0136	.0171	.0212	10
16		.0005	.0008	.0011	.0015	.0021	.0029	.0039	.0052	.0068	.0088	9
17		.0001	.0002	.0003	.0004	.0006	.0009	.0012	.0017	.0023	.0031	8
18		.0000	.0000	.0001	.0001	.0001	.0002	.0003	.0005	.0007	.0009	7
19		.0000	.0000	.0000	.0000	.0000	.0000	.0001	.0001	.0002	.0002	6
		.69	.68	.67	.66	.65	.64	.63	.62	.61	.60	P R

P	P	.41	.42	.43	.44	.45	.46	.47	.48	.49	.50	R
2		.0003	.0002	.0001	.0001	.0001	.0000	.0000	.0000	.0000	.0000	23
3		.0014	.0011	.0008	.0006	.0004	.0003	.0002	.0001	.0001	.0001	22
4		.0055	.0042	.0032	.0024	.0018	.0014	.0010	.0007	.0005	.0004	21
5		.0161	.0129	.0102	.0081	.0063	.0049	.0037	.0028	.0021	.0016	20
6		.0372	.0311	.0257	.0211	.0172	.0138	.0110	.0087	.0068	.0053	19
7		.0703	.0611	.0527	.0450	.0381	.0319	.0265	.0218	.0178	.0143	18
8		.1099	.0996	.0895	.0796	.0701	.0612	.0529	.0453	.0384	.0322	17
9		.1442	.1363	.1275	.1181	.1084	.0985	.0886	.0790	.0697	.0609	16
10		.1603	.1579	.1539	.1485	.1419	.1342	.1257	.1166	.1071	.0974	15
11		.1519	.1559	.1583	.1591	.1583	.1559	.1521	.1468	.1404	.1328	14
12		.1232	.1317	.1393	.1458	.1511	.1550	.1573	.1581	.1573	.1550	13
13		.0856	.0954	.1051	.1146	.1236	.1320	.1395	.1460	.1512	.1550	12
14		.0510	.0592	.0680	.0772	.0867	.0964	.1060	.1155	.1245	.1328	11
15		.0260	.0314	.0376	.0445	.0520	.0602	.0690	.0782	.0877	.0974	10
16		.0113	.0142	.0177	.0218	.0266	.0321	.0382	.0451	.0527	.0609	9
17		.0042	.0055	.0071	.0091	.0115	.0145	.0179	.0220	.0268	.0322	8
18		.0013	.0018	.0024	.0032	.0042	.0055	.0071	.0090	.0114	.0143	7
19		.0003	.0005	.0007	.0009	.0013	.0017	.0023	.0031	.0040	.0053	6
20		.0001	.0001	.0001	.0002	.0003	.0004	.0006	.0009	.0012	.0016	5
21		.0000	.0000	.0000	.0000	.0001	.0001	.0001	.0002	.0003	.0004	4
22		.0000	.0000	.0000	.0000	.0000	.0000	.0000	.0000	.0000	.0001	3
		.59	.58	.57	.56	.55	.54	.53	.52	.51	.50	P R

N = 50

R	P	.01	.02	.03	.04	.05	.06	.07	.08	.09	.10	R
0		.6050	.3642	.2181	.1299	.0769	.0453	.0266	.0155	.0090	.0052	50
1		.3056	.3716	.3372	.2706	.2025	.1447	.0999	.0672	.0443	.0286	49
2		.0756	.1858	.2555	.2762	.2611	.2262	.1843	.1433	.1073	.0779	48
3		.0122	.0607	.1264	.1842	.2199	.2311	.2219	.1993	.1698	.1386	47
4		.0015	.0145	.0459	.0902	.1360	.1733	.1963	.2037	.1973	.1809	46
5		.0001	.0027	.0131	.0346	.0658	.1018	.1359	.1629	.1795	.1849	45
6		.0000	.0004	.0030	.0108	.0260	.0487	.0767	.1063	.1332	.1541	44
7		.0000	.0001	.0006	.0028	.0086	.0195	.0363	.0581	.0828	.1076	43
8		.0000	.0000	.0001	.0006	.0024	.0067	.0147	.0271	.0440	.0643	42
9		.0000	.0000	.0000	.0001	.0006	.0020	.0052	.0110	.0203	.0333	41
10		.0000	.0000	.0000	.0000	.0001	.0005	.0016	.0039	.0082	.0152	40
11		.0000	.0000	.0000	.0000	.0000	.0001	.0004	.0012	.0030	.0061	39
12		.0000	.0000	.0000	.0000	.0000	.0000	.0001	.0004	.0010	.0022	38
13		.0000	.0000	.0000	.0000	.0000	.0000	.0000	.0001	.0003	.0007	37
14		.0000	.0000	.0000	.0000	.0000	.0000	.0000	.0000	.0001	.0002	36
15		.0000	.0000	.0000	.0000	.0000	.0000	.0000	.0000	.0000	.0001	35
		.99	.98	.97	.96	.95	.94	.93	.92	.91	.90	P R

TABLE C Binomial Distribution: Individual Terms $p(r) = c\binom{n}{r} p^r q^{n-r}$ (*Continued*)

R	P	.11	.12	.13	.14	.15	.16	.17	.18	.19	.20	
0		.0029	.0017	.0009	.0005	.0003	.0002	.0001	.0000	.0000	.0000	50
1		.0182	.0114	.0071	.0043	.0026	.0016	.0009	.0005	.0003	.0002	49
2		.0552	.0382	.0259	.0172	.0113	.0073	.0046	.0029	.0018	.0011	48
3		.1091	.0853	.0619	.0449	.0319	.0222	.0151	.0102	.0067	.0044	47
4		.1584	.1354	.1086	.0858	.0661	.0496	.0364	.0262	.0185	.0128	46
5		.1801	.1674	.1493	.1286	.1072	.0869	.0687	.0530	.0400	.0295	45
6		.1670	.1712	.1674	.1570	.1419	.1242	.1055	.0872	.0703	.0554	44
7		.1297	.1467	.1572	.1606	.1575	.1487	.1358	.1203	.1037	.0870	43
8		.0862	.1075	.1263	.1406	.1493	.1523	.1495	.1420	.1307	.1169	42
9		.0497	.0684	.0880	.1068	.1230	.1353	.1429	.1454	.1431	.1364	41
10		.0252	.0383	.0539	.0713	.0890	.1057	.1200	.1309	.1376	.1398	40
11		.0113	.0190	.0293	.0422	.0571	.0732	.0894	.1045	.1174	.1271	39
12		.0045	.0084	.0142	.0223	.0328	.0453	.0595	.0745	.0695	.1033	38
13		.0016	.0034	.0062	.0106	.0169	.0252	.0356	.0478	.0613	.0755	37
14		.0005	.0012	.0025	.0046	.0079	.0127	.0193	.0277	.0380	.0499	36
15		.0002	.0004	.0009	.0018	.0033	.0058	.0095	.0146	.0214	.0299	35
16		.0000	.0001	.0003	.0006	.0013	.0024	.0042	.0070	.0110	.0164	34
17		.0000	.0000	.0001	.0002	.0005	.0009	.0017	.0031	.0052	.0082	33
18		.0000	.0000	.0000	.0001	.0001	.0003	.0007	.0012	.0022	.0037	32
19		.0000	.0000	.0000	.0000	.0000	.0001	.0002	.0005	.0009	.0016	31
20		.0000	.0000	.0000	.0000	.0000	.0000	.0001	.0002	.0003	.0006	30
21		.0000	.0000	.0000	.0000	.0000	.0000	.0000	.0000	.0001	.0002	29
22		.0000	.0000	.0000	.0000	.0000	.0000	.0000	.0000	.0000	.0001	23
		.89	.88	.87	.86	.85	.84	.83	.82	.81	.80	P R

R	P	.21	.22	.23	.24	.25	.26	.27	.28	.29	.30	
1		.0001	.0001	.0000	.0000	.0000	.0000	.0000	.0000	.0000	.0000	49
2		.0007	.0004	.0002	.0001	.0001	.0000	.0000	.0000	.0000	.0000	48
3		.0028	.0018	.0011	.0007	.0004	.0002	.0001	.0001	.0000	.0000	47
4		.0088	.0059	.0039	.0025	.0015	.0010	.0006	.0004	.0002	.0001	46
5		.0214	.0152	.0106	.0073	.0049	.0033	.0022	.0014	.0009	.0006	45
6		.0427	.0322	.0238	.0173	.0123	.0087	.0060	.0040	.0027	.0018	44
7		.0713	.0571	.0447	.0344	.0259	.0191	.0139	.0099	.0069	.0048	43
8		.1019	.0865	.0718	.0583	.0463	.0361	.0276	.0207	.0152	.0110	42
9		.1263	.1139	.1001	.0859	.0721	.0592	.0476	.0375	.0290	.0220	41
10		.1377	.1317	.1226	.1113	.0985	.0852	.0721	.0598	.0485	.0386	40
11		.1331	.1351	.1332	.1278	.1194	.1089	.0970	.0845	.0721	.0602	39
12		.1150	.1258	.1293	.1311	.1294	.1244	.1166	.1068	.0957	.0838	38
13		.0894	.1021	.1129	.1210	.1261	.1277	.1261	.1215	.1142	.1050	37
14		.0628	.0761	.0891	.1010	.1110	.1186	.1233	.1248	.1233	.1189	36
15		.0400	.0515	.0639	.0766	.0888	.1000	.1094	.1165	.1209	.1223	35
16		.0233	.0318	.0417	.0529	.0648	.0769	.0885	.0991	.1080	.1147	34
17		.0124	.0179	.0249	.0334	.0432	.0540	.0655	.0771	.0882	.0983	33
18		.0060	.0093	.0137	.0193	.0264	.0349	.0444	.0550	.0661	.0772	32
19		.0027	.0044	.0069	.0103	.0148	.0205	.0277	.0360	.0454	.0558	31
20		.0011	.0019	.0032	.0050	.0077	.0112	.0159	.0217	.0288	.0370	30
21		.0004	.0008	.0014	.0023	.0036	.0056	.0084	.0121	.0168	.0227	29
22		.0001	.0003	.0005	.0009	.0016	.0026	.0041	.0062	.0090	.0128	28
23		.0000	.0001	.0002	.0004	.0007	.0011	.0018	.0029	.0045	.0067	27
24		.0000	.0000	.0001	.0001	.0002	.0004	.0008	.0013	.0021	.0032	26
25		.0000	.0000	.0000	.0000	.0001	.0002	.0003	.0005	.0009	.0014	25
26		.0000	.0000	.0000	.0000	.0000	.0001	.0001	.0002	.0003	.0006	24
27		.0000	.0000	.0000	.0000	.0000	.0000	.0000	.0001	.0001	.0002	23
28		.0000	.0000	.0000	.0000	.0000	.0000	.0000	.0000	.0000	.0001	22
		.79	.78	.77	.76	.75	.74	.73	.72	.71	.70	P R

R	P	.31	.32	.33	.34	.35	.36	.37	.38	.39	.40	
4		.0001	.0000	.0000	.0000	.0000	.0000	.0000	.0000	.0000	.0000	46
5		.0003	.0002	.0001	.0001	.0000	.0000	.0000	.0000	.0000	.0000	45
6		.0011	.0007	.0005	.0003	.0002	.0001	.0001	.0000	.0000	.0000	44
7		.0032	.0022	.0014	.0009	.0006	.0004	.0002	.0001	.0001	.0000	43
8		.0078	.0055	.0037	.0025	.0017	.0011	.0007	.0004	.0003	.0002	42
9		.0164	.0120	.0086	.0061	.0042	.0029	.0019	.0013	.0008	.0005	41
10		.0301	.0231	.0174	.0128	.0093	.0066	.0046	.0032	.0022	.0014	40
11		.0493	.0395	.0311	.0240	.0182	.0136	.0099	.0071	.0050	.0035	39
12		.0719	.0604	.0498	.0402	.0319	.0248	.0189	.0142	.0105	.0076	38
13		.0944	.0831	.0717	.0606	.0502	.0408	.0325	.0255	.0195	.0147	37
14		.1121	.1034	.0933	.0825	.0714	.0607	.0505	.0412	.0330	.0260	36
15		.1209	.1168	.1103	.1020	.0923	.0819	.0712	.0606	.0507	.0415	35
16		.1188	.1202	.1189	.1149	.1088	.1008	.0914	.0813	.0709	.0606	34
17		.1068	.1152	.1171	.1184	.1171	.1133	.1074	.0997	.0906	.0808	33
18		.0880	.0976	.1057	.1118	.1156	.1169	.1156	.1120	.1062	.0987	32
19		.0666	.0774	.0877	.0970	.1048	.1107	.1144	.1156	.1144	.1109	31
20		.0463	.0564	.0670	.0775	.0875	.0965	.1041	.1098	.1134	.1146	30
21		.0297	.0379	.0471	.0570	.0673	.0776	.0874	.0962	.1035	.1091	29
22		.0176	.0235	.0306	.0387	.0478	.0575	.0676	.0777	.0873	.0959	28
23		.0096	.0135	.0183	.0243	.0313	.0394	.0484	.0580	.0679	.0778	27
24		.0049	.0071	.0102	.0141	.0190	.0249	.0319	.0400	.0489	.0584	26
25		.0023	.0035	.0052	.0075	.0105	.0146	.0195	.0255	.0325	.0405	25
26		.0010	.0016	.0025	.0037	.0055	.0079	.0110	.0150	.0200	.0259	24
27		.0004	.0007	.0011	.0017	.0025	.0039	.0058	.0082	.0113	.0154	23
28		.0001	.0003	.0004	.0007	.0012	.0018	.0028	.0041	.0060	.0084	22
29		.0000	.0001	.0002	.0003	.0005	.0008	.0012	.0019	.0029	.0043	21
30		.0000	.0000	.0001	.0001	.0002	.0003	.0005	.0008	.0013	.0020	20
31		.0000	.0000	.0000	.0000	.0001	.0001	.0002	.0003	.0005	.0009	19
32		.0000	.0000	.0000	.0000	.0000	.0000	.0001	.0001	.0002	.0003	18
33		.0000	.0000	.0000	.0000	.0000	.0000	.0000	.0000	.0001	.0001	17
		.69	.68	.67	.66	.65	.64	.63	.62	.61	.60	P R

TABLE C Binomial Distribution: Individual Terms $p(r) = c\binom{n}{r} p^r q^{n-r}$ (*Continued*)

R	P	.41	.42	.43	.44	.45	.46	.47	.48	.49	.50	
8		.0001	.0001	.0000	.0000	.0000	.0000	.0000	.0000	.0000	.0000	42
9		.0003	.0002	.0001	.0001	.0000	.0000	.0000	.0000	.0000	.0000	41
10		.0009	.0006	.0004	.0002	.0001	.0001	.0001	.0000	.0000	.0000	40
11		.0024	.0016	.0010	.0007	.0004	.0003	.0002	.0001	.0001	.0000	39
12		.0054	.0037	.0026	.0017	.0011	.0007	.0005	.0003	.0002	.0001	38
13		.0109	.0079	.0057	.0040	.0027	.0018	.0012	.0008	.0005	.0003	37
14		.0200	.0152	.0113	.0082	.0059	.0041	.0029	.0019	.0013	.0008	36
15		.0334	.0264	.0204	.0155	.0116	.0085	.0061	.0043	.0030	.0020	35
16		.0508	.0418	.0337	.0267	.0207	.0158	.0118	.0086	.0062	.0044	34
17		.0706	.0605	.0508	.0419	.0339	.0269	.0209	.0159	.0119	.0087	33
18		.0899	.0803	.0703	.0604	.0508	.0420	.0340	.0270	.0210	.0160	32
19		.1053	.0979	.0893	.0799	.0700	.0602	.0597	.0419	.0340	.0270	31
20		.1134	.1099	.1044	.0973	.0888	.0795	.0697	.0600	.0506	.0419	30
21		.1126	.1157	.1126	.1092	.1038	.0967	.0884	.0791	.0695	.0598	29
22		.1031	.1086	.1119	.1131	.1119	.1086	.1033	.0963	.0880	.0788	28
23		.0872	.0957	.1028	.1082	.1115	.1126	.1115	.1082	.1029	.0960	27
24		.0682	.0780	.0872	.0956	.1025	.1079	.1112	.1124	.1112	.1080	26
25		.0493	.0587	.0684	.0781	.0873	.0955	.1026	.1079	.1112	.1123	25
26		.0329	.0409	.0497	.0590	.0687	.0783	.0875	.0957	.1027	.1080	24
27		.0203	.0263	.0333	.0412	.0500	.0593	.0690	.0786	.0877	.0960	23
28		.0116	.0157	.0206	.0266	.0335	.0415	.0502	.0596	.0692	.0788	22
29		.0061	.0086	.0118	.0159	.0203	.0268	.0338	.0417	.0504	.0598	21
30		.0030	.0044	.0062	.0087	.0113	.0160	.0210	.0270	.0339	.0419	20
31		.0013	.0020	.0030	.0044	.0063	.0088	.0120	.0161	.0210	.0270	19
32		.0006	.0009	.0014	.0021	.0031	.0044	.0063	.0088	.0120	.0160	18
33		.0002	.0003	.0006	.0009	.0014	.0021	.0031	.0044	.0063	.0087	17
34		.0001	.0001	.0002	.0003	.0005	.0009	.0014	.0020	.0030	.0044	16
35		.0000	.0000	.0001	.0001	.0002	.0003	.0005	.0009	.0013	.0020	15
36		.0000	.0000	.0000	.0000	.0001	.0001	.0002	.0003	.0005	.0008	14
37		.0000	.0000	.0000	.0000	.0000	.0000	.0001	.0001	.0002	.0003	13
38		.0000	.0000	.0000	.0000	.0000	.0000	.0000	.0000	.0001	.0001	12
		.59	.58	.57	.56	.55	.54	.53	.52	.51	.50	P R

N = 100

R	P	.01	.02	.03	.04	.05	.06	.07	.08	.09	.10	
0		.3660	.1326	.0476	.0169	.0059	.0021	.0007	.0002	.0001	.0000	100
1		.3697	.2707	.1471	.0703	.0312	.0131	.0053	.0021	.0008	.0003	99
2		.1849	.2734	.2252	.1450	.0812	.0414	.0198	.0090	.0039	.0016	98
3		.0610	.1823	.2275	.1973	.1396	.0864	.0486	.0254	.0125	.0059	97
4		.0149	.0902	.1706	.1994	.1781	.1338	.0888	.0536	.0301	.0159	96
5		.0029	.0353	.1013	.1595	.1800	.1639	.1283	.0895	.0571	.0339	95
6		.0005	.0114	.0496	.1052	.1500	.1657	.1529	.1233	.0895	.0596	94
7		.0001	.0031	.0206	.0589	.1060	.1420	.1545	.1440	.1188	.0889	93
8		.0000	.0007	.0074	.0285	.0649	.1054	.1352	.1455	.1366	.1148	92
9		.0000	.0002	.0023	.0121	.0349	.0687	.1040	.1293	.1381	.1304	91
10		.0000	.0000	.0007	.0046	.0167	.0399	.0712	.1024	.1243	.1319	90
11		.0000	.0000	.0002	.0016	.0072	.0209	.0439	.0728	.1006	.1199	89
12		.0000	.0000	.0000	.0005	.0028	.0099	.0245	.0470	.0738	.0988	88
13		.0000	.0000	.0000	.0001	.0010	.0043	.0125	.0276	.0494	.0743	87
14		.0000	.0000	.0000	.0000	.0003	.0017	.0058	.0149	.0304	.0513	86
15		.0000	.0000	.0000	.0000	.0001	.0006	.0025	.0074	.0172	.0327	85
16		.0000	.0000	.0000	.0000	.0000	.0002	.0010	.0034	.0090	.0193	84
17		.0000	.0000	.0000	.0000	.0000	.0001	.0004	.0015	.0044	.0106	83
18		.0000	.0000	.0000	.0000	.0000	.0000	.0001	.0006	.0020	.0054	82
19		.0000	.0000	.0000	.0000	.0000	.0000	.0000	.0002	.0009	.0026	81
20		.0000	.0000	.0000	.0000	.0000	.0000	.0000	.0001	.0003	.0012	80
21		.0000	.0000	.0000	.0000	.0000	.0000	.0000	.0000	.0001	.0005	79
22		.0000	.0000	.0000	.0000	.0000	.0000	.0000	.0000	.0000	.0002	78
23		.0000	.0000	.0000	.0000	.0000	.0000	.0000	.0000	.0000	.0001	77
		.99	.98	.97	.96	.95	.94	.93	.92	.91	.90	P R

R	P	.11	.12	.13	.14	.15	.16	.17	.18	.19	.20	
1		.0001	.0000	.0000	.0000	.0000	.0000	.0000	.0000	.0000	.0000	99
2		.0007	.0003	.0001	.0000	.0000	.0000	.0000	.0000	.0000	.0000	98
3		.0027	.0012	.0005	.0002	.0001	.0000	.0000	.0000	.0000	.0000	97
4		.0080	.0038	.0018	.0008	.0003	.0001	.0001	.0000	.0000	.0000	96
5		.0189	.0100	.0050	.0024	.0011	.0005	.0002	.0001	.0000	.0000	95
6		.0369	.0215	.0119	.0063	.0031	.0015	.0007	.0003	.0001	.0001	94
7		.0613	.0394	.0238	.0137	.0075	.0039	.0020	.0009	.0004	.0002	93
8		.0881	.0625	.0414	.0259	.0153	.0086	.0047	.0024	.0012	.0006	92
9		.1112	.0871	.0632	.0430	.0276	.0168	.0098	.0054	.0029	.0015	91
10		.1251	.1080	.0860	.0637	.0444	.0292	.0182	.0108	.0062	.0034	90
11		.1265	.1205	.1051	.0849	.0640	.0454	.0305	.0194	.0118	.0069	89
12		.1160	.1219	.1165	.1025	.0838	.0642	.0463	.0316	.0206	.0128	88
13		.0970	.1125	.1179	.1130	.1001	.0827	.0642	.0470	.0327	.0216	87
14		.0745	.0954	.1094	.1143	.1098	.0979	.0817	.0641	.0476	.0335	86
15		.0528	.0745	.0938	.1067	.1111	.1070	.0960	.0807	.0640	.0481	85
16		.0347	.0540	.0744	.0922	.1041	.1082	.1044	.0941	.0798	.0638	84
17		.0212	.0364	.0549	.0742	.0908	.1019	.1057	.1021	.0924	.0789	83
18		.0121	.0229	.0379	.0557	.0739	.0895	.0998	.1033	.1000	.0909	82
19		.0064	.0135	.0244	.0391	.0563	.0735	.0882	.0979	.1012	.0981	81
20		.0032	.0074	.0148	.0258	.0402	.0567	.0732	.0870	.0962	.0993	80
21		.0015	.0039	.0084	.0160	.0270	.0412	.0571	.0728	.0859	.0946	79
22		.0007	.0019	.0045	.0094	.0171	.0282	.0420	.0574	.0724	.0849	78

TABLE C Binomial Distribution: Individual Terms $p(r) = c\binom{n}{r} p^r q^{n-r}$ (*Continued*)

r											r
23	.0003	.0009	.0023	.0052	.0103	.0192	.0292	.0427	.0576	.0720	77
24	.0001	.0004	.0011	.0027	.0058	.0111	.0192	.0301	.0433	.0577	76
25	.0000	.0002	.0005	.0013	.0031	.0064	.0119	.0201	.0309	.0439	75
26	.0000	.0001	.0002	.0006	.0015	.0035	.0071	.0127	.0209	.0316	74
27	.0000	.0000	.0001	.0003	.0008	.0018	.0040	.0076	.0134	.0217	73
28	.0000	.0000	.0000	.0001	.0004	.0009	.0021	.0044	.0082	.0141	72
29	.0000	.0000	.0000	.0000	.0002	.0004	.0011	.0024	.0048	.0088	71
30	.0000	.0000	.0000	.0000	.0001	.0002	.0005	.0012	.0027	.0052	70
31	.0000	.0000	.0000	.0000	.0000	.0001	.0002	.0006	.0014	.0029	69
32	.0000	.0000	.0000	.0000	.0000	.0000	.0001	.0003	.0007	.0016	68
33	.0000	.0000	.0000	.0000	.0000	.0000	.0000	.0001	.0003	.0008	67
34	.0000	.0000	.0000	.0000	.0000	.0000	.0000	.0001	.0002	.0004	66
35	.0000	.0000	.0000	.0000	.0000	.0000	.0000	.0000	.0001	.0002	65
36	.0000	.0000	.0000	.0000	.0000	.0000	.0000	.0000	.0000	.0001	64
	.89	.88	.87	.86	.85	.84	.83	.82	.81	.80	P r

R	P	.21	.22	.23	.24	.25	.26	.27	.28	.29	.30	
7		.0001	.0000	.0000	.0000	.0000	.0000	.0000	.0000	.0000	.0000	93
8		.0003	.0001	.0001	.0000	.0000	.0000	.0000	.0000	.0000	.0000	92
9		.0007	.0003	.0002	.0001	.0000	.0000	.0000	.0000	.0000	.0000	91
10		.0018	.0009	.0004	.0002	.0001	.0000	.0000	.0000	.0000	.0000	90
11		.0038	.0021	.0011	.0005	.0003	.0001	.0001	.0000	.0000	.0000	89
12		.0076	.0043	.0024	.0012	.0006	.0003	.0001	.0001	.0000	.0000	88
13		.0136	.0082	.0048	.0027	.0014	.0007	.0004	.0002	.0001	.0000	87
14		.0225	.0144	.0089	.0052	.0030	.0016	.0009	.0004	.0002	.0001	86
15		.0343	.0253	.0152	.0095	.0057	.0033	.0018	.0010	.0005	.0002	85
16		.0484	.0350	.0241	.0159	.0100	.0061	.0035	.0020	.0011	.0006	84
17		.0636	.0487	.0356	.0248	.0165	.0106	.0065	.0038	.0022	.0012	83
18		.0780	.0634	.0490	.0361	.0254	.0171	.0111	.0069	.0041	.0024	82
19		.0895	.0772	.0631	.0492	.0365	.0259	.0177	.0115	.0072	.0044	81
20		.0963	.0881	.0764	.0629	.0493	.0369	.0254	.0182	.0120	.0076	80
21		.0975	.0947	.0869	.0756	.0626	.0494	.0373	.0269	.0186	.0124	79
22		.0931	.0959	.0932	.0858	.0749	.0623	.0495	.0376	.0273	.0190	78
23		.0839	.0917	.0944	.0919	.0847	.0743	.0621	.0495	.0378	.0277	77
24		.0716	.0850	.0905	.0931	.0905	.0837	.0735	.0618	.0496	.0380	76
25		.0578	.0712	.0822	.0893	.0913	.0894	.0828	.0731	.0615	.0496	75
26		.0444	.0579	.0708	.0814	.0883	.0906	.0833	.0810	.0725	.0613	74
27		.0323	.0448	.0580	.0704	.0806	.0873	.0896	.0873	.0812	.0720	73
28		.0224	.0329	.0451	.0580	.0701	.0799	.0864	.0886	.0864	.0804	72
29		.0148	.0231	.0335	.0455	.0580	.0697	.0793	.0855	.0855	.0856	71
30		.0093	.0154	.0237	.0340	.0458	.0580	.0694	.0787	.0847	.0868	70
31		.0056	.0098	.0160	.0242	.0344	.0460	.0580	.0691	.0781	.0840	69
32		.0032	.0060	.0103	.0165	.0243	.0349	.0462	.0579	.0688	.0776	68
33		.0018	.0035	.0063	.0107	.0170	.0252	.0352	.0464	.0579	.0685	67
34		.0009	.0019	.0037	.0067	.0112	.0175	.0257	.0358	.0466	.0579	66
35		.0005	.0010	.0021	.0040	.0070	.0116	.0179	.0261	.0359	.0468	65
36		.0002	.0005	.0011	.0023	.0042	.0073	.0120	.0183	.0265	.0362	64
37		.0001	.0003	.0006	.0012	.0024	.0045	.0077	.0123	.0187	.0268	63
38		.0000	.0001	.0003	.0006	.0013	.0026	.0047	.0079	.0127	.0191	62
39		.0000	.0001	.0001	.0003	.0007	.0015	.0028	.0049	.0082	.0130	61
40		.0000	.0000	.0001	.0002	.0004	.0008	.0016	.0029	.0051	.0085	60
41		.0000	.0000	.0000	.0001	.0002	.0004	.0009	.0017	.0031	.0053	59
42		.0000	.0000	.0000	.0000	.0001	.0002	.0004	.0009	.0018	.0032	58
43		.0000	.0000	.0000	.0000	.0000	.0001	.0002	.0005	.0010	.0019	57
44		.0000	.0000	.0000	.0000	.0000	.0000	.0001	.0002	.0005	.0010	56
45		.0000	.0000	.0000	.0000	.0000	.0000	.0000	.0001	.0003	.0005	55
46		.0000	.0000	.0000	.0000	.0000	.0000	.0000	.0001	.0001	.0003	54
47		.0000	.0000	.0000	.0000	.0000	.0000	.0000	.0000	.0001	.0001	53
48		.0000	.0000	.0000	.0000	.0000	.0000	.0000	.0000	.0000	.0001	52
		.79	.78	.77	.76	.75	.74	.73	.72	.71	.70	P r

R	P	.31	.32	.33	.34	.35	.36	.37	.38	.39	.40	
15		.0001	.0001	.0000	.0000	.0000	.0000	.0000	.0000	.0000	.0000	85
16		.0003	.0001	.0001	.0000	.0000	.0000	.0000	.0000	.0000	.0000	84
17		.0006	.0003	.0002	.0001	.0000	.0000	.0000	.0000	.0000	.0000	83
18		.0013	.0007	.0004	.0002	.0001	.0000	.0000	.0000	.0000	.0000	82
19		.0025	.0014	.0008	.0004	.0002	.0001	.0000	.0000	.0000	.0000	81
20		.0046	.0027	.0015	.0008	.0004	.0002	.0001	.0001	.0000	.0000	80
21		.0079	.0049	.0029	.0016	.0009	.0005	.0002	.0001	.0001	.0000	79
22		.0127	.0082	.0051	.0030	.0017	.0010	.0005	.0003	.0001	.0001	78
23		.0194	.0131	.0085	.0053	.0032	.0018	.0010	.0006	.0003	.0001	77
24		.0280	.0198	.0134	.0088	.0055	.0033	.0019	.0011	.0006	.0003	76
25		.0382	.0283	.0201	.0137	.0090	.0057	.0035	.0020	.0012	.0006	75
26		.0496	.0384	.0286	.0204	.0140	.0092	.0059	.0036	.0021	.0012	74
27		.0610	.0495	.0386	.0288	.0207	.0143	.0095	.0060	.0037	.0022	73
28		.0715	.0608	.0495	.0387	.0290	.0209	.0145	.0097	.0062	.0038	72
29		.0797	.0710	.0605	.0495	.0388	.0292	.0211	.0147	.0098	.0063	71
30		.0848	.0791	.0706	.0603	.0494	.0389	.0294	.0213	.0149	.0100	70
31		.0860	.0840	.0785	.0702	.0601	.0494	.0399	.0295	.0215	.0151	69
32		.0833	.0853	.0834	.0779	.0698	.0599	.0493	.0390	.0296	.0217	68
33		.0771	.0827	.0846	.0827	.0774	.0694	.0597	.0493	.0390	.0297	67
34		.0683	.0767	.0821	.0840	.0821	.0769	.0689	.0595	.0492	.0391	66
35		.0578	.0680	.0763	.0816	.0834	.0816	.0765	.0688	.0593	.0491	65
36		.0469	.0578	.0678	.0759	.0811	.0829	.0811	.0761	.0685	.0591	64
37		.0365	.0471	.0577	.0676	.0755	.0806	.0824	.0807	.0757	.0682	63
38		.0272	.0367	.0472	.0577	.0674	.0752	.0802	.0820	.0803	.0754	62
39		.0194	.0275	.0369	.0473	.0577	.0672	.0749	.0799	.0816	.0799	61
40		.0133	.0197	.0277	.0372	.0474	.0577	.0671	.0746	.0795	.0812	60

TABLE C Binomial Distribution: Individual Terms $p(r) = c\binom{n}{r} p^r q^{n-r}$ (*Continued*)

	.69	.68	.67	.66	.65	.64	.63	.62	.61	.60	
41	.0087	.0156	.0200	.0280	.0373	.0475	.0577	.0670	.0744	.0792	59
42	.0055	.0090	.0138	.0203	.0282	.0375	.0476	.0576	.0668	.0742	58
43	.0033	.0057	.0092	.0141	.0205	.0285	.0377	.0477	.0576	.0667	57
44	.0019	.0035	.0059	.0094	.0143	.0207	.0287	.0378	.0477	.0576	56
45	.0011	.0020	.0036	.0060	.0096	.0145	.0210	.0289	.0380	.0478	55
46	.0006	.0011	.0021	.0037	.0062	.0098	.0147	.0212	.0290	.0381	54
47	.0003	.0006	.0012	.0022	.0038	.0063	.0099	.0149	.0213	.0292	53
48	.0001	.0003	.0007	.0012	.0023	.0039	.0064	.0101	.0151	.0215	52
49	.0001	.0002	.0003	.0007	.0013	.0023	.0040	.0066	.0102	.0152	51
50	.0000	.0001	.0002	.0004	.0007	.0013	.0024	.0041	.0067	.0103	50
51	.0000	.0000	.0001	.0002	.0004	.0007	.0014	.0025	.0042	.0068	49
52	.0000	.0000	.0000	.0001	.0002	.0004	.0008	.0014	.0025	.0042	48
53	.0000	.0000	.0000	.0000	.0001	.0002	.0004	.0008	.0015	.0026	47
54	.0000	.0000	.0000	.0000	.0000	.0001	.0002	.0004	.0008	.0015	46
55	.0000	.0000	.0000	.0000	.0000	.0000	.0001	.0002	.0004	.0008	45
56	.0000	.0000	.0000	.0000	.0000	.0000	.0000	.0001	.0002	.0004	44
57	.0000	.0000	.0000	.0000	.0000	.0000	.0000	.0001	.0001	.0002	43
58	.0000	.0000	.0000	.0000	.0000	.0000	.0000	.0000	.0001	.0001	42
59	.0000	.0000	.0000	.0000	.0001	.0000	.0000	.0000	.0000	.0001	41
	.69	.68	.67	.66	.65	.64	.63	.62	.61	.60	P R

P	P	.41	.42	.43	.44	.45	.46	.47	.48	.49	.50	
23		.0001	.0000	.0000	.0000	.0000	.0000	.0000	.0000	.0000	.0000	77
24		.0002	.0001	.0000	.0000	.0000	.0000	.0000	.0000	.0000	.0000	76
25		.0003	.0002	.0001	.0000	.0000	.0000	.0000	.0000	.0000	.0000	75
26		.0007	.0003	.0002	.0001	.0000	.0000	.0000	.0000	.0000	.0000	74
27		.0013	.0007	.0004	.0002	.0001	.0000	.0000	.0000	.0000	.0000	73
28		.0023	.0013	.0007	.0004	.0002	.0001	.0000	.0000	.0000	.0000	72
29		.0039	.0024	.0014	.0008	.0004	.0002	.0001	.0000	.0000	.0000	71
30		.0065	.0040	.0024	.0014	.0008	.0004	.0002	.0001	.0001	.0000	70
31		.0102	.0066	.0041	.0025	.0014	.0008	.0004	.0002	.0001	.0001	69
32		.0152	.0103	.0067	.0042	.0025	.0015	.0008	.0004	.0002	.0001	68
33		.0218	.0154	.0104	.0068	.0043	.0026	.0015	.0008	.0004	.0002	67
34		.0298	.0219	.0155	.0105	.0069	.0043	.0026	.0015	.0009	.0005	66
35		.0391	.0299	.0220	.0156	.0106	.0069	.0044	.0026	.0015	.0009	65
36		.0491	.0391	.0300	.0221	.0157	.0107	.0070	.0044	.0027	.0016	64
37		.0590	.0490	.0391	.0300	.0222	.0157	.0107	.0070	.0044	.0027	63
38		.0680	.0588	.0489	.0391	.0301	.0222	.0158	.0108	.0071	.0045	62
39		.0751	.0677	.0587	.0489	.0391	.0301	.0223	.0158	.0108	.0071	61
40		.0796	.0748	.0675	.0586	.0438	.0391	.0301	.0223	.0159	.0108	60
41		.0809	.0733	.0745	.0673	.0534	.0487	.0391	.0301	.0223	.0159	59
42		.0790	.0806	.0790	.0743	.0672	.0583	.0487	.0390	.0301	.0223	58
43		.0740	.0787	.0804	.0788	.0741	.0670	.0582	.0486	.0390	.0301	57
44		.0666	.0739	.0785	.0802	.0786	.0739	.0669	.0581	.0485	.0390	56
45		.0576	.0656	.0737	.0784	.0800	.0784	.0738	.0668	.0580	.0485	55
46		.0479	.0576	.0665	.0736	.0782	.0798	.0783	.0737	.0667	.0580	54
47		.0382	.0480	.0576	.0665	.0736	.0781	.0797	.0781	.0736	.0666	53
48		.0293	.0383	.0480	.0577	.0665	.0735	.0781	.0797	.0781	.0735	52
49		.0216	.0235	.0384	.0481	.0577	.0664	.0735	.0780	.0796	.0780	51
50		.0153	.0218	.0296	.0385	.0482	.0577	.0665	.0735	.0780	.0796	50
51		.0104	.0155	.0219	.0297	.0386	.0482	.0578	.0665	.0735	.0780	49
52		.0068	.0105	.0156	.0220	.0298	.0387	.0483	.0578	.0665	.0735	48
53		.0043	.0069	.0106	.0156	.0221	.0299	.0388	.0483	.0579	.0666	47
54		.0026	.0044	.0070	.0107	.0157	.0221	.0299	.0388	.0484	.0580	46
55		.0015	.0026	.0044	.0070	.0108	.0158	.0222	.0300	.0389	.0485	45
56		.0008	.0015	.0027	.0044	.0071	.0108	.0158	.0222	.0300	.0390	44
57		.0005	.0009	.0016	.0027	.0045	.0071	.0108	.0158	.0223	.0301	43
58		.0002	.0005	.0009	.0016	.0027	.0045	.0071	.0108	.0159	.0223	42
59		.0001	.0002	.0005	.0009	.0016	.0027	.0045	.0071	.0109	.0159	41
60		.0001	.0001	.0002	.0005	.0009	.0016	.0027	.0045	.0071	.0108	40
61		.0000	.0001	.0001	.0002	.0005	.0009	.0016	.0027	.0045	.0071	39
62		.0000	.0000	.0001	.0001	.0002	.0005	.0009	.0016	.0027	.0045	38
63		.0000	.0000	.0000	.0001	.0001	.0002	.0005	.0009	.0016	.0027	37
64		.0000	.0000	.0000	.0000	.0001	.0001	.0002	.0005	.0009	.0016	36
65		.0000	.0000	.0000	.0000	.0000	.0001	.0001	.0002	.0005	.0009	35
66		.0000	.0000	.0000	.0000	.0000	.0000	.0001	.0001	.0002	.0005	34
67		.0000	.0000	.0000	.0000	.0000	.0000	.0000	.0001	.0001	.0002	33
68		.0000	.0000	.0000	.0000	.0000	.0000	.0000	.0000	.0001	.0001	32
69		.0000	.0000	.0000	.0000	.0000	.0000	.0000	.0000	.0000	.0001	31
		.59	.58	.57	.56	.55	.54	.53	.52	.51	.50	P R

Table D Poisson probabilities

					λ					
k	0.005	0.01	0.02	0.03	0.04	0.05	0.06	0.07	0.08	0.09
0	0.9950	0.9900	0.9802	0.9704	0.9608	0.9512	0.9418	0.9324	0.9231	0.9139
1	0.0050	0.0099	0.0192	0.0291	0.0384	0.0476	0.0565	0.0653	0.0738	0.0823
2	0.0000	0.0000	0.0002	0.0004	0.0008	0.0012	0.0017	0.0023	0.0030	0.0037
3	0.0000	0.0000	0.0000	0.0000	0.0000	0.0000	0.0000	0.0001	0.0001	0.0001

					λ					
k	0.1	0.2	0.3	0.4	0.5	0.6	0.7	0.8	0.9	1.0
0	0.9048	0.8187	0.7408	0.6703	0.6065	0.5488	0.4966	0.4493	0.4066	0.3679
1	0.0905	0.1637	0.2222	0.2681	0.3033	0.3293	0.3476	0.3595	0.3659	0.3679
2	0.0045	0.0164	0.0333	0.0536	0.0758	0.0988	0.1217	0.1438	0.1647	0.1839
3	0.0002	0.0011	0.0033	0.0072	0.0126	0.0198	0.0284	0.0383	0.0494	0.0613
4	0.0000	0.0001	0.0002	0.0007	0.0016	0.0030	0.0050	0.0077	0.0111	0.0153
5	0.0000	0.0000	0.0000	0.0001	0.0002	0.0004	0.0007	0.0012	0.0020	0.0031
6	0.0000	0.0000	0.0000	0.0000	0.0000	0.0000	0.0001	0.0002	0.0003	0.0005
7	0.0000	0.0000	0.0000	0.0000	0.0000	0.0000	0.0000	0.0000	0.0000	0.0001

					λ					
k	1.1	1.2	1.3	1.4	1.5	1.6	1.7	1.8	1.9	2.0
0	0.3329	0.3012	0.2725	0.2466	0.2231	0.2019	0.1827	0.1653	0.1496	0.1353
1	0.3662	0.3614	0.3543	0.3452	0.3347	0.3230	0.3106	0.2975	0.2842	0.2707
2	0.2014	0.2169	0.2303	0.2417	0.2510	0.2584	0.2640	0.2678	0.2700	0.2707
3	0.0738	0.0867	0.0998	0.1128	0.1255	0.1378	0.1496	0.1607	0.1710	0.1804
4	0.0203	0.0260	0.0324	0.0395	0.0471	0.0551	0.0636	0.0723	0.0812	0.0902
5	0.0045	0.0062	0.0084	0.0111	0.0141	0.0176	0.0216	0.0260	0.0309	0.0361
6	0.0008	0.0012	0.0018	0.0026	0.0035	0.0047	0.0061	0.0078	0.0098	0.0120
7	0.0001	0.0002	0.0003	0.0005	0.0008	0.0011	0.0015	0.0020	0.0027	0.0034
8	0.0000	0.0000	0.0001	0.0001	0.0001	0.0002	0.0003	0.0005	0.0006	0.0009
9	0.0000	0.0000	0.0000	0.0000	0.0000	0.0000	0.0001	0.0001	0.0001	0.0002

					λ					
k	2.1	2.2	2.3	2.4	2.5	2.6	2.7	2.8	2.9	3.0
0	0.1225	0.1108	0.1003	0.0907	0.0821	0.0743	0.0672	0.0608	0.0550	0.0498
1	0.2572	0.2438	0.2306	0.2177	0.2052	0.1931	0.1815	0.1703	0.1596	0.1494
2	0.2700	0.2681	0.2652	0.2613	0.2565	0.2510	0.2450	0.2384	0.2314	0.2240
3	0.1890	0.1966	0.2033	0.2090	0.2138	0.2176	0.2205	0.2225	0.2237	0.2240
4	0.0992	0.1082	0.1169	0.1254	0.1336	0.1414	0.1488	0.1557	0.1622	0.1680
5	0.0417	0.0476	0.0538	0.0602	0.0668	0.0735	0.0804	0.0872	0.0940	0.1008
6	0.0146	0.0174	0.0206	0.0241	0.0278	0.0319	0.0362	0.0407	0.0455	0.0504
7	0.0044	0.0055	0.0068	0.0083	0.0099	0.0118	0.0139	0.0163	0.0188	0.0216
8	0.0011	0.0015	0.0019	0.0025	0.0031	0.0038	0.0047	0.0057	0.0068	0.0081
9	0.0003	0.0004	0.0005	0.0007	0.0009	0.0011	0.0014	0.0018	0.0022	0.0027
10	0.0001	0.0001	0.0001	0.0002	0.0002	0.0003	0.0004	0.0005	0.0006	0.0008
11	0.0000	0.0000	0.0000	0.0000	0.0000	0.0001	0.0001	0.0001	0.0002	0.0002
12	0.0000	0.0000	0.0000	0.0000	0.0000	0.0000	0.0000	0.0000	0.0000	0.0001

					λ					
k	3.1	3.2	3.3	3.4	3.5	3.6	3.7	3.8	3.9	4.0
0	0.0450	0.0408	0.0369	0.0334	0.0302	0.0273	0.0247	0.0224	0.0202	0.0183
1	0.1397	0.1304	0.1217	0.1135	0.1057	0.0984	0.0915	0.0850	0.0789	0.0733
2	0.2165	0.2087	0.2008	0.1929	0.1850	0.1771	0.1692	0.1615	0.1539	0.1465
3	0.2237	0.2226	0.2209	0.2186	0.2158	0.2125	0.2087	0.2046	0.2001	0.1954
4	0.1734	0.1781	0.1823	0.1858	0.1888	0.1912	0.1931	0.1944	0.1951	0.1954
5	0.1075	0.1140	0.1203	0.1264	0.1322	0.1377	0.1429	0.1477	0.1522	0.1563
6	0.0555	0.0608	0.0662	0.0716	0.0771	0.0826	0.0881	0.0936	0.0989	0.1042
7	0.0246	0.0278	0.0312	0.0348	0.0385	0.0425	0.0466	0.0508	0.0551	0.0595
8	0.0095	0.0111	0.0129	0.0148	0.0169	0.0191	0.0215	0.0241	0.0269	0.0298
9	0.0033	0.0040	0.0047	0.0056	0.0066	0.0076	0.0089	0.0102	0.0116	0.0132

Table D Poisson probabilities *(Continued)*

10	0.0010	0.0013	0.0016	0.0019	0.0023	0.0028	0.0033	0.0039	0.0045	0.0053
11	0.0003	0.0004	0.0005	0.0006	0.0007	0.0009	0.0011	0.0013	0.0016	0.0019
12	0.0001	0.0001	0.0001	0.0002	0.0002	0.0003	0.0003	0.0004	0.0005	0.0006
13	0.0000	0.0000	0.0000	0.0000	0.0001	0.0001	0.0001	0.0001	0.0002	0.0002
14	0.0000	0.0000	0.0000	0.0000	0.0000	0.0000	0.0000	0.0000	0.0000	0.0001

λ

k	4.1	4.2	4.3	4.4	4.5	4.6	4.7	4.8	4.9	5.0
0	0.0166	0.0150	0.0136	0.0123	0.0111	0.0101	0.0091	0.0082	0.0074	0.0067
1	0.0679	0.0630	0.0583	0.0540	0.0500	0.0462	0.0427	0.0395	0.0365	0.0337
2	0.1393	0.1323	0.1254	0.1188	0.1125	0.1063	0.1005	0.0948	0.0894	0.0842
3	0.1904	0.1852	0.1798	0.1743	0.1687	0.1631	0.1574	0.1517	0.1460	0.1404
4	0.1951	0.1944	0.1933	0.1917	0.1898	0.1875	0.1849	0.1820	0.1789	0.1755
5	0.1600	0.1633	0.1662	0.1687	0.1708	0.1725	0.1738	0.1747	0.1753	0.1755
6	0.1093	0.1143	0.1191	0.1237	0.1281	0.1323	0.1362	0.1398	0.1432	0.1462
7	0.0640	0.0686	0.0732	0.0778	0.0824	0.0869	0.0914	0.0959	0.1002	0.1044
8	0.0328	0.0360	0.0393	0.0428	0.0463	0.0500	0.0537	0.0575	0.0614	0.0653
9	0.0150	0.0168	0.0188	0.0209	0.0232	0.0255	0.0280	0.0307	0.0334	0.0363
10	0.0061	0.0071	0.0081	0.0092	0.0104	0.0118	0.0132	0.0147	0.0164	0.0181
11	0.0023	0.0027	0.0032	0.0037	0.0043	0.0049	0.0056	0.0064	0.0073	0.0082
12	0.0008	0.0009	0.0011	0.0014	0.0016	0.0019	0.0022	0.0026	0.0030	0.0034
13	0.0002	0.0003	0.0004	0.0005	0.0006	0.0007	0.0008	0.0009	0.0011	0.0013
14	0.0001	0.0001	0.0001	0.0001	0.0002	0.0002	0.0003	0.0003	0.0004	0.0005
15	0.0000	0.0000	0.0000	0.0000	0.0001	0.0001	0.0001	0.0001	0.0001	0.0002

λ

k	5.1	5.2	5.3	5.4	5.5	5.6	5.7	5.8	5.9	6.0
0	0.0061	0.0055	0.0050	0.0045	0.0041	0.0037	0.0033	0.0030	0.0027	0.0025
1	0.0311	0.0287	0.0265	0.0244	0.0225	0.0207	0.0191	0.0176	0.0162	0.0149
2	0.0793	0.0746	0.0701	0.0659	0.0618	0.0580	0.0544	0.0509	0.0477	0.0446
3	0.1348	0.1293	0.1239	0.1185	0.1133	0.1082	0.1033	0.0985	0.0938	0.0892
4	0.1719	0.1681	0.1641	0.1600	0.1558	0.1515	0.1472	0.1428	0.1383	0.1339
5	0.1753	0.1748	0.1740	0.1728	0.1714	0.1697	0.1678	0.1656	0.1632	0.1606
6	0.1490	0.1515	0.1537	0.1555	0.1571	0.1584	0.1594	0.1601	0.1605	0.1606
7	0.1086	0.1125	0.1163	0.1200	0.1234	0.1267	0.1298	0.1326	0.1353	0.1377
8	0.0692	0.0731	0.0771	0.0810	0.0849	0.0887	0.0925	0.0962	0.0998	0.1033
9	0.0392	0.0423	0.0454	0.0486	0.0519	0.0552	0.0586	0.0620	0.0654	0.0688
10	0.0200	0.0220	0.0241	0.0262	0.0285	0.0309	0.0334	0.0359	0.0386	0.0413
11	0.0093	0.0104	0.0116	0.0129	0.0143	0.0157	0.0173	0.0190	0.0207	0.0225
12	0.0039	0.0045	0.0051	0.0058	0.0065	0.0073	0.0082	0.0092	0.0102	0.0113
13	0.0015	0.0018	0.0021	0.0024	0.0028	0.0032	0.0036	0.0041	0.0046	0.0052
14	0.0006	0.0007	0.0008	0.0009	0.0011	0.0013	0.0015	0.0017	0.0019	0.0022
15	0.0002	0.0002	0.0003	0.0003	0.0004	0.0005	0.0006	0.0007	0.0008	0.0009
16	0.0001	0.0001	0.0001	0.0001	0.0001	0.0002	0.0002	0.0002	0.0003	0.0003
17	0.0000	0.0000	0.0000	0.0000	0.0000	0.0001	0.0001	0.0001	0.0001	0.0001

λ

k	6.1	6.2	6.3	6.4	6.5	6.6	6.7	6.8	6.9	7.0
0	0.0022	0.0020	0.0018	0.0017	0.0015	0.0014	0.0012	0.0011	0.0010	0.0009
1	0.0137	0.0126	0.0116	0.0106	0.0098	0.0090	0.0082	0.0076	0.0070	0.0064
2	0.0417	0.0390	0.0364	0.0340	0.0318	0.0296	0.0276	0.0258	0.0240	0.0223
3	0.0848	0.0806	0.0765	0.0726	0.0688	0.0652	0.0617	0.0584	0.0552	0.0521
4	0.1294	0.1249	0.1205	0.1162	0.1118	0.1076	0.1034	0.0992	0.0952	0.0912
5	0.1579	0.1549	0.1519	0.1487	0.1454	0.1420	0.1385	0.1349	0.1314	0.1277
6	0.1605	0.1601	0.1595	0.1586	0.1575	0.1562	0.1546	0.1529	0.1511	0.1490
7	0.1399	0.1418	0.1435	0.1450	0.1462	0.1472	0.1480	0.1486	0.1489	0.1490
8	0.1066	0.1099	0.1130	0.1160	0.1188	0.1215	0.1240	0.1263	0.1284	0.1304
9	0.0723	0.0757	0.0791	0.0825	0.0858	0.0891	0.0923	0.0954	0.0985	0.1014

Table D Poisson probabilities *(Continued)*

k										
10	0.0441	0.0469	0.0498	0.0528	0.0558	0.0588	0.0618	0.0649	0.0679	0.0710
11	0.0245	0.0265	0.0285	0.0307	0.0330	0.0353	0.0377	0.0401	0.0426	0.0452
12	0.0124	0.0137	0.0150	0.0164	0.0179	0.0194	0.0210	0.0227	0.0245	0.0264
13	0.0058	0.0065	0.0073	0.0081	0.0089	0.0098	0.0108	0.0119	0.0130	0.0142
14	0.0025	0.0029	0.0033	0.0037	0.0041	0.0046	0.0052	0.0058	0.0064	0.0071
15	0.0010	0.0012	0.0014	0.0016	0.0018	0.0020	0.0023	0.0026	0.0029	0.0033
16	0.0004	0.0005	0.0005	0.0006	0.0007	0.0008	0.0010	0.0011	0.0013	0.0014
17	0.0001	0.0002	0.0002	0.0002	0.0003	0.0003	0.0004	0.0004	0.0005	0.0006
18	0.0000	0.0001	0.0001	0.0001	0.0001	0.0001	0.0001	0.0002	0.0002	0.0002
19	0.0000	0.0000	0.0000	0.0000	0.0000	0.0000	0.0000	0.0001	0.0001	0.0001

λ

k	7.1	7.2	7.3	7.4	7.5	7.6	7.7	7.8	7.9	8.0
0	0.0008	0.0007	0.0007	0.0006	0.0006	0.0005	0.0005	0.0004	0.0004	0.0003
1	0.0059	0.0054	0.0049	0.0045	0.0041	0.0038	0.0035	0.0032	0.0029	0.0027
2	0.0208	0.0194	0.0180	0.0167	0.0156	0.0145	0.0134	0.0125	0.0116	0.0107
3	0.0492	0.0464	0.0438	0.0413	0.0389	0.0366	0.0345	0.0324	0.0305	0.0286
4	0.0874	0.0836	0.0799	0.0764	0.0729	0.0696	0.0663	0.0632	0.0602	0.0573
5	0.1241	0.1204	0.1167	0.1130	0.1094	0.1057	0.1021	0.0986	0.0951	0.0916
6	0.1468	0.1445	0.1420	0.1394	0.1367	0.1339	0.1311	0.1282	0.1252	0.1221
7	0.1489	0.1486	0.1481	0.1474	0.1465	0.1454	0.1442	0.1428	0.1413	0.1396
8	0.1321	0.1337	0.1351	0.1363	0.1373	0.1382	0.1388	0.1392	0.1395	0.1396
9	0.1042	0.1070	0.1096	0.1121	0.1144	0.1167	0.1187	0.1207	0.1224	0.1241
10	0.0740	0.0770	0.0800	0.0829	0.0858	0.0887	0.0914	0.0941	0.0967	0.0993
11	0.0478	0.0504	0.0531	0.0558	0.0585	0.0613	0.0640	0.0667	0.0695	0.0722
12	0.0283	0.0303	0.0323	0.0344	0.0366	0.0388	0.0411	0.0434	0.0457	0.0481
13	0.0154	0.0168	0.0181	0.0196	0.0211	0.0227	0.0243	0.0260	0.0278	0.0296
14	0.0078	0.0086	0.0095	0.0104	0.0113	0.0123	0.0134	0.0145	0.0157	0.0169
15	0.0037	0.0041	0.0046	0.0051	0.0057	0.0062	0.0069	0.0075	0.0083	0.0090
16	0.0016	0.0019	0.0021	0.0024	0.0026	0.0030	0.0033	0.0037	0.0041	0.0045
17	0.0007	0.0008	0.0009	0.0010	0.0012	0.0013	0.0015	0.0017	0.0019	0.0021
18	0.0003	0.0003	0.0004	0.0004	0.0005	0.0006	0.0006	0.0007	0.0008	0.0009
19	0.0001	0.0001	0.0001	0.0002	0.0002	0.0002	0.0003	0.0003	0.0003	0.0004
20	0.0000	0.0000	0.0001	0.0001	0.0001	0.0001	0.0001	0.0001	0.0001	0.0002
21	0.0000	0.0000	0.0000	0.0000	0.0000	0.0000	0.0000	0.0000	0.0001	0.0001

λ

k	8.1	8.2	8.3	8.4	8.5	8.6	8.7	8.8	8.9	9.0
0	0.0003	0.0003	0.0002	0.0002	0.0002	0.0002	0.0002	0.0002	0.0001	0.0001
1	0.0025	0.0023	0.0021	0.0019	0.0017	0.0016	0.0014	0.0013	0.0012	0.0011
2	0.0100	0.0092	0.0086	0.0079	0.0074	0.0068	0.0063	0.0058	0.0054	0.0050
3	0.0269	0.0252	0.0237	0.0222	0.0208	0.0195	0.0183	0.0171	0.0160	0.0150
4	0.0544	0.0517	0.0491	0.0466	0.0443	0.0420	0.0398	0.0377	0.0357	0.0337
5	0.0882	0.0849	0.0816	0.0784	0.0752	0.0722	0.0692	0.0663	0.0635	0.0607
6	0.1191	0.1160	0.1128	0.1097	0.1066	0.1034	0.1003	0.0972	0.0941	0.0911
7	0.1378	0.1358	0.1338	0.1317	0.1294	0.1271	0.1247	0.1222	0.1197	0.1171
8	0.1395	0.1392	0.1388	0.1382	0.1375	0.1366	0.1356	0.1344	0.1332	0.1318
9	0.1256	0.1269	0.1280	0.1290	0.1299	0.1306	0.1311	0.1315	0.1317	0.1318
10	0.1017	0.1040	0.1063	0.1084	0.1104	0.1123	0.1140	0.1157	0.1172	0.1186
11	0.0749	0.0776	0.0802	0.0828	0.0853	0.0878	0.0902	0.0925	0.0948	0.0970
12	0.0505	0.0530	0.0555	0.0579	0.0604	0.0629	0.0654	0.0679	0.0703	0.0728
13	0.0315	0.0334	0.0354	0.0374	0.0395	0.0416	0.0438	0.0459	0.0481	0.0504
14	0.0182	0.0196	0.0210	0.0225	0.0240	0.0256	0.0272	0.0289	0.0306	0.0324
15	0.0098	0.0107	0.0116	0.0126	0.0136	0.0147	0.0158	0.0169	0.0182	0.0194
16	0.0050	0.0055	0.0060	0.0066	0.0072	0.0079	0.0086	0.0093	0.0101	0.0109
17	0.0024	0.0026	0.0029	0.0033	0.0036	0.0040	0.0044	0.0048	0.0053	0.0058
18	0.0011	0.0012	0.0014	0.0015	0.0017	0.0019	0.0021	0.0024	0.0026	0.0029
19	0.0005	0.0005	0.0006	0.0007	0.0008	0.0009	0.0010	0.0011	0.0012	0.0014

Table D Poisson probabilities *(Continued)*

k										
20	0.0002	0.0002	0.0002	0.0003	0.0003	0.0004	0.0004	0.0005	0.0005	0.0006
21	0.0001	0.0001	0.0001	0.0001	0.0001	0.0002	0.0002	0.0002	0.0002	0.0003
22	0.0000	0.0000	0.0000	0.0000	0.0001	0.0001	0.0001	0.0001	0.0001	0.0001

λ

k	9.1	9.2	9.3	9.4	9.5	9.6	9.7	9.8	9.9	10.0
0	0.0001	0.0001	0.0001	0.0001	0.0001	0.0001	0.0001	0.0001	0.0001	0.0000
1	0.0010	0.0009	0.0009	0.0008	0.0007	0.0007	0.0006	0.0005	0.0005	0.0005
2	0.0046	0.0043	0.0040	0.0037	0.0034	0.0031	0.0029	0.0027	0.0025	0.0023
3	0.0140	0.0131	0.0123	0.0115	0.0107	0.0100	0.0093	0.0087	0.0081	0.0076
4	0.0319	0.0302	0.0285	0.0269	0.0254	0.0240	0.0226	0.0213	0.0201	0.0189
5	0.0581	0.0555	0.0530	0.0506	0.0483	0.0460	0.0439	0.0418	0.0398	0.0378
6	0.0881	0.0851	0.0822	0.0793	0.0764	0.0736	0.0709	0.0682	0.0656	0.0631
7	0.1145	0.1118	0.1091	0.1064	0.1037	0.1010	0.0982	0.0955	0.0928	0.0901
8	0.1302	0.1286	0.1269	0.1251	0.1232	0.1212	0.1191	0.1170	0.1148	0.1126
9	0.1317	0.1315	0.1311	0.1306	0.1300	0.1293	0.1284	0.1274	0.1263	0.1251
10	0.1198	0.1210	0.1219	0.1228	0.1235	0.1241	0.1245	0.1249	0.1250	0.1251
11	0.0991	0.1012	0.1031	0.1049	0.1067	0.1083	0.1098	0.1112	0.1125	0.1137
12	0.0752	0.0776	0.0799	0.0822	0.0844	0.0866	0.0888	0.0908	0.0928	0.0948
13	0.0526	0.0549	0.0572	0.0594	0.0617	0.0640	0.0662	0.0685	0.0707	0.0729
14	0.0342	0.0361	0.0380	0.0399	0.0419	0.0439	0.0459	0.0479	0.0500	0.0521
15	0.0208	0.0221	0.0235	0.0250	0.0265	0.0281	0.0297	0.0313	0.0330	0.0347
16	0.0118	0.0127	0.0137	0.0147	0.0157	0.0168	0.0180	0.0192	0.0204	0.0217
17	0.0063	0.0069	0.0075	0.0081	0.0088	0.0095	0.0103	0.0111	0.0119	0.0128
18	0.0032	0.0035	0.0039	0.0042	0.0046	0.0051	0.0055	0.0060	0.0065	0.0071
19	0.0015	0.0017	0.0019	0.0021	0.0023	0.0026	0.0028	0.0031	0.0034	0.0037
20	0.0007	0.0008	0.0009	0.0010	0.0011	0.0012	0.0014	0.0015	0.0017	0.0019
21	0.0003	0.0003	0.0004	0.0004	0.0005	0.0006	0.0006	0.0007	0.0008	0.0009
22	0.0001	0.0001	0.0002	0.0002	0.0002	0.0002	0.0003	0.0003	0.0004	0.0004
23	0.0000	0.0001	0.0001	0.0001	0.0001	0.0001	0.0001	0.0001	0.0002	0.0002
24	0.0000	0.0000	0.0000	0.0000	0.0000	0.0000	0.0000	0.0001	0.0001	0.0001

Table E Values of e^{-x}

$$e^{-1.37} = 0.25411$$

x	e^{-x} (value)	x	e^{-x} (value)	x	e^{-x} (value)	x	e^{-x} (value)
0.00	1.00000	0.50	0.60653	1.00	0.36788	1.50	0.22313
0.01	0.99005	0.51	0.60050	1.01	0.36422	1.51	0.22091
0.02	0.98020	0.52	0.59452	1.02	0.36060	1.52	0.21871
0.03	0.97045	0.53	0.58860	1.03	0.35701	1.53	0.21654
0.04	0.96079	0.54	0.58275	1.04	0.35345	1.54	0.21438
0.05	0.95123	0.55	0.57695	1.05	0.34994	1.55	0.21225
0.06	0.94176	0.56	0.57121	1.06	0.34646	1.56	0.21014
0.07	0.93239	0.57	0.56553	1.07	0.34301	1.57	0.20805
0.08	0.92312	0.58	0.55990	1.08	0.33960	1.58	0.20598
0.09	0.91393	0.59	0.55433	1.09	0.33622	1.59	0.20393
0.10	0.90484	0.60	0.54881	1.10	0.33287	1.60	0.20190
0.11	0.89583	0.61	0.54335	1.11	0.32956	1.61	0.19989
0.12	0.88692	0.62	0.53794	1.12	0.32628	1.62	0.19790
0.13	0.87809	0.63	0.53259	1.13	0.32303	1.63	0.19593
0.14	0.86936	0.64	0.52729	1.14	0.31982	1.64	0.19398
0.15	0.86071	0.65	0.52205	1.15	0.31664	1.65	0.19205
0.16	0.85214	0.66	0.51685	1.16	0.31349	1.66	0.19014
0.17	0.84366	0.67	0.51171	1.17	0.31037	1.67	0.18825
0.18	0.83527	0.68	0.50662	1.18	0.30728	1.68	0.18637
0.19	0.82696	0.69	0.50158	1.19	0.30422	1.69	0.18452
0.20	0.81873	0.70	0.49659	1.20	0.30119	1.70	0.18268
0.21	0.81058	0.71	0.49164	1.21	0.29820	1.71	0.18087
0.22	0.80252	0.72	0.48675	1.22	0.29523	1.72	0.17907
0.23	0.79453	0.73	0.48191	1.23	0.29229	1.73	0.17728
0.24	0.78663	0.74	0.47711	1.24	0.28938	1.74	0.17552
0.25	0.77880	0.75	0.47237	1.25	0.28650	1.75	0.17377
0.26	0.77105	0.76	0.46767	1.26	0.28365	1.76	0.17204
0.27	0.76338	0.77	0.46301	1.27	0.28083	1.77	0.17033
0.28	0.75578	0.78	0.45841	1.28	0.27804	1.78	0.16864
0.29	0.74826	0.79	0.45384	1.29	0.27527	1.79	0.16696
0.30	0.74082	0.80	0.44933	1.30	0.27253	1.80	0.16530
0.31	0.73345	0.81	0.44486	1.31	0.26982	1.81	0.16365
0.32	0.72615	0.82	0.44043	1.32	0.26714	1.82	0.16203
0.33	0.71892	0.83	0.43605	1.33	0.26448	1.83	0.16041
0.34	0.71177	0.84	0.43171	1.34	0.26185	1.84	0.15882
0.35	0.70469	0.85	0.42741	1.35	0.25924	1.85	0.15724
0.36	0.69768	0.86	0.42316	1.36	0.25666	1.86	0.15567
0.37	0.69073	0.87	0.41895	1.37	0.25411	1.87	0.15412
0.38	0.68386	0.88	0.41478	1.38	0.25158	1.88	0.15259
0.39	0.67706	0.89	0.41066	1.39	0.24908	1.89	0.15107
0.40	0.67032	0.90	0.40657	1.40	0.24660	1.90	0.14957
0.41	0.66365	0.91	0.40252	1.41	0.24414	1.91	0.14808
0.42	0.65705	0.92	0.39852	1.42	0.24171	1.92	0.14661
0.43	0.65051	0.93	0.39455	1.43	0.23931	1.93	0.14515
0.44	0.64404	0.94	0.39063	1.44	0.23693	1.94	0.14370
0.45	0.63763	0.95	0.38674	1.45	0.23457	1.95	0.14227
0.46	0.63128	0.96	0.38289	1.46	0.23224	1.96	0.14086
0.47	0.62500	0.97	0.37908	1.47	0.22993	1.97	0.13946
0.48	0.61878	0.98	0.37531	1.48	0.22764	1.98	0.13807
0.49	0.61263	0.99	0.37158	1.49	0.22537	1.99	0.13670

Table E Values of e^{-x}

$e^{-1.37} = 0.25411$ *(Continued)*

x	e^{-x} (value)	x	e^{-x} (value)	x	e^{-x} (value)	x	e^{-x} (value)
2.00	0.13534	2.40	0.09072	2.80	0.06081	4.00	0.01832
2.01	0.13399	2.41	0.08982	2.81	0.06020	4.10	0.01657
2.02	0.13266	2.42	0.08892	2.82	0.05961	4.20	0.01500
2.03	0.13134	2.43	0.08804	2.83	0.05901	4.30	0.01357
2.04	0.13003	2.44	0.08716	2.84	0.05843	4.40	0.01227
2.05	0.12873	2.45	0.08629	2.85	0.05784	4.50	0.01111
2.06	0.12745	2.46	0.08543	2.86	0.05727	4.60	0.01005
2.07	0.12619	2.47	0.08458	2.87	0.05670	4.70	0.00910
2.08	0.12493	2.48	0.08374	2.88	0.05613	4.80	0.00823
2.09	0.12369	2.49	0.08291	2.89	0.05558	4.90	0.00745
2.10	0.12246	2.50	0.08208	2.90	0.05502	5.00	0.00674
2.11	0.12124	2.51	0.08127	2.91	0.05448	5.10	0.00610
2.12	0.12003	2.52	0.08046	2.92	0.05393	5.20	0.00552
2.13	0.11884	2.53	0.07966	2.93	0.05340	5.30	0.00499
2.14	0.11765	2.54	0.07887	2.94	0.05287	5.40	0.00452
2.15	0.11648	2.55	0.07808	2.95	0.05234	5.50	0.00409
2.16	0.11533	2.56	0.07730	2.96	0.05182	5.60	0.00370
2.17	0.11418	2.57	0.07654	2.97	0.05130	5.70	0.00335
2.18	0.11304	2.58	0.07577	2.98	0.05079	5.80	0.00303
2.19	0.11192	2.59	0.07502	2.99	0.05029	5.90	0.00274
2.20	0.11080	2.60	0.07427	3.00	0.04979	6.00	0.00248
2.21	0.10970	2.61	0.07353	3.05	0.04736	6.25	0.00193
2.22	0.10861	2.62	0.07280	3.10	0.04505	6.50	0.00150
2.23	0.10753	2.63	0.07208	3.15	0.04285	6.75	0.00117
2.24	0.10646	2.64	0.07136	3.20	0.04076	7.00	0.00091
2.25	0.10540	2.65	0.07065	3.25	0.03877	7.50	0.00055
2.26	0.10435	2.66	0.06995	3.30	0.03688	8.00	0.00034
2.27	0.10331	2.67	0.06925	3.35	0.03508	8.50	0.00020
2.28	0.10228	2.68	0.06856	3.40	0.03337	9.00	0.00012
2.29	0.10127	2.69	0.06788	3.45	0.03175	9.50	0.00007
2.30	0.10026	2.70	0.06721	3.50	0.03020	10.00	0.00005
2.31	0.09926	2.71	0.06654	3.55	0.02872		
2.32	0.09827	2.72	0.06587	3.60	0.02732		
2.33	0.09730	2.73	0.06522	3.65	0.02599		
2.34	0.09633	2.74	0.06457	3.70	0.02472		
2.35	0.09537	2.75	0.06393	3.75	0.02352		
2.36	0.09442	2.76	0.06329	3.80	0.02237		
2.37	0.09348	2.77	0.06266	3.85	0.02128		
2.38	0.09255	2.78	0.06204	3.90	0.02024		
2.39	0.09163	2.79	0.06142	3.95	0.01925		

Table F Areas under the normal curve

Each entry in the table indicates the proportion of the total area under the normal curve contained in the segment bounded by a perpendicular raised at the mean and a perpendicular raised at a distance of x standard deviation units.

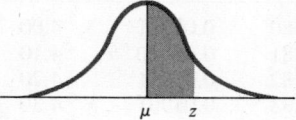

To illustrate: 43.57 percent of the area under a normal curve lies between the maximum ordinate and a point 1.52 standard deviation units away.

x	0.00	0.01	0.02	0.03	0.04	0.05	0.06	0.07	0.08	0.09
0.0	0.0000	0.0040	0.0080	0.0120	0.0160	0.0199	0.0239	0.0279	0.0319	0.0359
0.1	0.0398	0.0438	0.0478	0.0517	0.0557	0.0596	0.0636	0.0675	0.0714	0.0753
0.2	0.0793	0.0832	0.0871	0.0910	0.0948	0.0987	0.1026	0.1064	0.1103	0.1141
0.3	0.1179	0.1217	0.1255	0.1293	0.1331	0.1368	0.1406	0.1443	0.1480	0.1517
0.4	0.1554	0.1591	0.1628	0.1664	0.1700	0.1736	0.1772	0.1808	0.1844	0.1879
0.5	0.1915	0.1950	0.1985	0.2019	0.2054	0.2088	0.2123	0.2157	0.2190	0.2224
0.6	0.2257	0.2291	0.2324	0.2357	0.2389	0.2422	0.2454	0.2486	0.2518	0.2549
0.7	0.2580	0.2612	0.2642	0.2673	0.2704	0.2734	0.2764	0.2794	0.2823	0.2852
0.8	0.2881	0.2910	0.2939	0.2967	0.2995	0.3023	0.3051	0.3078	0.3106	0.3133
0.9	0.3159	0.3186	0.3212	0.3238	0.3264	0.3289	0.3315	0.3340	0.3365	0.3389
1.0	0.3413	0.3438	0.3461	0.3485	0.3508	0.3531	0.3554	0.3577	0.3599	0.3621
1.1	0.3643	0.3665	0.3686	0.3708	0.3729	0.3749	0.3770	0.3790	0.3810	0.3830
1.2	0.3849	0.3869	0.3888	0.3907	0.3925	0.3944	0.3962	0.3980	0.3997	0.4015
1.3	0.4032	0.4049	0.4066	0.4082	0.4099	0.4115	0.4131	0.4147	0.4162	0.4177
1.4	0.4192	0.4207	0.4222	0.4236	0.4251	0.4265	0.4279	0.4292	0.4306	0.4319
1.5	0.4332	0.4345	0.4357	0.4370	0.4382	0.4394	0.4406	0.4418	0.4429	0.4441
1.6	0.4452	0.4463	0.4474	0.4484	0.4495	0.4505	0.4515	0.4525	0.4535	0.4545
1.7	0.4554	0.4564	0.4573	0.4582	0.4591	0.4599	0.4608	0.4616	0.4625	0.4633
1.8	0.4641	0.4649	0.4656	0.4664	0.4671	0.4678	0.4686	0.4693	0.4699	0.4706
1.9	0.4713	0.4719	0.4726	0.4732	0.4738	0.4744	0.4750	0.4756	0.4761	0.4767
2.0	0.4772	0.4778	0.4783	0.4788	0.4793	0.4798	0.4803	0.4808	0.4812	0.4817
2.1	0.4821	0.4826	0.4830	0.4834	0.4838	0.4842	0.4846	0.4850	0.4854	0.4857
2.2	0.4861	0.4864	0.4868	0.4871	0.4875	0.4878	0.4881	0.4884	0.4887	0.4890
2.3	0.4893	0.4896	0.4898	0.4901	0.4904	0.4906	0.4909	0.4911	0.4913	0.4916
2.4	0.4918	0.4920	0.4922	0.4925	0.4927	0.4929	0.4931	0.4932	0.4934	0.4936
2.5	0.4938	0.4940	0.4941	0.4943	0.4945	0.4946	0.4948	0.4949	0.4951	0.4952
2.6	0.4953	0.4955	0.4956	0.4957	0.4959	0.4960	0.4961	0.4962	0.4963	0.4964
2.7	0.4965	0.4966	0.4967	0.4968	0.4969	0.4970	0.4971	0.4972	0.4973	0.4974
2.8	0.4974	0.4975	0.4976	0.4977	0.4977	0.4978	0.4979	0.4979	0.4980	0.4981
2.9	0.4981	0.4982	0.4982	0.4983	0.4984	0.4984	0.4985	0.4985	0.4986	0.4986
3.0	0.4986	0.4987	0.4987	0.4988	0.4988	0.4989	0.4989	0.4989	0.4990	0.4990
3.1	0.4990	0.4991	0.4991	0.4991	0.4992	0.4992	0.4992	0.4992	0.4993	0.4993
3.2	0.4993	0.4993	0.4994	0.4994	0.4994	0.4994	0.4994	0.4995	0.4995	0.4995
3.3	0.4995	0.4995	0.4995	0.4996	0.4996	0.4996	0.4996	0.4996	0.4996	0.4997
3.4	0.4997	0.4997	0.4997	0.4997	0.4997	0.4997	0.4997	0.4997	0.4998	0.4998
3.5	0.4998	0.4998	0.4998	0.4998	0.4998	0.4998	0.4998	0.4998	0.4998	0.4998
3.6	0.4998	0.4998	0.4999	0.4999	0.4999	0.4999	0.4999	0.4999	0.4999	0.4999
3.7	0.4999	0.4999	0.4999	0.4999	0.4999	0.4999	0.4999	0.4999	0.4999	0.4999
3.8	0.4999	0.4999	0.4999	0.4999	0.4999	0.4999	0.4999	0.5000	0.5000	0.5000
3.9	0.5000	0.5000	0.5000	0.5000	0.5000	0.5000	0.5000	0.5000	0.5000	0.5000

Table G Random numbers

2271	2572	8665	3272	9033	8256	2822	3646	7599	0270
3025	0788	5311	7792	1837	4739	4552	3234	5572	9885
3382	6151	1011	3778	9951	7709	8060	2258	8536	2290
7870	5799	6032	9043	4526	8100	1957	9539	5370	0046
1697	0002	2340	6959	1915	1626	1297	1533	6572	3835
3395	3381	1862	3250	8614	5683	6757	5628	2551	6971
6081	6526	3028	2338	5702	8819	3679	4829	9909	4712
3470	9879	2935	1141	6398	6387	5634	9589	3212	7963
0432	8641	5020	6612	1038	1547	0948	4278	0020	6509
4995	5596	8286	8377	8567	8237	3520	8244	5694	3326
8246	6718	3851	5870	1216	2107	1387	1621	5509	5772
7825	8727	2849	3501	3551	1001	0123	7873	5926	6078
6258	2450	2962	1183	3666	4156	4454	8239	4551	2920
3235	5783	2701	2378	7460	3398	1223	4688	3674	7872
2525	9008	5997	0885	1053	2340	7066	5328	6412	5054
5852	9739	1457	8999	2789	9068	9829	1336	3148	7875
0440	3769	7864	4029	4494	9829	1339	4910	1303	9161
0820	4641	2375	2542	4093	5364	1145	2848	2792	0431
7114	2842	8554	6881	6377	9427	8216	1193	8042	8449
6558	9301	9096	0577	8520	5923	4717	0188	8545	8745
0345	9937	5569	0279	8951	6183	7787	7808	5149	2185
7430	2074	9427	8422	4082	5629	2971	9456	0649	7981
8030	7345	3389	4739	5911	1022	9189	2565	1982	8577
6272	6718	3849	4715	3156	2823	4174	8733	5600	7702
4894	9847	5611	4763	8755	3388	5114	3274	6681	3657
2676	5984	6806	2692	4012	0934	2436	0869	9557	2490
9305	2074	9378	7670	8284	7431	7361	2912	2251	7395
5138	2461	7213	1905	7775	9881	8782	6272	0632	4418
2452	4200	8674	9202	0812	3986	1143	7343	2264	9072
8882	3033	8746	7390	8609	1144	2531	6944	8869	1570
1087	9336	8020	9166	4472	8293	2904	7949	3165	7400
5666	2841	8134	9588	2915	4116	2802	6917	3993	8764
9790	2228	9702	1690	7170	7511	1937	0723	4505	7155
3250	8860	3294	2684	6572	3415	5750	8726	2647	6596
5450	3922	0950	0890	6434	2306	2781	1066	3681	2404
5765	0765	7311	5270	5910	7009	0240	7435	4568	6484
8408	1939	0599	5347	2160	7376	4696	6969	0787	3838
8460	7658	6906	9177	1492	4680	3719	3456	8681	6736
4198	7244	3849	4819	1008	6781	3388	5253	7041	6712
9872	4441	6712	9614	2736	5533	9062	2534	0855	7946
6485	0487	0004	5563	1481	1546	8245	6116	6920	0990
2064	0512	9509	0341	8131	7778	8609	9417	1216	4189
9927	8987	5321	3125	9992	9449	5951	5872	2057	5731
4918	9690	6121	8770	6053	6931	7252	5409	1869	4229
8099	5821	3899	2685	6781	3178	0096	2986	8878	8991
1901	4974	1262	6810	4673	8772	6616	2632	7891	9970
8273	6675	4925	3924	2274	3860	1662	7480	8674	4503
2878	8213	3170	5126	0434	9481	7029	8688	4027	3340
6088	1182	3242	0835	1765	8819	3462	9820	5759	4189
5773	6600	5306	0354	8295	0148	6608	9064	3421	8570
2742	6731	3741	4890	8818	3208	3171	5755	2301	5517
9112	8964	7544	8932	1281	2355	5563	1638	7331	2387
0462	1288	6055	6983	0294	5271	4846	1094	4234	5361
4616	3678	1168	1044	6235	6534	5263	9716	3890	6999
4976	3383	9718	1256	9764	0393	7132	6904	2607	4733

Table G Random numbers *(Continued)*

9852	1453	9473	2914	4262	7582	9644	7969	3711	8334
7535	0446	3178	0206	5295	2991	5923	3215	8824	9362
7469	6789	4044	7141	1840	6288	7101	0746	2695	7019
1886	1976	1856	9117	7133	7428	9700	0223	3753	0230
6268	0733	1136	7479	7881	0953	5499	1158	4729	1442
2097	9561	2830	0500	3695	4707	4083	8432	0611	7583
3735	1763	8090	1457	8894	0426	0433	9165	4050	1432
9296	5868	7340	8139	7202	0465	9028	7987	3800	4555
8527	5288	8545	9063	3106	4875	5945	3058	7682	5313
7866	9022	0844	5307	0612	8102	8529	7529	6106	5733
9954	5248	6636	4988	3463	2461	5451	4552	3026	1554
2613	8184	5800	6939	5684	7008	0950	0891	6856	0228
2668	4852	8006	8893	6596	9808	3623	9518	3102	6927
3002	2998	5357	6634	7099	2299	0198	4755	0795	3344
8422	4435	0134	4587	4886	2413	2774	6136	8993	2721
3690	2492	7171	7720	6509	7549	2330	5733	4730	1759
0813	6790	6858	1489	2669	3743	1901	4971	8280	1127
6477	5289	4092	4223	6454	7632	7577	2816	9202	0811
0772	2160	7236	0812	4195	5589	0830	8261	9232	5944
5692	9870	3583	8997	1533	6466	8830	7271	3809	3291
2080	3828	7880	0586	8482	7811	6807	3309	2729	8636
1039	3382	7600	1077	4455	8806	1822	1669	7501	5622
7227	0104	4141	1521	9104	5563	1392	8238	4882	0861
8506	6348	4612	8252	1062	1757	0964	2983	2244	0253
5086	0303	7423	3298	3979	2831	2257	1508	7642	3981
0092	1629	0377	3590	2209	4839	6332	1490	3092	2590
0935	5565	2315	8030	7651	5189	0075	9353	1921	9572
2605	3973	8204	4143	2677	0034	8601	3340	8383	2818
7277	9889	0390	5579	4620	5650	0310	2082	4664	0494
5484	3900	3485	0741	9069	5920	4326	7704	6525	2503
6905	7127	5933	1137	7583	6450	5658	7678	3444	5163
8387	5323	3753	1859	6043	0294	5110	6340	9137	6137
4094	4957	0163	9717	4118	4276	9465	8820	4127	2954
4951	3781	5101	1815	7068	6379	7252	1086	8919	5677
9047	0199	5068	7447	1664	9278	1708	3625	2864	1351
7274	9512	0074	6677	8676	0222	3335	1976	1645	3914
9192	4011	0255	5458	6942	8043	6201	1587	0972	3391
0554	1690	6333	1931	9433	2661	8690	2313	6999	3436
9231	5627	1815	7171	8036	1832	2031	6298	6073	7121
3995	9677	7765	3194	3222	4191	2734	4469	8617	6694
2402	6250	9362	7373	4757	1716	1942	0417	5921	7456
5295	7385	5474	2123	7035	9983	5192	1840	6176	3818
5177	1191	2106	3351	5057	0967	4538	1246	3374	3286
7315	3365	7203	1231	0546	6612	1038	1425	2709	5780
5775	7517	8974	3961	2183	5295	3096	8536	9442	3139
5500	2276	6307	2346	1285	7000	5306	0414	3383	2137
3251	8902	8843	2112	8567	8131	8116	5270	5994	7445
4675	1435	2192	0874	2897	0262	5092	5541	4014	2086
3543	6130	4247	4859	2660	7852	9096	0578	0097	4746
3521	8772	6612	0721	3899	2999	1263	7017	8057	4983
5573	9396	3464	1702	9204	3389	5678	2589	0288	4633
7478	7569	7551	3380	2152	5411	2647	7242	2800	6183
3339	2854	9691	9562	3252	9848	6030	8472	2266	1270
5505	8474	3167	8552	5409	1556	4247	4652	2953	5394
6381	2086	5457	7703	2758	2963	8167	6712	9820	5654

Table G Random numbers *(Continued)*

6975	5239	0762	5846	2431	0543	4956	8787	9651	2605
7185	4019	7332	2820	4853	8636	9505	6575	0365	6648
4510	1658	5615	2194	1901	4975	1895	4383	0415	3771
7752	0105	4769	2994	7445	0781	4960	4253	9451	6518
4834	4043	6591	3646	8918	4603	1970	9145	7615	3905
8866	6036	9755	4508	9061	2080	3406	9856	1298	6281
6622	4612	2030	7299	8414	8822	5176	9443	6054	6462
9094	8973	3335	2183	5192	1630	0959	8143	9182	8012
5618	6445	2983	0375	2540	2735	4901	5515	4787	7058
2705	2693	1944	8074	2015	3261	5529	7193	5401	9531
1797	4334	3293	2632	3770	1675	9363	7795	3331	8995
9448	5174	5869	0448	8613	4400	6938	5161	8691	2838
3461	1304	9682	8577	4449	1896	8328	1698	7138	1141
7092	5007	5596	8522	2580	4495	4728	8948	4434	2438
5533	4294	0939	4050	1225	6414	5895	0148	7053	5935
7852	8988	5951	4919	7404	2426	4450	2358	3082	4561
8313	8456	9892	0981	6736	8021	6226	5573	1664	9489
1158	2241	9861	7588	2669	5480	9160	4267	1690	7278
9338	7226	0025	8844	8181	5565	2418	9394	0837	3106
7711	1336	3251	8902	8425	5766	3262	5848	3545	7073
2656	1863	3884	6516	6922	1808	1896	8853	0964	3089
7980	9370	2850	3818	7281	8352	9637	0618	2430	6525
1409	7865	5908	4296	1888	2792	4014	1667	1295	0814
7657	6630	5000	1493	5459	5869	0315	8134	9587	2184
2863	5450	1329	8787	8795	4604	2615	0075	1433	7707
3988	2042	2906	8995	0818	9288	1650	0803	8319	2533
4551	2815	8941	4893	8612	4844	0042	3890	7068	8512
5772	4732	2829	3931	9540	6256	5420	2179	9448	5489
9150	1435	3817	8975	4276	9569	0175	6663	0045	5549
5764	7914	8280	1337	3779	8197	9105	5985	1054	2866
5895	0044	5021	3846	7599	0398	5212	9509	0134	4656
6857	1174	8085	6503	5355	3027	1708	3626	7059	0167
2538	2669	3746	3270	1214	9983	8434	1344	1160	3292
9983	1387	1410	8891	2523	8705	9190	2986	7654	5142
5061	9529	2922	2199	8310	6954	8090	5371	0672	6281
9999	4226	2815	8817	5606	5190	0495	7867	9968	5951
9078	5936	2393	7875	6871	3163	9203	2863	5693	9973
4823	2291	8925	6306	1717	0320	2549	3107	5488	0303
1232	1384	5698	9313	3501	3238	7227	0220	6118	7655
7694	6484	0279	8528	7214	1750	0577	8418	0698	5403
9207	6903	9703	2028	3460	0778	3795	0698	3974	8522
1886	2080	3719	3602	3896	1214	9862	1969	6782	9237
6963	4197	6405	8683	7573	0842	9306	2596	7404	9999
1797	2315	5434	0787	3809	9129	4511	0708	2181	9119
6534	5578	4158	6256	3721	7515	3905	1905	7153	3552
2325	4238	8861	6098	8837	7690	0497	8848	6601	1553
6598	4628	1023	9747	4860	3437	7414	7609	9938	8335
4592	5016	4434	7133	7218	4602	1690	7914	8819	3600
1765	8822	5278	2324	3715	0431	7780	4955	9683	8998
6139	3275	7731	3351	5306	0323	5387	3901	4151	2922
3911	8334	5465	6647	8773	7456	9954	5141	3573	5570
6840	0366	6962	3462	1724	6661	7221	6074	9262	3461
5572	8838	8132	9398	0737	7125	7388	7686	9814	1760
2337	5303	3720	3917	7238	9925	7940	7818	1676	9780
3138	6014	4909	1143	7551	3380	2713	7649	2784	0175

Table G   Random numbers *(Continued)*

1921	9046	6300	7460	8271	5547	6701	0098	8559	0490
6735	5991	6857	0756	1943	7657	6399	1052	1816	7694
6089	1958	8748	8310	6953	7775	9274	1579	2548	7028
1767	9764	0628	5037	1549	9048	0644	3850	5555	6104
1748	3753	1385	0381	9998	4075	7729	2301	5381	3510
7589	4195	5691	2192	0770	0900	2608	5048	5751	9322
9717	9635	7972	1198	6177	0745	2199	8205	4559	9159
9063	3318	6412	4632	6904	2713	7858	3089	7970	4058
0915	5751	9112	1040	0561	9128	4285	5149	3002	2868
6456	8457	3348	4322	6493	8656	7532	1384	6198	1236
8700	0874	3291	7324	4471	5375	6769	2113	0950	0682
8050	3921	0741	8814	4262	7477	7042	7027	4275	2143
6663	2373	0845	8119	6862	7477	7148	9840	7154	4495
3648	5449	3166	7921	3125	9884	5739	1775	4927	5394
4238	9071	4578	2589	0076	6636	7915	8735	4907	6195
6015	5642	4642	2904	7841	4004	7299	8310	6953	7986
1088	7727	1060	9589	3315	9763	6349	5029	6314	9714
2027	6143	5814	9032	9856	1146	5864	6943	8234	0896
1929	8697	3478	5184	6468	9513	6333	1699	7348	6139
7651	4561	1036	9359	6992	8957	0203	8199	9733	7220
1629	0688	9118	7121	7870	5877	2438	1767	9065	4264
5650	0473	0543	5273	5716	1335	0448	8490	6438	4090
0604	0231	1836	4004	5457	8963	7017	8226	9427	8318
0602	9868	2575	8785	8517	8975	4276	9677	2064	0599
5913	4536	4377	5975	3607	6362	6467	5009	1142	7238
9111	0238	9718	4707	3767	0311	2967	1839	5788	9242
7787	7707	4838	5910	3042	4640	2299	4400	4446	0423
5731	3738	3526	2380	8557	8789	9577	1122	8310	7057
4827	9203	2758	2946	5915	0771	1426	2919	8014	4156
1653	8923	5386	3684	7891	5200	9884	5545	2742	7205
3561	7178	2486	1353	6928	9463	7229	1545	7616	4224
8301	8225	7398	4084	8697	9670	7404	9893	1467	3681
5658	7221	6074	8944	6446	7141	2731	6380	4621	2703
4819	1007	6258	4628	0705	0757	4546	8906	7679	4180
9239	0523	3369	6234	6115	1524	8575	3502	3971	9376

From Donald B. Owen, *Handbook of Statistical Tables*, Reading, Mass.: Addison-Wesley, 1962. Courtesy of U.S. Atomic Energy Commission.

Table H Percentage points of the *t* distribution

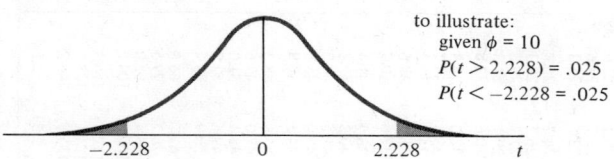

to illustrate:
given $\phi = 10$
$P(t > 2.228) = .025$
$P(t < -2.228 = .025$

ϕ \ α	0.25	0.20	0.15	0.10	0.05	0.025	0.01	0.005	0.0005
1	1.000	1.376	1.963	3.078	6.314	12.706	31.821	63.657	636.619
2	0.816	1.061	1.386	1.886	2.920	4.303	6.965	9.925	31.598
3	0.765	0.978	1.250	1.638	2.353	3.182	4.541	5.841	12.941
4	0.741	0.941	1.190	1.533	2.132	2.776	3.747	4.604	8.610
5	0.727	0.920	1.156	1.476	2.015	2.571	3.365	4.032	6.859
6	0.718	0.906	1.134	1.440	1.943	2.447	3.143	3.707	5.959
7	0.711	0.896	1.119	1.415	1.895	2.365	2.998	3.499	5.405
8	0.706	0.889	1.108	1.397	1.860	2.306	2.896	3.355	5.041
9	0.703	0.883	1.100	1.383	1.833	2.262	2.821	3.250	4.781
10	0.700	0.879	1.093	1.372	1.812	2.228	2.764	3.169	4.587
11	0.697	0.876	1.088	1.363	1.796	2.201	2.718	3.106	4.437
12	0.695	0.873	1.083	1.356	1.782	2.179	2.681	3.055	4.318
13	0.694	0.870	1.079	1.350	1.771	2.160	2.650	3.012	4.221
14	0.692	0.868	1.076	1.345	1.761	2.145	2.624	2.977	4.140
15	0.691	0.866	1.074	1.341	1.753	2.131	2.602	2.947	4.073
16	0.690	0.865	1.071	1.337	1.746	2.120	2.583	2.921	4.015
17	0.689	0.863	1.069	1.333	1.740	2.110	2.567	2.898	3.965
18	0.688	0.862	1.067	1.330	1.734	2.101	2.552	2.878	3.922
19	0.688	0.861	1.066	1.328	1.729	2.093	2.539	2.861	3.883
20	0.687	0.860	1.064	1.325	1.725	2.086	2.528	2.845	3.850
21	0.686	0.859	1.063	1.323	1.721	2.080	2.518	2.831	3.819
22	0.686	0.858	1.061	1.321	1.717	2.074	2.508	2.819	3.792
23	0.685	0.858	1.060	1.319	1.714	2.069	2.500	2.807	3.767
24	0.685	0.857	1.059	1.318	1.711	2.064	2.492	2.397	3.745
25	0.684	0.856	1.058	1.316	1.708	2.060	2.485	2.787	3.725
26	0.684	0.856	1.058	1.315	1.706	2.056	2.479	2.779	3.707
27	0.684	0.855	1.057	1.314	1.703	2.052	2.473	2.771	3.690
28	0.683	0.855	1.056	1.313	1.701	2.048	2.467	2.763	3.674
29	0.683	0.854	1.055	1.311	1.699	2.045	2.462	2.756	3.659
30	0.683	0.854	1.055	1.310	1.697	2.042	2.457	2.750	3.646
40	0.681	0.851	1.050	1.303	1.684	2.021	2.423	2.704	3.551
60	0.679	0.848	1.046	1.296	1.671	2.000	2.390	2.660	3.460
120	0.677	0.845	1.041	1.289	1.658	1.980	2.358	2.617	3.373
∞	0.674	0.842	1.036	1.282	1.645	1.960	2.326	2.576	3.291

Abridged from Table III of R. A. Fisher and F. Yates, *Statistical Tables for Biological, Agricultural and Medical Research*, published by Oliver & Boyd, Edinburgh, and by permission of the authors and publishers.

example
for $\phi = 8$ degrees
of freedom:

$P[\chi^2 > 13.36]$
$= .10$

Table I Percentage points of the χ^2 distribution

ϕ \ P	0.995	0.99	0.975	0.95	0.90	0.75	0.50	0.25	0.10	0.05	0.025	0.01	0.005
1	0.0^4393	0.0^3157	0.0^3982	0.0^393	0.0158	0.102	0.455	1.323	2.71	3.84	5.02	6.63	7.88
2	0.0100	0.0201	0.0506	0.103	0.211	0.575	1.386	2.77	4.61	5.99	7.38	9.21	10.60
3	0.0717	0.115	0.216	0.352	0.584	1.213	2.37	4.11	6.25	7.81	9.35	11.34	12.84
4	0.207	0.297	0.484	0.711	1.064	1.923	3.36	5.39	7.78	9.49	11.14	13.28	14.86
5	0.412	0.554	0.831	1.145	1.610	2.67	4.35	6.63	9.24	11.07	12.83	15.09	16.75
6	0.676	0.872	1.237	1.635	2.20	3.45	5.35	7.84	10.64	12.59	14.45	16.81	18.55
7	0.989	1.239	1.690	2.17	2.83	4.25	6.35	9.04	12.02	14.07	16.01	18.48	20.3
8	1.344	1.646	2.18	2.73	3.49	5.07	7.34	10.22	13.36	15.51	17.53	20.1	22.0
9	1.735	2.09	2.70	3.33	4.17	5.90	8.34	11.39	14.68	16.92	19.02	21.7	23.6
10	2.16	2.56	3.25	3.94	4.87	6.74	9.34	12.55	15.99	18.31	20.5	23.2	25.2
11	2.60	3.05	3.82	4.57	5.58	7.58	10.34	13.70	17.28	19.68	21.9	24.7	26.8
12	3.07	3.57	4.40	5.23	6.30	8.44	11.34	14.85	18.55	21.0	23.3	26.2	28.3
13	3.57	4.11	5.01	5.89	7.04	9.30	12.34	15.98	19.81	22.4	24.7	27.7	29.8
14	4.07	4.66	5.63	6.57	7.79	10.17	13.34	17.12	21.1	23.7	26.1	29.1	31.3
15	4.60	5.23	6.26	7.26	8.55	11.04	14.34	18.25	22.3	25.0	27.5	30.6	32.8
16	5.14	5.81	6.91	7.96	9.31	11.91	15.34	19.37	23.5	26.3	28.8	32.0	34.3
17	5.70	6.41	7.56	8.67	10.09	12.79	16.34	20.5	24.8	27.6	30.2	33.4	35.7
18	6.26	7.01	8.23	9.39	10.86	13.68	17.34	21.6	26.0	28.9	31.5	34.8	37.2
19	6.84	7.63	8.91	10.12	11.65	14.56	18.34	22.7	27.2	30.1	32.9	36.2	38.6
20	7.43	8.26	9.59	10.85	12.44	15.45	19.34	23.8	28.4	31.4	34.2	37.6	40.0
21	8.03	8.90	10.28	11.59	13.24	16.34	20.3	24.9	29.6	32.7	35.5	38.9	41.4
22	8.64	9.54	10.98	12.31	14.04	17.24	21.3	26.0	30.8	33.9	36.8	40.3	42.8
23	9.26	10.20	11.69	13.09	14.85	18.14	22.3	27.1	32.0	35.2	38.1	41.6	44.2
24	9.89	10.86	12.40	13.85	15.66	19.04	23.3	28.2	33.2	36.4	39.4	43.0	45.6
25	10.52	11.52	13.12	14.61	16.47	19.94	24.3	29.3	34.4	37.7	40.6	44.3	46.9
26	11.16	12.20	13.84	15.38	17.29	20.8	25.3	30.4	35.6	38.9	41.9	45.6	48.3
27	11.81	12.88	14.57	16.15	18.11	21.7	26.3	31.5	36.7	40.1	43.2	47.0	49.6
28	12.46	13.56	15.31	16.93	18.94	22.7	27.3	32.6	37.9	41.3	44.5	48.3	51.0
29	13.12	14.26	16.05	17.71	19.77	23.6	28.3	33.7	39.1	42.6	45.7	49.6	52.3
30	13.79	14.95	16.79	18.49	20.6	24.5	29.3	34.8	40.3	43.8	47.0	50.9	53.7
40	20.7	22.2	24.4	26.5	29.1	33.7	39.3	45.6	51.8	55.8	59.3	63.7	66.8
50	28.0	29.7	32.4	34.8	37.7	42.9	49.3	56.3	63.2	67.5	71.4	76.2	79.5
60	35.5	37.5	40.5	43.2	46.5	52.3	59.3	67.0	74.4	79.1	83.3	88.4	92.0
70	43.3	45.4	48.8	51.7	55.3	61.7	69.3	77.6	85.5	90.5	95.0	100.4	104.2
80	51.2	53.5	57.2	60.4	64.3	71.1	79.3	88.1	96.6	101.9	106.6	112.3	116.3
90	59.2	61.8	65.6	69.1	73.3	80.6	89.3	98.6	107.6	113.1	118.1	124.1	128.3
100	67.3	70.1	74.2	77.9	82.4	90.1	99.3	109.1	118.5	124.3	129.6	135.8	140.2
x_α	-2.58	-2.33	-1.96	-1.64	-1.28	-0.674	0.000	0.674	1.282	1.645	1.960	2.33	2.58

Abridged from "Table of percentage points of the χ^2 distribution" by Catherine M. Thompson, *Biometrika*, Vol. 32 (1941), pp. 187–191, and is reprinted here by permission of the author and editor of *Biometrika*.

Table J The F Distribution

Degrees of freedom in the numerator are recorded at the top of the table, and the degrees of freedom in the denominator are indicated at the sides. The first listed or smaller value is the value on the F scale to the right of which lies 0.05 of the area under the curve. The second listed or larger value is the value on the F scale to the right of which lies 0.01 of the area under the curve.

(handwritten in margin: d = degrees of freedom; $z - 1 = 1$; $B = 3$)

ϕ_n degrees of freedom (numerator)

ϕ_d	1	2	3	4	5	6	8	10	12	16	20	30	40	50	100	ϕ_d
1	161 / 4,052	200 / 4,999	216 / 5,403	225 / 5,625	230 / 5,764	234 / 5,859	239 / 5,981	242 / 6,056	244 / 6,106	246 / 6,169	248 / 6,208	250 / 6,258	251 / 6,286	252 / 6,302	253 / 6,334	1
2	18.51 / 98.49	19.00 / 99.00	19.16 / 99.17	19.25 / 99.25	19.30 / 99.30	19.33 / 99.33	19.37 / 99.36	19.39 / 99.40	19.41 / 99.42	19.43 / 99.44	19.44 / 99.45	19.46 / 99.47	19.47 / 99.48	19.47 / 99.48	19.49 / 99.49	2
3	10.13 / 34.12	9.55 / 30.82	9.28 / 29.46	9.12 / 28.71	9.01 / 28.24	8.94 / 27.91	8.84 / 27.49	8.78 / 27.23	8.74 / 27.05	8.69 / 26.83	8.66 / 26.69	8.62 / 26.50	8.60 / 26.41	8.58 / 26.35	8.56 / 26.23	3
4	7.71 / 21.20	6.94 / 18.00	6.59 / 16.69	6.39 / 15.98	6.26 / 15.52	6.16 / 15.21	6.04 / 14.80	5.96 / 14.54	5.91 / 14.37	5.84 / 14.15	5.80 / 14.02	5.74 / 13.83	5.71 / 13.74	5.70 / 13.69	5.66 / 13.57	4
5	6.61 / 16.26	5.79 / 13.27	5.41 / 12.06	5.19 / 11.39	5.05 / 10.97	4.95 / 10.67	4.82 / 10.27	4.74 / 10.05	4.68 / 9.89	4.60 / 9.68	4.56 / 9.55	4.50 / 9.38	4.46 / 9.29	4.44 / 9.24	4.40 / 9.13	5
6	5.99 / 13.74	5.14 / 10.92	4.76 / 9.78	4.53 / 9.15	4.39 / 8.75	4.28 / 8.47	4.15 / 8.10	4.06 / 7.87	4.00 / 7.72	3.92 / 7.52	3.87 / 7.39	3.81 / 7.23	3.77 / 7.14	3.75 / 7.09	3.71 / 6.99	6
7	5.59 / 12.25	4.74 / 9.55	4.35 / 8.45	4.12 / 7.85	3.97 / 7.46	3.87 / 7.19	3.73 / 6.84	3.63 / 6.62	3.57 / 6.47	3.49 / 6.27	3.44 / 6.15	3.38 / 5.98	3.34 / 5.90	3.32 / 5.85	3.28 / 5.75	7
8	5.32 / 11.26	4.46 / 8.65	4.07 / 7.59	3.84 / 7.01	3.69 / 6.63	3.58 / 6.37	3.44 / 6.03	3.34 / 5.82	3.28 / 5.67	3.20 / 5.48	3.15 / 5.36	3.08 / 5.20	3.05 / 5.11	3.03 / 5.06	2.98 / 4.96	8
9	5.12 / 10.56	4.26 / 8.02	3.86 / 6.99	3.63 / 6.42	3.48 / 6.06	3.37 / 5.80	3.23 / 5.47	3.13 / 5.26	3.07 / 5.11	2.98 / 4.92	2.93 / 4.80	2.86 / 4.64	2.82 / 4.56	2.80 / 4.51	2.76 / 4.41	9
10	4.96 / 10.04	4.10 / 7.56	3.71 / 6.55	3.48 / 5.99	3.33 / 5.64	3.22 / 5.39	3.07 / 5.06	2.97 / 4.85	2.91 / 4.71	2.82 / 4.52	2.77 / 4.41	2.70 / 4.25	2.67 / 4.17	2.64 / 4.12	2.59 / 4.01	10
11	4.84 / 9.65	3.98 / 7.20	3.59 / 6.22	3.36 / 5.67	3.20 / 5.32	3.09 / 5.07	2.95 / 4.74	2.86 / 4.54	2.79 / 4.40	2.70 / 4.21	2.65 / 4.10	2.57 / 3.94	2.53 / 3.86	2.50 / 3.80	2.45 / 3.70	11
12	4.75 / 9.33	3.88 / 6.93	3.49 / 5.95	3.26 / 5.41	3.11 / 5.06	3.00 / 4.82	2.85 / 4.50	2.76 / 4.30	2.69 / 4.16	2.60 / 3.98	2.54 / 3.86	2.46 / 3.70	2.42 / 3.61	2.40 / 3.56	2.35 / 3.46	12
13	4.67 / 9.07	3.80 / 6.70	3.41 / 5.74	3.18 / 5.20	3.02 / 4.86	2.92 / 4.62	2.77 / 4.30	2.67 / 4.10	2.60 / 3.96	2.51 / 3.78	2.46 / 3.67	2.38 / 3.51	2.34 / 3.42	2.32 / 3.37	2.26 / 3.27	13
14	4.60 / 8.86	3.74 / 6.51	3.34 / 5.56	3.11 / 5.03	2.96 / 4.69	2.85 / 4.46	2.70 / 4.14	2.60 / 3.94	2.53 / 3.80	2.44 / 3.62	2.39 / 3.51	2.31 / 3.34	2.27 / 3.26	2.24 / 3.21	2.19 / 3.11	14
15	4.54 / 8.68	3.68 / 6.36	3.29 / 5.42	3.06 / 4.89	2.90 / 4.56	2.79 / 4.32	2.64 / 4.00	2.55 / 3.80	2.48 / 3.67	2.39 / 3.48	2.33 / 3.36	2.25 / 3.20	2.21 / 3.12	2.18 / 3.07	2.12 / 2.97	15
16	4.49 / 8.53	3.63 / 6.23	3.24 / 5.29	3.01 / 4.77	2.85 / 4.44	2.74 / 4.20	2.59 / 3.89	2.49 / 3.69	2.42 / 3.55	2.33 / 3.37	2.28 / 3.25	2.20 / 3.10	2.16 / 3.01	2.13 / 2.96	2.07 / 2.86	16

Table J The F Distribution (Continued)

ϕ_n degrees of freedom (numerator)

ϕ_d	1	2	3	4	5	6	8	10	12	16	20	30	40	50	100	ϕ_d
17	4.45 / 8.40	3.59 / 6.11	3.20 / 5.18	2.96 / 4.67	2.81 / 4.34	2.70 / 4.10	2.55 / 3.79	2.45 / 3.59	2.38 / 3.45	2.29 / 3.27	2.23 / 3.16	2.15 / 3.00	2.11 / 2.92	2.08 / 2.86	2.02 / 2.76	17
18	4.41 / 8.28	3.55 / 6.01	3.16 / 5.09	2.93 / 4.58	2.77 / 4.25	2.66 / 4.01	2.51 / 3.71	2.41 / 3.51	2.34 / 3.37	2.25 / 3.19	2.19 / 3.07	2.11 / 2.91	2.07 / 2.83	2.04 / 2.78	1.98 / 2.68	18
19	4.38 / 8.18	3.52 / 5.93	3.13 / 5.01	2.90 / 4.50	2.74 / 4.17	2.63 / 3.94	2.48 / 3.63	2.38 / 3.43	2.31 / 3.30	2.21 / 3.12	2.15 / 3.00	2.07 / 2.84	2.02 / 2.76	2.00 / 2.70	1.94 / 2.60	19
20	4.35 / 8.10	3.49 / 5.85	3.10 / 4.94	2.87 / 4.43	2.71 / 4.10	2.60 / 3.87	2.45 / 3.56	2.35 / 3.37	2.28 / 3.23	2.18 / 3.05	2.12 / 2.94	2.04 / 2.77	1.99 / 2.69	1.96 / 2.63	1.90 / 2.53	20
25	4.24 / 7.77	3.38 / 5.57	2.99 / 4.68	2.76 / 4.18	2.60 / 3.86	2.49 / 3.63	2.34 / 3.32	2.24 / 3.13	2.16 / 2.99	2.06 / 2.81	2.00 / 2.70	1.92 / 2.54	1.87 / 2.45	1.84 / 2.40	1.77 / 2.29	25
30	4.17 / 7.56	3.32 / 5.39	2.92 / 4.51	2.69 / 4.02	2.53 / 3.70	2.42 / 3.47	2.27 / 3.17	2.16 / 2.98	2.09 / 2.84	1.99 / 2.66	1.93 / 2.55	1.84 / 2.38	1.79 / 2.29	1.76 / 2.24	1.69 / 2.13	30
40	4.08 / 7.31	3.23 / 5.18	2.84 / 4.31	2.61 / 3.83	2.45 / 3.51	2.34 / 3.29	2.18 / 2.99	2.07 / 2.80	2.00 / 2.66	1.90 / 2.49	1.84 / 2.37	1.74 / 2.20	1.69 / 2.11	1.66 / 2.05	1.59 / 1.94	40
50	4.03 / 7.17	3.18 / 5.06	2.79 / 4.20	2.56 / 3.72	2.40 / 3.41	2.29 / 3.18	2.13 / 2.88	2.02 / 2.70	1.95 / 2.56	1.85 / 2.39	1.78 / 2.26	1.69 / 2.10	1.63 / 2.00	1.60 / 1.94	1.52 / 1.82	50
60	4.00 / 7.08	3.15 / 4.98	2.76 / 4.13	2.52 / 3.65	2.37 / 3.34	2.25 / 3.12	2.10 / 2.82	1.99 / 2.63	1.92 / 2.50	1.81 / 2.32	1.75 / 2.20	1.65 / 2.03	1.59 / 1.93	1.56 / 1.87	1.48 / 1.74	60
80	3.96 / 6.96	3.11 / 4.88	2.72 / 4.04	2.48 / 3.56	2.33 / 3.25	2.21 / 3.04	2.05 / 2.74	1.95 / 2.55	1.88 / 2.41	1.77 / 2.24	1.70 / 2.11	1.60 / 1.94	1.54 / 1.84	1.51 / 1.78	1.42 / 1.65	80
100	3.94 / 6.90	3.09 / 4.82	2.70 / 3.98	2.46 / 3.51	2.30 / 3.20	2.19 / 2.99	2.03 / 2.69	1.92 / 2.51	1.85 / 2.36	1.75 / 2.19	1.68 / 2.06	1.57 / 1.89	1.51 / 1.79	1.48 / 1.73	1.39 / 1.59	100
150	3.91 / 6.81	3.06 / 4.75	2.67 / 3.91	2.43 / 3.44	2.27 / 3.14	2.16 / 2.92	2.00 / 2.62	1.89 / 2.44	1.82 / 2.30	1.71 / 2.12	1.64 / 2.00	1.54 / 1.83	1.47 / 1.72	1.44 / 1.66	1.34 / 1.51	150
200	3.89 / 6.76	3.04 / 4.71	2.65 / 3.88	2.41 / 3.41	2.26 / 3.11	2.14 / 2.90	1.98 / 2.60	1.87 / 2.41	1.80 / 2.28	1.69 / 2.09	1.62 / 1.97	1.52 / 1.79	1.45 / 1.69	1.42 / 1.62	1.32 / 1.48	200
400	3.86 / 6.70	3.02 / 4.66	2.62 / 3.83	2.39 / 3.36	2.23 / 3.06	2.12 / 2.85	1.96 / 2.55	1.85 / 2.37	1.78 / 2.23	1.67 / 2.04	1.60 / 1.92	1.49 / 1.74	1.42 / 1.64	1.38 / 1.57	1.28 / 1.42	400
1000	3.85 / 6.66	3.00 / 4.62	2.61 / 3.80	2.38 / 3.34	2.22 / 3.04	2.10 / 2.82	1.95 / 2.53	1.84 / 2.34	1.76 / 2.20	1.65 / 2.01	1.58 / 1.89	1.47 / 1.71	1.41 / 1.61	1.36 / 1.54	1.26 / 1.38	1000
∞	3.84 / 6.64	2.99 / 4.60	2.60 / 3.78	2.37 / 3.32	2.21 / 3.02	2.09 / 2.80	1.94 / 2.51	1.83 / 2.32	1.75 / 2.18	1.64 / 1.99	1.57 / 1.87	1.46 / 1.69	1.40 / 1.59	1.35 / 1.52	1.24 / 1.36	∞

Reprinted by permission from *Statistical Methods*, 6th edition, by George W. Snedecor and William G. Cochran, Copyright © 1967 by The Iowa State University Press, Ames, Iowa.

Table K Table of values of *r* for values of *z**

z*	0.00	0.01	0.02	0.03	0.04	0.05	0.06	0.07	0.08	0.09
0.0	0.0000	0.0100	0.0200	0.0300	0.0400	0.0500	0.0599	0.0699	0.0798	0.0898
0.1	0.0997	0.1096	0.1194	0.1293	0.1391	0.1489	0.1587	0.1684	0.1781	0.1878
0.2	0.1974	0.2070	0.2165	0.2260	0.2355	0.2449	0.2543	0.2636	0.2729	0.2821
0.3	0.2913	0.3004	0.3095	0.3185	0.3275	0.3364	0.3452	0.3540	0.3627	0.3714
0.4	0.3800	0.3885	0.3969	0.4053	0.4136	0.4219	0.4301	0.4382	0.4462	0.4542
0.5	0.4621	0.4700	0.4777	0.4854	0.4930	0.5005	0.5080	0.5154	0.5227	0.5299
0.6	0.5370	0.5441	0.5511	0.5581	0.5649	0.5717	0.5784	0.5850	0.5915	0.5980
0.7	0.6044	0.6107	0.6169	0.6231	0.6291	0.6352	0.6411	0.6469	0.6527	0.6584
0.8	0.6640	0.6696	0.6751	0.6805	0.6858	0.6911	0.6963	0.7014	0.7064	0.7114
0.9	0.7163	0.7211	0.7259	0.7306	0.7352	0.7398	0.7443	0.7487	0.7531	0.7574
1.0	0.7616	0.7658	0.7699	0.7739	0.7779	0.7818	0.7857	0.7895	0.7932	0.7969
1.1	0.8005	0.8041	0.8076	0.8110	0.8144	0.8178	0.8210	0.8243	0.8275	0.8306
1.2	0.8337	0.8367	0.8397	0.8426	0.8455	0.8483	0.8511	0.8538	0.8565	0.8591
1.3	0.8617	0.8643	0.8668	0.8693	0.8717	0.8741	0.8764	0.8787	0.8810	0.8832
1.4	0.8854	0.8875	0.8896	0.8917	0.8937	0.8957	0.8977	0.8996	0.9015	0.9033
1.5	0.9052	0.9069	0.9087	0.9104	0.9121	0.9138	0.9154	0.9170	0.9186	0.9202
1.6	0.9217	0.9232	0.9246	0.9261	0.9275	0.9289	0.9302	0.9316	0.9329	0.9342
1.7	0.9354	0.9367	0.9379	0.9391	0.9402	0.9414	0.9425	0.9436	0.9447	0.9458
1.8	0.9468	0.9478	0.9498	0.9488	0.9508	0.9518	0.9527	0.9536	0.9545	0.9554
1.9	0.9562	0.9571	0.9579	0.9587	0.9595	0.9603	0.9611	0.9619	0.9626	0.9633
2.0	0.9640	0.9647	0.9654	0.9661	0.9668	0.9674	0.9680	0.9687	0.9693	0.9699
2.1	0.9705	0.9710	0.9716	0.9722	0.9727	0.9732	0.9738	0.9743	0.9748	0.9753
2.2	0.9757	0.9762	0.9767	0.9771	0.9776	0.9780	0.9785	0.9789	0.9793	0.9797
2.3	0.9801	0.9805	0.9809	0.9812	0.9816	0.9820	0.9823	0.9827	0.9830	0.9834
2.4	0.9837	0.9840	0.9843	0.9846	0.9849	0.9852	0.9855	0.9858	0.9861	0.9863
2.5	0.9866	0.9869	0.9871	0.9874	0.9876	0.9879	0.9881	0.9884	0.9886	0.9888
2.6	0.9890	0.9892	0.9895	0.9897	0.9899	0.9901	0.9903	0.9905	0.9906	0.9908
2.7	0.9910	0.9912	0.9914	0.9915	0.9917	0.9919	0.9920	0.9922	0.9923	0.9925
2.8	0.9926	0.9928	0.9929	0.9931	0.9932	0.9933	0.9935	0.9936	0.9937	0.9938
2.9	0.9940	0.9941	0.9942	0.9943	0.9944	0.9945	0.9946	0.9947	0.9949	0.9950
3.0	0.9951									
4.0	0.9993									
5.0	0.9999									

Reprinted from Table V-B of R. A. Fisher, *Statistical Methods for Research Workers*, published by Oliver & Boyd, Edinburgh, and by permission of the authors and publishers.

INDEX